T0270892

Theories and Analyses of Beams and Axisymmetric Circular Plates

Theories and Analyses of Beams and Axisymmetric Circular Plates

J. N. Reddy

CRC Press
Taylor & Francis Group
Boca Raton London New York

CRC Press is an imprint of the
Taylor & Francis Group, an **informa** business

First edition published 2022
by CRC Press
6000 Broken Sound Parkway NW, Suite 300, Boca Raton, FL 33487-2742

and by CRC Press
4 Park Square, Milton Park, Abingdon, Oxon, OX14 4RN

CRC Press is an imprint of Taylor & Francis Group, LLC

ISBN: 978-1-032-14739-0 (hbk)
ISBN: 978-1-032-14741-3 (pbk)
ISBN: 978-1-003-24084-6 (ebk)

DOI: 10.1201/9781003240846

Typeset in CMR10
by KnowledgeWorks Global Ltd.

Publisher's note: This book has been prepared from camera-ready copy provided by the authors.

To

My loving wife,

Aruna Reddy

The author has been very fortunate to have a companion and loving wife, who selflessly took care of his needs with care and affection. She does not know how much the author benefited professionally because of her time, kindness, and generosity. The author is more grateful to her than any words can express. He is ever thankful for everything she did, and hopes that her life is filled with peace and happiness even after he leaves this world.

Contents

Preface xvii

List of symbols used xix

About the Author... xxiii

Chapter 1   Mechanics Preliminaries... 1

    1.1 General Comments... 1
    1.2 Beams and Plates... 2
    1.3 Vectors and Tensors 4
        1.3.1 Vectors and Coordinate Systems 4
        1.3.2 Summation Convention 5
        1.3.3 Stress Vector and Stress Tensor 6
        1.3.4 The Gradient Operator 9
    1.4 Review of the Equations of Solid Mechanics.............. 13
        1.4.1 Green–Lagrange Strain Tensor 13
        1.4.2 The Second Piola–Kirchhoff Stress Tensor 17
        1.4.3 Equations of Motion.. 18
        1.4.4 Stress–Strain Relations.................................... 18
    1.5 Functionally Graded Structures 19
        1.5.1 Background... 19
        1.5.2 Mori–Tanaka Scheme..................................... 20
        1.5.3 Voigt Scheme: Rule of Mixtures 21
        1.5.4 Exponential Model 21
        1.5.5 Power-Law Model.. 22
    1.6 Modified Couple Stress Effects................................... 23
        1.6.1 Background... 23
        1.6.2 The Strain Energy Functional 24
    1.7 Chapter Summary... 24

Chapter 2   Energy Principles and Variational Methods.............. 27

    2.1 Concepts of Work and Energy 27
        2.1.1 Historical Background 27
        2.1.2 Objectives of the Chapter 28
        2.1.3 Concept of Work Done 29
    2.2 Strain Energy and Complementary Strain Energy 31

2.3 Total Potential Energy and Total Complementary Energy 35
2.4 Virtual Work 36
    2.4.1 Virtual Displacements 36
    2.4.2 Virtual Forces 39
2.5 Calculus of Variations and Duality Pairs 39
    2.5.1 The Variational Operator 39
    2.5.2 Functionals and Their Variations 41
    2.5.3 Fundamental Lemma of Variational Calculus ... 42
    2.5.4 Extremum of a Functional 43
    2.5.5 The Euler Equations and Duality Pairs 43
    2.5.6 Natural and Essential Boundary Conditions 46
2.6 The Principle of Virtual Displacements 49
2.7 Principle of Minimum Total Potential Energy 51
2.8 Hamilton's Principle 54
    2.8.1 Preliminary Comments 54
    2.8.2 Statement of the Principle 55
    2.8.3 Euler–Lagrange Equations 56
2.9 Chapter Summary 58

Chapter 3   **The Classical Beam Theory** 61

3.1 Introductory Comments 61
3.2 Kinematics 62
3.3 Equations of Motion 63
    3.3.1 Preliminary Comments 63
    3.3.2 Vector Approach 64
    3.3.3 Energy Approach 66
3.4 Governing Equations in Terms of Displacements 69
    3.4.1 Material Constitutive Relations 69
    3.4.2 Uniaxial Stress–Strain Relations 69
    3.4.3 Material Gradation through the Beam Height . 70
    3.4.4 Beam Constitutive Equations 70
    3.4.5 Equations of Motion 72
        3.4.5.1 The general case (with FGM, VKN, and MCS) 72
        3.4.5.2 Homogeneous beams with VKN and MCS 73
        3.4.5.3 Linearized FGM beams with MCS 73
        3.4.5.4 Linearized homogeneous beams with MCS 74
3.5 Equations in Terms of Displacements and Bending Moment 74
    3.5.1 Preliminary Comments 74

3.5.2   General Case with FGM, MCS, and VKN 75
3.5.3   Special Cases 76
    3.5.3.1   Homogeneous beams with VKN
          and MCS... 76
    3.5.3.2   Linearized FGM beams with MCS 77
    3.5.3.3   Linearized homogeneous beams
          with MCS 77
3.6   Cylindrical Bending of FGM Rectangular Plates........ 77
3.6.1   Cylindrical Bending.. 77
3.6.2   Governing Equations in Terms of Stress
    Resultants.. 78
3.6.3   Governing Equations in Terms of
    Displacements... 79
3.7   Exact Solutions 80
3.7.1   Bending Solutions... 80
3.7.2   Buckling and Natural Vibrations..................... 92
    3.7.2.1   Buckling solutions............................. 92
    3.7.2.2   Natural frequencies 97
3.8   The Navier Solutions 102
3.8.1   The General Procedure 102
3.8.2   Navier's Solution of Equations of Motion....... 103
3.8.3   Bending Solutions... 105
3.8.4   Natural Vibrations 108
3.8.5   Transient Analysis 109
3.9   Energy and Variational Methods 113
3.9.1   Introduction.. 113
3.9.2   The Ritz Method... 114
    3.9.2.1   Background and model problem 114
    3.9.2.2   The Ritz approximation 115
    3.9.2.3   Requirements on the approxima-
          tion functions................................. 118
3.9.3   The Weighted-Residual Methods.................... 141
3.10   Chapter Summary.. 146

Chapter 4   The First-Order Shear Deformation Beam Theory 151

4.1   Introductory Comments... 151
4.2   Displacements and Strains 152
4.3   Equations of Motion 152
4.3.1   Vector Approach.. 152
4.3.2   Energy Approach... 154
4.4   Governing Equations in Terms of Displacements....... 156
4.4.1   Beam Constitutive Equations......................... 156
4.4.2   Equations of Motion for the General Case 157

4.4.3 Equations of Motion without the Couple
       Stress and Thermal Effects............................. 158
4.4.4 Equations of Motion for Homogeneous
       Beams 158
4.4.5 Linearized Equations of Motion for FGM
       Beams... 159
4.4.6 Linearized Equations for Homogeneous
       Beams... 159
4.5 Mixed Formulation of the TBT 160
4.6 Exact Solutions 162
     4.6.1 Bending Solutions.. 162
     4.6.2 Buckling Solutions... 174
     4.6.3 Natural Vibration.. 176
4.7 Relations between CBT and TBT 177
     4.7.1 Background.. 177
     4.7.2 Bending Relations between the CBT
            and TBT... 178
            4.7.2.1 Summary of equations of the CBT .. 178
            4.7.2.2 Summary of equations of the TBT .. 178
            4.7.2.3 Relationships by similarity and
                      load equivalence 179
     4.7.3 Bending Relationships for FGM Beams with
            the Couple Stress Effect 187
            4.7.3.1 Summary of equations of CBT and
                      TBT... 187
            4.7.3.2 General relationships 187
            4.7.3.3 Specialized relationships 189
     4.7.4 Buckling Relationships 200
            4.7.4.1 Summary of governing equations..... 200
     4.7.5 Frequency Relationships 202
            4.7.5.1 Governing equations of the CBT 202
            4.7.5.2 Governing equations of the TBT 202
            4.7.5.3 Relationship................................... 203
4.8 The Navier Solutions 203
     4.8.1 General Solution.. 203
     4.8.2 Bending Solution 205
     4.8.3 Natural Vibrations 207
4.9 Solutions by Variational Methods 213
4.10 Chapter Summary.. 218

Chapter 5   Third-Order Beam Theories.................................. 221

5.1 Introduction 221
     5.1.1 Why a Third-Order Theory? 221
     5.1.2 Present Study 222

5.2   A General Third-Order Theory.................................. 223
    5.2.1   Kinematics... 223
    5.2.2   Equations of Motion 225
    5.2.3   Equations of Motion without Couple Stress
         Effects.. 228
    5.2.4   Constitutive Relations 228
5.3   A Third-Order Theory with Vanishing Shear
Stress on the Top and Bottom Faces 230
    5.3.1   The General Case 230
    5.3.2   The Reddy Third-Order Beam Theory
         (RBT)... 232
         5.3.2.1   Kinematics....................................... 232
         5.3.2.2   Equations of motion using
               Hamilton's principle........................ 233
         5.3.2.3   Constitutive relations 235
         5.3.2.4   Equations of motion in terms of
               the generalized displacements: the
               general case..................................... 238
         5.3.2.5   Equations of motion in terms of
               the generalized displacements: the
               linear case 239
    5.3.3   Levinson's Third-Order Beam Theory
         (LBT) 241
         5.3.3.1   Equations of motion 241
         5.3.3.2   Equations of motion in terms of
               the displacements............................ 242
         5.3.3.3   Equations of motion for the linear
               case 243
         5.3.3.4   Equations of equilibrium for the
               linear case 243
         5.3.3.5   Linearized equations without the
               couple stress effect 244
5.4   Exact Solutions for Bending 244
    5.4.1   The Reddy Beam Theory 244
    5.4.2   The Simplified RBT 248
         5.4.2.1   FGM beams 248
         5.4.2.2   Homogeneous beams........................ 250
    5.4.3   The Levinson Beam Theory 251
         5.4.3.1   FGM beams 251
         5.4.3.2   Homogeneous beams........................ 253
5.5   Bending Relationships for the RBT 259
    5.5.1   Preliminary Comments................................... 259
    5.5.2   Summary of Equations 259
    5.5.3   General Relationships..................................... 260

5.5.4   Bending Relationships for the
Simplified RBT..263

5.5.5   Relationships between the LBT
and the CBT 265

5.5.6   Numerical Examples...................................267

5.5.7   Buckling Relationships 273

5.5.7.1   Summary of equations of the CBT..273

5.5.7.2   Summary of equations of the RBT..274

5.6   Navier Solutions..280

5.6.1   The Reddy Beam Theory (RBT) 280

5.6.1.1   Bending analysis...........................284

5.6.1.2   Natural vibration...........................285

5.6.2   The Levinson Beam Theory (LBT)...............285

5.6.3   Numerical Results......................................286

5.7   Solutions by Variational Methods...........................292

5.8   Chapter Summary...298

Chapter 6   Classical Theory of Circular Plates 303

6.1   General Relations..303

6.1.1   Preliminary Comments...............................303

6.1.2   Kinematic Relations 304

6.1.2.1   Modified Green–Lagrange strains 304

6.1.2.2   Curvature tensor............................304

6.1.3   Stress–Strain Relations..............................305

6.1.4   Strain Energy Functional 305

6.2   Governing Equations of the CPT.............................306

6.2.1   Displacements and Strains..........................306

6.2.2   Equations of Motion 306

6.2.3   Isotropic Constitutive Relations 310

6.2.4   Displacement Formulation of the CPT..........311

6.2.5   Mixed Formulation of the CPT 312

6.3   Solutions for Homogeneous Plates in Bending..........314

6.3.1   Governing Equations 314

6.3.2   Exact Solutions..315

6.3.3   Numerical Examples....................................316

6.4   Bending Solutions for FGM Plates 323

6.4.1   Governing Equations 323

6.4.2   Exact Solutions..324

6.5   Buckling and Natural Vibration 332

6.5.1   Buckling Solutions.....................................332

6.5.2   Natural Frequencies....................................335

6.6   Variational Solutions...340

6.6.1   Introductory Comments 340

    6.6.2   Variational Statement 340
    6.6.3   The Ritz Method 341
    6.6.4   The Galerkin Method 344
    6.6.5   Natural Frequencies and Buckling Loads 348
        6.6.5.1   Variational statement 348
6.7   Chapter Summary 352

Chapter 7   First-Order Theory of Circular Plates 355

7.1   Governing Equations 355
    7.1.1   Displacements and Strains 355
    7.1.2   Equations of Motion 356
    7.1.3   Plate Constitutive Relations 357
    7.1.4   Equations of Motion in Terms of the
        Displacements 358
        7.1.4.1   The general case 358
        7.1.4.2   Nonlinear equations of equilibrium .. 359
        7.1.4.3   Linear equations of equilibrium
                without couple stress 360
        7.1.4.4   Linear equations of equilibrium
                without couple stress and FGM 360
7.2   Exact Solutions of Isotropic Circular Plates 360
7.3   Exact Solutions for FGM Circular Plates 365
    7.3.1   Governing Equations 365
    7.3.2   Exact Solutions 366
    7.3.3   Examples 368
7.4   Bending Relationships between CPT and FST 372
    7.4.1   Summary of the Governing Equations 372
    7.4.2   Relationships 372
    7.4.3   Examples 374
7.5   Bending Relationships for Functionally Graded
   Circular Plates 379
    7.5.1   Introduction 379
    7.5.2   Summary of Equations 379
    7.5.3   Relationships between the CPT and FST 381
7.6   Chapter Summary 392

Chapter 8   Third-Order Theory of Circular Plates 395

8.1   Governing Equations 395
    8.1.1   Preliminary Comments 395
    8.1.2   Displacements and Strains 395
    8.1.3   Equations of Motion 396
    8.1.4   Plate Constitutive Equations 398
8.2   Exact Solutions of the TST 400

8.3   Relationships between CPT and TST........................ 403
      8.3.1   Bending Relationships 403
              8.3.1.1   Classical plate theory (CPT) 403
              8.3.1.2   Third-order shear deformation plate
                        theory (TST) 404
      8.3.2   Relationships 405
      8.3.3   Buckling Relationships 409
              8.3.3.1   Governing equations 409
              8.3.3.2   Relationship between CPT
                        and FST.. 410
              8.3.3.3   Relationship between CPT
                        and TST 411
8.4   Chapter Summary.. 414

Chapter 9   Finite Element Analysis of Beams 415

9.1   Introduction 415
      9.1.1   The Finite Element Method 415
      9.1.2   Interpolation Functions 417
      9.1.3   Present Study... 422
9.2   Displacement Model of the CBT............................ 423
      9.2.1   Governing Equations and Variational
              Statements.. 423
      9.2.2   Finite Element Model............................... 425
9.3   Mixed Finite Element Model of the CBT 427
      9.3.1   Variational Statements 427
      9.3.2   Finite Element Model............................... 429
9.4   Displacement Finite Element Model of the TBT....... 430
      9.4.1   Governing Equations and Variational
              Statements.. 430
      9.4.2   The Finite Element Model 431
9.5   Mixed Finite Element Model of the TBT 433
      9.5.1   Governing Equations and Variational
              Statements.. 433
      9.5.2   Finite Element Model............................... 435
9.6   Displacement Finite Element Model of the RBT....... 435
      9.6.1   Governing Equations 435
      9.6.2   Weak Forms... 436
      9.6.3   Finite Element Model............................... 437
9.7   Time Approximation (Full Discretization)................ 439
      9.7.1   Introduction... 439
      9.7.2   Newmark's Method 439
      9.7.3   Fully Discretized Equations...................... 440
9.8   Solution of Nonlinear Algebraic Equations 442

9.8.1   Preliminary Comments 442
9.8.2   Direct Iteration Procedure 443
9.8.3   Newton's Iteration Procedure 444
9.8.4   Load Increments 450
9.9   Tangent Stiffness Coefficients 451
9.9.1   Definition of Tangent Stiffness Coefficients 451
9.9.2   The Displacement Model of the CBT 451
9.9.3   The Mixed Model of the CBT 453
9.9.4   The Displacement Model of the TBT 453
9.9.5   The Mixed Model of the TBT 454
9.9.6   The Displacement Model of the RBT 454
9.10 Post-Computations 455
9.10.1 General Comments 455
9.10.2 CBT Finite Element Models 455
9.10.3 TBT Finite Element Models 456
9.10.4 RBT Displacement Model 457
9.11 Numerical Results 458
9.11.1 Geometry and Boundary Conditions 458
9.11.2 Material Constitution 458
9.11.3 Examples 459
9.12 Chapter Summary 482

Chapter 10   Finite Element Analysis of Circular Plates 487

10.1 Introductory Remarks 487
10.2 Displacement Model of the CPT 488
10.2.1 Weak Forms 488
10.2.2 Finite Element Model 490
10.3 Mixed Model of the CPT 491
10.3.1 Weak Forms 491
10.3.2 Finite Element Model 492
10.4 Displacement Model of the FST 494
10.4.1 Weak Forms 494
10.4.2 Finite Element Model 496
10.5 Displacement Model of the TST 498
10.5.1 Variational Statements 498
10.5.2 Finite Element Model 501
10.6 Tangent Stiffness Coefficients 504
10.6.1 Preliminary Comments 504
10.6.2 The Displacement Model of the CPT 504
10.6.3 The Mixed Model of the CPT 504
10.6.4 The Displacement Model of the FST 505
10.6.5 The Displacement Model of the TST 505
10.7 Numerical Results 506

10.7.1 Preliminary Comments.................................... 506
10.7.2 Linear Analysis... 507
10.7.3 Nonlinear Analysis without Couple Stress
        Effect 510
10.7.4 Nonlinear Analysis with Couple Stress
        Effect 517
10.8 Chapter Summary.. 522

References 525

Papers with Collaborators 535

Answers 541

Index.. 549

Preface

The motivation for composing this book has come from the need to fill the gap in the literature and provide a comprehensive treatment of the classical and shear deformation theories of beams and axisymmetric circular plates in one volume. The book is a compendium of all related works by the author and his colleagues on the subject over his lifetime. The book contains detailed derivations of the governing equations, analytical solutions, variational solutions, and numerical solutions (FEM) of the classical and shear deformation theories of beams and axisymmetric circular plates. The readers and users will benefit to have such a comprehensive book available in the literature as a reference for finding the governing equations, analytical and numerical solutions of bending, vibration, and buckling for problems with various boundary conditions.

In this present book, classical and shear deformation theories are presented, accounting for through-thickness variation of two-constituent functionally graded material, modified couple stress (i.e., strain gradient), and the von Kármán nonlinearity. Analytical solutions of the linear theories and finite element analysis of linear and nonlinear theories are included.

Chapter 1 is devoted to a brief review of mechanics preliminaries that include vectors and tensors, summation convention, governing equations of solid mechanics, an introduction to functionally graded materials (FGMs), and the modified couple stress model. A reader familiar with these may skip this chapter but it is recommended that a casual walk through the chapter is beneficial to see the notation used. Chapter 2 deals with the concepts of work and energy, strain energy, and virtual work, and elements of the calculus of variations and variational principles of solid and structural mechanics. These ideas are useful in the development of variationally consistent theories of beams and plates and their solution by direct variational methods such as the Ritz and Galerkin methods.

The main thesis of the book begins with Chapter 3, which presents a detailed discussion of the classical beam theory (CBT), including kinematics, constitutive models, and governing equations of motion. The governing equations of plate strips (i.e., cylindrical bending of plates) are also discussed. The chapter also contains analytical and numerical solutions of the linearized equations. Analytical solutions include solutions by direct integration as well as the Navier method. The Ritz and other variational methods are introduced in this chapter and illustrated by their applications to CBT. Chapters 4 and 5 follow the same sequence of developments for first-order (TBT) and third-order (RBT) theories, respectively, of beams. A major feature of these chapters is the development of the algebraic relationships between the solutions of the

TBT and CBT and RBT and CBT. That is, if one has the analytical solution of a beam problem using the CBT, the relationships allow one to obtain the solutions of the same problem by the TBT and RBT models.

Chapters 6–8 are dedicated to the classical, first-order, and third-order theories of axisymmetric circular plates, following the same sequence of steps (i.e., derivation of equations, analytical and variational solutions, and relationships between classical and shear deformation theories).

Finally, finite element formulations and numerical solutions of beams and axisymmetric circular plates, respectively, are presented in Chapters 9 and 10. These chapters contain extensive theoretical results in the form of weak-form development, finite element models, tangent stiffness coefficient derivations, and numerical results for linear and nonlinear analysis of beams and circular plates. To keep the size of the book within reasonable limits, numerical results in these chapters are limited to static bending analysis.

The major feature of the book is the comprehensive treatment (within the scope of the book) of the subject matter. The readers never have to consult another source to follow the developments; although many references are provided mostly to acknowledge the developments, it is not necessary to read them to follow what is presented here. Of course, having a background in mechanics of materials, elasticity, and a first course on the finite element method would help. Historical notes are included in several places to make it interesting and derive some level of appreciation for those who have contributed to the subject matter covered herein. Few exercise problems are also included, but extensions and applications of the theories developed herein are possible. For such tasks, this book is an excellent reference to researchers.

As already stated, many of the results included herein were obtained during the course of the author's lifetime in collaboration with his students, postdocs, and colleagues around the world. For the reason of missing some, the names (a large number) of these individuals are not listed here. Instead, a list of papers coauthored with them on topics related to the subject matter are included at the back of the book. *The author is very appreciative of the friendship and collaboration of all these colleagues over the years.* The author is pleased to acknowledge the help of Dr. Eugenio Ruocco, Dr. Praneeth Nampally, Mr. Ho Yong Shin, and Ms. Alekhya Banki with the proofreading of the manuscript prior to its publication. A book of this nature, full of mathematical statements, is bound to have some typos and errors. The author requests the readers to send any comments and corrections to *jnreddy@tamu.edu*.

J. N. Reddy
College Station, Texas
http://mechanics.tamu.edu

Anyone who has never made a mistake has never tried anything new.
Albert Einstein

List of symbols used

The meaning of various symbols used in the book for some important quantities is defined in the following table. The list is not exhaustive (c_i and K_i are constants used at various places).

Symbol	Meaning
a	Outer radius of a circular or annular plate
a_{ij}	Coefficients of matrix $[A] = \mathbf{A}$
A	Area of cross section of a beam
$A_{xx}, A_{xz}, \ldots$	Axial and shear stiffness coefficients
b	Inner radius of an annular plate; width of a beam cross section
B_{xx}	Bending-stretching coupling stiffness
c_f	Modulus of elastic foundation per unit length
c_v, c_p	Specific heat at constant volume and pressure, respectively
$d\Gamma$	Surface element
dA	Area element ($dA = dxdy$)
$d\Omega$	Area element ($d\Omega = dxdy$) or volume element ($d\Omega = dxdydz$)
D_{xx}, D_{xz}	Bending and higher-order shear stiffness coefficients
$\bar{D}_{xx}$	Effective stiffness coefficient, $\bar{D}_{xx} = D_{xx} - \alpha F_{xx}$,
D^e_{xx}	Effective stiffness coefficient, $D^e_{xx} = D_{xx} + A_{xy}$, A_{xy} being the stiffness coefficient due to couple stress; also, $\hat{D}_{xx} = \bar{D}_{xx} - \alpha \bar{F}_{xx}$
D^*_{xx}	Effective stiffness coefficient, $D^*_{xx} = D_{xx}A_{xx} - B_{xx}B_{xx}$
$\hat{\mathbf{e}}_i$	Basis vector in the x_i-direction
$(\hat{\mathbf{e}}_r, \hat{\mathbf{e}}_\theta, \hat{\mathbf{e}}_z)$	Basis vectors in the (r, θ, z) system
$(\hat{\mathbf{e}}_x, \hat{\mathbf{e}}_y, \hat{\mathbf{e}}_z)$	Basis vectors in the (x, y, z) system
$(\hat{\mathbf{e}}_1, \hat{\mathbf{e}}_2, \hat{\mathbf{e}}_3)$	Basis vectors in the (x_1, x_2, x_3) system
E	Modulus of elasticity
E_1, E_2	Moduli of elasticity of a functionally graded structure or an orthotropic material
$E_{xx}, E_{yy}, \ldots$	Green strain components in rectangular Cartsian system; E_{xx} are the higher-order stiffness coefficient
$E_{rr}, E_{\theta\theta}, \ldots$	Green strain components in cyndrical coordinate system
$\mathbf{E}$	Green–Lagrange strain tensor
$\mathbf{f}$	Body force vector
f_x, f_y, f_z	Body force components in the x, y, and z directions
$F_{xx}, F_{rr}, \ldots$	Higher-order stress resultants
$F^\alpha_i, \mathbf{F}^\alpha$	Finite element force vectors
$\mathbf{F}$	Deformation gradient, $\mathbf{F} = (\boldsymbol{\nabla}\mathbf{x})^{\mathrm{T}}$
G	Shear modulus
h	Height of a beam or thickness of a plate; length of a finite element
$H_{xx}, H_{rr}, \ldots$	Higher-order stress resultants
I	Second moment of area, $I = bh^3/12$
$\mathbf{I}$	Unit second-order tensor
J	Determinant of $\mathbf{J}$ (Jacobian)
J_n	Bessel function of the first kind and of the nth order
$\mathbf{J}$	Jacobian (of transformation) matrix

Symbol	Meaning
k	Extensional spring constant
k_R	Rotational spring constant
K	Kinetic energy; bulk modulus
K_s	Shear correction coefficient
$\mathbf{K}$	Finite element stiffness matrix
$K_{ij}^{\alpha\beta}, \mathbf{K}^{\alpha\beta}$	Finite element stiffness submatrices
l, ℓ	Material length scale used couple stress model
L	Length of a beam
$\mathbf{m}$	Couple stress tensor
$\mathbf{M}$	Finite element mass matrix
$M_{ij}^{\alpha\beta}, \mathbf{M}^{\alpha\beta}$	Finite element mass submatrices
$\mathcal{M}_{xy}$	Couple stress
$M_{xx}, M_{rr}, \ldots$	Bending stress resultants
n	Index/exponent used in power-law model
$\hat{\mathbf{n}}$	Unit normal vector in the current configuration
n_i	ith component of the unit normal vector $\hat{\mathbf{n}}$
(n_x, n_y, n_z)	Components of the unit normal vector $\hat{\mathbf{n}}$
$N_{xx}, N_{rr}, \ldots$	Stretching stress resultants
$P_{xx}, P_{rr}, \ldots$	Higher-order stress resultants
q	Distributed transverse load per unit length
Q_{ij}	Plane stress-reduced elastic coefficients
r	Radial coordinate in the cylindrical polar system; $r = |\mathbf{r}|$
$\mathbf{r}$	Position vector in cylindrical coordinates, $\mathbf{x}$
(r, θ, z)	Cylindrical coordinate system
R	Outer radius of a circular plate
t	Time
$\mathbf{t}$	Stress vector; traction vector
$\mathbf{t}_i$	Stress vector on x_i-plane, $\mathbf{t}_i = \sigma_{ij}\hat{\mathbf{e}}_j$
T	Temperature
u	Axial displacement
$\mathbf{u}$	Displacement vector
u_r, u_θ, u_z	Components of a displacement vector $\mathbf{u}$ in a cylindrical coordinate system
u_x, u_y, u_z	Components of a displacement vector $\mathbf{u}$ in a rectangular Cartesian coordinate system
U	Strain energy of a body
U_0	Strain energy density of a body
v	Velocity, $v = |\mathbf{v}|$
(v_1, v_2, v_3)	Components of velocity vector $\mathbf{v}$ in (x_1, x_2, x_3) system
(v_r, v_θ, v_z)	Components of velocity vector $\mathbf{v}$ in (r, θ, z) system
$\mathbf{v}$	Velocity vector, $\mathbf{v} = \frac{D\mathbf{x}}{Dt}$
$\mathbf{v}_n$	Velocity vector normal to the plane (whose normal is $\hat{\mathbf{n}}$)
V	Potential energy due to external loads; shear force
V_1, V_2	Material volume fractions for functionally graded material
V_{eff}	Effective shear force
V_E	Work done by external forces $(= W_E)$
w	Transverse displacement component
W_E	External work done by forces
W_I	Internal work stored in the body
$\mathbf{x}$	Position vector in the current configuration
(x, y, z)	Rectangular Cartesian coordinates
(x_1, x_2, x_3)	Rectangular Cartesian coordinates (spatial)
(X_1, X_2, X_3)	Rectangular Cartesian coordinates (material)
Y_n	Bessel function of the second kind and of the nth order

Greek symbols

Symbol	Meaning
α	Angle; parameter in time approximations scheme; also, $\alpha = 4/3h^2$, h being the total height of the beam or plate
α_T	Coefficient of thermal expansion
β	Heat transfer coefficient (also other uses); also $\beta = 3\alpha = 4/h^2$
γ	Parameter in a time approximation scheme
Γ	Total boundary
δ	Dirac delta; variational symbol
δ_{ij}	Components of the unit tensor, $\mathbf{I}$ (Kronecker delta)
$\Delta, \boldsymbol{\Delta}$	Increment; generalized displacement vector
ϵ	Tolerance specified for nonlinear convergence
ε_{ij}	Infinitesimal strain components
ε_{ijk}	Alternating symbol
ζ, η	Natural (normalized) coordinate
θ	Angular coordinate in the cylindrical and spherical coordinate systems
λ	Lamé constant; eigenvalue
μ	Lamé constant
ν, ν_{ij}	Poisson's ratio; Poisson's ratios for an orthotropic material
ξ	Natural (normalized) coordinate
Π	Total potential energy
ρ	Mass density
$\boldsymbol{\sigma}$	Stress tensor
σ_{ij}	Components of the stress tensor in the rectangular coordinate system (x_1, x_2, x_3)
$\sigma_{rr}, \sigma_{\theta\theta}, \sigma_{r\theta}, \cdots$	Components of the stress tensor $\boldsymbol{\sigma}$ in the cylindrical coordinate system (r, θ, z)
τ	Shear stress
$\boldsymbol{\tau}$	Viscous stress tensor
ϕ	Angular coordinate in the spherical coordinate system
ϕ_i	Hermite cubic interpolation functions
ϕ_x, ϕ_r	Rotation functions
$\chi, \boldsymbol{\chi}$	Curvature and curvature tensor
ψ	Warping function; stream function
ψ_i	Lagrange interpolation functions
ω	Angular velocity
$\boldsymbol{\omega}$	Rotation vector
Ω	Domain of a problem; natural frequency
$\boldsymbol{\Omega}$	Spin tensor or skew symmetric part of the velocity
ω_i	Components of vorticity vector $\boldsymbol{\omega}$ in the rectangular coordinate system

Other symbols

Symbol	Meaning
∇	Gradient operator (with respect to $\mathbf{X}$)
∇_x	Gradient operator with respect to $\mathbf{x}$
∇^2	Laplace operator, $\nabla^2 = \nabla \cdot \nabla$
$[\,]$	Matrix of components of the enclosed tensor
$\{\,\}$	Column of components of the enclosed vector
$\cdot$	Symbol for the dot product or scalar product
$\times$	Symbol for the cross product or vector product

Table 1

Conversion factors

s = second; lb = pound; in = inch; ft = foot; hp = horse power;
kg = kilogram (= 10^3 grams); m = meter; mm = millimeter (10^{-3} m);
N = Newton; W = Watt; Pa = Pascal = N/m^2;
kN = 10^3 N; MN = 10^6 N; MPa = 10^6 Pa; GPa = 10^9 Pa

Quantity	US customary unit	SI equivalent
Mass	lb (mass)	0.4536 kg
Length	in	25.4 mm
	ft	0.3048 m
Density	lb/in^3	27.68×10^3 kg/m^3
Force	lb (force)	4.448 N
Pressure or stress	lb/in^2 (psi)	6.895 kN/m^2
Moment or torque	lb in	0.1130 Nm
Power	ft lb/s	1.356 W
	hp (550 ft lb/s)	745.7 W

Note:

Quotes by various people included in this book were found at different web sites; for example, visit:

http://naturalscience.com/dsqhome.html,

http://thinkexist.com/quotes/david˙hilbert/, and *http://www.yalescientific. org/.*

The historical notes included in various footnotes can be found at different websites, especially Wikipedia, https://en.wikipedia.org/.

This author is motivated to include the quotes for their wit and wisdom. The author cannot vouch for the accuracy of the quotes or the historical notes. The reason for the inclusion of the historical notes is to remind the readers that we are "standing on the shoulders" of many giants before us.

A few words of caution about Wikipedia: Most of the references cited there belong to the authors who contributed to the subject matter, and they are neither authoritative nor original contributions to the subject; some selfish authors tried to promote their own work at the expense of not giving credit to the original contributors. In addition, the readers should be very careful in accepting what is found there as technically accurate. It is advised that the readers consult the papers and books by well-known researchers on the technical topic/subject.

If you are not willing to learn, no one can help you. If you are determined to learn, no one can stop you.   Zig Ziglar

About the Author

J. N. Reddy, the O'Donnell Foundation Chair IV Professor in J. Mike Walker '66 Department of Mechanical Engineering at Texas A&M University, is a highly cited researcher, author of 24 textbooks and over 750 journal papers, and a leader in the applied mechanics field for nearly 50 years. He is known worldwide for his significant contributions to the field of applied mechanics through the authorship of widely used textbooks on mechanics of materials, continuum mechanics, linear and nonlinear finite element analyses, energy principles and variational methods, and composite materials and structures. His pioneering works on the development of shear deformation theories of beams, plates, and shells (that bear his name in the literature as *the Reddy third-order plate theory* and *the Reddy layerwise theory*), and nonlocal and non-classical continuum mechanics have had a major impact, and have led to new research developments and applications. Some of the ideas on shear deformation theories and penalty finite element models of fluid flows have been implemented into commercial finite element computer programs like Abaqus, NISA, and HyperXtrude (Altair).

Recent honors and awards include: the 2019 Timoshenko Medal from the American Society of Mechanical Engineers, 2018 Theodore von Kármán Medal from the Engineering Mechanics Institute of the American Society of Civil Engineers, the 2017 John von Neumann Medal from the U.S. Association of Computational Mechanics, the 2016 Prager Medal from the Society of Engineering Science, the 2016 Thomson Reuters IP and Science's Web of Science Highly Cited Researchers – Most Influential Minds, and the 2016 ASME Medal from the American Society of Mechanical Engineers. He is a member of the US National Academy of Engineering and foreign fellow of the Canadian Academy of Engineering, the Chinese Academy of Engineering, the Brazilian National Academy of Engineering, the Indian National Academy of Engineering, the Royal Academy of Engineering of Spain, the European Academy of Sciences, and the Academia Scientiarum et Artium Europaea (the European Academy of Sciences and Arts). For additional details, visit http://mechanics.tamu.edu/.

Teachers must make selfless efforts to make their own hard-earned knowledge and expertise accessible to motivated students. They must derive personal reward, not from demonstrating to students that they are experts, but from being able to explain complex ideas in a way that they can comprehend and ultimately master the ideas and utilize them in their professional life. I also believe that any measure of success must include giving back to the community.

There is no "complete" mathematical model of anything we study. We only try to "improve" on what we already know. Junuthula Narasimha (J. N.) Reddy

1 Mechanics Preliminaries

Minds are like parachutes. They only function when they are open.
James Dewar

1.1 GENERAL COMMENTS

Engineers of all types contribute to science and technology for the benefit of mankind. They construct mathematical models, develop analytical and numerical approaches and methodologies, and design and manufacture various types of devices, systems, or processes. Mathematical models, engineering experiments, and numerical simulations constitute the three main pillars of scientific activity. Engineering analysis is an aid to designing systems for specific functionalities, and they involve (1) mathematical model development, (2) data acquisition by measurements, (3) numerical simulations, and (4) validation of the results in light of any experimental evidence. The most challenging task for engineers is to identify a suitable mathematical model of the system's behavior. It is in this connection this book is composed to provide interested readers with the theories and analyses of beams and circular plates. That is, we develop appropriate mathematical models (i.e., governing equations) for bending, buckling, natural vibration, and transient (to a limited extent) analyses of beams and axisymmetric circular plates. The book contains an up-to-date, relatively complete treatment of these specialized topics.

It is important to understand that all models, mathematical or experimental, are required to satisfy the laws of physics; beyond that, they are only approximate representations of the actual system or process. *There is no exact model of anything we study, and we only build on what we know to make them better for the intended purpose of the study.* In particular, continuum mechanics is not an exact science; as it stands now, it is not complete, and it will never be complete as we explore new phenomena. However, continuum mechanics is responsible for many advances in science and engineering, and we continue to build on it and make it better. Thus, the theories and analyses presented in this book for beams and axisymmetric circular plates form a basis for future developments.

This chapter is devoted to a review of preliminaries from engineering mechanics. The review includes: vectors and tensors, the definitions of the Green–Lagrange strain tensor, infinitesimal strain tensor, measures of stress, equations of elasticity, and stress–strain relations for plane stress problems, an introduction to functionally graded materials, and an introduction to the modified couple stress concept. These preliminaries are used in the coming

DOI: 10.1201/9781003240846-1

chapters to develop the theories of beams and axisymmetric circular plates. Readers familiar with these may skip this chapter, but it is advised that they browse through the chapter to understand the notation used.

1.2   BEAMS AND PLATES

Beams are structural members that have a ratio of length-to-cross-sectional dimensions very large, say, 10 to 100 or more and subjected to forces, both along and transverse to the length and moments that tend to rotate them about an axis perpendicular to their length. When all applied loads are along the length only, they are called *bars* (i.e., bars experience only tensile or compressive stresses and strains and no bending deformation). *Cables* (or ropes) may be viewed as a very flexible form of bars, which can only take tension and not compression. *Plates* are a two-dimensional version of beams, with plate in-plane dimensions much larger in order of magnitude than the thickness. Thus, plates are thin bodies subjected to forces, in the plane as well as in the direction normal to the plane and bending moments about either axis in the plane. Geometrically, plates can be used in different shapes: circular, rectangular, triangular, rhombic, or polygonal. Ancient Egyptians, Greeks, Indus valley civilizations, and Romans used beams and plates of various shapes in their temples, monumental buildings, and tombs. Because of their geometry and loads applied, the beams and plates are stretched and bent (by design, in infinitesimally small magnitudes) from their original shapes. Such members are known as *structural elements* and their study constitutes *structural mechanics*, which is a subset of solid mechanics. The difference between structural elements and three-dimensional solid bodies, such as solid blocks and spheres that have no restrictions on their geometric make up, is that the latter may change their original geometry, but they may not show significant "bending" deformation.

All deformable solids can be analyzed for stress and deformation using the elasticity equations. However, the original geometry, induced deformation, and stress fields can be predicted, for most practical engineering problems involving structural elements, with simplified theories in the place of the three-dimensional elasticity theory. Beams (including frames), plates, and shells are analyzed using structural theories that are derived from three-dimensional elasticity theory by making certain simplifying assumptions concerning the deformation (kinematics) and stress states in these members. The development of such theories dates back to Leonardo da Vinci[1], Galileo Galilei[2],

---

[1]In his "Codice Atlantico," Leonardo Da Vinci (1452–1519) made the first attempt known to us to correlate bending deflection and geometry for a beam.

[2]The book "Discorsi e dimostrazioni matematiche intorno a due nuove scienze attinenti la mecanica e i moti locali" by Galileo Galilei (1564–1642) is considered to be the first book devoted to structural mechanics.

Jacob Bernoulli[3], and Leonhard Euler[4]. The first one is *the Euler–Bernoulli beam theory*, a theory that is covered in all undergraduate mechanics of materials books. In the Euler–Bernoulli beam theory, the transverse shear strain is neglected, making the beam infinitely rigid in the transverse direction. The second one is popularly known as *the Timoshenko beam theory* [1, 2][5] which accounts for the transverse shear strain (γ_{xz}). In a recent paper, Elishakoff [3] pointed out that the beam theory that incorporates both the rotary inertia and shear deformation as is known presently, with shear correction coefficient included, should be referred to as the *Timoshenko–Ehrenfest beam theory* because the original paper published by Timoshenko had a coauthor by name Paul Ehrenfest. In view of the fact that many people have contributed to the development of shear deformation theories, Reddy [4] coined the phrase *first-order shear deformation theory*. Unfortunately, most people do not read original papers they cite, and errors in giving the due credit are propagated from one writing to the next (consult the article by Reddy and Srinivasa [5] for some misattributions and misnomers in mechanics).

All modern developments are dedicated to refinements to the above stated theories, by expanding the displacements in terms of higher-order terms and accounting for other non-classical continuum mechanics aspects (e.g., stress and strain gradient effects and material length scales). For example, a general higher-order theory is of the form

$$\mathbf{u} = u_x\,\hat{\mathbf{e}}_x + u_y\,\hat{\mathbf{e}}_y + u_z\,\hat{\mathbf{e}}_z, \tag{1.2.1}$$

where

$$u_x = \sum_{i=0}^{m} z^i \phi_x^{(i)}(x,t), \quad u_y = 0, \quad u_z = \sum_{i=0}^{p} z^i \psi_z^{(i)}(x,t). \tag{1.2.2}$$

Here $\phi_x^{(0)} = u$ and $\psi_z^{(0)} = w$ denote the midplane displacements along the x and z directions, respectively, and $\phi_x^{(i)}$ and $\psi_x^{(i)}$ are the higher-order terms, which can be mathematically interpreted as higher-order generalized displacements with the meaning

$$\phi_x^{(i)} = \frac{1}{(i)!}\left(\frac{\partial^i u_1}{\partial z^i}\right)_{z=0}, \quad \psi_z^{(i)} = \frac{1}{(i)!}\left(\frac{\partial^i u_3}{\partial z^i}\right)_{z=0}. \tag{1.2.3}$$

---

[3] Jacob Bernoulli (1655–1705) was one of the many prominent Swiss mathematicians in the Bernoulli family. Jacob Bernoulli (1655–1705), along with his brother Johann Bernoulli (1667–1748), was one of the founders of calculus of variations. Jacob Bernoulli and Daniel Bernoulli (1700–1782) (son of Johann Bernoulli) are credited for initiating a beam theory.

[4] Leonhard Euler (1707–1783) was a pioneering Swiss mathematician and physicist who put forward the theory in 1750: "Methodus inveniendi lineas curvas maximi minimive proprietate gaudentes," *Leonhardi Euleri Opera Omnia Ser.* I, 14, 1744.

[5] Stephan Prokofyevich Timoshenko (1878–1972) was a Ukrainian, Russian, and American engineer and academician, who is considered to be the father of modern engineering mechanics.

For a general third-order beam theory, we have $m = 3$ and $p = 2$ in Eq. (1.2.2). The third-order beam theory of Reddy, derived from this third-order plate theory (see Reddy [6]–[9] and Heyliger and Reddy [10]), adopts a displacement field that is a special case of Eq. (1.2.1) and imposes zero transverse shear stress conditions on the bounding planes (i.e., top and bottom faces) of the beam to express the variables introduced with the higher order terms in terms of the variables that appear in the Euler–Bernoulli and Timoshenko beam theories.

In the remaining part of this chapter, we review some mathematical preliminaries involving calculus of vectors and tensors, and equations of solid mechanics that are useful in the sequel. The topic of vectors and tensors is in itself a major subject, and books are devoted to its treatment. Here we assume that the readers are sufficiently familiar with the subject, and we only review some useful concepts. The equations of solid mechanics include the strain–displacement relations, equations of motion in terms of stresses, and stress–strain relations. Other mechanics preliminaries needed in this book, such as the energy and variational principles (including the principles of virtual displacements and the minimum total potential energy, and Hamilton's principle), are presented in Chapter 2. The principle of virtual displacements plays a major role in the development of the governing equations of higher-order beam and plate theories presented in this book.

1.3 VECTORS AND TENSORS

1.3.1 VECTORS AND COORDINATE SYSTEMS

The elementary notion of a vector as being one with "magnitude" and "direction" is a geometric concept and applies to directed line segments. In the broader context, vectors can be quantities, such as functions and matrices, and satisfy the rules of vector addition and multiplication of a vector by a scalar. The terms "magnitude" and "direction" take different meaning in different contexts. Engineering examples of vectors are provided by displacements, velocities, forces, heat flux, and so on; and they are endowed with a direction and a magnitude. Note that entities like speed and temperature are scalars (i.e., they have only magnitudes but no directions). Stress as a measure of force per unit area is a vector (stress vector) whereas representation of a stress tensor requires not only a stress vector but also the specification of the area on which it acts. In written or typed material, a vector or tensor is denoted by a boldface letter, $\mathbf{A}$, such as used in this book, and its magnitude is denoted by $|\mathbf{A}|$ or just A.

We begin with an orthonormal Cartesian coordinate system (x_1, x_2, x_3) with the following orthonormal basis vectors:

$$\{\hat{\mathbf{e}}_x, \hat{\mathbf{e}}_y, \hat{\mathbf{e}}_z\} \qquad \text{or} \qquad \{\hat{\mathbf{e}}_1, \hat{\mathbf{e}}_2, \hat{\mathbf{e}}_3\}. \tag{1.3.1}$$

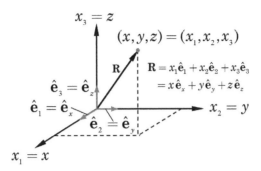

Figure 1.3.1: Rectangular Cartesian coordinates.

The associated Cartesian coordinates are denoted by $(x, y, z) = (x_1, x_2, x_3)$. The familiar rectangular Cartesian coordinate system is shown in Fig. 1.3.1. We shall always use right-handed coordinate systems.

We can represent any vector $\mathbf{A}$ in three-dimensional space as a linear combination of the orthonormal basis as

$$\mathbf{A} = A_1\hat{\mathbf{e}}_1 + A_2\hat{\mathbf{e}}_2 + A_3\hat{\mathbf{e}}_3. \tag{1.3.2}$$

The vectors $A_1\hat{\mathbf{e}}_1$, $A_2\hat{\mathbf{e}}_2$, and $A_3\hat{\mathbf{e}}_3$ are called the *vector components* of $\mathbf{A}$, and A_1, A_2, and A_3 are called *scalar components* of $\mathbf{A}$ associated with the basis $(\hat{\mathbf{e}}_1, \hat{\mathbf{e}}_2, \hat{\mathbf{e}}_3)$. Also, we use the notation $\mathbf{A} = (A_1, A_2, A_3)$ to denote a vector by its components. Thus, the position vector $\mathbf{R}$ can be written as $\mathbf{R} = x_1\hat{\mathbf{e}}_1 + x_2\hat{\mathbf{e}}_2 + x_3\hat{\mathbf{e}}_3$.

1.3.2 SUMMATION CONVENTION

Equation (1.3.2) can be expressed as

$$\mathbf{A} = \sum_{i=1}^{3} A_i\,\hat{\mathbf{e}}_i, \tag{1.3.3}$$

which can be shortened, by omitting the summation symbol, and understanding that summation over the range of the index is implied when an index is repeated, to

$$\mathbf{A} = A_i\,\hat{\mathbf{e}}_i. \tag{1.3.4}$$

The repeated index is called *dummy index* and thus can be replaced by *any other symbol that has not already been used*. Thus we can also write

$$\mathbf{A} = A_i\,\hat{\mathbf{e}}_i = A_m\,\hat{\mathbf{e}}_m, \text{ and so on.} \tag{1.3.5}$$

It is convenient at this time to introduce the Kronecker delta δ_{ij} and alternating symbol ε_{ijk} for representing the dot product and cross product of two orthonormal vectors in a right-handed basis system. We define the dot product $\hat{\mathbf{e}}_i \cdot \hat{\mathbf{e}}_j$ between the orthonormal basis vectors of a right-handed system as

$$\hat{\mathbf{e}}_i \cdot \hat{\mathbf{e}}_j \equiv \delta_{ij} = \begin{cases} 1, \text{ if } i = j, \text{ for any fixed value of } i, j \\ 0, \text{ if } i \neq j, \text{ for any fixed value of } i, j, \end{cases} \quad (1.3.6)$$

where δ_{ij} is called the *Kronecker delta symbol*. Similarly, we define the cross product $\hat{\mathbf{e}}_i \times \hat{\mathbf{e}}_j$ for a right-handed system as

$$\hat{\mathbf{e}}_i \times \hat{\mathbf{e}}_j \equiv \varepsilon_{ijk}\hat{\mathbf{e}}_k, \quad (1.3.7)$$

where

$$\varepsilon_{ijk} = \begin{cases} 1, \text{ if } i, j, k \text{ are in cyclic order} \\ \quad \text{and not repeated } (i \neq j \neq k), \\ -1, \text{ if } i, j, k \text{ are not in cyclic order} \\ \quad \text{and not repeated } (i \neq j \neq k), \\ 0, \text{ if any of } i, j, k \text{ are repeated.} \end{cases} \quad (1.3.8)$$

The symbol ε_{ijk} is called the *alternating symbol* or *permutation symbol*.

In an orthonormal basis, the scalar product $\mathbf{A} \cdot \mathbf{B}$ and vector product $\mathbf{A} \times \mathbf{B}$ can be expressed in the index form using the Kronecker delta symbol δ_{ij} and alternating symbol ε_{ijk} as

$$\mathbf{A} \cdot \mathbf{B} = (A_i\hat{\mathbf{e}}_i) \cdot (B_j\hat{\mathbf{e}}_j) = A_iB_j\delta_{ij} = A_iB_i, \quad (1.3.9)$$
$$\mathbf{A} \times \mathbf{B} = (A_i\hat{\mathbf{e}}_i) \times (B_j\hat{\mathbf{e}}_j) = A_iB_j\varepsilon_{ijk}\hat{\mathbf{e}}_k. \quad (1.3.10)$$

The Kronecker delta and the permutation symbol are related by the identity, known as the *ε-δ identity*:

$$\varepsilon_{ijk}\varepsilon_{imn} = \delta_{jm}\delta_{kn} - \delta_{jn}\delta_{km}. \quad (1.3.11)$$

Then the length of a vector in an orthonormal basis can be expressed as $A = \sqrt{\mathbf{A} \cdot \mathbf{A}} = \sqrt{A_iA_i} = \sqrt{A_1^2 + A_2^2 + A_3^2}$. Similarly, we have $R^2 = x_ix_i$.

1.3.3 STRESS VECTOR AND STRESS TENSOR

Consider the equilibrium of an element of a continuum acted upon by forces. The surface force acting on a small element of area in a continuous medium depends not only on the magnitude of the area but also upon the orientation of the area. It is customary to denote the direction of a plane area by means of a unit vector drawn normal to that plane (see Fig. 1.3.2). To fix the direction of the normal, we assign a *sense of travel* along the contour of the boundary of the plane area in question. The direction of the normal is taken by convention as that in which a right-handed screw advances as it is rotated according to the sense of travel along the boundary curve or contour (see Fig. 1.3.2). Let the unit normal vector be given by $\hat{\mathbf{n}}$. Then the area can be denoted by $\mathbf{s} = s\hat{\mathbf{n}}$.

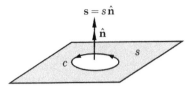

Figure 1.3.2: Plane area as a vector. Unit normal vector and sense of travel are shown.

If we denote by $\Delta\mathbf{F}(\hat{\mathbf{n}})$ the force on an elemental area $\hat{\mathbf{n}}\Delta s = \Delta\mathbf{s}$ located at the position $\mathbf{r}$ (see Fig. 1.3.3), the *stress vector* is defined as

$$\mathbf{t}(\hat{\mathbf{n}}) = \lim_{\Delta s \to 0} \frac{\Delta\mathbf{F}(\hat{\mathbf{n}})}{\Delta s}. \qquad (1.3.12)$$

We see that the stress vector is a point function of the unit normal $\hat{\mathbf{n}}$, which denotes the orientation of the surface Δs. The component of $\mathbf{t}$ that is in the direction of $\hat{\mathbf{n}}$ is called the *normal stress*. The component of $\mathbf{t}$ that is normal to $\hat{\mathbf{n}}$ (or in the plane) is called a *shear* stress.

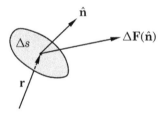

Figure 1.3.3: Force on an area element.

At a fixed point $\mathbf{r} = \mathbf{x}$ for each given unit vector $\hat{\mathbf{n}}$, there is a stress vector $\mathbf{t}(\hat{\mathbf{n}})$ acting on the plane normal to $\hat{\mathbf{n}}$. To establish a relationship between $\mathbf{t}$ and $\hat{\mathbf{n}}$ and introduce the stress tensor, we now set up an infinitesimal tetrahedron in Cartesian coordinates, as shown in Fig. 1.3.4.

If $-\mathbf{t}_1, -\mathbf{t}_2, -\mathbf{t}_3$, and $\mathbf{t}$ denote the stress vectors in the outward directions on the faces of the infinitesimal tetrahedron whose areas are Δs_1, Δs_2, Δs_3, and Δs, respectively, we have by Newton's second law for the mass inside the tetrahedron:

$$\mathbf{t}\Delta s - \mathbf{t}_1\Delta s_1 - \mathbf{t}_2\Delta s_2 - \mathbf{t}_3\Delta s_3 + \rho\Delta v\mathbf{f} = \rho\Delta v\mathbf{a}, \qquad (1.3.13)$$

where Δv is the volume of the tetrahedron, ρ is the density, $\mathbf{f}$ is the body

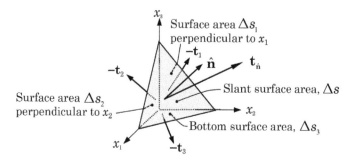

Figure 1.3.4: Tetrahedral element in Cartesian coordinates.

force per unit mass, and $\mathbf{a}$ is the acceleration. Since the total vector area of a closed surface is zero, we have

$$\Delta s \hat{\mathbf{n}} - \Delta s_1 \hat{\mathbf{e}}_1 - \Delta s_2 \hat{\mathbf{e}}_2 - \Delta s_3 \hat{\mathbf{e}}_3 = \mathbf{0}. \tag{1.3.14}$$

It follows that

$$\Delta s_1 = (\hat{\mathbf{n}} \cdot \hat{\mathbf{e}}_1)\Delta s, \quad \Delta s_2 = (\hat{\mathbf{n}} \cdot \hat{\mathbf{e}}_2)\Delta s, \quad \Delta s_3 = (\hat{\mathbf{n}} \cdot \hat{\mathbf{e}}_3)\Delta s. \tag{1.3.15}$$

The volume of the element Δv can be expressed as

$$\Delta v = \frac{\Delta h}{3}\Delta s, \tag{1.3.16}$$

where Δh is the perpendicular distance from the origin to the slant face.

Substitution of Eqs. (1.3.15) and (1.3.16) in Eq. (1.3.13) and dividing throughout by Δs reduces it to

$$\mathbf{t} = (\hat{\mathbf{n}} \cdot \hat{\mathbf{e}}_1)\mathbf{t}_1 + (\hat{\mathbf{n}} \cdot \hat{\mathbf{e}}_2)\mathbf{t}_2 + (\hat{\mathbf{n}} \cdot \hat{\mathbf{e}}_3)\mathbf{t}_3 + \rho\frac{\Delta h}{3}(\mathbf{a} - \mathbf{f}).$$

In the limit when the tetrahedron shrunk to a point (to obtain the relation at a point), $\Delta h \to 0$, we are left with

$$\mathbf{t} = (\hat{\mathbf{n}} \cdot \hat{\mathbf{e}}_1)\mathbf{t}_1 + (\hat{\mathbf{n}} \cdot \hat{\mathbf{e}}_2)\mathbf{t}_2 + (\hat{\mathbf{n}} \cdot \hat{\mathbf{e}}_3)\mathbf{t}_3 = (\hat{\mathbf{n}} \cdot \hat{\mathbf{e}}_i)\mathbf{t}_i, \tag{1.3.17}$$

which can be displayed as

$$\mathbf{t} = \hat{\mathbf{n}} \cdot (\hat{\mathbf{e}}_1\mathbf{t}_1 + \hat{\mathbf{e}}_2\mathbf{t}_2 + \hat{\mathbf{e}}_3\mathbf{t}_3). \tag{1.3.18}$$

The terms in the parenthesis are to be treated as a dyad, called *stress dyad* or *stress tensor* $\boldsymbol{\sigma}$:

$$\boldsymbol{\sigma} \equiv \hat{\mathbf{e}}_1\mathbf{t}_1 + \hat{\mathbf{e}}_2\mathbf{t}_2 + \hat{\mathbf{e}}_3\mathbf{t}_3. \tag{1.3.19}$$

The stress tensor is a point property of the medium that is independent of the unit normal vector $\hat{\mathbf{n}}$. Thus, we have[6]

$$\mathbf{t}(\hat{\mathbf{n}}) = \hat{\mathbf{n}} \cdot \boldsymbol{\sigma} \quad (t_i = n_j \sigma_{ji}) \tag{1.3.20}$$

and the dependence of $\mathbf{t}$ on $\hat{\mathbf{n}}$ has been explicitly displayed. Equation (1.3.20) is known as *Cauchy's formula*, and $\boldsymbol{\sigma}$ is termed the *Cauchy stress tensor*.

It is useful to resolve the stress vectors $\mathbf{t}_1, \mathbf{t}_2$, and $\mathbf{t}_3$ into their orthogonal components. We have

$$\mathbf{t}_i = \sigma_{i1}\hat{\mathbf{e}}_1 + \sigma_{i2}\hat{\mathbf{e}}_2 + \sigma_{i3}\hat{\mathbf{e}}_3 = \sigma_{ij}\hat{\mathbf{e}}_j \tag{1.3.21}$$

for $i = 1, 2, 3$. Hence, the Cauchy stress tensor can be expressed in the rectangular Cartesian system using the summation notation as

$$\boldsymbol{\sigma} = \hat{\mathbf{e}}_i \mathbf{t}_i = \sigma_{ij}\hat{\mathbf{e}}_i\hat{\mathbf{e}}_j. \tag{1.3.22}$$

The component σ_{ij} represents the stress (force per unit area at a point) on a plane perpendicular to the ith coordinate and in the jth coordinate direction (see Fig. 1.3.5). The stress vector $\mathbf{t}$ represents the vectorial stress on a plane whose normal coincides with $\hat{\mathbf{n}}$.

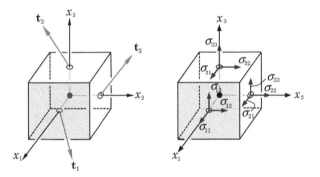

Figure 1.3.5: Definition of stress components in Cartesian rectangular coordinates.

1.3.4 THE GRADIENT OPERATOR

Let us denote a scalar field by $\phi = \phi(\mathbf{x}) = \phi(x_1, x_2, x_3)$, where $\mathbf{x} = (x_1, x_2, x_3)$ is a position vector of a typical point in space. The differential change in ϕ is given by

$$d\phi = \frac{\partial \phi}{\partial x_1} dx_1 + \frac{\partial \phi}{\partial x_2} dx_2 + \frac{\partial \phi}{\partial x_3} dx_3. \tag{1.3.23}$$

---

[6]In some books, $\boldsymbol{\sigma}$ is defined to be the transpose of that defined in Eq. (1.3.19); see Reddy [11]. This is because Eq. (1.3.17) can be expressed, in view of the fact that $\hat{\mathbf{n}} \cdot \hat{\mathbf{e}}_i$ is a scalar quantity that can be placed on the other side of the vector $\mathbf{t}_i$, making Eq. (1.3.18) to become $\mathbf{t} = (\mathbf{t}_1\hat{\mathbf{e}}_1 + \mathbf{t}_2\hat{\mathbf{e}}_2 + \mathbf{t}_3\hat{\mathbf{e}}_3) \cdot \hat{\mathbf{n}}$ and $\mathbf{t}(\hat{\mathbf{n}}) = \boldsymbol{\sigma} \cdot \hat{\mathbf{n}} = \hat{\mathbf{n}} \cdot \boldsymbol{\sigma}^{\mathrm{T}}$.

The differentials dx_1, dx_2, and dx_3 are components of $d\mathbf{x}$. Since $\hat{\mathbf{e}}_i \cdot \hat{\mathbf{e}}_j = \delta_{ij}$, we can write

$$d\phi = \hat{\mathbf{e}}_1 \frac{\partial \phi}{\partial x_1} \cdot \hat{\mathbf{e}}_1 \, dx_1 + \hat{\mathbf{e}}_2 \frac{\partial \phi}{\partial x_2} \cdot \hat{\mathbf{e}}_2 \, dx_1 + \hat{\mathbf{e}}_3 \frac{\partial \phi}{\partial x_3} \cdot \hat{\mathbf{e}}_3 \, dx_3$$

$$= (dx_1 \, \hat{\mathbf{e}}_1 + dx_2 \, \hat{\mathbf{e}}_2 + dx_3 \, \hat{\mathbf{e}}_3) \cdot \left(\hat{\mathbf{e}}_1 \frac{\partial \phi}{\partial x_1} + \hat{\mathbf{e}}_2 \frac{\partial \phi}{\partial x_2} + \hat{\mathbf{e}}_3 \frac{\partial \phi}{\partial x_3} \right)$$

$$= d\mathbf{x} \cdot \left(\hat{\mathbf{e}}_1 \frac{\partial \phi}{\partial x_1} + \hat{\mathbf{e}}_2 \frac{\partial \phi}{\partial x_2} + \hat{\mathbf{e}}_3 \frac{\partial \phi}{\partial x_3} \right). \tag{1.3.24}$$

Let us now denote the magnitude of $d\mathbf{x}$ by $ds \equiv |d\mathbf{x}|$. Then $\hat{\mathbf{e}} = d\mathbf{x}/ds$ is a unit vector in the direction of $d\mathbf{x}$, and we have

$$\left(\frac{d\phi}{ds} \right)_{\hat{\mathbf{e}}} = \hat{\mathbf{e}} \cdot \left(\hat{\mathbf{e}}_1 \frac{\partial \phi}{\partial x_1} + \hat{\mathbf{e}}_2 \frac{\partial \phi}{\partial x_2} + \hat{\mathbf{e}}_3 \frac{\partial \phi}{\partial x_3} \right). \tag{1.3.25}$$

The derivative $(d\phi/ds)$ is called the *directional derivative* of ϕ, and it is the *rate of change* of ϕ with respect to distance. Because the magnitude of this vector is equal to the maximum value (by being along the vector $d\mathbf{x}$) of the directional derivative, it is called the *gradient vector* and is denoted by $\boldsymbol{\nabla}\phi$:

$$\text{grad } \phi = \boldsymbol{\nabla}\phi \equiv \hat{\mathbf{e}}_1 \frac{\partial \phi}{\partial x_1} + \hat{\mathbf{e}}_2 \frac{\partial \phi}{\partial x_2} + \hat{\mathbf{e}}_3 \frac{\partial \phi}{\partial x_3}. \tag{1.3.26}$$

It is important to note that whereas the gradient operator $\boldsymbol{\nabla}$ has some of the properties of a vector, it does not have them all, because it is an operator. For instance, $\boldsymbol{\nabla} \cdot \mathbf{A}$ is a scalar, called the divergence of vector $\mathbf{A}$, whereas $\mathbf{A} \cdot \boldsymbol{\nabla}$ is a scalar *differential operator*. Thus, the del operator does not commute in this sense. The dot product of del operator with a vector is called the *divergence of a vector* and is denoted by

$$\boldsymbol{\nabla} \cdot \mathbf{A} \equiv \text{div} \, \mathbf{A} = \frac{\partial A_i}{\partial x_i}. \tag{1.3.27}$$

If we take the divergence of the gradient vector, we have

$$\text{div}(\text{grad } \phi) \equiv \boldsymbol{\nabla} \cdot \boldsymbol{\nabla}\phi = (\boldsymbol{\nabla} \cdot \boldsymbol{\nabla})\phi = \nabla^2 \phi. \tag{1.3.28}$$

The notation $\nabla^2 = \boldsymbol{\nabla} \cdot \boldsymbol{\nabla}$ is called the *Laplace operator*. In the Cartesian rectangular coordinate system, this reduces to the form

$$\nabla^2 \phi = \frac{\partial^2 \phi}{\partial x^2} + \frac{\partial^2 \phi}{\partial y^2} + \frac{\partial^2 \phi}{\partial z^2} = \frac{\partial^2 \phi}{\partial x_i \partial x_i}. \tag{1.3.29}$$

The curl of a vector is defined as the del operator operating on a vector by means of the cross product [the ith component of $(\boldsymbol{\nabla} \times \mathbf{A})$ is $\frac{\partial A_k}{\partial x_j} \varepsilon_{jki}$]:

$$\text{curl } \mathbf{A} = \boldsymbol{\nabla} \times \mathbf{A} = \hat{\mathbf{e}}_j \frac{\partial}{\partial x_j} \times \hat{\mathbf{e}}_k A_k = \frac{\partial A_k}{\partial x_j} \left(\hat{\mathbf{e}}_j \times \hat{\mathbf{e}}_k \right) = \frac{\partial A_k}{\partial x_j} \varepsilon_{jki} \, \hat{\mathbf{e}}_i. \tag{1.3.30}$$

We also note that the gradient of a vector, $\boldsymbol{\nabla}\mathbf{A}$, is a dyad (i.e., second-order tensor) because it has two base vectors to represent it: $\boldsymbol{\nabla}\mathbf{A} = \hat{\mathbf{e}}_j \frac{\partial A_i}{\partial x_j} \hat{\mathbf{e}}_i = \frac{\partial A_i}{\partial x_j} \hat{\mathbf{e}}_j \hat{\mathbf{e}}_i$. One should make note of the order of the base vectors[7]. The transpose of $\boldsymbol{\nabla}\mathbf{A}$ is to interchange the basis vectors $(\boldsymbol{\nabla}\mathbf{A})^{\mathrm{T}} = \frac{\partial A_i}{\partial x_j} \hat{\mathbf{e}}_i \hat{\mathbf{e}}_j$.

Useful expressions for the integrals of the gradient, divergence, and curl of a vector can be established between volume integrals and surface integrals[8]. Let Ω denote a region in space surrounded by the closed surface Γ. Let $d\Gamma$ be a differential element of surface and $\hat{\mathbf{n}}$ the unit outward normal, and let $d\Omega$ be a differential volume element. The following integral relations between volume and surface integrals (or between area integrals and line integrals) are proven to be useful in the coming chapters. In three dimensions, these relations involve the gradient, curl, and divergence of field variables. The specific forms are presented here.

Gradient theorem

$$\int_\Omega \boldsymbol{\nabla}\phi \, d\Omega = \oint_\Gamma \hat{\mathbf{n}}\phi \, d\Gamma \quad \left[\int_\Omega \hat{\mathbf{e}}_i \frac{\partial \phi}{\partial x_i} \, d\Omega = \oint_\Gamma \hat{\mathbf{e}}_i n_i \phi \, d\Gamma \right]. \qquad (1.3.31)$$

Curl theorem (also known as Kelvin–Stokes' theorem[9])

$$\int_\Omega \boldsymbol{\nabla}\times\mathbf{A} \, d\Omega = \oint_\Gamma \hat{\mathbf{n}} \times \mathbf{A} \, d\Gamma \quad \left[\int_\Omega \varepsilon_{ijk}\hat{\mathbf{e}}_k \frac{\partial A_j}{\partial x_i} \, d\Omega = \oint_\Gamma \varepsilon_{ijk}\hat{\mathbf{e}}_k n_i A_j \, d\Gamma \right]. \qquad (1.3.32)$$

Divergence theorem (also known as Green-Gauss's theorem[10]),

$$\int_\Omega \boldsymbol{\nabla}\cdot\mathbf{A} \, d\Omega = \oint_\Gamma \hat{\mathbf{n}} \cdot \mathbf{A} \, d\Gamma \quad \left[\int_\Omega \frac{\partial A_i}{\partial x_i} \, d\Omega = \oint_\Gamma n_i A_i \, d\Gamma \right]. \qquad (1.3.33)$$

The three theorems can be expressed in a single equation as

$$\int_\Omega \boldsymbol{\nabla} * \mathbf{F} \, d\Omega = \oint_\Gamma \hat{\mathbf{n}} * \mathbf{F} \, d\Gamma, \qquad (1.3.34)$$

---

[7] In some books the gradient operator $\boldsymbol{\nabla}$ is defined, different from that in Eq. (1.3.26), as one with the backward operation: $\boldsymbol{\nabla}\mathbf{A} = (\partial \mathbf{A}/\partial x_j)\hat{\mathbf{e}}_j = (\partial A_i/\partial x_j)\,\hat{\mathbf{e}}_i\,\hat{\mathbf{e}}_j$.

[8] The notion of surface integrals was introduced by Joseph-Louis Lagrange (1736–1813) in 1760 and again in 1811 in the second edition of his *Mécanique Analytique* in more general terms. He discovered the divergence theorem in 1762.

[9] Named after Lord Kelvin (1824–1907) and George Stokes (1819–1903).

[10] Carl Friedrich Gauss (1777–1855) used surface integrals while working on the gravitational attraction of an elliptical spheroid in 1813, when he proved special cases of the divergence theorem. George Green (1793–1841) proved special cases of the theorem in 1828 in "An Essay on the Application of Mathematical Analysis to the Theories of Electricity and Magnetism."

where * is a gradient, curl, or divergence operation, and the field variable is necessarily be a vector or tensor field when * denotes curl or divergence operation.

The forms of a typical vector and its gradient, curl, and divergence in the cylindrical coordinate system (see Fig. 1.3.6) are presented here for a ready reference when we study circular plates.

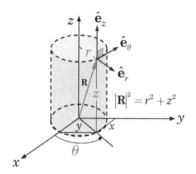

Figure 1.3.6: Cylindrical coordinate system.

Cylindrical coordinate system (r, θ, z)

$$\text{Position vector: } \mathbf{R} = r\,\hat{\mathbf{e}}_r + z\,\hat{\mathbf{e}}_z = x\hat{\mathbf{e}}_x + y\hat{\mathbf{e}}_y + z\,\hat{\mathbf{e}}_z, \tag{1.3.35}$$

$$\text{Relation to } (x, y, z): \quad x = r\cos\theta, \quad y = r\sin\theta, \quad z = z, \tag{1.3.36}$$

$$\hat{\mathbf{e}}_r = \cos\theta\,\hat{\mathbf{e}}_x + \sin\theta\,\hat{\mathbf{e}}_y, \quad \hat{\mathbf{e}}_\theta = -\sin\theta\,\hat{\mathbf{e}}_x + \cos\theta\,\hat{\mathbf{e}}_y, \quad \hat{\mathbf{e}}_z = \hat{\mathbf{e}}_z, \tag{1.3.37}$$

$$\frac{\partial \hat{\mathbf{e}}_r}{\partial \theta} = -\sin\theta\,\hat{\mathbf{e}}_x + \cos\theta\,\hat{\mathbf{e}}_y = \hat{\mathbf{e}}_\theta,$$

$$\frac{\partial \hat{\mathbf{e}}_\theta}{\partial \theta} = -(\cos\theta\,\hat{\mathbf{e}}_x + \sin\theta\,\hat{\mathbf{e}}_y) = -\hat{\mathbf{e}}_r. \tag{1.3.38}$$

All other derivatives of the base vectors of the cylindrical coordinate system are zero. A typical vector $\mathbf{u}$ (such as the displacement), which is a function of the coordinates, can be expressed in the cylindrical coordinate system in terms of its components (u_r, u_θ, u_z) as

$$\mathbf{u} = u_r\,\hat{\mathbf{e}}_r + u_\theta\,\hat{\mathbf{e}}_\theta + u_z\,\hat{\mathbf{e}}_z. \tag{1.3.39}$$

Then ∇ and its various operations on $\mathbf{u}$ are given by

$$\nabla = \hat{\mathbf{e}}_r\frac{\partial}{\partial r} + \frac{1}{r}\hat{\mathbf{e}}_\theta\frac{\partial}{\partial \theta} + \hat{\mathbf{e}}_z\frac{\partial}{\partial z}, \quad \nabla^2 = \frac{1}{r}\left[\frac{\partial}{\partial r}\left(r\frac{\partial}{\partial r}\right) + \frac{1}{r}\frac{\partial^2}{\partial \theta^2} + r\frac{\partial^2}{\partial z^2}\right], \tag{1.3.40}$$

$$\nabla \cdot \mathbf{u} = \frac{1}{r}\left[\frac{\partial(ru_r)}{\partial r} + \frac{\partial u_\theta}{\partial \theta} + r\frac{\partial u_z}{\partial z}\right], \tag{1.3.41}$$

$$\nabla \times \mathbf{u} = \left(\frac{1}{r}\frac{\partial u_z}{\partial \theta} - \frac{\partial u_\theta}{\partial z}\right)\hat{\mathbf{e}}_r + \left(\frac{\partial u_r}{\partial z} - \frac{\partial u_z}{\partial r}\right)\hat{\mathbf{e}}_\theta + \frac{1}{r}\left[\frac{\partial(ru_\theta)}{\partial r} - \frac{\partial u_r}{\partial \theta}\right]\hat{\mathbf{e}}_z,$$

$$(1.3.42)$$

$$\nabla \mathbf{u} = \frac{\partial u_r}{\partial r}\,\hat{\mathbf{e}}_r\hat{\mathbf{e}}_r + \frac{\partial u_\theta}{\partial r}\,\hat{\mathbf{e}}_r\hat{\mathbf{e}}_\theta + \frac{1}{r}\left(\frac{\partial u_r}{\partial \theta} - u_\theta\right)\hat{\mathbf{e}}_\theta\hat{\mathbf{e}}_r + \frac{\partial u_z}{\partial r}\,\hat{\mathbf{e}}_r\hat{\mathbf{e}}_z + \frac{\partial u_r}{\partial z}\,\hat{\mathbf{e}}_z\hat{\mathbf{e}}_r$$

$$+ \frac{1}{r}\left(u_r + \frac{\partial u_\theta}{\partial \theta}\right)\hat{\mathbf{e}}_\theta\hat{\mathbf{e}}_\theta + \frac{1}{r}\frac{\partial u_z}{\partial \theta}\,\hat{\mathbf{e}}_\theta\hat{\mathbf{e}}_z + \frac{\partial u_\theta}{\partial z}\,\hat{\mathbf{e}}_z\hat{\mathbf{e}}_\theta + \frac{\partial u_z}{\partial z}\,\hat{\mathbf{e}}_z\hat{\mathbf{e}}_z.$$

$$(1.3.43)$$

1.4 REVIEW OF THE EQUATIONS OF SOLID MECHANICS

1.4.1 GREEN–LAGRANGE STRAIN TENSOR

For most part, the measure of strain in solid mechanics is the Green–Lagrange strain tensor[11] defined by (see Reddy [11])

$$\mathbf{E} = \tfrac{1}{2}\left[(\nabla\mathbf{u}) + (\nabla\mathbf{u})^{\mathrm{T}} + (\nabla\mathbf{u})\cdot(\nabla\mathbf{u})^{\mathrm{T}}\right], \qquad (1.4.1a)$$

where $\mathbf{u}(\mathbf{X}, t)$ is the displacement vector of a material particle occupying location $\mathbf{X}$ in the reference configuration (and the same material particle occupies a location $\mathbf{x} = \mathbf{X} + \mathbf{u}$ in the deformed body), and ∇ is the gradient operator with respect to $\mathbf{X}$:

$$\nabla = \hat{\mathbf{E}}_1\frac{\partial}{\partial X_1} + \hat{\mathbf{E}}_2\frac{\partial}{\partial X_2} + \hat{\mathbf{E}}_3\frac{\partial}{\partial X_3} = \hat{\mathbf{E}}_i\frac{\partial}{\partial X_i}, \qquad (1.4.1b)$$

where $(\hat{\mathbf{E}}_1, \hat{\mathbf{E}}_2, \hat{\mathbf{E}}_3)$ are the unit base vectors in the coordinate system (X_1, X_2, X_3). Clearly, the last term in Eqs. (1.4.1a) is nonlinear in the displacement gradients. In terms of the displacement components (u_1, u_2, u_3) referred to the rectangular coordinates (X_1, X_2, X_3), we have (see Fig. 1.4.1)

$$E_{ij} = \tfrac{1}{2}\left(\frac{\partial u_i}{\partial X_j} + \frac{\partial u_j}{\partial X_i} + \frac{\partial u_m}{\partial X_i}\frac{\partial u_m}{\partial X_j}\right), \qquad (1.4.2)$$

where the summation on repeated (or dummy) index m over the range of $m = 1, 2, 3$ is implied.

---

[11]There are several measures of strains. The most commonly used strain measures are: the Cauchy–Green deformation tensor, $\mathbf{C} = \mathbf{F}^{\mathrm{T}}\cdot\mathbf{F}$; the Green–Lagrange strain tensor, $2\mathbf{E} = \mathbf{C} - \mathbf{I}$; and the Euler–Almansi strain tensor, $2\mathbf{e} = \mathbf{I} - \mathbf{F}^{-\mathrm{T}}\cdot\mathbf{F}^{-1}$. Here $\mathbf{F}$ denotes the deformation gradient, $\mathbf{F}^{\mathrm{T}} = \nabla\mathbf{x}$, and $\mathbf{I}$ is the unit second order tensor (see Reddy [11] for details). George Green (1793–1841) was a British mathematical physicist and well-known for Cauchy–Green tensor and Green's theorem. Joseph-Louis Lagrange (1736–1813) was an Italian mathematician and astronomer, later naturalized French. He made significant contributions to analysis, number theory, and classical and celestial mechanics. Leonhard Euler was succeeded by Lagrange as the director of mathematics at the Prussian Academy of Sciences in Berlin.

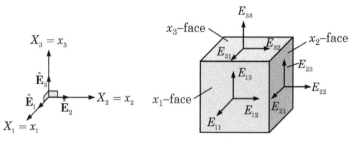

E_{ij} − Strain in the jth direction on the ith face

Figure 1.4.1: Notation used for the Green strain components in the rectangular Cartesian coordinate system.

In expanded notation, the Green strain tensor components referred to the rectangular Cartesian coordinate system (X_1, X_2, X_3) in the reference (unde-formed) configuration in terms of the displacement components (u_1, u_2, u_3) are given by

$$E_{11} = \frac{\partial u_1}{\partial X_1} + \frac{1}{2}\left[\left(\frac{\partial u_1}{\partial X_1}\right)^2 + \left(\frac{\partial u_2}{\partial X_1}\right)^2 + \left(\frac{\partial u_3}{\partial X_1}\right)^2\right],$$

$$E_{22} = \frac{\partial u_2}{\partial X_2} + \frac{1}{2}\left[\left(\frac{\partial u_1}{\partial X_2}\right)^2 + \left(\frac{\partial u_2}{\partial X_2}\right)^2 + \left(\frac{\partial u_3}{\partial X_2}\right)^2\right], \qquad (1.4.3a)$$

$$E_{33} = \frac{\partial u_3}{\partial X_3} + \frac{1}{2}\left[\left(\frac{\partial u_1}{\partial X_3}\right)^2 + \left(\frac{\partial u_2}{\partial X_3}\right)^2 + \left(\frac{\partial u_3}{\partial X_3}\right)^2\right],$$

$$2E_{12} = \frac{\partial u_1}{\partial X_2} + \frac{\partial u_2}{\partial X_1} + \frac{\partial u_1}{\partial X_1}\frac{\partial u_1}{\partial X_2} + \frac{\partial u_2}{\partial X_1}\frac{\partial u_2}{\partial X_2} + \frac{\partial u_3}{\partial X_1}\frac{\partial u_3}{\partial X_2},$$

$$2E_{13} = \frac{\partial u_1}{\partial X_3} + \frac{\partial u_3}{\partial X_1} + \frac{\partial u_1}{\partial X_1}\frac{\partial u_1}{\partial X_3} + \frac{\partial u_2}{\partial X_1}\frac{\partial u_2}{\partial X_3} + \frac{\partial u_3}{\partial X_1}\frac{\partial u_3}{\partial X_3}, \qquad (1.4.3b)$$

$$2E_{23} = \frac{\partial u_2}{\partial X_3} + \frac{\partial u_3}{\partial X_2} + \frac{\partial u_1}{\partial X_2}\frac{\partial u_1}{\partial X_3} + \frac{\partial u_2}{\partial X_2}\frac{\partial u_2}{\partial X_3} + \frac{\partial u_3}{\partial X_2}\frac{\partial u_3}{\partial X_3}.$$

The components E_{11}, E_{22}, and E_{33} are the normal (i.e., extensional) strains, and E_{12}, E_{23}, and E_{13} are the shear strains.

By definition, the Green–Lagrange strain tensor is symmetric, $E_{ij} = E_{ji}$. It is the measure often used in the large deformation analysis. It is a strain measure that is "energetically conjugate" to the second Piola–Kirchhoff stress tensor introduced in Section 1.4.2. As we shall see shortly, we will consider a special case of **E** that is suitable for small strains but accounts for moderately large rotations, as experienced in beams and plates.

The Green–Lagrange strain tensor components in the cylindrical coordinate system $(r = X_1, \theta = X_2, z = X_3$; see Fig. 1.4.2) are given by

$$E_{rr} = \frac{\partial u_r}{\partial r} + \tfrac{1}{2}\left[\left(\frac{\partial u_r}{\partial r}\right)^2 + \left(\frac{\partial u_\theta}{\partial r}\right)^2 + \left(\frac{\partial u_z}{\partial r}\right)^2\right],$$

$$E_{\theta\theta} = \frac{u_r}{r} + \frac{1}{r}\frac{\partial u_\theta}{\partial \theta} + \frac{1}{2}\left[\left(\frac{1}{r}\frac{\partial u_r}{\partial \theta}\right)^2 + \left(\frac{1}{r}\frac{\partial u_\theta}{\partial \theta}\right)^2 + \left(\frac{1}{r}\frac{\partial u_z}{\partial \theta}\right)^2\right.$$
$$\left. - \frac{2}{r^2}u_\theta\frac{\partial u_r}{\partial \theta} + \frac{2}{r^2}u_r\frac{\partial u_\theta}{\partial \theta} + \left(\frac{u_\theta}{r}\right)^2 + \left(\frac{u_r}{r}\right)^2\right], \qquad (1.4.4a)$$

$$E_{zz} = \frac{\partial u_z}{\partial z} + \tfrac{1}{2}\left[\left(\frac{\partial u_r}{\partial z}\right)^2 + \left(\frac{\partial u_\theta}{\partial z}\right)^2 + \left(\frac{\partial u_z}{\partial z}\right)^2\right],$$

$$2E_{r\theta} = \frac{1}{r}\frac{\partial u_r}{\partial \theta} + \frac{\partial u_\theta}{\partial r} - \frac{u_\theta}{r} + \frac{1}{r}\frac{\partial u_r}{\partial \theta}\frac{\partial u_r}{\partial r} + \frac{1}{r}\frac{\partial u_\theta}{\partial r}\frac{\partial u_\theta}{\partial \theta}$$
$$+ \frac{1}{r}\frac{\partial u_z}{\partial r}\frac{\partial u_z}{\partial \theta} + \frac{u_r}{r}\frac{\partial u_\theta}{\partial r} - \frac{u_\theta}{r}\frac{\partial u_r}{\partial r},$$

$$2E_{rz} = \left(\frac{\partial u_r}{\partial z} + \frac{\partial u_z}{\partial r} + \frac{\partial u_r}{\partial r}\frac{\partial u_r}{\partial z} + \frac{\partial u_\theta}{\partial r}\frac{\partial u_\theta}{\partial z} + \frac{\partial u_z}{\partial r}\frac{\partial u_z}{\partial z}\right), \qquad (1.4.4b)$$

$$2E_{\theta z} = \frac{\partial u_\theta}{\partial z} + \frac{1}{r}\frac{\partial u_z}{\partial \theta} + \frac{1}{r}\frac{\partial u_r}{\partial \theta}\frac{\partial u_r}{\partial z} + \frac{1}{r}\frac{\partial u_\theta}{\partial \theta}\frac{\partial u_\theta}{\partial z}$$
$$+ \frac{1}{r}\frac{\partial u_z}{\partial \theta}\frac{\partial u_z}{\partial z} - \frac{u_\theta}{r}\frac{\partial u_r}{\partial z} + \frac{u_r}{r}\frac{\partial u_\theta}{\partial z}.$$

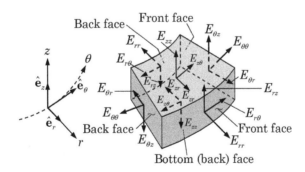

Figure 1.4.2: Notation used for the Green strain components on a volume element in the cylindrical coordinate system (r, θ, z); see Fig. 1.3.6. As indicated in Fig. 1.4.1, $E_{\xi\eta}$ is the strain on the plane perpendicular to ξ-coordinate and in the η-coordinate direction, where ξ and η take on the symbols r, θ, and z. The notation shown here for strains follows that of the stress components shown in Fig. 1.3.5.

When the displacement gradients are small (say less than 1%), that is, $|\nabla \mathbf{u}| \ll 1$,

$$\frac{\partial u_i}{\partial X_j} \ll 1, \quad \left(\frac{\partial u_i}{\partial X_j}\right)^2 \approx 0, \text{ for any } i \text{ and } j,$$

we may neglect the nonlinear terms in the definition of the Green strain tensor $\mathbf{E}$ and obtain the linearized strain tensor ε, called the *infinitesimal strain tensor* $(X_i \approx x_i)$:

$$\varepsilon = \tfrac{1}{2}\left[(\nabla \mathbf{u}) + (\nabla \mathbf{u})^{\mathrm{T}}\right]; \quad \varepsilon_{ij} = \tfrac{1}{2}\left(\frac{\partial u_i}{\partial x_j} + \frac{\partial u_j}{\partial x_i}\right). \quad (1.4.5)$$

In expanded form in the rectangular coordinate system (x, y, z), the infinitesimal strain tensor components are

$$\varepsilon_{xx} = \frac{\partial u_x}{\partial x}, \quad \varepsilon_{yy} = \frac{\partial u_y}{\partial y}, \quad \varepsilon_{zz} = \frac{\partial u_z}{\partial z},$$

$$2\varepsilon_{xy} = \frac{\partial u_x}{\partial y} + \frac{\partial u_y}{\partial x}, \quad 2\varepsilon_{xz} = \frac{\partial u_x}{\partial z} + \frac{\partial u_z}{\partial x}, \quad 2\varepsilon_{yz} = \frac{\partial u_y}{\partial z} + \frac{\partial u_z}{\partial y}. \quad (1.4.6)$$

If one presumes that the strains are small and but rotations about the y-axis of the material lines transverse to the x-axis are moderately large, that is, the squares and products of $\partial u_z/\partial x$ and $\partial u_z/\partial y$ are not negligible but squares and products of $\partial u_x/\partial x$, $\partial u_y/\partial y$, and $\partial u_z/\partial z$ are negligible, the strains resulting from the Green strain tensor components are known as the Föppl–von Kármán strains[12]

$$\varepsilon_{xx} = \frac{\partial u_x}{\partial x} + \frac{1}{2}\left(\frac{\partial u_z}{\partial x}\right)^2, \quad \varepsilon_{yy} = \frac{\partial u_y}{\partial y} + \frac{1}{2}\left(\frac{\partial u_z}{\partial y}\right)^2, \quad \varepsilon_{zz} = \frac{\partial u_z}{\partial z}, \quad (1.4.7a)$$

$$2\varepsilon_{xy} = \frac{\partial u_x}{\partial y} + \frac{\partial u_y}{\partial x} + \frac{\partial u_z}{\partial x}\frac{\partial u_z}{\partial y}, \quad 2\varepsilon_{xz} = \frac{\partial u_x}{\partial z} + \frac{\partial u_z}{\partial x},$$

$$2\varepsilon_{yz} = \frac{\partial u_y}{\partial z} + \frac{\partial u_z}{\partial y}. \quad (1.4.7b)$$

---

[12]They are named after August Föppl and Theodore von Kármán. August Otto Föppl (1854–1924) was a professor of Technical Mechanics and Graphical Statics at the Technical University of Munich, Germany. Theodore von Kármán (1881–1963) was a Hungarian-American mathematician, aerospace engineer, and physicist. He received his doctorate under the guidance of Ludwig Prandtl at the University of Göttingen, Germany in 1908. He was invited to the United States by Robert A. Millikan to advise California Institute of Technology (Caltech) engineers on the design of a wind tunnel. In 1930, he accepted the directorship of the Guggenheim Aeronautical Laboratory at the California Institute of Technology (GALCIT). His contributions include: theories of non-elastic buckling and supersonic aerodynamics. He made additional contributions to elasticity, vibration, heat transfer, and crystallography.

In the cylindrical coordinates, the von Kármán nonlinear strains are

$$\varepsilon_{rr} = \frac{\partial u_r}{\partial r} + \frac{1}{2}\left(\frac{\partial u_z}{\partial r}\right)^2, \qquad \varepsilon_{\theta\theta} = \frac{u_r}{r} + \frac{1}{r}\frac{\partial u_\theta}{\partial \theta} + \frac{1}{2}\left(\frac{1}{r}\frac{\partial u_z}{\partial \theta}\right)^2,$$

$$\varepsilon_{zz} = \frac{\partial u_z}{\partial z} + \frac{1}{2}\left(\frac{\partial u_z}{\partial z}\right)^2, \qquad 2\varepsilon_{r\theta} = \frac{1}{r}\frac{\partial u_r}{\partial \theta} + \frac{\partial u_\theta}{\partial r} - \frac{u_\theta}{r} + \frac{1}{r}\frac{\partial u_z}{\partial r}\frac{\partial u_z}{\partial \theta},$$

$$2\varepsilon_{rz} = \frac{\partial u_r}{\partial z} + \frac{\partial u_z}{\partial r}, \qquad 2\varepsilon_{\theta z} = \frac{\partial u_\theta}{\partial z} + \frac{1}{r}\frac{\partial u_z}{\partial \theta} + \frac{1}{r}\frac{\partial u_r}{\partial \theta}\frac{\partial u_r}{\partial z}. \qquad (1.4.8)$$

1.4.2 THE SECOND PIOLA–KIRCHHOFF STRESS TENSOR

The Cauchy[13] stress tensor $\boldsymbol{\sigma}$ (sometimes called "true stress") introduced in Eqs. (1.3.19) and (1.3.22) is the most natural and physical measure of the state of stress at a point in the deformed body and measured as the force in the deformed body per unit area of the deformed body. Since the geometry of the deformed body is not known (and yet to be determined), the governing equations must be written in terms of the known reference configuration[14], say, configuration at $t = 0$. This need gives rise to come up with a measure of stress that can be calculated using the known reference configuration. One such measure is the second Piola–Kirchhoff stress tensor[15], which is a measure of the *transformed internal force (from the deformed to the undeformed body) per undeformed area*. It is a mathematical entity introduced for the convenience of calculating stresses in a deformed solid.

The Green strain tensor $\mathbf{E}$ can be shown to be the dual (or energetically conjugate) to the second Piola–Kirchhoff stress tensor $\mathbf{S}$ (see Reddy [11]) in the sense that the strain energy density stored in an elastic body is equal to the product of $\mathbf{E}$ and $\mathbf{S}$ and it is invariant (i.e., independent of the coordinate system used). The second Piola–Kirchhoff stress tensor $\mathbf{S}$ and the Cauchy stress tensor $\boldsymbol{\sigma}$ are related according to

$$\mathbf{S}^{\mathrm{T}} = J\mathbf{F}^{-1}\cdot\boldsymbol{\sigma}^{\mathrm{T}}\cdot\mathbf{F}^{-\mathrm{T}}, \qquad \boldsymbol{\sigma}^{\mathrm{T}} = \frac{1}{J}\mathbf{F}\cdot\mathbf{S}^{\mathrm{T}}\cdot\mathbf{F}^{\mathrm{T}}. \qquad (1.4.9)$$

where $\mathbf{F}$ is the deformation gradient defined by [11]

$$\mathbf{F} = (\boldsymbol{\nabla}\mathbf{x})^{\mathrm{T}} = \mathbf{I} + (\boldsymbol{\nabla}\mathbf{u})^{\mathrm{T}}. \qquad (1.4.10)$$

---

[13] Baron Augustin-Louis Cauchy (1789–1857) was a French mathematician, engineer, and physicist who made pioneering contributions to mathematical analysis and continuum mechanics.

[14] We shall use the term *configuration* to mean the simultaneous position of all material points of a body for any fixed time.

[15] Gustav Robert Kirchhoff (1824–1887) was a German physicist who contributed to the fundamental understanding of electrical circuits, spectroscopy, black-body radiation by heated objects, and theoretical mechanics.

and J is the determinant of $\mathbf{F}$, called the *Jacobian of the motion*. The deformation gradient $\mathbf{F}$, in general, involves both stretch and rotation. In Eq. (1.4.10), $\mathbf{I}$ denotes the second-order identity tensor (i.e., $\mathbf{I} = \delta_{ij}\,\hat{\mathbf{E}}_i\hat{\mathbf{E}}_j$; see Fig. 1.4.1).

1.4.3   EQUATIONS OF MOTION

The principle of balance of linear momentum as applied to a deformed solid continuum and expressed in terms of the Cauchy stress tensor $\boldsymbol{\sigma}$ gives

$$\boldsymbol{\nabla}_x \cdot \boldsymbol{\sigma} + \mathbf{f} = \rho\frac{\partial^2\mathbf{u}}{\partial t^2}, \tag{1.4.11}$$

where $\boldsymbol{\nabla}_x$ is the gradient operator with respect to the spatial coordinate $\mathbf{x}$ occupied by the material particle X which was at location $\mathbf{X}$ in the undeformed body (i.e., the displacement vector is $\mathbf{u} = \mathbf{x} - \mathbf{X}$); $\mathbf{f}$ is the body force vector measured per unit deformed volume; and ρ is the mass per unit deformed volume. Equation (1.4.10) is not useful for the analysis of large deformation because there the measure of the stress is the second Piola–Kirchhoff stress tensor, $\mathbf{S}$. Therefore, we express the equation of motion in terms of the second Piola–Kirchhoff stress tensor $\mathbf{S}$ as

$$\boldsymbol{\nabla} \cdot [\mathbf{S} \cdot (\mathbf{I} + \boldsymbol{\nabla}\mathbf{u})] + \hat{\mathbf{f}} = \rho_0\frac{\partial^2\mathbf{u}}{\partial t^2}, \tag{1.4.12}$$

where $\boldsymbol{\nabla}$ is the gradient operator with respect to the material coordinate $\mathbf{X}$, ρ_0 is the mass density measured in the undeformed body and $\hat{\mathbf{f}}$ is the body force per unit volume in the undeformed body. Clearly, the equations of motion expressed in terms of the second Piola–Kirchhoff stress tensor are nonlinear, and this (spatial) nonlinearity is in addition to any nonlinearity that may come from the strain–displacement relations and constitutive relations.

1.4.4   STRESS–STRAIN RELATIONS

Due to the smallness of the thickness dimension in beams and plates, the normal stress in the thickness direction, namely, σ_{zz}, is assumed to be small and negligible compared to the in-plane stresses. More importantly, in the case of an orthotropic material with different moduli in the material 1 and 2 directions (i.e., planes of material symmetry), the shear stresses are assumed to be only function of their respective shear strains $\sigma_{ij} = G_{ij}2\varepsilon_{ij}$ (no sum on repeated subscripts) for $i \neq j = 1, 2, 3$. Then the 3-D constitutive equations resulting from the application of Hooke's law[16] must be modified to account for

---

[16]Robert Hooke (1635–1703) was an English scientist and architect and recently called "England's Leonardo." Hooke's law states that the force (F) needed to elongate or compress a spring by some distance (x) is linearly proportional to the distance, $F = kx$, where k is the proportionality constant which is characteristic of the spring stiffness.

this fact. The stress–strain relations obtained are termed plane-stress-reduced constitutive equations, which are adopted for beams, plates, and shells, whose thickness is very small compared to the other dimensions.

Here, we assume that the beam or plate material is characterized as orthotropic with respect to the (x, y, z) system (i.e., the material coordinates coincide with the coordinates used to describe the governing equations). Then we have

$$
\begin{Bmatrix} \sigma_{xx} \\ \sigma_{yy} \\ \sigma_{xy} \end{Bmatrix} = \begin{bmatrix} Q_{11} & Q_{12} & 0 \\ Q_{12} & Q_{22} & 0 \\ 0 & 0 & Q_{66} \end{bmatrix} \begin{Bmatrix} \varepsilon_{xx} - \alpha_x \, \Delta T \\ \varepsilon_{yy} - \alpha_y \, \Delta T \\ 2\varepsilon_{xy} \end{Bmatrix},
$$

$$
\begin{Bmatrix} \sigma_{yz} \\ \sigma_{xz} \end{Bmatrix} = \begin{bmatrix} Q_{44} & 0 \\ 0 & Q_{55} \end{bmatrix} \begin{Bmatrix} 2\varepsilon_{yz} \\ 2\varepsilon_{xz} \end{Bmatrix},
$$

(1.4.13)

where Q_{ij} are the *plane stress-reduced elastic stiffness coefficients*; α_x and α_y are the coefficients of thermal expansion along the x and y directions, respectively; and $\Delta T = T - T_0$ is the temperature increment from a reference state T_0. The elastic coefficients Q_{ij} are related to the six independent engineering constants $(E_1, E_2, \nu_{12}, G_{12}, G_{13}, G_{23})$ as follows:

$$
Q_{11} = \frac{E_1}{1 - \nu_{12}\nu_{21}}, \quad Q_{12} = \frac{\nu_{12}E_2}{1 - \nu_{12}\nu_{21}}, \quad Q_{22} = \frac{E_2}{1 - \nu_{12}\nu_{21}},
$$

$$
Q_{66} = G_{12}, \quad Q_{44} = G_{23}, \quad Q_{55} = G_{13}.
$$

(1.4.14)

Note that ν_{21} is computed from the following reciprocal relationship implied by the symmetry of elasticity tensor (see Reddy [8] for details):

$$
\nu_{21} = \nu_{12} \, \frac{E_2}{E_1}.
$$

(1.4.15)

1.5 FUNCTIONALLY GRADED STRUCTURES

1.5.1 BACKGROUND

Functionally graded materials (FGMs) are characterized by the variation in composition of two or more materials gradually over surface or volume, resulting in a composite material that has desired properties. An FGM can be designed for specific functionality and application. Most structures found in nature – from sea shells, trees and plants, to organs of living bodies – are multi-material graded structures, formed over millions of years, to satisfy certain functionalities. In the modern times, the man-made FGMs were proposed (see [12] and [13]) as thermal barrier materials for applications in space planes, space structures, nuclear reactors, turbine rotors, flywheels, and gears, to name only a few. As conceived and manufactured today, these materials are isotropic and non-homogeneous. In general, all the multi-phase materials, in which the material properties are varied gradually in a predetermined manner, fall into the category of functionally gradient materials. As stated before, the functionally gradient material characteristics are present in most structures found in nature, and perhaps, a better understanding of the highly complex form of materials in nature will help us in synthesizing new materials (the science of so called "biomimetics"). Such property enhancements endow FGMs with material properties such as resilience to fracture. FGMs promise attractive applications in a wide variety of wear coating and thermal shielding problems such as gears, cams, cutting tools, high temperature chambers, furnace liners, turbines, micro-electronics, and space structures.

A large number journal papers dealing with functionally graded beams and plates have appeared in the literature and a critical review of these papers is not a focus of this introduction to FGM structures [14] (also see, e.g., [15]–[41] and references therein). A majority of these works considered two-constituent FGM structures, and typically the material variation is considered through the thickness of beams, plates, and shell structures. The works of Praveen and Reddy [19] and Reddy [22] have also considered the von Kármán nonlinearity in functionally graded plates.

With the progress of technology and fast growth of the use of nanostructures, FGMs have found potential applications in micro and nano scales in the form of shape memory alloy thin films [42], atomic force microscopes (AFMs) [43], electrically actuated actuators [44], and micro switches [45], to name a few. The von Kármán nonlinearity may have significant contribution to the response of micro- and nano-scale devices such as biosensors and AFMs [46].

A typical FGM represents a particulate composite with a prescribed distribution of volume fractions of constituent phases. In the case of beams, plates, and shells, the material properties are assumed to vary continuously through the thickness. The effective properties of macroscopic homogeneous beams, plates, and shells are derived from the microscopic heterogeneous material distributions using homogenization techniques [14, 47, 48]. Several models, like the rule of mixtures [19, 22], Hashin–Shtrikman type bounds [49], Mori–Tanaka scheme [48, 50], and self-consistent schemes [51] are available in the literature for determination of the bounds for the effective properties. Voigt scheme and the Mori–Tanaka scheme [50] have been generally used for the study of FGM plates and structures by researchers [35, 52].

1.5.2 MORI–TANAKA SCHEME

For those parts of the graded microstructure that have a well-defined continuous matrix and discontinuous reinforcement, the overall properties and local fields can be closely predicted by Mori–Tanaka estimates. The assumption of spherical particles embedded in a matrix is considered. The primary matrix phase is assumed to be reinforced by spherical particles of secondary phase.

Mori and Tanaka [48, 50] derived a method to calculate the average internal stress in the matrix of a material. This has been reformulated by Benveniste [53] for use in the computation of the effective properties of composite materials. According to the Mori–Tanaka scheme, the effective elastic properties of the FGM can be expressed as

$$\frac{K - K_1}{K_2 - K_1} = \frac{1 - V_1}{1 + V_1 \frac{K_2 - K_1}{K_1 + \frac{4}{3}G_1}}, \quad \frac{G - G_1}{G_2 - G_1} = \frac{1 - V_1}{1 + V_1 \frac{G_2 - G_1}{G_1 + f_1}}, \tag{1.5.1}$$

where

$$f_1 = \frac{G_1(9K_1 + 8G_1)}{6(K_1 + 2G_1)}, \quad V_2 = 1 - V_1, \tag{1.5.2}$$

in which K and G are bulk modulus and shear modulus, respectively, and V is the volume fraction of the material (the subscript 1 and 2 refer to materials 1 and 2, respectively). The bulk modulus K and shear modulus G are related to Young's modulus E and Poisson's ratio ν, by the following equations:

$$E = \frac{9KG}{3K + G}, \quad \nu = \frac{3K - 2G}{2(3K + G)}. \tag{1.5.3}$$

1.5.3   VOIGT SCHEME: RULE OF MIXTURES

There are two rule of mixture models to describe the effective mechanical properties of a composite comprising two elastically isotropic constituent phases: the Voigt and Reuss models [54]. The Voigt model corresponds to axial loads and the Reuss model to transverse loads.

Voigt scheme has been adopted in most analysis of FGM structures [18, 22, 25, 35, 37, 38, 39, 41]. The advantage of the Voigt model is the simplicity of implementation and the ease of computation. According to Voigt scheme, the effective property P of the composite of two phases is the weighted average of the properties of the constituent phases:

$$P(z) = P_1 V_1(z) + P_2 V_2(z), \quad V_2(z) = 1 - V_1(z), \tag{1.5.4}$$

where P_1 and P_2 represent the constituent material properties (e.g., modulus, conductivity, and so on) of materials 1 and 2, respectively; and, V_1 and V_2 represent the volume fractions of materials 1 and 2, respectively, which may vary with respect to thickness coordinate z.

1.5.4   EXPONENTIAL MODEL

The exponential model, which is often employed in fracture studies, is based on the formula (see [29, 30])

$$P(z) = P_1 \exp\left[-\alpha\left(\frac{1}{2} - \frac{z}{h}\right)\right], \quad \alpha = \log\left(\frac{P_1}{P_2}\right). \tag{1.5.5}$$

1.5.5   POWER-LAW MODEL

The variation of properties through the thickness is considered to be either exponential (called E-FGM), as given in Eq. (1.5.5), or based on a power series (called P-FGM), as presented in Eqs. (1.5.4) and (1.5.6), which covers most of the existing analytical models.

The volume fractions of materials 1 and 2, V_1 and V_2 can be expressed in the form of power law as (see Fig. 1.5.1)

$$V_1(z) = \left(\frac{1}{2} + \frac{z}{h}\right)^n, \quad V_2(z) = 1 - V_1(z). \tag{1.5.6}$$

where n is the volume fraction exponent (termed here as the *power-law index*). Then the property P as a function of the thickness coordinate z is given by Eq. (1.5.4). Fig. 1.5.2 shows the variation of the volume fraction of ceramic,

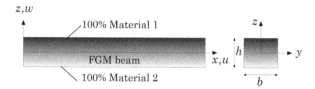

Figure 1.5.1: Geometry of a through-thickness functionally graded beam.

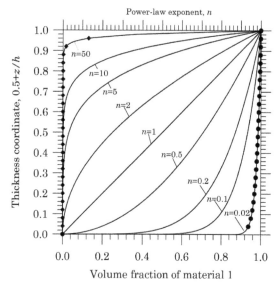

Figure 1.5.2: Volume fraction of material 1, V_1, through the beam thickness for various values of power-law index, n.

V_1, through the beam thickness for various values of the power-law index n. Note that the volume fraction $V_1(z)$ decreases with increasing value of n. The Power law is most popular because of its simplicity and algebraic nature.

Equations (1.5.4) and (1.5.6) can be combined to express a typical material property variation through the beam height or plate thickness, h, as

$$P(z) = (P_1 - P_2)\, V_1(z) + P_2, \quad V_1(z) = \left(\frac{1}{2} + \frac{z}{h}\right)^n . \tag{1.5.7}$$

1.6 MODIFIED COUPLE STRESS EFFECTS

1.6.1 BACKGROUND

The increasing demand for safe, lightweight, and environmentally acceptable structures has increased the need to investigate new structural configurations, including cellular or architected beams and plates. Such structures offer higher load-bearing capacity compared to their conventional counter parts. Computational models that take into account all architectural details are prohibitively expensive, requiring nonlocal continuum theories which account for the structural details, without homogenizing the structure, are needed. The *modified couple stress theory* of Mindlin [55], Koiter [56], and Toupin [57], and the *strain gradient theory* of [58]–[61] provide examples of such nonlocal theories. A more complete review of the early developments can be found in the paper of Srinivasa and Reddy [62]. The strain gradient theory is a more general form of the modified couple stress theory and the relationship between the modified couple stress theory and the strain gradient theory can be found in the work of Reddy and Srinivasa [63]. In recent years a number of attempts have been made to bring microstructural length scales into the continuum description of beams and plates. Such models are useful in determining the structural response of micro and nano devices made of a variety of new materials that require the consideration of small material length scales over which the neighboring secondary constituents interact, especially when the spatial resolution is comparable to the size of the secondary constituents. Examples of such materials are provided by nematic elastomers, carbon nanotube composites [64], and CNT-reinforced environment-resistant coatings [65].

Microstructure-dependent theories are developed for the Bernoulli-Euler beam by Park and Gao [66, 67], for the shear deformable beams and plates by Ma, Gao, and Reddy [68]–[70], for the third-order theory of plates for bending and vibration by Aghababaei and Reddy [71], and for vibration and buckling of shear deformable beams by Araujo dos Santos and Reddy [72, 73]. In the last two decades, Reddy and his colleagues [68]–[85] have published a large number of papers dealing with linear and nonlinear bending of classical and first- and third-order shear deformable beams and plates using the modified couple stress theory. Some of these works have accounted for the the von Kármán nonlinearity and functionally graded (through the thickness) materials. The von Kármán nonlinearity may have significant contribution to the response of beam-like elements used in micro- and nano-scale devices such as, biosensors and AFMs [46, 86].

1.6.2   THE STRAIN ENERGY FUNCTIONAL

Let $\mathbf{u}$ denote the displacement vector of an arbitrary point in the beam or plate. The rotation vector $\boldsymbol{\omega}$ is defined as

$$\boldsymbol{\omega} = \tfrac{1}{2}\left(\boldsymbol{\nabla} \times \mathbf{u}\right). \tag{1.6.1}$$

Physically, $\boldsymbol{\omega}$ denotes the macro-rotation at a point of the continuum. The curvature tensor $\boldsymbol{\chi}$, which represents the rate of change of the rotation, is defined as (assumed to be small):

$$\boldsymbol{\chi} = \frac{1}{2}\left[\boldsymbol{\nabla}\boldsymbol{\omega} + (\boldsymbol{\nabla}\boldsymbol{\omega})^{\mathrm{T}}\right]. \tag{1.6.2}$$

The modified couple stress theory is based on the hypothesis that the rate of change of macro-rotations cause additional stresses, called *couple stresses*, in the continuum. The modified couple stress tensor $\mathbf{m}$ is related to the curvature tensor $\boldsymbol{\chi}$ through the constitutive relations [55]:

$$\mathbf{m} = 2G\ell^2\,\boldsymbol{\chi}, \tag{1.6.3}$$

where ℓ is the length scale parameter (sometimes denoted by l) and G is the shear modulus.

According to the modified couple stress theory, the strain energy potential of an elastic beam of length a or circular plate of radius a can be expressed as (see Section 2.2 for the concept of strain energy)

$$U = \frac{1}{2}\int_A \left[\int_0^a \left(\boldsymbol{\sigma} : \boldsymbol{\varepsilon} + \mathbf{m} : \boldsymbol{\chi}\right)dx\right] dA, \tag{1.6.4}$$

where A is the area of cross section (for beam we set $dA = dydz$ and for circular plate we take $dx = dr$ and $dA = rd\theta dz$), $\boldsymbol{\sigma}$ is the Cauchy stress tensor, $\boldsymbol{\varepsilon}$ is the simplified Green–Lagrange strain tensor, $\mathbf{m}$ is the deviatoric part of the symmetric couple stress tensor, and $\boldsymbol{\chi}$ is the symmetric curvature tensor defined in Eq. (1.6.2). In the coming chapters, these relations will be specialized to various beam and plate theories.

1.7   CHAPTER SUMMARY

In this chapter, beginning with a short discussion of vectors and tensors and the introduction of the Cauchy stress vector and Cauchy stress tensor, measures of Green strain tensor, infinitesimal strain tensor and the von Kármán strain tensor components are reviewed. The definition of the second Piola–Kirchhoff stress tensor is introduced, but for small strains (as is the case with the present study), it is indistinguishable from the Cauchy stress tensor. Then the equations of motion of a deformable solid are presented, and stress–strain relations for a linear elastic material are summarized.

In addition, an introduction to two-constituent functionally graded materials is presented and various models of material gradation are reviewed. Then the modified couple stress theory is briefly visited, and the pertinent equations are summarized. Overall, the contents of this chapter will be utilized in the coming chapters.

There are other nonlocal models [87, 88, 89, 92]. Among them the stress gradient model (especially the differential model) of Eringen [89]–[93] has been used to study beams and circular plates, and the topic is not included in this book. Interested readers may consult [94]–[101] and references therein.

SUGGESTED EXERCISES

1.1 Establish the relations in Eqs. (1.3.41)–(1.3.43).

1.2 Verify the relations in Eqs. (1.4.4a) and (1.4.4b)

1.3 Establish the equation of motion in Eq. (1.4.12):

$$\nabla \cdot [\mathbf{S} \cdot (\mathbf{I} + \nabla \mathbf{u})] + \mathbf{f} = \rho_0 \frac{\partial^2 \mathbf{u}}{\partial t^2},$$

where ∇ is the gradient operator with respect to the material coordinate $\mathbf{X}$, ρ_0 is the mass density measured in the undeformed body and $\hat{\mathbf{f}}$ is the body force per unit volume of the undeformed body.

1.4 Establish the relations in Eq. (1.4.14) beginning with the strain–stress relations (Hooke's law) for the plane stress case (i.e., $\sigma_3 \equiv \sigma_{33} = 0$):

$$\begin{Bmatrix} \varepsilon_1 \\ \varepsilon_2 \\ \varepsilon_6 \end{Bmatrix} = \begin{bmatrix} \frac{1}{E_1} & -\frac{\nu_{21}}{E_2} & 0 \\ -\frac{\nu_{12}}{E_1} & \frac{1}{E_2} & 0 \\ 0 & 0 & \frac{1}{G_{12}} \end{bmatrix} \begin{Bmatrix} \sigma_1 \\ \sigma_2 \\ \sigma_6 \end{Bmatrix}.$$

1.5 If the displacement vector is given by

$$\mathbf{u}(x_1, x_2, x_3) = [u(x_1) + x_3\, \phi(x_1)]\, \hat{\mathbf{e}}_1 + w(x_1)\, \hat{\mathbf{e}}_3,$$

determine the components of ω and χ.

If your theory is found to be against the second law of thermodynamics, I give you no hope; there is nothing for it but to collapse in deepest humiliation. Arthur Eddington

2 Energy Principles and Variational Methods

The science of today is the technology of tomorrow. Edward Teller

2.1 CONCEPTS OF WORK AND ENERGY

2.1.1 HISTORICAL BACKGROUND

We begin this chapter with comments on the history of the concepts of work, energy, and variational theory. Archimedes (287–212 B.C.), a Greek mathematician, physicist, and engineer, is regarded as the first to use work arguments in his study of levers. The most primitive ideas of variational theory are present in the writings of the Greek philosopher Aristotle (384–322 B.C.), revived by Italian polymath Galileo Galilei (1564–1642), and finally formulated into a principle of least time by the French mathematician Fermat (1607–1665). The phrase *virtual velocities* was used by the Swiss mathematician Johann (or Jean) Bernoulli (1667–1748) in his 1717 letter to Varignon (1654–1722). The development of early variational calculus had to await the works of Newton (1642–1727) and Leibniz (1646–1716). The earliest applications of such variational ideas included the classical *isoperimetric problem* of finding among closed curves of given length the one that encloses the greatest area, and Newton's problem of determining the solid of revolution of "minimum resistance." In 1696, Jean Bernoulli proposed the problem of the *brachistochrone*: among all curves connecting two points, find the curve traversed in the shortest time by a particle under the influence of gravity. It was a challenge to the mathematicians of their day to solve the problem using the rudimentary tools of analysis available to them or whatever new ones they could develop. Solutions were presented by some of the greatest mathematicians of the time: Leibniz, Jean Bernoulli's older brother Jacques Bernoulli, L'Hopital, and Newton.

A general method for solving variational problems was given by the Swiss genius Leonhard Euler (1707–1783) in 1732 when he presented a "general solution of the isoperimetric problem." Pierre Louis Maupertuis (1698–1759), a French mathematician and philosopher, is credited to have put forward the principle of least action in his *Mémoires de l'Académie des Sciences* in 1740. It was in Euler's 1732 work and subsequent publication of the principle of least action in his book *Methodus inveniendi lineas curvas ...* in 1744 that variational concepts found a permanent home in mechanics. He developed ideas surrounding the principle of minimum total potential energy in his work

DOI: 10.1201/9781003240846-2

on the *elastica*, and he demonstrated the relationship between variational equations and those governing the flexure and buckling of thin rods.

A great impetus to the development of variational mechanics began in the writings of Lagrange (1736–1813), first in his correspondence with Euler. Euler worked intensely in developing Lagrange's method but delayed publishing his results until Lagrange's works were published in 1760. Lagrange used D'Alembert's principle to convert dynamics to statics and then used the principle of virtual displacements to derive his famous equations governing the laws of dynamics in terms of kinetic and potential energy. Euler's work, together with Lagrange's *Mécanique analytique* of 1788, laid down the basis for the variational theory of dynamical systems. Further generalizations appeared in the fundamental work of Sir William Rowan Hamilton (1805–1865), an Irish mathematician, in 1834. Collectively, all these works have had a monumental impact on virtually every branch of mechanics.

2.1.2 OBJECTIVES OF THE CHAPTER

In this chapter, we introduce the principles of virtual work and their special cases, which are useful in finding point displacements and forces as well as for determining continuous solutions of beams and circular plates by variational methods and the finite element method. Toward this end, we first introduce the concepts of *work done* and *energy stored* in deformable solids subjected to external forces. Then the related energy principles are introduced and illustrated with the help of extensional deformation of rods or bars. These concepts and principles will be used in the coming chapters to (1) derive the governing equations, (2) find point displacements and forces, (3) determine continuous solutions using classical variational methods (e.g., the Ritz and Galerkin methods), and (4) formulate finite element models. A vast majority of the material included here is based on Chapters 3–5 of the author's book [9], and the readers are advised to consult it for additional details.

Work done by forces in moving through the displacements induced in the body is of special interest and use in solid and structural mechanics, as will be seen shortly. Work done by a force (or moment) is defined as the product of the force (or moment) and the displacement (or rotation) *in the direction* of the force (or moment). The work done is a scalar (i.e., a real number) with units of N-m (Newton meters). When the forces and displacements are functions of position and time, we take integrals of the product over the space and time, producing a scalar value; and it is positive whenever both displacement (rotation) and force (moment) have the same direction and negative if they are in the opposite directions.

In general, all bodies are deformable under the action of forces. A body is said to be *rigid*, when the range of forces applied do not cause considerable geometric changes. Deformable bodies and rigid bodies are engineering representation of changes or lack thereof, respectively, in the geometry. When the body returns to its original shape after the forces causing the deformation are removed, we say that the body has undergone *elastic* deformation; otherwise (i.e., leaving some permanent geometric changes or strains) the body is said to have undergone inelastic or plastic deformation. Deformations are measured using certain strain measures and displacements, which were introduced in Section 1.4.

2.1.3   CONCEPT OF WORK DONE

In a deformable body, work is done by externally applied forces and internally generated forces (which are often measured as stresses). The work done by external forces, denoted with W_E, is work put into the system, whereas the work done by internal forces (stresses), W_I, is the work stored in the body. Sometimes, W_E is called the potential energy due to the applied loads and is denoted by V_E. The internal forces generated inside the body due to the application of the external loads move through the internal displacements. In general, the work done by internal forces in moving through their respective displacements is *not* equal to the work done by (or potential energy due to) external forces in moving through their respective displacements (i.e., $-W_E \neq W_I$). In fact, the Clapeyron's[1] theorem states that "the internal energy due to the deformation of a linearly elastic body in equilibrium under external loads is equal to half the work done by the external forces, provided these forces remain fixed during the deformation: $W_I = 0.5\,W_E$."

To gain further understanding of work done by external forces and work done by internal forces, consider a spring–mass system in static equilibrium [see Fig. 2.1.1(a)]. Suppose that the mass m is placed slowly (to eliminate dynamic effects) at the end of a *linear elastic* spring. When it reaches equilibrium state, the spring will elongate by an amount e_0, measured from its undeformed state. In this case, the force due to gravity is equal to $F = mg$. Clearly, F is independent of the extension e_0 in the spring, and F does not change during the course of the extension e going from 0 to its final value e_0 [see Fig. 2.1.1(b)]. The work done by F in moving through de is $F\,de$. Then the total external work done by F is

$$W_E = \int_0^{e_0} F\,de = F\,e_0. \tag{2.1.1}$$

The force induced in the spring, F_s, is proportional to the displacement in the spring. Therefore, the value of F_s goes from zero when $e = 0$ to its final value F_s^f when $e = e_0$ [see Fig. 2.1.1(c)]. The work done by F_s in moving through

---

[1] Benoît Paul Émile Clapeyron (1799-1864) was a French engineer and physicist and one of the founders of thermodynamics.

de is $F_s\, de$. The total work done by F_s is

$$W_I = \int_0^{e_0} F_s(e)\, de. \tag{2.1.2}$$

Since the spring is assumed to be linearly elastic (not a limitation of what is being derived) with spring constant k, we have $F_s = ke$, and the total work done by the spring force (internal) is

$$W_I = \int_0^{e_0} ke\, de = \tfrac{1}{2}\, ke_0^2 = \tfrac{1}{2} F\, e_0. \tag{2.1.3}$$

Thus, Clapeyron's theorem is verified. In this case, the internal work done is due to the straining of the spring.

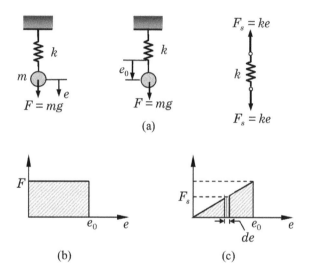

Figure 2.1.1: A linear spring–mass system in equilibrium. (a) Elongation due to weight mg. (b) External work done. (c) Internal work done.

Energy stored in a deformable body is defined as the capacity of the body to perform work. For example, the internal work done is stored as elastic energy, known as the strain energy. In the case of a linear elastic spring, the internal work done is nothing but the strain energy stored in the spring [see Eq. (2.1.3)], which is available to restore the spring back to its original state when the external force causing the deformation is removed. Thus, energy is a measure of the capacity of all forces that can be associated with body to perform work. This understanding can be extended to axial deformation (i.e., stretching) of elastic bars.

2.2 STRAIN ENERGY AND COMPLEMENTARY STRAIN ENERGY

Consider axial deformation of a bar of area of cross section A. The schematic of the elongation of an element of length dx of the bar under the action of axial stress σ_{xx} is shown in Fig. 2.2.1(a). We presume that Poisson's effect is neglected (i.e., contraction perpendicular to the direction of the load is negligible). Note that the element is in static equilibrium, and we wish to determine the work done by the internal force associated with stress σ_{xx}. Suppose that the element is deformed slowly so that axial strain varies from 0 to its final value ε_{xx}. At any instant during the strain variation from ε_{xx} to $\varepsilon_{xx} + d\varepsilon_{xx}$, we assume that σ_{xx} (due to ε_{xx}) is kept constant so that equilibrium is maintained. Then the work done by $A\sigma_{xx}$ in moving through the displacement $d\varepsilon_{xx}\, dx$ is $A\sigma_{xx}\, d\varepsilon_{xx}\, dx = \sigma_{xx}\, d\varepsilon_{xx}\, A\, dx = dU_0\, A\, dx$, where dU_0 denotes the work done per unit volume in portion dx of the bar.

Referring to the stress–strain diagram in Fig. 2.2.1(b), dU_0 represents the elemental area *under* the stress–strain curve. The elemental area in the complement (in the rectangle formed by ε_{xx} and $d\sigma_{xx}$) is

$$dU_0^* = \varepsilon_{xx}\, d\sigma_{xx}.$$

The total area under the curve is obtained by integrating from zero to the final value of the strain (during which the stress changes according to its relation to the strain):

$$U_0 = \int_0^{\varepsilon_{xx}} \sigma_{xx}\, d\varepsilon_{xx}.$$

The quantity U_0 is known as the *strain energy density*.

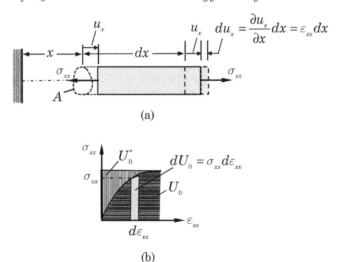

(a)

(b)

Figure 2.2.1: Computation of strain energy for a uniaxially loaded member. (a) Work done by force $\sigma_{xx}\,A$. (b) Definitions of strain energy density and complementary strain energy density.

Similarly, the total area above the stress–strain curve in Fig. 2.2.1(b) is (during which the strain changes according to its relation to the stress)

$$U_0^* = \int_0^{\sigma_{xx}} \varepsilon_{xx}\, d\sigma_{xx},$$

which is termed the *complementary strain energy density*. Note that the strain energy density is expressed in terms of strains, while the complementary strain energy is expressed in terms of stresses. In general, $U_0^* \neq U_0$. They are equal in value when the stress–strain relations are linear (but they are still expressed in terms of different quantities).

The internal work done (or strain energy stored) by $A\sigma_{xx}$ over the whole element of length dx during the entire deformation is

$$dU = \int_0^{\varepsilon_{xx}} \sigma_{xx}\, d\varepsilon_{xx}\, A dx = U_0\, A\, dx,$$

and the total work done or strain energy stored in the entire body is obtained by integrating over the length of the bar:

$$U = \int_0^L A U_0\, dx.$$

This is the internal strain energy stored in the body and is called the *strain energy*. Similarly, the *complementary strain energy* is given by

$$U^* = \int_0^L A U_0^*\, dx.$$

The discussion can be extended to three dimensions. The strain energy per unit volume or the strain energy density is

$$U_0 = \int_0^{\varepsilon_{ij}} \sigma_{ij}\, d\varepsilon_{ij}. \tag{2.2.1}$$

Then the total internal work done by forces due to the stresses σ_{ij} is given by the integral of U_0 over the volume of the body:

$$U = \int_\Omega U_0\, d\Omega, \tag{2.2.2}$$

where $d\Omega$ denotes the volume element and U represents the strain energy stored in the body, and it is known as the *strain energy* of the body.

The complementary strain energy density, U_0^*, can be computed in a similar way from

$$U_0^* = \int_0^{\sigma_{ij}} \varepsilon_{ij}\, d\sigma_{ij}. \tag{2.2.3}$$

The *complementary strain energy* U^* of an elastic body is defined by

$$U^* = \int_\Omega U_0^* \, d\Omega. \tag{2.2.4}$$

For linear elastic solids (i.e., when the material obeys Hooke's law), both the strain energy and complementary strain energy can be expressed as

$$U = \frac{1}{2} \int_\Omega \varepsilon : \sigma(\varepsilon) \, d\Omega, \quad U^* = \frac{1}{2} \int_\Omega \varepsilon(\sigma) : \sigma \, d\Omega, \tag{2.2.5}$$

where ":" denotes the "double-dot product" between two second-order tensors, with the following meaning for rectangular Cartesian components:

$$\varepsilon : \sigma = \varepsilon_{ij} \sigma_{ij}.$$

We note that, although the value of U and U^* is the same for linear elastic solids, U is expressed in terms of strains (or displacements), while U^* is expressed in terms of stresses (or forces).

In general, the internal work done W_I and internal complementary work done W_I^* in a deformable solid consist of energies due to stress and deformation caused by mechanical, thermal, magnetic, and other stimuli. Thus, the strain energy U and complementary strain energy U^* are only a part of W_I and W_I^*, respectively. *In the present study, unless stated otherwise, we only consider energies due to mechanical forces only.* Therefore, we have

$$W_I = U, \quad W_I^* = U^*. \tag{2.2.6}$$

Example 2.2.1 ─────────────────────────────────────

Determine the total (a) strain energy and (b) complementary strain energy of the pin-connected structure shown in Fig. 2.2.2(a), assuming linear stress–strain relation.

Solution: First we note that U is expressed in terms of strains (and displacements) and U^* is expressed in terms of stresses (and forces). Thus, we must find the strains and stresses in each member of the structure.

(a) In order to compute the strain in each of the two members, we consider the deformed geometry shown in Fig. 2.2.2(b). We have

$$
\begin{aligned}
\varepsilon^{(1)} = \frac{B\bar{O}}{BO} - 1 &= \left[\frac{(a-u)^2 + (b+v)^2}{a^2 + b^2} \right]^{\frac{1}{2}} - 1 \\
&= \left[\frac{a^2 + b^2 + u^2 + v^2 + 2(bv - au)}{a^2 + b^2} \right]^{\frac{1}{2}} - 1 \\
&\approx \left[1 + 2\frac{(bv - au)}{a^2 + b^2} \right]^{\frac{1}{2}} - 1 \approx \frac{(bv - au)}{L^2},
\end{aligned}
\tag{1}
$$

$$\varepsilon^{(2)} = \frac{C\bar{O}}{CO} - 1 = \left[\frac{(a-u)^2 + v^2}{a^2}\right]^{\frac{1}{2}} - 1$$

$$= \left[\frac{a^2 + u^2 + v^2 - 2au}{a^2}\right]^{\frac{1}{2}} - 1 \approx \left[1 - 2\frac{u}{a}\right]^{\frac{1}{2}} - 1 \approx -\frac{u}{a}, \tag{2}$$

where $L^2 = a^2 + b^2$.

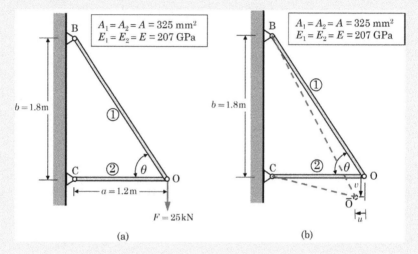

Figure 2.2.2: A truss structure with kinematics of deformation.

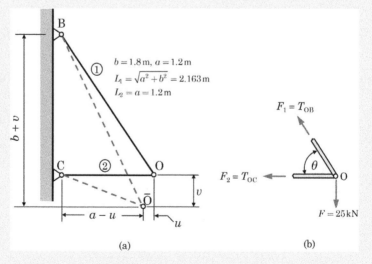

Figure 2.2.3: A truss structure with equilibrium of internal forces.

The total strain energy of the structure is

$$
U = \frac{EA}{2} \left[\left(\varepsilon^{(1)} \right)^2 L_1 + \left(\varepsilon^{(2)} \right)^2 L_2 \right]
$$

$$
= \frac{EA}{2} \left[\left(-\frac{u}{L_2} \right)^2 L_2 + \left(\frac{bv - au}{L_1^2} \right)^2 L_1 \right]
$$

$$
= \frac{EA}{2} \left[\left(\frac{u^2}{a} \right) + \left(\frac{(bv - au)^2}{(a^2 + b^2)^{3/2}} \right) \right]. \tag{3}
$$

(b) To compute the complementary strain energy, we must first compute the forces. From the free-body-diagram of joint O shown in Fig. 2.2.3(b), the axial forces in each member are

$$
F_1 = F_{OB} = \frac{F}{\sin \theta} = F \times \frac{\sqrt{a^2 + b^2}}{b} \quad (= 30.046\,\text{kN}),
$$

$$
F_2 = F_{OC} = -F \cot \theta = -\frac{Fa}{b} \quad (= -16.667\,\text{kN}). \tag{4}
$$

Then the stress $\sigma^{(i)}$ in ith member is calculated as (with $\sigma^{(i)} = F_i/A_i$)

$$
\sigma^{(1)} = \frac{F_1}{A_1} = \frac{F}{A} \times \frac{\sqrt{a^2 + b^2}}{b} \quad (= 92.449\,\text{MPa}),
$$

$$
\sigma^{(2)} = \frac{F_2}{A_2} = -\frac{Fa}{Ab} \quad (= -51.283\,\text{MPa}). \tag{5}
$$

where the total complementary energy is

$$
U^* = \frac{A}{2E} \left[\left(\sigma^{(1)} \right)^2 L_1 + \left(\sigma^{(2)} \right)^2 L_2 \right] = \frac{F^2}{2AEb^2} \left[\left(a^2 + b^2 \right)^{\frac{3}{2}} + a^3 \right]. \tag{6}
$$

2.3 TOTAL POTENTIAL ENERGY AND TOTAL COMPLEMENTARY ENERGY

As discussed in the previous sections, strain energy U is the internal energy stored in the body due to the deformation of the body, and it is the same as the work done by internal forces in moving through their respective displacements. The strain energy is expressed in terms of strains or displacements. This requires the use of constitutive relations to express the internal stresses in terms of strains. The complementary strain energy U^* is also the energy stored in the body due to the deformation of the body, and it is the same as the work done by internal forces in moving through their respective displacements. However, the complementary strain energy is expressed in terms of stresses or forces with the help of constitutive equations. *Unless stated otherwise, in the following discussion, linear stress–strain relations are assumed.*

The work done on the body by external forces is denoted by W_E. When W_E involves only the work done by external forces that are independent of the deformation, we call it potential due to external loads and denote it by V_E (since we used V to denote the transverse shear force in the case of beams, a subscript is used to distinguish between the shear force and potential energy due to external loads). The sum of U and V_E is termed the *total potential energy*, and it is denoted by $\Pi = U + V_E$:

$$\Pi(\mathbf{u}) = U + V_E = \frac{1}{2}\int_\Omega \boldsymbol{\sigma}(\boldsymbol{\varepsilon}) : \boldsymbol{\varepsilon}\, d\Omega - \left[\int_\Omega \mathbf{f}\cdot\mathbf{u}\, d\Omega + \oint_\Gamma \mathbf{t}\cdot\mathbf{u}\, d\Gamma\right], \quad (2.3.1)$$

where Ω denotes the volume occupied by the body and Γ is its total boundary. Similarly, the sum of U^* and V_E is termed the *total complementary energy*:

$$\Pi^*(\boldsymbol{\sigma}) = U^* + V_E = \frac{1}{2}\int_\Omega \boldsymbol{\sigma} : \boldsymbol{\varepsilon}(\boldsymbol{\sigma})\, d\Omega - \left[\int_\Omega \mathbf{f}\cdot\mathbf{u}\, d\Omega + \oint_\Gamma \mathbf{t}\cdot\mathbf{u}\, d\Gamma\right]. \quad (2.3.2)$$

For example, the total potential energy of a bar of length L, modulus E, and area of cross section A, subjected to distributed axial load $f(x)$ (measured per unit length) and point load P at $x = L$ is given by

$$\Pi(u) = \frac{1}{2}\int_\Omega \sigma_{xx}\varepsilon_{xx} d\Omega - \left[\int_0^L f(x)\,u(x)\,dx + P\,u(L)\right]$$
$$= \frac{1}{2}\int_0^L E(x)A(x)\left(\frac{du}{dx}\right)^2 dx - \left[\int_0^L f(x)\,u(x)\,dx + P\,u(L)\right], \quad (2.3.3)$$

where we have used the following kinematic and constitutive relations:

$$\varepsilon_{xx}(x) = \frac{du}{dx}, \quad \sigma_{xx}(x) = E(x)\,\varepsilon_{xx}(x), \quad (2.3.4)$$

and evaluated the volume integral (because all variables are only functions of x) as

$$\int_\Omega (\cdot)\, d\Omega = \int_0^L \int_A (\cdot)\, dA\, dx = \int_0^L (\cdot)\, A(x)\, dx. \quad (2.3.5)$$

The complementary energy will be simply in terms of σ_{xx}.

2.4 VIRTUAL WORK

2.4.1 VIRTUAL DISPLACEMENTS

From purely geometrical considerations, a given mechanical system can take many possible configurations. The set of configurations that satisfy the geometric constraints (e.g., geometric boundary conditions) of the system is called the *set of admissible configurations*. Of all admissible configurations, only one of them corresponds to the equilibrium configuration under the applied loads, and it is this configuration that also satisfies Newton's second law (i.e., equilibrium of forces and moments). The admissible configurations are restricted to a neighborhood of the true configuration so that they are obtained from infinitesimal *variations* of the true configuration (i.e., infinitesimal movement of the material points). During such variations, the geometric constraints of the system are not violated, and all applied forces are fixed at their actual equilibrium values. When a mechanical system experiences such variations in its equilibrium configuration, it is said to undergo *virtual displacements*. These displacements need not have any relationship with the actual displacements, which might occur due to a change in the loads and/or boundary conditions. The displacements are called virtual because they are *imagined* to take place (i.e., hypothetical) with the actual loads acting at their fixed values.

As an example, consider a cantilever beam with fixed end at $x = 0$; the geometric boundary conditions are:[2] $u(0) = \hat{u}$, $w(0) = \hat{w}$, and $dw/dx = \hat{\theta}$ at $x = 0$, where u is the axial displacement, w is the transverse deflection, and $\theta \equiv -(dw/dx)$ is the slope. The set of all functions $w(x)$ and $u(x)$ that satisfy the geometric boundary conditions is called *the space of admissible configurations*. In the present problem, the space of admissible configurations consists of pairs of elements $\{(u_i, w_i)\}$ of the form

$$u_1(x) = \hat{u} + a_1 x, \qquad\qquad w_1(x) = \hat{w} + \hat{\theta}x + b_1 x^2,$$
$$u_2(x) = \hat{u} + a_1 x + a_2 x^2, \quad w_2(x) = \hat{w} + \hat{\theta}x + b_1 x^2 + b_2 x^3,$$

where a_i and b_i are constants. The pair (u, w) that also satisfies, in addition to the geometric boundary conditions, the equilibrium equations and force boundary conditions of the problem is the equilibrium solution. The virtual displacements, $\delta u(x)$ and $\delta w(x)$, must necessarily be of the form

$$\delta u_1 = a_1 x, \;\; \delta w_1 = b_1 x^2; \quad \delta u_2 = a_1 x + a_2 x^2, \;\; \delta w_2 = b_1 x^2 + b_2 x^3, \quad \text{and so on,}$$

which satisfy the homogeneous form of the specified geometric boundary conditions:

$$\delta w(0) = 0, \quad \left.\frac{d\delta w}{dx}\right|_{x=0} = 0, \quad \delta u(0) = 0. \qquad (2.4.1)$$

Thus, the virtual displacements at the boundary points at which the geometric conditions are specified (independent of the specified values) are necessarily zero.

The work done by the actual forces through a virtual displacement of the actual configuration is called *virtual work* done by actual forces. If we denote the virtual displacement by $\delta \mathbf{u}$, the virtual work done by a constant force $\mathbf{F}$ is given by

$$\delta W = \mathbf{F} \cdot \delta \mathbf{u}. \qquad (2.4.2)$$

---

[2]In this case, the specified geometric boundary conditions are all zero: $\hat{u} = \hat{w} = \hat{\theta} = 0$.

The virtual work done by actual forces in moving through virtual displacements in a deformable body consists of two parts: virtual work done by internal forces, δW_I, and virtual work done by external forces, δW_E. These may be computed as discussed next.

Consider a deformable body of volume Ω and closed surface area Γ. Suppose that the body is subjected to a body force $\mathbf{f}(\mathbf{x})$ per unit volume of the body, a specified surface force $\mathbf{t}(s)$ per unit area on a portion Γ_σ of the boundary, and specified displacement $\mathbf{u}$ on the remaining portion Γ_u of the boundary such that $\Gamma = \Gamma_u \cup \Gamma_\sigma$ and $\Gamma_u \cap \Gamma_\sigma$ is empty. For this case, the virtual displacement $\delta\mathbf{u}(\mathbf{x})$ is any function, small in magnitude (to keep the system in equilibrium), that satisfies the requirement

$$\delta\mathbf{u} = \mathbf{0} \quad \text{on} \ \ \Gamma_u.$$

The virtual work done by actual forces $\mathbf{f}$ and $\mathbf{t}$ in moving through the virtual displacement $\delta\mathbf{u}$ is

$$\delta W_E = -\left(\int_\Omega \mathbf{f} \cdot \delta\mathbf{u} \, d\Omega + \int_{\Gamma_\sigma} \mathbf{t} \cdot \delta\mathbf{u} \, d\Gamma \right). \tag{2.4.3}$$

The negative sign in front of the expression indicates that work is performed *on* the body.

As a result of the application of the loads, the body develops internal forces in the form of stresses. These stresses also perform work when the body is given a virtual displacement. In this study, we shall be concerned mainly with only *ideal systems*. An ideal system is one in which no work is dissipated by friction, heat, or other means. We assume that the virtual displacement $\delta\mathbf{u}$ is applied slowly from zero to its final value. Associated with the virtual displacement $\delta\mathbf{u}$ is the virtual strain $\delta\boldsymbol{\varepsilon}$, which can be computed according to the strain–displacement relation (assuming small strains, i.e., for the linear case):

$$\delta\boldsymbol{\varepsilon} = \tfrac{1}{2}\left[\boldsymbol{\nabla}(\delta\mathbf{u}) + (\boldsymbol{\nabla}(\delta\mathbf{u}))^{\mathsf{T}} \right]. \tag{2.4.4}$$

The internal virtual work stored, that is, the work done by the force due to actual stress $\boldsymbol{\sigma}$ in moving through the virtual displacement associated with the virtual strain $\delta\boldsymbol{\varepsilon}$ in the body *per unit volume*, analogous to the calculation of the actual strain energy density, is the virtual strain energy density, δU_0:

$$\delta U_0 = \int_0^{\delta\varepsilon} \boldsymbol{\sigma} : d(\delta\boldsymbol{\varepsilon}) = \int_0^{\delta\varepsilon_{ij}} \sigma_{ij} \, d(\delta\varepsilon_{ij}) = \sigma_{ij} \, \delta\varepsilon_{ij}. \tag{2.4.5}$$

We note that the result in Eq. (2.4.5) is arrived without the use of a constitutive equation, because $\boldsymbol{\sigma}$ is not a function of $\delta\boldsymbol{\varepsilon}$. The total internal virtual work stored in the body is denoted by δW_I, and it is equal to

$$\delta W_I = \int_\Omega \delta U_0 \, d\Omega = \int_\Omega \sigma_{ij} \, \delta\varepsilon_{ij} \, d\Omega. \tag{2.4.6}$$

2.4.2 VIRTUAL FORCES

Analogous to virtual displacements, we can also think of virtual forces. If a body is imagined to be subjected to a set of *self-equilibrating force system* $\delta\mathbf{F}$, the virtual work done by the virtual forces in moving through actual displacements $\mathbf{u}$ is given by

$$\delta W^* = \delta\mathbf{F} \cdot \mathbf{u},$$

and it is called *complementary virtual work*. The complementary internal virtual work and complementary external virtual work for a deformable body can be expressed, following the ideas already presented, as

$$\delta W_E^* = -\left(\int_\Omega \delta\mathbf{f} \cdot \mathbf{u}\,d\Omega + \int_\Gamma \delta\mathbf{t} \cdot \mathbf{u}\,d\Gamma \right), \tag{2.4.7}$$

$$\delta W_I^* = \int_\Omega \delta U_0^*\,d\Omega = \int_\Omega \boldsymbol{\varepsilon} : \delta\boldsymbol{\sigma}\,d\Omega, \tag{2.4.8}$$

where U_0^* is the complementary strain energy density

$$\delta U_0^* = \int_0^{\delta\sigma} \boldsymbol{\varepsilon} : d(\delta\boldsymbol{\sigma}) = \boldsymbol{\varepsilon} : \delta\boldsymbol{\sigma} = \varepsilon_{ij}\,\delta\sigma_{ij}. \tag{2.4.9}$$

Note that in selecting a virtual force system, one must make sure that the virtual forces are in equilibrium among themselves.

2.5 CALCULUS OF VARIATIONS AND DUALITY PAIRS

2.5.1 THE VARIATIONAL OPERATOR

The delta operator δ used in conjunction with virtual quantities has special importance in variational methods. The operator is called the *variational operator* because it is used to denote a variation (or change) in a given quantity. In this section, we discuss certain operational properties of δ and elements of variational calculus. With these tools in hand, we can study energy and variational principles of solid and structural mechanics.

Let $\mathbf{u}(\mathbf{x})$ be the true displacement (i.e., the one corresponding to equilibrium) at a point $\mathbf{x}$ in the deformed configuration of a body Ω, and suppose that $\mathbf{u} = \hat{\mathbf{u}}$ on the boundary Γ_u of the total boundary Γ. Then an admissible configuration is of the form

$$\bar{\mathbf{u}} = \mathbf{u} + \alpha\mathbf{v} \tag{2.5.1}$$

everywhere in the body, where $\mathbf{v}$ is an arbitrary function that satisfies the homogeneous geometric boundary condition of the system:

$$\mathbf{v} = \mathbf{0} \quad \text{on } \Gamma_u. \tag{2.5.2}$$

Here $\alpha \mathbf{v}$ represents a variation of the actual displacement vector $\mathbf{u}$. It should be understood that *the variations are small enough* (i.e., α is small) *not to disturb the equilibrium of the system, and the variation is consistent with the geometric constraint of the system*. Equation (2.5.1) defines a set of varied configurations; an infinite number of configurations $\bar{\mathbf{u}}$ can be generated for a fixed $\mathbf{v}$ by assigning values to α. All of these configurations satisfy the specified geometric boundary conditions on boundary Γ_u, and therefore they constitute the set of admissible configurations for the system. For any $\mathbf{v}$, all configurations reduce to the actual one when α is set to zero. Therefore, for any *fixed* point $\mathbf{x}$, $\alpha \mathbf{v}$ can be viewed as a change or *variation* in the actual displacement $\mathbf{u}$. This variation is often denoted by $\delta \mathbf{u}$:

$$\delta \mathbf{u} = \alpha \, \mathbf{v} \tag{2.5.3}$$

and $\delta \mathbf{u}$ is called the *first variation* of $\mathbf{u}$. We can show that a differential operator ∇ and the variational operator δ can be interchanged without changing the value:

$$\delta(\nabla \mathbf{u}) = \alpha(\nabla \mathbf{v}) = \nabla(\alpha \mathbf{v}) = \nabla(\delta \mathbf{u}). \tag{2.5.4}$$

For simplicity of discussion, consider a function of the dependent variable $u(x)$ and its derivative $u' \equiv du/dx$ in a one-dimensional problem:

$$F = F(x, u, u'). \tag{2.5.5}$$

For a fixed x, the change in F associated with a variation in u (and hence u') is

$$\Delta F = F(x, u + \alpha v, u' + \alpha v') - F(x, u, u')$$

or

$$\begin{aligned}
\Delta F = {} & F(x, u, u') + \frac{\partial F}{\partial u}\alpha v + \frac{\partial F}{\partial u'}\alpha v' + \frac{(\alpha v)^2}{2!}\frac{\partial^2 F}{\partial u^2} \\
& + \frac{2(\alpha v)(\alpha v')}{2!}\frac{\partial^2 F}{\partial u \partial u'} + \cdots - F(x, u, u') \\
= {} & \frac{\partial F}{\partial u}\alpha v + \frac{\partial F}{\partial u'}\alpha v' + O(\alpha^2),
\end{aligned} \tag{2.5.6}$$

where $O(\alpha^2)$ denotes terms of order α^2 and higher. The first total variation of $F(x, u, u')$ is defined by

$$\delta F \equiv \alpha \left[\lim_{\alpha \to 0} \frac{\Delta F}{\alpha}\right] = \frac{\partial F}{\partial u}\delta u + \frac{\partial F}{\partial u'}\delta u'. \tag{2.5.7}$$

There is an analogy between the first variation δF of F and the total differential dF of F. The total differential of F is

$$dF = \frac{\partial F}{\partial x}dx + \frac{\partial F}{\partial u}du + \frac{\partial F}{\partial u'}du'. \tag{2.5.8}$$

Since x is fixed during the variation of u to $u + \delta u$, we have $dx = 0$, and the analogy between δF in Eq. (2.5.7) and dF in Eq. (2.5.8) becomes apparent. That is, the variational operator δ acts like a differential operator with respect to the dependent variables. Indeed, the laws of variation of sums, products, ratios, powers, and so forth are completely analogous to the corresponding laws of differentiation.

If $G = G(u, v, w)$ is a function of several dependent variables u, v, and w (and possibly their derivatives), the total variation of G is the sum of partial variations with respect to u, v, and w:

$$\delta G = \delta_u G + \delta_v G + \delta_w G, \tag{2.5.9}$$

where, for example, δ_u denotes the partial variation with respect to u. The variational operator can be interchanged with differential (as already shown) and integral operators:

$$\delta\left(\frac{du}{dx}\right) = \frac{d}{dx}(\delta u), \quad \delta\left(\int_0^a u\, dx\right) = \int_0^a \delta u\, dx. \tag{2.5.10}$$

All of the discussion can be extended to functions F that depend on more than one dependent variable in two or three dimensions.

2.5.2 FUNCTIONALS AND THEIR VARIATIONS

In the study of variational formulations of continuum problems, we encounter integrals of functions of dependent variables that are themselves functions of other parameters, such as position and time. Examples of such integral statements are provided by the strain energy and complementary strain energy expressions of preceding sections [see, in particular, Eq. (2.3.1)]. Such integral expressions whose value is a real number are termed functionals.

A functional $\mathcal{F}$ is said to be *linear* if

$$\mathcal{F}(\alpha u + \beta v) = \alpha \mathcal{F}(u) + \beta \mathcal{F}(v) \tag{2.5.12}$$

for all real numbers α and β and dependent variables u, v. A *quadratic functional* $Q(u)$ is a functional that satisfies the relation

$$Q(\alpha u) = \alpha^2 Q(u) \tag{2.5.13}$$

for all real numbers α and the dependent variable u.

The first variation of a functional of u (and possibly its derivatives) can be calculated as follows: Let $I(u)$ denote the integral defined in the interval (a, b):

$$I(u) = \int_a^b F(x, u, u')\, dx, \tag{2.5.14}$$

where F is a function, in general, of x, u, and du/dx. When the limits a and b are independent of u, the first variation of the functional $I(u)$ is

$$\delta I(u; \delta u) = \delta \int_a^b F(x, u, u') \, dx = \int_a^b \delta F \, dx = \int_a^b \left(\frac{\partial F}{\partial u} \delta u + \frac{\partial F}{\partial u'} \delta u' \right) dx.$$
$$(2.5.15)$$

When the limits of integration, a and b, depend on the dependent variable, that is, $a = a(u)$ and $b = b(u)$, the Leibniz rule is used to calculate the first variation:

$$\delta I(u; v) = \alpha \left[\frac{d}{d\alpha} \int_{a(u+\alpha v)}^{b(u+\alpha v)} F(x, u + \alpha v, u' + \alpha v') dx \right]_{\alpha=0}. \qquad (2.5.16)$$

Using the Leibniz rule, we obtain

$$\delta I(u; \delta u) = \int_a^b \delta F(x, u, u') dx + F(b, u(b), u'(b))\delta b - F(a, u(a), u'(a))\delta a.$$
$$(2.5.17)$$

Thus, the variational operator can be used to compute the variation of any functional.

2.5.3   FUNDAMENTAL LEMMA OF VARIATIONAL CALCULUS

The *fundamental lemma of calculus of variations* is useful in obtaining differential equations from variational principles involving integral statements. The lemma can be stated as follows:

Lemma 2.1: *For any integrable function G, if the statement*

$$\int_a^b G(x) \cdot \eta(x) \, dx = 0 \qquad (2.5.18)$$

holds for any arbitrary continuous function $\eta(x)$ defined in the interval (a, b), then it follows that

$$G(x) = 0, \quad x \in (a, b).$$

Proof: Since $\eta(x)$ is arbitrary, it can be replaced by G. We have

$$\int_a^b G^2(x) \, dx = 0. \qquad (2.5.19)$$

Since an integral of a positive function, G^2, is positive, the aforementioned statement holds only if $G = 0$.

A more general statement of the fundamental lemma is as follows: If η is an arbitrary function of x in $a < x < b$ and $\eta(a)$ is arbitrary at $x = a$, then the statement

$$\int_a^b G(x) \, \eta(x) \, dx + B(a)\eta(a) = 0 \qquad (2.5.20)$$

implies that

$$G(x) = 0 \quad a < x < b \quad \text{and} \quad B(a) = 0. \tag{2.5.21}$$

The proof is very simple. Since $\eta(x)$ is independent of $\eta(a)$, one can set $\eta(a) = 0$ and $\eta(x) \neq 0$, and obtain $G(x) = 0$ for $a < x < b$. Similarly, taking $\eta(x) = 0$ and $\eta(x) \neq 0$, one can obtain $B(a) = 0$.

2.5.4   EXTREMUM OF A FUNCTIONAL

In the study of problems by variational principles and methods, we seek the extremum of integrals of functions, $F = F(x, u(x), u'(x))$, or functionals. Consider a simple, but typical, variational problem. Find a function $u = u(x)$ such that $u(a) = u_a$ and $u(b) = u_b$, and

$$I(u) = \int_a^b F(x, u(x), u'(x)) \, dx \tag{2.5.22}$$

is a minimum.

In analyzing the problem, we are not interested in all functions u, but only in those functions that satisfy the stated boundary (or end) conditions. The set of all such functions is called, for obvious reasons, the *set of competing functions* (or set of admissible functions). We shall denote the set by $\mathcal{C}$. The problem is to seek an element u from $\mathcal{C}$, which renders $I(u)$ a minimum. If $u \in \mathcal{C}$, then $(u + \alpha v) \in \mathcal{C}$ for every v satisfying the conditions $v(a) = v(b) = 0$. The space of all such elements is called the *space of admissible variations*, as already stated earlier.

Analogous to the necessary condition for the minimum of a function, a necessary condition for a functional $I(u)$ to be a minimum at u is that its first variation be zero,

$$\delta I(u) = 0. \tag{2.5.23}$$

Similarly, the sufficient condition for a functional to assume a relative minimum (maximum) is that the second variation $\delta^2 I(u)$ is greater (less) than zero.

2.5.5   THE EULER EQUATIONS AND DUALITY PAIRS

We now return to the problem of determining the minimum of a functional

$$I(u) = \int_a^b F(x, u, u') \, dx \tag{2.5.24}$$

subject to the end conditions:

$$u(a) = u_a, \quad u(b) = u_b. \tag{2.5.25}$$

It is clear that any candidate for the minimizing functional should satisfy the end conditions in Eq. (2.5.25) and be sufficiently differentiable (twice in the present case, as we shall see shortly). The set of all such functions is the set of admissible functions or competing functions for the present case. Functions from the admissible set can be viewed as smooth (i.e., differentiable twice) functions passing through points (a, u_a) and (b, u_b). Clearly, any element $\bar{u} \in \mathcal{C}$ (the set of competing functions) has the form

$$\bar{u} = u + \alpha\, v \equiv u + \delta u, \tag{2.5.26}$$

where α is a small number and v is a sufficiently differentiable function that satisfies the *homogeneous form* of the end conditions (because $\bar{u}$ must satisfy the specified end conditions) in Eq. (2.5.25):

$$\delta u(a) = 0, \quad \delta u(b) = 0 \tag{2.5.27}$$

and u is the function that minimizes the functional in Eq. (2.5.24). The set of all functions v or δu is the set of admissible variations, $\mathcal{H}$. Now assuming that, for each admissible function $\bar{u}$, $F(x, \bar{u}, \bar{u}')$ exists and is continuously differentiable with respect to its arguments and $I(\bar{u})$ takes one and only one real value, we seek the particular function $u(x)$ that makes the integral a minimum.

The necessary condition for $I(u)$ to attain a minimum is [see Eq. (2.5.23)]

$$0 = \delta I = \int_a^b \left(\frac{\partial F}{\partial u} \delta u + \frac{\partial F}{\partial u'} \delta u' \right) dx, \tag{2.5.28}$$

The statement in Eq. (2.5.28) is of no use unless we can deduce some result from the statement. The integrand has two terms, one with δu and the other with $\delta u'$, which are not independent of each other (the second one is a derivative of the first). If we can transfer the differentiation from $\delta u'$ to its coefficient, then the two terms in the integrand can be combined into one expression with δu multiplying it, and we can use the fundamental lemma of variational calculus to deduce that this expression, because δu is arbitrary, is necessarily zero. Integrating the second term[3] in the last equation by parts to transfer differentiation from $\delta u'$ to $\partial F/\partial u'$, we obtain

$$0 = \int_a^b \delta u \left[\frac{\partial F}{\partial u} - \frac{d}{dx}\left(\frac{\partial F}{\partial u'} \right) \right] dx + \left[\frac{\partial F}{\partial u'} \delta u \right]_a^b. \tag{2.5.29}$$

The boundary expression in Eq. (2.5.29) is of significance in defining the duality. The *primary variable* of the formulation is identified as u by removing

---

[3]The integration by parts involves recognizing the identity:

$$\frac{d}{dx}\left[G(x, u, u')\delta u \right] = G'\, \delta u + G\, \delta u' \quad \text{or} \quad G\, \delta u' = (G\, \delta u)' - G'\, \delta u \quad \text{with} \quad G = \frac{\partial F}{\partial u'}.$$

the delta symbol δ from δu in the boundary expression. The form of δu in the boundary expression is without a derivative. If there were a derivative, say $\delta u'$, then the primary variable would have been u' (which is the case when the functional $I(u)$ contains higher derivatives of the dependent variable u). The coefficient of δu in the boundary expression, namely, $\partial F/\partial u'$, is called the *secondary variable*. This definition is unambiguous, and the names – primary and secondary variables are just names (the names could have been interchanged). The duality pair $(u, \partial F/\partial u')$ represents the cause and effect (i.e., one causes the other). The idea of a duality pairs is fundamental in mechanics, and it plays a critical role in the specification of boundary conditions at a point, as will be apparent in the coming chapters.

The boundary term in Eq. (2.5.29) vanishes because δu is zero at $x = a$ and $x = b$ [see Eq. (2.5.27)]. The fact that δu is arbitrary inside the internal (a, b) and yet Eq. (2.5.29) holds implies, by the fundamental lemma of the calculus of variations, the expression in the square brackets be zero identically:

$$\frac{\partial F}{\partial u} - \frac{d}{dx}\left(\frac{\partial F}{\partial u'}\right) = 0 \quad \text{in } a < x < b. \tag{2.5.30}$$

Equation (2.5.30) is called the *Euler equation* of the functional in Eq. (2.5.24). Of all the admissible functions, the one that satisfies (i.e., the solution of) Eq. (2.5.30) is the true minimizer of the functional $I(u)$.

Next, consider the problem of finding the pair (u, v) defined on a two-dimensional region Ω (note that $d\Omega = dxdy$ and $d\Gamma = ds$) such that the following functional is to be minimized:

$$I(u, v) = \int_\Omega F(x, y, u, v, u_{,x}, v_{,x}, u_{,y}, v_{,y}) \, d\Omega, \tag{2.5.31}$$

where $u_{,x} = \partial u/\partial x$, $u_{,y} = \partial u/\partial y$, and so on. For the moment we assume that u and v are specified on the boundary Γ of Ω. The vanishing of the first variation of $I(u, v)$ is written as

$$\delta I(u, v) = \delta_u I(u, v) + \delta_v I(u, v) = 0.$$

Here δ_u and δ_v denote partial variations with respect to u and v, respectively. We have

$$\delta I(u, v; \delta u, \delta v) = \int_\Omega \left(\frac{\partial F}{\partial u}\delta u + \frac{\partial F}{\partial u_{,x}}\delta u_{,x} + \frac{\partial F}{\partial u_{,y}}\delta u_{,y} \right.$$
$$\left. + \frac{\partial F}{\partial v}\delta v + \frac{\partial F}{\partial v_{,x}}\delta v_{,x} + \frac{\partial F}{\partial v_{,y}}\delta v_{,y} \right) d\Omega. \tag{2.5.32}$$

The next step in the development involves the use of integration by parts or the gradient theorem on the second, third, fifth, and sixth terms in

Eq. (2.5.32). Consider the second term. We have

$$
\int_\Omega \frac{\partial F}{\partial u_{,x}} \frac{\partial \delta u}{\partial x}\, d\Omega = \int_\Omega \left[\frac{\partial}{\partial x}\left(\frac{\partial F}{\partial u_{,x}} \delta u \right) - \frac{\partial}{\partial x}\left(\frac{\partial F}{\partial u_{,x}} \right) \delta u \right] d\Omega
$$

$$
= \oint_\Gamma \frac{\partial F}{\partial u_{,x}} \delta u\, n_x\, d\Gamma - \int_\Omega \frac{\partial}{\partial x}\left(\frac{\partial F}{\partial u_{,x}} \right) \delta u\, d\Omega. \qquad (2.5.33)
$$

Using a similar procedure on the other terms and collecting the coefficients of δu and δv separately, we obtain

$$
\begin{aligned}
0 = \int_\Omega \Bigg\{ &\left[\frac{\partial F}{\partial u} - \frac{\partial}{\partial x}\left(\frac{\partial F}{\partial u_{,x}} \right) - \frac{\partial}{\partial y}\left(\frac{\partial F}{\partial u_{,y}} \right) \right] \delta u \\
+ &\left[\frac{\partial F}{\partial v} - \frac{\partial}{\partial x}\left(\frac{\partial F}{\partial v_{,x}} \right) - \frac{\partial}{\partial y}\left(\frac{\partial F}{\partial v_{,y}} \right) \right] \delta v \Bigg\} d\Omega \\
+ &\oint_\Gamma \left[\left(\frac{\partial F}{\partial u_{,x}} n_x + \frac{\partial F}{\partial u_{,y}} n_y \right) \delta u + \left(\frac{\partial F}{\partial v_{,x}} n_x + \frac{\partial F}{\partial v_{,y}} n_y \right) \delta v \right] d\Gamma. \quad (2.5.34)
\end{aligned}
$$

Now examining the boundary expressions and noting that δu and δv are independent variations, we identify u and v as the primary variables and

$$
\frac{\partial F}{\partial u_{,x}} n_x + \frac{\partial F}{\partial u_{,y}} n_y, \qquad \frac{\partial F}{\partial v_{,x}} n_x + \frac{\partial F}{\partial v_{,y}} n_y
$$

as the secondary variables. Thus, there are two duality pairs:

$$
\left(u,\ \frac{\partial F}{\partial u_{,x}} n_x + \frac{\partial F}{\partial u_{,y}} n_y \right), \quad \left(v,\ \frac{\partial F}{\partial v_{,x}} n_x + \frac{\partial F}{\partial v_{,y}} n_y \right).
$$

Since (u, v) are specified on Γ, $\delta u = \delta v = 0$ and the boundary expressions in Eq. (2.5.34) vanish. Then, since δu and δv are arbitrary and independent of each other in Ω, the fundamental lemma yields the following Euler equations:

$$
\delta u: \qquad \frac{\partial F}{\partial u} - \frac{\partial}{\partial x}\left(\frac{\partial F}{\partial u_{,x}} \right) - \frac{\partial}{\partial y}\left(\frac{\partial F}{\partial u_{,y}} \right) = 0, \qquad (2.5.35)
$$

$$
\delta v: \qquad \frac{\partial F}{\partial v} - \frac{\partial}{\partial x}\left(\frac{\partial F}{\partial v_{,x}} \right) - \frac{\partial}{\partial y}\left(\frac{\partial F}{\partial v_{,y}} \right) = 0. \qquad (2.5.36)
$$

2.5.6 NATURAL AND ESSENTIAL BOUNDARY CONDITIONS

First, consider the problem of minimizing the functional in Eq. (2.5.24) subject to no end conditions (hence $\delta u \neq 0$ at $x = a$ and $x = b$). The necessary condition for I to attain minimum yields, as before, the result in Eq. (2.5.29). Now suppose that $\frac{\partial F}{\partial u'}$ and δu are selected such that

$$
\frac{\partial F}{\partial u'} \delta u = 0 \quad \text{for } x = a \text{ and } x = b. \qquad (2.5.37)
$$

Then using the fundamental lemma of the calculus of variations, we obtain the Euler equation in Eq. (2.5.30).

From Eq. (2.5.37), it is clear that the *duality pair* is

$$\left(u, \ \frac{\partial F}{\partial u'} \right).$$

Equation (2.5.37) is satisfied identically for any of the following combinations:

$$
\begin{aligned}
&(1) && \delta u(a) = 0, && \delta u(b) = 0. \\[2mm]
&(2) && \delta v(a) = 0, && \frac{\partial F}{\partial u'}(b) = 0. \\[2mm]
&(3) && \frac{\partial F}{\partial u'}(a) = 0, && \delta v(b) = 0. \\[2mm]
&(4) && \frac{\partial F}{\partial u'}(a) = 0, && \frac{\partial F}{\partial u'}(b) = 0.
\end{aligned}
\qquad (2.5.38)
$$

The requirement that $\delta u = 0$ at an end point is equivalent to the requirement that u be specified (to be some value, including zero) at that point. The end conditions in Eq. (2.5.38) are classified into two types: *essential boundary conditions* are those in which the primary variable u is specified at the boundary, and *natural boundary conditions* are those in which the secondary variables are specified. Thus, we have the following types of boundary conditions:

Essential boundary conditions:

$$\delta u = 0 \ \text{ or specify } \ u = \hat{u} \ \text{ on the boundary.} \qquad (2.5.39)$$

Natural boundary conditions:

$$\frac{\partial F}{\partial u'} = 0 \text{ on the boundary.} \qquad (2.5.40)$$

Essential boundary conditions are also known as *Dirichlet* or *geometric* boundary conditions, and natural boundary conditions are known as *Neumann* or *force* boundary conditions. In general, *only one element of each of the duality pairs may be specified at any point of the boundary*. When neither element of a duality pair is specified but a relation is known between the elements of the pair, the boundary conditions are known as the *mixed boundary conditions*. In a given problem, only one of the four combinations given in Eq. (2.5.38) can be specified. Problems in which all of the boundary conditions are of essential type are called *Dirichlet boundary-value problems*, and those in which all of the boundary conditions are of natural type are called *Neumann boundary-value problems*. *Mixed boundary-value problems* are those in which both essential and natural boundary conditions are specified.

Example 2.5.1

Consider a linear elastic bar of length L, area of cross section A, and Young's modulus E. Assume that it is fixed at the left end, $x = 0$, supported axially by a linear elastic spring with spring constant k at $x = L$. The bar is subjected to distributed axial load $f(x)$ and a point load P at $x = L$, as shown in Fig. 2.5.1. Supposing that the total potential energy is a minimum at equilibrium, derive the equation of equilibrium (i.e., the Euler equation) and natural boundary conditions for the problem. Assume that the strains are small and Hooke's law holds.

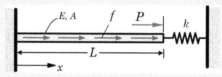

Figure 2.5.1: A bar fixed at the left end and spring-supported at the right end.

Solution: The total potential energy functional for the stretching of a linear elastic bar is

$$\Pi(u) = \int_0^L \left[\frac{EA}{2} \left(\frac{du}{dx} \right)^2 \right] dx + \frac{k}{2} [u(L)]^2 - \int_0^L fu \, dx - Pu(L), \qquad (1)$$

where u denotes the axial displacement of the bar. The first term in $\Pi(u)$ represents the strain energy U stored in the bar, the second term corresponds to the strain energy stored in the spring at $x = L$, and the third and fourth terms denote the external work done V_E on the bar by the distributed load f and point load P, respectively.

In order that Π be a minimum, its first variation must be zero: $\delta \Pi = 0$. The first variation is

$$\delta \Pi(u) = \int_0^L \left(EA \frac{du}{dx} \frac{d\delta u}{dx} - f \delta u \right) dx + ku(L) \, \delta u(L) - P \delta u(L), \qquad (2)$$

where δu is arbitrary in $0 < x < L$ and at $x = L$ but satisfies the condition $\delta u(0) = 0$. To use the fundamental lemma of variational calculus, we must relieve δu of any differentiation. Integrating the first term by parts, we obtain

$$\delta \Pi(u; \delta u) = \int_0^L \left[-\frac{d}{dx} \left(EA \frac{du}{dx} \right) - f \right] \delta u \, dx + \left[EA \frac{du}{dx} \delta u \right]_0^L + [ku(L) - P] \, \delta u(L)$$

$$= \int_0^L \delta u \left[-\frac{d}{dx} \left(EA \frac{du}{dx} \right) - f \right] dx + \delta u(L) \left[\left(EA \frac{du}{dx} \right)_{x=L} + ku(L) - P \right]$$

$$- \delta u(0) \left(EA \frac{du}{dx} \right)_{x=0}. \qquad (3)$$

The last term is zero because $\delta u(0) = 0$. Also, from the boundary term

$$\delta u(0) \cdot \left(EA \frac{du}{dx} \right)_{x=0},$$

we immediately identify that u as the primary variable and $N \equiv EA(du/dx)$, which is the axial force, as the secondary variable. Thus, one may specify either u or N at a boundary point, or one may have a relation between u and N at the boundary point (which is the case at $x = L$).

Since $\delta\Pi = 0$, setting the coefficients of δu in $(0, L)$, and δu at $x = L$ from Eq. (3) to zero separately, we obtain the Euler equation and the natural (or force) boundary condition of the problem:

Euler equation:

$$-\frac{d}{dx}\left(EA\frac{du}{dx}\right) - f = 0, \quad 0 < x < L. \tag{4}$$

Natural boundary condition:

$$EA\frac{du}{dx} + ku - P = 0 \text{ at } x = L. \tag{5}$$

We note that the natural boundary condition is of the mixed kind. Thus, the solution u of Eqs. (4) and (5) that satisfies $u(0) = 0$ is the minimizer of the energy functional $\Pi(u)$ in Eq. (1). Equation (4) can be obtained directly from Eq. (2.5.30) by substituting $(u' = du/dx)$

$$F(x, u, u') = \frac{EA}{2}\left(\frac{du}{dx}\right)^2 - fu, \quad \frac{\partial F}{\partial u} = -f, \quad \frac{\partial F}{\partial u'} = EA\frac{du}{dx}. \tag{6}$$

2.6   THE PRINCIPLE OF VIRTUAL DISPLACEMENTS

Consider a continuum occupying the volume Ω, each point occupied by a material particle, and is in equilibrium under the action of body forces $\mathbf{f}$ and surface tractions $\mathbf{t}$. Suppose that displacements are specified to be $\hat{\mathbf{u}}$, over portion Γ_u of the total boundary Γ of Ω; on the remaining boundary $\Gamma - \Gamma_u \equiv \Gamma_\sigma$, the tractions are specified to be $\hat{\mathbf{t}}$. The boundary portions Γ_u and Γ_σ are disjoint (i.e., do not overlap), and their union is the total boundary, Γ. Let $\mathbf{u} = (u_1, u_2, u_3)$ be the displacement vector corresponding to the equilibrium configuration of the body, and let σ_{ij} and ε_{ij} be the associated stress and strain components, respectively, referred to rectangular Cartesian system (x_1, x_2, x_3). Throughout this discussion, we assume that the strains are infinitesimal and rotations are possibly moderate so that no distinction between the Cauchy stress tensor $\boldsymbol{\sigma}$ and second Piola–Kirchhoff stress tensor $\mathbf{S}$ and between the infinitesimal strain tensor $\boldsymbol{\varepsilon}$ and the Green–Lagrange strain tensor $\mathbf{E}$ is made. We make no assumption concerning the constitutive behavior of the material body at the moment.

The set of admissible configurations is defined by sufficiently differentiable displacement fields that satisfy the geometric boundary conditions: $\mathbf{u} = \hat{\mathbf{u}}$ on Γ_u. Of all such admissible configurations, the actual one corresponds to the equilibrium configuration with the prescribed loads. In order to determine the displacement field $\mathbf{u}$ corresponding to the equilibrium configuration, we

let the body experience a virtual displacement $\delta\mathbf{u}$ from the equilibrium configuration. The virtual displacements are arbitrary, continuous with continuous
derivatives as dictated by the strain energy, and satisfy the homogeneous form
of the specified geometric boundary conditions, $\delta\mathbf{u} = \mathbf{0}$ on Γ_u. The principle
of virtual displacements states that *a continuous body is in equilibrium if and
only if the virtual work done by all forces, internal and external, acting on the
body is zero in a virtual displacement:*

$$\delta W = \delta W_I + \delta W_E = 0, \tag{2.6.1}$$

where δW_I is the virtual work due to the internal forces and δW_E is the virtual
work due to the external forces. For problems involving mechanical stimuli,
the internal virtual work is the same as the virtual strain energy, $\delta U = \delta W_I$.

The principle of virtual work is independent of any constitutive law. The
principle may be used to derive the equilibrium equations of deformable solids
in terms of stresses or stress resultants (i.e., forces), as will be shown in Chapters 3–8.

Next, we derive the stress equilibrium equations of three-dimensional elasticity using the principle of virtual displacements. We begin with the principle
of virtual displacements, $\delta W = 0$:

$$
\begin{aligned}
0 &= \int_\Omega \sigma_{ij}\,\delta\varepsilon_{ij}\,d\Omega - \left[\int_\Omega \mathbf{f}\cdot\delta\mathbf{u}\,d\Omega + \int_{\Gamma_\sigma} \hat{\mathbf{t}}\cdot\delta\mathbf{u}\,d\Gamma\right] \\
&= \tfrac{1}{2}\int_\Omega [\sigma_{ij}\,(\delta u_{i,j} + \delta u_{j,i})]\,d\Omega - \left[\int_\Omega f_i\delta u_i\,d\Omega + \int_{\Gamma_\sigma} \hat{t}_i\delta u_i\,d\Gamma\right] \\
&= -\int_\Omega (\sigma_{ji,j} + f_i)\,\delta u_i\,d\Omega - \int_{\Gamma_\sigma} \hat{t}_i\,\delta u_i\,d\Gamma + \int_\Gamma \sigma_{ji}n_j\delta u_i\,d\Gamma, \tag{2.6.2}
\end{aligned}
$$

where integration by parts (or the Green–Gauss theorem) is used to arrive at
the last step. Since $\delta u_i = 0$ on Γ_u, the last boundary integral reduces to one
on Γ_σ. Thus, we have

$$0 = -\int_\Omega (\sigma_{ji,j} + f_i)\,\delta u_i\,d\Omega + \int_{\Gamma_\sigma} (\sigma_{ij}n_j - \hat{t}_i)\,\delta u_i\,d\Gamma. \tag{2.6.3}$$

Since δu_i is arbitrary in Ω and independently on Γ_σ, we obtain the Euler
equations

$$\sigma_{ji,j} + f_i = 0 \quad\text{or}\quad \nabla\cdot\boldsymbol{\sigma} + \mathbf{f} = \mathbf{0} \quad\text{in } \Omega \tag{2.6.4}$$

and

$$\sigma_{ji}n_j - \hat{t}_i = 0 \quad\text{or}\quad \hat{\mathbf{n}}\cdot\boldsymbol{\sigma} - \hat{\mathbf{t}} = \mathbf{0} \quad\text{on } \Gamma_\sigma. \tag{2.6.5}$$

Thus, the principle of virtual displacements yields the stress equilibrium equation and stress boundary condition as the Euler equations. Next, we consider
a specific example from theory of beams.

2.7 PRINCIPLE OF MINIMUM TOTAL POTENTIAL ENERGY

The principle of virtual work discussed in the previous section is applicable to any continuous body with arbitrary constitutive behavior (e.g., linear or nonlinear elastic materials). The principle of minimum total potential energy is obtained as a special case from the principle of virtual displacements when the constitutive relations can be obtained from a potential function. Here, we restrict our discussion to materials that admit existence of a strain energy potential such that the stress is derivable from it. Such materials are termed *hyperelastic*.

For elastic bodies (in the absence of temperature variations), there exists a strain energy density U_0 such that

$$\sigma_{ij} = \frac{\partial U_0}{\partial \varepsilon_{ij}}. \tag{2.7.1}$$

The strain energy density U_0 is a function of strains at a point and is assumed to be positive definite (i.e., $U_0(\varepsilon_{ij}) > 0$ when $\varepsilon_{ij} \neq 0$ and $U_0 = 0$ if and only if $\varepsilon_{ij} = 0$). The statement of the principle of virtual displacements, $\delta W = 0$, can be expressed in terms of the strain energy density U_0 as

$$0 = \delta W = \int_\Omega \sigma_{ij}\, \delta\varepsilon_{ij}\, d\Omega - \left[\int_\Omega \mathbf{f} \cdot \delta\mathbf{u}\, d\Omega + \int_{\Gamma_\sigma} \hat{\mathbf{t}} \cdot \delta\mathbf{u}\, d\Gamma \right] \tag{2.7.2}$$

$$= \int_\Omega \frac{\partial U_0}{\partial \varepsilon_{ij}} \delta\varepsilon_{ij}\, d\Omega + \delta V_E$$

$$= \int_\Omega \delta U_0\, d\Omega + \delta V_E = \delta(U + V_E) \equiv \delta\Pi, \tag{2.7.3}$$

where V_E is the potential energy of external loads and U is the strain energy:

$$V_E = - \left[\int_\Omega \mathbf{f} \cdot \mathbf{u}\, d\Omega + \int_{\Gamma_\sigma} \hat{\mathbf{t}} \cdot \mathbf{u}\, d\Gamma \right], \quad U = \int_\Omega U_0\, d\Omega. \tag{2.7.4}$$

Note that we have used a constitutive equation, Eq. (2.7.1), in arriving at Eq. (2.7.3). As already defined in Section 2.3, the sum $U + V_E \equiv \Pi$ is called the *total potential energy*, and the statement

$$\delta\Pi \equiv \delta(U + V_E) = 0 \tag{2.7.5}$$

is known as the *principle of minimum total potential energy*. It is a statement of the fact that the energy of the system is the minimum only at its equilibrium configuration $\mathbf{u}$:

$$\Pi(\mathbf{u} + \alpha\mathbf{v}) \geq \Pi(\mathbf{u}) \quad \text{for all scalars } \alpha \text{ and admissible variations } \mathbf{v}. \tag{2.7.6}$$

The equality holds only for $\alpha = 0$.

Next, we show that the principle of minimum total potential energy, $\delta\Pi = 0$, as applied to an isotropic linear elastic solid in three dimensions yields the equations of equilibrium and force boundary conditions in terms of displacements. Suppose that the body occupying the volume Ω is subjected to body force $\mathbf{f}$ (measured per unit volume), prescribed surface traction $\hat{\mathbf{t}}$ (measured per unit area) on portion Γ_σ, and specified displacements $\hat{\mathbf{u}}$ on portion Γ_u of the total surface Γ of Ω.

The total potential energy for the problem at hand is given by [see Eq. (2.3.1)]

$$
\begin{aligned}
\Pi(\mathbf{u}) &= \int_\Omega \left[\tfrac{1}{2}\boldsymbol{\sigma}(\boldsymbol{\varepsilon}) : \boldsymbol{\varepsilon} - \mathbf{f} \cdot \mathbf{u}\right] d\Omega - \int_{\Gamma_\sigma} \hat{\mathbf{t}} \cdot \mathbf{u}\, d\Gamma \\
&= \int_\Omega \left(\tfrac{1}{2}\sigma_{ij}\varepsilon_{ij} - f_i u_i\right) d\Omega - \int_{\Gamma_\sigma} \hat{t}_i u_i\, d\Gamma,
\end{aligned}
\tag{2.7.7}
$$

where u_i denote the components of the displacement vector $\mathbf{u}$, σ_{ij} are the components of the stress tensor $\boldsymbol{\sigma}$, ε_{ij} are the components of the strain tensor $\boldsymbol{\varepsilon}$, f_i are the components of the body force vector $\mathbf{f}$, and t_i are the components of the stress vector (often called traction vector) $\mathbf{t}$ – all referred to a rectangular Cartesian coordinate system. The stress vector $\mathbf{t}$ is related to the stress tensor $\boldsymbol{\sigma}$ by Cauchy's formula [see Eq. (1.3.20)]:

$$
\mathbf{t} = \hat{\mathbf{n}} \cdot \boldsymbol{\sigma}.
\tag{2.7.8}
$$

Here $\hat{\mathbf{n}}$ denotes the unit normal vector to the surface Γ of the domain Ω. The first term under the volume integral in Eq. (2.7.7) represents the strain energy density of the elastic body, the second term represents the work done by the body force $\mathbf{f}$, and the surface integral denotes the work done by the specified traction $\hat{\mathbf{t}}$.

In order to express the strain energy in terms of the displacements, we must employ the stress–strain relations and strain–displacement equations. For an isotropic body, the stress–strain relations are given by

$$
\boldsymbol{\sigma} = 2\mu\,\boldsymbol{\varepsilon} + \lambda\,\mathrm{tr}\,(\boldsymbol{\varepsilon})\,\mathbf{I} \quad (\sigma_{ij} = 2\mu\varepsilon_{ij} + \lambda\delta_{ij}\varepsilon_{kk}),
\tag{2.7.9}
$$

where μ and λ are the Lamé constants and $\mathrm{tr}\,(\boldsymbol{\varepsilon})$ denotes the trace of the strain tensor (i.e., sum of the diagonal elements, $\varepsilon_{kk} = \varepsilon_{11} + \varepsilon_{22} + \varepsilon_{33}$). Hence,

$$
\boldsymbol{\sigma} : \boldsymbol{\varepsilon} = \sigma_{ij}\varepsilon_{ij} = 2\mu\varepsilon_{ij}\varepsilon_{ij} + \lambda\varepsilon_{ii}\varepsilon_{kk},
\tag{2.7.10}
$$

wherein sum on repeated indices is assumed.

The strain–displacement relations of the linearized elasticity are

$$
\varepsilon_{ij} = \tfrac{1}{2}\left(u_{i,j} + u_{j,i}\right),
\tag{2.7.11}
$$

where $u_{i,j} = (\partial u_i/\partial x_j)$. Substituting Eqs. (2.7.9) and (2.7.11) into Eq. (2.7.7), we obtain the final expression for the total potential energy in terms of the displacements:

$$\Pi(\mathbf{u}) = \int_\Omega \left[\frac{\mu}{4} (u_{i,j} + u_{j,i})(u_{i,j} + u_{j,i}) + \frac{\lambda}{2} u_{i,i} u_{k,k} - f_i u_i \right] d\Omega - \int_{\Gamma_\sigma} \hat{t}_i u_i \, d\Gamma.$$

(2.7.12)

where $d\Omega = dx_1 dx_2 dx_3$ is the volume element and $d\Gamma$ is the surface element.

Now we use the principle of minimum total potential energy, $\delta\Pi = 0$, and derive the Euler equations associated with the total potential energy functional Π in Eq. (2.7.12). Since the body is subjected to specified displacements on the portion Γ_u of the surface, the first variation of the displacement vector is zero on Γ_u:

$$\mathbf{u} = \hat{\mathbf{u}} \quad \text{and} \quad \delta\mathbf{u} = \mathbf{0} \quad \text{on } \Gamma_u.$$

(2.7.13)

Setting the first variation of Π to zero, we obtain

$$0 = \int_\Omega \left[\frac{\mu}{2} (\delta u_{i,j} + \delta u_{j,i})(u_{i,j} + u_{j,i}) + \lambda \delta u_{i,i} u_{k,k} - f_i \delta u_i \right] d\Omega - \int_{\Gamma_\sigma} \hat{t}_i \delta u_i \, d\Gamma,$$

(2.7.14)

wherein the product rule of variation is used and similar terms are combined. Using integration by parts (or the Green–Gauss theorem), we obtain

$$\int_\Omega \delta u_{i,j} (u_{i,j} + u_{j,i}) \, d\Omega = - \int_\Omega \delta u_i (u_{i,j} + u_{j,i})_{,j} \, d\Omega + \oint_\Gamma \delta u_i (u_{i,j} + u_{j,i}) n_j \, d\Gamma,$$

where n_j denotes the jth direction cosine of the unit normal to the surface. Hence, Eq. (2.7.14) becomes

$$0 = \int_\Omega \left[-\frac{\mu}{2} (u_{i,j} + u_{j,i})_{,j} \delta u_i - \frac{\mu}{2} (u_{i,j} + u_{j,i})_{,i} \delta u_j - \lambda u_{k,ki} \delta u_i - f_i \delta u_i \right] d\Omega$$

$$+ \oint_\Gamma \left[\frac{\mu}{2} (u_{i,j} + u_{j,i})(n_j \delta u_i + n_i \delta u_j) + \lambda u_{k,k} n_i \delta u_i \right] d\Gamma - \int_{\Gamma_\sigma} \delta u_i \hat{t}_i \, d\Gamma$$

$$= \int_\Omega \left[-\mu (u_{i,j} + u_{j,i})_{,j} - \lambda u_{k,ki} - f_i \right] \delta u_i \, d\Omega$$

$$+ \oint_\Gamma \left[\mu (u_{i,j} + u_{j,i}) + \lambda u_{k,k} \delta_{ij} \right] n_j \delta u_i \, d\Gamma - \int_{\Gamma_\sigma} \delta u_i \hat{t}_i \, d\Gamma.$$

(2.7.15)

In arriving at the last step, change of dummy indices is made to combine terms. Recognizing that the expression inside the square brackets of the closed surface integral is nothing but σ_{ij} and $\sigma_{ij} n_j = t_i$ by Cauchy's formula, we can write

$$\oint_\Gamma t_i \delta u_i \, d\Gamma = \int_{\Gamma_u} t_i \delta u_i \, d\Gamma + \int_{\Gamma_\sigma} t_i \delta u_i \, d\Gamma = \int_{\Gamma_\sigma} t_i \delta u_i \, d\Gamma.$$

The integral over Γ_u is zero by virtue of Eq. (2.7.13). Hence, Eq. (2.7.15) becomes

$$0 = \int_\Omega \left[-\mu (u_{i,j} + u_{j,i})_{,j} - \lambda u_{k,ki} - f_i \right] \delta u_i \, d\Omega + \int_{\Gamma_\sigma} (t_i - \hat{t}_i) \delta u_i \, d\Gamma.$$

(2.7.16)

Using the fundamental lemma of calculus of variations, we set the coefficients of δu_i in Ω and δu_i on Γ_σ to zero separately and obtain the Euler equations

$$\mu\big(u_{i,jj} + u_{j,ij}\big) + \lambda u_{k,ki} + f_i = 0 \text{ in } \Omega \quad \text{and} \quad t_i - \hat{t}_i = 0 \text{ on } \Gamma_\sigma, \quad (2.7.17)$$

for $i = 1, 2, 3$. Equation (2.7.17) is the well-known Navier equation of elasticity (contains three equations), which can be expressed in vector form as

$$\mu\nabla^2\mathbf{u} + (\lambda + \mu)\nabla(\nabla \cdot \mathbf{u}) + \mathbf{f} = \mathbf{0} \text{ in } \Omega, \quad \text{and} \quad \mathbf{t} - \hat{\mathbf{t}} = \mathbf{0} \text{ on } \Gamma_\sigma. \quad (2.7.18)$$

The principle of virtual displacements and the principle of minimum total potential energy give, when applied to an elastic body, the equilibrium equations as the Euler equations. The main difference between the two principles is that the principle of virtual displacements gives the equilibrium equations in terms of stresses or stress resultants, whereas the principle of minimum total potential energy gives them in terms of the displacements because, in the latter, constitutive and kinematic relations are used to replace the stresses (or stress resultants) in terms of the displacements.

The minimum character of the total potential energy functional $\Pi(u)$ can be easily established (see Reddy [9]):

$$\Pi(\mathbf{u}+\alpha\mathbf{v}) = \Pi(\mathbf{u})+\alpha^2 \int_\Omega \Big[\frac{\mu}{4}\left(v_{i,j} + v_{j,i}\right)\left(v_{i,j} + v_{j,i}\right)+\frac{\lambda}{2}v_{i,i}v_{k,k}\Big]d\Omega. \quad (2.7.19)$$

Since the coefficient of α^2 in Eq. (2.7.19) is a positive-definite quantity (being quadratic), it follows that

$$\Pi(\mathbf{u} + \alpha\mathbf{v}) \geq \Pi(\mathbf{u}) \qquad\qquad (2.7.20)$$

for any arbitrary real number α and vector $\mathbf{v}$ that vanishes on Γ_u (i.e., it is an element from the space of admissible variations).

2.8 HAMILTON'S PRINCIPLE

2.8.1 PRELIMINARY COMMENTS

The principle of virtual displacements discussed in Section 2.6 can be generalized to dynamical systems, and the generalization is known as Hamilton's principle of least action[4]. The principle assumes that the system under consideration is characterized by two energy functions: a *kinetic energy* K and

---

[4]Sir William Rowan Hamilton (1805–1865) was an Irish mathematician, who made important contributions to optics and classical mechanics. His reformulation of Newtonian mechanics, now called *Hamiltonian mechanics*, is considered central to the study of classical field theories and to the development of quantum mechanics. The Cayley–Hamilton theorem in linear algebra is also named after Arthur Cayley (1821–1895) and Hamilton.

a *potential energy* Π. For *continuous* systems (i.e., systems whose exact response cannot be described by a finite number of generalized coordinates), the energies can be expressed in terms of the dependent variables, such as the displacements and stress resultants.

2.8.2 STATEMENT OF THE PRINCIPLE

To derive Hamilton's principle, we begin with the assumption that the actual displacement (or motion) $\mathbf{u} = \mathbf{u}(\mathbf{x}, t)$ of a material particle in position $\mathbf{x}$ in the body may be *varied*, consistent with kinematic (essential) boundary conditions, to $\mathbf{u} + \delta\mathbf{u}$, where $\delta\mathbf{u}$ is the admissible variation (or virtual displacement) of the displacement. We suppose that the varied path differs from the actual path except at initial and final times, t_1 and t_2, respectively. Thus, an admissible displacement variation $\delta\mathbf{u}$ satisfies the conditions

$$\delta\mathbf{u} = \mathbf{0} \text{ on } \Gamma_u \text{ for all } t,$$
$$\delta\mathbf{u}(\mathbf{x}, t_1) = \delta\mathbf{u}(\mathbf{x}, t_2) = \mathbf{0} \text{ for all } \mathbf{x},$$

(2.8.1)

where Γ_u denotes the portion of the boundary of the body where the displacement vector $\mathbf{u}$ is specified.

The *work done on the body* at time t by the *resultant force* in moving through the virtual displacement $\delta\mathbf{u}$ is given by

$$\int_\Omega \mathbf{f} \cdot \delta\mathbf{u} \, d\Omega + \int_{\Gamma_\sigma} \hat{\mathbf{t}} \cdot \delta\mathbf{u} \, d\Gamma - \int_\Omega \boldsymbol{\sigma} : \delta\boldsymbol{\varepsilon} \, d\Omega,$$

(2.8.2)

where $\mathbf{f}$ is the body force vector, $\hat{\mathbf{t}}$ is the specified surface traction vector, and $\boldsymbol{\sigma}$ and $\boldsymbol{\varepsilon}$ are the stress and strain tensors. The last term in Eq. (2.8.2) represents the *virtual work* of internal forces *stored in the body*. The strains $\delta\boldsymbol{\varepsilon}$ are assumed to be compatible, in the sense that the strain–displacement relations in Eq. (1.4.5) are satisfied. The work done by the inertia force $\rho \partial^2 \mathbf{u} / \partial t^2$ in moving through the virtual displacement $\delta\mathbf{u}$ is given by

$$\int_\Omega \rho \frac{\partial^2 \mathbf{u}}{\partial t^2} \cdot \delta\mathbf{u} \, d\Omega$$

(2.8.3)

where $\rho = \rho(\mathbf{x})$ is the mass density of the medium. We have the result

$$\int_{t_1}^{t_2} \left\{ \int_\Omega \rho \frac{\partial^2 \mathbf{u}}{\partial t^2} \cdot \delta\mathbf{u} \, d\Omega - \left[\int_\Omega (\mathbf{f} \cdot \delta\mathbf{u} - \boldsymbol{\sigma} : \delta\boldsymbol{\varepsilon}) \, d\Omega + \int_{\Gamma_\sigma} \hat{\mathbf{t}} \cdot \delta\mathbf{u} \, d\Gamma \right] \right\} dt = 0,$$

or

$$\int_{t_1}^{t_2} \left[\int_\Omega \rho \dot{\mathbf{u}} \cdot \delta\dot{\mathbf{u}} \, d\Omega + \int_\Omega (\mathbf{f} \cdot \delta\mathbf{u} - \boldsymbol{\sigma} : \delta\boldsymbol{\varepsilon}) \, d\Omega + \int_{\Gamma_\sigma} \hat{\mathbf{t}} \cdot \delta\mathbf{u} \, d\Gamma \right] dt = 0, \quad \dot{\mathbf{u}} = \frac{\partial\mathbf{u}}{\partial t}.$$

(2.8.4)

In arriving at the expression in Eq. (2.8.4), integration by parts is used on the first term; the integrated terms vanish because of the initial and final conditions in Eq. (2.8.1). Equation (2.8.4) is known as *the general form of Hamilton's principle for a continuous medium* (conservative or not, and elastic or not).

We recall from the previous discussions that for an ideal elastic body, the forces $\mathbf{f}$ and $\mathbf{t}$ are conservative (i.e., they are derivable from a potential function V_E),

$$\delta V_E = -\left(\int_\Omega \mathbf{f} \cdot \delta \mathbf{u} \, d\Omega + \int_{\Gamma_\sigma} \hat{\mathbf{t}} \cdot \delta \mathbf{u} \, d\Gamma \right), \tag{2.8.5}$$

and that there exists a strain energy density potential $U_0 = U_0(\varepsilon_{ij})$ such that

$$\boldsymbol{\sigma} = \frac{\partial U_0}{\partial \boldsymbol{\varepsilon}} \quad \left(\sigma_{ij} = \frac{\partial U_0}{\partial \varepsilon_{ij}} \right) \quad \text{or} \quad \boldsymbol{\sigma} : \delta \boldsymbol{\varepsilon} = \delta U_0. \tag{2.8.6}$$

Substituting Eqs. (2.8.5) and (2.8.6) into Eq. (2.8.4), we obtain the statement of Hamilton's principle for an elastic (linear or nonlinear) medium:

$$\int_{t_1}^{t_2} [\delta K - (\delta U + \delta V_E)] dt = 0, \tag{2.8.7}$$

where K and U are the kinetic and strain energies, respectively,

$$\delta K = \int_\Omega \rho \, \delta \dot{\mathbf{u}} \cdot \dot{\mathbf{u}} \, d\Omega, \quad \delta U = \int_\Omega \delta U_0 \, d\Omega = \int_\Omega \boldsymbol{\sigma} : \delta \boldsymbol{\varepsilon} \, d\Omega. \tag{2.8.8}$$

For conservative systems, it is possible to write $\delta K - (\delta U + \delta V_E)$ as $\delta[K - (U + V_E)]$. The expression $L \equiv K - (U + V_E)$ is called the Lagrangian. We recall that the sum of the strain energy and potential energy of external forces, $U + V_E$, is called the total potential energy, Π, of the body, and we have $L = K - \Pi$.

In the following section, we present the Euler–Lagrange equations of motion resulting from the application of Hamilton's principle to continuous systems. When applied to material particle or a rigid body ($U = 0$), Hamilton's principle yields Newton's second law of motion, $\mathbf{F} = m\mathbf{a}$, where m is the mass and $\mathbf{a}$ is the acceleration vector.

2.8.3   EULER–LAGRANGE EQUATIONS

The Euler–Lagrange equations associated with the Lagrangian, $L = K - \Pi$, can be obtained from Eq. (2.8.7), with δV_E and $(\delta K, \delta U)$ given by Eqs. (2.8.5) and (2.8.8), respectively, as follows:

$$
\begin{aligned}
0 &= \int_{t_1}^{t_2} \int_\Omega (\rho \, \delta \dot{\mathbf{u}} \cdot \dot{\mathbf{u}} - \boldsymbol{\sigma} : \delta \boldsymbol{\varepsilon} + \mathbf{f} \cdot \delta \mathbf{u}) \, d\Omega \, dt + \int_{t_1}^{t_2} \int_{\Gamma_\sigma} \hat{\mathbf{t}} \cdot \delta \mathbf{u} \, d\Gamma \, dt \\
&= \int_{t_1}^{t_2} \left[\int_\Omega \left(\rho \frac{\partial^2 \mathbf{u}}{\partial t^2} - \nabla \cdot \boldsymbol{\sigma} - \mathbf{f} \right) \cdot \delta \mathbf{u} \, d\Omega + \int_{\Gamma_\sigma} (\mathbf{t} - \hat{\mathbf{t}}) \cdot \delta \mathbf{u} \, d\Gamma \right] dt, \tag{2.8.9}
\end{aligned}
$$

where $\boldsymbol{\sigma} : \delta\boldsymbol{\varepsilon} = \sigma_{ij}\,\delta\varepsilon_{ij} = \sigma_{ij}\,\delta u_{i,j}$, integration by parts (or the divergence theorem), and the fact that $\delta\mathbf{u} = \mathbf{0}$ on Γ_u are used in arriving at Eq. (2.8.9) from Eq. (2.8.7). Because $\delta\mathbf{u}$ is arbitrary for $\mathbf{x}$ in Ω and also on Γ_σ for all t, $t_1 < t < t_2$, it follows that

$$\rho\frac{\partial^2 \mathbf{u}}{\partial t^2} - \nabla\cdot\boldsymbol{\sigma} - \mathbf{f} = \mathbf{0} \quad \text{in } \Omega; \quad \mathbf{t} - \hat{\mathbf{t}} = \mathbf{0} \quad \text{on } \Gamma_\sigma. \tag{2.8.10}$$

The two equations in Eq. (2.8.10) are the Euler–Lagrange equations for a linearized elastic body. The first equation is the equation of motion and the second equation is the natural (or force) boundary condition.

Example 2.8.1

Consider the axial motion of an elastic bar of length L, area of cross section A, modulus of elasticity E, and mass density ρ, and subjected to distributed force f per unit length and an end load P. Determine the equations of motion of the bar when (a) no nonconservative forces are applied and (b) when the bar also experiences a nonconservative (viscous damping) force proportional to the velocity,

$$F^* = -\mu\frac{\partial u}{\partial t}, \tag{1}$$

where μ is the damping coefficient (a constant).

Solution: (a) The kinetic and total potential energies of the system are

$$K = \int_\Omega \frac{\rho}{2}\left(\frac{\partial u}{\partial t}\right)^2 d\Omega = \int_0^L \frac{\rho A}{2}\left(\frac{\partial u}{\partial t}\right)^2 dx, \tag{2}$$

$$\Pi(u) = \int_0^L \frac{A}{2}\sigma_{xx}\varepsilon_{xx}\,dx - \int_0^L fu\,dx - Pu(L), \tag{3}$$

wherein u, σ_{xx}, and ε_{xx} are assumed to be functions of x only, and

$$u(0,t) = 0 \quad \text{(bar is fixed at } x = 0), \tag{4}$$

$$\varepsilon_{xx} = \frac{\partial u}{\partial x} \quad \text{(strain–displacement relation)}. \tag{5}$$

Substituting for K and Π from Eqs. (2) and (3) into Eq. (2.8.7), we obtain

$$
\begin{aligned}
0 &= \int_{t_1}^{t_2}\left\{\int_0^L\left[A\rho\frac{\partial u}{\partial t}\frac{\partial \delta u}{\partial t} - A\sigma_{xx}\delta\left(\frac{\partial u}{\partial x}\right) + f\delta u\right]dx + P\delta u(L)\right\}dt \\
&= \int_0^L\left[\int_{t_1}^{t_2} -\frac{\partial}{\partial t}\left(\rho A\frac{\partial u}{\partial t}\right)\delta u\,dt + \rho A\frac{\partial u}{\partial t}\delta u\Big|_{t_1}^{t_2}\right]dx \\
&\quad + \int_{t_1}^{t_2}\left\{\int_0^L\left[\frac{\partial}{\partial x}(A\sigma_{xx}) + f\right]\delta u\,dx - (A\sigma_{xx}\delta u)\Big|_0^L + P\delta u(L)\right\}dt \\
&= -\int_{t_1}^{t_2}\left\{\int_0^L\left[\frac{\partial}{\partial t}\left(\rho A\frac{\partial u}{\partial t}\right) - \frac{\partial}{\partial x}(A\sigma_{xx}) - f\right]\delta u\,dx\right. \\
&\quad \left. - (A\sigma_{xx} - P)\big|_{x=L}\delta u(L)\right\}dt, \tag{6}
\end{aligned}
$$

where $\delta u(0,t) = 0$ and $\delta u(x,t_1) = \delta u(x,t_2) = 0$ are used to simplify the expression. The Euler–Lagrange equations are obtained by setting the coefficients of δu in $(0,L)$ and at $x = L$ to zero separately:

$$\frac{\partial}{\partial t}\left(\rho A \frac{\partial u}{\partial t}\right) - \frac{\partial}{\partial x}\left(A\sigma_{xx}\right) - f = 0, \quad 0 < x < L, \tag{7}$$

$$\left(A\sigma_{xx}\right)\big|_{x=L} - P = 0, \tag{8}$$

for all $t, t_1 < t < t_2$. For linear elastic materials, we have $\sigma_{xx} = E\varepsilon_{xx} = E(\partial u/\partial x)$, and Eqs. (7) and (8) can be expressed in terms of the displacement u as

$$\frac{\partial}{\partial t}\left(\rho A \frac{\partial u}{\partial t}\right) - \frac{\partial}{\partial x}\left(EA \frac{\partial u}{\partial x}\right) - f = 0, \quad 0 < x < L, \tag{9}$$

$$\left(AE \frac{\partial u}{\partial x}\right)\Big|_{x=L} - P = 0. \tag{10}$$

In Eqs. (7)–(10), the area of cross section A, density ρ, and modulus E can be, in general, function of position x and time t.

(b) Now suppose that the bar also experiences a nonconservative (e.g., viscous damping) force given in Eq. (1). Then the Euler–Lagrange equations from Eq. (2.8.7), which is in terms of the varied quantities, are given by

$$\frac{\partial}{\partial t}\left(\rho A \frac{\partial u}{\partial t}\right) - \frac{\partial}{\partial x}\left(EA \frac{\partial u}{\partial x}\right) - f + \mu \frac{\partial u}{\partial t} = 0, \quad 0 < x < L, \tag{11}$$

$$AE \frac{\partial u}{\partial x}\Big|_{x=L} - P = 0. \tag{12}$$

2.9 CHAPTER SUMMARY

In this chapter, the concepts of work and energy, expressions for strain energy of an elastic body, and the principle of virtual displacement, the principle of the minimum total potential energy, and Hamilton's principle are presented for derivation of the equations of motion or equilibrium of a continuous system. Starting with an assumed displacement field, one can compute strains (and virtual strains) in terms of the displacement functions introduced through the displacement expansions (and virtual displacements). Then the principle of virtual displacements or Hamilton's principle can be used to derive the Euler–Lagrange equations for any beam theory, as will be discussed in the coming chapters.

SUGGESTED EXERCISES

2.1 Consider the double pendulum shown in Fig. P2.1 at some instant of time. Write the potential energy (only due to gravity) of the system assuming that the energy is zero when $\theta_1 = \theta_2 = 0$. Neglect the weight of the pendulum bars and assume that they are rigid.

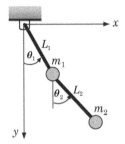

Figure P2.1

2.2 Determine the strain energy and external work done for the bar with an end nonlinear elastic spring with the force-deflection relationship $F_s = ke^n$, where e is the elongation in the spring (see Fig. P2.2).

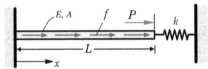

Figure P2.2

2.3 Determine the (a) strain energy and (b) complementary strain energy of the truss shown in Fig. P2.3. Assume elastic behavior of the form $\sigma = K\sqrt{\varepsilon}$. Express your answer in terms of displacements in the former case and applied loads in the latter case. Assume that the area of cross section of each member is A.

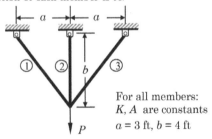

For all members:
K, A are constants
$a = 3$ ft, $b = 4$ ft

Figure P2.3

2.4 Establish the inequality in Eq. (2.7.20):

$$\Pi(\mathbf{u} + \alpha\mathbf{v}) \geq \Pi(\mathbf{u})$$

where Π is the total potential energy and $\mathbf{u}$ is the actual (i.e., one that satisfies the equations of equilibrium) displacement vector, $\mathbf{v}$ is an arbitrary vector such that it vanishes on the boundary where $\mathbf{u}$ is specified, and α is a real number.

2.5 Verify Eqs. (11) and (12) of **Example 2.8.1**.

Science is a wonderful thing if one does not have to earn one's living at it.

Albert Einstein

3 The Classical Beam Theory

Science is organized knowledge. Wisdom is organized life. Immanuel Kant

3.1 INTRODUCTORY COMMENTS

Because of the special geometric shapes of solid bodies (i.e., one or two dimensions being very large compared to the other two dimensions or one dimension), one may develop simplified continuum theories, compared to three-dimensional elasticity theory, and obtain very good engineering accuracy in predicting stresses and displacements. Such theories, termed structural theories, have been developed using one of the following two approaches.

1. Theories based on the assumed displacement expansions in terms of the powers of the thickness coordinate and unknown functions,

$$u_i(x, y, z, t) = \sum_{j=0}^{m} (z)^j \, \varphi_i^{(j)}, \quad i = 1, 2, \ldots, 3, \qquad (3.1.1)$$

where u_i ($i = 1, 2, 3$) are the *total* displacements of a point (x, y, z) in the structure and $\varphi_i^{(j)}$ are the "displacement" functions to be determined. The principle of virtual displacements or Hamilton's principle is used to determine the governing differential equations in terms of $\varphi_i^{(j)}$.

2. Theories based on the assumed stress expansions

$$\sigma_i(x, y, z, t) = \sum_{j=0}^{m} (z)^j \, \psi_i^{(j)}, \quad i = 1, 2, \ldots, 6, \qquad (3.1.2)$$

where σ_i are the stress components $\sigma_1 = \sigma_{xx}$, $\sigma_2 = \sigma_{yy}$, $\sigma_3 = \sigma_{zz}$, $\sigma_4 = \sigma_{yz}$, $\sigma_5 = \sigma_{xz}$, and $\sigma_6 = \sigma_{xy}$, and $\psi_i^{(j)}$ are the "stress functions," which are to be determined using the principle of complementary virtual work.

Here, the x-axis is taken along the length of the beam such that it passes through the geometric centroid of the beam; and the z-axis is taken vertically upward (i.e., transverse to the axis of a beam), with the y-axis into the plane of the page. In this book we only consider bending about the y-axis. In general, not all $\varphi_i^{(j)}$ or $\psi_i^{(j)}$ are linearly independent of each other (e.g., some may be the derivatives of the other). In this book, we only consider displacement-based theories.

DOI: 10.1201/9781003240846-3

61

The most commonly used displacement-based theories of beams are: (a) classical beam theory (CBT) (i.e., the theory in which transverse shear strain is assumed to be zero), (b) the first-order shear deformation beam theory, which accounts for transverse shear strains as a constant through the beam height, and (c) the third-order shear deformation beam theory, which accommodates quadratic variation of the transverse shear strains through the thickness. The commonly used names for these theories are the Euler–Bernoulli beam theory or CBT, the Timoshenko beam theory (TBT), and the Reddy third-order beam theory (RBT).

This chapter is dedicated to the development of the classical theory of straight beams, accounting for the von Kármán nonlinear strain, material gradation through the beam height, and modified couple stress effect. These theories as applied to beams, subjected to external forces that are only functions of x and time t, are governed by differential equations that are function of one spatial coordinate x and time t.

3.2 KINEMATICS

The displacement field of the CBT is constructed assuming that transverse lines perpendicular to the beam axis (x) remain: (1) straight, (2) inextensible, and (3) perpendicular to the tangents of the deflected x-axis (see Fig. 3.2.1). These three assumptions together are known as the *Euler–Bernoulli hypothesis*. Let the displacement vector be denoted as $\mathbf{u} = u_x\hat{\mathbf{e}}_x + u_y\hat{\mathbf{e}}_y + u_z\hat{\mathbf{e}}_z$, with (u_x, u_y, u_z) being the components referred to the (x, y, z) coordinates. For bending in the xz-plane (i.e., bending about the y axis), the Euler–Bernoulli hypothesis results in the following displacement field:

$$\mathbf{u}(x, z, t) = [u(x,t) + z\theta_x(x,t)]\,\hat{\mathbf{e}}_x + w(x,t)\hat{\mathbf{e}}_z, \quad \theta_x \equiv -\frac{\partial w}{\partial x}, \qquad (3.2.1)$$

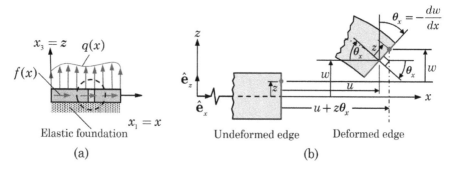

Figure 3.2.1: Kinematics of deformation of the classical beam theory (CBT).

where $(\hat{\mathbf{e}}_x, \hat{\mathbf{e}}_z)$ are the unit basis vectors along the (x, z) coordinates and (u, w) denote the axial and transverse displacements, respectively, of a point on the neutral axis (i.e., x axis) of the beam at time t. We note that $u_1 = u_x = u(x, t) + z\theta_x(x, t)$, $u_2 = u_y = 0$, and $u_3 = u_z = w$.

We assume that the axial strain $(\partial u_x / \partial x)$ and the curvature $(\partial^2 w / \partial x^2)$ are of order ϵ, and $(\partial w / \partial x)$ is of order $\sqrt{\epsilon}$, where $\epsilon \ll 1$ is a small parameter (compared to the ratio of the beam height to the length). Then the nonzero von Kármán strain tensor components [see Eqs. (1.4.7a) and (1.4.7b)] referred to the rectangular Cartesian system (x, y, z) are

$$\varepsilon_{xx} = \frac{\partial u_x}{\partial x} + \frac{1}{2}\left(\frac{\partial u_z}{\partial x}\right)^2, \quad \varepsilon_{xz} = \frac{1}{2}\left(\frac{\partial u_x}{\partial z} + \frac{\partial u_z}{\partial x}\right). \tag{3.2.2}$$

The underlined term in Eq. (3.2.2) is nonlinear. Equation (3.2.2) holds for any suitable expansion of the displacement fields u_x and u_z, including that in Eq. (3.2.1). Next, we specialize the strains in Eq. (3.2.2) for the displacement field in Eq. (3.2.1) of the Euler–Bernoulli beam theory. We note that the geometric nonlinearity appearing in Eq. (3.2.2) involves only the slope, $\theta = -\partial w / \partial x$.

Based on the displacement field in Eq. (3.2.1), the only nonzero strain in the present case is the axial strain ε_{xx} (i.e., the transverse shear strain $\varepsilon_{xz} = 0$). Using Eq. (3.2.2), we obtain:

$$\varepsilon_{xx}(x, z, t) = \frac{\partial u}{\partial x} + \frac{1}{2}\left(\frac{\partial w}{\partial x}\right)^2 + z\left(-\frac{\partial^2 w}{\partial x^2}\right) \equiv \varepsilon_{xx}^{(0)} + z\,\varepsilon_{xx}^{(1)}, \tag{3.2.3a}$$

$$\varepsilon_{xx}^{(0)}(x, t) = \frac{\partial u}{\partial x} + \frac{1}{2}\left(\frac{\partial w}{\partial x}\right)^2, \quad \varepsilon_{xx}^{(1)}(x, t) = -\frac{\partial^2 w}{\partial x^2}. \tag{3.2.3b}$$

Here, $\varepsilon_{xx}^{(0)}$ is known as the extensional (or membrane) strain and while $\varepsilon_{xx}^{(1)}$ is known as the bending (or flexural) strain.

3.3   EQUATIONS OF MOTION

3.3.1   PRELIMINARY COMMENTS

The displacement expansion in Eq. (3.2.1) contains two unknowns, namely, u and w. In deriving the governing equations, we assume that the applied forces f_x and f_z are only functions of x and t. The equations of motion governing these two displacement functions (u and w) can be determined using one of the two approaches: (a) the vector approach followed in continuum mechanics [see Eqs. (1.3.13)–(1.3.17)] or (b) the energy approach using Hamilton's principle (or the principle of virtual displacements). The vector approach typically requires identifying an infinitesimal element taken from the continuum with suitable forces displayed on it, while the energy approach requires the construction of the energy functional for the system. In general, these approaches do not necessarily result in the same set of governing equations.

This is particularly true for higher-order theories discussed in Chapters 5 and 8. Furthermore, the vector approach requires modification of the continuum mechanics equations to account for higher-order kinematics and couple stress effects; the approach also does not give any information about the boundary conditions. For the sake of completeness and information, here we illustrate both approaches.

3.3.2   VECTOR APPROACH

The vector form of the equations of motion governing a 2-D solid continuum are given by Eq. (1.4.12) (without the couple stress effect), which is expanded here in terms of the displacements and stresses in the rectangular Cartesian system (x, y, z), as

$$\frac{\partial}{\partial x}\left(\sigma_{xx} + \frac{\partial u_x}{\partial x}\sigma_{xx} + \frac{\partial u_x}{\partial z}\sigma_{xz}\right) + \frac{\partial}{\partial z}\left(\sigma_{xz} + \frac{\partial u_x}{\partial x}\sigma_{xz}\right) + f_x(x) = \rho\frac{\partial^2 u_x}{\partial t^2},$$
(3.3.1a)

$$\frac{\partial}{\partial x}\left(\sigma_{xz} + \frac{\partial u_z}{\partial x}\sigma_{xx}\right) + \frac{\partial}{\partial z}\left(\frac{\partial u_z}{\partial x}\sigma_{xz}\right) + f_z(x) = \rho\frac{\partial^2 u_z}{\partial t^2},$$
(3.3.1b)

where the second Piola–Kirchhoff stress components (S_{XX}, S_{XZ}) are assumed to be indistinguishable from the Cauchy stress components $(\sigma_{xx}, \sigma_{xz})$ for small strain theories presented in this book, f_x and f_z are the distributed forces measured per unit volume, and ρ is the mass density. If the beam is assumed to be resting on a linear elastic foundation with foundation modulus k_f, we must add $-k_f w$ as the reaction force of the foundation [i.e., replace f_z in Eq. (3.3.1b) with $f_z - k_f w$].

Equations (3.3.1a) and (3.3.1b) are valid for any displacement components (u_x, u_z), including those in Eq. (3.2.1). We now express these equations in terms of area-integrated stresses and the displacement unknowns (u, w) (a reduction of equations from 3D to 1D).

Integration of Eqs. (3.3.1a) and (3.3.1b) over the beam area of cross section A yields (note that u and w are only functions of x)

$$\frac{\partial}{\partial x}\left(N_{xx} + \frac{\partial u}{\partial x}N_{xx} - \frac{\partial^2 w}{\partial x^2}M_{xx} - \frac{\partial w}{\partial x}N_{xz}\right)$$
$$-\frac{\partial^2 w}{\partial x^2}N_{xz} + f = m_0\frac{\partial^2 u}{\partial t^2} - m_1\frac{\partial^3 w}{\partial x\partial t^2},$$
(3.3.2a)

$$\frac{\partial}{\partial x}\left(N_{xz} + \frac{\partial w}{\partial x}N_{xx}\right) - c_f w + q = m_0\frac{\partial^2 w}{\partial t^2},$$
(3.3.2b)

where N_{xx} is the axial force, N_{xz} is the transverse shear force (often denoted with Q_x or V in other books), and M_{xx} is the bending moment about the y

axis in the clockwise direction when the z-axis is positive upwards (all three quantities are resulting from stresses and thus termed *stress resultants*):

$$N_{xx} = \int_A \sigma_{xx}\, dA, \quad N_{xz} = \int_A \sigma_{xz}\, dA, \quad M_{xx} = \int_A z\sigma_{xx}\, dA, \tag{3.3.3}$$

(m_0, m_1, m_2) are the mass inertias defined by

$$m_0 = \int_A \rho\, dA, \quad m_1 = \int_A z\rho\, dA, \quad m_2 = \int_A z^2 \rho\, dA, \tag{3.3.4}$$

and $f = Af_x$ and $q = Af_z$ are the axial and transverse distributed forces measured per unit length, and $c_f = Ak_f$ is the foundation modulus measured as force per unit length per unit displacement. In arriving at the results in Eqs. (3.3.2a) and (3.3.2b), we have assumed that the transverse shear stress σ_{xz} is zero at the top and bottom of the beam (actually, in the CBT, it is zero if computed using the shear stress–strain relation, $\sigma_{xz} = 2G\varepsilon_{xz}$ because $\varepsilon_{xz} = 0$).

Next, multiply Eq. (3.3.1a) with z, integrate over the beam area of cross section, and use the fact that $\sigma_{xz} = 0$ at the top and bottom surfaces and obtain (several algebraic manipulations are involved before arriving at the result)

$$\frac{\partial}{\partial x}\left(M_{xx} + \frac{\partial u}{\partial x} M_{xx} - \frac{\partial^2 w}{\partial x^2} P_{xx} - \frac{\partial w}{\partial x} R_{xz} \right)$$
$$- N_{xz} - \frac{\partial u}{\partial x} N_{xz} + \frac{\partial^2 w}{\partial x^2} R_{xz} = m_1 \frac{\partial^2 u}{\partial t^2} - m_2 \frac{\partial^3 w}{\partial x \partial t^2}, \tag{3.3.5}$$

where (P_{xx}, R_{xx}) are the higher-order stress resultants

$$P_{xx} = \int_A \sigma_{xx} z^2\, dA, \quad R_{xz} = \int_A \sigma_{xz} z\, dA. \tag{3.3.6}$$

Note that area integrals of zf_x, zf_z, and $zc_f w$ vanish because they are only functions of x and because the x-axis passes through the geometric centroid of the beam cross section.

Now we use the order-of-magnitude assumption of various quantities discussed prior to Eq. (3.2.2), and simplify (by neglecting the terms that are negligibly small) Eqs. (3.3.2a), (3.3.2b), and (3.3.5). In particular, we assume that the following nonlinear expressions are negligible compared to the linear and the nonlinear terms introduced due to the von Kármán assumption:

$$\frac{\partial u}{\partial x} N_{xx}, \quad \frac{\partial u}{\partial x} N_{xz}, \quad \frac{\partial u}{\partial x} M_{xx},$$
$$\frac{\partial^2 w}{\partial x^2} M_{xx}, \quad \frac{\partial^2 w}{\partial x^2} P_{xx}, \quad \frac{\partial^2 w}{\partial x^2} N_{xz}, \quad \frac{\partial^2 w}{\partial x^2} R_{xz}. \tag{3.3.7}$$

We retain the expression $\frac{\partial w}{\partial x} N_{xx}$ only when we account for the von Kármán nonlinear strains. Thus, we obtain

$$\frac{\partial N_{xx}}{\partial x} + f = m_0 \frac{\partial^2 u}{\partial t^2} - m_1 \frac{\partial^3 w}{\partial x \partial t^2}, \qquad (3.3.8)$$

$$\frac{\partial N_{xz}}{\partial x} + \frac{\partial}{\partial x}\left(\frac{\partial w}{\partial x} N_{xx}\right) - c_f w + q = m_0 \frac{\partial^2 w}{\partial t^2}, \qquad (3.3.9)$$

$$\frac{\partial M_{xx}}{\partial x} - N_{xz} = m_1 \frac{\partial^2 u}{\partial t^2} - m_2 \frac{\partial^3 w}{\partial x \partial t^2}, \qquad (3.3.10)$$

There are three governing equations, namely, Eqs. (3.3.8)–(3.3.10). However, we have a problem with these equations when we try to express the stress resultants (N_{xx}, N_{xz}, M_{xx}) appearing in these equations in terms of the displacements (u, w) by invoking constitutive relations (yet to be discussed); the transverse shear force N_{xz}, if determined using the definition in Eq. (3.3.3), will be zero because $\varepsilon_{xz} = 0$ and hence $\sigma_{xz} = 2G\varepsilon_{xz} = 0$. Most importantly, the transverse shear force N_{xz} cannot be zero, because it is the force developed internal to the beam to balance the externally applied loads (e.g., q) that tend to shear the beam. To circumvent this situation, we compute N_{xz} from equilibrium using Eq. (3.3.10) and substitute into Eq. (3.3.9). Then the final two equations of motion associated with CBT are

$$\frac{\partial N_{xx}}{\partial x} + f = m_0 \frac{\partial^2 u}{\partial t^2} - m_1 \frac{\partial^3 w}{\partial x \partial t^2}, \qquad (3.3.11)$$

$$\frac{\partial^2 M_{xx}}{\partial x^2} + \frac{\partial}{\partial x}\left(\frac{\partial w}{\partial x} N_{xx}\right) - c_f w + q = m_0 \frac{\partial^2 w}{\partial t^2} + m_1 \frac{\partial^3 u}{\partial x \partial t^2}$$

$$- m_2 \frac{\partial^4 w}{\partial x^2 \partial t^2}. \qquad (3.3.12)$$

3.3.3 ENERGY APPROACH

Here, the equations of motion of the CBT are obtained as the Euler–Lagrange equations, which include the natural boundary conditions, by using Hamilton's principle [see Eq. (2.8.8)]

$$0 = \int_{t_1}^{t_2} [\delta K - (\delta U + \delta V_E)]\, dt, \qquad (3.3.13)$$

where δ denotes the variational operator, t_1 and t_2 denote the initial and final times (which are arbitrary), respectively, δK the virtual kinetic energy, δU is the virtual strain energy stored in the body (or the virtual work done by internal forces in moving through their respective virtual displacements), and δV_E is the virtual work done by external applied forces. Hamilton's principle stipulates that the virtual displacements vanish at $t = t_1$ and $t = t_2$, which is an important consideration in obtaining the equations of motion. Substituting the expressions for δK, δU, and δV_E, we have [the expressions for u_x and u_z are known from Eq. (3.2.1) in terms of u and w]

$$0 = \int_{t_1}^{t_2} \int_0^L \int_A \left(\rho_0 \frac{\partial \delta u_x}{\partial t} \frac{\partial u_x}{\partial t} + \rho_0 \frac{\partial \delta u_z}{\partial t} \frac{\partial u_z}{\partial t} - \delta \varepsilon_{xx} \sigma_{xx} \right) dA dx dt$$
$$+ \int_{t_1}^{t_2} \int_0^L \left[\delta u \, f + \delta w \, (q - c_f w) \right] dx dt, \tag{3.3.14a}$$

where L is the length of the beam, $(\delta u, \delta w)$ are the virtual displacements associated with (u, w), and c_f is the modulus of the elastic foundation on which the beam rests. Substituting for u_x and u_y, we obtain

$$0 = \int_{t_1}^{t_2} \int_0^L \int_A \left[\rho_0 \left(\frac{\partial \delta u}{\partial t} - z \frac{\partial^2 \delta w}{\partial x \partial t} \right) \left(\frac{\partial u}{\partial t} - z \frac{\partial^2 w}{\partial x \partial t} \right) + \rho_0 \frac{\partial \delta w}{\partial t} \frac{\partial w}{\partial t} \right.$$
$$\left. - \left(\frac{\partial \delta u}{\partial x} + \frac{\partial \delta w}{\partial x} \frac{\partial w}{\partial x} - z \frac{\partial^2 \delta w}{\partial x^2} \right) \sigma_{xx} \right] dA dx dt$$
$$+ \int_{t_1}^{t_2} \int_0^L \left[\delta u \, f + \delta w \, (q - c_f w) \right] dx dt. \tag{3.3.14b}$$

It is important to note that u and w are functions of x and t, but ε_{xx} and, therefore, σ_{xx} are functions of x, z, and t.

Carrying out the area integration, we arrive at the expression

$$0 = \int_{t_1}^{t_2} \int_0^L \left[m_0 \frac{\partial \delta u}{\partial t} \frac{\partial u}{\partial t} - m_1 \left(\frac{\partial \delta u}{\partial t} \frac{\partial^2 w}{\partial x \partial t} + \frac{\partial^2 \delta w}{\partial x \partial t} \frac{\partial u}{\partial t} \right) + m_2 \frac{\partial^2 \delta w}{\partial x \partial t} \frac{\partial^2 w}{\partial x \partial t} \right.$$
$$+ m_0 \frac{\partial \delta w}{\partial t} \frac{\partial w}{\partial t} - N_{xx} \left(\frac{\partial \delta u}{\partial x} + \frac{\partial w}{\partial x} \frac{\partial \delta w}{\partial x} \right) + M_{xx} \frac{\partial^2 \delta w}{\partial x^2}$$
$$\left. - c_f \delta w w + f \, \delta u + q \, \delta w \right] dx dt, \tag{3.3.15}$$

where (m_0, m_1, m_2) are the mass inertias defined in Eq. (3.3.4) and N_{xx} and M_{xx} are the stress resultants defined in Eq. (3.3.3). The Euler–Lagrange equations obtained from Eq. (3.3.15) are (we must relieve δu and δw of any differentiation with respect to x and/or t to be able to use the fundamental lemma of the calculus of variations; see Sections 2.5–2.8 for the details)

$$\frac{\partial N_{xx}}{\partial x} + f = m_0 \frac{\partial^2 u}{\partial t^2} - m_1 \frac{\partial^3 w}{\partial x \partial t^2}, \tag{3.3.16}$$

$$\frac{\partial^2 M_{xx}}{\partial x^2} + \frac{\partial}{\partial x} \left(\frac{\partial w}{\partial x} N_{xx} \right) - c_f w + q = m_0 \frac{\partial^2 w}{\partial t^2} + m_1 \frac{\partial^3 u}{\partial x \partial t^2} - m_2 \frac{\partial^4 w}{\partial x^2 \partial t^2}, \tag{3.3.17}$$

which are the same as those in Eqs. (3.3.11) and (3.3.12), derived using the vector approach.

The energy approach also yields three duality pairs, which indicate the quantities that may be specified on the boundary. The duality pairs are (the first element of each pair is the primary variable and the second element is the secondary variable)

$$(u, N_{xx}), \quad (w, V_{\text{eff}}), \quad \left(-\frac{\partial w}{\partial x}, M_{xx}\right). \tag{3.3.18}$$

It is clear that the axial displacement u is dual to the axial force N_{xx}, the transverse displacement w is dual to the effective shear force

$$V_{\text{eff}} \equiv \frac{\partial M_{xx}}{\partial x} + N_{xx}\frac{\partial w}{\partial x} + m_2\frac{\partial^3 w}{\partial x \partial t^2} - m_1\frac{\partial^2 u}{\partial t^2}, \tag{3.3.19}$$

and the rotation $-\partial w/\partial x$ is dual to the bending moment M_{xx}. It is interesting to note that the effective shear force has time derivative terms, which play a role in dynamics problems. It is rather difficult to see these terms in the boundary expressions in the vector approach.

The energy approach can also be used to derive the governing equations of beams accounting for modified couple stress effects. We simply add a term due to the modified couple stress to the virtual strain energy in Eq. (3.3.14a) [see Section 1.6 and Eqs. (1.6.1)–(1.6.3); also, see Reddy [74]]:

$$0 = \int_{t_1}^{t_2} \int_0^L \int_A \left(\rho_0 \frac{\partial \delta u_1}{\partial t}\frac{\partial u_1}{\partial t} + \rho_0 \frac{\partial \delta u_3}{\partial t}\frac{\partial u_3}{\partial t} - \delta\varepsilon_{xx}\,\sigma_{xx} - 2\mathcal{M}_{xy}\,\delta\chi_{xy}\right) dA\,dx\,dt$$

$$+ \int_{t_1}^{t_2}\int_0^L [\delta u\,f + \delta w\,(q - c_f w)]\,dx\,dt, \tag{3.3.20}$$

where $\mathcal{M}_{xy}$ is the couple stress induced by the difference between rates of rotations and χ_{xy} is the curvature component (dual to $\mathcal{M}_{xy}$)

$$\omega_y = \frac{1}{2}\left(\frac{\partial u_1}{\partial z} - \frac{\partial u_3}{\partial x}\right) = -\frac{\partial w}{\partial x}, \quad \chi_{xy} = \frac{1}{2}\frac{\partial\omega_y}{\partial x} = -\frac{1}{2}\frac{\partial^2 w}{\partial x^2}. \tag{3.3.21}$$

The governing equations of motion become (the only change is to replace M_{xx} with $\bar{M}_{xx} = M_{xx} + P_{xy}$ in the governing equations and boundary conditions)

$$\frac{\partial N_{xx}}{\partial x} + f = m_0\frac{\partial^2 u}{\partial t^2} - m_1\frac{\partial^3 w}{\partial x \partial t^2}, \tag{3.3.22}$$

$$\frac{\partial^2 \bar{M}_{xx}}{\partial x^2} + \frac{\partial}{\partial x}\left(\frac{\partial w}{\partial x}N_{xx}\right) - c_f w + q = m_0\frac{\partial^2 w}{\partial t^2} + m_1\frac{\partial^3 u}{\partial x \partial t^2} - m_2\frac{\partial^4 w}{\partial x^2 \partial t^2}, \tag{3.3.23}$$

where

$$\bar{M}_{xx} = M_{xx} + P_{xy}, \quad P_{xy} = \int_A \mathcal{M}_{xy}\,dA. \tag{3.3.24}$$

3.4   GOVERNING EQUATIONS IN TERMS OF DISPLACEMENTS

3.4.1   MATERIAL CONSTITUTIVE RELATIONS

For an isotropic, linear elastic material, the 3-D stress–strain relations are

$$\sigma_{ij} = 2\mu\,\varepsilon_{ij} + \lambda\,\delta_{ij}\,\varepsilon_{kk} - \alpha_T(3\lambda + 2\mu)(T - T_0)\delta_{ij},$$
$$\mathcal{M}_{ij} = 2\mu\,\ell^2\,\chi_{ij}, \tag{3.4.1}$$

where $\mu = G$ and λ are the Lamé parameters

$$\lambda = \frac{E\nu}{(1+\nu)(1-2\nu)}, \quad \mu = \frac{E}{2(1+\nu)} \tag{3.4.2}$$

with E being Young's modulus and ν being Poisson's ratio, α_T is the coefficient of thermal expansion, and $\Delta T = T_0 - T$ is the temperature increment from a reference temperature, T_0, and ℓ is the material length scale parameter, which is the square root of the ratio of the modulus of curvature to the modulus of shear, and it is a property measuring the effect of the couple stress.

3.4.2   UNIAXIAL STRESS–STRAIN RELATIONS

The constitutive relations in Eq. (3.4.1) can be simplified for the one-dimensional case by setting $\varepsilon_{22} = \varepsilon_{33} = 0$:

$$\sigma_{xx} = E(z)\left[\eta\varepsilon_{xx}(x,z) - \zeta\alpha_T\Delta T(x,z)\right], \quad \Delta T = T - T_0 \tag{3.4.3a}$$

where

$$\eta = \frac{(1-\nu)}{(1+\nu)(1-2\nu)}, \quad \zeta = \frac{1}{1-2\nu} \quad \nu \neq 0.5. \tag{3.4.3b}$$

However, these equations are not valid for $\nu = 0.5$ (i.e., for incompressible material), and they are not meaningful for one-dimensional problems because it amounts to using an equivalent Young's modulus of $\bar{E} = E\,\eta$ for mechanical deformation and $\bar{E} = E\,\zeta$ for thermal deformation. The values of η and ζ for $\nu = 0.3$ are $\eta = 1.346$ and $\zeta = 2.5$. Thus, using the constitutive relation (3.4.3a) in one-dimensional beam problem is to increase the value of E to $E\eta$, which is the reason why the deflections are reduced by the factor of η. Thus, in effect, the material is assumed to be stiffer than it is. Here, we use the uniaxial stress–strain relation

$$\sigma_{xx} = E(\varepsilon_{xx} - \alpha_T\Delta T). \tag{3.4.4}$$

One may recall that the stress–strain relations in Eq. (3.4.1) are obtained by inverting strain–stress relations, which are derived using the principle of superposition: when an infinitesimal block of material is subjected to normal stresses in three perpendicular directions simultaneously, the normal strain produced in a coordinate direction is the sum of the individual strains produced in that direction by the three normal stresses acting simultaneously. Since in the equilibrium consideration we only consider axial stress σ_{xx} and transverse shear stress σ_{xz}, the axial strain ε_{xx} is only due to σ_{xx}: $\varepsilon_{xx} = \sigma_{xx}/E$. Thus, it is reasonable (and conservative from design point of view) to use Eq. (3.4.4).

3.4.3 MATERIAL GRADATION THROUGH THE BEAM HEIGHT

Here, we assume that the beam is graded with two material combination through the beam height according to the relation [see Eq. (1.5.7)]

$$E(z) = (E_1 - E_2)\,V_1(z) + E_2, \quad V_1(z) = \left(\frac{1}{2} + \frac{z}{h}\right)^n, \qquad (3.4.5a)$$

where E_1 and E_2 are Young's moduli of the two materials used, n is the index that dictates the volume fraction $V_1(z)$ of material 1 and $V_2(z) = 1 - V_1(z)$ is the volume fraction of material 2 in the FGM. We assume that Poisson's ratio ν is a constant for the FGM material. Similarly, the thermal conductivity $k(z)$ and the mass density $\rho_0(z)$ are also assumed to vary similar to Eq. (3.4.5a); for example, the conductivity varies as

$$k(z) = (k_1 - k_2)\,V_1(z) + k_2, \quad V_1(z) = \left(\frac{1}{2} + \frac{z}{h}\right)^n. \qquad (3.4.5b)$$

3.4.4 BEAM CONSTITUTIVE EQUATIONS

The stress resultants defined by [see Eqs. (3.3.3) and (3.3.24)]

$$N_{xx} = \int_A \sigma_{xx}\, dA = A_{xx}\left[\frac{\partial u}{\partial x} + \frac{1}{2}\left(\frac{\partial w}{\partial x}\right)^2\right] - B_{xx}\frac{\partial^2 w}{\partial x^2} - N_{xx}^T, \qquad (3.4.6a)$$

$$M_{xx} = \int_A z\sigma_{xx}\, dA = B_{xx}\left[\frac{\partial u}{\partial x} + \frac{1}{2}\left(\frac{\partial w}{\partial x}\right)^2\right] - D_{xx}\frac{\partial^2 w}{\partial x^2} - M_{xx}^T, \qquad (3.4.6b)$$

$$P_{xy} = \int_A \mathcal{M}_{xy}\, dA = -A_{xy}\frac{\partial^2 w}{\partial x^2}, \qquad (3.4.6c)$$

$$N_{xx}^T = \int_A \alpha_T(z)\,E(z)\,\Delta T(z)\, dA, \quad M_{xx}^T = \int_A z\,\alpha_T(z)\,E(z)\,\Delta T(z)\, dA, \qquad (3.4.6d)$$

where N_{xx}^T and M_{xx}^T are the thermal stress resultants and A_{xx}, B_{xx}, D_{xx}, and A_{xy} are the extensional, extensional-bending, bending, and couple stress shear stiffness coefficients. They can be readily evaluated in terms of the constituent material properties (i.e., moduli) and the power-law exponent n (see Appendix at the end of this chapter for the evaluation of the integrals of $V_1(z)\,z^p$ for

$p = 1, 2, \ldots, 6)$ for a rectangular section beam. They are listed here for ready reference:

$$A_{xx} = \int_A E(z)\, dA = E_2 A_0 \frac{M+n}{1+n}$$

$$B_{xx} = \int_A E(z)z\, dA = E_2 B_0 \frac{n(M-1)}{2(1+n)(2+n)},$$

$$D_{xx} = \int_A E(z)z^2\, dA = E_2 I_0 \left[\frac{(6+3n+3n^2)M + (8n+3n^2+n^3)}{(1+n)(2+n)(3+n)}\right],$$

$$A_{xy} = \frac{\ell^2}{2(1+\nu)} \int_A E(z)\, dA = \ell^2 \frac{E_2 A_0}{2(1+\nu)} \frac{M+n}{1+n}, \qquad (3.4.7)$$

$$m_0 = \int_A \rho(z)\, dA = \frac{A_0}{1+n}(\rho_1 + n\rho_2),$$

$$m_1 = \int_A \rho(z)z\, dA = \frac{B_0 n}{2(1+n)(2+n)}(\rho_1 - \rho_2),$$

$$m_2 = \int_A \rho(z)z^2\, dA = I_0 \left[\frac{(6+3n+3n^2)\rho_1 + (8n+3n^2+n^3)\rho_2}{(1+n)(2+n)(3+n)}\right],$$

$$A_0 = bh, \quad B_0 = bh^2, \quad I_0 = \frac{bh^3}{12}, \quad M = \frac{E_1}{E_2},$$

where $A_0 = bh$ is the area of cross section (b is the width and h is the height of a rectangular cross section beam), $I_0 = bh^3/12$ is the second moment of area, and M is the modulus ratio. Note that Eq. (3.4.5a) can be recast in the form

$$E(z) = E_2\left[(M-1)V_1(z) + 1\right], \quad V_1(z) = \left(\frac{1}{2} + \frac{z}{h}\right)^n, \quad M = \frac{E_1}{E_2}.$$

Although other material variations can be used with any beam and plate theory, in this book we will exclusively consider the model in the above expression.

As an example of evaluation of the beam stiffness coefficients, consider

$$A_{xx} = b \int_{-\frac{h}{2}}^{\frac{h}{2}} E(z)\, dz = E_2 b \left[(M-1) \int_{-\frac{h}{2}}^{\frac{h}{2}} \left(\frac{1}{2} + \frac{z}{h}\right)^n dz + \int_{-\frac{h}{2}}^{\frac{h}{2}} dz\right]$$

$$= E_2 b \left[(M-1)\frac{h}{(1+n)} + h\right] = E_2 A_0 \frac{M+n}{1+n}.$$

For beams with $E = E(x)$ (i.e., $n = 0$ or $E_1 = E_2 = E$ and $\rho_1 = \rho_2 = \rho$), we have $A_{xx} = EA_0$, $B_{xx} = 0$, $D_{xx} = EI_0$, $A_{xy} = GA_0\ell^2$, $m_0 = \rho A_0$, $m_1 = 0$, and $m_2 = \rho I_0$.

Figure 3.4.1(a) contains the variation of the non-dimensional axial stiffness $\bar{A}_{xx} = A_{xx}/E_2 A_0$, $A_0 = bh$ and bending stiffness $\bar{D}_{xx} = D_{xx}/E_2 I_0$, $I_0 =$

$bh^3/12$ as functions of the volume fraction index n for various values of the modulus ratio $M = (E_1/E_2) \geq 1$, and Fig. 3.4.1(b) contains similar plots for the non-dimensional extensional-bending coupling stiffness $\bar{B}_{xx} = B_{xx}/E_2 B_0$ ($B_0 = bh^2$). It is clear that both $\bar{A}_{xx}$ and $\bar{D}_{xx}$ are the maximum at $n = 0$ and decrease with increasing values of n. However, $\bar{B}_{xx}$ is zero at $n = 0$, increases to a maximum at $n = \sqrt{2}$, and then decreases with increasing value of n. Thus, beams with nonzero B_{xx} will have a response that is not monotonic with respect to n.

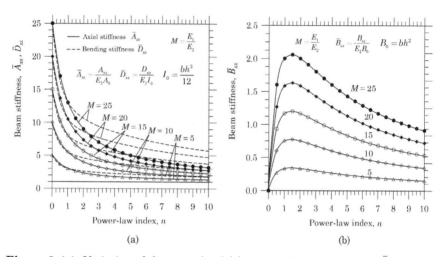

(a)   (b)

Figure 3.4.1: Variation of the normalized (a) axial stiffness coefficients $\bar{A}_{xx}(n)$ and $\bar{D}_{xx}(n)$; and (b) coupling stiffness coefficients $\bar{B}_{xx}(n)$ as functions of the power-law index exponent, n, for various values of the modulus ratio, $M = E_1/E_2$.

3.4.5   EQUATIONS OF MOTION

3.4.5.1   The general case (with FGM, VKN, and MCS)

The equations of motion of CBT, Eqs. (3.3.22) and (3.3.23), are valid for functionally graded material (FGM) beams with the von Kármán nonlinearity (VKN) and modified couple stress (MCS) terms. They can be expressed in terms of the displacements u and w using the beam constitutive relations from Eqs. (3.4.6a)–(3.4.6c). The resulting coupled (even when nonlinearity is not accounted for) differential equations are second order in u and fourth order in w:

$$-\frac{\partial}{\partial x}\left\{A_{xx}\left[\frac{\partial u}{\partial x} + \frac{1}{2}\left(\frac{\partial w}{\partial x}\right)^2\right] + B_{xx}\left(-\frac{\partial^2 w}{\partial x^2}\right) - N_{xx}^T\right\}$$

$$+ m_0 \frac{\partial^2 u}{\partial t^2} - m_1 \frac{\partial^3 w}{\partial t^2 \partial x} - f = 0, \qquad (3.4.8)$$

$$-\frac{\partial^2}{\partial x^2}\left\{B_{xx}\left[\frac{\partial u}{\partial x}+\frac{1}{2}\left(\frac{\partial w}{\partial x}\right)^2\right]+(D_{xx}+A_{xy})\left(-\frac{\partial^2 w}{\partial x^2}\right)-M_{xx}^T\right\}$$

$$-\frac{\partial}{\partial x}\left\{A_{xx}\frac{\partial w}{\partial x}\left[\frac{\partial u}{\partial x}+\frac{1}{2}\left(\frac{\partial w}{\partial x}\right)^2\right]-B_{xx}\frac{\partial w}{\partial x}\frac{\partial^2 w}{\partial x^2}-\frac{\partial w}{\partial x}N_{xx}^T\right\}$$

$$+c_f w+m_0\frac{\partial^2 w}{\partial t^2}+m_1\frac{\partial^3 u}{\partial t^2\partial x}-m_2\frac{\partial^4 w}{\partial t^2\partial x^2}-q=0. \tag{3.4.9}$$

The equations of equilibrium are obtained by simply setting all time derivative terms to zero.

3.4.5.2 Homogeneous beams with VKN and MCS

For beams with $E = E(x)$, we obtain $B_{xx} = 0$ and $m_1 = 0$. Then Eqs. (3.4.8) and (3.4.9) take the form

$$m_0\frac{\partial^2 u}{\partial t^2}-\frac{\partial}{\partial x}\left\{A_{xx}\left[\frac{\partial u}{\partial x}+\frac{1}{2}\left(\frac{\partial w}{\partial x}\right)^2\right]-N_{xx}^T\right\}-f=0, \tag{3.4.10}$$

$$m_0\frac{\partial^2 w}{\partial t^2}-m_2\frac{\partial^4 w}{\partial t^2\partial x^2}+\frac{\partial^2}{\partial x^2}\left[(D_{xx}+A_{xy})\frac{\partial^2 w}{\partial x^2}+M_{xx}^T\right]$$

$$-\frac{\partial}{\partial x}\left\{A_{xx}\frac{\partial w}{\partial x}\left[\frac{\partial u}{\partial x}+\frac{1}{2}\left(\frac{\partial w}{\partial x}\right)^2\right]-\frac{\partial w}{\partial x}N_{xx}^T\right\}+c_f w-q=0. \tag{3.4.11}$$

These two equations are coupled only because of the nonlinearity.

3.4.5.3 Linearized FGM beams with MCS

When we neglect the VKN, Eqs. (3.4.8) and (3.4.9) reduce to

$$-\frac{\partial}{\partial x}\left(A_{xx}\frac{\partial u}{\partial x}-B_{xx}\frac{\partial^2 w}{\partial x^2}-N_{xx}^T\right)+m_0\frac{\partial^2 u}{\partial t^2}-m_1\frac{\partial^3 w}{\partial t^2\partial x}-f=0, \tag{3.4.12}$$

$$-\frac{\partial^2}{\partial x^2}\left[B_{xx}\frac{\partial u}{\partial x}-(D_{xx}+A_{xy})\frac{\partial^2 w}{\partial x^2}-M_{xx}^T\right]+\frac{\partial}{\partial x}\left(\frac{\partial w}{\partial x}N_{xx}^T\right)$$

$$+c_f w+m_0\frac{\partial^2 w}{\partial t^2}+m_1\frac{\partial^3 u}{\partial t^2\partial x}-m_2\frac{\partial^4 w}{\partial t^2\partial x^2}-q=0. \tag{3.4.13}$$

Equations (3.4.12) and (3.4.13) are still coupled because of $B_{xx}\neq 0$ and $m_1\neq 0$ (second order in u and fourth order in w). The associated equations of static equilibrium are

$$-\frac{d}{dx}\left(A_{xx}\frac{du}{dx}-B_{xx}\frac{d^2 w}{dx^2}-N_{xx}^T\right)-f=0, \tag{3.4.14}$$

$$-\frac{d^2}{dx^2}\left[B_{xx}\frac{du}{dx}-(D_{xx}+A_{xy})\frac{d^2w}{dx^2}-M_{xx}^T\right]$$

$$+\frac{d}{dx}\left(\frac{dw}{dx}N_{xx}^T\right)+c_fw-q=0. \tag{3.4.15}$$

3.4.5.4   Linearized homogeneous beams with MCS

For linearized beams for which the material properties such as E and ρ_0 are at the most functions of x only, we set $B_{xx}=0$ and $m_1=0$ in Eqs. (3.4.12) and (3.4.13) and obtain

$$-\frac{\partial}{\partial x}\left(A_{xx}\frac{\partial u}{\partial x}-N_{xx}^T\right)+m_0\frac{\partial^2 u}{\partial t^2}-f=0, \tag{3.4.16}$$

$$\frac{\partial^2}{\partial x^2}\left[(D_{xx}+A_{xy})\frac{\partial^2 w}{\partial x^2}+M_{xx}^T\right]+\frac{\partial}{\partial x}\left(\frac{\partial w}{\partial x}N_{xx}^T\right)$$

$$+c_fw+m_0\frac{\partial^2 w}{\partial t^2}-m_2\frac{\partial^4 w}{\partial t^2\partial x^2}-q=0. \tag{3.4.17}$$

We set $A_{xy}=0$ if we do not wish to include the couple stress effect.

Equation (3.4.17) is uncoupled from Eq. (3.4.16), and one can solve each differential equation independent of the other. In particular, for a linear beam with $E=E(x)$ (i.e., $B_{xx}=0$ and $m_1=0$) under isothermal conditions (i.e., $\Delta T=0$), Eq. (3.4.17) reduces to

$$(EI_0+GA_0\ell^2)\frac{\partial^4 w}{\partial x^4}+c_f\,w+m_0\frac{\partial^2 w}{\partial t^2}-m_2\frac{\partial^4 w}{\partial t^2\partial x^2}=q(x,t), \tag{3.4.18}$$

which accounts for the rotatory inertia (m_2) and material length scale (ℓ).

It is clear that the length scale ℓ introduced due to the inclusion of the couple stress has the effect of increasing the bending stiffness, with equivalent bending stiffness being $(EI_0)_{equiv}=EI_0+GA_0\ell^2$. Thus, the effect of couple stress is to reduce the deflection (i.e., it has a stiffening effect).

The equilibrium equation is obtained from Eq. (3.4.18) by setting the time derivative terms to zero:

$$(EI_0+GA_0\ell^2)\frac{d^4 w}{dx^4}+c_f\,w=q(x). \tag{3.4.19}$$

The duality pairs for Eqs. (3.4.18) and (3.4.19) are $(w,N_{xz}=dM_{xx}/dx)$ and $(-dw/dx,M_{xx})$.

3.5   EQUATIONS IN TERMS OF DISPLACEMENTS AND BENDING MOMENT

3.5.1   PRELIMINARY COMMENTS

The equation governing the bending of beams is a fourth-order differential equation. Finite element models based on the fourth-order equation will require Hermite cubic interpolation of the deflection w (as will be discussed in detail in Chapter 9). To reduce the order (so that a lower order approximation of w can be used), we can reformulate the governing equations in terms of the displacements (u, w) and the bending moment M_{xx} (effective bending moment $\bar{M}_{xx} = M_{xx} + P_{xy}$ when the modified couple stress effect is included). Of course, while the order is reduced, we increase the number of unknowns. However, introducing the bending moment as a primary variable has the benefit of computing stresses at the nodes of a finite element mesh, as will be discussed in the sequel. Such formulations are termed *mixed formulation* because the displacement variables are mixed with force variables. Here, we consider only the isothermal case (i.e., set $N_{xx}^T = 0$ and $M_{xx}^T = 0$).

3.5.2 GENERAL CASE WITH FGM, MCS, AND VKN

Toward this end, we first rewrite Eqs. (3.4.6a) and (3.4.6b) in the form (i.e., strains $\varepsilon_{xx}^{(0)}$ and $\varepsilon_{xx}^{(1)}$ in terms of the stress resultants N_{xx} and $\bar{M}_{xx}$)

$$
\left\{ \begin{matrix} N_{xx} \\ \bar{M}_{xx} \end{matrix} \right\} = \begin{bmatrix} A_{xx} & B_{xx} \\ B_{xx} & \hat{D}_{xx} \end{bmatrix} \left\{ \begin{matrix} \varepsilon_{xx}^{(0)} \\ \varepsilon_{xx}^{(1)} \end{matrix} \right\},
\tag{3.5.1a}
$$

where $\hat{D}_{xx} = D_{xx} + A_{xy}$. Inverting the relations, we obtain

$$
\left\{ \begin{matrix} \varepsilon_{xx}^{(0)} \\ \varepsilon_{xx}^{(1)} \end{matrix} \right\} = \frac{1}{\hat{D}_{xx}^*} \begin{bmatrix} \hat{D}_{xx} & -B_{xx} \\ -B_{xx} & A_{xx} \end{bmatrix} \left\{ \begin{matrix} N_{xx} \\ \bar{M}_{xx} \end{matrix} \right\},
\tag{3.5.1b}
$$

where [and recall the definitions of $\varepsilon_{xx}^{(0)}$ and $\varepsilon_{xx}^{(1)}$ from Eq. (3.2.3b)]

$$
\hat{D}_{xx}^* = A_{xx}\hat{D}_{xx} - B_{xx}^2, \quad \varepsilon_{xx}^{(0)} = \frac{\partial u}{\partial x} + \frac{1}{2} \left(\frac{\partial w}{\partial x} \right)^2, \quad \varepsilon_{xx}^{(1)} = -\frac{\partial^2 w}{\partial x^2}.
\tag{3.5.1c}
$$

Solving the first equation of (3.5.1b) for N_{xx} in terms of $(u, w, \bar{M}_{xx})$ by replacing $\varepsilon_{xx}^{(0)}$ in terms of u and w, we obtain

$$
N_{xx} = \frac{\hat{D}_{xx}^*}{\hat{D}_{xx}} \left[\frac{\partial u}{\partial x} + \frac{1}{2} \left(\frac{\partial w}{\partial x} \right)^2 \right] + \frac{B_{xx}}{\hat{D}_{xx}} \bar{M}_{xx}.
\tag{3.5.2a}
$$

We note that $A_{xx} > 0$, $\hat{D}_{xx} > 0$, and $\hat{D}_{xx}^* > 0$. Using Eq. (3.5.2a) for N_{xx} in the second equation of (3.5.1b), we obtain

$$
-\frac{\partial^2 w}{\partial x^2} = -\bar{B}_{xx} \left[\frac{\partial u}{\partial x} + \frac{1}{2} \left(\frac{\partial w}{\partial x} \right)^2 \right] + \frac{1}{\hat{D}_{xx}} \bar{M}_{xx},
\tag{3.5.2b}
$$

where ($\bar{A}_{xx}$ is used in Eqs. (3.5.3) and (3.5.4))

$$
\bar{A}_{xx} = \frac{\hat{D}_{xx}^*}{\hat{D}_{xx}} = \frac{A_{xx}\hat{D}_{xx} - B_{xx}B_{xx}}{\hat{D}_{xx}}, \quad \bar{B}_{xx} = \frac{B_{xx}}{\hat{D}_{xx}}.
\tag{3.5.2c}
$$

Equations (3.3.22) and (3.3.23), with N_{xx} replaced by Eq. (3.5.2a), and Eq. (3.5.2b) constitute the governing equations for the mixed formulation in terms of the displacements (u, w), and bending moment $\bar{M}_{xx}$.

Writing Eqs. (3.3.22), (3.3.23), and (3.5.2b) in terms of $(u, w, \bar{M}_{xx})$, we have

$$-\frac{\partial}{\partial x}\left\{\bar{A}_{xx}\left[\frac{\partial u}{\partial x}+\frac{1}{2}\left(\frac{\partial w}{\partial x}\right)^2\right]+\bar{B}_{xx}\,\bar{M}_{xx}\right\}$$

$$+m_0\frac{\partial^2 u}{\partial t^2}-m_1\frac{\partial^3 w}{\partial t^2\partial x}=f, \qquad (3.5.3)$$

$$-\frac{\partial^2 \bar{M}_{xx}}{\partial x^2}-\frac{\partial}{\partial x}\left\{\bar{A}_{xx}\frac{\partial w}{\partial x}\left[\frac{\partial u}{\partial x}+\frac{1}{2}\left(\frac{\partial w}{\partial x}\right)^2\right]+\bar{B}_{xx}\frac{\partial w}{\partial x}\,\bar{M}_{xx}\right\}$$

$$+m_0\frac{\partial^2 w}{\partial t^2}+m_1\frac{\partial^3 u}{\partial t^2\partial x}-m_2\frac{\partial^4 w}{\partial t^2\partial x^2}+c_f w=q, \qquad (3.5.4)$$

$$-\frac{\partial^2 w}{\partial x^2}-\frac{1}{\hat{D}_{xx}}\bar{M}_{xx}+\bar{B}_{xx}\left[\frac{\partial u}{\partial x}+\frac{1}{2}\left(\frac{\partial w}{\partial x}\right)^2\right]=0. \qquad (3.5.5)$$

As we shall see in Chapter 9, the finite element model based on the mixed formulation admits linear or higher order Lagrange interpolation of all field variables. The model allows the computation of the bending moment, hence the bending stress, at the nodes of the finite element mesh.

3.5.3 SPECIAL CASES

3.5.3.1 Homogeneous beams with VKN and MCS

For beams with E independent of z, we have $\bar{B}_{xx}=0$, $m_1=0$, and $\bar{A}_{xx}=A_{xx}$ in Eqs. (3.5.3)–(3.5.5) and obtain

$$-\frac{\partial}{\partial x}\left\{A_{xx}\left[\frac{\partial u}{\partial x}+\frac{1}{2}\left(\frac{\partial w}{\partial x}\right)^2\right]\right\}+m_0\frac{\partial^2 u}{\partial t^2}=f, \qquad (3.5.6)$$

$$-\frac{\partial^2 \bar{M}_{xx}}{\partial x^2}-\frac{\partial}{\partial x}\left\{A_{xx}\frac{\partial w}{\partial x}\left[\frac{\partial u}{\partial x}+\frac{1}{2}\left(\frac{\partial w}{\partial x}\right)^2\right]\right\}$$

$$+m_0\frac{\partial^2 w}{\partial t^2}-m_2\frac{\partial^4 w}{\partial t^2\partial x^2}+c_f w=q, \qquad (3.5.7)$$

$$-\frac{\partial^2 w}{\partial x^2}-\frac{1}{\hat{D}_{xx}}\bar{M}_{xx}=0. \qquad (3.5.8)$$

3.5.3.2 Linearized FGM beams with MCS

In this case, we set the nonlinear terms to zero and obtain

$$-\frac{\partial}{\partial x}\left(\bar{A}_{xx}\frac{\partial u}{\partial x} + \bar{B}_{xx}\bar{M}_{xx}\right) + m_0\frac{\partial^2 u}{\partial t^2} - m_1\frac{\partial^3 w}{\partial t^2\partial x} = f, \tag{3.5.9}$$

$$-\frac{\partial^2 \bar{M}_{xx}}{\partial x^2} + m_0\frac{\partial^2 w}{\partial t^2} + m_1\frac{\partial^3 u}{\partial t^2\partial x} - m_2\frac{\partial^4 w}{\partial t^2\partial x^2} + c_f w = q, \tag{3.5.10}$$

$$-\frac{\partial^2 w}{\partial x^2} - \frac{1}{\hat{D}_{xx}}\bar{M}_{xx} + \bar{B}_{xx}\frac{\partial u}{\partial x} = 0. \tag{3.5.11}$$

When the modified couple stress effect is omitted, Eqs. (3.5.1a)–(3.5.11) hold with $\bar{M}_{xx}$ replaced with M_{xx} and $\hat{D}_{xx}$ with D_{xx}.

3.5.3.3 Linearized homogeneous beams with MCS

Here we set $\bar{B}_{xx} = 0$, $m_1 = 0$, and $\bar{A}_{xx} = A_{xx}$, and obtain

$$-\frac{\partial}{\partial x}\left(A_{xx}\frac{\partial u}{\partial x}\right) + m_0\frac{\partial^2 u}{\partial t^2} = f, \tag{3.5.12}$$

$$-\frac{\partial^2 \bar{M}_{xx}}{\partial x^2} + m_0\frac{\partial^2 w}{\partial t^2} - m_2\frac{\partial^4 w}{\partial t^2\partial x^2} + c_f w = q, \tag{3.5.13}$$

$$-\frac{\partial^2 w}{\partial x^2} - \frac{1}{\hat{D}_{xx}}\bar{M}_{xx} = 0. \tag{3.5.14}$$

Clearly, Eq. (3.5.12) governing the axial displacement u is decoupled from Eqs. (3.5.13) and (3.5.14), which are coupled among w and M_{xx} (i.e., bending variables).

3.6 CYLINDRICAL BENDING OF FGM RECTANGULAR PLATES

3.6.1 CYLINDRICAL BENDING

Consider an isotropic rectangular plate of in-plane dimensions a and b and thickness h. Let the x and y coordinates be parallel to the edges of the plate, with z being the thickness coordinate, as shown in Fig. 3.6.1. Let (u, v, w) denote the displacements along the coordinates (x, y, z). Suppose that the plate is relatively long in the y-direction compared to the dimension along the x-direction, the boundary conditions and loads do not change with respect to the y coordinate, and the edges parallel to the x-axis are free (i.e., not subjected to any geometric or force boundary conditions). The transverse load q (measured per unit area) is only a function of x, and edges parallel to the y-axis each may have a specified geometric (e.g., simply supported, clamped, or free) or force boundary condition (e.g., distributed load, $q = q(x)$, constant shear force, or bending moment) which does not change with y. In such cases, the plate bends into a cylindrical surface, with the transverse deflection w

being only a function of x, and $\varepsilon_{yy} = 0$ and $\varepsilon_{xy} = 0$; thus, the deformation can be described using one-dimensional equations. In such cases, the plate is said to undergo *cylindrical bending*, and it is sufficient to consider a strip of the plate between any two lines parallel to the x-axis. In this section, we formulate the equations governing the plate strips under cylindrical bending.

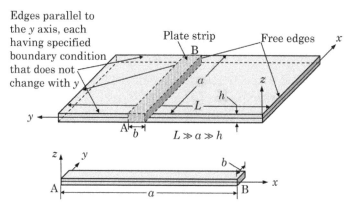

Figure 3.6.1: Cylindrical bending of a rectangular plate with edges $y = 0, b$ free; and edges $x = 0, a$, each having a boundary condition that does not vary with y.

3.6.2   GOVERNING EQUATIONS IN TERMS OF STRESS RESULTANTS

The governing equations of motion according to the classical plate theory (CPT) are given by

$$\frac{\partial N_{xx}^p}{\partial x} + f = m_0 \frac{\partial^2 u}{\partial t^2} - m_1 \frac{\partial^3 w}{\partial x \partial t^2}, \qquad (3.6.1)$$

$$\frac{\partial N_{xy}^p}{\partial x} = m_0 \frac{\partial^2 v}{\partial t^2}, \qquad (3.6.2)$$

$$\frac{\partial^2 \bar{M}_{xx}^p}{\partial x^4} + \frac{\partial}{\partial x}\left(N_{xx}^p \frac{\partial w}{\partial x}\right) - c_f w + q = m_0 \frac{\partial^2 w}{\partial t^2} - m_2 \frac{\partial^4 w}{\partial x^2 \partial t^2} + m_1 \frac{\partial^3 u}{\partial x \partial t^2}, \qquad (3.6.3)$$

where $\bar{M}_{xx}^p = M_{xx}^p + P_{xy}^p$ and $(N_{xx}^p, N_{xy}^p, M_{xx}^p, P_{xy}^p)$ are known [cf. Eqs. (3.4.6a)–(3.4.6c)] in terms of the strains as

$$N_{xx}^p = \int_{-\frac{h}{2}}^{\frac{h}{2}} \sigma_{xx}\, dz = A_{xx}^p \varepsilon_{xx}^{(0)} + B_{xx}^p \varepsilon_{xx}^{(1)} - N_{xx}^T,$$

$$M_{xx}^p = \int_{-\frac{h}{2}}^{\frac{h}{2}} z\, \sigma_{xx}\, dz = B_{xx}^p \varepsilon_{xx}^{(0)} + D_{xx}^p \varepsilon_{xx}^{(1)} - M_{xx}^T, \qquad (3.6.4a)$$

$$(N_{xx}^T, M_{xx}^T) = \frac{1}{(1-\nu)} \int_{-\frac{h}{2}}^{\frac{h}{2}} (1, z)\alpha_T(z)\, E(z)\Delta T(z)\, dz,$$

(3.6.4b)

$$N_{xy}^p = \int_{-\frac{h}{2}}^{\frac{h}{2}} \sigma_{xy}\, dz = A_{xy}^p \frac{\partial v}{\partial x}, \quad P_{xy}^p = \int_{-\frac{h}{2}}^{\frac{h}{2}} z\, \mathcal{M}_{xy}\, dz = S_{xy}^p \varepsilon_{xx}^{(1)},$$

$$(A_{xx}^p, B_{xx}^p, D_{xx}^p) = \frac{1}{(1-\nu^2)} \int_{-\frac{h}{2}}^{\frac{h}{2}} E(z)(1, z, z^2)\, dz,$$

(3.6.5)

$$A_{xy}^p = \frac{1}{(1+\nu)} \int_{-\frac{h}{2}}^{\frac{h}{2}} E(z)\, dz, \quad S_{xy}^p = \frac{\ell^2}{(1+\nu)} \int_{-\frac{h}{2}}^{\frac{h}{2}} E(z)\, dz.$$

Here, the superscript p on the stress resultants refers to the plate strips, α_T denotes the thermal coefficient of expansion, and ν denotes Poisson's ratio. The plane-stress–reduced constitutive relations are adopted (see **Problem 1.4**). It is clear from Eq. (3.6.4b) that the displacement $v(x, t)$ appears only in Eq. (3.6.2), which is decoupled from Eqs. (3.6.1) and (3.6.3) even in the general case. Therefore, Eq. (3.6.2) is not considered further (v is zero as there is no force in the y direction). Equations (3.6.1) and (3.6.3) governing the motion of plate strips are identical to Eqs. (3.3.22) and (3.3.23) for beams, with (N_{xx}, M_{xx}, P_{xy}) replaced by $(N_{xx}^p, M_{xx}^p, P_{xy}^p)$ for plate strips.

3.6.3 GOVERNING EQUATIONS IN TERMS OF DISPLACEMENTS

Using the relations in Eqs. (3.6.4a) and (3.6.4b), the equations of motion (3.6.1) and (3.6.3) can be written in terms of the displacements u and w as

$$-\frac{\partial}{\partial x}\left\{ A_{xx}^p \left[\frac{\partial u}{\partial x} + \frac{1}{2}\left(\frac{\partial w}{\partial x}\right)^2 \right] + B_{xx}^p\left(-\frac{\partial^2 w}{\partial x^2}\right) - N_{xx}^T \right\}$$
$$+ m_0 \frac{\partial^2 u}{\partial t^2} - m_1 \frac{\partial^3 w}{\partial t^2 \partial x} - f = 0,$$

(3.6.6)

$$-\frac{\partial^2}{\partial x^2}\left\{ B_{xx}^p \left[\frac{\partial u}{\partial x} + \frac{1}{2}\left(\frac{\partial w}{\partial x}\right)^2 \right] + (D_{xx}^p + S_{xy}^p)\left(-\frac{\partial^2 w}{\partial x^2}\right) - M_{xx}^T \right\}$$
$$-\frac{\partial}{\partial x}\left\{ A_{xx}^p \frac{\partial w}{\partial x}\left[\frac{\partial u}{\partial x} + \frac{1}{2}\left(\frac{\partial w}{\partial x}\right)^2 \right] - B_{xx}^p \frac{\partial w}{\partial x}\frac{\partial^2 w}{\partial x^2} - \frac{\partial w}{\partial x}N_{xx}^T \right\}$$
$$+ c_f w + m_0 \frac{\partial^2 w}{\partial t^2} + m_1 \frac{\partial^3 u}{\partial t^2 \partial x} - m_2 \frac{\partial^4 w}{\partial t^2 \partial x^2} - q = 0.$$

(3.6.7)

Because of the similarity of the plate strip (cylindrical bending) equations to the beam equations, the special cases presented in the preceding sections are valid for plate strips with the appropriate change (or scaling) of the material stiffness coefficients (i.e., change A_{xx} to $A_{xx}^p = A_{xx}/(1-\nu^2)$ and so on).

3.7 EXACT SOLUTIONS

3.7.1 BENDING SOLUTIONS

Here, we present exact solutions to the *linear* (i.e., omit the nonlinear terms) equations of equilibrium (i.e., all quantities are independent of time t) without axial force and elastic foundation. The exact solutions of the CBT presented in this section are also valid for plate strips with a change of the coefficients.

By omitting all time derivative terms and setting $f = 0$ and $c_f = 0$ in Eqs. (3.4.14) and (3.4.15), we obtain

$$-\frac{d}{dx}\left(A_{xx}\frac{du}{dx} - B_{xx}\frac{d^2w}{dx^2} - N_{xx}^T\right) = 0, \qquad (3.7.1)$$

$$-\frac{d^2}{dx^2}\left[B_{xx}\frac{du}{dx} - (D_{xx} + A_{xy})\frac{d^2w}{dx^2} - M_{xx}^T\right] = q. \qquad (3.7.2)$$

We further assume that the beam stiffness coefficients as well as the thermal forces are all constant. The assumption that $c_f = 0$ simplifies the solution procedure; otherwise, the solution cannot be obtained by simple integration (see **Example 3.7.4** for the case $c_f = 0$).

Equations (3.7.1) and (3.7.2), when expressed in terms of the stress resultants N_{xx} and $\bar{M}_{xx}$ take the form (and integrating once)

$$\frac{dN_{xx}}{dx} = 0 \;\;\rightarrow\;\; N_{xx} = K_1, \qquad (3.7.3a)$$

$$\frac{d^2\bar{M}_{xx}}{dx^2} = -q \;\;\rightarrow\;\; \frac{d\bar{M}_{xx}}{dx} = -\int q(x)\,dx + K_2. \qquad (3.7.3b)$$

Integrating Eq. (3.7.3b), we obtain

$$\bar{M}_{xx} = -\int^x\left(\int^\xi q(\eta)d\eta\right)d\xi + K_2 x + K_3, \qquad (3.7.4)$$

where K_i $(i = 1, 2, 3)$ are constants of integration, which can be determined using the boundary conditions, as will be explained shortly. Here, ξ and η denote dummy coordinates of integration (ultimately, the resulting expression will be in terms of x only). They are introduced to keep track of the sequence of integration; of course, one may replace these dummy variables with x and understand that integration is carried out one at a time.

When the stress resultants N_{xx} and $\bar{M}_{xx}$ in Eqs. (3.7.3a) and (3.7.4) are expressed in terms of the displacements u and w using Eqs. (3.4.6a) and (3.4.6b) (by omitting the nonlinear terms) and writing in matrix form, we obtain

$$\begin{bmatrix} A_{xx} & B_{xx} \\ B_{xx} & \hat{D}_{xx} \end{bmatrix} \left\{ \begin{array}{c} \frac{du}{dx} \\ -\frac{d^2w}{dx^2} \end{array} \right\} = \left\{ \begin{array}{c} K_1 + N_{xx}^T \\ F(x) + M_{xx}^T, \end{array} \right\}, \qquad (3.7.5a)$$

where

$$F(x) = -\int^x \int^\xi q(\eta)\,d\eta\,d\xi + K_2 x + K_3. \tag{3.7.5b}$$

Solving for du/dx and $-d^2w/dx^2$, we obtain

$$\frac{du}{dx} = \frac{1}{\hat{D}^*_{xx}}\left[\hat{D}_{xx}\left(K_1 + N^T_{xx}\right) - B_{xx}\left(F(x) + M^T_{xx}\right)\right], \tag{3.7.6}$$

$$-\frac{d^2w}{dx^2} = \frac{1}{\hat{D}^*_{xx}}\left[-B_{xx}\left(K_1 + N^T_{xx}\right) + A_{xx}\left(F(x) + M^T_{xx}\right)\right], \tag{3.7.7}$$

where

$$\hat{D}^*_{xx} = A_{xx}\hat{D}_{xx} - B^2_{xx}, \quad \hat{D}_{xx} = D_{xx} + A_{xy}. \tag{3.7.8}$$

Integrating Eqs. (3.7.6) and (3.7.7), we obtain

$$u(x) = \frac{\hat{D}_{xx}}{\hat{D}^*_{xx}}\left(K_1 + N^T_{xx}\right)x - \frac{B_{xx}}{\hat{D}^*_{xx}}M^T_{xx}\,x$$
$$+ \frac{B_{xx}}{\hat{D}^*_{xx}}\int^x\int^\xi\int^\eta q(\zeta)\,d\zeta\,d\eta\,d\xi$$
$$- \frac{B_{xx}}{\hat{D}^*_{xx}}\left(K_2\frac{x^2}{2} + K_3 x + K_4\right), \tag{3.7.9}$$

$$-\frac{dw}{dx} = \left[\frac{A_{xx}}{\hat{D}^*_{xx}}M^T_{xx} - \frac{B_{xx}}{\hat{D}^*_{xx}}\left(K_1 + N^T_{xx}\right)\right]x$$
$$- \frac{A_{xx}}{\hat{D}^*_{xx}}\int^x\int^\xi\int^\eta q(\zeta)\,d\zeta\,d\eta\,d\xi$$
$$+ \frac{A_{xx}}{\hat{D}^*_{xx}}\left(K_2\frac{x^2}{2} + K_3 x + K_5\right), \tag{3.7.10}$$

$$w(x) = -\left[\frac{A_{xx}}{\hat{D}^*_{xx}}M^T_{xx} - \frac{B_{xx}}{\hat{D}^*_{xx}}\left(K_1 + N^T_{xx}\right)\right]\frac{x^2}{2}$$
$$- \frac{A_{xx}}{\hat{D}^*_{xx}}\left(K_2\frac{x^3}{6} + K_3\frac{x^2}{2} + K_5 x + K_6\right)$$
$$+ \frac{A_{xx}}{\hat{D}^*_{xx}}\int^x\int^\xi\int^\eta\int^\zeta q(\mu)\,d\mu\,d\zeta\,d\eta d\xi. \tag{3.7.11}$$

Once $q(x)$ is specified, one can easily evaluate the integrals. For example, when $q = q_0$, a constant (i.e., uniform distributed load), we have

$$\int q(x)dx = q_0 x, \quad \int^x\int^\xi q(\eta)\,d\eta\,d\xi = \frac{q_0 x^2}{2},$$

$$\int^x\int^\xi\int^\eta q(\zeta)\,d\zeta\,d\eta d\xi = \frac{q_0 x^3}{6}, \quad \int^x\int^\xi\int^\eta\int^\zeta q(\mu)\,d\mu\,d\zeta\,d\eta\,d\xi = \frac{q_0 x^4}{24}.$$

The six constants of integration are determined using six boundary conditions[1], three at each end of the beam (i.e., one element of the each of the three duality pairs at each point; see Eq. (3.3.18) and Fig. 3.7.1): (u, N_{xx}), $(w, V_{\text{eff}} = d\bar{M}_{xx}/dx)$, and $(dw/dx, \bar{M}_{xx})$. The stress resultants N_{xx} and $\bar{M}_{xx}$ can be computed using Eqs. (3.4.6a) and (3.4.6b).

Type of support	Essential BCs (geometric BCs)	Natural BCs (force BCs)
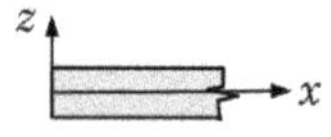 Free	None	$N_{xx} = 0$ $V_{\text{eff}} = 0$ $\bar{M}_{xx} = 0$
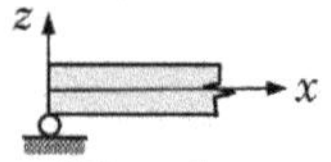 Hinged	$w = 0$	$N_{xx} = 0$ $\bar{M}_{xx} = 0$
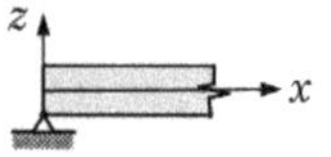 Pinned	$u = 0$ $w = 0$	$\bar{M}_{xx} = 0$
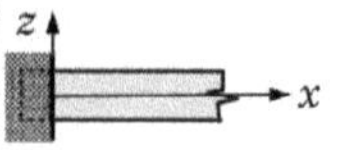 Clamped/Fixed	$u = 0$ $w = 0$ $\dfrac{\partial w}{\partial x} = 0$	None

Figure 3.7.1: Typical boundary conditions for the classical beam theory (CBT). We note that one element in each of the three pairs, or a relation between the two elements of the same pair, must be specified: (u, N_{xx}), $(w, V_{\text{eff}} = d\bar{M}_{xx}/dx)$, and $(dw/dx, \bar{M}_{xx})$.

Example 3.7.1

Determine the exact solution of an FGM beam with two different kinds of "simply-supported" boundary conditions: (a) pinned at the left end (to eliminate the rigid axial motion) and hinged at the right end and (b) pinned at both ends. The beam is subjected to uniform distributed load (UDL) of intensity $q(x) = q_0$.

Solution: (a) The boundary conditions for the pinned–hinged beam are

$$u(0) = 0, \quad w(0) = 0, \quad \bar{M}_{xx}(0) = 0, \quad N_{xx}(L) = 0, \quad w(L) = 0, \quad \bar{M}_{xx}(L) = 0. \quad (3.7.12)$$

In view of Eqs. (3.7.3a), (3.7.4), (3.7.9), and (3.7.11), the condition $N_{xx}(L) = 0$ gives $K_1 = 0$; conditions $\bar{M}_{xx}(0) = \bar{M}_{xx}(L)$ give $K_3 = 0$ and $K_2 = q_0 L/2$; $u(0) = 0$

---

[1]A hinged–hinged beam will have translational rigid body motion because the axial displacement is not specified at any point; a pinned–pinned beam will develop axial strain even without axial force, which will be considered in the nonlinear analysis.

gives $K_4 = 0$; and $w(0)$ gives $K_6 = 0$. The only constant to be determined is K_5. From Eq. (3.7.11), the condition $w(L) = 0$ gives

$$K_5 = -\left(M_{xx}^T - \frac{B_{xx}}{A_{xx}} N_{xx}^T\right) \frac{L}{2} - \frac{q_0 L^3}{24}. \qquad (3.7.13)$$

Therefore, the exact solutions for a pinned–hinged FGM beam are [the stress resultants N_{xx} and $\bar{M}_{xx}$ are computed using Eqs. (3.4.6a) and (3.4.6b); $\xi = x/L$]:

$$u(\xi) = \frac{B_{xx}}{\hat{D}_{xx}^*} \frac{q_0 L^3}{12} \left(-3\xi^2 + 2\xi^3\right) + \frac{L}{\hat{D}_{xx}^*} \left(\hat{D}_{xx} N_{xx}^T - B_{xx} M_{xx}^T\right) \xi, \qquad (3.7.14a)$$

$$-\frac{dw}{dx} = -\frac{A_{xx}}{\hat{D}_{xx}^*} \frac{q_0 L^3}{24} \left(1 - 6\xi^2 + 4\xi^3\right)$$
$$- \frac{L}{2\hat{D}_{xx}^*} \left(A_{xx} M_{xx}^T - B_{xx} N_{xx}^T\right) (1 - 2\xi), \qquad (3.7.14b)$$

$$w(\xi) = \frac{A_{xx}}{\hat{D}_{xx}^*} \frac{q_0 L^4}{24} \left(\xi - 2\xi^3 + \xi^4\right)$$
$$+ \frac{L^2}{2\hat{D}_{xx}^*} \left(A_{xx} M_{xx}^T - B_{xx} N_{xx}^T\right) \left(\xi - \xi^2\right), \qquad (3.7.14c)$$

$$\bar{M}_{xx}(\xi) = \frac{q_0 L^2}{2} \left(\xi - \xi^2\right), \quad N_{xx} = -N_{xx}^T, \qquad (3.7.14d)$$

$$N_{xz}(\xi) = \frac{dM_{xx}}{dx} = \frac{q_0 L}{2} (1 - 2\xi). \qquad (3.7.14e)$$

We note that the stress resultants $(N_{xx}, N_{xz}, \bar{M}_{xx})$ are independent of the material properties, as expected.

For homogeneous beams (i.e., $B_{xx} = 0$ and $\hat{D}_{xx}^* = \hat{D}_{xx} A_{xx}$), the displacements are given by

$$u(\xi) = \frac{1}{A_{xx}} N_{xx}^T x, \qquad (3.7.15a)$$

$$-\frac{dw}{dx} = -\frac{L}{2\hat{D}_{xx}} M_{xx}^T (1 - 2\xi) - \frac{1}{\hat{D}_{xx}} \frac{q_0 L^3}{24} \left(1 - 6\xi^2 + 4\xi^3\right), \qquad (3.7.15b)$$

$$w(\xi) = \frac{L^2}{2\hat{D}_{xx}} M_{xx}^T \left(\xi - \xi^2\right) + \frac{1}{\hat{D}_{xx}} \frac{q_0 L^4}{24} \left(\xi - 2\xi^3 + \xi^4\right). \qquad (3.7.15c)$$

(b) For the pinned–pinned beam, boundary condition $N_{xx}(L) = 0$ in Eq. (3.7.12) is replaced with $u(L) = 0$. Hence, except for K_1 and K_5, all other constants determined before (i.e., $K_2 = q_0 L/2$, $K_3 = 0$, $K_4 = 0$, and $K_6 = 0$) are valid for this case also. The boundary conditions $u(L) = 0$ and $w(L) = 0$ give

$$K_1 = \frac{B_{xx}}{\hat{D}_{xx}} \left(M_{xx}^T + \frac{q_0 L^2}{12}\right) - N_{xx}^T,$$
$$K_5 = -\frac{\hat{D}_{xx}^*}{\hat{D}_{xx} A_{xx}} \left(\frac{1}{2} M_{xx}^T L + \frac{q_0 L^3}{24}\right). \qquad (3.7.16)$$

Since the expressions for $\bar{M}_{xx}$ and N_{xz} do not contain K_1 and K_5, they are the same for pinned–hinged beams and pinned–pinned beams. Hence, the expressions for u, w, $-dw/dx$, and N_{xx} for the pinned–pinned FGM beam are (we recall that $A_{xx}\hat{D}_{xx} - B_{xx}B_{xx} = \hat{D}_{xx}^*$):

$$u(\xi) = \frac{B_{xx}}{\hat{D}_{xx}^*}\frac{q_0 L^3}{12}\left(\xi - 3\xi^2 + 2\xi^3\right), \tag{3.7.17a}$$

$$w(\xi) = \frac{B_{xx}B_{xx}}{\hat{D}_{xx}^*\hat{D}_{xx}}\frac{q_0 L^4}{24}\xi^2 - \frac{1}{\hat{D}_{xx}}\frac{q_0 L^4}{24}\xi$$

$$- \frac{A_{xx}}{\hat{D}_{xx}^*}\frac{q_0 L^4}{24}\left(2\xi^3 - \xi^4\right) + \frac{L^2}{2\hat{D}_{xx}}M_{xx}^T\left(\xi - \xi^2\right)$$

$$= \frac{A_{xx}}{\hat{D}_{xx}^*}\frac{q_0 L^4}{24}\left(\xi^2 - 2\xi^3 + \xi^4\right) + \frac{1}{\hat{D}_{xx}}\frac{q_0 L^4}{24}\left(\xi - \xi^2\right)$$

$$+ \frac{L^2}{2\hat{D}_{xx}}M_{xx}^T\left(\xi - \xi^2\right), \tag{3.7.17b}$$

$$-\frac{dw}{dx} = -\frac{A_{xx}}{\hat{D}_{xx}^*}\frac{q_0 L^3}{12}\left(\xi - 3\xi^2 + 2\xi^3\right) - \frac{1}{\hat{D}_{xx}}\frac{q_0 L^3}{24}\left(1 - 2\xi\right)$$

$$- \frac{L}{2\hat{D}_{xx}}M_{xx}^T\left(1 - 2\xi\right), \tag{3.7.17c}$$

$$N_{xx} = \frac{B_{xx}}{\hat{D}_{xx}}\left(M_{xx}^T + \frac{q_0 L^2}{12}\right) - N_{xx}^T. \tag{3.7.17d}$$

The solution for the homogeneous pinned–pinned beams is

$$u(\xi) = 0, \tag{3.7.18a}$$

$$w(\xi) = \frac{1}{\hat{D}_{xx}}\frac{q_0 L^4}{24}\left(\xi - 2\xi^3 + \xi^4\right) + \frac{L^2}{2\hat{D}_{xx}}M_{xx}^T\left(\xi - \xi^2\right), \tag{3.7.18b}$$

$$-\frac{dw}{dx} = -\frac{1}{\hat{D}_{xx}}\frac{q_0 L^3}{24}\left(1 - 6\xi^2 + 4\xi^3\right) - \frac{L}{2\hat{D}_{xx}}M_{xx}^T\left(1 - 2\xi\right). \tag{3.7.18c}$$

Here we consider pinned-pinned functionally graded beams of length $L = 100$ in (254 cm), height $h = 1$ in (2.54 cm), and width $b = 1$ in (2.54 cm). and subjected to uniform distributed load of intensity q_0 lb/in ($q_0 = 1$ lb/in $= 175$ N/m). The FGM beam is made of two materials with the following values of the moduli and Poisson's ratio:

$$E_1 = 30 \times 10^6 \text{ psi (207 GPa)}, \quad E_2 = 10 \times 10^6 \text{ psi (69 GPa)}, \quad \nu = 0.3.$$

We shall investigate the parametric effects of the volume fraction index, n, and boundary conditions on the transverse deflections and bending moment.

Figure 3.7.2 contains plots of $u(x)$ vs. x/L and $w(x)$ vs. x/L, while Fig. 3.7.3 contains plots of slope $-(dw/dx)(x)$ vs. x/L and bending moment $M_{xx}(x)$ vs. x/L. The axial displacement $u(x)$ exists only because of the coupling coefficient B_{xx} (and $f = 0$), and the way B_{xx} varies with n [see Fig. 3.4.1(b)] is reflected in the variation of $u(x)$ with n. In particular, $u(x)$ increases with increasing values of n for $n < 5$ but the magnitude of $u(x)$ decreases with increasing

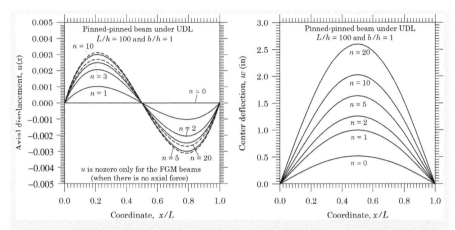

Figure 3.7.2: Plots of displacements $u(x)$ and $w(x)$ versus x/L for pinned–pinned FGM beams under uniform distributed transverse load.

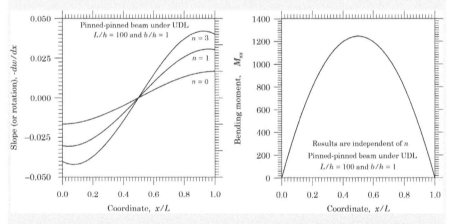

Figure 3.7.3: Plots of slope and bending moment $-(dw/dx)(x)$ and $M_{xx}(x)$ versus x/L for pinned–pinned FGM beams under uniform distributed transverse load.

values of n. On the other hand, the deflection and slope increase their magnitudes with the increasing values of n as the bending stiffness D_{xx} dominates bending.

Figure 3.7.4 contains variations of the maximum displacements $u(0.25L)$ and $w(0.5L)$ with the volume fraction index n. It is clear from the plots that the displacement u increases with n for $n \leq 5$ and then decreases with the increasing values of n, as dictated by the variation of B_{xx} with n. Both $w(0.5L)$ and $-(dw/dx)(L)$ (not shown here) increase with n but there are two parts, the first part exhibits rapid increase with n followed by slow increase due to the interplay between B_{xx} and D_{xx} in the bending response.

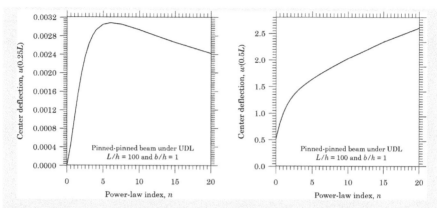

Figure 3.7.4: Plots of maximum displacements $u(0.25L)$ and $w(0.5L)$ versus n for pinned–pinned FGM beams under uniform distributed transverse load.

Example 3.7.2

Determine the exact solution of an FGM beam clamped at both ends and is subjected to uniform distributed transverse load of intensity $q(x) = q_0$. Plot the solution $w(x)$, $-dw/dx$, and $M(x)$ as a functions of x using the data from **Example 3.7.1**, $E_1 = 30 \times 10^6$ psi (207 GPa) , $E_2 = 10 \times 10^6$ psi (69 GPa), and $\nu = 0.3$.

Solution: The boundary conditions for this clamped–clamped beam are

$$u = 0, \quad w = 0, \quad \frac{dw}{dx} = 0, \quad \text{at } x = 0, L. \tag{3.7.19}$$

The condition $u(0) = 0$ gives $K_4 = 0$, $w(0) = 0$ gives $K_6 = 0$, and (dw/dx) at $x = 0$ gives $K_5 = 0$. Applying the boundary conditions at $x = L$, we obtain the following algebraic equations from Eqs. (3.7.9)–(3.7.11):

$$0 = \hat{D}_{xx} K_1 L + B_{xx}\frac{q_0 L^3}{6} - B_{xx}\left(K_2\frac{L^2}{2} + K_3 L\right),$$

$$0 = -B_{xx} K_1 L + A_{xx}\left(K_2\frac{L^2}{2} + K_3 L\right) - A_{xx}\frac{q_0 L^3}{6}, \tag{3.7.20}$$

$$0 = B_{xx} K_1 \frac{L^2}{2} - A_{xx}\left(K_2\frac{L^3}{6} + K_3\frac{L^2}{2}\right) + A_{xx}\frac{q_0 L^4}{24}.$$

From the last two equations, by eliminating K_1, we obtain $K_2 = q_0 L/2$. Substituting this value into the first two equations of (3.7.20), we obtain

$$0 = \hat{D}_{xx} K_1 L - B_{xx}\left(\frac{q_0 L^3}{12} + K_3 L\right),$$

$$0 = -B_{xx} K_1 L + A_{xx}\left(\frac{q_0 L^3}{12} + K_3 L\right), \tag{3.7.21a}$$

which yield

$$K_1 = 0, \quad K_2 = \frac{q_0 L}{2}, \quad K_3 = -\frac{q_0 L^2}{12}. \tag{3.7.21b}$$

Then, the exact solutions for a clamped FGM beam (the couple stress effect is included through $\hat{D}_{xx}^*$) are ($N_{xx} = 0$)

$$u(\xi) = \frac{B_{xx}}{\hat{D}_{xx}^*} \frac{q_0 L^3}{12} \left(\xi - 3\xi^2 + 2\xi^3 \right), \tag{3.7.22}$$

$$\frac{dw}{dx} = \frac{A_{xx}}{\hat{D}_{xx}^*} \frac{q_0 L^3}{12} \left(-\xi + 3\xi^2 - 2\xi^3 \right), \tag{3.7.23}$$

$$w(\xi) = \frac{A_{xx}}{\hat{D}_{xx}^*} \frac{q_0 L^4}{24} \left(\xi^2 - 2\xi^3 + \xi^4 \right), \tag{3.7.24}$$

$$M_{xx}(\xi) = -\frac{q_0 L^2}{12} \left(1 - 6\xi + 6\xi^2 \right), \tag{3.7.25}$$

$$N_{xz} = \frac{dM_{xx}}{dx} = \frac{q_0 L}{2} \left(1 - 2\xi \right). \tag{3.7.26}$$

We note that the bending moment and shear force are independent of material properties. The effect of B_{xx} on the displacements is through $\hat{D}_{xx}^* = \hat{D}_{xx} A_{xx} - B_{xx} B_{xx}$.

The variation of $u(x)$ for the clamped–clamped beam is the same as that of the pinned–pinned beam. Figure 3.7.5 contains plots of $w(x)$ vs. x/L and $-(dw/dx)(x)$ vs. x/L while Fig. 3.7.6 contains the bending moment $M_{xx}(x)$ vs. x/L. The data used here is the same as that used for pinned–pinned beams. The results obtained show the same trends as those discussed for the pinned–pinned beams.

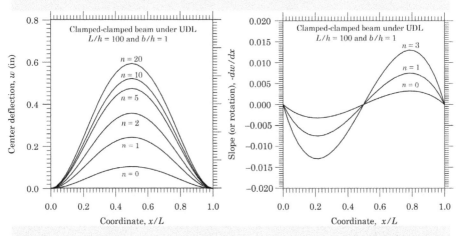

Figure 3.7.5: Plots of displacements $w(x)$ versus x/L and the maximum slope $-(dw/dx)$ versus x/L for clamped–clamped FGM beams under uniform distributed transverse load for various values of the power-law exponent n. The material properties used are: $E_1 = 30 \times 10^6$ psi (207 GPa) , $E_2 = 10 \times 10^6$ psi (69 GPa), and $\nu = 0.3$.

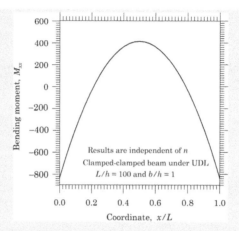

Figure 3.7.6: Plots of bending moment $M_{xx}(x)$ versus x/L for clamped–clamped FGM beams under uniform distributed transverse load.

Example 3.7.3

Determine the exact solution of a beam clamped at the left end, vertically spring–supported at the right end, and subjected to uniform distributed transverse load of intensity $q(x) = q_0$, as shown in Fig. 3.7.7(a). Assume that the spring is linearly elastic, with a spring constant of k (N/m). Specialize the results to (i) cantilevered beam shown in Fig. 3.7.7(b) and (ii) propped cantilever beam shown in Fig. 3.7.7(c).

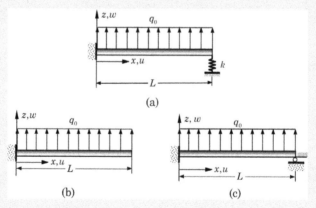

Figure 3.7.7: (a) A uniform loaded beam clamped at the left end $x = 0$ and spring–supported at the right end $x = L$. (b) A cantilevered beam. (c) A propped cantilevered beam.

Solution: Although temperature effect is not needed in this problem, we develop the solution accounting for the thermal resultants. The boundary conditions at the clamped end, $x = 0$, are

$$u = 0, \quad w = 0, \quad \frac{dw}{dx} = 0, \tag{3.7.27}$$

and at the right end, $x = L$, we have

$$N_{xx} = 0, \quad \frac{d\bar{M}_{xx}}{dx} + kw = 0, \quad \bar{M}_{xx} = 0. \tag{3.7.28}$$

As before, $u(0) = 0$ gives $K_4 = 0$, $w(0) = 0$ gives $K_6 = 0$, and (dw/dx) at $x = 0$ gives $K_5 = 0$. From Eq. (3.7.3a), $N_{xx}(L) = 0$ gives $K_1 = 0$ and $\bar{M}_{xx}(L) = 0$ gives $K_2 L + K_3 = q_0 L^2/2$. Finally, the mixed boundary condition $(d\bar{M}_{xx}/dx) + kw = 0$ at $x = L$ yields

$$
\begin{aligned}
0 &= -q_0 L + K_2 + \frac{kL^2}{2\hat{D}_{xx}^*}\left[-\left(A_{xx}\, M_{xx}^T - B_{xx}\, N_{xx}^T \right) - A_{xx}\left(K_2 \frac{L}{3} + K_3 \right) \right. \\
&\qquad\qquad \left. + A_{xx}\frac{q_0 L^2}{12} \right] \\
&= -q_0 L + K_2 + \frac{kL^2}{2\hat{D}_{xx}^*}\left[-\left(A_{xx}\, M_{xx}^T - B_{xx}\, N_{xx}^T \right) + A_{xx}\left(K_2 \frac{2L}{3} - \frac{5q_0 L^2}{12} \right) \right] \\
&= -q_0 L + \hat{c}_1\, K_2 - \hat{c}_2\, k,
\end{aligned}
\tag{3.7.29a}
$$

where

$$\hat{c}_1 = 1 + \frac{A_{xx}}{3\hat{D}_{xx}^*}kL^3, \quad \hat{c}_2 = \frac{L^2}{2\hat{D}_{xx}^*}\left[\left(A_{xx}\, M_{xx}^T - B_{xx}\, N_{xx}^T \right) + A_{xx}\frac{5q_0 L^2}{12} \right]. \tag{3.7.29b}$$

Then we have

$$K_2 = \frac{1}{\hat{c}_1}\left(\hat{c}_2\, k + q_0 L \right), \quad K_3 = \frac{q_0 L^2}{2} - K_2 L. \tag{3.7.29c}$$

Thus, the exact solutions for an FGM beam (with MCS) clamped at $x = 0$ and supported at $x = L$ by a linearly elastic spring are

$$
\begin{aligned}
u(\xi) &= \frac{L}{\hat{D}_{xx}^*}\left(\hat{D}_{xx}\, N_{xx}^T - B_{xx}\, M_{xx}^T \right)\xi \\
&\quad - \frac{B_{xx}}{\hat{D}_{xx}^*}\left[\frac{q_0 L^3}{6}\left(3\xi - \xi^3 \right) + \frac{K_2 L^2}{2}\left(2\xi - \xi^2 \right) \right],
\end{aligned}
\tag{3.7.30}
$$

$$
\begin{aligned}
w(\xi) &= -\frac{L^2}{2\hat{D}_{xx}^*}\left(A_{xx}\, M_{xx}^T - B_{xx}\, N_{xx}^T \right)\xi^2 \\
&\quad - \frac{A_{xx}}{\hat{D}_{xx}^*}\left[\frac{q_0 L^4}{24}\left(6\xi^2 - \xi^4 \right) - \frac{K_2 L^3}{6}\left(3\xi^2 - \xi^4 \right) \right],
\end{aligned}
\tag{3.7.31}
$$

$$
\begin{aligned}
-\frac{dw}{dx} &= \frac{L}{\hat{D}_{xx}^*}\left(A_{xx}\, M_{xx}^T - B_{xx}\, N_{xx}^T \right)\xi \\
&\quad + \frac{A_{xx}}{\hat{D}_{xx}^*}\left[\frac{q_0 L^3}{6}\left(3\xi - \xi^3 \right) - \frac{K_2 L^2}{2}\left(2\xi - \xi^2 \right) \right],
\end{aligned}
\tag{3.7.32}
$$

where K_2 is defined by Eqs. (3.7.29a) and (3.7.29b). The solution in Eqs. (3.7.30)–(3.7.32) have two special cases, as explained next.

(a) *Cantilevered beam* [see Fig. 3.7.7(b)]. For a cantilevered beam fixed at $x = 0$ and free at $x = L$ and subjected to uniform distributed load, we set $k = 0$ and obtain $\hat{c}_1 = 1$ and $K_2 = q_0 L$. Then the solution becomes

$$u(\xi) = \frac{L}{\hat{D}_{xx}^*} \left(\hat{D}_{xx} N_{xx}^T - B_{xx} M_{xx}^T \right) \xi + \frac{B_{xx}}{\hat{D}_{xx}^*} \frac{q_0 L^3}{6} \left(3\xi - 3\xi^2 + \xi^3 \right), \quad (3.7.33)$$

$$w(\xi) = -\frac{L^2}{2\hat{D}_{xx}^*} \left(A_{xx} M_{xx}^T - B_{xx} N_{xx}^T \right) \xi^2 + \frac{A_{xx}}{\hat{D}_{xx}^*} \frac{q_0 L^4}{24} \left(6\xi^2 - 4\xi^3 + \xi^4 \right),$$
$$(3.7.34)$$

$$-\frac{dw}{dx} = \frac{L}{\hat{D}_{xx}^*} \left(A_{xx} M_{xx}^T - B_{xx} N_{xx}^T \right) \xi - \frac{A_{xx}}{\hat{D}_{xx}^*} \frac{q_0 L^3}{6} \left(3\xi - 3\xi^2 + \xi^3 \right), \quad (3.7.35)$$

$$M_{xx}(\xi) = -\frac{q_0 L^2}{2} \left(1 - 2\xi + \xi^2 \right), \quad N_{xz} = q_0 L \left(1 - \xi \right), \quad N_{xx} = 0. \quad (3.7.36)$$

(b) *Propped Cantilever* [see Fig. 3.7.7(c)]. For a beam clamped at $x = 0$ and hinged at $x = L$, we set $1/k = 0$ (i.e. $k \to \infty$) and obtain K_2 from the second line of Eq. (3.7.29a) as

$$K_2 L = \left[\left(M_{xx}^T - \frac{B_{xx}}{A_{xx}} N_{xx}^T \right) \frac{3}{2} + \frac{5q_0 L^2}{8} \right]. \quad (3.7.37)$$

Then solution in Eqs. (3.7.30)–(3.7.32) becomes

$$u(\xi) = \frac{L}{4\hat{D}_{xx}^*} \frac{B_{xx}}{A_{xx}} \left(A_{xx} M_{xx}^T - B_{xx} N_{xx}^T \right) \left(2\xi - 3\xi^2 \right) + \frac{L}{A_{xx}} N_{xx}^T \xi$$
$$+ \frac{B_{xx}}{\hat{D}_{xx}^*} \frac{q_0 L^3}{48} \left(6\xi - 15\xi^2 + 8\xi^3 \right), \quad (3.7.38)$$

$$w(\xi) = \frac{L^2}{4\hat{D}_{xx}^*} \left(A_{xx} M_{xx}^T - B_{xx} N_{xx}^T \right) \left(\xi^2 - \xi^3 \right)$$
$$+ \frac{A_{xx}}{\hat{D}_{xx}^*} \frac{q_0 L^4}{48} \left(3\xi^2 - 5\xi^3 + 2\xi^4 \right), \quad (3.7.39)$$

$$-\frac{dw}{dx} = -\frac{L}{4\hat{D}_{xx}^*} \left(A_{xx} M_{xx}^T - B_{xx} N_{xx}^T \right) \left(2\xi - 3\xi^2 \right)$$
$$- \frac{A_{xx}}{\hat{D}_{xx}^*} \frac{q_0 L^3}{48} \left(6\xi - 15\xi^2 + 8\xi^3 \right), \quad (3.7.40)$$

$$M_{xx}(\xi) = -\frac{1}{2A_{xx}} \left(1 - 3\xi \right) \left(A_{xx} M_{xx}^T - B_{xx} N_{xx}^T \right)$$
$$- \frac{q_0 L^2}{24} \left(3 - 15\xi + 12\xi^2 \right), \quad (3.7.41)$$

$$N_{xz}(\xi) = -\frac{3}{2L A_{xx}} \left(A_{xx} M_{xx}^T - B_{xx} N_{xx}^T \right) - \frac{q_0 L}{8} \left(-5 + 8\xi \right), \quad N_{xx} = 0. \quad (3.7.42)$$

The results obtained in Examples 3.7.1–3.7.3 can be specialized to beams with $E = E(x)$ by setting $B_{xx} = 0$, which reduces $\hat{D}_{xx}^* = A_{xx}\hat{D}_{xx}$; for beams without the modified couple stress, we set $\hat{D}_{xx} = D_{xx}$.

Example 3.7.4 ——————————————————————————————————

Determine the exact bending solutions of an isotropic beam (i.e., $B_{xx} = 0$) of length L, bending stiffness D_{xx}, resting on the entire length on an elastic foundation with foundation modulus $c_f = k$, simply supported at both ends, and subjected to uniform distributed load, q_0.

Solution: For this case, Eqs. (3.7.1) and (3.7.2) are decoupled. Therefore, we can simply solve Eq. (3.7.2). The general solution of

$$D_{xx}^e \frac{d^4 w}{dx^4} + kw = q_0 \tag{3.7.43}$$

is

$$w(x) = \left(c_1 \sinh \frac{2\beta x}{L} + c_2 \cosh \frac{2\beta x}{L} \right) \sin \frac{2\beta x}{L}$$
$$+ \left(c_3 \sinh \frac{2\beta x}{L} + c_4 \cosh \frac{2\beta x}{L} \right) \cos \frac{2\beta x}{L} + \frac{q_0}{k}, \tag{3.7.44a}$$

where $D_{xx}^e = D_{xx} + A_{xy}$ and

$$\beta^4 = \frac{kL^4}{64 D_{xx}^e}, \tag{3.7.44b}$$

and c_1, c_2, c_3, and c_4 are constants of integration to be determined using the boundary conditions.

For a beam with both edges simply supported, the deflection is symmetrical with respect to the center of the beam. Suppose that the origin of the coordinate system is taken at the center of the beam. Then the slope (dw/dx) and shear force $(V = dM_{xx}/dx)$ at the center, $x = L/2$, must be zero. The boundary conditions are

$$\frac{dw}{dx} = \frac{d^3 w}{dx^3} = 0 \text{ at } x = 0; \quad w = \frac{d^2 w}{dx^2} = 0 \text{ at } x = L/2. \tag{3.7.45}$$

The first two boundary conditions give $c_2 = c_3 = 0$. The next two boundary conditions yield

$$c_1 \sin \beta \sinh \beta + c_4 \cos \beta \cosh \beta + \frac{q_0}{k} = 0, \quad c_1 \cos \beta \cosh \beta + c_4 \sin \beta \sinh \beta = 0,$$

from which we obtain

$$c_1 = -\frac{2q_0}{k} \left(\frac{\sin \beta \sinh \beta}{\cos 2\beta + \cosh 2\beta} \right), \quad c_4 = -\frac{2q_0}{k} \left(\frac{\cos \beta \cosh \beta}{\cos 2\beta + \cosh 2\beta} \right).$$

Then the deflection in Eq. (3.7.44a) takes the form

$$w(x) = \frac{q_0 L^4}{64 D_{xx}^e \beta^4} \left[1 - \left(\frac{2 \sin \beta \sinh \beta}{\cos 2\beta + \cosh 2\beta} \right) \sin \frac{2\beta x}{L} \sinh \frac{2\beta x}{L} \right.$$
$$\left. - \left(\frac{2 \cos \beta \cosh \beta}{\cos 2\beta + \cosh 2\beta} \right) \cos \frac{2\beta x}{L} \cosh \frac{2\beta x}{L} \right]. \tag{3.7.46}$$

The maximum deflection occurs at the center of the beam (i.e., at $x = 0$) and it is given by

$$w_{max} = w(0) = \frac{5q_0L^4}{384D_{xx}^e}\varphi_1(\beta), \quad \varphi_1(\beta) = \frac{6}{5\beta^4}\left(1 - \frac{2\cos\beta\cosh\beta}{\cos 2\beta + \cosh 2\beta}\right). \quad (3.7.47)$$

The rotation at $x = L/2$ is given by

$$\theta_{max} = \left.\frac{dw}{dx}\right|_{x=L/2} = \frac{q_0L^3}{24D_{xx}^e}\varphi_2(\beta), \quad \varphi_2(\beta) = \frac{3}{4\beta^3}\left(\frac{2\sinh 2\beta - \sin 2\beta}{\cos 2\beta + \cosh 2\beta}\right). \quad (3.7.48)$$

The maximum bending moment also occurs at $x = 0$, and it is given by

$$M_{max} = \left[-D_{xx}^e\frac{d^2w}{dx^2}\right]_{x=0} = \frac{q_0L^2}{8}\varphi_3(\beta), \quad \varphi_3(\beta) = \frac{2}{\beta^2}\left(\frac{\sinh\beta\sin\beta}{\cos 2\beta + \cosh 2\beta}\right).$$
$$(3.7.49)$$

3.7.2 BUCKLING AND NATURAL VIBRATIONS

3.7.2.1 Buckling solutions

A straight beam subjected to compressive load $N_{xx} = -N_{xx}^0$ along the centroidal axis (x) shortens the beam as the load increases from zero to a certain magnitude. However, due to a small eccentricity of the axial compressive load from the centroidal axis, a large lateral (perpendicular to the axis of the beam) deflection may occur, and the beam is said to be *unstable*. The onset of this instability is called *buckling*. Figure 3.7.8 shows typical buckling modes for different boundary conditions. The magnitude of the smallest compressive

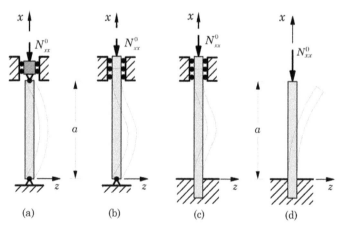

Figure 3.7.8: Buckling of beams under various boundary conditions: (a) hinged–hinged, (b) hinged–clamped, (c) clamped–clamped, and (d) clamped–free boundary conditions.

axial load at which the beam becomes unstable is termed the *critical buckling load*. If the load is increased beyond this critical buckling load, it results in a large deflection and the beam seeks another equilibrium configuration. Thus, the load at which a beam becomes unstable is of practical importance in design. This section is devoted to the determination of the critical buckling loads by exact analysis of the beams equations based on the CBT.

The governing equation of a homogeneous isotropic beam under the applied in-plane compressive load N_{xx}^0 (after the onset of buckling) can be obtained either from Eq. (3.3.17) or from Eq. (3.3.23) by replacing N_{xx} with $N_{xx} = -N_{xx}^0$ and omitting the inertia terms, thermal resultants, distributed transverse load $q(x)$, and axial load $f(x)$. Equation (3.3.17) is a special case of Eq. (3.3.23) as the latter contains the effect of the couple stress.

The derivation of the governing equation for buckling of beams by vector approach requires the use of the deformed configuration, as shown in Fig. 3.7.9. However, as noted before, the vector approach does not provide the boundary conditions.

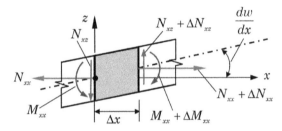

Figure 3.7.9: A typical element from the deformed beam, with internal forces and moments generated, to derive equations of equilibrium using the vector approach for buckling analysis.

Summing the forces in the x- and z-directions and moments about the y-axis, we obtain

$$-N_{xx} + (N_{xx} + \Delta N_{xx}) = 0, \qquad (3.7.50a)$$

$$-V_x + (V_x + \Delta V_x) = 0, \qquad (3.7.50b)$$

$$-M_{xx} + (M_{xx} + \Delta M_{xx}) - V_x \Delta x + N_{xx} \Delta x \frac{dw}{dx} = 0. \qquad (3.7.50c)$$

Taking the limit $\Delta x \to 0$, we obtain the following three equations:

$$\frac{dN_{xx}}{dx} = 0, \quad \frac{dN_{xz}}{dx} = 0, \quad \frac{dM_{xx}}{dx} - N_{xz} + N_{xx} \frac{dw}{dx} = 0. \qquad (3.7.51)$$

Presuming that the axial equation of equilibrium is trivially satisfied and combining the remaining two equations [i.e., substituting for N_{xz} from the third equation into the second equation of Eq. (3.7.51)], we arrive at

$$\frac{d^2 M_{xx}}{dx^2} + \frac{d}{dx}\left(N_{xx}\frac{dw}{dx}\right) = 0 \rightarrow \frac{d^2 M_{xx}}{dx^2} + N_{xx}\frac{d^2 w}{dx^2} = 0. \qquad (3.7.52)$$

Accounting for the couple stress, we have [set $B_{xx} = 0$, $c_f = 0$, $N_{xx} = -N_{xx}^0$, and the time derivative terms to zero in Eq. (3.4.13)]

$$(D_{xx} + A_{xy})\frac{d^4 w}{dx^4} + N_{xx}^0 \frac{d^2 w}{dx^2} = 0, \qquad (3.7.53)$$

where w denotes the transverse deflection measured from the pre-buckling equilibrium state, D_{xx} is the bending stiffness, and A_{xy} stiffness due to the couple stress effect [see Eq. (3.4.7) for the definition].

Integrating Eq. (3.7.53) twice with respect to x, we obtain

$$D_{xx}^e \frac{d^2 w}{dx^2} + N_{xx}^0 w = K_1 x + K_2, \quad D_{xx}^e = D_{xx} + A_{xy}, \qquad (3.7.54)$$

or

$$\frac{d^2 w}{dx^2} + \lambda^2 w = \frac{1}{D_{xx}^e}(K_1 x + K_2). \qquad (3.7.55a)$$

where K_1 and K_2 are constants, and

$$\lambda^2 = \frac{N_{xx}^0}{D_{xx}^e} \quad \text{or} \quad N_{xx}^0 = D_{xx}^e \lambda^2. \qquad (3.7.55b)$$

The solution of the second-order equation in Eq. (3.7.55a) has two parts: homogeneous solution and particular solution. The homogeneous and particular solutions of the second-order equation (3.7.55a) are given, respectively, by

$$w_h(x) = c_1 \sin \lambda x + c_2 \cos \lambda x, \quad w_p(x) = \frac{1}{D_{xx}^e \lambda^2}(K_1 x + K_2). \qquad (3.7.56)$$

Hence, the complete solution is

$$w(x) = c_1 \sin \lambda x + c_2 \cos \lambda x + c_3 x + c_4, \qquad (3.7.57)$$

where $c_3 = K_1/D_{xx}^e \lambda^2$ and $c_4 = K_2/D_{xx}^e \lambda^2$. The four constants c_1, c_2, c_3, and c_4 are determined (not all four constant can be zero) using four boundary conditions of the problem (one of the boundary conditions yields a relation of λ). Once λ is known, the buckling load can be determined using Eq. (3.7.55b). Various types of boundary conditions for beams are listed below (the boundary condition on M_{xx} is expressed in terms of the second derivative of w).

Free:

$$M_{xx} = 0 \quad \text{and} \quad N_{xz} - N_{xx}^0 \frac{d^2 w}{dx^2} = 0$$

or

$$\frac{d^2w}{dx^2} = 0 \quad \text{and} \quad D_{xx}^e \frac{d^3w}{dx^3} - N_{xx}^0 \frac{d^2w}{dx^2} = 0. \qquad (3.7.58)$$

Hinged (or) Simply Supported:

$$w = 0, \quad M_{xx} = 0 \quad \left(\text{or} \quad \frac{d^2w}{dx^2} = 0 \right). \qquad (3.7.59)$$

Fixed (or) Clamped:

$$w = 0, \quad \frac{dw}{dx} = 0. \qquad (3.7.60)$$

Elastically Simply Supported:

$$w = s_k F_0, \quad M_{xx} = -D_{xx} \frac{d^2w}{dx^2} = 0, \qquad (3.7.61)$$

where F_0 is the vertical reacting force of the elastic support and s_k is the inverse of the elastic (spring) constant. When $s_k = 0$ (i.e., the support is rigid), we recover the conventional simply supported boundary condition. When s_k is very large, the boundary condition approaches that of a free edge.

Elastically Clamped:

$$w = 0, \quad \frac{dw}{dx} = -k_R M_0, \qquad (3.7.62)$$

where M_0 is the reacting moment (clockwise) of the elastic support and k_R is the rotational spring constant. When $k_R = 0$ (i.e., the restraint is rigid), we recover the conventional clamped boundary condition. On the other hand, if k_R is very large (i.e., the restraint is very flexible), the condition approaches that of a simply supported case. We now consider couple of examples next.

Example 3.7.5 ──

Determine the critical buckling load of a simply supported beam of length L.

Solution: For a simply supported beam, the boundary conditions in Eq. (3.7.59) at $x = 0, L$ gives

$$w(0) = 0: \quad c_2 + c_4 = 0 \quad \text{or} \quad c_4 = -c_2,$$

$$\left. \frac{d^2w}{dx^2} \right|_{x=0} = 0: \quad -\lambda^2 c_2 = 0, \quad \text{or} \quad c_2 = 0,$$

$$w(L) = 0: \quad c_1 \sin \lambda L + c_3 L = 0, \qquad (3.7.63)$$

$$\left. \frac{d^2w}{dx^2} \right|_{x=L} = 0: \quad -\lambda^2 c_1 \sin \lambda L = 0.$$

From the above equations, it follows that

$$c_1 \sin \lambda L = 0, \quad c_2 = 0, \quad c_3 = 0, \quad c_4 = 0. \qquad (3.7.64)$$

The first equation implies that either $c_1 = 0$ or (and) $\sin \lambda L = 0$. If $c_1 = 0$, then the buckling deflection is zero, implying that the beam did not begin to buckle. For nonzero deflection w, we must have

$$\sin \lambda L = 0 \quad \text{which implies} \quad \lambda_n L = L \sqrt{\frac{N_{xx}^0}{D_{xx}^e}} = n\pi \equiv e_n,$$

or

$$N_{xx}^0 = D_{xx}^e \left(\frac{n\pi}{L}\right)^2 = D_{xx}^e \lambda_n^2. \tag{3.7.65}$$

The critical buckling load (i.e., the smallest value of N_{xx}^0 at which the plate buckles) occurs when $n = 1$:

$$N_{cr} = \frac{D_{xx}^e \pi^2}{L^2} = 9.8696 \frac{D_{xx}^e}{L^2}. \tag{3.7.66}$$

We note that the effect of couple stress is to increase the buckling load and $A_{xy} > 0$ adds to the bending stiffness, D_{xx}.

Example 3.7.6

Determine the critical buckling load of a clamped beam of length L.

Solution: For a clamped beam, the boundary conditions in Eq. (3.7.60) at $x = 0, L$ yields

$$w(0) = 0 : \qquad\qquad c_2 + c_4 = 0 \quad \text{or} \quad c_4 = -c_2,$$

$$\left.\frac{dw}{dx}\right|_{x=0} = 0 : \qquad\qquad \lambda c_1 + c_3 = 0 \quad \text{or} \quad c_3 = -\lambda c_1,$$

$$w(L) = 0 : \quad c_1 \sin \lambda L + c_2 \cos \lambda L + c_3 L + c_4 = 0, \tag{3.7.67}$$

$$\left.\frac{dw}{dx}\right|_{x=L} = 0 : \quad \lambda \left(c_1 \cos \lambda L - c_2 \sin \lambda L\right) + c_3 = 0.$$

Using the first two equations, c_3 and c_4 can be eliminated from the last two equations to obtain

$$c_1 \left(\sin \lambda L - \lambda L\right) + c_2 \left(\cos \lambda L - 1\right) = 0, \quad c_1 \left(\cos \lambda L - 1\right) - c_2 \sin \lambda L = 0. \tag{3.7.68}$$

For nontrivial solution (i.e., for nonzero values of c_2 and c_3), we require that the determinant of the above pair of equations be zero, giving

$$\lambda L \sin \lambda L + 2 \cos \lambda L - 2 = 0. \tag{3.7.69}$$

This nonlinear (transcendental) equation can be solved by an iterative method (e.g., Newton's) for various roots of the equation. The smallest root of this equation is $\lambda L = 2\pi$, and the critical buckling load becomes

$$N_{cr} = \frac{4D_{xx}^e \pi^2}{L^2} = 39.4784 \frac{D_{xx}^e}{L^2}. \tag{3.7.70}$$

Thus, the buckling load of a clamped beam is four times that of a simply supported beam because of the clamped support provides greater stiffness.

For beams with different boundary conditions, the values of the constants c_1–c_4, characteristic equation for buckling loads, and roots of the characteristic equation are presented in Table 3.7.1 (from Reddy [8, 113]).

Table 3.7.1: Values of the constants and eigenvalues for buckling of beams with various boundary conditions $[\lambda^2 \equiv N_{xx}^0/D_{xx}^e = (e_n/L)^2]$.

Beam type	Constants†	Characteristic Equation* and values of $e_n = \lambda_n L$
Hinged-Hinged	$c_1 \neq 0$ $c_2 = c_3 = c_4 = 0$	$\sin e_n = 0, \; e_n = n\pi$
Fixed-Fixed	$c_1 = 1/(\sin e_n - e_n)$ $c_3 = -1/\lambda_n, \; c_2 = -c_4$ $c_2 = 1/(\cos e_n - 1)$	$e_n \sin e_n = 2(1 - \cos e_n)$ $e_n = 2\pi, 8.987, 4\pi, \cdots$
Fixed-Free	$c_1 = c_3 = 0$ $c_2 = -c_4 \neq 0$	$\cos e_n = 0$ $e_n = (2n - 1)\pi/2$
Free-Free	$c_1 = c_3 = 0$ $c_2 \neq 0, \; c_4 \neq 0$	$\sin e_n = 0$ $e_n = n\pi$
Hinged-Fixed	$c_1 = 1/(e_n \cos e_n)$ $c_3 = -1, \; c_2 = c_4 = 0$	$\tan e_n = e_n$ $e_n = 4.493, 7.725, \cdots$

† $w(x) = c_1 \sin \lambda x + c_2 \cos \lambda x + c_3 x + c_4$.
*For critical buckling load, only the first (minimum) value of $e_1 = \lambda_1 L$ is needed.

3.7.2.2   Natural frequencies

For vibration in the absence of any applied transverse mechanical and thermal loads, Eq. (3.7.53) is modified to account for the inertia terms:

$$D_{xx}^e \frac{\partial^4 w}{\partial x^4} - \hat{N}_{xx} \frac{\partial^2 w}{\partial x^2} = m_2 \frac{\partial^4 w}{\partial x^2 \partial t^2} - m_0 \frac{\partial^2 w}{\partial t^2}, \qquad (3.7.71)$$

where $\hat{N}_{xx}$ is the *applied* (tensile) axial force.

For natural vibration, we assume the solution of Eq. (3.7.53) to be periodic

$$w(x, t) = w_0(x) \cos \omega t, \qquad (3.7.72)$$

where ω is the natural frequency of vibration and $w_0(x)$ is the mode shape. Substituting for w from Eq. (3.7.72) into Eq. (3.7.71) and cancelling the factor $\cos \omega t$ (because the result should hold for any time t), we obtain

$$D_{xx}^e \frac{d^4 w_0}{dx^4} - \hat{N}_{xx} \frac{d^2 w_0}{dx^2} = m_0 \omega^2 w_0 - m_2 \omega^2 \frac{d^2 w_0}{dx^2}. \qquad (3.7.73)$$

Equation (3.7.73) can be expressed in the following general form:

$$a\frac{d^4 w_0}{dx^4} + b\frac{d^2 w_0}{dx^2} - cw_0 = 0, \tag{3.7.74a}$$

where

$$a = D^e_{xx}, \quad b = \omega^2 m_2 - \hat{N}_{xx}, \quad c = \omega^2 m_0. \tag{3.7.74b}$$

We assume solution of the homogeneous equation in (3.7.74a) in the form

$$w_0(x) = Ae^{rx}, \tag{3.7.75}$$

and substitute it into Eq. (3.7.74a) to obtain

$$ar^4 + br^2 - c = 0 \quad \text{or} \quad as^2 + bs - c = 0 \quad (s = r^2). \tag{3.7.76}$$

The roots of the above equation are

$$s_1 = \frac{1}{2a}\left(-b - \sqrt{b^2 + 4ac}\right) \equiv -\lambda^2, \quad s_2 = \frac{1}{2a}\left(-b + \sqrt{b^2 + 4ac}\right) \equiv \mu^2. \tag{3.7.77}$$

Hence, the solution can be expressed as

$$w_0(x) = c_1 \sin \lambda x + c_2 \cos \lambda x + c_3 \sinh \mu x + c_4 \cosh \mu x, \tag{3.7.78}$$

where

$$\lambda = \sqrt{\frac{1}{2a}\left(b + \sqrt{b^2 + 4ac}\right)}, \quad \mu = \sqrt{\frac{1}{2a}\left(-b + \sqrt{b^2 + 4ac}\right)}, \tag{3.7.79}$$

and c_1, c_2, c_3, and c_4 are integration constants, which are to be determined using the boundary conditions. In reality, the four boundary conditions (two at each end of the beam) are used to determine λ and only three of the four constants.

The frequency of vibration is determined from λ (or μ) as follows. From Eq. (3.7.79) we have

$$\left(2a\lambda^2 - b\right)^2 = b^2 + 4ac \quad \text{or} \quad a\lambda^4 - b\lambda^2 - c = 0,$$

and using the definitions of a, b, and c, we obtain

$$D^e_{xx}\lambda^4 - \left(\omega^2 m_2 - \hat{N}_{xx}\right)\lambda^2 - \omega^2 m_0 = 0, \tag{3.7.80}$$

from which we have

$$\omega^2 = \frac{D^e_{xx}\lambda^4 + \hat{N}_{xx}\lambda^2}{m_0 + \lambda^2 m_2}. \tag{3.7.81}$$

If the applied in-plane force, $\hat{N}_{xx}$, is zero, the natural frequency of vibration, with rotary inertia included, is given by

$$\omega^2 = \frac{D^e_{xx}\lambda^4}{m_0 + \lambda^2 m_2} = \frac{D^e_{xx}}{m_0}\left(1 - \frac{\lambda^2 m_2}{m_0 + \lambda^2 m_2}\right)\lambda^4. \tag{3.7.82}$$

In addition, if rotary inertia m_2 is neglected, we have

$$\omega = \lambda^2 \sqrt{\frac{D_{xx}^e}{m_0}}. \tag{3.7.83}$$

We now consider couple of examples (refer to **Examples 3.7.5** and **3.7.6** for details of using the boundary conditions to determine the characteristic equations).

Example 3.7.7 ————————————————————————

Determine the fundamental frequency of a simply supported beam of length L.

Solution: For a simply supported beam, we have

$$w_0 = 0, \text{ and } M_{xx} = 0 \text{ or } \frac{d^2 w_0}{dx^2} = 0 \text{ at } x = 0, L. \tag{3.7.84}$$

From Eq. (3.7.78) we have

$$\frac{d^2 w_0}{dx^2} = -\lambda^2 \left(c_1 \sin \lambda x + c_2 \cos \lambda x\right) + \mu^2 \left(c_3 \sinh \mu x + c_4 \cosh \mu x\right). \tag{3.7.85}$$

Using the boundary conditions, we obtain $c_2 + c_4 = 0$ and $-\lambda^2 c_2 + \mu^2 c_4 = 0$, implying $c_2 = c_4 = 0$; and $c_1 \sin \lambda L + c_3 \sinh \mu L = 0$ and $-\lambda^2 c_1 \sin \lambda L + \mu^2 c_3 \sinh \mu L = 0$. These yield (because $\sinh x \neq 0$ when $x \neq 0$, $-\lambda^2 + \mu^2 \neq 0$, and $\lambda^2 + \mu^2 \neq 0$)

$$c_1 \sin \lambda L = 0, \quad c_2 = c_3 = c_4 = 0.$$

For nontrivial solution, $c_1 \neq 0$, giving $\lambda = \frac{n\pi}{L}$; then from Eq. (3.7.82) it follows that

$$\omega_n = \left(\frac{n\pi}{L}\right)^2 \left[\left(\frac{D_{xx}^e}{m_0}\right) \left(1 + \frac{\hat{N}_{xx}}{(\frac{n\pi}{L})^2 D_{xx}^e}\right) \left(\frac{1}{1 + (\frac{n\pi}{L})^2 \frac{m_2}{m_0}}\right)\right]^{\frac{1}{2}}. \tag{3.7.86}$$

If the rotary inertia, m_2, is neglected, we obtain

$$\omega_n = \left(\frac{n\pi}{L}\right)^2 \left[\left(\frac{D_{xx}^e}{m_0}\right) \left(1 + \frac{\hat{N}_{xx}}{(\frac{n\pi}{L})^2 D_{xx}^e}\right)\right]^{\frac{1}{2}}. \tag{3.7.87}$$

Thus, the effect of the in-plane tensile force $\hat{N}_{xx}$ is to increase the natural frequencies. If we have a very flexible beam, say a cable with large tension, the second term under the radical in Eq. (3.7.87) becomes very large in comparison with unity; if n is not large, we have

$$\omega_n \approx \frac{n\pi}{L} \sqrt{\frac{\hat{N}_{xx}}{m_0}}, \tag{3.7.88}$$

which are the natural frequencies of a cable in tension. We also note from Eq. (3.7.87) that frequencies of natural vibration decrease when a compressive force instead of a tensile force is applied.

When $\hat{N}_{xx} = 0$, we obtain from Eq. (3.7.86)

$$\omega_n = \left(\frac{n\pi}{L}\right)^2 \left[\frac{D_{xx}^e}{m_0}\left(\frac{m_0}{m_0 + (\frac{n\pi}{L})^2 m_2}\right)\right]^{\frac{1}{2}}$$

$$= \left(\frac{n\pi}{L}\right)^2 \left[\frac{D_{xx}^e}{m_0}\left(1 - \frac{m_2(\frac{n\pi}{L})^2}{(\frac{n\pi}{L})^2 m_2 + m_0}\right)\right]^{\frac{1}{2}}. \tag{3.7.89}$$

Thus, both the principal inertia m_0 and the rotatory inertia m_2 have the effect of decreasing the magnitude of frequencies of natural vibration; the bending stiffness D_{xx}^e (which includes the couples stress effect) has the effect of increasing the magnitude of natural frequencies. If the rotatory inertia is neglected (i.e., $m_2 = 0$), we obtain

$$\omega_n = \left(\frac{n\pi}{L}\right)^2 \sqrt{\frac{D_{xx}}{m_0}}. \tag{3.7.90}$$

Example 3.7.8

Determine the fundamental frequency of a clamped beam of length L.

Solution: For a clamped beam, the boundary conditions are

$$w_0 = 0, \quad \frac{dw_0}{dx} = 0 \quad \text{at} \quad x = 0, L. \tag{3.7.91}$$

From Eq. (3.7.78), we have

$$\frac{dw_0}{dx} = \lambda\left(c_1 \cos \lambda x - c_2 \sin \lambda x\right) + \mu\left(c_3 \cosh \mu x + c_4 \sinh \mu x\right).$$

Using the boundary conditions at $x = 0$ gives $c_2 + c_4 = 0$ and $\lambda c_1 + \mu c_3 = 0$. The boundary conditions at $x = L$ gives

$$0 = c_1 \sin \lambda L + c_2 \cos \lambda L + c_3 \sinh \mu L + c_4 \cosh \mu L,$$
$$0 = \lambda\left(c_1 \cos \lambda L - c_2 \sin \lambda L\right) + \mu\left(c_3 \cosh \mu L + c_4 \sinh \mu L\right).$$

Using $c_4 = -c_2$ and $c_3 = -(\lambda/\mu)c_4$, the last two relations can be expressed solely in terms of the constants c_1 and c_2:

$$\begin{bmatrix} \sin \lambda L - \left(\frac{\lambda}{\mu}\right)\sinh \mu L & \cos \lambda L - \cosh \mu L \\ \cos \lambda L - \cosh \mu L & -\sin \lambda L - \left(\frac{\mu}{\lambda}\right)\sinh \mu L \end{bmatrix} \begin{Bmatrix} c_1 \\ c_2 \end{Bmatrix} = \begin{Bmatrix} 0 \\ 0 \end{Bmatrix}. \tag{3.7.92}$$

For nonzero c_1 and c_2, we require the determinant of the coefficient matrix of Eq. (3.7.92) to vanish. This leads to the characteristic polynomial

$$-2 + 2\cos \lambda L \cosh \mu L + \left(\frac{\lambda}{\mu} - \frac{\mu}{\lambda}\right)\sin \lambda L \sinh \mu L = 0. \tag{3.7.93}$$

The transcendental equation needs to be solved iteratively for λ and μ, in conjunction with

$$\lambda^2 = \mu^2 + \frac{b}{a} = \mu^2 + \frac{m_2\omega^2 - \hat{N}_{xx}}{D_{xx}^e}. \tag{3.7.94}$$

For natural vibration without rotatory inertia, Eq. (3.7.93) takes the simpler form

$$\cos \lambda L \cosh \lambda L - 1 = 0. \qquad (3.7.95)$$

The roots of Eq. (3.7.95) are

$$\lambda_1 L = 4.730, \quad \lambda_2 L = 7.853, \quad \cdots, \quad \lambda_n L \approx \left(n + \frac{1}{2}\right)\pi.$$

The previous two examples involving simply supported (or pinned–hinged) and clamped (or fixed–fixed) beams illustrate the procedure for determining the natural frequencies. For beams with other boundary conditions, the values of the constants c_1–c_4, characteristic equation for vibration, and roots (i.e., natural frequencies) of the characteristic equation are presented in Table 3.7.2 (from Reddy [8, 113]).

Table 3.7.2: Values of the constants and eigenvalues for natural vibration of beams with various boundary conditions ($\lambda_n^4 \equiv \omega_n^2 m_0/D_{xx}^e = (e_n/L)^4$). Rotary inertia is not included.

Beam type	Constants[†]	Characteristic Equation and values of $e_n \equiv \lambda_n L$
Hinged-Hinged	$c_1 \neq 0$	$\sin e_n = 0, \; e_n = n\pi$
	$c_2 = c_3 = c_4 = 0$	
Fixed-Fixed	$c_1 = -c_3 = 1/p_n$	$\cos e_n \cosh e_n - 1 = 0$
	$-c_2 = c_4 = 1/q_n$	$e_n = 4.730, 7.853, \cdots$
	$p_n = (\sin e_n - \sinh e_n)$,	$q_n = (\cos e_n - \cosh e_n)$
Fixed-Free	$c_1 = -c_3 = 1/p_n$	$\cos e_n \cosh e_n + 1 = 0$
	$-c_2 = c_4 = 1/q_n$	$e_n = 1.875, 4.694, \cdots$
	$p_n = (\sin e_n + \sinh e_n)$	$q_n = (\cos e_n + \cosh e_n)$
Free-Free	$c_1 = c_3 = 1/p_n$	$\cos e_n \cosh e_n - 1 = 0$
	$c_2 = c_4 = -1/q_n$	$e_n = 4.730, 7.853, \cdots$
	$p_n = (\sin e_n - \sinh e_n)$	$q_n = (\cos e_n - \cosh e_n)$
Hinged-Fixed	$c_1 = 1/\sin e_n$	$\tan e_n = \tanh e_n$
	$c_2 = c_4 = 0$	$e_n = 3.927, 7.069, \cdots$
	$c_3 = 1/\sinh e_n$	
Hinged-Free	$c_1 = 1/\sin e_n$	$\tan e_n = \tanh e_n$
	$c_2 = c_4 = 0$	$e_n = 3.927, 7.069, \cdots$
	$c_3 = -1/\sinh e_n$	

[†] $W(x) = c_1 \sin \lambda x + c_2 \cos \lambda x + c_3 \sinh \mu x + c_4 \cosh \mu x.$

3.8   THE NAVIER SOLUTIONS

3.8.1   THE GENERAL PROCEDURE

Analytical solutions of linear differential equations for certain boundary conditions can be obtained using Navier's method[2]. In Navier's method of solution, the unknown variables (u, w) and applied loads (f, q) are expanded as linear combinations of unknown coefficients (U_n, W_n) and known functions $(f_n(x), g_n(x))$ in the form:

$$u = \sum_{n=1}^{N} U_n \, F_n(x), \quad w = \sum_{n=1}^{N} W_n \, F_n(x).$$

The choice of the functions used $(F_n(x), G_n(x))$ is restricted to those which meet the differentiability conditions imposed by the differential equations as well as the specified boundary conditions on the variables. In most cases, these functions $(F_n(x), G_n(x))$ are selected to be trigonometric functions. The substitution of the expansions for the unknowns (u, w) into the governing equations will dictate the choice of the expansions used for the loads (f, q).

When the expansions for the dependent unknown variables and loads are substituted into the set of differential equations, *each* differential equation of the pair results in an algebraic equation of the form $\mathcal{F}(U_n, W_n)\, H(x) = 0$, where $\mathcal{F}$ is an expression containing only the coefficients U_n and W_n and $H(x)$ is a function of x, which may be equal to $F_n(x)$, $G_n(x)$, or their derivatives. Then we declare that $\mathcal{F} = 0$, since it must hold for all possible values of x (hence $H(x) \neq 0$). If this does not happen, then the differential equations are said not to admit the Navier solution.

To fix the ideas, we consider a single linear differential equation of the form

$$-\frac{d}{dx}\left(a\frac{du}{dx}\right) + b\frac{du}{dx} + c\,u = f, \quad 0 < x < L, \tag{3.8.1}$$

where u is the function to be determined, a, b, c, and f are known functions of position x. Equation (3.8.1) is subjected to boundary conditions of the type (one element of the duality pair at each boundary point)

$$u = \hat{u} \ \text{ and } \ a\frac{du}{dx} = \hat{q} \ \text{ at } x = 0 \text{ and } x = L. \tag{3.8.2}$$

In the Navier method, we seek the solution in the form

$$u(x) = \sum_{n=1}^{\infty} c_n \, \phi_n(x), \tag{3.8.3}$$

---

[2]L.M. H. Navier (1785–1836) was a French engineer and scientist, who is credited with developing the first correct differential equation of thin plates subjected to distributed lateral loads. His lecture notes *Résumé des Leçons de Méchanique* were published by École Polytechnique, Paris, in 1819.

where $\{\phi_n(x)\}$ is a linearly independent set of functions that is selected such that the specified boundary conditions are satisfied. If, upon the substitution of the expansion (3.8.3) into Eq. (3.8.1), an expression of the form

$$\sum_{n=1}^{\infty} S_n \psi_n(x) = 0, \quad 0 < x < L, \tag{3.8.4}$$

is obtained, then we say that the Navier solution exists; otherwise, we do not have the Navier solution with the choice of the functions for the problem at hand. In Eq. (3.8.4) S_n, $n = 1, 2, \ldots, \infty$, are expressions which contain linear combinations of the parameters $c_1, c_2, \ldots, c_n$ which are independent of x, and $\{\psi_n\}$ is a linearly independent set of functions of x (often derivatives of functions ϕ_n). Since the expression in Eq. (3.8.4) must hold for all x in $(0, L)$, we set $S_n = 0$ for $n = 1, 2, \ldots$. This set of equations can be solved to determine $c_1, c_2, \ldots,$, and then the solution is given by Eq. (3.8.3).

3.8.2 NAVIER'S SOLUTION OF EQUATIONS OF MOTION

Now we focus on beams. The linearized equations of motion from Eqs. (3.4.12) and (3.4.13) with constant stiffness coefficients $(A_{xx}, B_{xx}, D_{xx}, A_{xy})$, mass inertia coefficients (m_0, m_1, m_2), and the foundation modulus c_f are:

$$-A_{xx}\frac{\partial^2 u}{\partial x^2} + B_{xx}\frac{\partial^3 w}{\partial x^3} + \frac{\partial N_{xx}^T}{\partial x} + m_0\frac{\partial^2 u}{\partial t^2} - m_1\frac{\partial^3 w}{\partial t^2 \partial x} - f(x,t) = 0, \tag{3.8.5}$$

$$-B_{xx}\frac{\partial^3 u}{\partial x^3} + (D_{xx} + A_{xy})\frac{\partial^4 w}{\partial x^4} + \frac{\partial^2 M_{xx}^T}{\partial x^2} + c_f w$$

$$+ m_0\frac{\partial^2 w}{\partial t^2} + m_1\frac{\partial^3 u}{\partial t^2 \partial x} - m_2\frac{\partial^4 w}{\partial t^2 \partial x^2} - q(x,t) = 0. \tag{3.8.6}$$

The Navier solutions of Eqs. (3.8.5) and (3.8.6) for beams exist only when the boundary conditions are of the "simply supported" type (i.e., both ends are hinged, not restricting the axial displacement u at the ends), defined by

$$N_{xx} = 0, \quad w = 0, \quad M_{xx} = 0, \text{ at } x = 0 \text{ and } x = L, \tag{3.8.7}$$

where L is the length of the beam and the origin of the x-coordinate is taken at the left end of the beam. We note that all of the specified boundary conditions are homogeneous. The Navier solution procedure applied to time-dependent problems only results in ordinary differential equations in time for the coefficients of the expansions, as described next.

The boundary conditions in Eq. (3.8.7) are met by the following form of the displacements $u(x,t)$ and $w(x,t)$:

$$u(x,t) = \sum_{m=1}^{\infty} U_m(t)\cos\alpha_m x, \quad w(x,t) = \sum_{m=1}^{\infty} W_m(t)\sin\alpha_m x, \quad \alpha_m = \frac{m\pi}{L},$$

$$\tag{3.8.8}$$

where $U_m(t)$ and $W_m(t)$ are the coefficients to be determined such that Eqs. (3.8.5) and (3.8.6) are satisfied everywhere in the domain $(0, L)$ for any arbitrary time t. Substituting the expansions from Eq. (3.8.8) into Eqs. (3.8.5) and (3.8.6), we obtain

$$\sum_{m=1}^{\infty} \left(A_{xx} \alpha_m^2 U_m - B_{xx} \alpha_m^3 W_m + m_0 \frac{d^2 U_m}{dt^2} - m_1 \alpha_m \frac{d^2 W_m}{dt^2} \right) \cos \alpha_m x$$

$$+ \frac{\partial N_{xx}^T}{\partial x} - f(x, t) = 0, \qquad (3.8.9)$$

$$\sum_{m=1}^{\infty} \left[-B_{xx} \alpha_m^3 U_m + (D_{xx} + A_{xy}) \alpha_m^4 W_m + c_f W_m - m_1 \alpha_m \frac{d^2 U_m}{dt^2} \right.$$

$$\left. + \left(m_0 + m_2 \alpha_m^2 \right) \frac{d^2 W_m}{dt^2} \right] \sin \alpha_m x + \frac{\partial^2 M_{xx}^T}{\partial x^2} - q(x, t) = 0. \qquad (3.8.10)$$

Equations (3.8.9) and (3.8.10) suggest that the forces $f(x, t)$ be expanded in the cosine series and N_{xx}^T, M_{xx}^T, and $q(x, t)$ be expanded in sine series:

$$f(x, t) = \sum_{m=1}^{\infty} F_m \cos \alpha_m x, \quad q(x, t) = \sum_{m=1}^{\infty} Q_m \sin \alpha_m x,$$

$$M_{xx}^T(x, y) = \sum_{m=1}^{\infty} M_m^T \sin \alpha_m x, \quad N_{xx}^T(x, y) = \sum_{m=1}^{\infty} N_m^T \sin \alpha_m x, \qquad (3.8.11)$$

where the coefficients F_m, Q_m, N_m^T, and M_m^T are calculated from

$$Q_m(t) = \frac{2}{L} \int_0^L q(x, t) \sin \alpha_m x \, dx, \quad F_m(t) = \frac{2}{L} \int_0^L f(x, t) \cos \alpha_m x \, dx,$$

$$\qquad (3.8.12)$$

with similar expressions for N_m^T and M_m^T. For example, the coefficients Q_n for some typical loads are:

$$Q_m = \begin{cases} q_0 \ (m = 1), & \text{sinusoidal load of intensity } q_0, \\ \frac{4}{m\pi} q_0 \ (m = 1, 3, 5, \ldots), & \text{uniform load of intensity } q_0, \\ \frac{2}{L} Q_0 \sin \frac{m\pi}{2} \ (m = 1, 2, 3, \ldots), & \text{point load } Q_0 \text{ at the beam center.} \end{cases}$$

$$\qquad (3.8.13)$$

With the load expansions in Eq. (3.8.12), Eqs. (3.8.9) and (3.8.10) become

$$\sum_{m=1}^{\infty} \left(A_{xx} \alpha_m^2 U_m - B_{xx} \alpha_m^3 W_m + m_0 \frac{d^2 U_m}{dt^2} - m_1 \alpha_m \frac{d^2 W_m}{dt^2} \right.$$

$$\left. + \alpha_m N_m^T - F_m \right) \cos \alpha_m x = 0, \qquad (3.8.14)$$

$$\sum_{m=1}^{\infty} \left[-B_{xx}\alpha_m^3 U_m + (D_{xx} + A_{xy})\alpha_m^4 W_m + c_f W_m - m_1\alpha_m \frac{d^2 U_m}{dt^2} \right.$$

$$\left. + (m_0 + m_2\alpha_m^2) \frac{d^2 W_m}{dt^2} - \alpha_m^2 M_m^T - Q_m \right] \sin\alpha_m x = 0. \qquad (3.8.15)$$

Since the above equations must hold for any x and m, we obtain

$$\begin{bmatrix} A_{xx}\alpha_m^2 & -B_{xx}\alpha_m^3 \\ -B_{xx}\alpha_m^3 & (D_{xx} + A_{xy})\alpha_m^4 + c_f \end{bmatrix} \begin{Bmatrix} U_m \\ W_m \end{Bmatrix}$$

$$+ \begin{bmatrix} m_0 & -m_1\alpha_m \\ -m_1\alpha_m & (m_0 + m_2\alpha_m^2) \end{bmatrix} \begin{Bmatrix} \ddot{U}_m \\ \ddot{W}_m \end{Bmatrix} = \begin{Bmatrix} -\alpha_m N_m^T + F_m \\ \alpha_m^2 M_m^T + Q_m \end{Bmatrix}. \qquad (3.8.16)$$

The above equation can be expressed in symbolic form as

$$\mathbf{K\Delta} + \mathbf{M\ddot{\Delta}} = \mathbf{F}. \qquad (3.8.17)$$

Next we discuss several special cases of this equation. We note that Eq. (3.8.17) is valid only for hinged–hinged straight beams (FGM as well as homogeneous).

3.8.3 BENDING SOLUTIONS

For static bending, we set the time derivative terms in Eq. (3.8.16) to zero and obtain

$$\begin{bmatrix} A_{xx}\alpha_m^2 & -B_{xx}\alpha_m^3 \\ -B_{xx}\alpha_m^3 & (D_{xx} + A_{xy})\alpha_m^4 + c_f \end{bmatrix} \begin{Bmatrix} U_m \\ W_m \end{Bmatrix} = \begin{Bmatrix} -\alpha_m N_m^T + F_m \\ \alpha_m^2 M_m^T + Q_m \end{Bmatrix}. \qquad$$
$$(3.8.18)$$

Inversion of the above equation for each m, for any applied load parameters (F_m, Q_m, N_m^T, M_M^T), gives the coefficients (U_m, W_m):

$$U_m = \frac{1}{D_{xx}^m} \left[(D_{xx} + A_{xy})\alpha_m^4 + c_f \right] \left(-\alpha_m N_m^T + F_m \right)$$

$$+ \frac{1}{D_{xx}^m} \left[B_{xx}\alpha_m^3 \left(\alpha_m^2 M_m^T + Q_m \right) \right], \qquad (3.8.19)$$

$$W_m = \frac{1}{D_{xx}^m} \left[A_{xx}\alpha_m^2 \left(\alpha_m^2 M_m^T + Q_m \right) + B_{xx}\alpha_m^3 \left(-\alpha_m N_m^T + F_m \right) \right], \qquad (3.8.20)$$

where

$$D_{xx}^m = A_{xx}\alpha_m^2 \left[(D_{xx} + A_{xy})\alpha_m^4 + c_f \right] - B_{xx}^2\alpha_m^6. \qquad (3.8.21)$$

The solution $(u(x), w(x))$ is given by Eq. (3.8.8). The solution can be specialized to a number of cases. For homogeneous isotropic beams (i.e., when

$B_{xx} = 0$), the solution in Eqs. (3.8.19) and (3.8.20) becomes

$$U_m = \frac{1}{D_{xx}^m} \left[(D_{xx} + A_{xy}) \alpha_m^4 + c_f \right] \left(-\alpha_m \, N_m^T + F_m \right), \tag{3.8.22}$$

$$W_m = \frac{A_{xx}\alpha_m^2}{D_{xx}^m} \left(\alpha_m^2 \, M_m^T + Q_m \right), \tag{3.8.23}$$

$$D_{xx}^m = A_{xx}\alpha_m^2 \left[(D_{xx} + A_{xy}) \alpha_m^4 + c_f \right].$$

Thus, the axial displacement $u(x)$ and transverse deflection $w(x)$ are decoupled from the influence of the mechanical loads. In addition, if the couple stress effect and foundation modulus are not considered (i.e., $A_{xy} = 0$ and $c_f = 0$, we obtain ($A_{xx} = EA$, $D_{xx} = EI$, and $D_{xx}^m = A_{xx}D_{xx}\alpha_m^6$)

$$U_m = \frac{1}{\alpha_m^2 \, EA} \left(F_m - \alpha_m \, N_m^T \right),$$

$$W_m = \frac{1}{\alpha_m^4 \, EI} \left(Q_m + \alpha_m^2 \, M_m^T \right). \tag{3.8.24}$$

Further, when the thermal forces are zero, we obtain

$$U_m = \frac{F_m L^2}{m^2 \pi^2 EA}; \quad W_m = \frac{Q_m L^4}{m^4 \pi^4 EI}. \tag{3.8.25}$$

We note that the solution in Eq. (3.8.8) is analytical but an infinite series. Thus, one must choose the number of terms, M, in the series to obtain the solution, and the number of terms depend on the accuracy one desires. For distributed loads, typically $M = 10$ or so, while for point loads, it can be as large as $M = 50$ to 100. For the sinusoidal load, $M = 1$, and it is an exact solution.

Example 3.8.1 ————————————————————————————

Determine the analytical solution of a simply supported beam subjected to three types of loads: (a) point load, $q(x) = Q_0 \, \delta(x - x_0)$, at the center ($x_0 = 0.5L$), (b) uniform distributed load, $q(x) = q_0$, and (c) sinusoidal load, $q(x) = q_0 \sin \frac{\pi x}{L}$. Use the following material and geometric parameters of the beam with a rectangular cross section:

$$E_1 = 14.4 \text{ GPa}, \quad E_2 = E = 1.44 \text{ GPa}, \; \; c_f = 0,$$

$$h = 5 \times 17.6 \times 10^{-6} \text{ m}, \; \; b = 2h, \; L = 20h, \; q_0 = 1.0 \text{ N/m}. \tag{3.8.26}$$

Solution: Numerical results of the normalized center deflection, $\bar{w} = w(0.5L)(E_2 I/q_0 L^4) \times 10^2$, of a simply supported FGM beams under various types of loads and for different values of the power-law index n are presented in Table 3.8.1 for two different values of the material length scale parameter (appearing in the couple stress constitutive model), $\ell/h \equiv \gamma = 0$ and $\gamma = 0.1$. Clearly, as expected, the effect of the couple stress is to stiffen the beam. Figure 3.8.1 contains plots of dimensionless deflection, $\bar{w}(x) = w(0.5L)(E_2 I/q_0 L^4)$, of FGM beams under uniform $[q(x) = q_0]$ and sinusoidal $[q(x) = q_0 \sin(\pi x/L)]$ transverse loads as a function of the dimensionless distance along the length, x/L. Plots of dimensionless deflections of FGM beams under uniform distributed transverse load as a function of the power-law index n are shown in Fig. 3.8.2.

Table 3.8.1: Center deflections $\bar{w} \times 10^2$ of simply supported FGM beams under various types of loads and for different values of the power-law index[a] n and the ratio $\ell/h = \gamma$ $[\bar{w} = w(0.5L)(E_2 I/q_0 L^4)]$.

	Point load (100)[b]		Uniform load (9)		Sinusoidal load (1)	
n	$\gamma = 0$	$\gamma = 0.1$	$\gamma = 0$	$\gamma = 0.1$	$\gamma = 0$	$\gamma = 0.1$
0	0.2083	0.1996	0.1302	0.1248	0.1027	0.0984
1	0.4876	0.4617	0.3047	0.2886	0.2403	0.2275
2	0.7153	0.6750	0.4471	0.4219	0.3525	0.3326
3	0.8442	0.7985	0.5276	0.4991	0.4160	0.3935
5	0.9496	0.9048	0.5935	0.5655	0.4679	0.4459
10	1.0442	1.0044	0.6526	0.6277	0.5145	0.4949
50	1.4704	1.4192	0.9190	0.8870	0.7246	0.6993
100	1.6855	1.6233	1.0534	1.0146	0.8306	0.7999

[a]$n = 0$ corresponds to material 1; $n = \infty$ corresponds to material 2.
[b]The number in the parenthesis denotes the number of terms used to evaluate the series; note that for sinusoidal case the solution is obtained with one term.

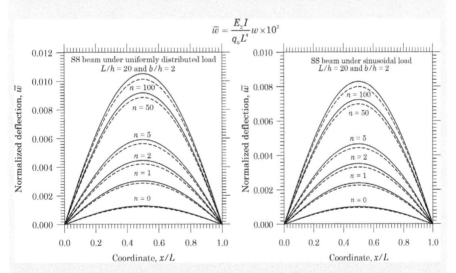

Figure 3.8.1: Plots of the normalized transverse deflection, $\bar{w}$ versus x/L for various values of n and ℓ/h for uniform and sinusoidal distributed load of intensity q_0.

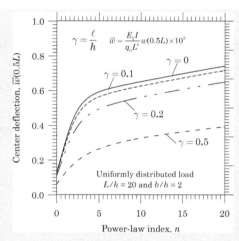

Figure 3.8.2: Plots of normalized transverse deflection, $\bar{w}$ versus n for various values of $\gamma = \ell/h$ for simply supported beams under uniform distributed transverse load of intensity q_0.

3.8.4 NATURAL VIBRATIONS

For natural (or free) vibration analysis, we set all force coefficients to zero and assume a periodic solution in the form

$$U_m(t) = U_m^0\, e^{-i\omega_m t}, \quad W_m(t) = W_m^0\, e^{-i\omega_m t}, \tag{3.8.27}$$

where ω_m is the frequency of natural vibration and $i = \sqrt{-1}$. Substitution of Eq. (3.8.27) into Eq. (3.8.17) yields the following eigenvalue problem:

$$\left(\mathbf{K} - \omega_m^2 \mathbf{M}\right) \boldsymbol{\Delta}^0 = \mathbf{0}, \quad \boldsymbol{\Delta}_m^0 = (U_m^0, W_m^0). \tag{3.8.28}$$

The solution of the eigenvalue problem in Eq. (3.8.28) gives the frequency ω_m and mode shapes $(U_m^0 \cos\alpha_m x,\ W_m^0 \sin\alpha_m x)$. The frequency ω_m and amplitudes of the mode shapes $(\boldsymbol{\Delta}_m^0 = U_m^0, W_m^0)$ are determined by setting the determinant of the coefficient matrix in Eq. (3.8.28) to zero:

$$\left|\mathbf{K} - \omega_m^2 \mathbf{M}\right| = 0, \tag{3.8.29}$$

which yields a quadratic equation for $\lambda_m = \omega_m^2$:

$$a\,\lambda_m^2 - b\,\lambda_m + c = 0, \tag{3.8.30a}$$

where

$$
\begin{aligned}
a &= m_0\left(m_0 + m_2\alpha_m^2\right) - m_1^2\alpha_m^2,\\
c &= A_{xx}\alpha_m^2\left[(D_{xx} + A_{xy})\alpha_m^4 + c_f\right] - B_{xx}^2\alpha_m^6.\\
b &= m_0\left[(D_{xx} + A_{xy})\alpha_m^4 + c_f\right] - 2m_1 B_{xx}\alpha_m^4\\
&\quad + A_{xx}\alpha_m^2\left(m_0 + m_2\alpha_m^2\right).
\end{aligned}
\tag{3.8.30b}
$$

For homogeneous beams without the couple stress effect and elastic foundation (i.e., $A_{xy} = 0$ and $c_f = 0$), Eqs. (3.8.30a) and (3.8.30b) yield (with $A_{xx} = EA$, $D_{xx} = EI$, $m_0 = \rho A$, $m_1 = 0$, $B_{xx} = 0$, and $m_2 = \rho I$) the following expressions for longitudinal (i.e., axial) and flexural frequencies:

$$\text{Axial:}\ \ \omega_m = \alpha_m \sqrt{\frac{A_{xx}}{m_0}}, \quad \text{Flexural:}\ \ \omega_m = \alpha_m^2 \sqrt{\frac{D_{xx}}{(m_0 + m_2 \alpha_m^2)}}. \quad (3.8.31)$$

Example 3.8.2

Determine the first three natural frequencies of a simply supported beam with the following material and geometric parameters and rectangular cross section:

$$h = 17.6 \times 10^{-6} \text{ m}, \quad \rho_1 = \rho_2 = 1.22 \times 10^3 \text{ kg/m},$$
$$E_1 = 14.4 \text{ GPa}, \quad E_2 = 1.44 \text{ GPa}, \quad \rho_1 = 12.2 \times 10^3 \text{ kg/m}, \quad (3.8.32)$$
$$\rho_2 = 1.22 \times 10^3 \text{ kg/m}, \quad h = 17.6 \times 10^{-6} \text{ m}.$$

Solution: Table 3.8.2 contains the results for homogeneous $(n = 0)$ and functionally graded $(n \neq 0)$ beams. The rotary inertia, m_2, is included in the present study. Figure 3.8.3 contains plots of dimensionless fundamental frequency versus the power-law index n for different values of the ratio, $\gamma = \ell/h$. It is clear that, due to the nature of variation of the bending–stretching coupling coefficient B_{xx} with n, the fundamental frequency also experiences variation inverse to the stiffness B_{xx}.

Table 3.8.2: First three flexural natural frequencies $\bar{\omega}_m$ $(m = 1, 2, 3)$ of homogeneous and FGM simply supported beams $(\bar{\omega}_m = \omega_m L^2/\sqrt{\rho_2 A_0/E_2})$.

	$\bar{\omega}_1$			$\bar{\omega}_2$			$\bar{\omega}_3$		
n	$\gamma = 0$	$\gamma = .1$	$\gamma = .2$	$\gamma = 0$	$\gamma = .1$	$\gamma = .2$	$\gamma = 0$	$\gamma = .1$	$\gamma = .2$
0	9.86	10.07	10.68	39.32	40.16	42.60	88.02	89.91	95.36
1	8.69	8.93	9.62	34.69	35.64	38.37	77.74	79.88	86.00
2	8.42	8.66	9.37	33.59	34.57	37.38	75.29	77.51	83.80
5	9.24	9.46	10.11	36.85	37.75	40.34	82.54	84.56	90.35
10	10.33	10.53	11.11	41.17	41.98	44.31	92.12	93.93	99.15

3.8.5 TRANSIENT ANALYSIS

To determine the transient response of the beam, we must solve the set of ordinary differential equations in time, Eq. (3.8.17). These equations can be solve either numerically or analytically. First we discuss the analytical approach by considering the special case of isotropic homogeneous beam (i.e., $B_{xx} = 0$). Then Eq. (3.8.17) reduces to

$$M_m \frac{d^2 W_m}{dt^2} + K_m W_m - Q_m = 0, \quad \text{for}\ \ m = 1, 2, \ldots. \quad (3.8.33)$$

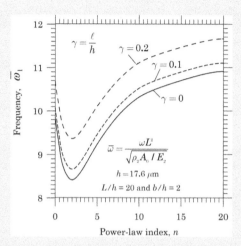

Figure 3.8.3: Plots of the normalized natural frequency, $\bar{\omega}_1$ versus n for different values of ℓ/h.

where

$$K_m = D_{xx}\alpha_m^4 , \quad M_m = m_0 + m_2\alpha_m^2. \tag{3.8.34}$$

Equation (3.8.33) represents a second-order ordinary differential equation in time for each m. The general solution of (3.8.33) is given by

$$W_m(t) = A_m^0 \sin \lambda_m t + B_m^0 \cos \lambda_m t + \frac{Q_m}{K_m} , \quad \lambda_m = \sqrt{\frac{K_m}{M_m}}, \tag{3.8.35}$$

where A_m^0 and B_m^0 are constants to be determined using the initial conditions of the problem. Since Eq. (3.8.33) is a second-order equation, we must specify $W_m(0)$ and $\dot{W}(0)$ (i.e., the displacement and velocity). The complete solution is then given by

$$w(x,t) = \sum_{m=1}^{\infty} \left[A_m^0 \sin \lambda_m t + B_m^0 \cos \lambda_m t + \frac{Q_m}{K_m} \right] \sin \alpha_m x, \tag{3.8.36}$$

where Q_m is as defined in Eq. (3.8.12).

Let the initial conditions on the displacement and velocity be

$$w(x,0) = d_0(x) = \sum_{m=1}^{\infty} D_m^0 \sin \alpha_m x, \quad \frac{\partial w}{\partial t}(x,0) = v_0 = \sum_{m=1}^{\infty} V_m^0 \sin \alpha_m x,$$
$$\tag{3.8.37}$$

where D_m^0 and V_m^0 have the same form as Q_m in Eq. (3.8.12). These conditions give

$$B_m^0 = D_m^0 - \frac{Q_m}{K_m}, \quad A_m^0 \lambda_m = V_m^0 - \frac{1}{K_m} \frac{dQ_m}{dt}. \tag{3.8.38}$$

Then the analytical solution becomes

$$w(x,t) = \sum_{m=1}^{\infty} \left[D_m^0 \cos \lambda_m t + \frac{V_m^0}{\lambda_m} \sin \lambda_m t \right.$$

$$\left. + \frac{Q_m}{K_m} (1 - \cos \lambda_m t) - \frac{1}{K_m \lambda_m} \frac{dQ_m}{dt} \sin \lambda_m t \right] \sin \alpha_m x. \tag{3.8.39}$$

When the initial deflection and velocity are zero, we obtain

$$w(x,t) = \sum_{m=1}^{\infty} \left[\frac{Q_m}{K_m} (1 - \cos \lambda_m t) - \frac{1}{K_m \lambda_m} \frac{dQ_m}{dt} \sin \lambda_m t \right] \sin \alpha_m x. \tag{3.8.40}$$

In particular, for a beam simply supported at both ends and subjected to a suddenly applied uniform load $q = q_0 H(t)$, Q_m is defined by Eq. (3.8.12) and the analytical solution becomes

$$w(x,t) = \frac{4q_0}{\pi} \sum_{m=1,3,\ldots}^{\infty} \frac{1}{m K_m} (1 - \cos \lambda_m t) \sin \alpha_m x, \quad \text{for } m \text{ odd.} \tag{3.8.41}$$

The second-order differential equation (3.8.33) can also be solved using the Laplace transform method, which proves to be useful when coupled differential equations are involved. Here, we describe the method as applied to Eq. (3.8.33).

The *Laplace transform* of a function $f(t)$ is defined by

$$\bar{f}(s) \equiv \mathcal{L}[f(t)] = \int_0^\infty e^{-st} f(t) dt. \tag{3.8.42}$$

The Laplace transforms of some typical functions are presented in Table 3.8.3.

To solve Eq. (3.8.33) by the Laplace transform method, let $\mathcal{L}[W_m(t)] = \bar{W}_m(s)$. Then Eq. (3.8.33) transforms to

$$M_m \left(s^2 \bar{W}_m(s) - s W_m(0) - \dot{W}_m(0) \right) + K_m \bar{W}_m(s) - \bar{F}_m(s) = 0, \tag{3.8.43}$$

or

$$\left(s^2 + \lambda_m^2 \right) \bar{W}_m(s) = \left(s W_m(0) + \dot{W}_m(0) + \frac{\bar{F}_m}{M_m} \right). \tag{3.8.44}$$

Now suppose that the applied load is uniformly distributed step load. Then

$$\bar{F}_m(s) = \left(\frac{4q_0}{a \alpha_m} \right) \frac{1}{s}. \tag{3.8.45}$$

Table 3.8.3: The Laplace transform pairs of typical functions.[†]

$f(t)$	$\bar{f}(s)$	$f(t)$	$\bar{f}(s)$
1	$\frac{1}{s}$	$\frac{df}{dt}$	$s\bar{f}(s) - f(0)$
$tf(t)$	$-\bar{f}'(s)$	$t^n f(t)$	$(-1)^n \bar{f}^{(n)}(s)$
e^{at}	$\frac{1}{s-a}$	$e^{at}f(t)$	$\bar{f}(s-a)$
te^{at}	$\frac{1}{(s-a)^2}$	$t^n e^{at}, \;\; n=0,1,2,\cdots$	$\frac{n!}{(s-a)^{n+1}}$
$\frac{d^2 f}{dt^2}$	$s^2\bar{f}(s) - sf(0) - f'(0)$	$\frac{1}{t}f(t)$	$\int_s^\infty f(\xi)\,d\xi$
$e^{at} - e^{bt}$	$\frac{a-b}{(s-a)(s-b)}$	$(ae^{at} - be^{bt})$	$\frac{s(a-b)}{(s-a)(s-b)}$
$\sin at$	$\frac{a}{s^2+a^2}$	$\cos at$	$\frac{s}{s^2+a^2}$
$\frac{1}{a^2}(1 - \cos at)$	$\frac{1}{s(s^2+a^2)}$	$\frac{1}{a^3}(at - \sin at)$	$\frac{1}{s^2(s^2+a^2)}$
$\frac{t}{2a}\sin at$	$\frac{s}{(s^2+a^2)^2}$	$\frac{1}{2a^3}(\sin at - at\cos at)$	$\frac{1}{(s^2+a^2)^2}$
$t\cos at$	$\frac{s^2-a^2}{(s^2+a^2)^2}$	$\frac{1}{2a}(\sin at + at\cos at)$	$\frac{s^2}{(s^2+a^2)^2}$
$\frac{1}{b}e^{at}\sin bt)$	$\frac{1}{(s-a)^2+b^2}$	$\frac{\cos at - \cos bt}{b^2-a^2}$	$\frac{s}{(s^2+a^2)(s^2+b^2)}$
$\sinh at$	$\frac{a}{s^2-a^2}$	$\cosh at$	$\frac{s}{s^2-a^2}$
$t^n f(t)$	$(-)^n \frac{d\bar{f}}{ds}$	$tf(t)$	$-\frac{d}{ds}\bar{f}(s)$
$\frac{1}{t}f(t)$	$\int_s^\infty \bar{f}(r)\,dr$	$\int_0^t f(\tau)\,d\tau$	$\frac{1}{s}\bar{f}(s)$
$e^{at}f(t)$	$\bar{f}(s-a)$	$f * g$	$\bar{f}(s)\bar{q}(s)$
$H(t-c)$	$\frac{e^{-cs}}{s}$	$\delta(t-c)$	e^{-cs}

[†] a and b are constants; H and δ are the Heaviside and Dirac delta functions; $f * g = \int_0^t f(\tau)g(t-\tau)d\tau$.

Substituting the above expression into Eq. (3.8.44) and using the inverse Laplace transform (see Table 3.8.3), we obtain

$$W_m(t) = \left[W_m(0)\cos\lambda_m t + \frac{\dot{W}_m(0)}{\lambda_m}\sin\lambda_m t \right] + \left(\frac{4q_0}{a\alpha_m K_m}\right)(1 - \cos\lambda_m t),$$

$$(3.8.46)$$

where we have used the identity

$$\frac{1}{s(s^2+\lambda_m^2)} = \frac{1}{\lambda_m^2}\left(\frac{1}{s} - \frac{s}{s^2+\lambda_m^2}\right).$$

For $W_m(0) = D_m^0$ and $\dot{W}_m(0) = V_m^0$, Eq. (3.8.46) becomes

$$W_m(t) = \left[D_m^0\cos\lambda_m t + \frac{V_m^0}{\lambda_n}\sin\lambda_m t \right] + \left(\frac{4q_0}{a\alpha_m K_m}\right)(1 - \cos\lambda_m t). \quad (3.8.47)$$

If the initial conditions are zero, then $D_m^0 = V_m^0 = 0$ and $W_m(t)$ becomes

$$W_m(t) = \left(\frac{4q_0}{a\alpha_m K_m}\right)(1 - \cos\lambda_m t)$$

and thus the solution obtained with the Laplace transform method is the same as the solution obtained earlier. The complete solution is given by

$$w(x,t) = \sum_{m=1}^{\infty} \left(\frac{4q_0}{a\alpha_m K_m} \right) (1 - \cos \lambda_m t) \sin \alpha_m x. \qquad (3.8.48)$$

The procedure used to solve Eq. (3.8.33) can be used to solve Eq. (3.8.17), which requires us to deal with a set of (matrix) differential equations.

3.9 ENERGY AND VARIATIONAL METHODS

3.9.1 INTRODUCTION

In this section, we introduce approximate methods that utilize the energy principles, such as the minimum total potential energy, the principle of virtual displacements, or Hamilton's principle, to determine solutions of the beam problems. These methods will be employed in the forthcoming chapters to solve problems introduced there. Therefore, first we introduce the methods in a general context and then use them to solve beam problems of this chapter. The methods described in this section seek solution of differential equations in terms of parameters that are determined by satisfying an energy principle. Such solution methods are called *direct variational methods,* because the approximate solutions are obtained directly by using the same energy principle that is used to derive the governing equations.

The assumed solutions are in the form of a finite series, $\sum_{i=1}^{N} c_i \phi_i(x)$, which is linear combination of *undetermined parameters* c_i with appropriately chosen functions, called *approximation functions* $\phi_i(x)$ that satisfy the boundary conditions (in some methods, only the specified essential boundary conditions), much like in the Navier solution method, except the series is a finite one. However, the methods to be discussed here are applicable to problems with any suitable boundary conditions, as long as we can find the approximation functions that satisfy the required continuity and boundary conditions, which will be detailed in the sequel. Since the solution of a continuum problem, in general, cannot be represented by a finite set of functions, error is introduced into the solution. Therefore, the solution obtained is an *approximation* of the true solution for the problem. As the number of linearly independent terms in the assumed solution is increased, the error in the approximation will be reduced, and the assumed solution converges to the desired solution. The parameters of the assumed solution are adjusted or *varied* to satisfy an energy principle, and hence all these methods are also termed *variational methods.*

The variational methods to be described here include the methods of Ritz, Galerkin, and Petrov–Galerkin [114, 115]. Examples of applications of these methods in this chapter are limited to bars and beams based on the CBT.

3.9.2   THE RITZ METHOD

3.9.2.1   Background and model problem

The Ritz method was proposed by Walther Ritz [114][3]. The method uses a functional (or variational) statement, such as the principle of minimum total potential energy (see Sections 2.3 and 2.7), of the problem to directly determine an approximate solution, without using the governing differential equation.

To describe the method, we consider the governing equation (expressed in terms of the displacement u) of the axial deformation of a bar of length a:

$$-\frac{d}{dx}\left(E(x)A(x)\frac{du}{dx}\right) + c(x)u(x) = f(x) \quad \text{in} \ \ 0 \le x \le a, \qquad (3.9.1)$$

where $u(x)$ is the axial displacement, $E(x)$ is Young's modulus (accounting for possible inhomogeneous material in the x coordinate), area of cross section $A(x)$ (accounting for possible variation of the cross sectional area with x), $c(x)$ is the resistance offered by any surrounding medium (e.g., when the bar is embedded in another medium), and $f(x)$ is the axially distributed force. Suppose that the boundary conditions are of the form

$$u(0) = 0, \quad \left[EA\frac{du}{dx} + ku\right]_{x=a} = P, \qquad (3.9.2)$$

where P is the point load at $x = a$ and k is the linearly elastic spring constant (see Fig. 3.9.1).

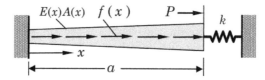

Figure 3.9.1: Axial deformation of a non-uniform bar fixed at the left end and supported axially by a linearly elastic spring at $x = a$.

---

[3]Walther Heinrich Wilhelm Ritz (1878–1909) was a Swiss theoretical physicist. He is most famous, in addition to the variational method named after him – the Ritz method – for his work with Johannes Rydberg on the Rydberg–Ritz combination principle. In 1915 Boris Grigoryevich Galerkin (1871–1945), a Soviet mathematician and engineer, proposed the idea of variational approximation. Before Galerkin, Ivan Grigoryevich Bubnov (1972–1919), a Russian marine engineer, developed a similar approach for the solution of differential equations, which is considered to be a variant of the Ritz method. The phrase "Rayleigh–Ritz method" is considered to be a misnomer because Rayleigh quotient is a special case of the Ritz method applied to an eigenvalue problem in one dimension. Also, one should know that the Galerkin method and the Ritz method are not the same in general; see Reddy [9] for additional discussion.

According to the principle of minimum total potential energy applied to a bar, the actual solution u_0 (from the linear vector space $\mathcal{U}$ of functions which vanish at $x = 0$) of Eqs. (3.9.1) and (3.9.2) minimizes the quadratic functional

$$\Pi(u) = \frac{1}{2} \int_0^a \left[E(x)A(x) \left(\frac{du}{dx} \right)^2 + c(x)u^2(x) \right] dx + \frac{1}{2}k[u(a)]^2$$
$$- \left[\int_0^a f(x)\,u(x)\,dx + P\,u(a) \right]. \tag{3.9.3}$$

Thus, the principle, $\delta\Pi(u) = 0$, is equivalent to Eq. (3.9.1) and the second condition in Eq. (3.9.2) (the first condition must be satisfied by every possible candidate for an approximate solution).

3.9.2.2 The Ritz approximation

We seek an approximation $U_N(x)$ of the solution $u(x)$ of Eq. (3.9.1) in the form

$$u(x) \approx U_N(x) = \sum_{i=1}^N c_i\phi_i(x) + \phi_0(x), \tag{3.9.4}$$

where $\phi_i(x)$ are functions from $\mathcal{U}$ (i.e., satisfy $\phi_i(0) = 0$), and c_i are parameters to be determined such that $U_N(x)$ minimizes $\Pi(U_N)$. The number of parameters, N, is preselected, and one expects to obtain increasingly accurate solution as N is increased (a property of the approximation known as the *convergence*). In minimizing Π, the parameters c_i $(i = 1, 2, \ldots, N)$ are adjusted or *varied*. This process of determining c_i by minimizing $\Pi(U_N)$ (or requiring $\delta\Pi(U_N) = 0$) is known as *the Ritz method,* and c_i are termed the *Ritz parameters.*

The particular form of the approximate solution in Eq. (3.9.4) resembles that of the exact solution to any differential equation: the homogeneous solution and the particular solution. That is, the first part $\sum_{j=1}^N c_j\phi_j(x)$ can be viewed as the homogeneous part and the second term $\phi_0(x)$ as the particular part.

The necessary condition for the minimum of Π is (see Section 2.5 for a study of the elements of calculus of variations)

$$0 = \delta\Pi = \int_0^a \left(EA\frac{du}{dx}\frac{d\delta u}{dx} + cu\,\delta u - f\,\delta u \right) dx$$
$$+ ku(a)\,\delta u(a) - P\,\delta u(a), \tag{3.9.5}$$

or

$$B(\delta u, u) = \ell(\delta u), \tag{3.9.6a}$$

where $B(\cdot, \cdot)$ is the bilinear functional and $\ell(\cdot)$ is the linear functional:

$$B(\delta u, u) = \int_0^a \left(EA \frac{du}{dx} \frac{d\delta u}{dx} + cu\,\delta u \right) dx + ku(a)\,\delta u(a), \qquad (3.9.6b)$$

$$\ell(\delta u) = \int_0^a f\delta u\,dx + P\delta u(a). \qquad (3.9.6c)$$

Writing Eq. (3.9.5) in the form of (3.9.6a) is known as the *variational problem*. As will be seen shortly, $B(\cdot, \cdot)$ results in the coefficient matrix of the set of algebraic equations and $\ell(\cdot)$ in the right-hand side vector of the equations.

So far, we have set up the minimization statement based on the principle of total potential energy. Now we apply the Ritz method by substituting the expression U_N for u either into the total potential energy functional $\Pi(u)$ of Eq. (3.9.3) or the variational problem in Eq. (3.9.6a); both yield the same result, while the former needs the additional step of minimizing Π, being a function of c_i, with respect to each of the c_i, $i = 1, 2, \ldots, N$. We shall discuss both here.

Substitution of U_N from Eq. (3.9.4) into the total potential energy functional Π in Eq. (3.9.3) yields Π as a function of the parameters $c_1, c_2, \ldots, c_N$:

$$\Pi(c_i) = \int_0^a \left[\frac{EA}{2} \left(\sum_{j=1}^N c_j \frac{d\phi_j}{dx} + \frac{d\phi_0}{dx} \right)^2 + \frac{c}{2} \left(\sum_{j=1}^N c_j\phi_j + \phi_0 \right)^2 \right.$$

$$\left. - f \left(\sum_{j=1}^N c_j\phi_j + \phi_0 \right) \right] dx + \frac{k}{2} \left(\sum_{j=1}^N c_j\phi_j(a) + \phi_0(a) \right)^2$$

$$- P \left(\sum_{j=1}^N c_j\phi_j(a) + \phi_0(a) \right). \qquad (3.9.7)$$

After the indicated integration with respect to x is carried out (since we know E, A, c, f, and ϕ_i, and ϕ_0 are known functions of x), Π is a function of $c_1, c_2, \ldots, c_N$. The minimum of a function of several variables is determined by setting the total derivative of Π (a necessary condition for the minimum) to zero:

$$0 = d\Pi = \frac{\partial \Pi}{\partial c_1} dc_1 + \frac{\partial \Pi}{\partial c_2} dc_2 + \cdots + \frac{\partial \Pi}{\partial c_N} dc_N. \qquad (3.9.8)$$

Since $\{c_1, c_2, \ldots, c_N\}$ is a linearly independent set, dc_1, dc_2, etc. are independent variations (e.g., we can take $dc_1 = 1$ and all other increments $dc_i = 0$ for $i \neq 1$ and obtain $\partial/\partial c_1 = 0$), we have

$$\frac{\partial \Pi}{\partial c_1} = 0, \quad \frac{\partial \Pi}{\partial c_2} = 0, \quad \cdots, \quad \frac{\partial \Pi}{\partial c_N} = 0. \qquad (3.9.9)$$

This gives a set of N linear algebraic equations among $c_1, c_2, \ldots, c_N$ (the algebraic equations are numbered according to the parameter number, i.e., the ith equation is that results from $\partial \Pi / \partial c_i = 0$):

$$0 = \frac{\partial \Pi}{\partial c_1} = \int_0^a \left[EA \frac{d\phi_1}{dx} \left(\sum_{j=1}^N c_j \frac{d\phi_j}{dx} + \frac{d\phi_0}{dx} \right) + c\phi_1 \left(\sum_{j=1}^N c_j \phi_j + \phi_0 \right) - f\phi_1 \right] dx$$

$$+ k\phi_1(a) \left(\sum_{j=1}^N c_j \phi_j(a) + \phi_0(a) \right) - P\phi_1(a), \qquad \text{(Eq. 1)}$$

$$0 = \frac{\partial \Pi}{\partial c_2} = \int_0^a \left[EA \frac{d\phi_2}{dx} \left(\sum_{j=1}^N c_j \frac{d\phi_j}{dx} + \frac{d\phi_0}{dx} \right) + c\phi_2 \left(\sum_{j=1}^N c_j \phi_j + \phi_0 \right) - f\phi_2 \right] dx$$

$$+ k\phi_2(a) \left(\sum_{j=1}^N c_j \phi_j(a) + \phi_0(a) \right) - P\phi_2(a), \qquad \text{(Eq. 2)}$$

$$\cdots\cdots\cdots \quad \cdots\cdots \quad \cdots\cdots \quad \cdots\cdots$$

$$0 = \frac{\partial \Pi}{\partial c_i} = \int_0^a \left[EA \frac{d\phi_i}{dx} \left(\sum_{j=1}^N c_j \frac{d\phi_j}{dx} + \frac{d\phi_0}{dx} \right) + c\phi_i \left(\sum_{j=1}^N c_j \phi_j + \phi_0 \right) - f\phi_i \right] dx$$

$$+ k\phi_i(a) \left(\sum_{j=1}^N c_j \phi_j(a) + \phi_0(a) \right) - P\phi_i(a), \qquad \text{(Eq. } i\text{)}$$

$$\cdots\cdots\cdots \quad \cdots\cdots \quad \cdots\cdots \quad \cdots\cdots$$

$$0 = \frac{\partial \Pi}{\partial c_N} = \int_0^a \left[EA \frac{d\phi_N}{dx} \left(\sum_{j=1}^N c_j \frac{d\phi_j}{dx} + \frac{d\phi_0}{dx} \right) + c\phi_N \left(\sum_{j=1}^N c_j \phi_j + \phi_0 \right) - f\phi_N \right] dx$$

$$+ k\phi_N(a) \left(\sum_{j=1}^N c_j \phi_j(a) + \phi_0(a) \right) - P\phi_N(a). \qquad \text{(Eq. } N\text{)}$$

In particular, the ith equation of the set of N equations is

$$0 = \sum_{j=1}^N \left\{ \int_0^a \left(EA \frac{d\phi_i}{dx} \frac{d\phi_j}{dx} + c\phi_i \phi_j \right) dx + k\phi_i(a)\phi_j(a) \right\} c_j$$

$$- \int_0^a \left(-EA \frac{d\phi_0}{dx} \frac{d\phi_i}{dx} - c\phi_0 \phi_i + f\phi_i \right) dx - P\phi_i(a) + k\phi_0(a)\phi_i(a)$$

$$\equiv \sum_{j=1}^N A_{ij} c_j - b_i, \qquad (3.9.10a)$$

where

$$A_{ij} = B(\phi_i, \phi_j) = \int_0^a \left(EA \frac{d\phi_i}{dx} \frac{d\phi_j}{dx} + c\phi_i\phi_j \right) dx + k\phi_i(a)\phi_j(a),$$

$$b_i = \ell(\phi_i) = -\int_0^a \left(EA \frac{d\phi_0}{dx} \frac{d\phi_i}{dx} + c\phi_0\phi_i \right) dx \qquad (3.9.10b)$$

$$- k\phi_0(a)\phi_i(a) + \int_0^a f\phi_i \, dx + P\phi_i(a).$$

In matrix form, the N linear equations can be expressed as

$$\mathbf{Ac} = \mathbf{b}. \qquad (3.9.11)$$

Equations (3.9.10a) and (3.9.10b) can also be arrived, more directly, by substituting the Ritz approximation (3.9.4) and its variation

$$\delta u \approx \delta U_N = \sum_{i=1}^N \delta c_i \, \phi_i(x) \qquad (3.9.12)$$

into the variational statement $\delta\Pi = 0$, instead of substituting Eq. (3.9.3) in Π and then taking its variation with respect to c_i. This results in

$$0 = \sum_{i=1}^N \delta c_i \left\{ \int_0^a \left[EA \frac{d\phi_i}{dx} \left(\sum_{j=1}^N c_j \frac{d\phi_j}{dx} + \frac{d\phi_0}{dx} \right) + c\phi_i \left(\sum_{j=1}^N c_j\phi_j + \phi_0 \right) - f\phi_i \right] dx \right.$$

$$\left. + k\phi_i(a) \left(\sum_{j=1}^N c_j\phi_j(a) + \phi_0(a) \right) - P\phi_i(a) \right\}. \qquad (3.9.13)$$

Since δc_i are arbitrary and linearly independent of each other, the expression in the curly brackets should vanish and we obtain the ith equation, Eq. (3.9.10a), of the set. We note that, in view of the fact that $\sum_{i=1}^N \delta c_i$ in Eq. (3.9.13) does not enter the final result, the ith equation of the system can be obtained directly from the variational problem in Eq. (3.9.6a) by substituting Eq. (3.9.4) for u and ϕ_i for δu. Thus, *one needs to have only a variational statement equivalent to the governing equations of the problem.* The existence of a quadratic functional is not necessary to use the Ritz method. In addition, the Ritz method can be used to solve a variational problem that is nonlinear, resulting in nonlinear algebraic equations among the c_i.

3.9.2.3   Requirements on the approximation functions

The most difficult part of using the Ritz method for the solution of differential equations is the selection of the *approximation functions* ϕ_0 and ϕ_i that meet the requirements. There exist no formulas for their derivation, but certain guidelines will help their construction. The fact that no systematic procedure to derive ϕ_0 and ϕ_i exists lead to the development of the finite element method, which most often uses the Ritz method to derive the element equations.

At this point, we recall that every differential equation has duality pairs, which are identified during the development of the variational statement (or the so-called weak form). The second-order equations have one duality pair while higher-order equations have multiple duality pairs. In structural mechanics, the duality pairs consist of a displacement variable and force variable. The displacement variable is denoted as the primary variable and the force variable is identified as the secondary variable. We also recall that the specification of a primary variable is termed the essential (or geometric) boundary condition and the specification of a secondary variable is called natural (or force) boundary condition. In the model problem discussed in this section, the displacement u is the primary variable and $N = EA(du/dx)$ is the force variable. The natural boundary conditions of the problem are included in the variational statement $\delta\Pi = 0$ or its equivalent. Hence, in the Ritz method, U_N *must satisfy only the specified essential boundary conditions.*

In order that U_N satisfies the essential boundary conditions *for any* c_i, it is convenient to choose the approximate solution in the form of Eq. (3.9.4) (now one may understand why ϕ_0 is introduced) and require $\phi_0(x)$ to satisfy the *actual* specified essential boundary conditions (i.e., the particular solution). For instance, if $u(x)$ is specified to be $u(0) = \hat{u}$, then $U_N(0) = \hat{u}$. If we require $\phi_0(x)$ to be the lowest–order function that meets the condition $\phi_0(0) = \hat{u}$, then Eq. (3.9.4) gives

$$U_N(0) = \sum_{i=1}^{N} c_i\phi_i(0) + \phi_0(0) = \sum_{i=1}^{N} c_i\phi_i(0) + \hat{u} \quad \rightarrow \quad \sum_{i=1}^{N} c_i\phi_i(0) = 0. \quad (3.9.14)$$

The above result further implies, for arbitrary and linearly independent set $\{c_i\}$ the requirement

$$\phi_i(0) = 0 \text{ for all } i = 1, 2, \ldots, N \qquad (3.9.15)$$

whenever $u(0)$ is specified. Stating in other words, $\phi_i(x)$ *is required to satisfy the "homogeneous form" of the specified essential boundary conditions.*

It is also of interest to ensure that the Ritz approximation $U_N(x)$ converges to the true solution $u(x)$ of the problem as the value of N is increased, both ϕ_i $(i = 1, 2, \ldots, N)$ and ϕ_0 must satisfy certain additional requirements. The concept of convergence of a sequence of solutions (e.g., $N = 1$, $N = 2$, and so on) involves a limit, say u_0, of the sequence. If the limit is not a part of the sequence $\{U_N\}_{N=1}^{\infty}$, then there is no hope of attaining the convergence. For example, if the true solution to a certain problem is of the form $u_0(x) = ax^2 + bx^3 + cx^5$, where a, b, and c are constants, then the sequence of approximations

$$U_1 = c_1x^3, \ U_2 = c_1x^3 + c_2x^4, \ \ldots, \ U_N = c_1x^3 + c_2x^4 + \cdots + c_Nx^{N+2}$$

will not converge to the true solution because the sequence does not contain the x^2 term that is present in the limit function u_0. The sequence is said to be incomplete. As a rule (to be on safe side), in selecting an approximate solution, one should include all terms up to the highest–order term in the expression used for U_N (even if individually ϕ_i are incomplete). If a certain term is not a part of the true solution, like the x^4 term, its (i.e., the term which is not a part of the limit) coefficient will turn out to be zero by the time all terms of the true solution are included in the approximation.

In summary, the requirements and guidelines on the selection of ϕ_0 and ϕ_i of the Ritz approximation are as follows:

1. ϕ_0 must satisfy the *specified* essential boundary conditions. It is identically zero if all of the specified essential boundary conditions of the problem are homogeneous. One must only select the lowest–order function that meets the specified nonzero essential boundary condition. For example, if the specified essential boundary condition is $u(0) = \hat{u} \neq 0$, then $\phi_0(x) = (1 + ax + bx^2 + \cdots +)\hat{u}$ satisfies the condition $\phi_0(0) = \hat{u}$ for any a, b, and so on. However, the lowest–order function is the constant term. Other terms should not be included in $\phi_0(x)$.
2. ϕ_i $(i = 1, 2, \ldots, N)$ must satisfy the following three conditions:
   (a) $\phi_i(x)$ be continuous, as required by the variational statement being used;
   (b) $\phi_i(x)$ satisfy the *homogeneous form* of the specified essential boundary conditions; and
   (c) the set $\{\phi_i\}$ be linearly independent and complete.   (3.9.16)

The properties listed in Eq. (3.9.16) provide guidelines for selecting the approximation functions $\phi_0(x)$ and $\phi_i(x)$. Apart from the guidelines, the selection of the coordinate functions is largely arbitrary (i.e., they can be algebraic, trigonometric, or other functions). For reasons of the simplicity of evaluating integrals of algebraic functions, often we prefer algebraic polynomials (like in the finite element method) for ϕ_0 and ϕ_i. As a general rule, the coordinate functions ϕ_i should be selected from an admissible set [i.e., those meeting the conditions in Eq. (3.9.16)], from the lowest order to a desirable order, without missing any intermediate terms (i.e., satisfying the completeness property). Also, ϕ_0 should be any lowest–order (including zero) that satisfied the specified essential boundary conditions of the problem; it has no continuity (differentiability) requirement.

Some general features of the Ritz method are noted here.

1. The Ritz method applies to all problems, linear or nonlinear, as long as one can cast the governing equations in the form of a variational statement [see Eqs. (3.9.6a) and (3.9.6b)]. The procedure to develop variational forms, also called *weak forms*, from governing differential equations arising in any field, can be found in [9, 155].

2. If the variational problem $B(v, u) = \ell(v)$ used in the Ritz approximation is such that $B(v, u)$ is symmetric in v and u, the coefficient matrix $\mathbf{A}$ of the resulting algebraic problem $(\mathbf{Ac} = \mathbf{b})$ is also symmetric and, therefore, only elements above or below the diagonal of the coefficient matrix $\mathbf{A}$ need to be computed.

3. If $B(v, u)$ is nonlinear in u, the resulting algebraic equations $\mathbf{A(c)c} = \mathbf{b}$ will also be nonlinear in the parameters c_i. To solve such nonlinear equations, a variety of numerical methods are available (e.g., Newton's method). Generally, there is more than one solution to the set of nonlinear equations, and one of them is selected as the solution based on some physical or mathematical criterion.

4. If the approximation functions ϕ_0 and $\phi_i(x)$ are selected to satisfy the requirements in Eq. (3.9.16), the assumed approximation $U_N(x)$, when substituted into the minimum total potential energy principle, normally converges to the actual solution $u(x)$ with an increase in the number of parameters (i.e., as $N \to \infty$). A mathematical proof of such an assertion is not given here, but interested readers can consult the references at the end of the book.

5. For increasing values of N, the previously computed coefficients of the algebraic equations in Eq. (3.9.10b) remain unchanged, provided the previously selected approximation functions are not changed; thus, one must compute only the new coefficients A_{ij} and b_i.

6. In general, the governing equation(s) and natural boundary conditions of the problem are satisfied only in the variational (or integral) sense, and not point wise, by the computed Ritz solution. Therefore, the solution(s) obtained from the Ritz approximation generally does not satisfy the governing equations in a point-wise sense.

7. Since a continuous system is approximated by a finite number of coordinates (or degrees of freedom), the approximate system is less flexible than the actual system (recall that $\Pi(u) < \Pi(U)$ for any U that is not equal to the exact solution u). Consequently, the displacements obtained by the Ritz method using the principle of minimum total potential energy converge to the exact displacement from below:

$$|\Pi(U_m)| < |\Pi(U_n)|, \quad \text{for } m > n,$$

where U_N denotes the N-parameter Ritz approximation of u obtained using the principle of minimum total potential energy. The inequality is used to indicate that U_m is closer to the exact solution than U_n when $m > n$.

Next, we illustrate the use of the Ritz method through a number of examples dealing with axially–loaded members and beams.

Example 3.9.1

Consider a variable cross-section bar with axial stiffness, which is governed by Eqs (3.9.1) and (3.9.2) with $EA = S[2-(x/L)]$, $c = 0$, and $f = f_0$; here L is the length of the bar and S is a constant. Develop the N-parameter Ritz solution using algebraic polynomials for $\phi_i(x)$ and obtain the numerical values of the Ritz parameters c_i.

Solution: The specified boundary conditions for the problem contain one essential, $u(0) = 0$ and the other one is natural $EA(du/dx) + ku = P$ at $x = L$, which is already accounted for in the total potential energy functional of Eq. (3.9.3). The specified essential boundary condition is homogeneous. Therefore, we select $\phi_0 = 0$. Consequently, all $\phi_i(x)$ must be such that $\phi_i(0) = 0$ for $i = 1, 2, \ldots, N$.

To construct $\phi_i(x)$, We begin with $\phi_1(x)$. The lowest order admissible and complete polynomial (i.e., has at least a nonzero first derivative and $\phi_1(0) = 0$) is $\phi_1(x) = x$ (any nonzero constant is absorbed into the parameter c_1). The second function in the set is $\phi_2(x) = x^2$ (one may also include x but it does not serve any purpose as it is already included in ϕ_1). Thus, the set $\{\phi_i\}_{i=1}^{N} = \{x^i\}_{i=1}^{N}$ is linearly independent set of functions that makes U_N a complete polynomial.

For the choice of algebraic polynomials, the N-parameter Ritz approximation for the bar problem is

$$U_N(x) = \sum_{i=1}^{N} c_i\phi_i(x), \quad \phi_i(x) = x^i. \tag{3.9.17}$$

Then the coefficients A_{ij} and b_i of Eq. (3.9.10b) for $EA = S[2 - (x/L)]$, $\phi_i(x) = x^i$, and $f = f_0$ (a constant) are

$$A_{ij} = S \int_0^L \left(2 - \frac{x}{L}\right) \frac{d\phi_i}{dx} \frac{d\phi_j}{dx}\, dx + k\phi_i(L)\phi_j(L)$$

$$= S \int_0^L ij \left(2 - \frac{x}{L}\right) x^{i+j-2}\, dx + k\,(L)^{i+j}$$

$$= S \frac{ij(1+i+j)}{(i+j-1)(i+j)} (L)^{i+j-1} + k\,(L)^{i+j}, \tag{3.9.18a}$$

$$b_i = -\int_0^L EA\frac{d\phi_0}{dx}\frac{d\phi_i}{dx}\, dx - k\phi_0(L)\phi_i(L) + \int_0^L f\phi_i\, dx + P\phi_i(L)$$

$$= \int_0^L f_0\phi_i\, dx + P\,\phi_i(L) = \frac{f_0}{i+1}(L)^{i+1} + P\,(L)^i. \tag{3.9.18b}$$

For $k = 0$ and $f_0 = 0$, we can express the Ritz equations as

$$\sum_{j=1}^{N} A_{ij}c_j - b_i = 0 \quad \text{or} \quad \sum_{j=1}^{N} \bar{A}_{ij}\bar{c}_j - \bar{b}_i = 0, \quad \bar{c}_j = \frac{SL^{j-1}}{P}c_j,$$

with

$$\bar{A}_{ij} = \frac{ij(1+i+j)}{(i+j-1)(i+j)}, \quad \bar{b}_i = 1.$$

Now we specialize the above equations for $N = 1, 2$, and 3.

One-parameter Ritz solution $(N = 1)$:

$$\bar{A}_{11} = \frac{3}{2}, \quad \bar{b}_1 = 1, \Rightarrow \bar{c}_1 = \frac{\bar{b}_1}{\bar{A}_{11}} = \frac{2}{3}.$$

The one-term Ritz solution is $(c_1 = \bar{c}_1 P/S)$

$$U_1(x) = \frac{2P}{3S}x = \frac{2PL}{3S}\frac{x}{L}.$$

Two-parameter Ritz solution $(N = 2)$:

$$\frac{1}{6}\begin{bmatrix} 9 & 8 \\ 8 & 10 \end{bmatrix}\begin{Bmatrix} \bar{c}_1 \\ \bar{c}_2 \end{Bmatrix} = \begin{Bmatrix} 1 \\ 1 \end{Bmatrix},$$

whose solution by Cramer's rule is

$$\bar{c}_1 = \frac{6}{13}, \quad \bar{c}_2 = \frac{3}{13}.$$

Hence, the two-parameter Ritz solution becomes

$$U_2(x) = c_1 x + c_2 x^2 = \frac{3PL}{13S}\left(2\frac{x}{L} + \frac{x^2}{L^2}\right).$$

Three-parameter Ritz solution $(N = 3)$:

$$\frac{1}{60}\begin{bmatrix} 90 & 80 & 75 \\ 80 & 100 & 108 \\ 75 & 108 & 126 \end{bmatrix}\begin{Bmatrix} \bar{c}_1 \\ \bar{c}_2 \\ \bar{c}_3 \end{Bmatrix} = \begin{Bmatrix} 1 \\ 1 \\ 1 \end{Bmatrix},$$

and the solution is

$$\bar{c}_1 = 50.7937 \times 10^{-2}, \quad \bar{c}_2 = 7.9365 \times 10^{-2}, \quad \bar{c}_3 = 10.5820 \times 10^{-2}.$$

Hence, the three-parameter Ritz solution becomes

$$U_3(x) = c_1 x + c_2 x^2 + c_3 x^3 = \frac{PL}{S}\left(50.7937\frac{x}{L} + 7.9365\frac{x^2}{L^2} + 10.5820\frac{x^3}{L^3}\right)10^{-2}.$$

The exact solution of Eqs (3.9.1) and (3.9.2) with $u(0) = 0$, $k = 0$, $EA = S[2 - (x/L)]$, and $f = f_0$ is

$$u_0(x) = \frac{f_0 L}{S}x + \frac{(f_0 L - P)L}{S}\log\left(1 - \frac{x}{2L}\right)$$
$$\approx \frac{f_0 L}{S}x + \frac{PL - f_0 L^2}{S}\left[\left(\frac{x}{2L}\right) + \frac{1}{2}\left(\frac{x}{2L}\right)^2 + \frac{1}{3}\left(\frac{x}{2L}\right)^3 + \frac{1}{4}\left(\frac{x}{2L}\right)^4 + \cdots\right].$$
$$(3.9.19)$$

Table 3.9.1 contains a comparison of the Ritz coefficients $\bar{c}_i$ for $N = 1, 2, \ldots, 8$ with the exact coefficients in Eq. (3.9.19). The first two Ritz coefficients $\bar{c}_i$ have converged to the exact ones for $N = 8$. However, the Ritz solution for $N < 8$ is reasonably close to the exact solution because coefficients of the higher-order terms contribute less and less to the solution, as can be seen from Table 3.9.2. In fact, the six-parameter solution matches with the exact solution up to the fourth decimal point for $x = 0, 0.1, 0.2, \ldots, 1$.

Table 3.9.1: The Ritz coefficients $(\bar{c}_i \times 10^2)$ for the axial deformation of an isotropic, linearly elastic, nonuniform bar fixed at one end and subjected to a point force P at the other end.

N	$\bar{c}_1$	$\bar{c}_2$	$\bar{c}_3$	$\bar{c}_4$	$\bar{c}_5$	$\bar{c}_6$	$\bar{c}_7$	$\bar{c}_8$
1	66.667							
2	46.154	23.077						
3	50.794	7.936	10.582					
4	49.844	14.019	0.000	5.452				
5	50.029	12.062	6.100	-1.872	2.994			
6	49.994	12.615	3.426	3.621	-2.506	1.713		
7	50.001	12.471	4.416	0.576	2.605	-1.764	1.008	
8	50.000	12.507	4.090	1.968	-0.517	2.045	-1.384	0.605
Exact	50.000	12.500	4.167	1.562	0.625	0.260	0.112	0.049

Table 3.9.2: Comparison of the Ritz solutions $(\bar{u} = u \times (S/PL))$ with the analytical solution of the bar problem described by Eqs. (3.9.1) and (3.9.2).

x/L	$N=1$	$N=2$	$N=3$	$N=4$	Exact
0.0	0.0000	0.0000	0.0000	0.0000	0.0000
0.1	0.0667	0.0485	0.0517	0.0513	0.0513
0.2	0.1333	0.1015	0.1056	0.1054	0.1054
0.3	0.2000	0.1592	0.1624	0.1626	0.1625
0.4	0.2667	0.2215	0.2226	0.2232	0.2231
0.5	0.3333	0.2885	0.2870	0.2877	0.2877
0.6	0.4000	0.3600	0.3562	0.3566	0.3567
0.7	0.4667	0.4361	0.4307	0.4307	0.4308
0.8	0.5333	0.5169	0.5113	0.5108	0.5108
0.9	0.6000	0.6023	0.5986	0.5979	0.5978
1.0	0.6667	0.6923	0.6931	0.6931	0.6931

Example 3.9.2

Consider a beam of length L and constant bending stiffness EI, loaded with uniform distributed transverse load of intensity q_0. Develop the N-parameter Ritz solution with algebraic polynomials using the CBT for beams with (a) simply supported (the pinned–hinged and pinned–pinned boundary conditions are the same for pure bending problems) and (b) clamped boundary conditions (see Fig. 3.9.2), and then compute the one-, two-, and three-parameter solutions, and compare with the exact solution of the problem.

Solution: Since the axial displacement u is decoupled from the bending deflection w, we do not include the strain energy due to stretching deformation. We first set up the variational problem using the principle of the minimum total potential energy (or the principle of virtual displacements). The total potential energy functional for this problem is constructed by assuming that the volume integration can be represented as area integration followed by line integration.

$$\int_V (\cdot)dV = \int_0^L \left(\int_A (\cdot)dA \right) dx.$$

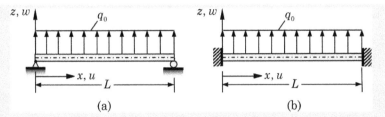

Figure 3.9.2: (a) Simply supported and (b) clamped beams under uniform distributed load.

The assumption is valid because z appears explicitly in the integrand, allowing integration with respect to z first.

The total potential energy of the beam is

$$\Pi(w) = \frac{1}{2} \int_0^L \int_A \sigma_{xx} \, \varepsilon_{xx} \, dA \, dx - \int_0^L q \, w \, dx$$

$$= -\frac{1}{2} \int_0^L M_{xx} \frac{d^2 w}{dx^2} \, dx - \int_0^L q \, w \, dx$$

$$= \frac{1}{2} \int_0^L EI \left(\frac{d^2 w}{dx^2} \right)^2 dx - \int_0^L q \, w \, dx, \qquad (3.9.20)$$

where we have made use of the following strain-0-displacement and constitutive relations in arriving at the last line:

$$\varepsilon_{xx} = -z \frac{d^2 w}{dx^2}, \quad \sigma_{xx} = E \varepsilon_{xx}, \quad M_{xx} = \int_A \sigma_{xx} \, z \, dA = -EI \frac{d^2 w}{dx^2}. \qquad (3.9.21)$$

Then the principle of minimum total potential energy for the problem can be expressed as

$$0 = \int_0^L \left(EI \frac{d^2 \delta w}{dx^2} \frac{d^2 w}{dx^2} - \delta w \, q_0 \right) dx \equiv B(\delta w, w) - \ell(\delta w), \qquad (3.9.22a)$$

where

$$B(\delta w, w) = \int_0^L EI \frac{d^2 \delta w}{dx^2} \frac{d^2 w}{dx^2} \, dx, \quad \ell(\delta w) = \int_0^L \delta w \, q \, dx. \qquad (3.9.22b)$$

Next, we construct the approximation functions $\phi_0(x)$ and $\phi_i(x)$. Towards this end, we note that the duality pairs of the CBT are: (w, N_{xz}) and $(dw/dx, M_{xx})$, $N_{xz} = dM_{xx}/dx$ and M_{xx} being the shear force and bending moment, respectively. We now consider the Ritz approximations for the two different boundary conditions.

(a) Simply supported beam. The boundary conditions for this case are:

$$w = 0, \quad M_{xx} = -EI \frac{d^2 w}{dx^2} = 0 \text{ at } x = 0, L.$$

Of the four boundary conditions, only the first two are the specified essential boundary conditions and they are homogeneous. Hence, we have $\phi_0 = 0$. To construct

ϕ_1, we note that ϕ_i must satisfy the (already) homogeneous boundary conditions: $\phi_i(0) = 0$ and $\phi_i(L) = 0$. Thus, ϕ_1 is of the form $\phi_1 = x(L-x)$. Then, the subsequent functions can be identified using this form as follows:

$$\phi_1 = x(L-x), \quad \phi_2 = x^2(L-x), \quad \phi_i = x^i(L-x), \dots, \phi_N = x^N(L-x). \quad (3.9.23)$$

Substituting the N-parameter Ritz approximation

$$w(x) \approx W_N(x) = \sum_{j=1}^{N} c_j \phi_j(x) + \phi_0(x) = \sum_{j=1}^{N} c_j x^j (L-x) \qquad (3.9.24)$$

for w and ϕ_i for δw into Eq. (3.9.22a), we obtain

$$0 = \int_0^L \left[EI\phi_i'' \left(\sum_{j=1}^{N} c_j \phi_j'' \right) - \phi_i q_0 \right] dx$$

$$= \sum_{j=1}^{N} A_{ij} c_j - b_i \quad \text{or} \quad \mathbf{Ac} = \mathbf{b}, \qquad (3.9.25a)$$

where

$$A_{ij} = \int_0^L EI \frac{d^2\phi_i}{dx^2} \frac{d^2\phi_j}{dx^2} \, dx$$

$$= EI \int_0^L ij \left[L(i-1)x^{i-2} - (i+1)x^{i-1} \right] \left[L(j-1)x^{j-2} - (j+1)x^{j-1} \right] dx$$

$$= ij \left[\frac{(i-1)(j-1)}{(i+j-3)} - \frac{2(ij-1)}{(i+j-2)} + \frac{(i+1)(j+1)}{(i+j-1)} \right] EI \, L^{i+j-1}, \qquad (3.9.25b)$$

$$b_i = \int_0^L q_0 \phi_i \, dx = \int_0^L q_0 \left[Lx^i - x^{i+1} \right] dx = \frac{q_0 L^{i+2}}{(i+1)(i+2)}. \qquad (3.9.25c)$$

Note that when the denominator in the expressions of A_{ij} is less than equal to zero, then that expression is not to be included in the evaluation of A_{ij}. We can specialize the results for $N = 1, 2$, and 3.

One-parameter Ritz solution($N = 1$) For this case, we have

$$A_{11} = 4EIL, \quad b_1 = \frac{q_0 L^3}{6} \implies c_1 = \frac{b_1}{A_{11}} = \frac{q_0 L^2}{24EI}.$$

The Ritz solution is

$$W_1(x) = \frac{q_0 L^4}{24EI} \left(\frac{x}{L} - \frac{x^2}{L^2} \right), \quad W_1(0.5L) = \frac{q_0 L^4}{96EI} = 0.01042 \frac{q_0 L^4}{EI}.$$

Two-parameter Ritz solution($N = 2$) For the two-parameter case, the previously computed coefficients A_{11} and b_1 are valid; we need to compute only the remaining coefficients, A_{12}, A_{21}, A_{22}, and b_2. We have

$$EIL \begin{bmatrix} 4 & 2L \\ 2L & 4L^2 \end{bmatrix} \begin{Bmatrix} c_1 \\ c_2 \end{Bmatrix} = \frac{q_0 L^3}{12} \begin{Bmatrix} 2 \\ L \end{Bmatrix}.$$

The solution of these equations is

$$c_1 = \frac{q_0 L^2}{24EI}, \quad c_2 = 0.$$

Thus, the two-parameter Ritz solution is the same as the one-parameter Ritz solution. Since $c_2 = 0$, one might jump to the conclusion that the Ritz approximation has already converged; but it is obvious (because this problem has algebraic exact solution which is a fourth-degree polynomial when $q = q_0 = $ constant) that it has not yet converged.

Three-parameter Ritz solution$(N = 3)$ Once again, the previously computed coefficients are valid here. With the additional coefficients computed, we have

$$EIL \begin{bmatrix} 4 & 2L & 2L^2 \\ 2L & 4L^2 & 4L^3 \\ 2L^2 & 4L^3 & 4.8L^4 \end{bmatrix} \begin{Bmatrix} c_1 \\ c_2 \\ c_3 \end{Bmatrix} = \frac{q_0 L^3}{12} \begin{Bmatrix} 2 \\ L \\ 0.6L^2 \end{Bmatrix},$$

whose solution is

$$c_1 = c_2 L = -c_3 L^2 = \frac{q_0 L^2}{24EI}.$$

The three-parameter Ritz solution becomes

$$W_3(x) = \frac{q_0 L^4}{24EI} \left[\left(\frac{x}{L} - \frac{x^2}{L^2} \right) + \left(\frac{x^2}{L^2} - \frac{x^3}{L^3} \right) - \left(\frac{x^3}{L^3} - \frac{x^4}{L^4} \right) \right]$$

$$= \frac{q_0 L^4}{24EI} \left(\frac{x}{L} - 2\frac{x^3}{L^3} + \frac{x^4}{L^4} \right).$$

This three-parameter Ritz solution coincides with the exact solution

$$w_0(x) = \frac{q_0 L^4}{24EI} \left(\frac{x}{L} - 2\frac{x^3}{L^3} + \frac{x^4}{L^4} \right). \tag{3.9.26}$$

This is not coincidental. The solution to a fourth-order differential equation with constant coefficients is a fourth-degree polynomial, and the three-parameter Ritz approximation with polynomial approximation functions is also a fourth-degree polynomial. If trigonometric approximation functions are used, one will not obtain the exact solution because a sine series representation of a constant is an infinite series, although the values obtained with two or three functions may be very close to exact.

The maximum deflection occurs at the center of the beam and it is

$$w(0.5L) = \frac{5}{384} \frac{q_0 L^4}{EI} = 0.01302 \frac{q_0 L^4}{EI}.$$

The maximum deflection predicted by the one-parameter Ritz approximation is in 20% error. Figure 3.9.3 shows a comparison of the Ritz solutions with the exact solution for this case. The one-parameter Ritz solution is in considerable error.

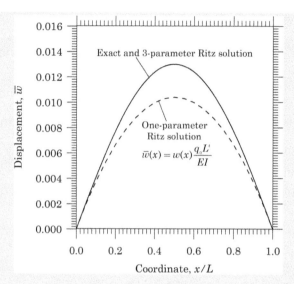

Figure 3.9.3: A comparison of the Ritz solutions $\bar{w} = w \times (q_0 L^4/EI)$ with the exact solution for the simply supported beam under uniform load.

(b) Clamped beam. The boundary conditions for the clamped case are:

$$w = 0, \quad \frac{dw}{dx} = 0 \text{ at } x = 0, L.$$

In this case all four specified boundary conditions are the essential type and they are homogeneous. Hence, we have $\phi_0 = 0$. To construct ϕ_1, we note that ϕ_i must satisfy the (already) homogeneous boundary conditions: $\phi_i(0) = 0$, $\phi_i(L) = 0$, $\phi_i'(0) = 0$, and $\phi_i'(L) = 0$, where the prime denotes the derivative with respect to x. Since there are four conditions, we select a five-parameter polynomial (if it were a four-parameter polynomial, all coefficients will be determined to be zero) for ϕ_1:

$$\phi_1(x) = \alpha_0 + \alpha_1 x + \alpha_2 x^2 + \alpha_3 x^3 + \alpha_4 x^4$$

and obtain $\alpha_0 = \alpha_1 = 0$, $\alpha_2 L^2 + \alpha_3 L^3 + \alpha_4 L^4 = 0$, and $2\alpha_2 L + 3\alpha_3 L^2 + 4\alpha_4 L^3 = 0$. Solving these two equations for α_2 and α_3 in terms of α_4, we obtain $\alpha_2 = \alpha_4 L^2$ and $\alpha_3 = -2\alpha_4 L$. We set $\alpha_4 = 1$ (because it gets absorbed into the coefficient c_1), and obtain

$$\phi_1(x) = x^2 L^2 - 2x^3 L + x^4 = x^2(L^2 - 2Lx + x^2) = x^2(L - x)^2.$$

Similarly, by inspection of ϕ_1, we can select $\phi_2(x) = x^3(L - x)^2$. Thus, the ith function becomes $\phi_i(x) = x^{i+1}(L - x)^2$.

Since the exact solution is a fourth-order polynomial (when $q = q_0 = $ constant), we expect that one-parameter Ritz solution would be exact. If we go for two- or three-parameter Ritz solutions, we begin to get $c_2 = 0$, $c_3 = 0$ and so on.

Substituting the one-parameter Ritz approximation $W_1(x) = c_1\phi_1(x)$ into Eq. (3.9.22a), we obtain, $A_{11}c_1 = F_1$ or $c_1 = F_1/A_{11}$, with

$$A_{11} = EI \int_0^L \left(\frac{d^2\phi_1}{dx^2}\right)^2 dx = \frac{4}{5}EIL^5, \quad F_1 = q_0 \int_0^L \phi_1 dx = \frac{q_0 L^5}{30}.$$

Then the one-parameter Ritz solution is given by

$$W_1(x) = \frac{q_0 L^4}{24EI} \frac{x^2}{L^2} \left(1 - \frac{x}{L}\right)^2, \tag{3.9.27}$$

which coincides with the exact solution. The maximum deflection of the clamped beam is $w(0.5L) = q_0 L^4/384EI$, which is one-fifth of the simply supported beam.

Example 3.9.3

Consider a functionally graded beam of length L and constant stiffness A_{xx}, B_{xx}, and D_{xx}, loaded with uniform distributed transverse load of intensity q_0. The beam is clamped at the left end and supported at the right end by a linear elastic spring with spring constant k [see Fig. 3.9.4(a)]. Develop the N-parameter Ritz solution using algebraic polynomials and the CBT.

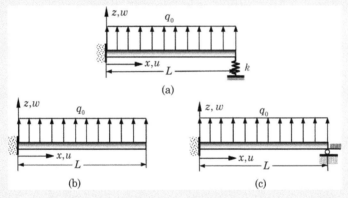

(a)

(b)   (c)

Figure 3.9.4: (a) A beam clamped at the left end and supported at the right end by a linear elastic spring with spring constant k, and subjected to uniform distributed load; (b) Cantilever beam ($k = 0$); and (c) propped cantilever beam ($k = \infty$).

Solution: For this problem, we have coupling between the stretching and bending deformations, bringing both axial and transverse displacements into play. The total potential energy for this problem is given by

$$\Pi(u, w) = \frac{1}{2} \int_0^L \int_A \sigma_{xx} \varepsilon_{xx} \, dA \, dx - \int_0^L q w \, dx + \frac{1}{2} k[w(L)]^2$$

$$= \frac{1}{2} \int_0^L \left(N_{xx} \frac{du}{dx} - M_{xx} \frac{d^2 w}{dx^2}\right) dx - \int_0^L q w \, dx + \frac{1}{2} k[w(L)]^2, \tag{3.9.28a}$$

or, in terms of the displacements,

$$\Pi(u, w) = \frac{1}{2} \int_0^L \left[\left(A_{xx} \frac{du}{dx} - B_{xx} \frac{d^2 w}{dx^2} \right) \frac{du}{dx} \right.$$

$$+ \left(B_{xx} \frac{du}{dx} - D_{xx} \frac{d^2 w}{dx^2} \right) \left(-\frac{d^2 w}{dx^2} \right) \right] dx$$

$$+ \tfrac{1}{2} k [w(L)]^2 - \int_0^L q\, w\, dx, \qquad (3.9.28b)$$

where we have made use of the following strain–displacement and constitutive relations in arriving at Eq. (3.9.28b):

$$\varepsilon_{xx} = \frac{du}{dx} - z \frac{d^2 w}{dx^2}, \quad \sigma_{xx} = E(z)\, \varepsilon_{xx},$$

$$N_{xx} = \int_A \sigma_{xx}\, dA = A_{xx} \frac{du}{dx} - B_{xx} \frac{d^2 w}{dx^2}, \qquad (3.9.29)$$

$$M_{xx} = \int_A \sigma_{xx}\, z\, dA = B_{xx} \frac{du}{dx} - D_{xx} \frac{d^2 w}{dx^2}.$$

Then the principle of minimum total potential energy for the problem can be expressed as

$$0 = \int_0^L \left[\left(A_{xx} \frac{du}{dx} - B_{xx} \frac{d^2 w}{dx^2} \right) \frac{d\delta u}{dx} + \left(B_{xx} \frac{du}{dx} - D_{xx} \frac{d^2 w}{dx^2} \right) \left(-\frac{d^2 \delta w}{dx^2} \right) \right] dx$$

$$+ k\, w(L)\, \delta w(L) - \int_0^L \delta w\, q\, dx. \qquad (3.9.30)$$

This statement is equivalent, because δu and δw are independent variations, to the following two statements (collect the expressions involving δu and δw separately):

$$0 = \int_0^L \left(A_{xx} \frac{du}{dx} - B_{xx} \frac{d^2 w}{dx^2} \right) \frac{d\delta u}{dx}\, dx \qquad (3.9.31)$$

$$0 = \int_0^L \left[\left(B_{xx} \frac{du}{dx} - D_{xx} \frac{d^2 w}{dx^2} \right) \left(-\frac{d^2 \delta w}{dx^2} \right) - \delta w\, q \right] dx + k\, w(L)\, \delta w(L). \qquad (3.9.32)$$

Toward finding the approximation functions, we note that the duality pairs of the CBT for FGM beams are: (u, N_{xx}), (w, N_{xz}), and $(dw/dx, M_{xx})$, where $N_{xz} = dM_{xx}/dx$ is the shear force. The boundary conditions for this case are:

$$\text{At } x = 0: \qquad u = w = \frac{dw}{dx} = 0,$$

$$\text{At } x = L: \quad N_{xx} = M_{xx} = \frac{dM_{xx}}{dx} + kw = 0. \qquad (3.9.33)$$

Of the six boundary conditions, the first three are specified geometric boundary conditions and they are homogeneous; the remaining are force boundary conditions.

Next, we construct the approximation functions for both $u(x)$ and $w(x)$. Since the specified boundary conditions on the primary variables $(u, w, dw/dx)$ are homogeneous, we have $\varphi_0 = 0$ and $\phi_0 = 0$. Suppose that the Ritz approximation of u and w are

$$u(x) \approx U_M(x) = \sum_{j=1}^{M} c_j \varphi_j(x), \quad w(x) \approx W_N(x) = \sum_{j=1}^{N} d_j \phi_j(x), \qquad (3.9.34)$$

where φ_j and ϕ_j must satisfy the requirement $\varphi_j = 0$, $\phi_j = 0$, and $d\phi_j/dx = 0$ at $x = 0$. Obviously, we can choose the following functions, which meet the requirements:

$$\varphi_j = x^j, \quad \phi_j = x^{j+1}, \quad j = 1, 2, \ldots, M \quad \text{or} \quad N. \qquad (3.9.35)$$

Substituting Eq. (3.9.34) for u and w, and $\varphi_i = x^i$ for δu and $\delta w = \phi_i = x^{i+1}$ into the variational statements in Eqs. (3.9.31) and (3.9.32), we obtain

$$0 = \sum_{j=1}^{M} \left(\int_0^L A_{xx} \frac{d\varphi_i}{dx} \frac{d\varphi_j}{dx} \, dx \right) c_j - \sum_{j=1}^{N} \left(\int_0^L B_{xx} \frac{d\varphi_i}{dx} \frac{d^2\phi_j}{dx^2} \, dx \right) d_j$$

$$= \sum_{j=1}^{M} A_{xx} \left(\frac{ij}{i+j-1} \right) (L)^{i+j-1} c_j - \sum_{j=1}^{N} B_{xx} \left(\frac{ij(1+j)}{i+j-1} \right) (L)^{i+j-1} d_j, \quad (3.9.36)$$

$$0 = -\sum_{j=1}^{M} \left(\int_0^L B_{xx} \frac{d^2\phi_i}{dx^2} \frac{d\varphi_j}{dx} \, dx \right) c_j + \sum_{j=1}^{N} \Bigg(\int_0^L D_{xx} \frac{d^2\phi_i}{dx^2} \frac{d^2\phi_j}{dx^2} \, dx$$

$$+ k\,\phi_i(L)\,\phi_j(L) \Bigg) d_j - \int_0^L \phi_i\, q_0\, dx$$

$$= -\sum_{j=1}^{M} B_{xx} \left(\frac{ij(1+i)}{i+j-1} \right) (L)^{i+j-1} c_j + \sum_{j=1}^{N} \Bigg[D_{xx} \left(\frac{ij(1+i)(1+j)}{i+j-1} \right)$$

$$+ kL^3 \Bigg] (L)^{i+j-1} d_j - q_0 \left(\frac{1}{i+2} \right) (L)^{i+2}. \qquad (3.9.37)$$

These equations can be solved for any chosen values of M and N.

To obtain numerical results, we assume $M = N$ and use the following geometric and material data:

$$L = 10.0 \text{ in.}, \quad b = 1 \text{ in.}, \quad h = 0.1 \text{ in.},$$

$$E_1 = 30 \times 10^6 \text{ psi}, \quad E_2 = 10 \times 10^6 \text{ psi}, \quad \nu = 0.3 \qquad (3.9.38)$$

The results (magnitudes are amplified) are normalized with respect to L and q_0. Figures 3.9.5 and 3.9.6 contain plots of the normalized displacements $\bar{u}$ and $\bar{w}$ versus x/L for the FGM beams with the power-law index $n = 0, 1$, and 2 for the case $k = 0$ (i.e., cantilever beam). The one-parameter Ritz solutions (being linear and quadratic for u and w, respectively) are in significant error, but two-parameter solutions are very close to the analytical solutions (for $n = 0$, the two-parameter solution is almost exact).

Plots of $\bar{w}$ versus x/L are presented in Figs. 3.9.7 and 3.9.8 for various values of the stiffness ratio $\mu = kL^3/D_{xx}$. The ratio $\mu \geq D_{xx}$ corresponds to the propped cantilever beam (the one-parameter Ritz solutions are negligibly small, as the chosen Ritz approximations do not satisfy the boundary condition $w(L) = 0$), and $\mu < D_{xx}$ corresponds to the spring-supported beams.

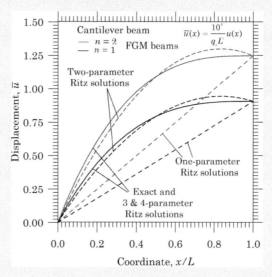

Figure 3.9.5: A comparison of the dimensionless Ritz solutions $\bar{u} = u \times (10^4/q_0 L^3)$ with the exact solution for the FGM cantilever beam under uniform load.

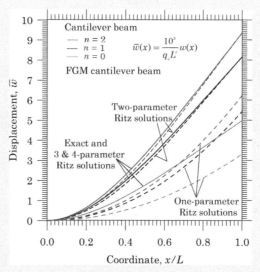

Figure 3.9.6: A comparison of the dimensionless Ritz solutions $\bar{w} = w \times (10^2/q_0 L^4)$ with the exact solution for the FGM cantilever beam ($k = 0$) under uniform load (note that the one-parameter Ritz solution for $w(x)$ is quadratic).

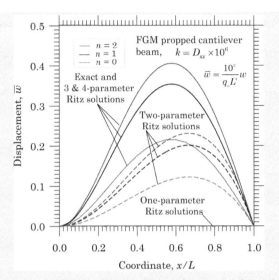

Figure 3.9.7: A comparison of the Ritz solutions $\bar{w} = w \times (10^2/q_0 L^4)$ with the exact solution for the FGM propped cantilever beam $(kL^3/D_{xx} = 10^6)$ under uniform load.

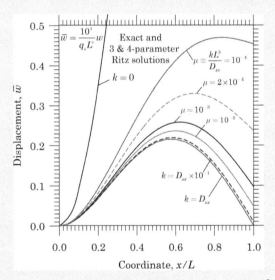

Figure 3.9.8: A comparison of the Ritz solutions $\bar{w} = w \times (10^2/q_0 L^4)$ with the exact solution for the spring-supported cantilever beam under uniform load; results are shown for a various ratios of $\mu = kL^3/D_{xx} < 10^6$.

Example 3.9.4

Consider a functionally graded beam of length L and constant stiffness A_{xx}, B_{xx}, and D_{xx}, loaded with distributed transverse load of $q(x)$. The beam is subjected to force boundary conditions on both ends of the beam (see Fig. 3.9.9). Develop the lowest order Ritz formulation using an algebraic polynomial and the classical beam theory (CBT).

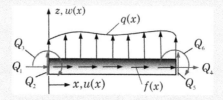

Figure 3.9.9: An FGM beam member with end loads and distributed loads.

Solution: This example is presented here to illustrate how the Ritz method can be used to formulate discrete equations in terms of the duality pairs (u, N_{xx}), (w, N_{xz}), and $(-dw/dx, M_{xx})$ at the ends of the beam member. This also serves as a prelude to the displacement-based finite element for the CBT.

The governing differential equations as per the CBT for the FGM beam are [see Eqs. (3.7.1) and (3.7.2), without the thermal forces N_{xx}^T and M_{xx}^T but including the couple stress effect]

$$-\frac{d}{dx}\left(A_{xx}\frac{du}{dx} - B_{xx}\frac{d^2w}{dx^2}\right) = f(x), \quad 0 < x < L, \tag{3.9.39}$$

$$-\frac{d^2}{dx^2}\left(B_{xx}\frac{du}{dx} - D_{xx}^e\frac{d^2w}{dx^2}\right) = q(x), \quad 0 < x < L, \tag{3.9.40}$$

where $D_{xx}^e = D_{xx} + A_{xy}$ and $(f(x), q(x))$ are the axial and transverse distributed forces (measured per unit length). We wish to seek the lowest order Ritz approximation of these two equations using algebraic polynomials.

From Eqs. (3.7.9) and (3.7.11), it is clear that the homogeneous solutions (i.e., when $f = 0$ and $q = 0$) for $u(x)$ and $w(x)$ are quadratic and cubic polynomials, respectively. In order to have the possibility of getting the exact solution, one must employ this degree of polynomials. However, since we plan to write the parameters in terms of the duality variables at the ends of the beam, we assume Ritz approximations of the type

$$u(x) \approx U_2(x) = c_1 + c_2 x, \tag{3.9.41}$$

$$w(x) \approx W_4(x) = d_1 + d_2 x + d_3 x^2 + d_4 x^3, \tag{3.9.42}$$

where the parameters c_i and d_i will expressed in terms of the generalized displacements at the ends of the beam (Δ_i is in the same direction as the Q_i; (Δ_i, Q_i),

$i = 1, 2, \ldots, 6$, are the duality pairs) as follows:

$$\Delta_1 \equiv U_2(0) = c_1, \quad \Delta_4 \equiv U_2(L) = c_1 + c_2 L \;\rightarrow\; c_2 = \frac{\Delta_4 - \Delta_1}{L}, \qquad (3.9.43)$$

$$\Delta_2 \equiv W_4(0) = d_1, \quad \Delta_3 \equiv -\frac{dW_4}{dx}\bigg|_{x=0} = -d_2,$$

$$\Delta_5 \equiv W_4(L) = d_1 + d_2 L + d_3 L^2 + d_4 L^3, \qquad (3.9.44)$$

$$\Delta_6 \equiv -\frac{dW_4}{dx}\bigg|_{x=L} = -d_2 - 2d_3 L - 3d_4 L^2,$$

Expressing d_3 and d_4 in terms of Δ_i, we obtain

$$d_3 = \frac{3}{L^2}\left(\Delta_5 - \Delta_2 + L\Delta_3\right) + \frac{1}{L}\left(\Delta_6 - \Delta_3\right),$$

$$d_4 = -\frac{1}{L^2}\left(\Delta_6 - \Delta_3\right) - \frac{2}{L^3}\left(\Delta_5 - \Delta_2 + L\Delta_3\right).$$

Then Eqs. (3.9.41) and (3.9.42) take the form

$$U_2(x) = \left(1 - \frac{x}{L}\right)\Delta_1 + \left(\frac{x}{L}\right)\Delta_4$$
$$\equiv \psi_1(x)\Delta_1 + \psi_2(x)\Delta_4, \qquad (3.9.45)$$

$$W_4(x) = \left(1 - 3\frac{x^2}{L^2} + 2\frac{x^3}{L^3}\right)\Delta_2 - L\frac{x}{L}\left(1 - \frac{x}{L}\right)^2\Delta_3$$
$$+ \left(3\frac{x^2}{L^2} - 2\frac{x^3}{L^3}\right)\Delta_5 + L\left(\frac{x^2}{L^2} - \frac{x^3}{L^3}\right)\Delta_6$$

$$\equiv \varphi_1(x)\Delta_2 + \varphi_2(x)\Delta_3 + \varphi_3(x)\Delta_5 + \varphi_4(x)\Delta_6, \qquad (3.9.46)$$

where ψ_i are the linear (Lagrange) interpolation functions and φ_i are the Hermite cubic interpolation functions:

$$\psi_1(x) = 1 - \frac{x}{L}, \quad \psi_2(x) = \frac{x}{L}, \qquad (3.9.47a)$$

$$\varphi_1(x) = 1 - 3\left(\frac{x}{L}\right)^2 + 2\left(\frac{x}{L}\right)^3,$$

$$\varphi_2(x) = -L\frac{x}{L}\left[1 - 2\left(\frac{x}{L}\right) + \left(\frac{x}{L}\right)^2\right],$$

$$\varphi_3(x) = \left(\frac{x}{L}\right)^2\left(3 - 2\frac{x}{L}\right),$$

$$\varphi_4(x) = L\left(\frac{x}{L}\right)^2\left(1 - \frac{x}{L}\right).$$

$$(3.9.47b)$$

The principle of the minimum total potential energy for this case is equivalent to the following two statements [see Eqs. (3.9.31) and (3.9.32)]:

$$0 = \int_0^L \left[\left(A_{xx}\frac{du}{dx} - B_{xx}\frac{d^2w}{dx^2}\right)\frac{d\delta u}{dx} - f\,\delta u\right] dx - \delta\Delta_1\,Q_1 - \delta\Delta_4\,Q_4, \qquad (3.9.48a)$$

$$0 = \int_0^L \left[\left(-B_{xx}\frac{du}{dx} + D_{xx}^e\frac{d^2w}{dx^2}\right)\frac{d^2\delta w}{dx^2} - q\,\delta w\right] dx$$
$$- \delta\Delta_2\,Q_2 - \delta\Delta_3\,Q_3 - \delta\Delta_5\,Q_5 - \delta\Delta_6\,Q_6, \qquad (3.9.48b)$$

where

$$Q_1 = -N_{xx}(0), \quad Q_2 = -\left.\frac{dM_{xx}}{dx}\right|_{x=0}, \quad Q_3 = -M_{xx}(0),$$

$$Q_4 = N_{xx}(L), \quad Q_5 = \left.\frac{dM_{xx}}{dx}\right|_{x=L}, \quad Q_6 = M_{xx}(L).$$

(3.9.49)

Substituting the expansions in Eqs. (3.9.45) for $u(x)$ and (3.9.46) for $w(x)$ in Eqs. (3.9.48a) and (3.9.48b) and carrying out the integration with respect to x, we obtain the following discrete equations:

$$\begin{bmatrix} \mathbf{K}^{11} & \mathbf{K}^{12} \\ \mathbf{K}^{21} & \mathbf{K}^{22} \end{bmatrix} \left[\begin{Bmatrix} \mathbf{\Delta}^1 \\ \mathbf{\Delta}^2 \end{Bmatrix} \right] = \begin{Bmatrix} \mathbf{F}^1 \\ \mathbf{F}^2 \end{Bmatrix},$$

(3.9.50a)

where (for $i, j = 1, 2$; and $k, p, q = 1, 2, 3, 4$)

$$K_{ij}^{11} = A_{xx} \int_0^L \frac{d\psi_i}{dx} \frac{d\psi_j}{dx} \, dx, \qquad K_{jk}^{12} = -B_{xx} \int_0^L \frac{d\psi_j}{dx} \frac{d^2\varphi_k}{dx^2} \, dx,$$

$$K_{kj}^{21} = -B_{xx} \int_0^L \frac{d^2\varphi_k}{dx^2} \frac{d\psi_j}{dx} \, dx = K_{jk}^{12}, \quad K_{pq}^{22} = D_{xx}^e \int_0^L \frac{d^2\varphi_p}{dx^2} \frac{d^2\varphi_q}{dx^2} \, dx, \quad (3.9.50b)$$

$$f_i = \int_0^L f(x)\psi_i(x) \, dx \ (i = 1, 2), \quad q_p = \int_0^L q(x)\varphi_p(x) \, dx.$$

For the choice of functions ψ_i and φ_i in Eqs. (3.9.47a) and (3.9.47b), the numerical values of the coefficients of various matrices are

$$\mathbf{K}^{11} = \frac{A_{xx}}{L} \begin{bmatrix} 1 & -1 \\ -1 & 1 \end{bmatrix}, \quad \mathbf{K}^{12} = -\frac{B_{xx}}{L} \begin{bmatrix} 0 & -1 & 0 & 1 \\ 0 & 1 & 0 & -1 \end{bmatrix},$$

(3.9.51a)

$$\mathbf{K}^{22} = \frac{2D_{xx}}{L^3} \begin{bmatrix} 6 & -3L & -6 & -3L \\ -3L & 2L^2 & 3L & L^2 \\ -6 & 3L & 6 & 3L \\ -3L & L^2 & 3L & 2L^2 \end{bmatrix}, \quad \mathbf{K}^{21} = -\frac{B_{xx}}{L} \begin{bmatrix} 0 & 0 \\ -1 & 1 \\ 0 & 0 \\ 1 & -1 \end{bmatrix}, \quad (3.9.51b)$$

$$\mathbf{\Delta}^1 = \begin{Bmatrix} \Delta_1 \\ \Delta_4 \end{Bmatrix}, \quad \mathbf{\Delta}^2 = \begin{Bmatrix} \Delta_2 \\ \Delta_3 \\ \Delta_5 \\ \Delta_6 \end{Bmatrix}, \quad \mathbf{F}^1 = \begin{Bmatrix} f_1 + Q_1 \\ f_2 + Q_4 \end{Bmatrix}, \quad \mathbf{F}^2 = \begin{Bmatrix} q_1 + Q_2 \\ q_2 + Q_3 \\ q_3 + Q_5 \\ q_4 + Q_6 \end{Bmatrix}.$$

(3.9.51c)

This completes the development of the Ritz equations for beams subjected to point generalized forces at the ends and distributed axial and transverse loads over the span. Next, we illustrate the use of the Ritz method to determine transient response of beams (see Section 3.8.5).

Example 3.9.5 ⎯⎯⎯⎯⎯⎯⎯⎯⎯⎯⎯⎯⎯⎯⎯⎯⎯⎯⎯⎯⎯⎯⎯⎯⎯⎯

Consider a functionally graded beam of length L and constant stiffness A_{xx}, B_{xx}, and D_{xx}, and loaded with axial load $f(x,t)$ and transverse load of $q(x,t)$. Formulate the Ritz discrete equations to determine the transient response of an FGM beam.

Solution: The variational statement of the problem at hand is provided by Hamilton's principle [see Eqs. (3.3.15) and (3.3.20)]:

$$
0 = \int_0^T \int_0^L \left[m_0 \left(\frac{\partial \delta u}{\partial t} \frac{\partial u}{\partial t} + \frac{\partial \delta w}{\partial t} \frac{\partial w}{\partial t} \right) - m_1 \left(\frac{\partial \delta u}{\partial t} \frac{\partial^2 w}{\partial x \partial t} + \frac{\partial^2 \delta w}{\partial x \partial t} \frac{\partial u}{\partial t} \right) \right.
$$
$$
- \frac{\partial \delta u}{\partial x} \left(A_{xx} \frac{\partial u}{\partial x} - B_{xx} \frac{\partial^2 w}{\partial x^2} \right) + \frac{\partial^2 \delta w}{\partial x^2} \left(B_{xx} \frac{\partial u}{\partial x} - D_{xx}^e \frac{\partial^2 w}{\partial x^2} \right)
$$
$$
\left. + m_2 \frac{\partial^2 \delta w}{\partial x \partial t} \frac{\partial^2 w}{\partial x \partial t} - c_f \delta w w + f \, \delta u + q \, \delta w \right] dx dt, \qquad (3.9.52)
$$

where $D_{xx}^e = D_{xx} + A_{xy}$, A_{xy} being the stiffness due to the couple stress [see Eq. (3.4.7)]. Since there are no specified nonzero generalized point forces in this problem, Eq. (3.9.52) is valid for an FGM beam with arbitrary geometric boundary conditions. The boundary conditions involve specifying one element of each of the following three pairs:

$$
(u, N_{xx}), \quad \left(w, \frac{d\bar{M}_{xx}}{dx} \right), \quad \left(\frac{dw}{dx}, \bar{M}_{xx} \right), \quad \text{where } \bar{M}_{xx} = M_{xx} + P_{xx}. \qquad (3.9.53)
$$

Let the (M,N)-parameter Ritz approximation is of (u, w) in the form:

$$
u(x, t) \approx U_M(x, t) = \sum_{j=1}^M U_j(t) \psi_j(x),
$$
$$
w(x, t) \approx W_N(x, t) = \sum_{j=1}^N W_j(t) \phi_j(x), \qquad (3.9.54)
$$

The functions $\phi_0(x)$ and ψ_0 are zero in the present case because the specified geometric boundary conditions are homogeneous.

The variational statement in Eq. (3.9.52) must be modified to account for the fact that the chosen Ritz approximation is one of "separation of variables" type; that is, time and space are separated, with the coefficients being functions of time t only and the approximation functions being functions of x only. Consequently, we relieve $(\delta u, \delta w)$ of any time derivatives so that the fundamental lemma of calculus of variations can be utilized to extract two variational statements from Eq. (3.9.52). Thus, we have

$$
0 = \int_0^T \int_0^L \left[\delta u \left(-m_0 \frac{\partial^2 u}{\partial t^2} + m_1 \frac{\partial^3 w}{\partial x \partial t^2} \right) - \frac{\partial \delta u}{\partial x} \left(A_{xx} \frac{\partial u}{\partial x} - B_{xx} \frac{\partial^2 w}{\partial x^2} \right) + f \, \delta u \right.
$$
$$
- m_0 \, \delta w \frac{\partial^2 w}{\partial t^2} + \frac{\partial \delta w}{\partial x} \left(-m_2 \frac{\partial^3 w}{\partial x \partial t^2} + m_1 \frac{\partial^2 u}{\partial t^2} \right)
$$
$$
\left. + \frac{\partial^2 \delta w}{\partial x^2} \left(B_{xx} \frac{\partial u}{\partial x} - D_{xx}^e \frac{\partial^2 w}{\partial x^2} \right) - c_f \delta w w + q \, \delta w \right] dx dt, \qquad (3.9.55)
$$

where we have made use of the requirement of Hamilton's principle that $\delta u(x, 0) = \delta u(x, T) = 0$ and $\delta w(x, 0) = \delta w(x, T) = 0$.

The variational statement in Eq. (3.9.55) is equivalent to the following two variational statements, obtained by collecting the terms involving δu and δw separately:

$$0 = \int_0^T \int_0^L \left[\delta u \left(-m_0 \frac{\partial^2 u}{\partial t^2} + m_1 \frac{\partial^3 w}{\partial x \partial t^2} \right) - \frac{\partial \delta u}{\partial x} \left(A_{xx} \frac{\partial u}{\partial x} - B_{xx} \frac{\partial^2 w}{\partial x^2} \right) + f \, \delta u \right] dx dt,$$

(3.9.56)

$$0 = \int_0^T \int_0^L \left[-m_0 \, \delta w \frac{\partial^2 w}{\partial t^2} + \frac{\partial \delta w}{\partial x} \left(-m_2 \frac{\partial^3 w}{\partial x \partial t^2} + m_1 \frac{\partial^2 u}{\partial t^2} \right) \right.$$
$$\left. + \frac{\partial^2 \delta w}{\partial x^2} \left(B_{xx} \frac{\partial u}{\partial x} - D_{xx}^e \frac{\partial^2 w}{\partial x^2} \right) - c_f \delta w w + q \, \delta w \right] dx dt.$$

(3.9.57)

Next, we substitute the approximations U_M and W_N from Eq. (3.9.54) for u and w, and $\delta u = \psi_i$ and $\delta w = \phi_i$ into the variational statements in Eqs. (3.9.56) and (3.9.57) and obtain

$$0 = \int_0^T \sum_{i=1}^M \delta U_i \left\{ \int_0^L \left[-m_0 \psi_i \left(\sum_{j=1}^M \psi_j \frac{d^2 U_j}{dt^2} \right) + m_1 \psi_i \left(\sum_{j=1}^N \frac{d\phi_j}{dx} \frac{d^2 W_j}{dt^2} \right) \right.\right.$$
$$\left.\left. - A_{xx} \frac{d\psi_i}{\partial x} \left(\sum_{j=1}^M \frac{d\psi_j}{dx} U_j \right) + B_{xx} \frac{d\psi_i}{\partial x} \left(\sum_{j=1}^N \frac{d^2 \phi_j}{dx^2} W_j \right) + f \psi_i \right] dx \right\} dt,$$

(3.9.58)

$$0 = \sum_{i=1}^N \delta W_i \left\{ \int_0^L \left[-m_0 \phi_i \left(\sum_{j=1}^N \phi_j \frac{d^2 W_j}{dt^2} \right) - m_2 \frac{d\phi_i}{dx} \left(\sum_{j=1}^N \frac{d\phi_j}{dx} \frac{d^2 W_j}{dt^2} \right) \right.\right.$$
$$+ m_1 \frac{d\phi_i}{dx} \left(\sum_{j=1}^M \psi_j \frac{d^2 U_j}{dt^2} \right) + B_{xx} \frac{d^2 \phi_i}{dx^2} \left(\sum_{j=1}^M \frac{d\psi_j}{dx} U_j \right)$$
$$\left.\left. - D_{xx}^e \frac{d^2 \phi_i}{dx^2} \left(\sum_{j=1}^N \frac{d^2 \phi_j}{dx^2} W_j \right) - c_f \phi_i \left(\sum_{j=1}^N \phi_j W_j \right) + q \, \phi_i \right] dx \right\} dt.$$

(3.9.59)

Since $\{\delta U_i\}$ and $\{\delta W_i\}$ are sets of independent variations, Eqs. (3.9.58) and (3.9.59) require that the expressions in the curl braces be zero independently, giving the equations:

$$0 = \sum_{j=1}^M \left[-\left(\int_0^L m_0 \, \psi_i \psi_j \, dx \right) \frac{d^2 U_j}{dt^2} - \left(\int_0^L A_{xx} \frac{d\psi_i}{dx} \frac{d\psi_j}{dx} dx \right) U_j \right]$$
$$+ \sum_{j=1}^N \left[\left(\int_0^L m_1 \psi_i \frac{d\phi_j}{dx} dx \right) \frac{d^2 W_j}{dt^2} + \left(\int_0^L B_{xx} \frac{d\psi_i}{dx} \frac{d^2 \phi_j}{dx^2} dx \right) W_j \right] + \int_0^a f \psi_i \, dx.$$

(3.9.60)

$$0 = \sum_{j=1}^{N} \left\{ \left(\int_0^L B_{xx} \frac{d^2\phi_i}{dx^2} \frac{d\psi_j}{dx} dx \right) U_j + \left(\int_0^L m_1 \frac{d\phi_i}{dx} \psi_j \, dx \right) \frac{d^2 U_j}{dt^2} \right\}$$

$$- \sum_{j=1}^{M} \left\{ \left[\int_0^L \left(m_0 \phi_i \phi_j + m_2 \frac{d\phi_i}{dx} \frac{d\phi_j}{dx} \right) dx \right] \frac{d^2 W_j}{dt^2} \right.$$

$$\left. + \left[\int_0^L \left(D_{xx}^e \frac{d^2\phi_i}{dx^2} \frac{d^2\phi_j}{dx^2} + c_f \phi_i \phi_j \right) dx \right] W_j \right\} + \int_0^L q \phi_i \, dx, \quad (3.9.61)$$

or

$$0 = \sum_{j=1}^{M} M_{ij}^{11} \frac{d^2 U_j}{dt^2} + \sum_{j=1}^{N} M_{ij}^{12} \frac{d^2 W_j}{dt^2} + \sum_{j=1}^{M} K_{ij}^{11} U_j + \sum_{j=1}^{N} K_{ij}^{12} W_j - F_i^1,$$

$$\qquad\qquad\qquad\qquad\qquad\qquad\qquad\qquad\qquad\qquad (3.9.62)$$

$$0 = \sum_{j=1}^{M} M_{ij}^{21} \frac{d^2 U_j}{dt^2} + \sum_{j=1}^{N} M_{ij}^{22} \frac{d^2 W_j}{dt^2} + \sum_{j=1}^{M} K_{ij}^{21} U_j + \sum_{j=1}^{N} K_{ij}^{22} W_j - F_i^2,$$

where

$$M_{ij}^{11} = \int_0^L m_0 \psi_i \psi_j \, dx, \qquad M_{ij}^{12} = -\int_0^a m_1 \psi_i \frac{d\phi_j}{dx} dx,$$

$$M_{ij}^{21} = -\int_0^L m_1 \frac{d\phi_i}{dx} \psi_j \, dx, \qquad M_{ij}^{22} = \int_0^L \left(m_0 \phi_i \phi_j + m_2 \frac{d\phi_i}{dx} \frac{d\phi_j}{dx} \right) dx,$$

$$K_{ij}^{11} = \int_0^L A_{xx} \frac{d\psi_i}{dx} \frac{d\psi_j}{dx} dx, \qquad K_{ij}^{12} = -\int_0^L B_{xx} \frac{d\psi_i}{\partial x} \frac{d^2\phi_j}{dx^2} dx, \qquad (3.9.63)$$

$$K_{ij}^{21} = -\int_0^L B_{xx} \frac{d^2\phi_i}{dx^2} \frac{d\psi_j}{dx} dx, \qquad K_{ij}^{22} = \int_0^L \left(D_{xx}^e \frac{d^2\phi_i}{dx^2} \frac{d^2\phi_j}{dx^2} + c_f \phi_i \phi_j \right) dx,$$

$$F_i^1 = \int_0^L f \psi_i \, dx, \quad F_i^2 = \int_0^L q \, \phi_i \, dx.$$

Equations (3.9.62) can be expressed in matrix form as

$$\begin{bmatrix} \mathbf{M}^{11} & \mathbf{M}^{12} \\ \mathbf{M}^{21} & \mathbf{M}^{22} \end{bmatrix} \left\{ \begin{matrix} \ddot{\mathbf{U}} \\ \ddot{\mathbf{W}} \end{matrix} \right\} + \begin{bmatrix} \mathbf{K}^{11} & \mathbf{K}^{12} \\ \mathbf{K}^{21} & \mathbf{K}^{22} \end{bmatrix} \left\{ \begin{matrix} \mathbf{U} \\ \mathbf{W} \end{matrix} \right\} = \left\{ \begin{matrix} \mathbf{F}^1 \\ \mathbf{F}^2 \end{matrix} \right\}, \qquad (3.9.64)$$

Once the functions ψ_i $(i = 1, 2, \ldots, M)$ and ϕ_i $(i = 1, 2, \ldots, N)$ are selected, one can evaluate the coefficients $M_{ij}^{\alpha\beta}$, $K_{ij}^{\alpha\beta}$, and F_i^α.

Equations (3.9.64) are second-order differential equations in time, and they can be solved either by a numerical method or using the Laplace transform method discussed in Section 3.8.5. Here we discuss the use of Taylor series-based approximation scheme, namely, the Newmark scheme, to formulate the final algebraic equations resulting from Eq. (3.9.63). First, we rewrite Eq. (3.9.64) in the form

$$\mathbf{M}\ddot{\boldsymbol{\Delta}} + \mathbf{K}\boldsymbol{\Delta} = \mathbf{F}, \quad 0 < t \leq T \qquad (3.9.65a)$$

subjected to the following initial conditions:

$$\boldsymbol{\Delta}(0) = \boldsymbol{\Delta}^0, \quad \dot{\boldsymbol{\Delta}}(0) = \boldsymbol{\Delta}^1. \qquad (3.9.65b)$$

Let us consider the following (α, γ)-*family of approximation*, where the function and its first time derivative are approximated as

$$\boldsymbol{\Delta}^{s+1} \approx \boldsymbol{\Delta}^s + \delta t \, \dot{\boldsymbol{\Delta}}^s + \frac{1}{2}(\delta t)^2 \left[(1-\gamma)\ddot{\boldsymbol{\Delta}}^s + \gamma \ddot{\boldsymbol{\Delta}}^{s+1} \right], \qquad (3.9.66\text{a})$$

$$\dot{\boldsymbol{\Delta}}^{s+1} \approx \dot{\boldsymbol{\Delta}}^s + a_2 \ddot{\boldsymbol{\Delta}}^s + a_1 \ddot{\boldsymbol{\Delta}}^{s+1}. \qquad (3.9.66\text{b})$$

Here α and γ are parameters that determine the stability and accuracy of the scheme. Equations (3.9.66a) and (3.9.66b) are Taylor's series expansions of $\boldsymbol{\Delta}^{s+1}$ and $\dot{\boldsymbol{\Delta}}^{s+1}$, respectively, about $t = t_s$. Here $\boldsymbol{\Delta}^s$ denotes the value of $\boldsymbol{\Delta}(t)$ at $t = t_s = s\,\delta t$, δt being the time step size.

The fully discretized form of Eq. (3.9.64) are obtained using the approximations introduced in Eqs. (3.9.66a) and (3.9.66b). First, we eliminate $\ddot{\boldsymbol{\Delta}}^{s+1}$ from Eqs. (3.9.66a) and (3.9.66b) and write the result for $\dot{\boldsymbol{\Delta}}^{s+1}$:

$$\dot{\boldsymbol{\Delta}}^{s+1} = a_6 \left(\boldsymbol{\Delta}^{s+1} - \boldsymbol{\Delta}^s \right) - a_7 \dot{\boldsymbol{\Delta}}^s - a_8 \ddot{\boldsymbol{\Delta}}^s, \qquad (3.9.67)$$

$$a_6 = \frac{2\alpha}{\gamma \Delta t}, \quad a_7 = \frac{2\alpha}{\gamma} - 1, \quad a_8 = \left(\frac{\alpha}{\gamma} - 1 \right) \Delta t. \qquad (3.9.68)$$

Now pre-multiplying Eq. (3.9.65a) with $\mathbf{M}$ and substituting for $\mathbf{M}\ddot{\boldsymbol{\Delta}}^{s+1}$ from Eq. (3.9.65a) (evaluated at $s+1$), we obtain

$$\left(\mathbf{M} + \tfrac{\gamma(\delta t)^2}{2}\mathbf{K} \right) \boldsymbol{\Delta}^{s+1} = \mathbf{M}\mathbf{b}^s + \tfrac{\gamma(\delta t)^2}{2}\mathbf{F}^{s+1}, \qquad (3.9.69\text{a})$$

where

$$\mathbf{b}^s = \boldsymbol{\Delta}^s + \delta t \, \dot{\boldsymbol{\Delta}}^s + \frac{1}{2}(1-\gamma)(\delta t)^2 \, \ddot{\boldsymbol{\Delta}}^s. \qquad (3.9.69\text{b})$$

Now, multiplying throughout with $2/[\gamma(\Delta t)^2]$ we arrive at

$$\left(\tfrac{2}{\gamma(\delta t)^2}\mathbf{M} + \mathbf{K} \right) \boldsymbol{\Delta}^{s+1} = \tfrac{2}{\gamma(\Delta t)^2}\mathbf{M}\mathbf{b}_s + \mathbf{F}^{s+1}. \qquad (3.9.69\text{c})$$

Using Eq. (3.9.67) for $\dot{\boldsymbol{\Delta}}^{s+1}$ in Eq. (3.9.69c) and collecting terms, we obtain the recursive relation:

$$\hat{\mathbf{K}}\mathbf{u}^{s+1} = \hat{\mathbf{F}}^{s,s+1}, \qquad (3.9.70\text{a})$$

where

$$\hat{\mathbf{K}} = \mathbf{K} + a_3 \mathbf{M}, \quad \hat{\mathbf{F}}^{s,s+1} = \mathbf{F}^{s+1} + \mathbf{M}\bar{\boldsymbol{\Delta}}^s, \quad \bar{\boldsymbol{\Delta}}^s = a_3 \boldsymbol{\Delta}^s + a_4 \dot{\boldsymbol{\Delta}}^s + a_5 \ddot{\boldsymbol{\Delta}}^s,$$

$$a_3 = \frac{2}{\gamma(\delta t)^2}, \quad a_4 = a_3 \, \delta t, \quad a_5 = \frac{1}{\gamma} - 1.$$

The two special cases of the (α, γ)-family of approximation are: (1) constant-average acceleration method $(\alpha = \gamma = 1/2)$, which is known as the Newmark scheme, and (2) linear acceleration method $(\alpha = 1/2$ and $\gamma = 1/3)$. For these two cases, the fully discretized equations in (3.9.70a) take the following explicit forms:

Constant-average acceleration scheme $(\alpha = \gamma = 1/2)$:

$$\left(\tfrac{4}{(\delta t)^2}\mathbf{M} + \mathbf{K} \right) \boldsymbol{\Delta}^{s+1} = \mathbf{F}^{s+1} + \tfrac{4}{(\delta t)^2}\mathbf{M}\boldsymbol{\Delta}^s + \tfrac{4}{\delta t}\mathbf{M}\dot{\boldsymbol{\Delta}}^s + \mathbf{M}\ddot{\boldsymbol{\Delta}}^s. \qquad (3.9.71)$$

Linear acceleration scheme ($\alpha = 1/2$ and $\gamma = 1/3$):

$$\left(\tfrac{6}{(\Delta t)^2}\mathbf{M} + \mathbf{K}\right)\mathbf{u}^{s+1} = \mathbf{F}^{s+1} + \tfrac{6}{(\delta t)^2}\mathbf{M}\mathbf{\Delta}^s + \tfrac{6}{\delta t}\mathbf{M}\dot{\mathbf{\Delta}}^s + 2\mathbf{M}\ddot{\mathbf{u}}^s. \tag{3.9.72}$$

For all schemes in which $\gamma \geq \alpha \geq 1/2$ are numerically stable (i.e., the error introduced during each time step does not grow indefinitely); for $\alpha \geq \tfrac{1}{2}$ and $\gamma < \alpha$, the stability requirement is

$$\Delta t \leq \Delta t_{cr} = \left[\tfrac{1}{2}\omega_{\max}^2(\alpha - \gamma)\right]^{-1/2}, \tag{3.9.73}$$

where $\omega_{\max}^2 = \lambda_{\max}$ is the maximum eigenvalue of the system in Eq. (3.9.65a). The stability characteristics of various schemes are as follows:

- $\alpha = \tfrac{1}{2}$, $\gamma = \tfrac{1}{2}$, the constant-average acceleration method (stable),
- $\alpha = \tfrac{1}{2}$, $\gamma = \tfrac{1}{3}$, the linear acceleration method (conditionally stable), (3.9.74)
- the central difference method (conditionally stable).

3.9.3   THE WEIGHTED-RESIDUAL METHODS

Here, we discuss the general method of weighted–residuals and its special cases. We begin with the model equation (3.9.1):

$$-\frac{d}{dx}\left(E(x)A(x)\frac{du}{dx}\right) + c(x)u(x) = f(x) \quad \text{in } 0 \leq x \leq a, \tag{3.9.75}$$

with the boundary conditions of the form

$$u(0) = 0, \quad \left[EA\frac{du}{dx} + ku\right]_{x=a} = P. \tag{3.9.76}$$

We seek an approximation of u in the same form as before:

$$u(x) \approx U_N(x) = \sum_{i=1}^{N} c_i\phi_i(x) + \phi_0(x). \tag{3.9.77}$$

Substituting U_N for u into Eq. (3.9.75), we obtain an expression that does not satisfy the original equation, leaving a *residual*:

$$\mathbb{R}_N \equiv -\frac{d}{dx}\left(E(x)A(x)\frac{dU_N}{dx}\right) + c(x)U_N(x) - f(x) \neq 0. \tag{3.9.78}$$

In the weighted-residual method, as the name suggests, we require $\mathbb{R}$ to be orthogonal (in the functions sense) to a set of N linearly independent set of functions, called the *weight functions* $\psi_i(x)$ in the following sense:

$$\int_0^a \psi_i \, \mathbb{R}_N(\mathbf{x}, \{c\}, \{\phi\}, f)dx = 0, \quad (i = 1, 2, \ldots, N). \tag{3.9.79}$$

Equation (3.9.79) provides N linearly independent relations among the c_i. Substituting R_N from Eq. (3.9.78) into Eq. (3.9.79), we obtain

$$0 = \int_0^a \psi_i \left[-\frac{d}{dx} \left(E(x)A(x)\frac{dU_N}{dx} \right) + c(x)U_N(x) - f \right] dx$$

$$= \sum_{j=1}^N A_{ij}c_j - b_i = 0 \Rightarrow \mathbf{Ac} = \mathbf{b}, \quad i = 1, 2, \ldots, N, \qquad (3.9.80)$$

where

$$\begin{aligned} A_{ij} &= \int_0^a \psi_i \left[-\frac{d}{dx} \left(E(x)A(x)\frac{d\phi_j}{dx} \right) + c(x)\phi_j \right] dx, \\ b_i &= \int_0^a \psi_i \left[\frac{d}{dx} \left(E(x)A(x)\frac{d\phi_0}{dx} \right) - c(x)\phi_0(x) + f \right] dx. \end{aligned} \qquad (3.9.81)$$

We note that the coefficient matrix $\mathbf{A}$ is unsymmetric.

Since the weighted-residual statement in Eq. (3.9.81) does not contain any boundary conditions of the problem, we must make sure that U_N satisfies *all* of the specified boundary conditions of the problem. This in turn places certain requirements on the approximation functions ϕ_0 and ϕ_j, which are different from those in the Ritz method. The approximation functions (ϕ_0, ϕ_j) and weight functions ψ_i in a weighted-residual method must satisfy the following requirements:

1. ϕ_0 is the lowest order function that satisfies *all* specified boundary conditions of the problem [of the type in Eq. (3.9.76)]; ϕ_0 is necessarily zero when the specified boundary conditions are all homogeneous.
2. $\phi_j (j = 1, 2, \ldots, N)$ must satisfy three conditions:
   (a) Each ϕ_j is *continuous* as required in the weighted-residual statement; that is, ϕ_j should be such that U_N yields a nonzero value of $R(U_N)$ (i.e., matrix $\mathbf{A}$ is invertible).
   (b) Each ϕ_j satisfies the *homogeneous form* of all specified (i.e., geometric as well as force) boundary conditions.
   (c) The set $\{\phi_j\}$ is *linearly independent* and *complete*.
3. ψ_i should be linearly independent.

There are two main differences between the approximation functions used in the Ritz method and those used in a weighted-residual method:

(1) *Continuity.* The approximation functions used in a weighted-residual method are required to have the same differentiability as in the differential equation, whereas those used in the Ritz method must be differentiable as required by the total potential energy functional. For example, in the case of a second-order differential equation, the total potential energy functional

contains only the first-order derivatives of the dependent variable. Thus, the approximation function ϕ_1 used in the weighted-residual method is required to be twice differentiable, whereas that used in the Ritz method has to be differentiable only once.

(2) *Boundary Conditions.* The approximation functions used in the weighted–residual method must satisfy the homogeneous form of the specified geometric as well as force boundary conditions, whereas those used in the Ritz method must satisfy the homogeneous form of only the specified essential boundary conditions (because the natural boundary conditions are already included in the total potential energy).

Both of these differences require ϕ_i used in a weighted-residual method to be of higher order than those used in the Ritz method. In an effort to satisfy the homogeneous form of all of the specified boundary conditions in a weighted-residual method, the order of the functions automatically go up.

Various special cases of the weighted-residual method differ from each other due to the choice of the weight function, ψ_i. Some of these are:

The Petrov–Galerkin method:	$\psi_i \neq \phi_i.$	
Galerkin's method:	$\psi_i = \phi_i.$	(3.9.82)
Least-squares method:	$\psi_i = \partial R/\partial c_i.$	
Collocation method:	$\psi_i = \delta(\mathbf{x} - \mathbf{x}_i).$	

Here $\delta(\cdot)$ denotes the Dirac delta function. Although the least-squares method is listed as a special case of the weighted-residual method here, it is based on the concept of minimizing the integral of the square of the equation residual due to the approximation. It can be viewed as a special case only in linear problems.

Sometimes there is a confusion that the Ritz method and the Galerkin method are the same. In the original form the two methods were proposed, they are not the same. The Ritz method is based on a weaker integral statement than the Galerkin method. However, these two methods can give the same set of final algebraic equations among the unknown coefficients c_i for the following three cases:

1. The specified boundary conditions of the problem are all the essential type, and therefore the requirements on ϕ_i and ϕ_0 in both methods are the same.
2. The problem has both essential and natural boundary conditions, but the approximate functions used in the Galerkin method are also used in the Ritz method.
3. The governing equations of the problems are even order, allowing integration by parts to reduce the Galerkin statement to that used in the Ritz method and using the same approximation functions in both cases. In this case, calling it a Galerkin method is not correct; it is a weak-form Galerkin method (or the Ritz method).

Example 3.9.6

Consider the simply supported (isotropic) beam problem of **Example 3.9.2**. Solve the problem using: (a) one-parameter Galerkin approximation with algebraic polynomials and (b) two-parameter collocation approximation with trigonometric functions.

Solution: (a) *Weighted–Residual Methods with Algebraic Polynomials.* Fist, we must determine the approximation functions needed for any weighted-residual method for this case. For the simply supported beam at hand, when the full beam is used, $\phi_0(x)$ is identically zero because all specified boundary conditions are homogeneous ($w = M_{xx} = 0$ at $x = 0, L$). Then $\phi_i(x)$ are determined such that the following conditions are met:

$$\phi_i(0) = \phi_i(L) = 0, \quad \left.\frac{d^2\phi_i}{dx^2}\right|_{x=0,L} = 0.$$

Since there are four conditions, We begin with a complete fourth-degree polynomial (with five constants) for $\phi_1(x)$:

$$\phi_1(x) = \alpha_0 + \alpha_1 x + \alpha_2 x^2 + \alpha_3 x^3 + \alpha_4 x^4$$

and find $\alpha_0 = 0, \alpha_2 = 0$ and

$$\alpha_1 + \alpha_3 L^2 + \alpha_4 L^3 = 0, \quad 6\alpha_3 + 12\alpha_4 L = 0,$$

from which we have

$$\alpha_3 = -2\alpha_4 L, \quad \alpha_1 = \alpha_4 L^3.$$

Thus, we have

$$\phi_1(x) = L^4\left(\frac{x}{L} - 2\frac{x^3}{L^3} + \frac{x^4}{L^4}\right). \tag{3.9.83}$$

Similar procedure can be used to determine $\phi_2(x)$ and so on. However, knowing the exact solution for a simply–supported beam under uniformly distributed load is a forth-order polynomial, we try one-parameter solutions using various methods.

(i) The Galerkin Method. Substituting $\phi_1(x)$ into the Galerkin weighted-integral statement for the problem, we obtain

$$0 = \int_0^L \phi_1(x)\left(EI\frac{d^4w}{dx^4} - q_0\right) dx = \int_0^L \phi_1(x)\left(24EIc_1 - q_0\right) dx \;\Rightarrow\; c_1 = \frac{q_0}{24EI}.$$

Thus, the one-parameter Galerkin solution is

$$W_{G1}(x) = \frac{q_0 L^4}{24EI}\left(\frac{x}{L} - 2\frac{x^3}{L^3} + \frac{x^4}{L^4}\right), \tag{3.9.84}$$

which coincides with the exact solution of the problem.

(ii) Other Weighted-Residual Methods. We note that the solution obtained here is independent of the particular weighted-residual method used because $A(W_1) - q_0$ is a constant. It can be shown that a one-parameter Ritz solution with ϕ_1 given by Eq. (3.9.83) also yields the same exact solution.

Half-Beam Model. If we exploit the solution symmetry about $x = L/2$, we can use one-half of the beam to solve the problem. In using the first half of the beam, $0 \leq x \leq L/2$, we must address the boundary conditions at $x = L/2$. The boundary conditions at the line of symmetry are that the slope is zero, $(dw/dx) = 0$, and the shear force is zero, $EI(d^3w/dx^3) = 0$. Hence, for this case, the approximation functions ϕ_i of the weighted-residual method must satisfy the conditions

$$\text{at } x = 0: \quad \phi_i = 0, \quad \frac{d^2\phi_i}{dx^2} = 0, \quad \text{and at } x = \frac{L}{2}: \quad \frac{d\phi_i}{dx} = 0, \quad \frac{d^3\phi_i}{dx^3} = 0.$$

Obviously, the function $\phi_1(x)$ of Eq. (3.9.82) satisfies the conditions above. Hence, we obtain the same exact solution using any of the weighted-residual methods.

(b) Collocation method with trigonometric functions. The functions $\phi_i(x) = \sin \frac{(2i-1)\pi x}{L}$ $(i = 1, 2, \ldots)$ satisfy the boundary conditions $(w = 0$ and $M_{xx} = 0$ at $x = 0, L)$ of the full beam. They also satisfy the boundary conditions of the half beam. The reason for selecting odd functions $\sin \pi x/L$, $\sin 3\pi x/L, \ldots$ is that the even functions will not contribute to the solution (also, the even functions will not satisfy the vanishing slope at $x = L/2$).

The two-parameter approximation is

$$W_2(x) = c_1 \sin \frac{\pi x}{L} + c_2 \sin \frac{3\pi x}{L}. \tag{3.9.85}$$

Due to the symmetry of the solution about $x = L/2$, it is sufficient to take the two points in the first half of the beam. The residual is

$$\mathbb{R}_N(x, c_1, c_2) = EI \frac{d^4 W_2}{dx^4} - q_0$$

$$= EI \left[c_1 \left(\frac{\pi}{L} \right)^4 \sin \frac{\pi x}{L} + c_2 \left(\frac{3\pi}{L} \right)^4 \sin \frac{3\pi x}{L} \right] - q_0.$$

Using collocation points at $x_1 = L/4$ and $x_2 = L/2$, we obtain

$$\mathbb{R}_N(L/4, c_1, c_2) = EI \left[c_1 \left(\frac{\pi}{L} \right)^4 \sin \frac{\pi}{4} + c_2 \left(\frac{3\pi}{L} \right)^4 \sin \frac{3\pi}{4} \right] - q_0 = 0,$$

$$\mathbb{R}_N(L/2, c_1, c_2) = EI \left[c_1 \left(\frac{\pi}{L} \right)^4 \sin \frac{\pi}{2} + c_2 \left(\frac{3\pi}{L} \right)^4 \sin \frac{3\pi}{2} \right] - q_0 = 0,$$

which yield

$$EI \left(\frac{\pi}{L} \right)^4 \begin{bmatrix} 1 & 81 \\ 1 & -81 \end{bmatrix} \begin{Bmatrix} c_1 \\ c_2 \end{Bmatrix} = q_0 \begin{Bmatrix} \sqrt{2} \\ 1 \end{Bmatrix}. \tag{3.9.86}$$

Solving these equations for c_1 and c_2, we obtain

$$c_1 = \frac{1 + \sqrt{2}}{2} \frac{q_0 L^4}{EI\pi^4}, \quad c_2 = -\frac{1 - \sqrt{2}}{162} \frac{q_0 L^4}{EI\pi^4}. \tag{3.9.87}$$

The two-parameter collocation solution becomes

$$W_2(x) = \frac{q_0 L^4}{162 EI\pi^4} \left[81(1 + \sqrt{2}) \sin \frac{\pi x}{L} + (-1 + \sqrt{2}) \sin \frac{3\pi x}{L} \right]. \tag{3.9.88}$$

The maximum deflection is

$$W_2(L/2) = \frac{82 + 80\sqrt{2}}{162}\frac{q_0 L^4}{EI\pi^4} = \frac{q_0 L^4}{80.8676EI} = \frac{4.7485}{384}\frac{q_0 L^4}{EI}, \tag{3.9.89}$$

which is about 5% lower compared with the exact value, $(5q_0 L^4/384EI) = (q_0 L^4/76.8EI)$.

If we have used $x = L/2$ and $x = 3L/4$, we would have obtained the same solution. We also note that we cannot use $x = L/4$ and $x = 3L/4$ because they both yield the same relations among c_1 and c_2. Thus, a judicious choice of the collocation points is necessary to obtain accurate solutions.

3.10 CHAPTER SUMMARY

This chapter is a most comprehensive treatment of straight beams, beginning with the development of governing equations to analytical and approximate solutions. In particular, the CBT, often known as the Euler–Bernoulli beam theory, is developed accounting for material variation through the beam height, the von Kármán nonlinearity, and the effect of the modified couple stress. In this theory, the transverse shear strain is neglected and therefore the transverse stress would be zero when computed using the constitutive relation $\sigma_{xz} = G\gamma_{xz}$. However, the transverse shear force and shear stress cannot be zero when beam are subjected to transverse forces. Therefore, they are computed using the equilibrium relation (as is known from a course on mechanics of materials):

$$N_{xz}(x, z) = \frac{M_{xx}(x)}{dx}, \quad \sigma_{xz} = \frac{V(x)Q(z)}{Ib}, \quad V = \frac{dM}{dx}, \tag{3.10.90}$$

where Q is the first moment of area and b is the width of the beam at z (the x-axis passes through the geometric centroid of the cross section). For a beam of cross-sectional dimensions $b \times h$ (h is the height), the first moment of area is given by

$$Q(z) = \frac{bh^2}{8}\left(1 - \frac{4z^2}{h^2}\right). \tag{3.10.91}$$

Exact solutions (by direct integration), the Navier solutions (which is is valid only for hinged–hinged FGM as well as homogeneous beams), and variational solutions of the theory for the linear case are presented. A number of numerical examples are presented to show the effect of geometry, boundary conditions, and material parameters. Interested readers may consult the books by Reddy [8, 9, 102, 113, 155], and references therein, for additional examples. Of course, the classical variational methods discussed in here have limitations when load or material variations are discontinuous. The finite element method discussed in Chapter 9 is a generalization of the Ritz method, which has no such limitations.

SUGGESTED EXERCISES

3.1 Derive the following linear equations of equilibrium by summing the forces and moments (i.e., using the vector approach) of a typical beam element, as shown in Fig. P3.1:

$$\frac{dN_{xx}}{dx} + f = 0,$$

$$\frac{dN_{xz}}{dx} + q = 0,$$

$$\frac{dM_{xx}}{dx} - N_{xz} = 0,$$

where N_{xx}, N_{xz}, and M_{xx} stress resultants defined in Eq. (3.3.3).

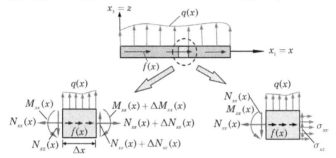

Figure P3.1

3.2 Verify the result in Eq. (3.3.5).

3.3 Using an inverse procedure, construct the Lagrangian, $L = K - (U + V_E)$, associated with Eqs. (3.5.3)–(3.5.5).

3.4 Determine the exact solution of the linearized equations, Eqs. (3.7.1) and (3.7.2), for a beam of length L with pinned–hinged boundary conditions and subjected to linearly varying distributed load, $q(x) = q_0(1 - x/L)$.

3.5 Repeat **Problem 3.4** for a cantilever beam.

3.6 Consider the system of beams shown in Fig. P3.6. The steel beam AB ($E_1 = 30 \times 10^6$ psi, $I_1 = 22$ in⁴.) is fixed at point B and connected at the other end A by a steel rod (diameter $d = 0.25$ in. and length 10 ft.) to the center of the wooden ($E_1.5 \times 10^6$ psi, $I_2 = 415$ in⁴.) beam DCE. The wooden beam carries a uniform load on intensity 420 lb/ft (acting down). Assuming that the steel rod (E_1) fits snugly between the two beams before the uniform load is placed on the beam DCE, determine: (a) force in the steel rod, (b) the maximum deflection at point C, and (c) the maximum bending moment and its location in beam DCE. *Hint:* Use the method of superposition to determine the force in the cable.

3.7 Consider a simply–supported (i.e., pinned–hinged) beam ABC shown in Fig. P4.6. The beam has a modulus of $E_1 = 200$ GPa, cross-sectional dimensions of width 164 mm and height 400 mm, and loaded with uniform load of intensity $q_0 = 30$ kN/m (acting down). Before the uniform distributed load is applied to the beam, there is a small gap of 0.02 m between the beam and the elastic post at point B at the center of the beam. Assuming that the modulus of elasticity of the post is the same as that of the beam and the diameter of the post is 40 mm, determine the: (a) force in the post at point B, (b) reactions at the supports A and C, and the (c) deflection $w(x)$ and its value at $x = 3$ m and at $x = 6$ m. *Hint:* Use the method of superposition to determine the force in the post.

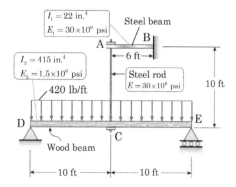

Figure P3.6

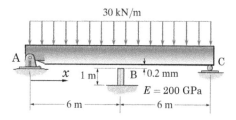

Figure P3.7

3.8 Determine the characteristic equation for buckling loads of a fixed (clamped)-hinged beam.

3.9 Determine the characteristic equation for natural frequencies of a fixed (clamped)-hinged beam.

CHAPTER 3 APPENDIX

The following integrals of $V_1(z)$ appearing in Eq. (3.4.7) are useful in evaluating the integrals for A_{xx}, B_{xx}, and so on for functionally graded beams and plates when the modulus of elasticity $E(z)$ is given by Eq. (3.4.5a):

$$\int_{-\frac{h}{2}}^{\frac{h}{2}} \left(\frac{1}{2} + \frac{z}{h} \right)^n dz = \frac{h}{1+n}, \tag{3A.1}$$

$$\int_{-\frac{h}{2}}^{\frac{h}{2}} \left(\frac{1}{2} + \frac{z}{h} \right)^n z\, dz = \frac{h^2 n}{2(1+n)(2+n)}, \tag{3A.2}$$

$$\int_{-\frac{h}{2}}^{\frac{h}{2}} \left(\frac{1}{2} + \frac{z}{h} \right)^n z^2\, dz = \frac{h^3(2+n+n^2)}{4(1+n)(2+n)(3+n)}, \tag{3A.3}$$

$$\int_{-\frac{h}{2}}^{\frac{h}{2}} \left(\frac{1}{2} + \frac{z}{h} \right)^n z^3\, dz = \frac{h^4 n(8+3n+n^2)}{8(1+n)(2+n)(3+n)(4+n)}, \tag{3A.4}$$

$$\int_{-\frac{h}{2}}^{\frac{h}{2}} \left(\frac{1}{2}+\frac{z}{h}\right)^n z^4\, dz = \frac{h^5(24+18n+23n^2+6n^3+n^4)}{16(1+n)(2+n)(3+n)(4+n)(5+n)}, \tag{3A.5}$$

$$\int_{-\frac{h}{2}}^{\frac{h}{2}} \left(\frac{1}{2}+\frac{z}{h}\right)^n z^5\, dz = \frac{h^6 n(184+110n+55n^2+10n^3+n^4)}{32(1+n)(2+n)(3+n)(4+n)(5+n)(6+n)}, \tag{3A.6}$$

$$\int_{-\frac{h}{2}}^{\frac{h}{2}} \left(\frac{1}{2}+\frac{z}{h}\right)^n z^6\, dz = \frac{h^7(720+660n+964n^2+405n^3+115n^4+15n^5+n^6)}{64(1+n)(2+n)(3+n)(4+n)(5+n)(6+n)(7+n)}. \tag{3A.7}$$

These integral values can be used to determine the structural stiffness coefficients A_{xx}, B_{xx}, and so on for an FGM beam with the material variation in Eq. (3.4.5a). We have

$$A_{xx} = b\int_{-\frac{h}{2}}^{\frac{h}{2}} E(z)\, dz = E_2 b\left[(M-1)\int_{-\frac{h}{2}}^{\frac{h}{2}}\left(\frac{1}{2}+\frac{z}{h}\right)^n dz + \int_{-\frac{h}{2}}^{\frac{h}{2}} dz\right]$$
$$= E_2 b\left[(M-1)\frac{h}{(1+n)}+h\right] = E_2 bh\frac{M+n}{1+n},$$

$$B_{xx} = b\int_{-\frac{h}{2}}^{\frac{h}{2}} z\,E(z)\, dz = E_2 b\left[(M-1)\int_{-\frac{h}{2}}^{\frac{h}{2}}\left(\frac{1}{2}+\frac{z}{h}\right)^n z\, dz + \int_{-\frac{h}{2}}^{\frac{h}{2}} z\, dz\right]$$
$$= E_2 b\left[(M-1)\frac{nh^2}{2(1+n)(2+n)}+0\right] = E_2\frac{bh^2}{2}\frac{(M-1)n}{(1+n)(2+n)},$$

$$D_{xx} = b\int_{-\frac{h}{2}}^{\frac{h}{2}} z^2\,E(z)\, dz = E_2 b\left[(M-1)\int_{-\frac{h}{2}}^{\frac{h}{2}}\left(\frac{1}{2}+\frac{z}{h}\right)^n z^2\, dz + \int_{-\frac{h}{2}}^{\frac{h}{2}} z^2\, dz\right]$$
$$= E_2 b\left[(M-1)\frac{h^3(2+n+n^2)}{4(1+n)(2+n)(3+n)}+\frac{h^3}{12}\right]$$
$$= E_2\frac{bh^3}{12}\left[\frac{3M(2+n+n^2)+8n+3n^2+n^3}{(1+n)(2+n)(3+n)}\right],$$

$$E_{xx} = b\int_{-\frac{h}{2}}^{\frac{h}{2}} z^3\,E(z)\, dz = E_2 b\left[(M-1)\int_{-\frac{h}{2}}^{\frac{h}{2}}\left(\frac{1}{2}+\frac{z}{h}\right)^n z^3\, dz + \int_{-\frac{h}{2}}^{\frac{h}{2}} z^3\, dz\right]$$
$$= E_2\frac{bh^4}{8}\left[\frac{(M-1)n(8+3n+n^2)}{(1+n)(2+n)(3+n)(4+n)}\right],$$

$$F_{xx} = b\int_{-\frac{h}{2}}^{\frac{h}{2}} z^4\,E(z)\, dz = E_2 b\left[(M-1)\int_{-\frac{h}{2}}^{\frac{h}{2}}\left(\frac{1}{2}+\frac{z}{h}\right)^n z^4\, dz + \int_{-\frac{h}{2}}^{\frac{h}{2}} z^4\, dz\right]$$
$$= E_2 b\left[(M-1)\frac{h^5(24+18n+23n^2+6n^3+n^4)}{16(1+n)(2+n)(3+n)(4+n)(5+n)}+\frac{h^5}{80}\right]$$
$$= E_2\frac{bh^5}{80}\left[\frac{5Mf_1+nf_2}{(1+n)(2+n)(3+n)(4+n)(5+n)}\right]$$

$$f_1 = (24+18n+23n^2+6n^3+n^4), \quad f_2 = (184+110n+55n^2+10n^3+n^4),$$

$$G_{xx} = b\int_{-\frac{h}{2}}^{\frac{h}{2}} z^5\,E(z)\, dz = E_2 b\left[(M-1)\int_{-\frac{h}{2}}^{\frac{h}{2}}\left(\frac{1}{2}+\frac{z}{h}\right)^n z^5\, dz + \int_{-\frac{h}{2}}^{\frac{h}{2}} z^5\, dz\right]$$
$$= E_2\frac{bh^6}{32}\left[\frac{(M-1)n(184+110n+55n^2+10n^3+n^4)}{(1+n)(2+n)(3+n)(4+n)(5+n)(6+n)}\right],$$

$$H_{xx} = b \int_{-\frac{h}{2}}^{\frac{h}{2}} z^6 E(z)\, dz = E_2 b \left[(M-1) \int_{-\frac{h}{2}}^{\frac{h}{2}} \left(\frac{1}{2} + \frac{z}{h} \right)^n z^5\, dz + \int_{-\frac{h}{2}}^{\frac{h}{2}} z^6\, dz \right]$$

$$= E_2 b \left[\frac{h^7 (M-1) g_1}{64(1+n)(2+n)(3+n)(4+n)(5+n)(6+n)(7+n)} + \frac{h^7}{448} \right]$$

$$= E_2 \frac{bh^7}{448} \left[\frac{7M g_1 + n g_2}{(1+n)(2+n)(3+n)(4+n)(5+n)(7+n)} \right],$$

$$g_1 = (720 + 660n + 964n^2 + 405n^3 + 115n^4 + 15n^5 + n^6),$$

$$g_2 = (8448 + 6384n + 3934n^2 + 1155n^3 + 217n^4 + 21n^5 + n^6). \tag{3A.8}$$

It is one thing for the human mind to extract from the phenomena of nature the laws which it has itself put into them; it may be a far harder thing to extract laws over which it has no control.
                                                                    Arthur Eddington

4 The First-Order Shear Deformation Beam Theory

Science is not only a disciple of reason but, also, one of romance and passion.

Stephen Hawking

4.1 INTRODUCTORY COMMENTS

In this chapter, we consider the first-order shear deformation theory, most commonly known as the *Timoshenko beam theory* (TBT)[1]. The TBT not only brings the transverse shear strain $\gamma_{xz} = 2\varepsilon_{xz}$ and shear stress σ_{xz} into the calculations, but the theory also removes the inconsistency of the classical beam theory, in the sense that the transverse shear force is nonzero even through the constitutive equation by virtue of the transverse shear strain being nonzero.

However, in the TBT, the transverse shear stress through the beam thickness is only represented as a constant, whereas the elasticity equations (as discussed in mechanics of materials books) show that the variation should be quadratic. To account for the inaccuracy in predicting the transverse shear force magnitude (not the shear stress distribution itself), *shear correction factor* (SCF) has been introduced (see [104]–[112], among many others). According to Timoshenko, the shear correction factor is the ratio of the average shear strain on a section to the shear strain at the geometric centroid of the cross section. The SCF, in general, is a function of the cross-sectional shape, material properties, boundary conditions, and so on. For rectangular sections, Timoshenko [104] proposed a SCF $K_s = 5(1 + \nu)/(6 + 5\nu)$, which takes the range of values, $(5/6) \le K_s \le (15/17)$ for $0 \le \nu \le 0.5$.

Following these preliminary comments, we proceed, in somewhat parallel fashion to the developments presented for the CBT in the preceding chapter, to develop the equations of motion and exact solutions, the Navier solution, and solutions by the Ritz and Galerkin methods for the linear case. The main difference between CBT and TBT is in the kinematics of the beam; once the displacement field is introduced, all other steps to derive the equations of motion are straightforward.

---

[1]See Section 1.1 for some historical comments.

DOI: 10.1201/9781003240846-4

4.2   DISPLACEMENTS AND STRAINS

The TBT is based on the displacement field (see Fig. 4.2.1)

$$\mathbf{u}(x, z, t) = [u(x, t) + z\phi_x(x, t)]\,\hat{\mathbf{e}}_x + w(x, t)\hat{\mathbf{e}}_z, \qquad (4.2.1)$$

where ϕ_x denotes the rotation (independent of the slope, $-\partial w/\partial x$) of the cross-sectional plane about the y-axis. In the TBT, the normality assumption of the CBT is relaxed and a constant state of transverse shear strain (and thus the constant shear stress computed from the constitutive equation) with respect to the transverse coordinate z is included. As stated earlier, the TBT requires a shear correction factor to compensate for the error due to this constant shear stress assumption.

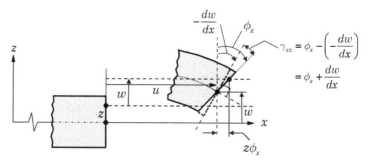

Figure 4.2.1: Kinematics of deformation of the Timoshenko beam theory.

The von Kármán nonlinear strains of the TBT are

$$\varepsilon_{xx}(x, z, t) = \frac{\partial u}{\partial x} + \frac{1}{2}\left(\frac{\partial w}{\partial x}\right)^2 + z\frac{\partial \phi_x}{\partial x} \equiv \varepsilon_{xx}^{(0)} + z\varepsilon_{xx}^{(1)}, \qquad (4.2.2a)$$

$$\varepsilon_{xx}^{(0)} = \frac{\partial u}{\partial x} + \frac{1}{2}\left(\frac{\partial w}{\partial x}\right)^2, \quad \varepsilon_{xx}^{(1)} = \frac{\partial \phi_x}{\partial x}, \qquad (4.2.2b)$$

$$\gamma_{xz}(x, t) = \phi_x + \frac{\partial w}{\partial x}. \qquad (4.2.2c)$$

We note that the strains involve only the first derivatives of the dependent unknowns (u, w, ϕ_x). In addition, the transverse shear strain γ_{xz} is only a function of x (i.e., constant through the beam height).

4.3   EQUATIONS OF MOTION

4.3.1   VECTOR APPROACH

Equations of motion, Eqs. (3.3.1a) and (3.3.1b) derived in Chapter 3, are also valid for the TBT as we have not invoked any displacement expansions. They are listed here for ready reference:

$$\frac{\partial}{\partial x}\left(\sigma_{xx} + \frac{\partial u_x}{\partial x}\sigma_{xx} + \frac{\partial u_x}{\partial z}\sigma_{xz}\right) + \frac{\partial}{\partial z}\left(\sigma_{xz} + \frac{\partial u_x}{\partial x}\sigma_{xz}\right) + f = \rho\frac{\partial^2 u_x}{\partial t^2},$$
(4.3.1a)

$$\frac{\partial}{\partial x}\left(\sigma_{xz} + \frac{\partial u_z}{\partial x}\sigma_{xx}\right) + \frac{\partial}{\partial z}\left(\frac{\partial u_z}{\partial x}\sigma_{xz}\right) + q = \rho\frac{\partial^2 u_z}{\partial t^2}.$$
(4.3.1b)

Integration of Eqs. (4.3.1a) and (4.3.1b) over the beam area of cross section, keeping the displacement expansion (4.2.1) in mind ($u_1 = u_x = u + z\phi_x$, $u_2 = u_y = 0$, and $u_3 = u_z = w$), yields

$$\frac{\partial}{\partial x}\left(N_{xx} + \frac{\partial u}{\partial x}N_{xx} + \frac{\partial \phi_x}{\partial x}M_{xx} + \phi_x N_{xz}\right)$$
$$+ \frac{\partial \phi_x}{\partial x}N_{xz} + f = m_0\frac{\partial^2 u}{\partial t^2} + m_1\frac{\partial^2 \phi_x}{\partial t^2},$$
(4.3.2)

$$\frac{\partial}{\partial x}\left(N_{xz} + \frac{\partial w}{\partial x}N_{xx}\right) + q = m_0\frac{\partial^2 w}{\partial t^2},$$
(4.3.3)

where (N_{xx}, N_{xz}, M_{xx}) are defined by [see Eq. (3.3.3)]:

$$N_{xx} = \int_A \sigma_{xx}\, dA, \quad N_{xz} = \int_A \sigma_{xz}\, dA, \quad M_{xx} = \int_A z\sigma_{xx}\, dA.$$
(4.3.4)

and (m_0, m_1, m_2) are the mass inertias defined by [see Eq. (3.3.4)]

$$m_0 = \int_A \rho\, dA, \quad m_1 = \int_A z\rho\, dA, \quad m_2 = \int_A z^2\rho\, dA.$$
(4.3.5)

Next, multiply Eq. (4.3.1a) with z and integrate over the beam area of cross section while keeping the displacement expansion (4.2.1) in mind [see Eq. (3.3.5)] and obtain

$$\frac{\partial}{\partial x}\left(M_{xx} + \frac{\partial u}{\partial x}M_{xx} + \frac{\partial \phi_x}{\partial x}P_{xx} + \phi_x R_{xz}\right)$$
$$- N_{xz} - \frac{\partial u}{\partial x}N_{xz} - \frac{\partial \phi_x}{\partial x}R_{xz} = m_1\frac{\partial^2 u}{\partial t^2} + m_2\frac{\partial^2 \phi_x}{\partial t^2},$$
(4.3.6)

where (P_{xx}, R_{xz}) are the higher-order stress resultants [see Eq. (3.3.6)]

$$P_{xx} = \int_A \sigma_{xx}z^2\, dA, \quad R_{xz} = \int_A \sigma_{xz}z\, dA.$$
(4.3.7)

Using the order-of-magnitude assumption of various quantities discussed prior to Eq. (3.2.2), and simplifying Eqs. (4.3.2), (4.3.3), and (4.3.6) (i.e., omit all the underline terms in these equations), we obtain

$$-\frac{\partial N_{xx}}{\partial x} + m_0 \frac{\partial^2 u}{\partial t^2} + m_1 \frac{\partial^2 \phi_x}{\partial t^2} - f = 0, \tag{4.3.8}$$

$$-\frac{\partial N_{xz}}{\partial x} - \frac{\partial}{\partial x}\left(N_{xx}\frac{\partial w}{\partial x}\right) + m_0 \frac{\partial^2 w}{\partial t^2} - q = 0, \tag{4.3.9}$$

$$-\frac{\partial M_{xx}}{\partial x} + N_{xz} + m_2 \frac{\partial^2 \phi_x}{\partial t^2} + m_1 \frac{\partial^2 u}{\partial t^2} = 0. \tag{4.3.10}$$

These equations are the same as those in Eqs. (3.3.8)–(3.3.10), except for the inertia terms (where ϕ_x replaced $-\partial w/\partial x$). Since there are three generalized displacements (u, w, ϕ_x), we need all three equations.

4.3.2   ENERGY APPROACH

Hamilton's principle for the TBT has the same form as that in Eq. (3.3.14a), except that we must add the virtual strain energy terms associated with the transverse stress σ_{xz}. To make it more complete, we also add the strain energy term due to the modified couple stress effect. The curvature in the TBT is given by [see Eqs. (1.6.1) and (1.6.2)]

$$\omega_y = \frac{1}{2}\left(\frac{\partial u_1}{\partial z} - \frac{\partial u_3}{\partial x}\right) = \frac{1}{2}\left(\phi_x - \frac{\partial w}{\partial x}\right),$$

$$\chi_{xy} = \frac{1}{2}\frac{\partial \omega_y}{\partial x} = \frac{1}{4}\left(\frac{\partial \phi_x}{\partial x} - \frac{\partial^2 w}{\partial x^2}\right). \tag{4.3.11}$$

The statement of Hamilton's principle is

$$0 = \int_{t_1}^{t_2}\int_0^L\int_A \left(\rho\frac{\partial \delta u_x}{\partial t}\frac{\partial u_x}{\partial t} + \rho\frac{\partial \delta u_z}{\partial t}\frac{\partial u_z}{\partial t} - \delta\varepsilon_{xx}\,\sigma_{xx}\right.$$

$$\left. - K_s\delta\gamma_{xz}\,\sigma_{xz} - 2\mathcal{M}_{xy}\,\delta\chi_{xy}\right)dA\,dx\,dt$$

$$+ \int_{t_1}^{t_2}\int_0^L [\delta u\,f + \delta w\,(q - c_f w)]\,dx\,dt, \tag{4.3.12}$$

where K_s is the shear correction factor.

We note that the shear correction factor only modifies the shear force computed and not the shear stress distribution. Thus, we use K_s to compute the transverse shear force N_{xz} in the TBT as follows:

$$N_{xz} = K_s \int_A \sigma_{xz}\,dz. \tag{4.3.13}$$

The factor K_s is computed such that the strain energy due to transverse shear stresses in Eq. (4.3.13) equals the strain energy due to the "true" transverse stresses predicted by the three-dimensional elasticity theory.

For example, consider a homogeneous beam with a rectangular cross section, with width b and height h. As derived in a book on mechanics of materials, the shear stress distribution through the thickness of the beam (at a distance x) is given by

$$\sigma_{xz}^C = \frac{3V(x)}{2bh}\left[1 - \left(\frac{2z}{h}\right)^2\right], \quad -\frac{h}{2} \le z \le \frac{h}{2}, \tag{4.3.14}$$

where $V(x)$ is the transverse force. The transverse shear stress in the TBT is a constant, $\sigma_{xz}^F = V(x)/A = V(x)/bh$. The strain energies due to transverse shear stresses in the two cases are

$$U_s^C = \frac{1}{2G}\int_A \left(\sigma_{xz}^C\right)^2 dA = \frac{3V^2}{5Gbh},$$
$$U_s^F = \frac{1}{2G}\int_A \left(\sigma_{xz}^F\right)^2 dA = \frac{V^2}{2Gbh}. \tag{4.3.15}$$

The shear correction factor is the ratio of U_s^F to U_s^C, which gives $K_s = 5/6$.

Substituting for the displacement expansion and virtual strains and integrating over the area of cross section, we obtain

$$0 = \int_{t_1}^{t_2}\int_0^L \int_A \left[\rho\left(\frac{\partial\delta u}{\partial t} + z\frac{\partial\delta\phi_x}{\partial t}\right)\left(\frac{\partial u}{\partial t} + z\frac{\partial\phi_x}{\partial t}\right) + \rho\frac{\partial\delta w}{\partial t}\frac{\partial w}{\partial t}\right.$$
$$- \left(\frac{\partial\delta u}{\partial x} + \frac{\partial\delta w}{\partial x}\frac{\partial w}{\partial x} + z\frac{\partial\delta\phi_x}{\partial x}\right)\sigma_{xx} - \left(\delta\phi_x + \frac{\partial\delta w}{\partial x}\right)\sigma_{xz}$$
$$\left. - \frac{1}{2}\left(\frac{\partial\delta\phi_x}{\partial x} - \frac{\partial^2\delta w}{\partial x^2}\right)M_{xy}\right]dA\,dx\,dt$$
$$+ \int_{t_1}^{t_2}\int_0^L \left[\delta u\,f + \delta w\,(q - c_f w)\right]dx\,dt \tag{4.3.16a}$$

$$= \int_{t_1}^{t_2}\int_0^L \left[m_0\frac{\partial\delta u}{\partial t}\frac{\partial u}{\partial t} + m_1\left(\frac{\partial\delta\phi_x}{\partial t}\frac{\partial u}{\partial t} + \frac{\partial\delta u}{\partial t}\frac{\partial\phi_x}{\partial t}\right) + m_2\frac{\partial\delta\phi_x}{\partial t}\frac{\partial\phi_x}{\partial t}\right.$$
$$+ m_0\frac{\partial\delta w}{\partial t}\frac{\partial w}{\partial t} - \left(\frac{\partial\delta u}{\partial x} + \frac{\partial\delta w}{\partial x}\frac{\partial w}{\partial x}\right)N_{xx} - \frac{\partial\delta\phi_x}{\partial x}M_{xx}$$
$$\left. - \delta\phi_x N_{xz} - \frac{\partial\delta w}{\partial x}N_{xz} - \frac{1}{2}\left(\frac{\partial\delta\phi_x}{\partial x} - \frac{\partial^2\delta w}{\partial x^2}\right)P_{xy}\right]dx\,dt$$
$$+ \int_{t_1}^{t_2}\int_0^L \left[\delta u\,f + \delta w\,(q - c_f w)\right]dx\,dt, \tag{4.3.16b}$$

where

$$N_{xx} = \int_A \sigma_{xx}\,dA, \quad N_{xz} = K_s\int_A \sigma_{xz}\,dA,$$
$$M_{xx} = \int_A z\sigma_{xx}\,dA, \quad P_{xy} = \int_A M_{xy}\,dA. \tag{4.3.17}$$

The Euler–Lagrange equations of the TBT are

$$-\frac{\partial N_{xx}}{\partial x} + m_0 \frac{\partial^2 u}{\partial t^2} + m_1 \frac{\partial^2 \phi_x}{\partial t^2} - f = 0, \qquad (4.3.18)$$

$$-\frac{\partial N_{xz}}{\partial x} - \frac{1}{2}\frac{\partial^2 P_{xy}}{\partial x^2} - \frac{\partial}{\partial x}\left(N_{xx}\frac{\partial w}{\partial x}\right) + c_f w + m_0 \frac{\partial^2 w}{\partial t^2} - q = 0, \qquad (4.3.19)$$

$$-\frac{\partial M_{xx}}{\partial x} - \frac{1}{2}\frac{\partial P_{xy}}{\partial x} + N_{xz} + m_2 \frac{\partial^2 \phi_x}{\partial t^2} + m_1 \frac{\partial^2 u}{\partial t^2} = 0. \qquad (4.3.20)$$

These equations are the same as those in Eqs. (4.3.8)–(4.3.10), except of the addition of foundation term $c_f w$ and the couple stress contributions P_{xy}.

The three duality pairs for the TBT are

$$(u, N_{xx}), \quad (w, V_{\text{eff}}), \quad (\phi_x, \bar{M}_{xx}). \qquad (4.3.21)$$

where the effective shear force and bending moments are

$$V_{\text{eff}} \equiv N_{xz} + N_{xx}\frac{\partial w}{\partial x} + \frac{1}{2}\frac{\partial P_{xy}}{\partial x}, \qquad (4.3.22a)$$

$$\bar{M}_{xx} = M_{xx} + \tfrac{1}{2}P_{xy}. \qquad (4.3.22b)$$

We note that the effective shear force in the TBT does not have any time derivative terms but it has the modified couple stress term.

The linearized equations of equilibrium can obtained from Eqs. (4.3.18)–(4.3.20) be setting the nonlinear and time-derivative terms to zero:

$$-\frac{dN_{xx}}{dx} - f = 0, \qquad (4.3.23)$$

$$-\frac{dN_{xz}}{dx} - \frac{1}{2}\frac{d^2 P_{xy}}{dx^2} + c_f w - q = 0, \qquad (4.3.24)$$

$$-\frac{dM_{xx}}{dx} - \frac{1}{2}\frac{dP_{xy}}{dx} + N_{xz} = 0. \qquad (4.3.25)$$

4.4 GOVERNING EQUATIONS IN TERMS OF DISPLACEMENTS

4.4.1 BEAM CONSTITUTIVE EQUATIONS

The material constitutive relations for beams were discussed in Section 3.4.1; see Eqs. (3.4.1) and (3.4.4): $\sigma_{xx} = E\left(\varepsilon_{xx} - \alpha_T \Delta T\right)$ and $M_{xy} = 2G\ell^2 \chi_{xy}$. Using these constitutive relations, we can express the stress resultants $(N_{xx}, M_{xx}, N_{xz}, P_{xy})$ in terms of the strains as

$$N_{xx} = \int_A \sigma_{xx}\, dA = A_{xx}\left[\frac{\partial u}{\partial x} + \frac{1}{2}\left(\frac{\partial w}{\partial x}\right)^2\right] + B_{xx}\frac{\partial \phi_x}{\partial x} - N_{xx}^T, \qquad (4.4.1a)$$

$$M_{xx} = \int_A z\sigma_{xx}\, dA = B_{xx}\left[\frac{\partial u}{\partial x} + \frac{1}{2}\left(\frac{\partial w}{\partial x}\right)^2\right] + D_{xx}\frac{\partial \phi_x}{\partial x} - M_{xx}^T, \qquad (4.4.1b)$$

$$P_{xy} = \int_A M_{xy}\, dA = \frac{1}{2}A_{xy}\left(\frac{\partial \phi_x}{\partial x} - \frac{\partial^2 w}{\partial x^2}\right), \tag{4.4.1c}$$

$$N_{xz} = K_s \int_A \sigma_{xz}\, dA = S_{xz}\left(\phi_x + \frac{\partial w}{\partial x}\right), \tag{4.4.1d}$$

where N_{xx}^T and M_{xx}^T are the thermal stress resultants [see Eq. (3.4.6d)] and A_{xx}, B_{xx}, D_{xx}, and A_{xy} are the extensional, extensional-bending, bending, and in-plane shear stiffness coefficients defined in Eq. (3.4.7), and S_{xz} is the shear stiffness

$$S_{xz} = K_s \int_A G(z)\, dA = \frac{K_s}{2(1+\nu)} \int_A E(z)\, dA \equiv K_s A_{xz}, \tag{4.4.2}$$

where K_s is the shear correction coefficient, G the shear modulus $[G(z) = E(z)/2(1+\nu)]$, and ν is Poisson's ratio, which is assumed to be a constant. It is clear from Eqs. (4.4.1a) and (4.4.1b) that the in-plane displacement u and the bending displacements (w, ϕ_x) are coupled by virtue of the coupling stiffness coefficient B_{xx} even when the nonlinearity is omitted.

4.4.2 EQUATIONS OF MOTION FOR THE GENERAL CASE

The equations of motion in Eqs. (4.3.18)–(4.3.20) can now be expressed in terms of the displacements u, w, and ϕ_x with the help of the beam constitutive relations in Eqs. (4.4.1a)–(4.4.1d) as (see [68, 74])

$$-\frac{\partial}{\partial x}\left\{A_{xx}\left[\frac{\partial u}{\partial x} + \frac{1}{2}\left(\frac{\partial w}{\partial x}\right)^2\right] + B_{xx}\frac{\partial \phi_x}{\partial x} - N_{xx}^T\right\}$$
$$+ m_0 \frac{\partial^2 u}{\partial t^2} + m_1 \frac{\partial^2 \phi_x}{\partial t^2} = f, \tag{4.4.3}$$

$$-\frac{\partial}{\partial x}\left[S_{xz}\left(\phi_x + \frac{\partial w}{\partial x}\right)\right] - \frac{1}{4}\frac{\partial^2}{\partial x^2}\left[A_{xy}\left(\frac{\partial \phi_x}{\partial x} - \frac{\partial^2 w}{\partial x^2}\right)\right]$$
$$-\frac{\partial}{\partial x}\left\{A_{xx}\frac{\partial w}{\partial x}\left[\frac{\partial u}{\partial x} + \frac{1}{2}\left(\frac{\partial w}{\partial x}\right)^2\right] + B_{xx}\frac{\partial w}{\partial x}\left(\frac{\partial \phi_x}{\partial x}\right) - \frac{\partial w}{\partial x}N_{xx}^T\right\}$$
$$+ c_f w - q + m_0 \frac{\partial^2 w}{\partial t^2} = 0, \tag{4.4.4}$$

$$-\frac{\partial}{\partial x}\left\{B_{xx}\left[\frac{\partial u}{\partial x} + \frac{1}{2}\left(\frac{\partial w}{\partial x}\right)^2\right] + D_{xx}\frac{\partial \phi_x}{\partial x} - M_{xx}^T\right\} + S_{xz}\left(\phi_x + \frac{\partial w}{\partial x}\right)$$
$$-\frac{1}{4}\frac{\partial}{\partial x}\left[A_{xy}\left(\frac{\partial \phi_x}{\partial x} - \frac{\partial^2 w}{\partial x^2}\right)\right] + m_1 \frac{\partial^2 u}{\partial t^2} + m_2 \frac{\partial^2 \phi_x}{\partial t^2} = 0. \tag{4.4.5}$$

Equations (4.4.3)–(4.4.5) are fully coupled equations (i.e., each equation contains all three variables) and must be solved simultaneously.

4.4.3 EQUATIONS OF MOTION WITHOUT THE COUPLE STRESS AND THERMAL EFFECTS

When we omit the couple stress effect and thermal resultants, the equations of motion in Eqs. (4.4.3)–(4.4.5) reduce to

$$-\frac{\partial}{\partial x}\left\{A_{xx}\left[\frac{\partial u}{\partial x}+\frac{1}{2}\left(\frac{\partial w}{\partial x}\right)^2\right]+B_{xx}\frac{\partial\phi_x}{\partial x}\right\}$$

$$+m_0\frac{\partial^2 u}{\partial t^2}+m_1\frac{\partial^2\phi_x}{\partial t^2}=f,\quad(4.4.6)$$

$$-\frac{\partial}{\partial x}\left\{A_{xx}\frac{\partial w}{\partial x}\left[\frac{\partial u}{\partial x}+\frac{1}{2}\left(\frac{\partial w}{\partial x}\right)^2\right]+B_{xx}\frac{\partial w}{\partial x}\left(\frac{\partial\phi_x}{\partial x}\right)\right\}$$

$$-\frac{\partial}{\partial x}\left[S_{xz}\left(\phi_x+\frac{\partial w}{\partial x}\right)\right]+c_f w-q+m_0\frac{\partial^2 w}{\partial t^2}=0,\quad(4.4.7)$$

$$-\frac{\partial}{\partial x}\left\{B_{xx}\left[\frac{\partial u}{\partial x}+\frac{1}{2}\left(\frac{\partial w}{\partial x}\right)^2\right]+D_{xx}\frac{\partial\phi_x}{\partial x}\right\}+S_{xz}\left(\phi_x+\frac{\partial w}{\partial x}\right)$$

$$+m_1\frac{\partial^2 u}{\partial t^2}+m_2\frac{\partial^2\phi_x}{\partial t^2}=0.\quad(4.4.8)$$

4.4.4 EQUATIONS OF MOTION FOR HOMOGENEOUS BEAMS

For beams for which E is not a function of z (i.e., homogeneous through the beam height), the above equations simplify to ($A_{xx}=EA_0$, $B_{xx}=0$, $D_{xx}=EI_0$, $A_{xy}=GA_0\ell^2$, $S_{xz}=GA_0$, and $m_1=0$)

$$-\frac{\partial}{\partial x}\left\{A_{xx}\left[\frac{\partial u}{\partial x}+\frac{1}{2}\left(\frac{\partial w}{\partial x}\right)^2\right]-N_{xx}^T\right\}+m_0\frac{\partial^2 u}{\partial t^2}=f,\quad(4.4.9)$$

$$-\frac{\partial}{\partial x}\left[S_{xz}\left(\phi_x+\frac{\partial w}{\partial x}\right)\right]-\frac{1}{4}\frac{\partial^2}{\partial x^2}\left[A_{xy}\left(\frac{\partial\phi_x}{\partial x}-\frac{\partial^2 w}{\partial x^2}\right)\right]$$

$$-\frac{\partial}{\partial x}\left\{A_{xx}\frac{\partial w}{\partial x}\left[\frac{\partial u}{\partial x}+\frac{1}{2}\left(\frac{\partial w}{\partial x}\right)^2\right]-\frac{\partial w}{\partial x}N_{xx}^T\right\}$$

$$+c_f w+m_0\frac{\partial^2 w}{\partial t^2}=q,\quad(4.4.10)$$

$$-\frac{\partial}{\partial x}\left(D_{xx}\frac{\partial\phi_x}{\partial x}-M_{xx}^T\right)+S_{xz}\left(\phi_x+\frac{\partial w}{\partial x}\right)$$

$$-\frac{1}{4}\frac{\partial}{\partial x}\left[A_{xy}\left(\frac{\partial\phi_x}{\partial x}-\frac{\partial^2 w}{\partial x^2}\right)\right]+m_2\frac{\partial^2\phi_x}{\partial t^2}=0.\quad(4.4.11)$$

All of the above three equations are coupled by virtue of the nonlinearity.

4.4.5 LINEARIZED EQUATIONS OF MOTION FOR FGM BEAMS

For FGM beams without the nonlinear terms, the equations of motion are

$$-\frac{\partial}{\partial x}\left(A_{xx}\frac{\partial u}{\partial x}+B_{xx}\frac{\partial\phi_x}{\partial x}-N_{xx}^T\right)+m_0\frac{\partial^2 u}{\partial t^2}+m_1\frac{\partial^2\phi_x}{\partial t^2}=f, \qquad (4.4.12)$$

$$-\frac{\partial}{\partial x}\left[S_{xz}\left(\phi_x+\frac{\partial w}{\partial x}\right)\right]-\frac{1}{4}\frac{\partial^2}{\partial x^2}\left[A_{xy}\left(\frac{\partial\phi_x}{\partial x}-\frac{\partial^2 w}{\partial x^2}\right)\right]$$
$$+c_f w+m_0\frac{\partial^2 w}{\partial t^2}=q, \qquad (4.4.13)$$

$$-\frac{\partial}{\partial x}\left(B_{xx}\frac{\partial u}{\partial x}+D_{xx}\frac{\partial\phi_x}{\partial x}-M_{xx}^T\right)+S_{xz}\left(\phi_x+\frac{\partial w}{\partial x}\right)$$
$$-\frac{1}{4}\frac{\partial}{\partial x}\left[A_{xy}\left(\frac{\partial\phi_x}{\partial x}-\frac{\partial^2 w}{\partial x^2}\right)\right]+m_1\frac{\partial^2 u}{\partial t^2}+m_2\frac{\partial^2\phi_x}{\partial t^2}=0. \qquad (4.4.14)$$

Again, the three equations are coupled only because of the coefficient B_{xx}.

4.4.6 LINEARIZED EQUATIONS FOR HOMOGENEOUS BEAMS

For beams with $E=E(x)$ and without accounting for the nonlinear terms, the equations of motion become

$$-\frac{\partial}{\partial x}\left(A_{xx}\frac{\partial u}{\partial x}-N_{xx}^T\right)+m_0\frac{\partial^2 u}{\partial t^2}=f, \qquad (4.4.15)$$

$$-\frac{\partial}{\partial x}\left[S_{xz}\left(\phi_x+\frac{\partial w}{\partial x}\right)\right]-\frac{1}{4}\frac{\partial^2}{\partial x^2}\left[A_{xy}\left(\frac{\partial\phi_x}{\partial x}-\frac{\partial^2 w}{\partial x^2}\right)\right]$$
$$+c_f w-q+m_0\frac{\partial^2 w}{\partial t^2}=0, \qquad (4.4.16)$$

$$-\frac{\partial}{\partial x}\left(D_{xx}\frac{\partial\phi_x}{\partial x}-M_{xx}^T\right)+S_{xz}\left(\phi_x+\frac{\partial w}{\partial x}\right)$$
$$-\frac{1}{4}\frac{\partial}{\partial x}\left[A_{xy}\left(\frac{\partial\phi_x}{\partial x}-\frac{\partial^2 w}{\partial x^2}\right)\right]+m_2\frac{\partial^2\phi_x}{\partial t^2}=0. \qquad (4.4.17)$$

The last two equations are decoupled from the first equation.

In the case in which the couple stress effect and foundation modulus are not included and the material coefficients are independent of x (i.e., homogeneous beams), Eqs. (4.4.16) and (4.4.17) can be combined into a single equation solely in terms of w (i.e., eliminate ϕ_x):

$$D_{xx}\frac{\partial^4 w}{\partial x^4}+m_0\frac{\partial^2 w}{\partial t^2}-\left(m_2+m_0\frac{D_{xx}}{S_{xz}}\right)\frac{\partial^4 w}{\partial t^2\partial x^2}+\frac{m_0 m_2}{S_{xz}}\frac{\partial^4 w}{\partial t^4}$$
$$+c_f w-q+\frac{D_{xx}}{S_{xz}}\frac{\partial^2 q}{\partial x^2}-\frac{m_2}{S_{xz}}\frac{\partial^2 q}{\partial t^2}+\frac{\partial^2 M_{xx}^T}{\partial x^2}=0. \qquad (4.4.18)$$

The equations of equilibrium without the couple stress effect are:

$$-\frac{d}{dx}\left(A_{xx}\frac{du}{dx} - N_{xx}^T\right) = f, \tag{4.4.19}$$

$$-\frac{d}{dx}\left[S_{xz}\left(\phi_x + \frac{dw}{dx}\right)\right] + c_f w = q, \tag{4.4.20}$$

$$-\frac{d}{dx}\left(D_{xx}\frac{d\phi_x}{dx} - M_{xx}^T\right) + S_{xz}\left(\phi_x + \frac{dw}{dx}\right) = 0. \tag{4.4.21}$$

4.5   MIXED FORMULATION OF THE TBT

Here we present the governing equations of TBT in terms of (u, w, M_{xx}) by eliminating ϕ_x from the equations. Because of the algebraic complexity, we limit the formulation for the case in which the modified couple stress terms and thermal effects are not accounted for. Further, as will be seen shortly, the formulation is not suitable for dynamics problems. Although we carry the time derivative terms in the derivation, it will be clear that they do not allow the elimination of ϕ_x.

First, we specialize Eqs. (4.3.18)–(4.3.20) for this case:

$$-\frac{\partial N_{xx}}{\partial x} + m_0\frac{\partial^2 u}{\partial t^2} + m_1\frac{\partial^2 \phi_x}{\partial t^2} = f, \tag{4.5.1}$$

$$-\frac{\partial N_{xz}}{\partial x} - \frac{\partial}{\partial x}\left(N_{xx}\frac{\partial w}{\partial x}\right) + c_f w + m_0\frac{\partial^2 w}{\partial t^2} = q, \tag{4.5.2}$$

$$-\frac{\partial M_{xx}}{\partial x} + N_{xz} + m_2\frac{\partial^2 \phi_x}{\partial t^2} + m_1\frac{\partial^2 u}{\partial t^2} = 0. \tag{4.5.3}$$

Substituting for N_{xz} from Eq. (4.5.3) into Eq. (4.5.2), the three equations are reduced to two equations:

$$-\frac{\partial N_{xx}}{\partial x} + m_0\frac{\partial^2 u}{\partial t^2} + m_1\frac{\partial^2 \phi_x}{\partial t^2} = f, \tag{4.5.4}$$

$$-\frac{\partial}{\partial x}\left(\frac{\partial M_{xx}}{\partial x}\right) - \frac{\partial}{\partial x}\left(N_{xx}\frac{\partial w}{\partial x}\right) + c_f w$$
$$+ m_0\frac{\partial^2 w}{\partial t^2} + \frac{\partial}{\partial x}\left(m_2\frac{\partial^2 \phi_x}{\partial t^2} + m_1\frac{\partial^2 u}{\partial t^2}\right) = q. \tag{4.5.5}$$

Writing Eqs. (4.4.1a) and (4.4.1b) in matrix form (without the thermal forces), we have

$$\left\{\begin{array}{c} N_{xx} \\ M_{xx} \end{array}\right\} = \left[\begin{array}{cc} A_{xx} & B_{xx} \\ B_{xx} & D_{xx} \end{array}\right]\left\{\begin{array}{c} \frac{\partial u}{\partial x} + \frac{1}{2}\left(\frac{\partial w}{\partial x}\right)^2 \\ \frac{\partial \phi_x}{\partial x} \end{array}\right\}. \tag{4.5.6}$$

We introduce the following stiffness coefficients for use in the coming developments:

$$D_{xx}^* = A_{xx}D_{xx} - B_{xx}^2, \quad \bar{A}_{xx} = \frac{D_{xx}^*}{D_{xx}}, \quad \bar{B}_{xx} = \frac{B_{xx}}{D_{xx}}. \tag{4.5.7}$$

Inverting Eq. (4.5.6), we obtain

$$\frac{\partial u}{\partial x} + \frac{1}{2}\left(\frac{\partial w}{\partial x}\right)^2 = \frac{1}{D_{xx}^*}\left(D_{xx}N_{xx} - B_{xx}M_{xx}\right), \tag{4.5.8a}$$

$$\frac{\partial \phi_x}{\partial x} = \frac{1}{D_{xx}^*}\left(-B_{xx}N_{xx} + A_{xx}M_{xx}\right). \tag{4.5.8b}$$

Solving Eq. (4.5.8a) for N_{xx} in terms of (u, w, M_{xx}), we obtain

$$N_{xx} = \bar{A}_{xx}\left[\frac{\partial u}{\partial x} + \frac{1}{2}\left(\frac{\partial w}{\partial x}\right)^2\right] + \bar{B}_{xx}M_{xx}. \tag{4.5.9}$$

Using Eq. (4.5.9) for N_{xx} in Eq. (4.5.8b), we obtain

$$\frac{\partial \phi_x}{\partial x} = -\bar{B}_{xx}\left[\frac{\partial u}{\partial x} + \frac{1}{2}\left(\frac{\partial w}{\partial x}\right)^2\right] + \frac{1}{D_{xx}}M_{xx}. \tag{4.5.10}$$

Next, substituting for N_{xz} from Eq. (4.4.1d) into Eq. (4.5.3), we obtain

$$S_{xz}\left(\phi_x + \frac{\partial w}{\partial x}\right) - \frac{\partial M_{xx}}{\partial x} + m_2\frac{\partial^2 \phi_x}{\partial t^2} + m_1\frac{\partial^2 u}{\partial t^2} = 0, \tag{4.5.11a}$$

or

$$S_{xz}\left(\frac{\partial \phi_x}{\partial x} + \frac{\partial^2 w}{\partial x^2}\right) - \frac{\partial}{\partial x}\left(\frac{\partial M_{xx}}{\partial x} - m_2\frac{\partial^2 \phi_x}{\partial t^2} - m_1\frac{\partial^2 u}{\partial t^2}\right) = 0. \tag{4.5.11b}$$

Now substituting for $\partial \phi_x/\partial x$ from Eq. (4.5.10) into Eq. (4.5.11b), we arrive at

$$-\frac{\partial^2 w}{\partial x^2} = -\frac{1}{S_{xz}}\frac{\partial}{\partial x}\left(\frac{\partial M_{xx}}{\partial x} - m_2\frac{\partial^2 \phi_x}{\partial t^2} - m_1\frac{\partial^2 u}{\partial t^2}\right)$$

$$-\bar{B}_{xx}\left[\frac{\partial u}{\partial x} + \frac{1}{2}\left(\frac{\partial w}{\partial x}\right)^2\right] + \frac{1}{D_{xx}}M_{xx}. \tag{4.5.12}$$

Equations (4.5.4) and (4.5.5), with N_{xx} replaced with Eq. (4.5.9), and Eq. (4.5.12) constitute the governing equations in terms of u, w, and M_{xx} for the mixed formulation of the TBT:

$$-\frac{\partial}{\partial x}\left\{\bar{A}_{xx}\left[\frac{\partial u}{\partial x} + \frac{1}{2}\left(\frac{\partial w}{\partial x}\right)^2\right] + \bar{B}_{xx}M_{xx}\right\}$$

$$+m_0\frac{\partial^2 u}{\partial t^2} + m_1\frac{\partial^2 \phi_x}{\partial t^2} = f, \tag{4.5.13}$$

$$-\frac{\partial^2 M_{xx}}{\partial x^2} - \frac{\partial}{\partial x}\left\{\bar{A}_{xx}\frac{\partial w}{\partial x}\left[\frac{\partial u}{\partial x} + \frac{1}{2}\left(\frac{\partial w}{\partial x}\right)^2\right] + \bar{B}_{xx}\frac{\partial w}{\partial x}M_{xx}\right\}$$

$$+c_f w + m_0\frac{\partial^2 w}{\partial t^2} + m_1\frac{\partial^3 u}{\partial t^2 \partial x} + m_2\frac{\partial^3 \phi_x}{\partial t^2 \partial x} = q, \tag{4.5.14}$$

$$-\frac{\partial}{\partial x}\left[\frac{\partial w}{\partial x} - \frac{1}{S_{xz}}\left(\frac{\partial M_{xx}}{\partial x} - m_2\frac{\partial^2\phi_x}{\partial t^2} - m_1\frac{\partial^2 u}{\partial t^2}\right)\right]$$

$$+\bar{B}_{xx}\left[\frac{\partial u}{\partial x} + \frac{1}{2}\left(\frac{\partial w}{\partial x}\right)^2\right] - \frac{1}{D_{xx}}M_{xx} = 0, \qquad (4.5.15)$$

where the effective rotation ϕ_x is [from Eq. (4.5.11a)]

$$\phi_x = -\frac{\partial w}{\partial x} + \frac{1}{S_{xz}}\left(\frac{\partial M_{xx}}{\partial x} - m_2\frac{\partial^2\phi_x}{\partial t^2} - m_1\frac{\partial^2 u}{\partial t^2}\right). \qquad (4.5.16)$$

We note that we did not succeed in getting rid of ϕ_x (to bring M_{xx} as the variable); ϕ_x appears in Eqs. (4.5.14)–(4.5.16) with a time derivative. Therefore, the mixed formulation presented here is useful only in the finite element analysis of static problems.

4.6   EXACT SOLUTIONS

4.6.1   BENDING SOLUTIONS

In this section we present exact solutions to the *linear* equations of equilibrium of FGM beams without the modified couple stress effect. This is because Eqs. (4.4.12)–(4.4.14) contain the fourth-order derivative of w, complicating the analytical solution procedure and requiring additional boundary conditions. The analytical solutions are obtained using direct integration of the governing equations, as was done in the case of the CBT (in fact, the steps followed here are exactly the same as those used for obtaining the exact solutions of the CBT in Section 3.7).

By omitting all time derivative, nonlinear, and couple stress terms, and setting $f = 0$ (i.e., no axial force) and $c_f = 0$ (foundation modulus) in Eqs. (4.4.3)–(4.4.5), we obtain

$$-\frac{d}{dx}\left[A_{xx}\left(\frac{du}{dx}\right) + B_{xx}\frac{d\phi_x}{dx} - N_{xx}^T\right] = 0, \qquad (4.6.1)$$

$$-\frac{d}{dx}\left[S_{xz}\left(\phi_x + \frac{dw}{dx}\right)\right] = q, \qquad (4.6.2)$$

$$-\frac{d}{dx}\left(B_{xx}\frac{du}{dx} + D_{xx}\frac{d\phi_x}{dx} - M_{xx}^T\right) + S_{xz}\left(\phi_x + \frac{dw}{dx}\right) = 0. \qquad (4.6.3)$$

Again, we further assume that the beam stiffness coefficients as well as the thermal forces are all constant.

Equations (4.6.1)–(4.6.3), when expressed in terms of the stress resultants N_{xx}, N_{xz}, and M_{xx} [see Eqs. (4.3.23)– (4.3.25) with $P_{xy} = 0$], take the following form:

$$\frac{dN_{xx}}{dx} = 0, \quad -\frac{dN_{xz}}{dx} - q = 0, \quad -\frac{dM_{xx}}{dx} + N_{xz} = 0. \qquad (4.6.4)$$

Substituting for N_{xz} from the third equation into the second equation, we obtain

$$\frac{dN_{xx}}{dx} = 0, \quad -\frac{d^2 M_{xx}}{dx^2} - q = 0. \tag{4.6.5}$$

These equations are similar to those in Eqs. (3.7.3a) and (3.7.3b), and after carrying out the integration once, we obtain

$$N_{xx} = c_1, \quad \frac{dM_{xx}}{dx} = -\int q(x)dx + c_2. \tag{4.6.6}$$

Integrating the second equation once again, we obtain

$$M_{xx} = -\int^x \int^\xi q(\eta)\,d\eta d\xi + c_2 x + c_3 \equiv F(x), \tag{4.6.7}$$

where c_i ($i = 1, 2, 3$) are constants of integration, similar to K_i in Eqs. (3.7.3a)–(3.7.4).

The left sides of Eqs. (4.6.6) and (4.6.7) can be expressed in terms of the displacements (u, ϕ_x) using Eqs. (4.4.1a) and (4.4.1b); we have

$$\begin{bmatrix} A_{xx} & B_{xx} \\ B_{xx} & D_{xx} \end{bmatrix} \begin{Bmatrix} \frac{du}{dx} \\ \frac{d\phi_x}{dx} \end{Bmatrix} = \begin{Bmatrix} c_1 + N_{xx}^T \\ F(x) + M_{xx}^T \end{Bmatrix}. \tag{4.6.8}$$

Solving for du/dx and $d\phi_x/dx$, we obtain

$$\frac{du}{dx} = \frac{1}{D_{xx}^*} \left[D_{xx} \left(c_1 + N_{xx}^T \right) - B_{xx} \left(F(x) + M_{xx}^T \right) \right], \tag{4.6.9}$$

$$\frac{d\phi_x}{dx} = \frac{1}{D_{xx}^*} \left[-B_{xx} \left(c_1 + N_{xx}^T \right) + A_{xx} \left(F(x) + M_{xx}^T \right) \right], \tag{4.6.10}$$

where

$$D_{xx}^* = A_{xx} D_{xx} - B_{xx}^2, \quad F(x) = -\int^x \int^\xi q(\eta)\,d\eta\,d\xi + c_2 x + c_3. \tag{4.6.11}$$

Integrating Eqs. (4.6.9) and (4.6.10) once, we obtain

$$u(x) = \left[\frac{D_{xx}}{D_{xx}^*} \left(c_1 + N_{xx}^T \right) - \frac{B_{xx}}{D_{xx}^*} M_{xx}^T \right] x + \frac{B_{xx}}{D_{xx}^*} \int^x \int^\xi \int^\eta q(\zeta)\,d\zeta\,d\eta\,d\xi$$
$$- \frac{B_{xx}}{D_{xx}^*} \left(c_2 \frac{x^2}{2} + c_3 x + c_4 \right), \tag{4.6.12}$$

$$\phi_x(x) = \left[\frac{A_{xx}}{D_{xx}^*} M_{xx}^T - \frac{B_{xx}}{D_{xx}^*} \left(c_1 + N_{xx}^T \right) \right] x - \frac{A_{xx}}{D_{xx}^*} \int^x \int^\xi \int^\eta q(\zeta)\,d\zeta\,d\eta d\xi$$
$$+ \frac{A_{xx}}{D_{xx}^*} \left(c_2 \frac{x^2}{2} + c_3 x + c_5 \right). \tag{4.6.13}$$

From Eqs. (4.6.4) and (4.6.6), we arrive at

$$N_{xz} = \frac{dM_{xx}}{dx} = -\int^x q(\xi)\, d\xi + c_2,$$
(4.6.14)

and using Eq. (4.4.1d), we obtain

$$\frac{dw}{dx} = \frac{1}{S_{xz}}\left(-\int^x q(\xi)\, d\xi + c_2\right) - \phi_x$$

$$= \frac{1}{S_{xz}}\left(-\int^x q(\xi)\, d\xi + c_2\right) - \left[\frac{A_{xx}}{D^*_{xx}}M^T_{xx} - \frac{B_{xx}}{D^*_{xx}}\left(c_1 + N^T_{xx}\right)\right]x$$

$$- \frac{A_{xx}}{D^*_{xx}}\left(-\int^x\int^\xi\int^\eta q(\zeta)\, d\zeta\, d\eta d\xi + c_2\frac{x^2}{2} + c_3 x + c_5\right),$$
(4.6.15)

or

$$w(x) = \frac{1}{S_{xz}}\left(-\int^x\int^\xi q(\eta)\, d\eta\, d\xi + c_2 x\right) - \left[\frac{A_{xx}}{D^*_{xx}}M^T_{xx} - \frac{B_{xx}}{D^*_{xx}}\left(c_1 + N^T_{xx}\right)\right]\frac{x^2}{2}$$

$$- \frac{A_{xx}}{D^*_{xx}}\left(-\int^x\int^\xi\int^\eta\int^\zeta q(\mu)\, d\mu\, d\zeta\, d\eta d\xi + c_2\frac{x^3}{6} + c_3\frac{x^2}{2} + c_5 x + c_6\right).$$
(4.6.16)

In Eq. (4.6.16) μ, ζ, η, and ξ are dummy coordinates to indicate the sequence of integration (ultimately, the resulting expression will be in terms of x only).

The six constants of integration are determined using six boundary conditions, three at each end of the beam. One (and only one) element of the each of three duality pairs at each boundary point must be known [see Eq. (4.3.21)]: (u, N_{xx}), (w, N_{xz}), and (ϕ_x, M_{xx}). We note that, in TBT, ϕ_x has replaced $-dw/dx$ as the primary variable, and it is dual to the bending moment M_{xx}. *One should not specify dw/dx in place of ϕ_x in the TBT.* The stress resultants $(N_{xx}, M_{xx}, P_{xy}, N_{xz})$ can be computed with the help of Eqs. (4.4.1a)–(4.4.1d).

Example 4.6.1

Determine the exact solution of a FGM beam with two alternative boundary conditions: (a) pinned at both ends and (b) pinned at the left end and hinged at the right end, and subjected to uniformly distributed transverse load of intensity $q(x) = q_0$. Tabulate the numerical values for u, w, and ϕ_x obtained with the CBT and the TBT. Also, investigate the effect of shear deformation by using various ratios of the length-to-height ratios, L/h. Use the following material properties in generating the numerical results:

$$E_1 = 30 \times 10^6 \text{ psi (207 GPa)}, \quad E_2 = 10 \times 10^6 \text{ psi (69 GPa)}, \quad \nu = 0.3, \quad K = \frac{5}{6}.$$

Solution: (a) The boundary conditions for the pinned–pinned beam are

$$u(0) = 0, \quad w(0) = 0, \quad M_{xx}(0) = 0, \quad u(L) = 0, \quad w(L) = 0, \quad M_{xx}(L) = 0. \quad (4.6.17)$$

The condition $u(0) = 0$ gives $c_4 = 0$, $w(0) = 0$ gives $c_6 = 0$, and $\bar{M}_{xx}(0) = 0$ gives $c_3 = 0$. Then $M_{xx}(L) = 0$ gives $c_2 = q_0 L/2$. Using the conditions $u(L) = 0$ and $w(L) = 0$ in Eqs. (4.6.12) and (4.6.16) yield the remaining two constants, c_1 and c_5, as

$$c_1 = \frac{B_{xx}}{D_{xx}} \left(M_{xx}^T + \frac{q_0 L^2}{12} \right) - N_{xx}^T,$$

$$c_5 = -\frac{D_{xx}^*}{D_{xx} A_{xx}} \left(\frac{1}{2} M_{xx}^T L + \frac{q_0 L^3}{24} \right). \quad (4.6.18)$$

Hence, the solution for the pinned–pinned FGM beam according to the TBT is [see Eqs. (3.7.20)–(3.7.23); $\xi = x/L$]

$$u(\xi) = \frac{B_{xx}}{D_{xx}^*} \frac{q_0 L^3}{12} \left(\xi - 3\xi^2 + 2\xi^3 \right), \quad (4.6.19a)$$

$$\phi_x(\xi) = -\frac{A_{xx}}{D_{xx}^*} \frac{q_0 L^3}{12} \left(\xi - 3\xi^2 + 2\xi^3 \right) - \frac{1}{D_{xx}} \frac{q_0 L^3}{24} \left(1 - 2\xi \right)$$

$$- \frac{L}{2 D_{xx}} M_{xx}^T \left(1 - 2\xi \right), \quad (4.6.19b)$$

$$w(\xi) = \frac{A_{xx}}{D_{xx}^*} \frac{q_0 L^4}{24} \left(\xi^2 - 2\xi^3 + \xi^4 \right) + \frac{1}{D_{xx}} \frac{q_0 L^4}{24} \left(\xi - \xi^2 \right)$$

$$+ \frac{1}{S_{xz}} \frac{q_0 L^2}{2} \left(\xi - \xi^2 \right) + \frac{L^2}{2 D_{xx}} M_{xx}^T \left(\xi - \xi^2 \right), \quad (4.6.19c)$$

$$N_{xx} = \frac{B_{xx}}{D_{xx}} \left(M_{xx}^T + \frac{q_0 L^2}{12} \right) - N_{xx}^T, \quad (4.6.19d)$$

$$M_{xx}(\xi) = \frac{q_0 L^2}{2} \left(\xi - \xi^2 \right), \quad N_{xz}(\xi) = \frac{q_0 L}{2} \left(1 - 2\xi \right). \quad (4.6.19e)$$

It is clear that *all functions except the transverse deflection are the same* as those predicted by the CBT. The transverse deflection has an additional positive term that *adds* to the value predicted by the CBT. Thus, the deflection w in the TBT is larger than that predicted by the CBT (i.e., the CBT underpredicts w).

For homogeneous beams (i.e., when $B_{xx} = 0$), we have $u(x) = 0$ and $N_{xx} = -N_{xx}^T$, and the expressions for ϕ_x and w become (M_{xx} and N_{xz} remain unchanged)

$$\phi_x(\xi) = -\frac{1}{D_{xx}} \frac{q_0 L^3}{24} \left(1 - 6\xi^2 + 4\xi^3 \right) - \frac{L}{2 D_{xx}} M_{xx}^T \left(1 - 2\xi \right), \quad (4.6.20a)$$

$$w(\xi) = \frac{1}{D_{xx}} \frac{q_0 L^4}{24} \left(\xi - 2\xi^3 + \xi^4 \right) + \frac{1}{S_{xz}} \frac{q_0 L^2}{2} \left(\xi - \xi^2 \right)$$

$$+ \frac{L^2}{2 D_{xx}} M_{xx}^T \left(\xi - \xi^2 \right). \quad (4.6.20b)$$

(b) For the pinned–hinged case, the boundary conditions remain the same as given in Eq. (4.6.17), except that the condition $u(L) = 0$ is replaced with $N_{xx}(L) = 0$.

The constants of integration become

$$c_1 = c_3 = c_4 = c_6 = 0, \quad c_2 = \frac{q_0 L}{2},$$

$$c_5 = -\frac{q_0 L^3}{24} - \frac{L}{2A_{xx}}\left(A_{xx}M_{xx}^T - B_{xx}N_{xx}^T\right). \tag{4.6.21}$$

The solutions for u, ϕ_x, and w in the case of the pinned–hinged beam become

$$u(\xi) = \frac{B_{xx}}{\hat{D}_{xx}^*}\frac{q_0 L^3}{12}\left(-3\xi^2 + 2\xi^3\right) + \frac{L}{D_{xx}^*}\left(D_{xx}N_{xx}^T - B_{xx}M_{xx}^T\right)\xi, \tag{4.6.22a}$$

$$\phi_x(\xi) = -\frac{A_{xx}}{D_{xx}^*}\frac{q_0 L^3}{24}\left(1 - 6\xi^2 + 4\xi^3\right)$$

$$- \frac{L}{2D_{xx}^*}\left(A_{xx}M_{xx}^T - B_{xx}N_{xx}^T\right)(1 - 2\xi), \tag{4.6.22b}$$

$$w(\xi) = \frac{A_{xx}}{D_{xx}^*}\frac{q_0 L^4}{24}\left(\xi - 2\xi^3 + \xi^4\right) + \frac{1}{S_{xz}}\frac{q_0 L^2}{2}\left(\xi - \xi^2\right)$$

$$+ \frac{L^2}{2D_{xx}^*}\left(A_{xx}M_{xx}^T - B_{xx}N_{xx}^T\right)\left(\xi - \xi^2\right). \tag{4.6.22c}$$

The axial force becomes $N_{xx} = 0$, and the bending moment M_{xx} and N_{xz} remain the same as those listed in Eq. (4.6.19e). These solutions match, except for the shear term, with those of the CBT.

For homogeneous beams, the solutions for u, ϕ_x, and w become

$$u(\xi) = \frac{L}{A_{xx}}N_{xx}^T\,\xi, \tag{4.6.23a}$$

$$\phi_x(\xi) = -\frac{1}{D_{xx}}\frac{q_0 L^3}{24}\left(1 - 3\xi^2 + 4\xi^3\right) - \frac{1}{2D_{xx}}M_{xx}^T(1 - 2\xi), \tag{4.6.23b}$$

$$w(\xi) = \frac{1}{D_{xx}}\frac{q_0 L^4}{24}\left(\xi - 2\xi^3 + \xi^4\right) + \frac{1}{S_{xz}}\frac{q_0 L^2}{2}\left(\xi - \xi^2\right)$$

$$+ \frac{L^2}{2D_{xx}}M_{xx}^T\left(\xi - \xi^2\right). \tag{4.6.23c}$$

The numerical results obtained by the TBT with the data used in **Example 3.7.1** are either the same or very close to those obtained using the CBT. Because of the fact that the beam considered is a thin beam ($L/h = 100$), the shear deformation effect is not seen. Table 4.6.1 shows the numerical results obtained with the two theories. The bending moment M_{xx} is independent of n and the theory.

Transverse deflections obtained with the CBT and TBT for three different length-to-height ratios, $L/h = 50$, $L/h = 20$, and $L/h = 10$, are presented in Table 4.6.2. It is clear that when the beam is moderately thick ($L/h = 20$) to very thick ($L/h = 10$), the CBT slightly underpredicts the deflections, although the difference between the two solutions is not be significant. The in-plane displacements are the same in both theories because both theories have the same in-plane equations.

In closing this example, we note that in the case of the beam hinged at both ends cannot be fully solved because the boundary conditions $N_{xx} = 0$ at $x = 0$ and $x = L$ both give the same condition, $c_1 = 0$.

Table 4.6.1: Numerical results obtained with the CBT and TBT for the displacements $\bar{u} = u(0.25L) \times 10^2$ and $w(0.5L)$ and slopes $\theta_x(L) = -(dw/dx)(L)$ and $\phi_x(L)$ of pinned–pinned FGM beams subjected to uniformly distributed load of intensity q_0 (all results are normalized with respect to q_0).

n	$\bar{u}$-CBT	$\bar{u}$-TBT	w-CBT	w-TBT	θ_x	ϕ_x
0.0	.00000	.00000	0.5208	0.5210	.01657	.01657
0.5	.04435	.04435	0.7984	0.7985	.02504	.02504
1.0	.09973	.09973	1.0014	1.0016	.03062	.03062
1.5	.15457	.15457	1.1502	1.1505	.03440	.03440
2.0	.20118	.20118	1.2635	1.2638	.03722	.03722
2.5	.23707	.23707	1.3528	1.3532	.03950	.03950
3.0	.26301	.26301	1.4261	1.4265	.04148	.04148
3.5	.28098	.28098	1.4887	1.4891	.04328	.04328
4.0	.29297	.29297	1.5440	1.5445	.04495	.04495
5.0	.30523	.30523	1.6415	1.6420	.04806	.04806
6.0	.30850	.30850	1.7286	1.7292	.05097	.05097
7.5	.30565	.30565	1.8487	1.8493	.05506	.05506
10.0	.29367	.29367	2.0305	2.0312	.06131	.06131
15.0	.26661	.26661	2.3443	2.3452	.07201	.07201
20.0	.24302	.24302	2.6047	2.6056	.08080	.08080

Table 4.6.2: Numerical results obtained with the classical (CBT) and the shear deformation (TBT) beam theories for the transverse deflections of a *pinned–pinned* FGM beam subjected to uniformly distributed load of intensity q_0 (the results are normalized with the load q_0); $\bar{w} = w(0.5L)\,10$ and $\hat{w} = w(0.5L) \times 10^2$.

n	$L/h = 50$		$L/h = 20$		$L/h = 10$	
	w-CBT	w-TBT	$\bar{w}$-CBT	$\bar{w}$-TBT	$\hat{w}$-CBT	$\hat{w}$-TBT
0.0	.06510	.06517	.04167	.04193	.05208	.05338
0.5	.09979	.09989	.06387	.06424	.07984	.08169
1.0	.12517	.12529	.08011	.08058	.10014	.10250
1.5	.14378	.14392	.09202	.11502	.11785	.03440
2.0	.15793	.15809	.10108	.10173	.12635	.12960
2.5	.16910	.10823	.10895	.16929	.13528	.13892
3.0	.17827	.17847	.11409	.11409	.14261	.14661
3.5	.18609	.18630	.11910	.11996	.14887	.15320
4.0	.19300	.19324	.12352	.12445	.15440	.15905
5.0	.20518	.20544	.13132	.13236	.16415	.16935
6.0	.21608	.21636	.13829	.13943	.17286	.17855
7.5	.23108	.23140	.14789	.14916	.18487	.19118
10.0	.25382	.25417	.16244	.16387	.20305	.21020
15.0	.29304	.29346	.18755	.18921	.23443	.24275
20.0	.32559	.32605	.20838	.21020	.26047	.26957

Example 4.6.2

Determine the exact solution of a clamped (at both ends) FGM beam under uniformly distributed transverse load of intensity $q(x) = q_0$. Obtain numerical results using the material properties from the previous example and tabulate the results.

Solution: The boundary conditions for beam clamped at both ends are

$$u = 0, \quad w = 0, \quad \phi_x = 0, \quad \text{at } x = 0, L. \tag{4.6.24}$$

The condition $u(0) = 0$ gives $c_4 = 0$, $w(0) = 0$ gives $c_6 = 0$, and $\phi_x(0)$ gives $c_5 = 0$. Applying the boundary conditions at $x = L$, we obtain the following algebraic equations from Eqs. (4.6.12), (4.6.13), and (4.6.16):

$$0 = \left[D_{xx}\left(c_1 + N_{xx}^T\right) - B_{xx} M_{xx}^T\right] L + B_{xx} \frac{q_0 L^3}{6} - B_{xx}\left(c_2 \frac{L^2}{2} + c_3 L\right),$$

$$0 = \left[A_{xx}M_{xx}^T - B_{xx}\left(c_1 + N_{xx}^T\right)\right] L + A_{xx}\left(c_2 \frac{L^2}{2} + c_3 L\right) - A_{xx}\frac{q_0 L^3}{6}, \tag{4.6.25}$$

$$0 = -\left[A_{xx}M_{xx}^T - B_{xx}\left(c_1 + N_{xx}^T\right)\right]\frac{L^2}{2} - A_{xx}\left(c_2\frac{L^3}{6} + c_3\frac{L^2}{2}\right) + A_{xx}\frac{q_0 L^4}{24}$$
$$+ \frac{D_{xx}^*}{S_{xz}}\left(-q_0\frac{L^2}{2} + c_2 L\right).$$

Multiplying Eq. (4.6.25)$_2$ [i.e., the second equation of (4.6.25)] with $L/2$ and adding to Eq. (4.6.25)$_3$, we obtain

$$0 = A_{xx}\left(c_2 - \frac{q_0 L}{2}\right)\frac{L^3}{12} + \frac{D_{xx}^*}{S_{xz}}\left(c_2 - \frac{q_0 L}{2}\right) L$$

from which we find that $c_2 = q_0 L/2$. Substituting this value into the first two equations of (4.6.25), we obtain

$$0 = \left[D_{xx}\left(c_1 + N_{xx}^T\right) - B_{xx} M_{xx}^T\right] L - B_{xx}\left(\frac{q_0 L^3}{12} + c_3 L\right),$$
$$0 = \left[A_{xx}M_{xx}^T - B_{xx}\left(c_1 + N_{xx}^T\right)\right] L + A_{xx}\left(\frac{q_0 L^3}{12} + c_3 L\right), \tag{4.6.26a}$$

which yield

$$c_1 + N_{xx}^T = 0, \quad c_3 = -M_{xx}^T - \frac{q_0 L^2}{12}. \tag{4.6.26b}$$

The exact solutions for a clamped–clamped FGM beam are (note that dw/dx is not equal to zero at the clamped end and thus it contributes to the shear stress there)

$$u(\xi) = \frac{B_{xx}}{D_{xx}^*}\frac{q_0 L^3}{12}\left(\xi - 3\xi^2 + 2\xi^3\right), \tag{4.6.27}$$

$$\phi_x(\xi) = \frac{A_{xx}}{D_{xx}^*}\frac{q_0 L^3}{12}\left(-\xi + 3\xi^2 - 2\xi^3\right), \tag{4.6.28}$$

$$w(\xi) = \frac{A_{xx}}{D_{xx}^*}\frac{q_0 L^4}{24}\left(\xi^2 - 2\xi^3 + \xi^4\right) + \frac{1}{S_{xz}}\frac{q_0 L^2}{2}\left(\xi - \xi^2\right), \tag{4.6.29}$$

$$M_{xx}(\xi) = -\frac{q_0 L^2}{12}\left(1 - 6\xi + 6\xi^2\right) - M_{xx}^T, \tag{4.6.30}$$

$$N_{xz}(\xi) = \frac{q_0 L}{2}\left(1 - 2\xi\right), \quad N_{xx} = -N_{xx}^T. \tag{4.6.31}$$

169

The numerical results obtained for $L/h = 100$ using the TBT are again either the same or very close to those obtained using the CBT. Table 4.6.3 shows the numerical results obtained with the two theories. Transverse deflections obtained with the CBT and TBT for three different length-to-height ratios, $L/h = 50, 20$, and 10 are presented in Table 4.6.4. The results are normalized with the load, q_0.

Table 4.6.3: Numerical results obtained with classical (CBT) and shear deformation (TBT) beam theories for the displacements $\bar{u} = u(0.25L) \times 10^2$ and $w(L)$ and slopes $\bar{\theta}_x(L) = -(dw/dx)(L) \times 10^2$ and $\bar{\phi}_x = \phi_x(L) \times 10^2$ of a clamped–clamped FGM beams under a uniformly distributed load.

n	$\bar{u}$-CBT	$\bar{u}$-TBT	w-CBT	w-TBT	$\bar{\theta}_x$	$\bar{\phi}_x$
0.0	.00000	.00000	.10417	.10430	.09460	.09460
0.5	.04435	.04435	.17246	.17265	.15663	.15663
1.0	.09973	.09973	.24379	.24403	.22141	.22141
1.5	.15457	.15457	.30724	.30752	.27903	.27903
2.0	.20118	.20118	.35765	.35798	.32482	.32482
2.5	.23707	.23707	.39511	.39548	.35884	.35884
3.0	.26301	.26301	.42212	.42252	.38337	.38337
3.5	.28098	.28098	.44154	.44197	.40100	.40100
4.0	.29297	.29297	.45573	.45619	.41389	.41389
5.0	.30523	.30523	.47481	.47533	.43122	.43122
6.0	.30850	.30850	.48751	.48808	.44276	.44276
7.5	.30565	.30565	.50187	.50250	.45580	.45580
10.0	.29367	.29367	.52209	.52280	.47416	.47416
15.0	.26661	.26661	.55956	.56039	.50819	.50819
20.0	.24302	.24302	.59405	.59496	.53952	.53952

Table 4.6.4: Numerical results obtained with the classical (CBT) and the shear deformation (TBT) beam theories for the transverse deflections of clamped–clamped FGM beams under a uniformly distributed load; $\bar{w} = w(0.5L)\,10$, $\hat{w} = w(0.5L) \times 10^2$, and $\tilde{w} = w(0.5L) \times 10^3$.

	$L/h = 50$		$L/h = 20$		$L/h = 10$	
n	$\bar{w}$-CBT	$\bar{w}$-TBT	$\hat{w}$-CBT	$\hat{w}$-TBT	$\tilde{w}$-CBT	$\tilde{w}$-TBT
0.0	.13021	.13086	.83333	.85933	.10417	.11717
0.5	.21558	.21651	.13797	.14168	.17246	.19103
1.0	.30474	.30592	.19504	.19976	.24379	.26743
1.5	.38405	.38546	.24579	.25144	.30724	.33550
2.0	.44707	.44869	.28612	.29262	.35765	.39015
2.5	.49389	.49571	.31609	.32337	.39511	.43151
3.0	.52765	.52965	.33770	.34570	.42212	.46212
3.5	.55192	.55409	.35323	.36190	.44154	.48487
4.0	.56966	.57198	.36458	.37387	.45573	.50216
5.0	.59351	.59611	.37984	.39024	.47481	.52681
6.0	.60938	.61223	.39001	.40138	.48751	.54438
7.5	.62733	.63049	.40149	.41412	.50187	.56501
10.0	.65261	.65619	.41767	.43197	.52209	.59359
15.0	.69945	.70361	.44765	.46429	.55956	.64276
20.0	.74256	.74711	.47524	.49344	.59405	.68505

Example 4.6.3

We wish to determine the exact solution of a beam clamped at the left end, vertically spring-supported at the right end, and subjected to uniformly distributed transverse load of intensity $q(x) = q_0$, as shown in Fig. 3.7.7(a). Assume that the spring is linearly elastic, with a spring constant of k (N/m). Specialize the results to (a) cantilevered beam shown in Fig. 3.7.7(b) and (b) propped cantilever beam shown in Fig. 3.7.7(c).

Solution: The boundary conditions at the clamped end, $x = 0$, are

$$u = 0, \quad w = 0, \quad \phi_x = 0, \tag{4.6.32a}$$

and at the right end, $x = L$, we have

$$N_{xx} = 0, \quad N_{xz} + kw = 0, \quad M_{xx} = 0. \tag{4.6.32b}$$

As before, using Eqs. (4.6.12), (4.6.13), and (4.6.16) for $u(x)$, $\phi_x(x)$, and $w(x)$, respectively, we obtain:

$$u(0) = 0 \;\rightarrow\; c_4 = 0, \quad \phi_x(L) = 0 \;\rightarrow\; c_5 = 0, \quad w(0) = 0 \;\rightarrow\; c_6 = 0.$$

From the first equation in Eq. (4.6.6), the condition $N_{xx}(L) = 0$ gives $c_1 = 0$; and $M_{xx}(L) = 0$ gives $c_2 L + c_3 = q_0 L^2/2$. Finally, the mixed boundary condition $N_{xz} + kw = 0$ at $x = L$ yields (with $c_3 = q_0 L^2/2 - c_3 L$) the result

$$
\begin{aligned}
0 = -q_0 L + c_2 + k &\left[\frac{1}{S_{xz}} \left(-\frac{q_0 L^2}{2} + c_2 L \right) - \left(\frac{A_{xx}}{D_{xx}^*} M_{xx}^T - \frac{B_{xx}}{D_{xx}^*} N_{xx}^T \right) \frac{L^2}{2} \right. \\
&\left. + \frac{A_{xx}}{D_{xx}^*} \left(c_2 \frac{L^3}{3} - \frac{5 q_0 L^4}{24} \right) \right] \equiv -q_0 L + \hat{c}_1 c_2 - k \hat{c}_2, \tag{4.6.33a}
\end{aligned}
$$

where

$$
\begin{aligned}
\hat{c}_1 &= 1 + kL \left(\frac{1}{S_{xz}} + \frac{A_{xx} L^2}{3 D_{xx}^*} \right), \\
\hat{c}_2 &= \frac{L^2}{2} \left[\left(\frac{A_{xx}}{D_{xx}^*} M_{xx}^T - \frac{B_{xx}}{D_{xx}^*} N_{xx}^T \right) + \frac{A_{xx}}{D_{xx}^*} \frac{5 q_0 L^2}{12} + \frac{1}{S_{xz}} q_0 \right].
\end{aligned}
\tag{4.6.33b}
$$

Thus, we have

$$c_1 = 0, \quad c_2 = \frac{1}{\hat{c}_1} (k \hat{c}_2 + q_0 L), \quad c_3 = \frac{q_0 L^2}{2} - c_2 L. \tag{4.6.34}$$

With all of the constants of integration known, the exact solutions for an FGM beam clamped at $x = 0$ and supported at $x = L$ by a linearly elastic spring are [expressed in terms of c_2, which is known from Eq. (4.6.34); $\xi = xL$]

$$u(\xi) = \frac{L}{D^*_{xx}} \left(D_{xx} N^T_{xx} - B_{xx} M^T_{xx} \right) \xi + \frac{B_{xx}}{D^*_{xx}} \left[\frac{q_0 L^3}{6} \left(\xi^3 - 3\xi \right) + \frac{c_2 L^2}{2} \left(2\xi - \xi^2 \right) \right],$$
(4.6.35)

$$\phi_x(\xi) = \frac{L}{D^*_{xx}} \left(A_{xx} M^T_{xx} - B_{xx} N^T_{xx} \right) \xi - \frac{A_{xx}}{D^*_{xx}} \left[\frac{q_0 L^3}{6} \left(\xi^3 - 3\xi \right) + \frac{c_2 L^2}{2} \left(2\xi - \xi^2 \right) \right],$$
(4.6.36)

$$w(\xi) = \frac{1}{S_{xz}} \left(-\frac{q_0 L^2}{2} \xi^2 + c_2 L \xi \right) - \frac{L^2}{2D^*_{xx}} \left(A_{xx} M^T_{xx} - B_{xx} N^T_{xx} \right) \xi^2$$
$$+ \frac{A_{xx}}{D^*_{xx}} \left[\frac{q_0 L^4}{24} \left(\xi^4 - 6\xi^2 \right) + \frac{c_2 L^3}{6} \left(3\xi^2 - \xi^3 \right) \right].$$
(4.6.37)

The solution in Eqs. (4.6.35)–(4.6.37) have two special cases: (a) $k = 0$ (fixed-free beam) and (b) $k \to \infty$ (fixed-hinged beam), as explained next.

(a) *Cantilevered beam* [see Fig. 3.7.7(b)]. For a cantilevered beam fixed at $x = 0$ and free at $x = L$, and subjected to uniformly distributed load, we set $k = 0$ and obtain $\hat{c}_1 = 1$ and $c_2 = q_0 L$. Then the solution becomes

$$u(\xi) = \frac{L}{D^*_{xx}} \left(D_{xx} N^T_{xx} - B_{xx} M^T_{xx} \right) \xi + \frac{B_{xx}}{D^*_{xx}} \frac{q_0 L^3}{6} \left(3\xi - 3\xi^2 + \xi^3 \right),$$
(4.6.38)

$$\phi_x(\xi) = \frac{L}{D^*_{xx}} \left(A_{xx} M^T_{xx} - B_{xx} N^T_{xx} \right) \xi - \frac{A_{xx}}{D^*_{xx}} \frac{q_0 L^3}{6} \left(3\xi - 3\xi^2 + \xi^3 \right),$$
(4.6.39)

$$w(\xi) = \frac{q_0 L^2}{2S_{xz}} \left(2\xi - \xi^2 \right) - \frac{L^2}{2D^*_{xx}} \left(A_{xx} M^T_{xx} - B_{xx} N^T_{xx} \right) \xi^2$$
$$+ \frac{A_{xx}}{D^*_{xx}} \frac{q_0 L^4}{24} \left(6\xi^2 - 4\xi^3 + \xi^4 \right),$$
(4.6.40)

$$M_{xx}(\xi) = -\frac{q_0 L^2}{2} \left(1 - 2\xi + \xi^2 \right), \quad N_{xz}(\xi) = q_0 L \left(1 - \xi \right).$$
(4.6.41)

(b) *Propped Cantilever* [see Fig. 3.7.7(c)]. For a beam clamped at $x = 0$ and hinged at $x = L$, we set $1/k = 0$ (i.e. $k \to \infty$) and obtain

$$c_2 L = \frac{1}{8(1 + 3\Omega)} \left[12 \left(M^T_{xx} - \frac{B_{xx}}{A_{xx}} N^T_{xx} \right) + 5q_0 L^2 + 12\Omega q_0 L^2 \right], \quad \Omega = \frac{D^*_{xx}}{A_{xx} S_{xz} L^2}.$$
(4.6.42)

Then the solution is known from Eqs. (4.6.35)–(4.6.37) with c_2 given by Eq. (4.6.42).

The analytical solution for a homogeneous isotropic beam on elastic foundation ($c_f \neq 0$) using the TBT is discussed next. The couple stress effect is not considered here. For this case, the equations of equilibrium for bending are decoupled with that of the axial deformation. The equations of interest are [see Eqs. (4.6.2) and (4.6.3); adding the foundation modulus term, $c_f w$]:

$$-\frac{d}{dx}\left[S_{xz}\left(\phi_x + \frac{dw}{dx}\right)\right] + c_f w - q = 0, \tag{4.6.43}$$

$$-\frac{d}{dx}\left(D_{xx}\frac{d\phi_x}{dx}\right) + S_{xz}\left(\phi_x + \frac{dw}{dx}\right) = 0. \tag{4.6.44}$$

Solving Eq. (4.6.43) for $d\phi_x/dx$ and substituting into the first derivative of Eq. (4.6.44), we obtain

$$\frac{d\phi_x}{dx} = -\frac{d^2w}{dx^2} + \frac{1}{S_{xz}}(c_f w - q), \tag{4.6.45}$$

$$-\frac{d^2}{dx^2}\left(-D_{xx}\frac{d^2w}{dx^2}\right) - \frac{1}{S_{xz}}\left(k\frac{d^2w}{dx^2} - \frac{d^2q}{dx^2}\right) + c_f w - q = 0. \tag{4.6.46}$$

Equation (4.6.46) can be expressed as

$$a\frac{d^4w}{dx^4} - b\frac{d^2w}{dx^2} + cw = F(x), \tag{4.6.47a}$$

where

$$a = D_{xx}, \quad b = \frac{c_f}{S_{xz}}, \quad c = c_f, \quad F(x) = q - \frac{1}{S_{xz}}\frac{d^2q}{dx^2}. \tag{4.6.47b}$$

Assume the solution of the homogeneous equation (4.6.47a) (i.e., with $F(x) = 0$) in the form $w_h(x) = Ae^{rx}$ and substitute it into Eq. (4.6.47a) and obtain

$$ar^4 - br^2 + c = 0 \quad \text{or} \quad as^2 - bs + c = 0 \ (s = r^2). \tag{4.6.48}$$

The roots of the above quadratic equation are

$$s_1 = \frac{1}{2a}\left(b - \sqrt{b^2 - 4ac}\right) \equiv -\lambda^2, \quad s_2 = \frac{1}{2a}\left(b + \sqrt{b^2 - 4ac}\right) \equiv \mu^2. \tag{4.6.49}$$

Hence, the homogeneous solution of Eq. (4.6.47a) can be expressed as

$$w_h(x) = c_1 \sin \lambda x + c_2 \cos \lambda x + c_3 \sinh \mu x + c_4 \cosh \mu x, \tag{4.6.50a}$$

where

$$\lambda = -\sqrt{\frac{1}{2a}\left(b - \sqrt{b^2 - 4ac}\right)},$$
$$\mu = \sqrt{\frac{1}{2a}\left(b + \sqrt{b^2 - 4ac}\right)}, \tag{4.6.50b}$$

and c_1, c_2, c_3, and c_4 are integration constants, which are to be determined using the boundary conditions (two at each edge of the beam). The particular solution, w_p, is determined using $F(x)$.

For example, for uniform load $q = q_0$, we have $F(x) = q_0$ and $w_p = q_0/c_f$ $(k = c_f)$, and the complete solution is

$$w(x) = c_1 \sin \lambda x + c_2 \cos \lambda x + c_3 \sinh \mu x + c_4 \cosh \mu x + \frac{q_0}{c_f}, \tag{4.6.51}$$

$$\frac{dw}{dx} = \lambda \left(c_1 \cos \lambda x - c_2 \sin \lambda x \right) + \mu \left(c_3 \cosh \mu x + c_4 \sinh \mu x \right), \tag{4.6.52}$$

$$\frac{d^2 w}{dx^2} = -\lambda^2 \left(c_1 \sin \lambda x + c_2 \cos \lambda x \right) + \mu^2 \left(c_3 \sinh \mu x + c_4 \cosh \mu x \right). \tag{4.6.53}$$

Then we can compute $d\phi_x/dx$ from Eq. (4.6.46) as

$$\begin{aligned}
\frac{d\phi_x}{dx} = {} & \left(\lambda^2 + \frac{c_f}{S_{xz}} \right) \left(c_1 \sin \lambda x + c_2 \cos \lambda x \right) \\
& - \left(\mu^2 - \frac{c_f}{S_{xz}} \right) \left(c_3 \sinh \mu x + c_4 \cosh \mu x \right).
\end{aligned} \tag{4.6.54}$$

The second derivative of ϕ_x is

$$\begin{aligned}
\frac{d^2 \phi_x}{dx^2} = {} & \lambda \left(\lambda^2 + \frac{c_f}{S_{xz}} \right) \left(c_1 \cos \lambda x - c_2 \sin \lambda x \right) \\
& - \mu \left(\mu^2 - \frac{c_f}{S_{xz}} \right) \left(c_3 \cosh \mu x + c_4 \sinh \mu x \right).
\end{aligned} \tag{4.6.55}$$

Then, from Eq. (4.6.44), we have $[\phi_x = (D_{xx}/S_{xz})(d^2\phi_x/dx^2) - (dw/dx)]$

$$\begin{aligned}
\phi_x(x) = {} & \lambda \frac{D_{xx}}{S_{xz}} \left(\lambda^2 + \frac{c_f}{S_{xz}} \right) \left(c_1 \cos \lambda x - c_2 \sin \lambda x \right) \\
& - \mu \frac{D_{xx}}{S_{xz}} \left(\mu^2 - \frac{c_f}{S_{xz}} \right) \left(c_3 \sinh \mu x + c_4 \cosh \mu x \right) \\
& - \lambda \left(c_1 \cos \lambda x - c_2 \sin \lambda x \right) - \mu \left(c_3 \cosh \mu x + c_4 \sinh \mu x \right). \tag{4.6.56}
\end{aligned}$$

For a simply supported beam on elastic foundation and loaded with uniformly distributed load, the boundary conditions are $w = M_{xx} = 0$ at $x = 0$ and $x = L$. These conditions give $(M_{xx} = D_{xx} d\phi_x/dx)$

$$c_2 + c_4 + \frac{q_0}{c_f} = 0,$$

$$\left(\lambda^2 + \frac{c_f}{S_{xz}} \right) c_2 - \left(\mu^2 - \frac{c_f}{S_{xz}} \right) c_4 = 0,$$

$$c_1 \sin \lambda L + c_2 \cos \lambda L + c_3 \sinh \mu L + c_4 \cosh \mu L + \frac{q_0}{c_f} = 0, \tag{4.6.57}$$

$$\left(\lambda^2 + \frac{c_f}{S_{xz}} \right) \left(c_1 \sin \lambda L + c_2 \cos \lambda L \right)$$

$$- \left(\mu^2 - \frac{c_f}{S_{xz}} \right) \left(c_3 \sinh \mu L + c_4 \cosh \mu L \right) = 0.$$

These equations can be used to determine all of the constants of integration.

4.6.2 BUCKLING SOLUTIONS

Here we consider buckling of homogeneous isotropic beams (with $c_f = 0$) using the TBT under the action of applied axial load $N_{xx} = -N_{xx}^0$. The governing equations are

$$-\frac{dM_{xx}}{dx} + N_{xz} = 0, \quad -\frac{dN_{xz}}{dx} + N_{xx}^0 \frac{d^2 w}{dx^2} = 0, \qquad (4.6.58)$$

where M_{xx} and N_{xz} are related to the deflection w and rotation function ϕ_x by

$$M_{xx} = D_{xx}\frac{d\phi_x}{dx}, \quad N_{xz} = S_{xz}\left(\phi_x + \frac{dw}{dx}\right). \qquad (4.6.59)$$

From the two equations in Eq. (4.6.58), we have

$$-\frac{d^2 M_{xx}}{dx^2} + N_{xx}^0 \frac{d^2 w}{dx^2} = 0, \quad \rightarrow \quad D_{xx}\frac{d^3 \phi_x}{dx^3} = N_{xx}^0 \frac{d^2 w}{dx^2}. \qquad (4.6.60)$$

Using Eqs. (4.6.58) and (4.6.59), we can establish the relation,

$$\frac{d\phi_x}{dx} = -\left(1 - \frac{N_{xx}^0}{S_{xz}}\right)\frac{d^2 w}{dx^2}. \qquad (4.6.61)$$

Using Eq. (4.6.61) in Eq. (4.6.60), we arrive at

$$D_{xx}\frac{d^4 w}{dx^4} + \frac{N_{xx}^0}{1 - \frac{N_{xx}^0}{S_{xz}}}\frac{d^2 w}{dx^2} = 0. \qquad (4.6.62)$$

Integrating the above equation twice with respect to x, we obtain

$$D_{xx}\frac{d^2 w}{dx^2} + \frac{N_{xx}^0}{1 - \frac{N_{xx}^0}{S_{xz}}}w = K_1 x + K_2, \qquad (4.6.63)$$

where K_1 and K_2 are the constants of integration. The homogeneous solution of Eq. (4.6.63) is determined from

$$D_{xx}\frac{d^2 w}{dx^2} + \frac{N_{xx}^0}{1 - \frac{N_{xx}^0}{S_{xz}}}w = 0 \quad \text{or} \quad \frac{d^2 w}{dx^2} + \lambda^2 w = 0, \qquad (4.6.64a)$$

where

$$\lambda^2 = \frac{N_{xx}^0}{D_{xx}\left(1 - \frac{N_{xx}^0}{S_{xz}}\right)} \quad \text{and} \quad N_{xx}^0 = \frac{D_{xx}\lambda^2}{\left(1 + \lambda^2 \frac{D_{xx}}{S_{xz}}\right)}. \qquad (4.6.64b)$$

The homogeneous and particular solutions of the second-order equation in Eq. (4.6.63) are given, respectively, by

$$w_h(x) = c_1 \sin \lambda x + c_2 \cos \lambda x, \quad w_p(x) = \frac{1}{D_{xx}\lambda^2}(K_1 x + K_2), \qquad (4.6.65)$$

which are exactly of the same form as those derived for the CBT in Eqs. (3.7.56) and (3.7.57), except for the definition of λ (also, D_{xx}^e is replaced with D_{xx}). Hence, the rest of the developments presented there apply here with λ defined as in Eq. (4.6.64b). The complete solution is given by

$$w(x) = c_1 \sin \lambda x + c_2 \cos \lambda x + c_3\, x + c_4, \qquad (4.6.66)$$

where $c_3 = K_1/D_{xx}\lambda^2$ and $c_4 = K_2/D_{xx}\lambda^2$. Three of the four constants c_1, c_2, c_3, c_4, and λ (or N_{xx}^0) are determined using (four) boundary conditions of the problem. Once λ is known, the buckling load can be determined using Eq. (4.6.64b).

Example 4.6.4

Determine the critical buckling load of a (a) simply supported and (b) clamped beams of length L using the TBT.

Solution: (a) *Simply supported beam.* We only need to have the value of λ to determine the buckling loads using the TBT. From **Example 3.7.5**, we have $\lambda L = n\pi$. Hence the buckling loads as per the TBT are given by

$$N_{xx}^0 = \frac{D_{xx}\lambda^2}{\left(1 + \lambda^2 \frac{D_{xx}}{S_{xz}}\right)} = \left[\frac{n\pi}{L}\right]^2 D_{xx} \left(1 + \left(\frac{n\pi}{L}\right)^2 \frac{D_{xx}}{S_{xz}}\right)^{-1}. \qquad (4.6.67)$$

It is clear that the buckling loads predicted by the TBT are smaller than those predicted by the CBT. In other words, the CBT over-predicts the buckling loads compared to the TBT. The critical buckling load is given by $(n = 1)$

$$N_{\text{crit}}^0 = 9.8696 \frac{D_{xx}}{L^2} \left(1 + 9.8696 \frac{D_{xx}}{S_{xz}L^2}\right)^{-1}. \qquad (4.6.68)$$

(b) *Clamped beam.* From **Example 3.7.6**, we have $\lambda L = 2n\pi$. Hence the buckling loads as per the TBT are given by

$$N_{xx}^0 = \left(\frac{2n\pi}{L}\right)^2 D_{xx} \left[1 + \left(\frac{2n\pi}{L}\right)^2 \frac{D_{xx}}{S_{xz}}\right]^{-1}. \qquad (4.6.69)$$

The critical buckling load is given by $(n = 1)$

$$N_{\text{crit}}^0 = 39.4784 \frac{D_{xx}}{L^2} \left(1 + 39.4784 \frac{D_{xx}}{S_{xz}L^2}\right)^{-1}. \qquad (4.6.70)$$

Buckling loads for other boundary conditions using the TBT can be easily found just by knowing the values of λ from Table 3.7.1.

4.6.3 NATURAL VIBRATION

For natural (or free) vibration in the absence of any applied transverse mechanical and thermal loads (and $c_f = 0$), the governing equations are [see Eqs. (4.4.16) and (4.4.17)]:

$$-\frac{\partial}{\partial x}\left[S_{xz}\left(\phi_x + \frac{\partial w}{\partial x}\right)\right] + m_0 \frac{\partial^2 w}{\partial t^2} = 0, \tag{4.6.71}$$

$$-\frac{\partial}{\partial x}\left(D_{xx}\frac{\partial \phi_x}{\partial x}\right) + S_{xz}\left(\phi_x + \frac{\partial w}{\partial x}\right) + m_2 \frac{\partial^2 \phi_x}{\partial t^2} = 0. \tag{4.6.72}$$

The solution of Eqs. (4.6.71) and (4.6.72) is assumed to be periodic

$$w(x,t) = W(x)\,e^{i\omega t}, \quad \phi_x(x,t) = \Phi(x)\,e^{i\omega t} \tag{4.6.73}$$

where ω is the natural frequency of vibration and (W, Φ) are the mode shapes. Substituting for (w, ϕ_x) from Eq. (4.6.73) into Eqs. (4.6.71) and (4.6.72), we obtain

$$-\frac{d}{dx}\left[S_{xz}\left(\Phi + \frac{dW}{dx}\right)\right] - m_0\,\omega^2\,W = 0, \tag{4.6.74}$$

$$-\frac{d}{dx}\left(D_{xx}\frac{d\Phi}{dx}\right) + S_{xz}\left(\Phi + \frac{dW}{dx}\right) - m_2\,\omega^2\,\Phi = 0. \tag{4.6.75}$$

We wish to rewrite these equations solely in terms of $W(x)$. Toward this end, we solve Eq. (4.6.74) for $d\Phi/dx$ in terms of W and its derivatives:

$$\frac{d\Phi}{dx} = -\frac{d^2 W}{dx^2} - \frac{m_0}{S_{xz}}\omega^2\,W. \tag{4.6.76}$$

Differentiating Eq. (4.6.75) once with respect to x and replacing the second expression with the help of Eq. (4.6.74), we obtain

$$D_{xx}\frac{d^3 \Phi}{dx^3} + m_0\,\omega^2\,W + m_2\,\omega^2\,\frac{d\Phi}{dx} = 0. \tag{4.6.77}$$

Now using Eq. (4.6.76), Eq. (4.6.77) can be solely expressed in terms of W and its derivatives as

$$D_{xx}\left(\frac{d^4 W}{dx^4} + \frac{m_0}{S_{xz}}\omega^2\frac{d^2 W}{dx^2}\right) - m_0\,\omega^2\,W + m_2\,\omega^2\left(\frac{d^2 W}{dx^2} + \frac{m_0}{S_{xz}}\omega^2\,W\right) = 0. \tag{4.6.78}$$

Equation (4.6.78) can be expressed in the following general form [cf. Eqs. (3.7.74a) and (3.7.74b)]:

$$a\frac{d^4 W}{dx^4} + b\frac{d^2 W}{dx^2} - cW = 0, \tag{4.6.79a}$$

where

$$a = D_{xx}, \quad b = \left(m_2 + m_0 \frac{D_{xx}}{S_{xz}}\right)\omega^2, \quad c = m_0\left(1 - \frac{m_2\omega^2}{S_{xz}}\right)\omega^2. \quad (4.6.79b)$$

We note that when we set $1/S_{xz} = 0$, we obtain Eqs. (3.7.74a) and (3.7.74b). Due to the similarity of Eqs. (4.6.79a) and (4.6.79b) to Eqs. (3.7.74a) and (3.7.74b), the following results from Eqs. (3.7.78) and (3.7.79) hold:

$$w_0(x) = c_1 \sin \lambda x + c_2 \cos \lambda x + c_3 \sinh \mu x + c_4, \cosh \mu x \quad (4.6.80a)$$

where

$$\lambda = \sqrt{\frac{1}{2a}\left(b + \sqrt{b^2 + 4ac}\right)}, \quad \mu = \sqrt{\frac{1}{2a}\left(-b + \sqrt{b^2 + 4ac}\right)}, \quad (4.6.80b)$$

and c_1, c_2, c_3, and c_4 are integration constants, which are to be determined using the boundary conditions. From the first expression in Eq. (4.6.80b), we obtain

$$\left(2a\lambda^2 - b\right)^2 = b^2 + 4ac \quad \text{or} \quad a\lambda^4 - b\lambda^2 - c = 0,$$

and using the definitions of a, b, and c from Eq. (4.6.79b), we obtain

$$D_{xx}\lambda^4 - \left(m_2 + \frac{m_0 D_{xx}}{S_{xz}}\right)\omega^2\lambda^2 - m_0\left(1 - \frac{m_2\omega^2}{S_{xz}}\right)\omega^2 = 0, \quad (4.6.81)$$

from which we obtain a quadratic equation for ω^2:

$$-\left(\frac{m_0 m_2}{S_{xz}}\right)\omega^4 + \left[m_0 + \left(m_2 + \frac{m_0 D_{xx}}{S_{xz}}\right)\lambda^2\right]\omega^2 - D_{xx}\lambda^4 = 0. \quad (4.6.82)$$

4.7   RELATIONS BETWEEN CBT AND TBT

4.7.1   BACKGROUND

Since the solutions of classical beam and plate theories are available in textbooks, it is desirable to have algebraic relationships that link the classical beam and plate solutions to those of the shear deformation theories. In the last two decades, Wang and his colleagues [116]–[121] (see Wang, Reddy, and Lee [120] for additional references) have developed such relationships connecting the solutions (e.g., deflections, shear forces, bending moments, buckling loads, and natural frequencies) of various shear deformation beam and plate theories to those of the respective classical beam and plate theories. These relationships provide an efficient and quick way to determine the solutions based on shear deformation theories. Such relationships were also developed for functionally graded beams with a microstructure-dependent length–scale parameter by Reddy and Arbind [122].

This section is concerned with the development of algebraic relationships between the bending solutions for deflections, slopes/rotations, bending moments, and shear forces of the linearized TBT for FGM beams in terms of the same quantities of the linearized CBT. These relationships enable the determination of various quantities of interest for any boundary conditions and loads using the TBT for FGM beams by knowing the corresponding quantities of the CBT for the same problem.

Following this introduction, the governing equations of the CBT and TBT are summarized. The equations are then used, using the similarity of the equations of the two theories and the load equivalence, to develop the relationships for bending deflections, slopes, and stress resultants of the two beam theories. The discussion is presented herein is limited to linear bending problems.

4.7.2   BENDING RELATIONS BETWEEN THE CBT AND TBT

4.7.2.1   Summary of equations of the CBT

The governing equations of equilibrium, expressed in terms of the stress resultants (N_{xx}, N_{xz}, M_{xx}), of the CBT are [see Eqs. (3.3.8)–(3.3.10)]

$$-\frac{dN_{xx}^C}{dx} = f, \quad -\frac{dN_{xz}^C}{dx} = q, \quad -\frac{dM_{xx}^C}{dx} + N_{xz}^C = 0, \qquad (4.7.1)$$

where (f, q) are the distributed axial and transversely distributed loads, respectively, and $(N_{xx}^C, M_{xx}^C, N_{xz}^C)$ are the axial force, bending moment, and shear force, respectively; here, the superscript "C" refers to the CBT. They are related to the generalized displacements $(u^C, w^C, \theta_x^C = -dw^C/dx)$ by [see Eqs. (3.4.6a) and (3.4.6b)]

$$N_{xx}^C = A_{xx}\frac{du^C}{dx} - B_{xx}\frac{d^2w^C}{dx^2} - N_{xx}^T,$$
$$M_{xx}^C = B_{xx}\frac{du^C}{dx} - D_{xx}\frac{d^2w^C}{dx^2} - M_{xx}^T, \qquad (4.7.2a)$$

or

$$\left\{\begin{array}{c} N_{xx}^C \\ M_{xx}^C \end{array}\right\} = \left[\begin{array}{cc} A_{xx} & B_{xx} \\ B_{xx} & D_{xx} \end{array}\right]\left\{\begin{array}{c} \frac{du^C}{dx} \\ -\frac{d^2w^C}{dx^2} \end{array}\right\} - \left\{\begin{array}{c} N_{xx}^T \\ M_{xx}^T \end{array}\right\}. \qquad (4.7.2b)$$

4.7.2.2   Summary of equations of the TBT

The governing equations of the TBT are the same as those for the CBT but the stress resultant–displacement relations are different. To distinguish the two sets of equations, we use the superscript "F" for the TBT:

$$-\frac{dN_{xx}^F}{dx} = f, \quad -\frac{dN_{xz}^F}{dx} = q, \quad -\frac{dM_{xx}^F}{dx} + N_{xz}^F = 0. \qquad (4.7.3)$$

The resultants $(N_{xx}^F, M_{xx}^F, N_{xz}^F)$ are related to the generalized displacements (u^F, w^F, ϕ_x^F) by

$$N_{xx}^F = A_{xx}\frac{du^F}{dx} + B_{xx}\frac{d\phi_x^F}{dx} - N_{xx}^T,$$

$$M_{xx}^F = B_{xx}\frac{du^F}{dx} + D_{xx}\frac{d\phi_x^F}{dx} - M_{xx}^T,$$

(4.7.4a)

or

$$\left\{\begin{array}{c} N_{xx}^F \\ M_{xx}^F \end{array}\right\} = \left[\begin{array}{cc} A_{xx} & B_{xx} \\ B_{xx} & D_{xx} \end{array}\right]\left\{\begin{array}{c} \frac{du^F}{dx} \\ \frac{d\phi_x^F}{dx} \end{array}\right\} - \left\{\begin{array}{c} N_{xx}^T \\ M_{xx}^T \end{array}\right\},$$

(4.7.4b)

and

$$N_{xz}^F = S_{xz}\left(\phi_x^F + \frac{dw^F}{dx}\right).$$

(4.7.4c)

Next, we use the similarity of the equations of the two theories and the *load equivalence* (i.e., f, q, and the beam stiffness coefficients A_{xx}, B_{xx}, and D_{xx} are the same in both theories) to develop the relationships between the solutions of the two theories.

4.7.2.3 Relationships by similarity and load equivalence

From the first equations of Eq. (4.7.1) and the first two equations of Eq. (4.7.3) that

$$\frac{dN_{xx}^F}{dx} = \frac{dN_{xx}^C}{dx} \rightarrow N_{xx}^F = N_{xx}^C + c_1,$$

(4.7.5)

$$\frac{dN_{xz}^F}{dx} = \frac{dN_{xz}^C}{dx} \rightarrow N_{xz}^F = N_{xz}^C + c_2,$$

(4.7.6)

where c_1 and c_2 are constants of integration. From the third equation in Eq. (4.7.3) and Eq. (4.7.6) we have

$$0 = -\frac{dM_{xx}^F}{dx} + N_{xz}^F = -\frac{dM_{xx}^F}{dx} + N_{xz}^C + c_2,$$

or, using the third equation in Eq. (4.7.1),

$$-\frac{dM_{xx}^F}{dx} + \frac{dM_{xx}^C}{dx} + c_2 = 0 \rightarrow M_{xx}^F = M_{xx}^C + c_2 x + c_3.$$

(4.7.7)

Using Eqs. (4.7.2b) and (4.7.4b) in Eqs. (4.7.5) and (4.7.7), we obtain

$$\left[\begin{array}{cc} A_{xx} & B_{xx} \\ B_{xx} & D_{xx} \end{array}\right]\left\{\begin{array}{c} \frac{du^F}{dx} \\ \frac{d\phi_x^F}{dx} \end{array}\right\} = \left[\begin{array}{cc} A_{xx} & B_{xx} \\ B_{xx} & D_{xx} \end{array}\right]\left\{\begin{array}{c} \frac{du^C}{dx} \\ -\frac{d^2 w^c}{dx^2} \end{array}\right\} + \left\{\begin{array}{c} c_1 \\ c_2\, x + c_3. \end{array}\right\}.$$

(4.7.8)

Solving for du^F/dx and $d\phi_x^F/dx$, we obtain

$$\frac{du^F}{dx} = \frac{du^C}{dx} + \frac{1}{D_{xx}^*}\left[c_1 D_{xx} - B_{xx}\left(c_2\, x + c_3\right)\right], \qquad (4.7.9)$$

$$\frac{d\phi_x^F}{dx} = -\frac{d^2w^C}{dx^2} + \frac{1}{D_{xx}^*}\left[-c_1 B_{xx} + A_{xx}\left(c_2\, x + c_3\right)\right]. \qquad (4.7.10)$$

Integrating the above equations, we obtain

$$u^F(x) = u^C(x) + \frac{1}{D_{xx}^*}\left[c_1 D_{xx}\, x - B_{xx}\left(c_2\,\frac{x^2}{2} + c_3\, x + c_4\right)\right], \qquad (4.7.11)$$

$$\phi_x^F(x) = -\frac{dw^C}{dx} + \frac{1}{D_{xx}^*}\left[-c_1 B_{xx}\, x + A_{xx}\left(c_2\,\frac{x^2}{2} + c_3\, x + c_5\right)\right]. \qquad (4.7.12)$$

Next, we use Eqs. (4.7.6), (4.7.4c), and (4.7.1) to obtain

$$S_{xz}\left(\phi_x^F + \frac{dw^F}{dx}\right) = N_{xz}^C + c_2 = \frac{dM_{xx}^C}{dx} + c_2. \qquad (4.7.13)$$

Solving for dw^F/dx,

$$\begin{aligned}
\frac{dw^F}{dx} &= -\phi_x^F + \frac{1}{S_{xz}}\left(\frac{dM_{xx}^C}{dx} + c_2\right) \\
&= \frac{dw^C}{dx} - \frac{1}{D_{xx}^*}\left[-c_1 B_{xx}\, x + A_{xx}\left(c_2\,\frac{x^2}{2} + c_3\, x + c_5\right)\right] \\
&\quad + \frac{1}{S_{xz}}\left(\frac{dM_{xx}^C}{dx} + c_2\right),
\end{aligned} \qquad (4.7.14)$$

and integrating once, we obtain

$$\begin{aligned}
w^F(x) &= w^C(x) + \frac{1}{D_{xx}^*}\left[c_1 B_{xx}\,\frac{x^2}{2} - A_{xx}\left(c_2\,\frac{x^3}{6} + c_3\,\frac{x^2}{2} + c_5\, x + c_6\right)\right] \\
&\quad + \frac{1}{S_{xz}}\left[M_{xx}^C(x) + c_2\, x\right].
\end{aligned} \qquad (4.7.15)$$

In summary, we have the following algebraic relationships between the generalized forces and generalized displacements between the classical and first-order shear deformation beam theories:

$$N_{xx}^F = N_{xx}^C + c_1, \qquad (4.7.16)$$

$$N_{xz}^F = N_{xz}^C + c_2, \qquad (4.7.17)$$

$$M_{xx}^F = M_{xx}^C + c_2 x + c_3, \qquad (4.7.18)$$

$$u^F(x) = u^C(x) + \frac{1}{D^*_{xx}}\left[c_1 D_{xx} x - B_{xx}\left(c_2 \frac{x^2}{2} + c_3 x + c_4\right)\right], \qquad (4.7.19)$$

$$\phi^F_x(x) = -\frac{dw^C}{dx} + \frac{1}{D^*_{xx}}\left[-c_1 B_{xx} x + A_{xx}\left(c_2 \frac{x^2}{2} + c_3 x + c_5\right)\right], \qquad (4.7.20)$$

$$w^F(x) = w^C(x) + \frac{1}{D^*_{xx}}\left[c_1 B_{xx}\frac{x^2}{2} - A_{xx}\left(c_2 \frac{x^3}{6} + c_3 \frac{x^2}{2} + c_5 x + c_6\right)\right]$$

$$+ \frac{1}{S_{xz}}\left[M^C_{xx}(x) + c_2 x\right], \qquad (4.7.21)$$

where c_i $(i = 1, 2, \ldots, 6)$ are constants of integration, which are to be determined using the boundary conditions of the particular beam problem being analyzed. For free (F), hinged (H), pinned (P), and clamped (C) ends, the boundary conditions are given by

F: $N^C_{xx} = N^F_{xx} = 0, \quad N^C_{xz} = N^F_{xz} = 0, \quad M^C_{xx} = M^F_{xx} = 0,$ $\qquad (4.7.22)$

H: $N^C_{xx} = N^F_{xx} = 0, \quad w^C = w^F = 0, \quad M^C_{xx} = M^F_{xx} = 0,$ $\qquad (4.7.23)$

P: $u^C = u^F = 0, \quad w^C = w^F = 0, \quad M^C_{xx} = M^F_{xx} = 0,$ $\qquad (4.7.24)$

C: $u^C = u^F = 0, \quad w^C = w^F = 0, \quad \dfrac{dw^C}{dx} = \dfrac{dw^F}{dx} = \phi^C_x = \phi^F_x = 0,$ $\quad (4.7.25)$

To this list we can add *elastically supported* cases:

$$N^C_{xx} = N^F_{xx} = 0, \quad N^C_{xz} = -k_E w^C, \quad N^F_{xz} = -k_E w^F, \quad M^C_{xx} = M^C_{xx} = 0.$$
$$(4.7.26)$$

where k_E is the spring constant of the elastic support (assumed to be linear). If a rotational spring is there at the support, the bending moment is then $M^C_{xx} = -k_T(dw^C/dx)$ and $M^F_{xx} = -k_T(dw^F/dx)$, where k_T is the torsional spring constant.

Table 4.7.1 (from Wang, Reddy, and Lee [120]) contains the bending relationships between the TBT and CBT for several standard boundary conditions. When one has the solution based on the CBT for any boundary conditions and load distribution $q(x)$, the corresponding solution based on the TBT can be obtained from Eqs. (4.7.16)–(4.7.21). Any point loads and moments will have to be introduced through the boundary conditions. If the point loads are in the interior of the beam span, one may use singularity functions to represent them as distributed loads.

It can be seen from Table 4.7.1 that the bending moments and shear forces are the same for statically determinate the TBT and CBT, that is, P-P and C-F beams (C-F means, for example, the left end is clamped and the right end is free). Also for these beams, the rotation in the TBT is equal to the slope of the CBT. For statically indeterminate C-H and C-C beams, the stress-resultants are not the same, because the compatibility equation involving the

Table 4.7.1: Generalized deflection and force relationships [120] between the TBT and CBT for pure bending of beams (F – free; H – hinged; C – clamped); $\mu = \Omega/(1 + 12\Omega)$ and $\Omega = D_{xx}/K_s S_{xz} L^2$.

B.C.	Relationships between TBT (F) and CBT (C)

H-H

$$w^F(x) = w^C(x) + \frac{\Omega L^2}{D_{xx}} M_{xx}^C(x)$$

$$\phi_x^F(x) = -\frac{dw^C}{dx}, \quad M_{xx}^F(x) = M_{xx}^C(x), \quad N_{xz}^F(x) = N_{xz}^C(x)$$

C-F

$$w^F(x) = w^C(x) + \frac{\Omega L^2}{D_{xx}}\left[M_{xx}^C(x) - M_{xx}^C(0)\right]$$

$$\phi_x^F(x) = -\frac{dw^C}{dx}, \quad M_{xx}^F(x) = M_{xx}^C(x), \quad N_{xz}^F(x) = N_{xz}^C(x)$$

C-H

$$w^F(x) = w^C(x) + \frac{\Omega L^2}{D_{xx}}\left[M_{xx}^C(x) - M_{xx}^C(0)\right]$$

$$+\frac{3\Omega L^2}{D_{xx}(1+3\Omega)}\frac{x}{L}\left(\Omega + \frac{x}{2L} - \frac{x^2}{6L^2}\right)M_{xx}^C(0)$$

$$\phi_x^F(x) = -\frac{dw^C}{dx} + \frac{3\Omega L}{D_{xx}(1+3\Omega)}\frac{x}{L}\left(1 - \frac{x}{2L}\right)M_{xx}^C(0)$$

$$M_{xx}^F(x) = M_{xx}^C(x) - \frac{3\Omega}{(1+3\Omega)}\left(1 - \frac{x}{L}\right)M_{xx}^C(0)$$

$$N_{xz}^F(x) = N_{xz}^C(x) + \frac{3\Omega}{(1+3\Omega)L}M_{xx}^C(0)$$

C-C

$$w^F(x) = w^C(x) + \frac{\Omega L^2}{D_{xx}}\left[M_{xx}^C(x) - M_{xx}^C(0)\right]$$

$$+\frac{3\mu L^3}{D_{xx}}\frac{x}{L}\left(\frac{2}{3}\frac{x^2}{L^2} - 4\Omega - \frac{x}{L}\right)\left[M_{xx}^C(L) - M_{xx}^C(0)\right]$$

$$\phi_x^F(x) = -\frac{dw^C}{dx} - \frac{6\mu L}{D_{xx}}\frac{x}{L}\left(\frac{x}{L} - 1\right)\left[M_{xx}^C(L) - M_{xx}^C(0)\right]$$

$$M_{xx}^F(x) = M_{xx}^C(x) - 6\mu\left(2\frac{x}{L} - 1\right)\left[M_{xx}^C(L) - M_{xx}^C(0)\right]$$

$$N_{xz}^F(x) = N_{xz}^E(x) - \frac{12\mu}{L}\left[M_{xx}^C(L) - M_{xx}^C(0)\right]$$

effect of transverse shear deformation is required for the solution. The deflection relationships show clearly (i.e., see the additional term in the expressions for the TBT) the effect of shear deformation. The shear-deflection component increases with increasing magnitude of the CBT moment (or the transverse load) and shear parameter.

We illustrate the procedure with several examples; the reader may find the solutions of the CBT for standard boundary conditions and loads in a book on mechanics of materials (e.g., Fenner and Reddy [123]).

Example 4.7.1

We wish to determine the exact TBT solutions for functionally graded beams with both ends *pinned*. The beam is subjected to uniformly distributed transverse load of intensity $q(x) = q_0$.

Solution: The boundary conditions $M_{xx}(0) = 0$, $u(0) = 0$, and $w(0) = 0$ give (respectively):

$$c_3 = 0, \quad c_4 = 0, \quad c_6 = \frac{D_{xx}^*}{A_{xx}S_{xz}}M_{xx}(0) = 0.$$

The boundary conditions $M_{xx}(L) = 0$, $u(L) = 0$, and $w(L) = 0$ give (using the above results):

$$c_2 = 0, \quad c_1 = 0, \quad c_5 = \frac{D_{xx}^*}{A_{xx}S_{xz}L}\left[M_{xx}(L) - M_{xx}(0)\right] = 0. \qquad (4.7.27)$$

Hence, the TBT solution becomes [see Eqs. (3.7.22)–(3.7.26) for the CBT]

$$N_{xx}^F(x) = N_{xx}^C(x) = 0, \qquad (4.7.28)$$

$$N_{xz}^F(x) = N_{xz}^C(x) = \frac{q_0 L}{2}\left(1 - 2\frac{x}{L}\right), \qquad (4.7.29)$$

$$M_{xx}^F(x) = M_{xx}^C(x) = \frac{q_0 L^2}{2}\left(\frac{x}{L} - \frac{x^2}{L^2}\right), \qquad (4.7.30)$$

$$u^F(x) = u^C(x) = \frac{B_{xx}}{D_{xx}^*}\frac{q_0 L^3}{12}\left(\frac{x}{L} - 3\frac{x^2}{L^2} + 2\frac{x^3}{L^3}\right), \qquad (4.7.31)$$

$$\phi_x^F(x) = -\frac{dw^C}{dx} = -\frac{1}{D_{xx}^*}\left(\frac{D_{xx}^*}{D_{xx}}M_{xx}^T\frac{L}{2} - \frac{B_{xx}^2}{D_{xx}}\frac{q_0 L^3}{24}\right)\left(1 - 2\frac{x}{L}\right)$$
$$- \frac{A_{xx}}{D_{xx}^*}\frac{q_0 L^3}{24}\left(1 - 6\frac{x^2}{L^2} + 4\frac{x^3}{L^3}\right), \qquad (4.7.32)$$

$$w^F(x) = w^C(x) + \frac{1}{S_{xz}}M_{xx}^C(x)$$
$$= -\frac{1}{D_{xx}^*}\left(\frac{D_{xx}^*}{D_{xx}}M_{xx}^T\frac{L^2}{2} - \frac{B_{xx}^2}{D_{xx}}\frac{q_0 L^4}{24}\right)\left(\frac{x}{L} - \frac{x^2}{L^2}\right)$$
$$+ \frac{A_{xx}}{D_{xx}^*}\frac{q_0 L^4}{24}\left(\frac{x}{L} - 2\frac{x^3}{L^3} + \frac{x^4}{L^4}\right) + \frac{q_0 L^2}{2S_{xz}}\left(\frac{x}{L} - \frac{x^2}{L^2}\right). \qquad (4.7.33)$$

These expressions are the same as those in **Example 4.6.1**.

Example 4.7.2

We wish to determine the exact TBT solutions for functionally graded beams with both ends *clamped*. The beam is subjected to uniformly distributed transverse load of intensity $q(x) = q_0$.

Solution: The boundary conditions at the left end $x = 0$ give:

$$u(0) = 0 \rightarrow c_4 = 0,$$
$$\phi_x(0) = 0 \rightarrow c_5 = 0, \qquad (4.7.34)$$
$$w(0) = 0 \rightarrow c_6 = \frac{D_{xx}^*}{A_{xx}S_{xz}}M_{xx}(0).$$

The boundary conditions at the right end $(x = L)$ of the beam, $u(L) = \phi_x(L) =$

$w(L) = 0$, give [using the results in Eq. (4.7.34)] the following relations:

$$0 = c_1 D_{xx} L - B_{xx} \left(c_2 \frac{L^2}{2} + c_3 L \right), \tag{4.7.35}$$

$$0 = -c_1 B_{xx} L + A_{xx} \left(c_2 \frac{L^2}{2} + c_3 L \right), \tag{4.7.36}$$

$$0 = \frac{1}{D_{xx}^*} \left[c_1 B_{xx} \frac{L^2}{2} - A_{xx} \left(c_2 \frac{L^3}{6} + c_3 \frac{L^2}{2} + c_6 \right) \right]$$
$$+ \frac{1}{S_{xz}} \left[M_{xx}^C(L) + c_2 L \right]. \tag{4.7.37}$$

Equations (4.7.35) and (4.7.36) yield

$$c_1 = 0, \quad c_2 \frac{L}{2} + c_3 = 0,$$

$$\left(\frac{A_{xx} L^3}{12 D_{xx}^*} + \frac{L}{S_{xz}} \right) c_2 = \frac{1}{S_{xz}} \left[M_{xx}^C(0) - M_{xx}^C(L) \right]. \tag{4.7.38}$$

We can write

$$\left(\frac{A_{xx} L^2}{12 D_{xx}^*} + \frac{1}{S_{xz}} \right) = \frac{1 + 12\Omega}{12 \Omega S_{xz}}, \quad \Omega = \frac{D_{xx}^*}{A_{xx} S_{xz} L^2} \tag{4.7.39}$$

so that

$$c_2 L = \frac{12\Omega}{(1 + 12\Omega)} \left[M_{xx}^C(0) - M_{xx}^C(L) \right]. \tag{4.7.40}$$

Hence, the solution for a clamped–clamped beam under *any* distributed load $q(x)$ becomes

$$N_{xx}^F(x) = N_{xx}^C(x) = 0, \tag{4.7.41}$$

$$N_{xz}^F(x) = N_{xz}^F(x) + c_2, \tag{4.7.42}$$

$$M_{xx}^F(x) = M_{xx}^C(x) + \frac{L}{2} \left(2\frac{x}{L} - 1 \right) c_2, \tag{4.7.43}$$

$$u^F(x) = u^C(x) - \frac{B_{xx}}{D_{xx}^*} \frac{L^2}{2} \left(\frac{x^2}{L^2} - \frac{x}{L} \right) c_2, \tag{4.7.44}$$

$$\phi_x^F(x) = -\frac{dw^C}{dx} + \frac{A_{xx}}{D_{xx}^*} \frac{L^2}{2} \left(\frac{x^2}{L^2} - \frac{x}{L} \right) c_2, \tag{4.7.45}$$

$$w^F(x) = w^C(x) + \frac{1}{S_{xz}} \left[M_{xx}^C(x) - M_{xx}^C(0) \right]$$
$$- \frac{A_{xx}}{D_{xx}^*} \frac{L^3}{12} \left(2\frac{x^3}{L^3} - 3\frac{x^2}{L^2} \right) c_2 + \frac{L}{S_{xz}} \frac{x}{L} c_2. \tag{4.7.46}$$

For uniformly distributed load $q(x) = q_0$, we have $M_{xx}^C(L) = M_{xx}^C = -q_0 L^2/2$. Hence, we have $c_2 = 0$ and $c_3 = 0$, and the expressions in Eqs. (4.7.42)–(4.7.46) become (the same as those in Example 4.6.2):

$$N_{xx}^F(x) = 0, \tag{4.7.47}$$

$$N_{xz}^F(x) = \frac{q_0 L}{2}\left(1 - \frac{x}{L}\right), \tag{4.7.48}$$

$$M_{xx}^F(x) = -\frac{q_0 L^2}{12}\left(1 - 6\frac{x}{L} + 6\frac{x^2}{L^2}\right), \tag{4.7.49}$$

$$u^F(x) = \frac{B_{xx}}{D_{xx}^*}\frac{q_0 L^3}{12}\left(\frac{x}{L} - 3\frac{x^2}{L^2} + 2\frac{x^3}{L^3}\right), \tag{4.7.50}$$

$$\phi_x^F(x) = \frac{A_{xx}}{D_{xx}^*}\frac{q_0 L^3}{12}\left(-\frac{x}{L} + 3\frac{x^2}{L^2} - 2\frac{x^3}{L^3}\right), \tag{4.7.51}$$

$$w^F(x) = \frac{A_{xx}}{D_{xx}^*}\frac{q_0 L^4}{24}\left(\frac{x^2}{L^2} - 2\frac{x^3}{L^3} + \frac{x^4}{L^4}\right) + \frac{1}{S_{xz}}\frac{q_0 L^2}{2}\left(\frac{x}{L} - \frac{x^2}{L^2}\right). \tag{4.7.52}$$

Example 4.7.3

We wish to determine the exact solution of a cantilever beam subjected to uniformly distributed transverse load of intensity $q(x) = q_0$ (N/m) and a point load F_0 (N) at $x = L$ (both positive upward). Assume that the spring is linearly elastic, with a spring constant of k (N/m).

Solution: The boundary conditions at the clamped end, $x = 0$, are

$$u^F = u^C = 0, \quad w^F = w^C = 0, \quad \phi_x^F = -\frac{dw^C}{dx} = 0, \tag{4.7.53a}$$

and at the right end, $x = L$, we have

$$N_{xx}^F = N_{xx}^C = 0, \quad N_{xz}^F = N_{xz}^C = F_0, \quad M_{xx}^F = M_{xx}^C = 0. \tag{4.7.53b}$$

The boundary conditions in Eq. (4.7.53a) yield

$$c_4 = 0, \quad c_5 = 0, \quad c_6 = \frac{D_{xx}^*}{A_{xx}S_{xz}}M_{xx}^C(0). \tag{4.7.54a}$$

The boundary conditions in Eq. (4.7.53b) give

$$c_1 = 0, \quad c_2 = 0, \quad c_3 = 0. \tag{4.7.54b}$$

The CBT solution for the problem is available from part (a) (i.e., $k = 0$) of **Example 3.7.3**, with a slight modification to account for the point load F_0: replace the expression for K_2 in Eqs. (3.7.30)–(3.7.32) with $K_2 = q_0 L + F_0$. We then have

$$D_{xx}^* u^C(x) = \left(D_{xx} N_{xx}^T - B_{xx} M_{xx}^T \right) x + B_{xx} \frac{q_0 L^3}{6} \left(3\frac{x}{L} - 3\frac{x^2}{L^2} + \frac{x^3}{L^3} \right)$$
$$+ B_{xx} \frac{F_0 L^2}{2} \frac{x}{L} \left(2 - \frac{x}{L} \right),$$

$$-D_{xx}^* \frac{dw^C}{dx} = \left(A_{xx} M_{xx}^T - B_{xx} N_{xx}^T \right) x - A_{xx} \frac{q_0 L^3}{6} \left(3\frac{x}{L} - 3\frac{x^2}{L^2} + \frac{x^3}{L^3} \right)$$
$$- A_{xx} \frac{F_0 L^2}{2} \frac{x}{L} \left(2 - \frac{x}{L} \right),$$

$$D_{xx}^* w^C(x) = -\frac{1}{2} \left(A_{xx} M_{xx}^T - B_{xx} N_{xx}^T \right) x^2 + A_{xx} \frac{q_0 L^4}{24} \left(6\frac{x^2}{L^2} - 4\frac{x^3}{L^3} + \frac{x^4}{L^4} \right)$$
$$+ A_{xx} \frac{F_0 L^3}{6} \frac{x^2}{L^2} \left(3 - \frac{x}{L} \right),$$

$$M_{xx}^C(x) = -\frac{q_0 L^2}{2} \left(1 - 2\frac{x}{L} + \frac{x^2}{L^2} \right) - F_0 L \left(1 - \frac{x}{L} \right),$$

$$N_{xz}^C = q_0 L \left(1 - \frac{x}{L} \right) + F_0.$$

Therefore, the TBT solution is

$$N_{xx}^F(x) = 0, \tag{4.7.55}$$

$$N_{xz}^F(x) = q_0 L \left(1 - \frac{x}{L} \right) + F_0, \tag{4.7.56}$$

$$M_{xx}^F(x) = -\frac{q_0 L^2}{2} \left(1 - 2\frac{x}{L} + \frac{x^2}{L^2} \right) - F_0 L \left(1 - \frac{x}{L} \right), \tag{4.7.57}$$

$$D_{xx}^* u^F(x) = B_{xx} \frac{q_0 L^3}{6} \left(3\frac{x}{L} - 3\frac{x^2}{L^2} + \frac{x^3}{L^3} \right) + B_{xx} \frac{F_0 L^2}{2} \frac{x}{L} \left(2 - \frac{x}{L} \right)$$
$$+ \left(D_{xx} N_{xx}^T - B_{xx} M_{xx}^T \right) x, \tag{4.7.58}$$

$$D_{xx}^* \phi_x^F(x) = -A_{xx} \frac{q_0 L^3}{6} \left(3\frac{x}{L} - 3\frac{x^2}{L^2} + \frac{x^3}{L^3} \right) - A_{xx} \frac{F_0 L^2}{2} \frac{x}{L} \left(2 - \frac{x}{L} \right)$$
$$+ L \left(A_{xx} M_{xx}^T - B_{xx} N_{xx}^T \right) \frac{x}{L}, \tag{4.7.59}$$

$$D_{xx}^* w^F(x) = A_{xx} \frac{q_0 L^4}{24} \left(6\frac{x^2}{L^2} - 4\frac{x^3}{L^3} + \frac{x^4}{L^4} \right) + \frac{D_{xx}^*}{S_{xz}} \frac{q_0 L^2}{2} \left(2\frac{x}{L} - \frac{x^2}{L^2} \right)$$
$$+ A_{xx} \frac{F_0 L^3}{6} \frac{x^2}{L^2} \left(3 - \frac{x}{L} \right) + \frac{D_{xx}^*}{S_{xz}} F_0 x$$
$$- \frac{L^2}{2} \left(A_{xx} M_{xx}^T - B_{xx} N_{xx}^T \right) \frac{x^2}{L^2}. \tag{4.7.60}$$

This completes the discussion of the bending relationships for FGM beams without the couple stress effect. Next, we derive the bending relationships for FGM beams with the couple stress effect.

4.7.3 BENDING RELATIONSHIPS FOR FGM BEAMS WITH THE COUPLE STRESS EFFECT

4.7.3.1 Summary of equations of CBT and TBT

Here, we establish relationships between the bending solutions (i.e., deflection, rotation, bending moment, and shear force) of the TBT for the microstructure-dependent FGM beams in terms of the corresponding quantities of the conventional CBT for beams of constant material and geometric properties. We take the distributed transverse load $q(x)$ and body couple $g(x)$ to be arbitrary. First, the governing equations of bending for the two theories are summarized here.

Conventional CBT: The governing equations of the CBT applied to homogeneous beams are [see Eq. (4.7.1)]

$$\frac{dN_{xx}^C}{dx} = -f, \quad \frac{dN_{xz}^C}{dx} = -q, \quad -\frac{dM_{xx}^C}{dx} + N_{xz}^C = 0, \qquad (4.7.61)$$

where

$$N_{xx}^C = A_{xx}\frac{du^C}{dx}, \quad M_{xx}^C = -D_{xx}\frac{d^2w^C}{dx^2}, \quad N_{xz}^C = -D_{xx}\frac{d^3w^C}{dx^3}. \qquad (4.7.62)$$

General TBT: The governing equations of the TBT applied to FGM beams with couple stress effect are [see Eqs. (4.3.18)–(4.3.20), with the addition of the couple g but omission of $c_f w$]

$$\frac{dN_{xx}^F}{dx} = -f, \quad \frac{d\bar{N}_{xz}^F}{dx} = -q, \quad -\frac{d\bar{M}_{xx}^F}{dx} + \bar{N}_{xz}^F = 0, \qquad (4.7.63)$$

where

$$\bar{M}_{xx}^F = M_{xx}^F + P_{xy}^F, \qquad \bar{N}_{xz}^F = N_{xz}^F + \frac{1}{2}\left(\frac{dP_{xy}^F}{dx} + g\right), \qquad (4.7.64)$$

$$N_{xx}^F = A_{xx}\frac{du^F}{dx} + B_{xx}\frac{d\phi_x^F}{dx}, \quad M_{xx}^F = B_{xx}\frac{du^F}{dx} + D_{xx}\frac{d\phi_x^F}{dx}, \qquad (4.7.65)$$

$$N_{xz}^F = S_{xz}\left(\phi_x^F + \frac{dw^F}{dx}\right), \quad P_{xy}^F = \frac{1}{2}A_{xy}\left(\frac{d\phi_x^F}{dx} - \frac{d^2w^F}{dx^2}\right), \qquad (4.7.66)$$

where a single superscript is used to denote the type of the theory, superscript C for the CBT and F for the TBT. The phrase "general TBT" refers to the TBT of microstructure-dependent FGM beams.

4.7.3.2 General relationships

From a comparison of Eq. (4.7.61) with Eq. (4.7.63), we have the following general relationships between the shear forces and bending moments of the two theories:

$$N_{xx}^F = N_{xx}^C + c_1, \quad \bar{N}_{xz}^F = N_{xz}^C + c_2, \quad \bar{M}_{xx}^F = M_{xx}^C + c_2 x + c_3, \quad (4.7.67)$$

where c_1, c_2, and c_3 are constants to be determined using boundary conditions of the particular beam problem according to the general TBT.

Next, substituting for N_{xz}^F and N_{xz}^C from Eqs. (4.7.62)) and (4.7.65) into (4.7.67)$_1$ [i.e., the first equation of (4.7.67)], we obtain

$$A_{xx}\frac{du^F}{dx} + B_{xx}\frac{d\phi_x^F}{dx} = A_{xx}\frac{du^C}{dx} + c_1. \quad (4.7.68)$$

Integrating once with respect to x gives

$$A_{xx}u^F + B_{xx}\phi_x^F = A_{xx}u^C + c_1 x + c_4. \quad (4.7.69)$$

Now substituting for M_{xx}^F, P_{xx}^F, and M_{xx}^C from Eqs. (4.7.62), (4.7.65), and (4.7.66) into Eq. (4.7.67)$_3$, we obtain

$$B_{xx}\frac{du^F}{dx} + \left(D_{xx} + \tfrac{1}{2}A_{xy}\right)\frac{d\phi_x^F}{dx} - \tfrac{1}{2}A_{xy}\frac{d^2 w^F}{dx^2} = -D_{xx}\frac{d^2 w^C}{dx^2} + c_2 x + c_3. \quad (4.7.70)$$

Substituting for du^F/dx from Eq. (4.7.68) into Eq. (4.7.70), we arrive at

$$\bar{D}_{xx}\frac{d\phi_x^F}{dx} = -B_{xx}\frac{du^C}{dx} + \tfrac{1}{2}A_{xy}\frac{d^2 w^F}{dx^2} - D_{xx}\frac{d^2 w^C}{dx^2} + c_2 x + c_3 - \frac{B_{xx}}{A_{xx}}c_1. \quad (4.7.71)$$

Integrating Eq. (4.7.71) once, we obtain

$$\bar{D}_{xx}\phi_x^F = -B_{xx}u^C + \tfrac{1}{2}A_{xy}\frac{dw^F}{dx} - D_{xx}\frac{dw^C}{dx} - \frac{B_{xx}}{A_{xx}}c_1 x + c_2\frac{x^2}{2} + c_3 x + c_5, \quad (4.7.72)$$

where

$$\bar{B}_{xx} = \frac{B_{xx}B_{xx}}{A_{xx}}, \quad \bar{D}_{xx} = D_{xx} + \tfrac{1}{2}A_{xy} - \bar{B}_{xx}. \quad (4.7.73)$$

From Eq. (4.7.67)$_2$ and making use of Eqs. (4.7.66) and (4.7.64), we obtain

$$S_{xz}\phi_x^F + S_{xz}\frac{dw^F}{dx} + \tfrac{1}{4}A_{xy}\frac{d^2\phi_x^F}{dx^2} - \tfrac{1}{4}A_{xy}\frac{d^3 w^F}{dx^3} = -D_{xx}\frac{d^3 w^C}{dx^3} + c_2 - \tfrac{1}{2}g. \quad (4.7.74)$$

Substituting for ϕ_x^F and $(d\phi_x^F/dx)$ from Eqs. (4.7.71) and (4.7.72) into Eq. (4.7.74), we arrive at the result

$$\bar{A}_{xz}\frac{dw^F}{dx} - \tfrac{1}{4}\bar{A}_{xy}\frac{d^3 w^F}{dx^3} = B_{xx}\left(S_{xz}u^C + \tfrac{1}{4}A_{xy}\frac{d^2 u^C}{dx^2}\right)$$

$$+ D_{xx}\left(S_{xz}\frac{dw^C}{dx} - \tilde{D}_{xx}\frac{d^3 w^C}{dx^3}\right) + \tilde{D}_{xx}c_2 - \tfrac{1}{2}\bar{D}_{xx}g$$

$$- S_{xz}\left(-\frac{B_{xx}}{A_{xx}}c_1 x + c_2\frac{x^2}{2} + c_3 x + c_5\right), \quad (4.7.75)$$

where

$$\bar{A}_{xz} = S_{xz}\left(A_{xy} + D_{xx} - \bar{B}_{xx}\right), \quad \bar{A}_{xy} = A_{xy}\left(D_{xx} - \bar{B}_{xx}\right),$$
$$\tilde{D}_{xx} = D_{xx} + \tfrac{1}{4}A_{xy} - \bar{B}_{xx}. \tag{4.7.76}$$

Integration once with respect to x yields

$$\bar{A}_{xz}w^F - \tfrac{1}{4}\bar{A}_{xy}\frac{d^2 w^F}{dx^2} = B_{xx}\left(S_{xz}\int u^C\, dx + \tfrac{1}{4}A_{xy}\frac{du^C}{dx}\right)$$
$$+ D_{xx}\left(S_{xz}w^C - \tilde{D}_{xx}\frac{d^2 w^C}{dx^2}\right)$$
$$- S_{xz}\left(-\frac{B_{xx}}{2A_{xx}}c_1 x^2 + c_2\frac{x^3}{6} + c_3\frac{x^2}{2} + c_5 x\right)$$
$$+ \tilde{D}_{xx}c_2 x - \tfrac{1}{2}\bar{D}_{xx}\int g\, dx + c_6. \tag{4.7.77}$$

4.7.3.3 Specialized relationships

Here we consider three cases: (1) functionally graded beams with the microstructural aspect neglected ($A_{xy} = 0$ and $g = 0$), (2) homogeneous beams with the microstructural aspect neglected ($A_{xy} = 0$, $g = 0$, and $B_{xx} = 0$), and (3) the general case in which the microstructural aspect as well as the functionally graded aspects are included ($A_{xy} \neq 0$ and $B_{xx} \neq 0$).

Functionally graded material beams without microstructural effect
For this case, we have

$$A_{xy} = 0, \quad \tilde{D}_{xx} = \bar{D}_{xx} = D_{xx} - \bar{B}_{xx}, \quad \bar{A}_{xz} = S_{xz}\bar{D}_{xx}. \tag{4.7.78}$$

Then the solution is given by

$$w^F(x) = \frac{B_{xx}}{\bar{D}_{xx}}\int u^C(x)\, dx + \frac{D_{xx}}{\bar{D}_{xx}}w^C(x) + \frac{1}{S_{xz}}M_{xx}^C(x) + \frac{1}{S_{xz}}c_2 x$$
$$+ \frac{c_6}{S_{xz}\bar{D}_{xx}} - \frac{1}{\bar{D}_{xx}}\left(-\frac{B_{xx}}{A_{xx}}c_1\frac{x^2}{2} + c_2\frac{x^3}{6} + c_3\frac{x^2}{2} + c_5 x\right). \tag{4.7.79}$$

Thus, there are only six constants of integration to be determined (because $P_{xy}^F = 0$). We have the following relationships between the TBT of FGM beams and the CBT of conventional homogeneous beams:

$$u^F(x) = u^C(x) + \frac{1}{A_{xx}}\left[-B_{xx}\phi_x^F(x) + c_1 x + c_4,\right] \tag{4.7.80}$$
$$w^F(x) = \frac{B_{xx}}{\bar{D}_{xx}}\int u^C(x)\, dx + \frac{D_{xx}}{\bar{D}_{xx}}w^C(x) + \frac{1}{S_{xz}}M_{xx}^C(x) + \frac{1}{S_{xz}}c_2 x + c_6$$
$$- \frac{1}{\bar{D}_{xx}}\left(-\frac{B_{xx}}{A_{xx}}c_1\frac{x^2}{2} + c_2\frac{x^3}{6} + c_3\frac{x^2}{2} + c_5 x\right), \tag{4.7.81}$$

$$\phi_x^F(x) = \frac{1}{\bar{D}_{xx}} \left(-D_{xx} \frac{dw^C}{dx} + c_2 \frac{x^2}{2} + c_3 x + c_5 \right)$$
$$- \frac{B_{xx}}{\bar{D}_{xx}} u^C - \frac{B_{xx}}{A_{xx} \bar{D}_{xx}} c_1 x, \tag{4.7.82}$$

$$N_{xx}^F(x) = N_{xx}^C(x) + c_1, \tag{4.7.83}$$

$$M_{xx}^F(x) = M_{xx}^C(x) + c_2 x + c_3, \tag{4.7.84}$$

$$N_{xz}^F(x) = N_{xz}^C(x) + c_2. \tag{4.7.85}$$

For homogeneous beams without microstructural effect, we have

$$A_{xy} = 0, \quad B_{xx} = 0, \quad \tilde{D}_{xx} = \bar{D}_{xx} = D_{xx}, \quad \bar{A}_{xz} = S_{xz} D_{xx}. \tag{4.7.86}$$

The relationships between the TBT and CBT solutions for homogeneous beams are

$$u^F(x) = u^C(x) + \frac{1}{A_{xx}} (c_1 x + c_4), \tag{4.7.87}$$

$$w^F(x) = w^C(x) + \frac{1}{S_{xz}} M_{xx}^C(x) - \frac{1}{D_{xx}} \left(c_2 \frac{x^3}{6} + c_3 \frac{x^2}{2} + c_5 x \right)$$
$$+ \frac{1}{S_{xz}} c_2 x + c_6, \tag{4.7.88}$$

$$\phi^F(x) = -\frac{dw^C}{dx} + \frac{1}{D_{xx}} \left(c_2 \frac{x^2}{2} + c_3 x + c_5 \right), \tag{4.7.89}$$

$$N_{xx}^F(x) = N_{xx}^C(x) + c_1, \tag{4.7.90}$$

$$M_{xx}^F(x) = M_{xx}^C(x) + c_2 x + c_3, \tag{4.7.91}$$

$$N_{xz}^F(x) = N^{xz}(x) + c_2. \tag{4.7.92}$$

The relationships in Eqs. (4.7.88), (4.7.89), (4.7.91), and (4.7.92) are the same as those derived in the book by Wang, Reddy, and Lee [120].

General beams. In this case, we must solve the second-order differential equation, Eq. (4.7.77), for w^F. Rewriting Eq. (4.7.77) as

$$\frac{d^2 w^F}{dx^2} - \alpha^2 w^F = p(x), \quad \alpha^2 = 4 \frac{\bar{A}_{xz}}{\bar{A}_{xy}}, \tag{4.7.93}$$

where

$$p(x) = -\frac{4}{\bar{A}_{xy}} \left[-S_{xz} \left(-\frac{B_{xx}}{2A_{xx}} c_1 x^2 + c_2 \frac{x^3}{6} + c_3 \frac{x^2}{2} + c_5 x \right) + \tilde{D}_{xx} c_2 x \right.$$
$$-0.5 \bar{D}_{xx} \int g \, dx + C_6 + B_{xx} \left(S_{xz} \int u^C \, dx + \tfrac{1}{4} A_{xy} \frac{du^C}{dx} \right)$$
$$\left. + D_{xx} \left(S_{xz} w^C - \tilde{D}_{xx} \frac{d^2 w^C}{dx^2} \right) \right]. \tag{4.7.94}$$

Then the homogeneous solution of Eq. (4.7.93) is $(A_{xy} \neq 0)$

$$w_h^F(x) = c_7 \cosh \alpha x + c_8 \sinh \alpha x, \qquad (4.7.95)$$

and the particular solution is

$$w_p^F(x) = \frac{1}{2\alpha} \left(e^{\alpha x} \int e^{-\alpha x} p(x)\, dx - e^{-\alpha x} \int e^{\alpha x} p(x)\, dx \right). \qquad (4.7.96)$$

Thus, we have the following relationships between the bending solution of the TBT and the CBT:

$$u^F(x) = u^C(x) + \frac{1}{A_{xx}} \left[-B_{xx}\phi_x^F(x) + c_1 x + c_4 \right], \qquad (4.7.97)$$

$$w^F(x) = w_h^F(x) + w_p^F(x), \qquad (4.7.98)$$

$$\phi^F(x) = \frac{1}{\bar{D}_{xx}} \left(-B_{xx} u^C + \tfrac{1}{2} A_{xy} \frac{dw^F}{dx} - D_{xx} \frac{dw^C}{dx} - \frac{B_{xx}}{A_{xx}} c_1 x \right.$$
$$\left. + c_2 \frac{x^2}{2} + c_3 x + c_5 \right), \qquad (4.7.99)$$

$$N_{xx}^F(x) = N_{xx}^C(x) + c_1, \qquad (4.7.100)$$

$$\bar{M}_{xx}^F(x) = M_{xx}^C(x) + c_2 x + c_3, \qquad (4.7.101)$$

$$\bar{N}_{xz}^F(x) = N_{xz}^C(x) + c_2. \qquad (4.7.102)$$

Example 4.7.4

Determine the TBT solution for a beam simply supported at the ends $x = 0, L$, and subjected to uniformly distributed transverse load $q = q_0$, zero axial load $f = 0$, and zero body couple $g = 0$.

Solution: The boundary conditions are

$$u^C(0) = u^F(0) = 0, \quad u^F(L) = u^C(L) = 0,$$

$$w^C = 0, \quad w^F = 0, \quad M_{xx}^C = 0, \quad M_{xx}^F = 0, \quad P_{xx}^F = 0 \text{ at } x = 0, L. \qquad (4.7.103)$$

The CBT solutions of a simply supported homogeneous beam are

$$w^C = -\frac{q_0}{24 D_{xx}} \left(L^3 x - 2L x^3 + x^4 \right),$$

$$\frac{dw^C}{dx} = -\frac{q_0}{24 D_{xx}} \left(L^3 - 6L x^2 + 4x^3 \right), \qquad (4.7.104)$$

$$\frac{d^2 w^C}{dx^2} = \frac{q_0}{2 D_{xx}} (Lx - x^2).$$

Using the boundary conditions on the stress resultants $\bar{N}_{xx}$ and $\bar{M}_{xx}$ from (4.7.103) in Eqs. (4.7.100) and (4.7.101), we conclude that

$$M_{xx}^F(0) = M_{xx}^C(0) = 0 \quad \Rightarrow \quad c_3 = 0,$$

$$M_{xx}^F(L) = M_{xx}^C(L) = 0 \quad \Rightarrow \quad c_2 = 0.$$

From Eq. (4.7.97) we have

$$u^F(0) = 0 \quad \Rightarrow \quad c_4 = B_{xx}\phi_x^F(0).$$

Using $u^C(L) = u^F(L) = 0$, from Eq. (4.7.80) we obtain the result

$$c_1 = \frac{B_{xx}}{L}\left[\phi_x^F(L) - \phi_x^F(0)\right].$$

Also, $w^C(0) = w^F(0) = 0$ and $M_{xx}^C(0) = 0$ give $c_6 = 0$.

Next, we rewrite the boundary conditions on M_{xx}^F and P^F in terms of the displacements

$$M_{xx}^F(L) = \left(B_{xx}\frac{du^F}{dx} + D_{xx}\frac{d\phi_x^F}{dx}\right)_{x=L} = 0,$$

$$M_{xx}^F(0) = \left(B_{xx}\frac{du^F}{dx} + D_{xx}\frac{d\phi_x^F}{dx}\right)_{x=0} = 0,$$

$$P_{xx}^F(0) = \tfrac{1}{2}A_{xy}\left(\frac{d\phi_x^F}{dx} - \frac{d^2w^F}{dx^2}\right)_{x=0} = 0,$$

$$P_{xx}^F(L) = \tfrac{1}{2}A_{xy}\left(\frac{d\phi_x^F}{dx} - \frac{d^2w^F}{dx^2}\right)_{x=L} = 0.$$

The last two equations, together with Eq. (4.7.70), imply

$$\frac{d\phi_x^F}{dx} = 0, \quad \text{at } x = L; \quad \frac{d^2w^F}{dx^2} = 0 \quad \text{at } x = 0, L.$$

Thus, we have determined five of the eight constants of integration, and we need to determine the remaining three constants, c_5, c_7, and c_8, using the following three conditions:

$$w^F(L) = 0, \quad \left(\frac{d\phi_x^F}{dx}\right)_{x=L} = 0, \quad \left(\frac{d^2w^F}{dx^2}\right)_{x=L} = 0.$$

To determine the particular solution $w_p^F(x)$ in Eq. (4.7.96), first we identify $p(x)$ of Eq. (4.7.94) for the problem at hand

$$p(x) = -\frac{4}{\bar{A}_{xy}}\left[S_{xz}D_{xx}w^C - \tilde{D}_{xx}D_{xx}\frac{d^2w^C}{dx^2} + S_{xz}c_5(L-x)\right.$$

$$\left. + S_{xz}\frac{B_{xx}}{A_{xx}}c_1\frac{x^2}{2} + c_6\right]$$

$$= \frac{4}{\bar{A}_{xy}}\left\{\frac{q_0}{24}\left[S_{xz}(L^3x - 2Lx^3 + x^4) + 12\tilde{D}_{xx}(Lx - x^2)\right]\right.$$

$$\left. + S_{xz}\frac{B_{xx}}{A_{xx}}c_1\frac{x^2}{2} + S_{xz}c_5(L-x) + c_6\right\}. \tag{4.7.105}$$

Then the particular solution is [see **Chapter 4 Appendix** for the evaluation of integrals in Eq. (4.7.96)]

$$w_p^F(x) = -\frac{q_0}{6\bar{A}_{xy}}\left[\left(\frac{24}{\alpha^6} + \frac{L^3x}{\alpha^2} - \frac{12Lx}{\alpha^4} + \frac{12x^2}{\alpha^4} - \frac{2Lx^3}{\alpha^2} + \frac{x^4}{\alpha^2}\right)S_{xz}\right.$$

$$\left. + 12\left(-\frac{2}{\alpha^4} + \frac{Lx}{\alpha^2} - \frac{x^2}{\alpha^2}\right)\tilde{D}_{xx}\right] + \frac{4S_{xz}}{\bar{A}_{xy}}\frac{L-x}{\alpha^2}c_5. \tag{4.7.106}$$

Applying the boundary conditions $w^F = w_h^F + w_p^F = 0$ at $x = 0, L$ and $(d^2 w^F / dx^2) = 0$ at $x = 0, L$, we arrive at the following four relations among c_1, c_5, c_7, and c_8:

$$c_7 - \frac{4q_0}{\bar{A}_{xy}} \left(\frac{1}{\alpha^6} S_{xz} - \frac{1}{\alpha^4} \tilde{D}_{xx} \right) + \frac{4 S_{xz} L}{\bar{A}_{xy} \alpha^2} c_5 = 0,$$

$$c_7 \cosh \alpha L + c_8 \sinh \alpha L - \frac{4q_0}{\bar{A}_{xy}} \left(\frac{1}{\alpha^6} S_{xz} - \frac{1}{\alpha^4} \tilde{D}_{xx} \right) = 0,$$

$$c_7 - \frac{4q_0}{\bar{A}_{xy}} \left(\frac{1}{\alpha^6} S_{xz} - \frac{1}{\alpha^4} \tilde{D}_{xx} \right) = 0.$$

The above equations yield

$$c_5 = 0, \quad c_7 = \frac{4q_0}{\bar{A}_{xy} \alpha^6} \left(S_{xz} - \alpha^2 \tilde{D}_{xx} \right), \quad c_8 = c_7 \left(\frac{1 - \cosh \alpha L}{\sinh \alpha L} \right).$$

With all constants of integration known, the complete solution for generalized displacements and forces is

$$u^F(x) = \frac{B_{xx}}{A_{xx}} \left[\phi_x^F(0) - \phi_x^F(x) \right], \tag{4.7.107}$$

$$w^F(x) = \frac{4q_0}{\bar{A}_{xy} \alpha^6} \left(S_{xz} - \alpha^2 \tilde{D}_{xx} \right) \frac{\sinh \alpha(L - x) + \sinh \alpha x}{\sinh \alpha L} + w_p^F(x), \tag{4.7.108}$$

$$\phi_x^F(x) = \frac{1}{\tilde{D}_{xx}} \left[\frac{1}{2} A_{xy} \frac{dw^F}{dx} + \frac{q_0 L^3}{24} \left(1 - 6 \frac{x^2}{L^2} + 4 \frac{x^3}{L^3} \right) \right], \tag{4.7.109}$$

$$M_{xx}^F(x) = -\frac{q_0 L^2}{2} \left(\frac{x}{L} - \frac{x^2}{L^2} \right) - \frac{1}{2} A_{xy} \left(\frac{d\phi_x^F}{dx} - \frac{d^2 w^F}{dx^2} \right), \tag{4.7.110}$$

$$N_{xz}^F(x) = -\frac{q_0 L}{2} \left(1 - 2 \frac{x}{L} \right) - \frac{1}{4} A_{xy} \left(\frac{d^2 \phi_x^F}{dx^2} - \frac{d^3 w^F}{dx^3} \right), \tag{4.7.111}$$

where w_p^F is given by equation (4.7.106) with $c_5 = 0$.

When the effect of the microstructure-dependent parameter is neglected (i.e., $A_{xy} = 0$ and $g = 0$), Eqs. (4.7.80)–(4.7.85) are valid with $u^C(x) = 0$, and we need to determine the six constants of integration using the following boundary conditions:

$$u^F = w^F = M_{xx}^F = u^C = w^C = M_{xx}^C = 0 \quad \text{at} \quad x = 0, L.$$

Using the boundary conditions $M_{xx}^F(0) = M_{xx}^F(L) = M_{xx}^C(0) = M_{xx}^C(L) = 0$, we obtain $c_2 = c_3 = 0$. Application of the boundary conditions $u^F(0) = 0$ and $u^F(L) = 0$ in Eq. (4.7.80) give $c_4 = B_{xx} \phi_x^F(0)$ and $c_1 L = B_{xx}[\phi_x^F(L) - \phi_x^F(0)]$. Using the condition $w^F(0) = 0$ in Eq. (4.7.81), in view of the fact that $w^C = 0$ and $M^C = 0$ (which in turn imply $d^2 w^C / dx^2 = 0$) at $x = 0$, gives $c_6 = 0$. The condition $w^F(L) = 0$, in view of the fact that $w^C = 0$ and $d^2 w^C / dx^2 = 0$ at $x = L$, gives

$$0 = -\frac{B_{xx}}{A_{xx}} c_1 \frac{L^2}{2} + c_5 L \quad \rightarrow \quad c_5 = \frac{B_{xx}}{A_{xx}} \frac{L}{2} c_1.$$

Thus, we have

$$c_2 = 0, \quad c_3 = 0,, \quad c_6 = 0, \quad c_1 L = B_{xx} \left[\phi_x^F(L) - \phi_x^F(0) \right]$$

$$c_4 = B_{xx} \phi_x^F(0), \quad c_5 = \frac{B_{xx}^2}{2 A_{xx}} \left[\phi_x^F(L) - \phi_x^F(0) \right],$$

where

$$\phi_x^F(x) = \frac{1}{\bar{D}_{xx}} \left[\frac{q_0 L^3}{24} \left(1 - 6\frac{x^2}{L^2} + 4\frac{x^3}{L^3} \right) - \frac{B_{xx}}{A_{xx}} c_1 x + c_5 \right]$$

so that

$$\phi_x^F(0) = \frac{1}{\bar{D}_{xx}} \left(\frac{q_0 L^3}{24} + c_5 \right), \quad \phi_x^F(L) = \frac{1}{\bar{D}_{xx}} \left(-\frac{q_0 L^3}{24} - \frac{B_{xx}}{A_{xx}} c_1 L + c_5 \right).$$

Thus, we obtain

$$c_1 = -\frac{q_0 L^2}{12} \frac{B_{xx}}{D_{xx}}, \quad c_4 = \frac{q_0 L^3}{24 D_{xx}}, \quad c_5 = -\frac{q_0 L^3}{24} \frac{\bar{B}_{xx}}{D_{xx}}.$$

The complete set of solutions for simply supported FGM beams are

$$u^F(x) = \frac{B_{xx}}{A_{xx} \bar{D}_{xx}} \frac{q_0 L^3}{24} \left[-\left(1 - 6\frac{x^2}{L^2} + 4\frac{x^3}{L^3} \right) + \frac{\bar{B}_{xx}}{D_{xx}} \left(1 - \frac{x}{L} \right) \right], \qquad (4.7.112)$$

$$w^F(x) = -\frac{q_0 L^4}{24 \bar{D}_{xx}} \left(\frac{x}{L} - 2\frac{x^3}{L^3} + \frac{x^4}{L^4} \right) - \frac{q_0 L^2}{2 S_{xz}} \frac{x}{L} \left(1 - \frac{x}{L} \right)$$
$$- \frac{1}{\bar{D}_{xx}} \left(\frac{B_{xx}}{A_{xx}} \frac{q_0 L^2}{24} \frac{B_{xx}}{D_{xx}} x^2 - \frac{q_0 L^3}{24} \frac{\bar{B}_{xx}}{D_{xx}} x \right), \qquad (4.7.113)$$

$$\phi^F(x) = -\frac{q_0 L^3}{24 \bar{D}_{xx}} \left[\left(1 - 6\frac{x^2}{L^2} + 4\frac{x^3}{L^3} \right) + \frac{\bar{B}_{xx}}{D_{xx}} \left(1 - \frac{x}{L} \right) \right], \qquad (4.7.114)$$

$$N_{xx}^F(x) = -\frac{q_0 L^2}{12} \frac{B_{xx}}{D_{xx}}, \qquad (4.7.115)$$

$$M_{xx}^F(x) = -\frac{q_0 L^2}{2} \frac{x}{L} \left(1 - \frac{x}{L} \right), \qquad (4.7.116)$$

$$N_{xz}^F(x) = -\frac{q_0 L}{2} \left(1 - 2\frac{x}{L} \right). \qquad (4.7.117)$$

For homogeneous beams (i.e., $A_{xy} = 0$ and $B_{xx} = 0$), we have $c_1 = c_2 = c_3 = c_4 = c_5 = c_6 = 0$ and the solution is $(u^F(x) = 0)$

$$w^F(x) = w^C(x) + \frac{1}{S_{xz}} M^C(x)$$
$$= -\frac{q_0 L^4}{24 D_{xx}} \frac{x}{L} \left(1 - 2\frac{x^2}{L^2} + \frac{x^3}{L^3} \right) - \frac{q_0 L^2}{2 S_{xz}} \frac{x}{L} \left(1 - \frac{x}{L} \right), \qquad (4.7.118)$$

$$\phi_x^F(x) = -\frac{dw^C}{dx} = \frac{q_0 L^3}{24 D_{xx}} \left(1 - 6\frac{x^2}{L^2} + 4\frac{x^3}{L^3} \right), \qquad (4.7.119)$$

$$M_{xx}^F(x) = M_{xx}^C(x) = -\frac{q_0 L^2}{2} \frac{x}{L} \left(1 - \frac{x}{L} \right), \qquad (4.7.120)$$

$$N_{xz}^F(x) = N_{xz}^C(x) = -\frac{q_0 L}{2} \left(1 - 2\frac{x}{L} \right). \qquad (4.7.121)$$

For uniformly distributed load of intensity q_0, the dimensionless transverse deflection, $\bar{w} = w(x)(EI/q_0 L^4)$, is plotted in Fig. 4.7.1 as a function of the normalized distance, $\bar{x} = x/L$ for two different ratios of the material length–scale parameter to the height of the beam, ℓ/h, and for three different values of the power-law index n. The following values are used in the numerical calculations [66, 67]:

$$E_1 = 14.4\,\text{GPa}, \quad E_2 = 1.44\,\text{GPa}, \quad \nu = 0.38, \quad h = 17.6\,\mu\text{m}, \quad b = 2h, \quad L = 20h\,. \tag{4.7.122}$$

Although explicit value of ℓ is not needed in presenting the numerical solutions, for a beam made of epoxy material, the value of ℓ is taken to be $\ell = 17.6$ mm.

For homogeneous beams with $L/h = 20$, the effect of shear deformation is not significant; however, for $\ell/h = 1$, the difference between the solutions of the TBT and the CBT homogeneous beams is significant. The effect of ℓ is to stiffen the beam. Figure 4.7.2 contains plots of the maximum transverse deflection (occur at the midspan) of simply supported beam as a function of the ℓ/h ratio for $n = 0, 1, 10$. The deflection decreases with the ratio ℓ/h and with decreasing values of n.

Example 4.7.5 ——————————————————

Determine the TBT solution of a beam clamped at both ends $x = 0, L$, using the relationships.

Solution: The boundary conditions are

$$u^C = u^F = 0, \quad w^C = 0, \quad w^F = 0, \quad \frac{dw^C}{dx} = 0,$$

$$\frac{dw^F}{dx} = 0, \quad \phi_x^F = 0 \text{ at } x = 0, L. \tag{4.7.123}$$

Again, we note that $u^C(x) = 0$ for a homogeneous beam. The boundary conditions give [see Eqs. (4.7.97) and (4.7.99)]

$$u^F(0) = 0, \quad \phi_x^F(0) = 0 \implies c_4 = 0, \quad c_5 = 0,$$
$$u^F(L) = 0, \quad \phi_x^F(L) = 0 \implies c_1 = 0, \quad c_2 L + 2c_3 = 0.$$

For uniform distributed load $q(x) = q_0$ and $f = 0$, the CBT solutions of a clamped homogeneous beam are

$$w^C = -\frac{q_0 L^4}{24 D_{xx}}\left(\frac{x^2}{L^2} - 2\frac{x^3}{L^3} + \frac{x^4}{L^4}\right),$$

$$\frac{dw^C}{dx} = -\frac{q_0 L^3}{24 D_{xx}}\left(2\frac{x}{L} - 6\frac{x^2}{L^2} + 4\frac{x^3}{L^3}\right), \tag{4.7.124}$$

$$\frac{d^2 w^C}{dx^2} = -\frac{q_0 L^2}{24 D_{xx}}\left(2 - 12\frac{x}{L} + 12\frac{x^2}{L^2}\right).$$

Then $p(x)$ of Eq. (4.7.94) becomes

$$p(x) = -\frac{4}{\bar{A}_{xy}}\left[-S_{xz}c_2\left(\frac{x^3}{6} - \frac{Lx^2}{4}\right) + \tilde{D}_{xx}c_2 x + \tilde{c}_6\right.$$
$$\left. + D_{xx}\left(S_{xz}w^C - \tilde{D}_{xx}\frac{d^2 w^C}{dx^2}\right)\right],$$

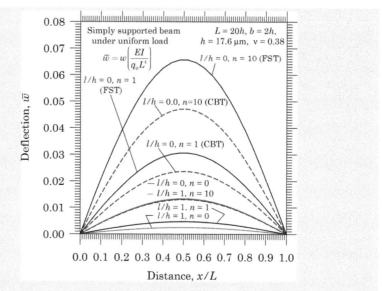

Figure 4.7.1: Deflection $\bar{w}(x)$ versus x/L for simply supported beam under uniformly distributed load.

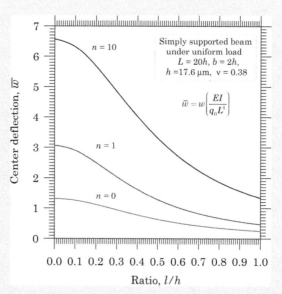

Figure 4.7.2: Maximum deflection $\bar{w}(L/2)$ versus ℓ/h for simply supported beam under uniformly distributed load.

or

$$p(x) = -\frac{4}{\bar{A}_{xy}}\left\{-S_{xz}c_2\left(\frac{x^3}{6} - \frac{Lx^2}{4}\right) + \tilde{D}_{xx}c_2 x + c_6\right.$$
$$\left. -\frac{q_0}{24}\left[S_{xz}(x^4 - 2Lx^3 + L^2x^2) - 2\tilde{D}_{xx}(6x^2 - 6Lx + L^2)\right]\right\}. \qquad (4.7.125)$$

Then the particular solution is (where $\alpha^2 = 4\bar{A}_{xz}/\bar{A}_{xy}$ and $A_{xy} \neq 0$)

$$w_p^F(x) = -\frac{4}{\bar{A}_{xy}}\left\{S_{xz}c_2\left(\frac{x^3}{6\alpha^2} + \frac{x}{\alpha^4} - \frac{Lx^2}{4\alpha^2} - \frac{L}{\alpha^4}\right) - \tilde{D}_{xx}c_2\frac{x}{\alpha^2} - c_6\frac{1}{\alpha^2}\right.$$
$$+\frac{q_0}{24}\left[S_{xz}\left(\frac{x^4}{\alpha^2} + \frac{12x^2}{\alpha^4} + \frac{24}{\alpha^6} - \frac{2Lx^3}{\alpha^2} - \frac{12Lx}{\alpha^4} + \frac{L^2x^2}{\alpha^2} + \frac{4L^2}{\alpha^4}\right)\right.$$
$$\left.\left. -2\tilde{D}_{xx}\left(\frac{6x^2}{\alpha^2} + \frac{24}{\alpha^4} - \frac{6Lx}{\alpha^2} + \frac{L^2}{\alpha^2}\right)\right]\right\}. \qquad (4.7.126)$$

The first derivative is

$$\frac{dw_p^F}{dx} = -\frac{4}{\bar{S}_{xy}}\left\{S_{xz}C_2\left(\frac{x^2}{2\alpha^2} + \frac{1}{\alpha^4} - \frac{Lx}{2\alpha^2}\right) - \tilde{D}_{xx}c_2\frac{1}{\alpha^2} + \frac{q_0}{24}\left[S_{xz}\left(\frac{4x^3}{\alpha^2} + \frac{24x}{\alpha^4}\right.\right.\right.$$
$$\left.\left.\left. -\frac{6Lx^2}{\alpha^2} - \frac{12L}{\alpha^4} + \frac{2L^2x}{\alpha^2}\right) - 2\tilde{D}_{xx}\left(\frac{12x}{\alpha^2} - \frac{6L}{\alpha^2}\right)\right]\right\}. \qquad (4.7.127)$$

The boundary conditions $w^F(0) = w^F(L) = 0$ and (dw^F/dx) at $x = 0, L$ give

$$0 = c_7 - \frac{4}{\bar{A}_{xy}\alpha^2}\left\{-\frac{S_{xz}L}{\alpha^2}c_2 - c_6 + \frac{q_0}{24}\left[S_{xz}\left(\frac{24}{\alpha^4} + \frac{4L^2}{\alpha^2}\right) - 2\tilde{D}_{xx}\left(\frac{24}{\alpha^2} + L^2\right)\right]\right\},$$

$$0 = c_7\cosh\alpha L + c_8\sinh\alpha L - \frac{4}{\bar{S}_{xy}\alpha^2}\left\{-S_{xz}c_2\frac{L^3}{12} - \tilde{D}_{xx}c_2 L - c_6,\right.$$
$$\left. +\frac{q_0}{24}\left[S_{xz}\left(\frac{24}{\alpha^4} + \frac{4L^2}{\alpha^2}\right) - 2\tilde{D}_{xx}\left(\frac{24}{\alpha^2} + L^2\right)\right]\right\},$$

$$0 = \alpha c_8 - \frac{4}{\bar{S}_{xy}\alpha^4}\left\{S_{xz}c_2 - \alpha^2\tilde{D}_{xx}c_2 + \frac{q_0 L}{2}\left(-S_{xz} + \alpha^2\tilde{D}_{xx}\right)\right\},$$

$$0 = \alpha\left(c_7\sinh\alpha L + c_8\cosh\alpha L\right) - \frac{4}{\bar{A}_{xy}\alpha^4}\left\{\left(S_{xz} - \alpha^2\tilde{D}_{xx}\right)c_2\right.$$
$$\left. +\frac{q_0 L}{2}\left(S_{xz} - \alpha^2\tilde{D}_{xx}\right)\right\}.$$

Let

$$K_1 = \frac{q_0}{24}\left[S_{xz}\left(\frac{24}{\alpha^4} + \frac{4L^2}{\alpha^2}\right) - 2\tilde{D}_{xx}\left(\frac{24}{\alpha^2} + L^2\right)\right],$$
$$K_2 = \left(S_{xz} - \alpha^2\tilde{D}_{xx}\right). \qquad (4.7.128)$$

Then we have the following four equations in terms of the four constants $(c_2, c_6, c_7,$

and c_8):

$$0 = c_7 - \frac{4}{\bar{A}_{xy}\alpha^2}\left(-S_{xz}c_2\frac{L}{\alpha^2} - c_6 + K_1\right),$$

$$0 = c_7\cosh\alpha L + c_8\sinh\alpha L - \frac{4}{\bar{S}_{xy}\alpha^2}\left[-\frac{L}{12}\left(S_{xz}L^2 + 12\tilde{D}_{xx}\right)c_2 - c_6 + K_1\right],$$

$$0 = \alpha c_8 - \frac{4}{\bar{A}_{xy}\alpha^4}\left(c_2 - \frac{q_0 L}{2}\right)K_2,$$

$$0 = \alpha\left(c_7\sinh\alpha L + c_8\cosh\alpha L\right) - \frac{4}{\bar{A}_{xy}\alpha^4}\left(c_2 + \frac{q_0 L}{2}\right)K_2.$$

The solution to these equations is

$$c_2 = 0, \quad c_3 = -\frac{L}{2}c_2 = 0, \quad c_6 = K_1 - \bar{A}_{xy}\alpha^2 c_7,$$
$$c_7 = \frac{2K_2 q_0 L}{\bar{A}_{xy}\alpha^5\sinh\alpha L}\left(\cosh\alpha L - 1\right), \quad c_8 = -\frac{2K_2 q_0 L}{\bar{A}_{xy}\alpha^5}. \tag{4.7.129}$$

The solution to the microstructure-dependent FGM beams is

$$u^F(x) = -\frac{B_{xx}}{A_{xx}}\phi_x^F(x), \quad N_{xx}^F(x) = N_{xx}^C(x), \tag{4.7.130}$$

$$w^F(x) = c_7\cosh\alpha x + c_8\sinh\alpha x + w_p^F(x), \tag{4.7.131}$$

$$\phi_x^F(x) = \frac{1}{\bar{D}_{xx}}\left(\tfrac{1}{2}A_{xy}\frac{dw^F}{dx} - D_{xx}\frac{dw^C}{dx} + c_2\frac{x^2}{2} + c_3 x\right), \tag{4.7.132}$$

$$M_{xx}^F(x) = \frac{1}{2}A_{xy}\left(\frac{d^2 w^F}{dx^2} - \frac{d\phi_x^F}{dx}\right) + M_{xx}^C(x) + c_2 x + c_3, \tag{4.7.133}$$

$$N_{xz}^F(x) = \tfrac{1}{4}A_{xy}\left(\frac{d^3 w^F}{dx^3} - \frac{d^2\phi_x^F}{dx^2}\right) + N_{xz}^C(x) + c_2. \tag{4.7.134}$$

For functionally graded beams without considering the effect of the couple stress (i.e., $A_{xy} = 0$ and $g = 0$), the relationships are much simpler than those of the general beam. Use of the boundary conditions from Eq. (4.7.121) in Eqs. (4.7.80)–(4.7.82) yields

$$c_1 = c_4 = c_5 = 0; \quad c_3 = -\frac{1}{2}c_2 L, \quad c_6 = -\frac{1}{S_{xz}}M^C(0),$$
$$c_2 = \frac{12\Omega}{(1 + 12\Omega)L}\left(M_{xx}^C(0) - M_{xx}^C(L)\right), \quad \Omega = \frac{D_{xx}}{S_{xz}L^2}.$$

The complete solution for simply supported FGM beams is

$$u^F(x) = -\frac{B_{xx}}{A_{xx}}\phi_x^F(x), \tag{4.7.135}$$

$$w^F(x) = \frac{D_{xx}}{\bar{D}_{xx}}w^C(x) + \frac{1}{\bar{D}_{xx}}\left[\Omega L^2\left(M_{xx}^C(x) - M_{xx}^C(0)\right)\right.$$
$$\left. + \frac{c_2}{12}\left(12\Omega x L^2 + 3x^2 L - 2x^3\right)\right], \tag{4.7.136}$$

$$\phi_x^F(x) = -\frac{dw^C}{dx} + \frac{c_2}{2}(x^2 - xL), \tag{4.7.137}$$

$$M_{xx}^F(x) = M_{xx}^C(x) + \frac{c_2}{2}(2x - L), \tag{4.7.138}$$

$$N_{xz}^F(x) = N_{xz}^C(x) + c_2, \tag{4.7.139}$$

where

$$M_{xx}^C(x) = \frac{q_0}{12}(6x^2 - 6Lx + L^2), \quad N_{xz}^C(x) = \frac{q_0}{2}(2x - L). \tag{4.7.140}$$

The above expressions are also valid for a homogeneous beam but with $u^F = 0$ (because $B_{xx} = 0$). Also note that $c_2 = 0$ whenever the load is such that $M_{xx}^C(0) = M_{xx}^C(L)$, as in the case of uniformly distributed load.

Numerical results are presented using the following data [the same as that in Eq. (4.7.122)]:

$$E_1 = 14.4 \text{ GPa}, \quad E_2 = 1.44 \text{ GPa}, \quad \nu = 0.38, \quad h = 17.6 \times 10^{-6} \text{ m}, \quad b = 2h, \quad L = 20h.$$

Figure 4.7.3 shows the plots of the dimensionless transverse deflection, $\bar{w} = w(x)(EI/q_0L^4)$, as a function of the normalized distance, $\bar{x} = x/L$ for two different ratios of the material length scale parameter to the height of the beam, ell/h, and for three different values of the power-law index n. The difference between the TBT solutions and the CBT solutions becomes smaller.

Figure 4.7.4 contains plots of the maximum transverse deflections of clamped beams as a function of the ℓ/h ratio for $n = 0, 1, 10$. Again, as expected, the deflection decreases with the ratio ℓ/h and with decreasing values of n.

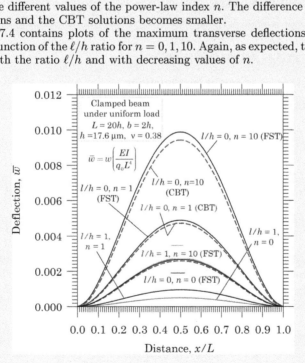

Figure 4.7.3: Deflection $\bar{w}(x)$ versus x/L for clamped (C–C) beam under uniformly distributed load.

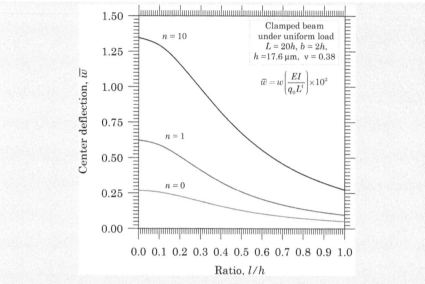

Figure 4.7.4: Maximum deflection $\bar{w}(L/2)$ versus ℓ/h for clamped beam under uniformly distributed load.

4.7.4 BUCKLING RELATIONSHIPS

4.7.4.1 Summary of governing equations

The buckling equations in terms of the stress resultants of the CBT are obtained from Eq. (4.6.58)

$$-\frac{dM_{xx}^C}{dx} + N_{xz}^C = 0, \quad -\frac{dN_{xz}^C}{dx} + N_{xx}^C \frac{d^2w^C}{dx^2} = 0, \tag{4.7.141}$$

where the superscript C refers to CBT. These equations are also valid for the TBT:

$$-\frac{dM_{xx}^F}{dx} + N_{xz}^F = 0, \quad -\frac{dN_{xz}^F}{dx} + N_{xx}^F \frac{d^2w^F}{dx^2} = 0. \tag{4.7.142}$$

The two equations in (4.7.141) can be combined into one by substituting the first equation for N_{xz} into the second equation in each case; and then using the stress resultant–displacement relation, $M_{xx}^C = -D_{xx}(d^2w^C/dx^2)$, to replace the bending moment, one arrives at (assuming that D_{xx} is a constant)

$$\frac{d^4w^C}{dx^4} + \frac{N_{xx}^C}{D_{xx}} \frac{d^2w^C}{dx^2} = 0. \tag{4.7.143}$$

Similarly, for TBT we have

$$D_{xx}\frac{d^2\phi_x^F}{dx^2} = S_{xz}\left(\phi_x^F + \frac{dw^F}{dx}\right), \tag{4.7.144}$$

$$N_{xx}^F\frac{d^2w^F}{dx^2} = S_{xz}\left(\frac{d\phi_x^F}{dx} + \frac{d^2w^F}{dx^2}\right). \tag{4.7.145}$$

By differentiating Eq. (4.7.144) and using Eq. (4.7.145), we obtain

$$D_{xx}\frac{d^3\phi_x^F}{dx^3} = N_{xx}^F\frac{d^2w^F}{dx^2}. \tag{4.7.146}$$

Equation (4.7.145) can be solved for $d\phi_x^F/dx$ as

$$\frac{d\phi_x^F}{dx} = -\left(1 - \frac{N_{xx}^F}{S_{xz}}\right)\frac{d^2w^F}{dx^2}. \tag{4.7.147}$$

Substitution of Eq. (4.7.147) into Eq. (4.7.146) gives the final result:

$$\frac{d^4w^F}{dx^4} + \frac{1}{D_{xx}}\left(\frac{N_{xx}^F}{1 - \frac{N_{xx}^F}{S_{xz}}}\right)\frac{d^2w^F}{dx^2} = 0. \tag{4.7.148}$$

By differentiating Eq. (4.7.144) and using Eq. (4.7.147), we can also obtain

$$\frac{d^3\phi_{xx}^F}{dx^3} + \frac{1}{D_{xx}}\left(\frac{N_{xx}^F}{1 - \frac{N_{xx}^F}{S_{xz}}}\right)\frac{d\phi_{xx}^F}{dx} = 0. \tag{4.7.149}$$

In view of the similarity of Eq. (4.7.148) to Eq. (4.7.143), and provided the boundary conditions are of the same form, it can be concluded that

$$N_{xx}^C = \frac{N_{xx}^F}{1 - \frac{N_{xx}^F}{S_{xz}}} \quad \text{or} \quad N_{xx}^F = \frac{N_{xx}^C}{1 + \frac{N_{xx}^C}{S_{xz}}} \tag{4.7.150}$$

and

$$\phi_x^F = -\left(1 - \frac{N_{xx}^F}{S_{xz}}\right)\frac{dw^F}{dx} + c_1, \tag{4.7.151}$$

$$w^F = c_2w^C + c_3x + c_4, \tag{4.7.152}$$

$$\phi_x^F = -\left(1 - \frac{N_{xx}^F}{S_{xz}}\right)\left(c_2\frac{dw^C}{dx} + c_3\right) + c_1, \tag{4.7.153}$$

where c_1, c_2, c_3, and c_4 are constants. **Example 4.6.4** already illustrates the use of the derived correspondence between the buckling load N_{xx}^C predicted by the CBT with the buckling load N_{xx}^F predicted by the TBT. The relationship in Eq. (4.7.150) does not hold for fixed–pinned beams because the boundary conditions do not match exactly.

4.7.5   FREQUENCY RELATIONSHIPS

4.7.5.1   Governing equations of the CBT

The equation of motion of the CBT is given by

$$\frac{\partial^2}{\partial x^2}\left(-D_{xx}\frac{\partial^2 w^C}{\partial x^2}\right) = m_0\frac{\partial^2 w^E}{\partial t^2} - m_2\frac{\partial^4 w^E}{\partial x^2 \partial t^2}. \tag{4.7.154}$$

For free vibration, we seek periodic solutions of the form $w^C(x,t) = W^C(x)e^{i\omega t}$, where $W^C(x)$ is the mode shape and ω is the frequency of natural vibration. Substitution of this expression into Eq. (4.7.154) gives

$$p\frac{d^4 W^E}{dx^4} + q\frac{d^2 W^E}{dx^2} - rW^E = 0, \tag{4.7.155}$$

where

$$p = D_{xx}, \quad q = m_2\omega^2, \quad r = m_0\omega^2. \tag{4.7.156}$$

If μ_1 and μ_2 are the roots of Eq. (4.7.155), where

$$(\mu_1)^2 = -\frac{q}{2p} - \frac{1}{2p}\sqrt{q^2 + 4pr}, \quad (\mu_2)^2 = -\frac{q}{2p} + \frac{1}{2p}\sqrt{q^2 + 4pr}, \tag{4.7.157}$$

then Eq. (4.7.155) can be written as

$$\left(\frac{d^2}{dx^2} + \mu_1\right)\left(\frac{d^2}{dx^2} + \mu_2\right)W^E = 0. \tag{4.7.158}$$

4.7.5.2   Governing equations of the TBT

The equations governing the natural vibration of the TBT are

$$\frac{d}{dx}\left[S_{xz}\left(\Phi_x^F + \frac{dW^F}{dx}\right)\right] + \omega^2 m_0 W^T = 0, \tag{4.7.159}$$

$$D_{xx}\frac{d^2\Phi_x^F}{dx^2} - S_{xz}\left(\Phi_x^F + \frac{dW^F}{dx}\right) + \omega^2 m_2\Phi_x^F = 0, \tag{4.7.160}$$

where W^F and Φ_x^F are deflection and rotation that define the mode shapes. Towards eliminating Φ_x^T from the two equations, we solve Eq. (4.7.159) for $d\Phi_x^T/dx$

$$\frac{d\Phi_x^F}{dx} = -\frac{m_0\omega^2}{S_{xz}}W^F - \frac{d^2 W^F}{dx^2}. \tag{4.7.161}$$

Differentiating Eq. (4.7.160) once and substituting for $d\Phi_x^T/dx$ and $d^3\Phi_x^T/dx^3$ from Eq. (4.7.161), we obtain

$$a\frac{d^4 W^F}{dx^4} + b\frac{d^2 W^T}{dx^2} - cW^T = 0, \tag{4.7.162}$$

where

$$a = D_{xx}, \quad b = \left(m_2 + m_0 \frac{D_{xx}}{S_{xz}} \right) w^2, \quad c = m_0 \left(1 - \frac{m_2}{S_{xz}} w^2 \right). \quad (4.7.163)$$

If λ_1 and λ_2 are the roots of Eq. (4.7.162), where

$$(\lambda_1)^2 = -\frac{b}{2a} - \frac{1}{2a}\sqrt{b^2 + 4ac}, \quad (\lambda_2)^2 = -\frac{b}{2a} + \frac{1}{2a}\sqrt{b^2 + 4ac}, \quad (4.7.164)$$

then Eq. (4.7.162) can be expressed as

$$\left(\frac{d^2}{dx^2} + \lambda_1 \right)\left(\frac{d^2}{dx^2} + \lambda_2 \right)W^F = 0. \quad (4.7.165)$$

4.7.5.3 Relationship

Comparing Eq. (4.7.165) with Eq. (4.7.158), and assuming that the mode shapes $W^C(x)$ and $W^F(x)$ are identical, we have the relationship (equating the plausible roots)

$$\mu_2 = \lambda_2 \quad \rightarrow \quad -\frac{q}{2p} + \frac{1}{2p}\sqrt{q^2 + 4pr} = -\frac{b}{2a} + \frac{1}{2a}\sqrt{b^2 + 4ac}. \quad (4.7.166)$$

4.8 THE NAVIER SOLUTIONS

4.8.1 GENERAL SOLUTION

Here, we consider Navier's solution to simply-supported (i.e., hinged at both ends) FGM beams using the linearized equations of motion from Eqs. (4.4.12)–(4.4.14) when the stiffness coefficients (A_{xx}, B_{xx}, D_{xx}, A_{xy}, and S_{xz}), mass inertias ($m_0.m_1, m_2$), and the foundation modulus c_f are constant (see Reddy [74]):

$$-\frac{\partial}{\partial x}\left(A_{xx}\frac{\partial u}{\partial x} + B_{xx}\frac{\partial \phi_x}{\partial x} - N_{xx}^T \right) + m_0\frac{\partial^2 u}{\partial t^2} + m_1\frac{\partial^2 \phi_x}{\partial t^2} = f(x,t), \quad (4.8.1)$$

$$-\frac{\partial}{\partial x}\left[S_{xz}\left(\phi_x + \frac{\partial w}{\partial x} \right) \right] - \frac{1}{4}\frac{\partial^2}{\partial x^2}\left[A_{xy}\left(\frac{\partial \phi_x}{\partial x} - \frac{\partial^2 w}{\partial x^2} \right) \right] + c_f w$$

$$- q(x,t) + m_0\frac{\partial^2 w}{\partial t^2} = 0, \quad (4.8.2)$$

$$-\frac{\partial}{\partial x}\left(B_{xx}\frac{\partial u}{\partial x} + D_{xx}\frac{\partial \phi_x}{\partial x} - M_{xx}^T \right) + S_{xz}\left(\phi_x + \frac{\partial w}{\partial x} \right)$$

$$- \frac{1}{4}\frac{\partial}{\partial x}\left[A_{xy}\left(\frac{\partial \phi_x}{\partial x} - \frac{\partial^2 w}{\partial x^2} \right) \right] + m_1\frac{\partial^2 u}{\partial t^2} + m_2\frac{\partial^2 \phi_x}{\partial t^2} = 0. \quad (4.8.3)$$

The simply supported boundary conditions for the TBT are:

$$N_{xx} = 0, \quad w = 0, \quad M_{xx} = 0, \text{ at } x = 0 \text{ and } x = L, \tag{4.8.4}$$

where L is the length of the beam and the origin of the x-coordinate is taken at the left end of the beam. Note that the boundary conditions are of those of a hinged–hinged beam. The boundary conditions in Eq. (4.8.4) are met by the following displacement expansions:

$$u(x,t) = \sum_{m=1}^{\infty} U_m(t) \cos \alpha_m x,$$

$$w(x,t) = \sum_{m=1}^{\infty} W_m(t) \sin \alpha_m x, \quad \alpha_m = \frac{m\pi}{L}, \tag{4.8.5}$$

$$\phi_x(x,t) = \sum_{m=1}^{\infty} S_m(t) \cos \alpha_m x,$$

where $U_m(t)$, $W_m(t)$, and $S_m(t)$ are the coefficients to be determined such that Eqs. (4.8.1)–(4.8.3) are satisfied everywhere in the domain $(0, L)$ for any arbitrary time t. The forces are expanded in the cosine and sine series as

$$f(x,t) = \sum_{m=1}^{\infty} F_m \cos \alpha_m x, \quad q(x,t) = \sum_{m=1}^{\infty} Q_m \sin \alpha_m x,$$

$$M_{xx}^T(x,y) = \sum_{m=1}^{\infty} M_m^T \sin \alpha_m x, \quad N_{xx}^T(x,y) = \sum_{m=1}^{\infty} N_m^T \sin \alpha_m x, \tag{4.8.6}$$

where the coefficients F_m, Q_m, N_m^T, and M_m^T are determined from

$$Q_m(t) = \frac{2}{L} \int_0^L q(x,t) \sin \alpha_m x \, dx,$$

$$F_m(t) = \frac{2}{L} \int_0^L f(x,t) \cos \alpha_m x \, dx, \tag{4.8.7}$$

with similar expressions for N_m^T and M_m^T.

Substituting the expansions from Eqs. (4.8.5) and (4.8.6) into Eqs. (4.8.1)–(4.8.3), and collecting the coefficients of U_m, W_m, and S_m separately, we obtain

$$0 = \sum_{m=1}^{\infty} \left(A_{xx} \alpha_m^2 U_m + B_{xx} \alpha_m^2 S_m + m_0 \frac{d^2 U_m}{dt^2} + m_1 \frac{d^2 S_m}{dt^2} \right.$$

$$\left. + \alpha_m N_m^T - F_m \right) \cos \alpha_m x, \tag{4.8.8}$$

$$0 = \sum_{m=1}^{\infty} \left[\left(S_{xz}\alpha_m - \tfrac{1}{4}A_{xy}\alpha_m^3 \right) S_m + \left(S_{xz}\alpha_m^2 + \tfrac{1}{4}A_{xy}\alpha_m^4 + c_f \right) W_m \right.$$

$$\left. + m_0 \frac{d^2 W_m}{dt^2} - Q_m \right] \sin \alpha_m x, \tag{4.8.9}$$

$$0 = \sum_{m=1}^{\infty} \left[B_{xx}\alpha_m^2 U_m + \left(D_{xx}\alpha_m^2 + \frac{1}{4}A_{xy}\alpha_m^2 + S_{xz} \right) S_m + \left(S_{xz}\alpha_m - \tfrac{1}{4}A_{xy}\alpha_m^3 \right) W_m \right.$$

$$\left. + m_1 \frac{d^2 U_m}{dt^2} + m_2 \frac{d^2 S_m}{dt^2} + \alpha_m M_m^T \right] \cos \alpha_m x. \tag{4.8.10}$$

Since the above equations must hold for any x and m, we obtain (the expressions in the square brackets)

$$\begin{bmatrix} A_{xx}\alpha_m^2 & 0 & B_{xx}\alpha_m^2 \\ 0 & S_{xz}\alpha_m^2 + \tfrac{1}{4}A_{xy}\alpha_m^4 + c_f & S_{xz}\alpha_m - \tfrac{1}{4}A_{xy}\alpha_m^3 \\ B_{xx}\alpha_m^2 & S_{xz}\alpha_m - \tfrac{1}{4}A_{xy}\alpha_m^3 & D_{xx}\alpha_m^2 + \tfrac{1}{4}A_{xy}\alpha_m^2 + S_{xz} \end{bmatrix} \begin{Bmatrix} U_m \\ W_m \\ S_m \end{Bmatrix}$$

$$+ \begin{bmatrix} m_0 & 0 & m_1 \\ 0 & m_0 & 0 \\ m_1 & 0 & m_2 \end{bmatrix} \begin{Bmatrix} \ddot{U}_m \\ \ddot{W}_m \\ \ddot{S}_m \end{Bmatrix} = \begin{Bmatrix} -\alpha_m N_m^T + F_m \\ Q_m \\ -\alpha_m M_m^T \end{Bmatrix}. \tag{4.8.11}$$

4.8.2 BENDING SOLUTION

For static bending, we set the time derivative terms in Eq. (4.8.11) to zero and obtain

$$\begin{bmatrix} A_{xx}\alpha_m^2 & 0 & B_{xx}\alpha_m^2 \\ 0 & S_{xz}\alpha_m^2 + \tfrac{1}{4}A_{xy}\alpha_m^4 + c_f & S_{xz}\alpha_m - \tfrac{1}{4}A_{xy}\alpha_m^3 \\ B_{xx}\alpha_m^2 & S_{xz}\alpha_m - \tfrac{1}{4}A_{xy}\alpha_m^3 & D_{xx}\alpha_m^2 + \tfrac{1}{4}A_{xy}\alpha_m^2 + S_{xz} \end{bmatrix} \begin{Bmatrix} U_m \\ W_m \\ S_m \end{Bmatrix}$$

$$= \begin{Bmatrix} -\alpha_m N_m^T + F_m \\ Q_m \\ -\alpha_m M_m^T \end{Bmatrix}. \tag{4.8.12}$$

The matrix equation in Eq. (4.8.12) can be inverted for each m for any applied load parameters (F_m, Q_m, N_m^T, M_M^T). For example, when $f = c_f = 0$ and thermal forces are not accounted for, Eq. (4.8.12) can be expressed as

$$K_{11}U_m + K_{13}S_m = 0,$$
$$K_{22}W_m + K_{23}S_m = Q_m, \tag{4.8.13}$$
$$K_{31}U_m + K_{32}W_m + K_{33}S_m = 0,$$

where [we recall that A_{xx}, B_{xx}, D_{xx}, and A_{xy} are defined in Eq. (3.4.7); S_{xz} defined in Eq. (4.4.2) is the same as A_{xy} with ℓ^2 replaced with K_s]

$$K_{11} = A_{xx}\left(\frac{m\pi}{L}\right)^2, \quad K_{13} = K_{31} = B_{xx}\left(\frac{m\pi}{L}\right)^2,$$

$$K_{22} = S_{xz}\left(\frac{m\pi}{L}\right)^2 + \tfrac{1}{4}A_{xy}\left(\frac{m\pi}{L}\right)^4 + c_f,$$

$$K_{23} = K_{32} = S_{xz}\left(\frac{m\pi}{L}\right) - \tfrac{1}{4}A_{xy}\left(\frac{m\pi}{L}\right)^3, \qquad (4.8.14)$$

$$K_{33} = S_{xz} + \left(D_{xx} + \tfrac{1}{4}A_{xy}\right)\left(\frac{m\pi}{L}\right)^2.$$

Equations (4.8.13) can be solved to obtain (U_m, W_m, S_m)

$$U_m = K_{13}K_{23}\left(Q_m/S_0\right), \quad W_m = \left(K_{11}K_{33} - K_{13}K_{13}\right)\left(Q_m/S_0\right),$$

$$S_m = -K_{11}K_{23}\left(Q_m/S_0\right), \quad S_0 = K_{11}\left(K_{22}K_{33} - K_{23}K_{23}\right) - K_{13}K_{13}K_{22}. \qquad (4.8.15)$$

and the solution $(u(x), w(x), \phi_x)$ is given by Eq. (4.8.5). The shear correction coefficient used here is based on the formula, $K_s = 5(1+\nu)/(6+5\nu)$.

Example 4.8.1 ————————————————————————————————

Determine the exact solution of a simply supported micro-beam subjected to three types of loads: (a) point load, $q(x) = Q_0\,\delta(x - x_0)$, at the center ($x_0 = 0.5L$), (b) uniformly distributed load, $q(x) = q_0$, and (c) sinusoidal load, $q(x) = q_0 \sin\frac{\pi x}{L}$. Use the following material and geometric parameters of micro-beams with rectangular cross section [74]:

$$E_1 = 14.4 \text{ GPa}, \quad E_2 = E = 1.44 \text{ GPa}, \, \quad \nu = 0.38, \quad c_f = 0,$$

$$h = 5 \times 17.6 \times 10^{-6} \text{ m}, \quad b = 2h, \quad L = 20h, \quad q_0 = 1.0 \text{ N/m}. \qquad (4.8.16)$$

Solution: Numerical results of the normalized center deflection $\bar{w}$ of a simply supported FGM beams under various types of loads (SSL = sinusoidal load, UDL = uniformly distributed load, and PTL = point load at the center of the beam) and for the two different beam theories, CBT and TBT, are presented in Table 4.8.1 as a function of ℓ/h ratio and the volume fraction index n. The effect of shear deformation on the deflection is not significant because of the length-to-height ratio is 20. In general, the effect of the length scale parameter ℓ is to make the beam behave stiffer by contributing to the bending stiffness.

Figure 4.8.1 contains plots of dimensionless deflection, $\bar{w}(x) = w(x)(D_{xx}/q_0 L^4) \times 10^2$, for the case of homogeneous isotropic beam (i.e., $n = 0$) under sinusoidal distributed transverse load (SSL) as a function of the dimensionless distance along the length, x/L, for various cases of including and not including the length scale parameter in the model. The difference between the results predicted by the CBT (see Section 3.8) and TBT is small and cannot be seen in these plots. In general, unless the beam length-to-height ratio is very small (of the order of 5), the effect of shear deformation is not significant on the deflections.

Figure 4.8.2 contains plots of dimensionless deflection, $\bar{w}(x) = w(x)(D_{xx}/q_0 L^4) \times$ 10^2, for functionally graded beams ($E_1 = 14.4$ GPa, and $E_2 = 1.44$ GPa) under sinusoidal distributed transverse load for both with and without the length scale parameter (ℓ). Through-thickness grading of the material in the beam helps to stiffen the beam so that the transverse deflections can be controlled.

Similar plots for the case of uniformly distributed load are presented in Figs. 4.8.3 and 4.8.4. Note that the dimensionless deflection plotted is $\bar{w}(x) = w(x)(D_{xx}/q_0 L^4) \times 10^3$. The dark symbol and blue line in Fig. 4.8.3 corresponds to the TBT solutions, which essentially coincide with the CBT solutions, indicating that for the choice of the geometric and material parameters used in this example the effect of shear deformation is negligible, and that TBT deflections are slightly larger than the CBT solutions. We note that the Navier solutions for the hinged-hinged FGM beams are different from exact solutions obtained by direct integration of the pinned-pinned FGM beams.

Table 4.8.1: Center deflections ($\bar{w} = w(0.5L)(D_{xx}/q_0 L^4) \times 10^3$ for UDL and SSL and $\bar{w} = w(0.5L)(D_{xx}/q_0 L^3) \times 10^4$ for PTL) of simply supported FGM beams under various types of loads and for different values of the power-law index n and the ratio ℓ/h (modified couple stress effect).

		Point load		Uniform load		Sinusoidal load	
n	ℓ/h	CBT	TBT	CBT	TBT	CBT	TBT
0	0.0	0.0367	0.0369	1.3021	1.3103	1.0266	1.0333
	0.2	0.0312	0.0315	1.1092	1.1162	0.8745	0.8802
	0.4	0.0216	0.0218	0.7679	0.7731	0.6054	0.6096
	0.6	0.0143	0.0144	0.5076	0.5116	0.4002	0.4034
	0.8	0.0097	0.0098	0.3442	0.3475	0.2714	0.2741
	1.0	0.0069	0.0070	0.2435	0.2464	0.1920	0.1943
1	0.0	0.0858	0.0863	3.0474	3.0624	2.4027	2.4148
	0.2	0.0701	0.0705	2.4900	2.5023	1.9632	1.9732
	0.4	0.0453	0.0456	1.6077	1.6165	1.2676	1.2747
	0.6	0.0285	0.0287	1.0108	1.0175	0.7970	0.8023
	0.8	0.0187	0.0189	0.6651	0.6707	0.5244	0.5289
	1.0	0.0130	0.0132	0.4620	0.4669	0.3642	0.3682
10	0.0	0.1838	0.1854	6.5261	6.5714	5.1454	5.1820
	0.2	0.1586	0.1600	5.6333	5.6726	4.4415	4.4733
	0.4	0.1125	0.1135	3.9941	4.0325	3.1491	3.1729
	0.6	0.0757	0.0765	2.6097	2.7122	2.1206	2.1309
	0.8	0.0520	0.0526	1.8458	1.8644	1.4552	1.4703
	1.0	0.0370	0.0376	1.3152	1.3315	1.0369	1.0502

4.8.3 NATURAL VIBRATIONS

For natural (or free) vibration analysis, we set all force coefficients to zero and assume a periodic solution in the form

$$U_m(t) = U_m^0 \, e^{-i\omega_m t}, \quad W_m(t) = W_m^0 \, e^{-i\omega_m t}, \quad S_m(t) = S_m^0 \, e^{-i\omega_m t}, \quad (4.8.17)$$

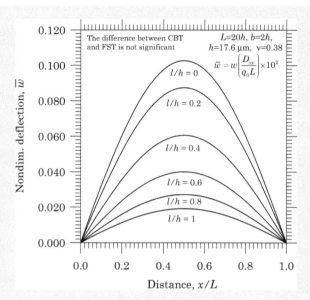

Figure 4.8.1: Plots of the transverse deflection $\bar{w}(x) = w(x)(D_{xx}/q_0 L^4) \times 10^2$ versus x/L for isotropic simply supported beams subjected to sinusoidal distributed transverse load.

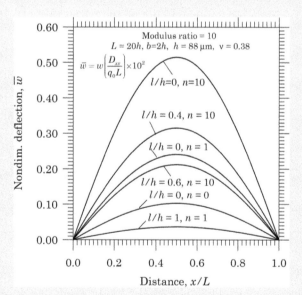

Figure 4.8.2: Plots of the transverse deflection $\bar{w}(x) = w(x)(D_{xx}/q_0 L^4) \times 10^2$ versus x/L for simply supported FGM beams with modified couple stress effect and subjected to sinusoidal distributed transverse load.

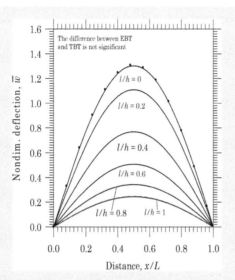

Figure 4.8.3: Plots of the transverse deflection $\bar{w}(x) = w(x)(D_{xx}/q_0 L^4) \times 10^3$ versus x/L for simply supported isotropic beams under uniformly distributed transverse load.

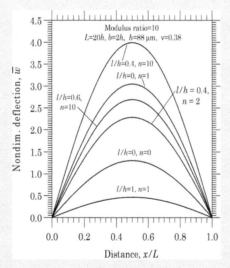

Figure 4.8.4: Plots of the transverse deflection $\bar{w}(x) = w(x)(D_{xx}/q_0 L^4) \times 10^3$ versus x/L for simply supported FGM beams with modified couple stress effect and subjected to uniformly distributed transverse load.

where ω_m is the frequency of natural vibration and $i = \sqrt{-1}$. Substitution of

Eq. (4.8.17) into Eq. (4.8.11) yields the following eigenvalue problem:

$$\left(K_{11} - \omega_m^2 M_{11}\right) U_m^0 + \left(K_{13} - \omega_m^2 M_{13}\right) S_m^0 = 0,$$
$$\left(K_{22} - \omega_m^2 M_{22}\right) W_m^0 + K_{23} S_m^0 = 0, \qquad (4.8.18)$$
$$\left(K_{31} - \omega_m^2 M_{31}\right) U_m^0 + K_{32} W_m^0 + \left(K_{33} - \omega_m^2 M_{33}\right) S_m^0 = 0,$$

where K_{ij} are defined in Eq. (4.8.14) and M_{ij} are

$$M_{11} = m_0, \quad M_{13} = M_{31} = m_1, \quad M_{22} = m_0, \quad M_{33} = m_2. \qquad (4.8.19)$$

In matrix form, we have

$$\begin{bmatrix} K_{11} - \omega_m^2 M_{11} & 0 & K_{13} - \omega_m^2 M_{13} \\ 0 & K_{22} - \omega_m^2 M_{22} & K_{23} \\ K_{31} - \omega_m^2 M_{31} & K_{32} & K_{33} - \omega_m^2 M_{33} \end{bmatrix} \begin{Bmatrix} U_m^0 \\ W_m^0 \\ S_m^0 \end{Bmatrix} = \begin{Bmatrix} 0 \\ 0 \\ 0 \end{Bmatrix}. \qquad (4.8.20)$$

Equation (4.8.20) defines an eigenvalue problem, where the eigenvalues are obtained by finding the roots of a quadratic equation in $\lambda_m = \omega_m^2$.

Example 4.8.2

Determine the first three natural frequencies of a simply supported micro-beam using the Navier solution approach. Use the same material and geometric properties of micro-beams with rectangular cross section as in **Example 3.8.2**, which are presented here for easy reference:

$$h = 17.6 \times 10^{-6} \text{ m}, \quad \rho_1 = \rho_2 = 1.22 \times 10^3 \text{ kg/m},$$
$$E_1 = 14.4 \text{ GPa}, \quad E_2 = 1.44 \text{ GPa}, \quad \rho_1 = 12.2 \times 10^3 \text{ kg/m}, \qquad (4.8.21)$$
$$\rho_2 = 1.22 \times 10^3 \text{ kg/m}, \quad h = 17.6 \times 10^{-6} \text{ m}.$$

Solution: The Navier solution approach requires solving the eigenvalue problem in Eq. (4.8.20) for each m, and determine the vibration frequency ω_m. The corresponding mode is provided by Eq. (4.8.5).

Table 4.8.2 contains the results for isotropic ($n = 0$) and functionally graded ($n \neq 0$) beams. The rotary inertia, m_2, is included in this study. The table contains the results for various cases of including or not including the length scale effect in homogeneous ($n = 0$) and functionally graded ($n \neq 0$) beams. Figures 4.8.5 and 4.8.6 contain plots of dimensionless fundamental frequencies versus the power-law index n for $\ell = 0$ and $\ell = h = 17.6 \times 10^{-6}$ m, respectively. It is clear that, due to the nature of variation of the bending–stretching coupling coefficient B_{xx} with n, the fundamental frequency also experiences associated variation. The effect of shear deformation is more significant when the micro-structure–dependent constitutive model is considered, which increases the magnitude of the frequency due to the stiffening effect.

Table 4.8.2: First three natural frequencies $\bar{\omega}_m$ $(m = 1, 2, 3)$ of homogeneous and functionally graded simply supported beams $(\bar{\omega}_m = \omega_m L^2 / \sqrt{\rho_2 A_0 / E_2})$; results were obtained using the Navier solution procedure.

n	ℓ/h	$\bar{\omega}_1$ CBT	$\bar{\omega}_1$ TBT	$\bar{\omega}_2$ CBT	$\bar{\omega}_2$ TBT	$\bar{\omega}_3$ CBT	$\bar{\omega}_3$ TBT
0	0.0	9.86	9.83	39.32	38.82	88.02	85.63
	0.2	10.68	10.65	42.60	42.06	95.36	92.78
	0.4	12.84	12.80	51.20	50.52	114.61	111.34
	0.6	15.79	15.73	62.97	62.01	140.97	136.39
	0.8	19.18	19.08	76.47	75.05	171.18	164.51
	1.0	22.80	22.66	90.92	88.84	203.54	193.82
1	0.0	8.69	8.67	34.69	34.29	77.74	75.79
	0.2	9.62	9.59	38.37	37.93	86.00	83.84
	0.4	11.97	11.93	47.75	47.16	107.72	104.15
	0.6	15.09	15.04	60.23	59.35	134.97	130.77
	0.8	18.61	18.52	74.25	72.91	166.39	160.69
	1.0	22.41	22.28	89.43	87.42	200.42	190.99
2	0.0	8.41	8.39	33.59	33.17	75.29	73.25
	0.2	9.37	9.34	37.38	36.91	83.80	81.53
	0.4	11.77	11.73	46.97	46.33	105.28	102.25
	0.6	14.93	14.87	59.61	58.68	133.62	129.20
	0.8	18.48	18.39	73.750	72.353	165.324	158.743
	1.0	22.22	22.08	88.68	86.61	198.79	189.12
5	0.0	9.24	9.20	36.85	36.29	82.54	79.83
	0.2	10.11	10.07	40.34	39.72	90.35	87.39
	0.4	12.37	12.32	49.34	48.54	110.52	106.74
	0.6	15.41	15.34	61.48	60.37	137.72	132.47
	0.8	18.86	18.76	75.26	73.65	168.57	161.10
	1.0	22.54	22.39	89.92	87.62	201.42	190.79
10	0.0	10.33	10.28	41.17	40.47	92.12	88.80
	0.2	11.12	11.07	44.31	43.56	99.15	95.58
	0.4	13.20	13.14	52.63	51.70	117.75	113.38
	0.6	16.09	16.00	64.13	62.88	143.49	137.66
	0.8	19.42	19.30	77.42	75.67	173.22	165.14
	1.0	23.09	22.92	92.04	89.57	205.95	194.63
20	0.0	8.42	8.39	33.59	33.17	75.29	73.25
	0.2	9.37	9.34	37.38	36.91	83.80	81.53
	0.4	11.77	11.73	46.967	46.33	105.28	102.25
	0.6	14.94	14.88	59.61	58.68	133.62	129.20
	0.8	18.48	18.39	73.75	72.35	165.32	158.74
	1.0	22.22	22.08	88.68	86.61	198.79	189.12

In summary, the effect of transverse shear deformation, independent of whether modified couple stress effect is included or not, is to reduce the magnitude of the natural frequencies. This is primarily due to the fact that in the TBT the beam transverse shear stiffness is finite as opposed to the CBT, where it is infinitely rigid.

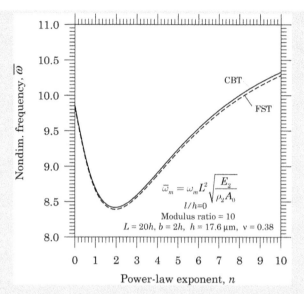

Figure 4.8.5: The effect of the power-law index on the fundamental natural frequency of a simply-supported FGM beam ($\ell = 0$).

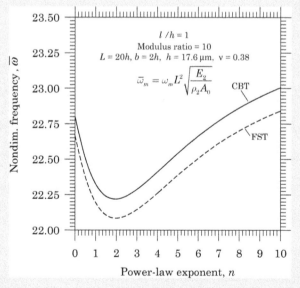

Figure 4.8.6: The effect of power-law index and the length scale parameter on the fundamental frequency ω_1 of a simply supported FGM beam ($\ell = h$).

4.9 SOLUTIONS BY VARIATIONAL METHODS

In this section, we use the Ritz and Galerkin methods introduced in Section 3.9 to solve the TBT beams. We shall use the same examples as those used in Section 3.9.

Example 4.9.1 ——————————————

Consider a beam of length L and constant bending stiffness EI, loaded with uniformly distributed transverse load of intensity q_0. Develop the N-parameter Ritz solution with algebraic polynomials using the TBT for (a) simply supported and (b) clamped boundary conditions (see Fig. 4.9.1), and then compute the one-, two-, and three-parameter solutions, and compare with the exact solution of the problem.

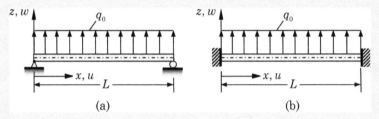

(a)   (b)

Figure 4.9.1: (a) Simply supported and (b) clamped beams under uniformly distributed load.

Solution: First, we need to set up the variational problem using the principle of the minimum total potential energy for the TBT. The total potential energy for this problem is given by (since the axial displacement u is decoupled from the bending variables w and ϕ_x, we do not include the strain energy due to stretching deformation; also, there are no nonzero forces which contribute to Π)

$$
\begin{aligned}
\Pi(w, \phi_x) &= \frac{1}{2} \int_0^L \int_A \left(\sigma_{xx} \varepsilon_{xx} + 2K_s \sigma_{xz} \varepsilon_{xz} \right) dA\, dx - \int_0^L q\, w\, dx \\
&= \frac{1}{2} \int_0^L \left[M_{xx} \frac{d\phi_x}{dx} + N_{xz} \left(\phi_x + \frac{dw}{dx} \right) \right] dx - \int_0^L q\, w\, dx \\
&= \frac{1}{2} \int_0^L \left[EI \left(\frac{d\phi_x}{dx} \right)^2 + GAK_s \left(\phi_x + \frac{dw}{dx} \right)^2 \right] dx - \int_0^L q\, w\, dx, \quad (4.9.1)
\end{aligned}
$$

where K_s is the shear correction coefficient, and the following strain-displacement and constitutive relations are used in arriving at the last line:

$$
\varepsilon_{xx} = z \frac{d\phi_x}{dx}, \quad 2\varepsilon_{xz} = \phi_x + \frac{dw}{dx}, \quad \sigma_{xx} = E\varepsilon_{xx}, \quad \sigma_{xz} = 2G\varepsilon_{xz}, \quad (4.9.2)
$$

$$
M_{xx} = \int_A \sigma_{xx}\, z\, dA = EI \frac{d\phi_x}{dx}, \quad N_{xz} = K_s \int_A \sigma_{xz}\, dA = GAK_s \left(\phi_x + \frac{dw}{dx} \right).
$$
$$
(4.9.3)
$$

Then the principle of minimum total potential energy, $\delta\Pi(w, \phi_x) = 0$, for the problem can be expressed as a pair of the following two integral statements:

$$0 = \int_0^L \left[GAK_s \frac{d\delta w}{dx} \left(\phi_x + \frac{dw}{dx} \right) - \delta w\, q \right] dx, \tag{4.9.4a}$$

$$0 = \int_0^L \left[EI \frac{d\delta\phi_x}{dx} \frac{d\phi_x}{dx} + GAK_s\, \delta\phi_x \left(\phi_x + \frac{dw}{dx} \right) \right] dx. \tag{4.9.4b}$$

First, we assume the Ritz approximations of w and ϕ_x are in the form

$$w(x) \approx W_M = \sum_{j=1}^{M} b_j \psi_j(x) + \psi_0(x), \quad \phi_x(x) \approx \Phi_N = \sum_{j=1}^{N} c_j \theta_j(x) + \theta_0(x), \tag{4.9.5}$$

where the approximation functions ψ_j and θ_j are independent of each other. However, The shear strain expression $2\varepsilon_{xz} = \phi_x + dw/dx$, for consistency, implies that w should be one order higher than ϕ_x (i.e., when W_M is quadratic, Φ_N should be linear, and so on). Next, we formulate the Ritz equations, postponing the construction of the approximation functions for the specific problem at hand to later. Substituting the Ritz approximations in Eq. (4.9.5) for w and ϕ_x and $\delta w = \psi_i$ and $\delta\phi_x = \theta_i$ into the variational statements in Eqs. (4.9.4a) and (4.9.4b), we obtain (assuming that the specified essential boundary conditions are homogeneous so that $\phi_0 = 0$ and $\psi_0 = 0$)

$$0 = \int_0^L \left[GAK_s \frac{d\psi_i}{dx} \left(\sum_{j=1}^{N} c_j \theta_j + \sum_{j=1}^{M} \frac{d\psi_j}{dx} b_j \right) - \psi_i q \right] dx$$

$$= \sum_{j=1}^{M} A_{ij} b_j + \sum_{j=1}^{N} B_{ij} c_j - F_i^1, \tag{4.9.6}$$

$$0 = \int_0^L \left[EI \frac{d\theta_i}{dx} \left(\sum_{j=1}^{N} \frac{d\theta_j}{dx} c_j \right) + GAK_s \theta_i \left(\sum_{j=1}^{N} \theta_j c_j + \sum_{j=1}^{M} \frac{d\psi_j}{dx} b_j \right) \right] dx$$

$$= \sum_{j=1}^{M} C_{ij} b_j + \sum_{j=1}^{N} D_{ij} c_j - F_i^2, \tag{4.9.7}$$

or in matrix form

$$\begin{bmatrix} \mathbf{A} & \mathbf{B} \\ \mathbf{C} & \mathbf{D} \end{bmatrix} \begin{Bmatrix} \mathbf{b} \\ \mathbf{c} \end{Bmatrix} = \begin{Bmatrix} \mathbf{F}^1 \\ \mathbf{F}^2 \end{Bmatrix}. \tag{4.9.8}$$

The coefficients A_{ij}, B_{ij}, and so on are given by

$$A_{ij} = \int_0^L GAK_s \frac{d\psi_i}{dx} \frac{d\psi_j}{dx}\, dx, \quad B_{ij} = \int_0^L GAK_s \frac{d\psi_i}{dx} \theta_j\, dx,$$

$$C_{ij} = \int_0^L GAK_s \theta_i \frac{d\psi_j}{dx}\, dx, \quad D_{ij} = \int_0^L \left(EI \frac{d\theta_i}{dx} \frac{d\theta_j}{dx} + GAK_s \theta_i \theta_j \right) dx, \tag{4.9.9}$$

$$F_i^1 = \int_0^L \psi_i q\, dx, \quad F_i^2 = 0.$$

(a) *Simply supported beam.* The boundary conditions for this case are:

$$w = 0, \quad M_{xx} = EI \frac{d\phi_x}{dx} = 0 \text{ at } x = 0, L.$$

Of the four boundary conditions, only the first two $(w(0) = w(L) = 0)$ are the specified geometric boundary conditions and they are homogeneous. Hence, we have $\psi_0 = \theta_0 = 0$. The ψ_i must satisfy the homogeneous boundary conditions: $\psi_i(0) = 0$ and $\psi_i(L) = 0$. Thus, ψ_1 is of the form $\psi_1 = x(L - x)$. On the other hand, ϕ_x is not specified in the problem. Hence, θ_i can be any function that is at lest once differentiable (and the set $\{\theta_i\}$ should be such that $\Phi_N(x)$ does not vanish at $x = 0$ or $x = L$). Then, the approximations of w and ϕ_x can be of the form:

$$W_M(x) = b_1\psi_1(x) + b_2\psi_2(x) + \cdots + b_M\psi_M(x), \quad \psi_i = x^i(L - x) \quad (M \geq 1);$$
(4.9.10a)

$$\Phi_N(x) = c_1\theta_1(x) + c_2\theta_2(x) + \cdots + c_N\theta_N(x), \quad \theta_i = x^{i-1} \quad (N \geq 2).$$
(4.9.10b)

Substituting the Ritz approximation functions into Eq. (4.9.8), we obtain

$$A_{ij} = GAK_sL^{i+j+1}\left[\frac{ij}{i+j-1} - \frac{i+j+2ij}{i+j} + \frac{(i+1)(j+1)}{i+j+1}\right] \quad (i,j = 1, 2, \ldots, M),$$

$$B_{ij} = GAK_sL^{i+j}\left[\frac{i}{i+j-1} - \frac{i+1}{i+j}\right] = C_{ji},$$

$$F_i^1 = q_0L^{i+2}\frac{1}{(i+1)(i+2)}, \quad (i = 1, 2, \ldots, M; j = 1, 2, \ldots, N),$$
(4.9.11)

$$D_{ij} = L^{i+j-3}\left[EI\frac{(i-1)(j-1)}{i+j-3} + GAK_sL^2\frac{1}{i+j-1}\right] \quad (i,j = 1, 2, \ldots, N), \quad F_i^2 = 0.$$

Note that when the denominator in the expressions of D_{ij} is less than equal to zero, then that expression is not to be included in the evaluation of D_{ij}. We can specialize the results for $(M = 1, N = 2)$, $(M = 2, N = 3)$, and so on.

For example, for $M = 1$ and $N = 2$ and $q = q_0$, the coefficients A_{ij}, B_{ij}, C_{ij}, D_{ij}, and F_i^1 are:

$$A_{11} = GAK_s\frac{L^3}{3}, \quad B_{11} = C_{11} = 0, \quad B_{12} = C_{21} = -GAK_s\frac{L^3}{6}, \quad D_{11} = GAK_s\,L,$$

$$D_{12} = D_{21} = GAK_s\frac{L^2}{2}, \quad D_{22} = EI\,L + GAK_s\frac{L^3}{3}, \quad F_1^1 = \frac{q_0L^3}{6}.$$
(4.9.12a)

The Ritz equations can be expressed in matrix form:

$$\begin{bmatrix} A_{11} & 0 & B_{12} \\ 0 & D_{11} & D_{12} \\ C_{21} & D_{21} & D_{22} \end{bmatrix} \begin{Bmatrix} b_1 \\ c_1 \\ c_2 \end{Bmatrix} = \frac{q_0L^3}{6} \begin{Bmatrix} 1 \\ 0 \\ 0 \end{Bmatrix}.$$
(4.9.12b)

The solution of these equations is

$$b_1 = \frac{q_0}{2GAK_s} + \frac{q_0L^2}{24EI}, \quad c_1 = -\frac{q_0L^3}{24EI}, \quad c_2 = \frac{q_0L^2}{12EI}.$$
(4.9.12c)

Then the $(M, N) = (1, 2)$-parameter Ritz solution is

$$W_1(x) = \left(\frac{q_0L^2}{2GAK_s} + \frac{q_0L^4}{24EI}\right)\frac{x}{L}\left(1 - \frac{x}{L}\right), \quad \Phi_2(x) = -\frac{q_0L^3}{24EI}\left(1 - 2\frac{x}{L}\right). \quad (4.9.13)$$

The exact solution is [see Eqs. (4.6.19b) and (4.6.19c), with $D^*_{xx} = A_{xx}D_{xx}$ and $D_{xx} = EI$]

$$w(x) = \frac{q_0 L^4}{24EI}\left(\frac{x}{L} - 2\frac{x^3}{L^3} + \frac{x^4}{L^4}\right) + \frac{1}{GAK_s}\frac{q_0 L^2}{2}\left(\frac{x}{L} - \frac{x^2}{L^2}\right), \qquad (4.9.14a)$$

$$\phi_x(x) = -\frac{q_0 L^3}{24EI}\left(1 - 6\frac{x^2}{L^2} + 4\frac{x^3}{L^3}\right), \qquad (4.9.14b)$$

The approximate and exact maximum values of the deflection are:

$$W_{\text{Ritz}} = \frac{4q_0 L^4}{384EI} + \frac{q_0 L^2}{8GAK_s} \quad w_{\text{exact}} = \frac{5q_0 L^4}{384EI} + \frac{q_0 L^2}{8GAK_s}. \qquad (4.9.15)$$

The Ritz solution is exact as far as the shear part and 20% in error for the bending part. For $M = 2$ and $N = 3$, the solution obtained is the same as that in the previous case with $b_2 = c_3 = 0$. However, for $M = 3$ and $N = 4$, we obtain the exact solution.

Figure 4.9.1 shows a comparison of the Ritz solutions for $w(x)$ with the exact solution for this case. The $(M, N) = (1, 2)$ Ritz solution is in considerable error (20%), while the $(M, N) = (3, 4)$ Ritz solution matches with the exact solution. Figure 4.9.2 contains a similar comparison of the rotation $\phi_x(x)$.

(b) Clamped beam. The boundary conditions for the clamped case are: $w = 0$, $\phi_x = 0$ at $x = 0, L$. In this case all four specified boundary conditions are the essential type and they are homogeneous. Hence, we have $\psi_0 = 0$ and $\theta_0 = 0$. Since the geometric boundary conditions on w and ϕ_x are the same, we can take $\psi_i = \theta_i$. From **Example 3.9.3**, we have $\psi_i(x) = x^i(L - x)$. Thus, we can use the Ritz approximations

$$W_M(x) = \sum_{j=1}^{M} b_j \psi_j(x), \quad \Phi_N(x) = \sum_{j=1}^{N} c_j \psi_j(x), \qquad (4.9.16)$$

with different number of terms for w and ϕ_x. Since the exact solution for w is a fourth-order polynomial and the exact solution for ϕ_x is a cubic polynomial (when $q = q_0 = $ constant), we expect that $(M, N) = (3, 2)$ Ritz solution would be exact.

Substituting the Ritz approximations from Eq. (4.9.16) into Eqs. (4.9.4a) and (4.9.4b), we obtain,

$$A_{ij} = GAK_s L^{i+j+1}\left[\frac{ij}{i+j-1} - \frac{i+j+2ij}{i+j} + \frac{(i+1)(j+1)}{i+j+1}\right] \ (i, j = 1, 2, \ldots, M),$$

$$B_{ij} = GAK_s L^{i+j+2}\left[\frac{i}{i+j} - \frac{2i+1}{i+j+1} + \frac{i+1}{1+j+2}\right] = C_{ji},$$

$$F_i^1 = q_0 L^{i+2}\frac{1}{(i+1)(i+2)}, \ (i = 1, 2, \ldots, M; j = 1, 2, \ldots, N), \qquad (4.9.17)$$

$$D_{ij} = EIL^{i+j+3}\left[\frac{ij}{i+j-1} - \frac{i+j+2ij}{i+j} + \frac{(i+1)(j+1)}{i+j+1}\right]$$

$$+ GAK_s\left[\frac{i}{i+j} - \frac{2i+1}{i+j+1} + \frac{i+1}{i+j+2}\right] (i, j = 1, 2, \ldots, N), \ F_i^2 = 0.$$

The cases $(M, N) = (1, 1), (2, 1), (2, 2)$ give results that are grossly in error. As expected the case $(M, N) = (3, 2)$ gives the exact solution.

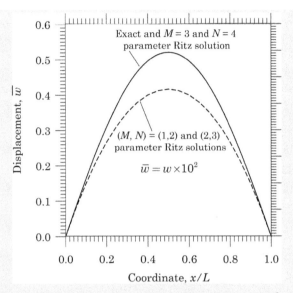

Figure 4.9.2: A comparison of the Ritz solutions $\bar{w} = w \times 10^2$ with the exact solution for the simply supported beam under uniform load.

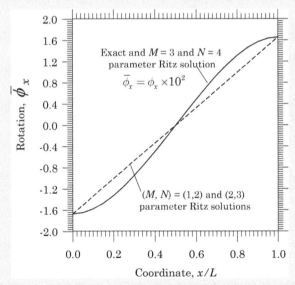

Figure 4.9.3: A comparison of the Ritz solutions $\bar{\phi}_x = \phi_x \times 10^2$ with the exact solution for the simply supported beam under uniform load.

4.10   CHAPTER SUMMARY

The shear deformation beam theory, often known as the TBT, is developed accounting for material variation through the thickness (i.e., FGM), the von Kármán nonlinearity and the effect of the modified couple stress. Algebraic relations between the TBT and the CBT are also presented. Exact solutions (by direct integration), the Navier solution, and variational solution of the theory for the linear case are presented. The nonlinear solutions by the finite element method are presented in Chapter 9.

SUGGESTED EXERCISES

4.1 Determine the exact solution for the case in which $A_{xy} \neq 0$ (i.e., account for the modified couple stress effect). That is, solve the following equations analytically:

$$-\frac{dN_{xx}}{dx} - f = 0,$$

$$-\frac{dN_{xz}}{dx} - \frac{1}{2}\frac{d^2 P_{xy}}{dx^2} - q = 0,$$

$$-\frac{dM_{xx}}{dx} - \frac{1}{2}\frac{dP_{xy}}{dx} + N_{xz} = 0,$$

where

$$N_{xx} = A_{xx}\frac{du}{dx} + B_{xx}\frac{d\phi_x}{dx} - N_{xx}^T, \quad M_{xx} = B_{xx}\frac{du}{dx} + D_{xx}\frac{d\phi_x}{dx} - M_{xx}^T,$$

$$P_{xy} = -\frac{1}{2}A_{xy}\left(\frac{d\phi_x}{dx} - \frac{d^2 w}{dx^2}\right), \quad N_{xz} = S_{xz}\left(\phi_x + \frac{dw}{dx}\right).$$

4.2 Assuming that E and ρ are constant and that the couple stress effect is not accounted for, eliminate ϕ_x from Eqs. (4.4.16) and (4.4.17) write a single equation in terms of w only [see Eq. (4.4.18)]:

$$D_{xx}\frac{\partial^4 w}{\partial x^4} + m_0\frac{\partial^2 w}{\partial t^2} - \left(m_2 + m_0\frac{D_{xx}}{S_{xz}}\right)\frac{\partial^4 w}{\partial t^2 \partial x^2} + \frac{m_0 m_2}{S_{xz}}\frac{\partial^4 w}{\partial t^4}$$

$$+ c_f w - q + \frac{D_{xx}}{S_{xz}}\frac{\partial^2 q}{\partial x^2} - \frac{m_2}{S_{xz}}\frac{\partial^2 q}{\partial t^2} + \frac{\partial^2 M_{xx}^T}{\partial x^2} = 0.$$

4.3 Determine the generalized displacements and generalized forces of a simply-supported beam subjected to a downward point load F_0 at the center using the relationships developed in **Example 4.7.1**. The bending moment in the CBT for this load is $M^C(x) = -F_0 x/2$ for $0 \leq x \leq L/2$ and $M^C(x) = F_0(x - L)/2$ for $L/2 \leq x \leq L$.

4.4 Determine the generalized displacements and generalized forces of a clamped beam subjected to a downward point load F_0 at the center using the relationships developed in **Example 4.7.2**. The bending moment in the CBT for this load is $M^C(x) = -F_0 x/2 + F_0 L/8$ for $0 \leq x \leq L/2$ and $M^C(x) = F_0(x - L)/2 + F_0 L/8$ for $L/2 \leq x \leq L$.

4.5 Determine the generalized displacements and generalized forces of a propped cantilever beam subjected to uniformly distributed load using the relationships in Eqs. (4.7.16)–(4.7.21) (i.e., determine the constants c_1 through c_6 using the boundary conditions). The solutions for the same problem by the CBT are available from Eqs. (3.7.43)–(3.7.47).

4.6 Solve **Problem 3.7** (see Fig. P4.6; diameter of the post is 40 mm) using the TBT; assume shear modulus of $G = 75$ GPa and shear correction coefficient of $K_s = 5/6$.

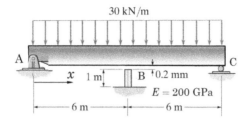

Figure P4.6

4.7 Show that the constants of integration c_1 through c_6 for the boundary conditions, C–F, F–C, C–H, H–C, and C–C, are as follows [see Eqs. (4.7.16)–(4.7.21)]:

(a) *Clamped-Free (C-F) Beams:* $c_1 = c_2 = c_3 = 0$, and $c_4 = M_{xx}^C(0)\Omega L^2$

(b) *Free–Clamped (F–C) Beams:*

$$c_1 = c_2 = c_3 = 0, \quad \text{and} \quad c_4 = M_{xx}^C(L)\Omega L^2$$

(c) *Clamped–Hinged (C–H) Beams:*

$$c_1 = \frac{3\Omega}{(1+3\Omega)L}M_{xx}^C(0), \quad c_2 = -c_1 L, \quad c_3 = 0, \quad c_4 = M_{xx}^C(0)\Omega L^2$$

(d) *Hinged–Clamped (H–C) Beams:*

$$c_1 = -\frac{3\Omega}{(1+3\Omega)L}M_{xx}^C(L), \quad c_2 = 0, \quad c_3 = -c_1 L, \quad c_4 = 0$$

(e) *Clamped–Clamped (C–C) Beams:*

$$c_1 = -\frac{12\Omega}{(1+12\Omega)L}\left[M_{xx}^C(L) - M_{xx}^C(0)\right]$$

$$c_2 = \frac{6\Omega}{(1+12\Omega)}\left[M_{xx}^C(L) - M_{xx}^C(0)\right]$$

$$c_3 = 0, \quad c_4 = M_{xx}^C(0)\Omega L^2$$

CHAPTER 4 APPENDIX

The following result can be found in any book containing tables of integrals:

$$\int x^m e^{\alpha x}\, dx = e^{\alpha x} \sum_{r=0}^{m}(-1)^r \frac{m!\, x^{m-r}}{(m-r)!\, \alpha^{r+1}},$$

For various values of m, we have the following results:

$$m = 0: \quad e^{-\alpha x}\int e^{\alpha x}\, dx = \frac{1}{\alpha},$$

$$m = 1: \quad e^{-\alpha x}\int x\, e^{\alpha x}\, dx = \frac{x}{\alpha} - \frac{1}{\alpha^2},$$

$$m = 2: \quad e^{-\alpha x} \int x^2 e^{\alpha x}\, dx = \frac{x^2}{\alpha} - \frac{2x}{\alpha^2} + \frac{2}{\alpha^3},$$

$$m = 3: \quad e^{-\alpha x} \int x^3 e^{\alpha x}\, dx = \frac{x^3}{\alpha} - \frac{3x^2}{\alpha^2} + \frac{6x}{\alpha^3} - \frac{6}{\alpha^4},$$

$$m = 4: \quad e^{-\alpha x} \int x^4 e^{\alpha x}\, dx = \frac{x^4}{\alpha} - \frac{4x^3}{\alpha^2} + \frac{12x^2}{\alpha^3} - \frac{24x}{\alpha^4} + \frac{24}{\alpha^5},$$

$$m = 5: \quad e^{-\alpha x} \int x^5 e^{\alpha x}\, dx = \frac{x^5}{\alpha} - \frac{5x^4}{\alpha^2} + \frac{20x^3}{\alpha^3} - \frac{60x^2}{\alpha^4} + \frac{120x}{\alpha^5} - \frac{120}{\alpha^6}.$$

By changing α to $-\alpha$, we can evaluate

$$e^{\alpha x} \int x^m e^{-\alpha x}\, dx$$

for $m = 0, 1, \cdots 5$. Then

$$\frac{1}{2\alpha}\left(e^{\alpha x}\int e^{-\alpha x}\,dx - e^{-\alpha x}\int e^{\alpha x}\,dx\right) = -\frac{1}{\alpha^2},$$

$$\frac{1}{2\alpha}\left(e^{\alpha x}\int x\,e^{-\alpha x}\,dx - e^{-\alpha x}\int x\,e^{\alpha x}\,dx\right) = -\frac{x}{\alpha^2},$$

$$\frac{1}{2\alpha}\left(e^{\alpha x}\int x^2\,e^{-\alpha x}\,dx - e^{-\alpha x}\int x^2\,e^{\alpha x}\,dx\right) = -\frac{x^2}{\alpha^2} - \frac{4}{\alpha^4},$$

$$\frac{1}{2\alpha}\left(e^{\alpha x}\int x^3\,e^{-\alpha x}\,dx - e^{-\alpha x}\int x^3\,e^{\alpha x}\,dx\right) = -\frac{x^3}{\alpha^2} - \frac{6x}{\alpha^4},$$

$$\frac{1}{2\alpha}\left(e^{\alpha x}\int x^4\,e^{-\alpha x}\,dx - e^{-\alpha x}\int x^4\,e^{\alpha x}\,dx\right) = -\frac{x^4}{\alpha^2} - \frac{12x^2}{\alpha^4} - \frac{24}{\alpha^6},$$

$$\frac{1}{2\alpha}\left(e^{\alpha x}\int x^5\,e^{-\alpha x}\,dx - e^{-\alpha x}\int x^5\,e^{\alpha x}\,dx\right) = -\frac{x^5}{\alpha^2} - \frac{20x^3}{\alpha^4} - \frac{120x}{\alpha^6}.$$

Anybody who has been seriously engaged in scientific work of any kind realizes that over the entrance to the gates of the temple of science are written the words: 'Ye must have faith.

Max Planck

5 Third-Order Beam Theories

Education is not the learning of many facts, but the training of the mind to think.
Albert Einstein

5.1 INTRODUCTION

5.1.1 WHY A THIRD-ORDER THEORY?

A "third-order" beam theory is one in which the displacement field is expanded up to and including the third-power of z, with each power of z having an independent dependent variable:

$$u_x(x, z) = z^0\, \varphi_x^{(0)} + z^1\, \varphi_x^{(1)} + z^2\, \varphi_x^{(2)} + z^3\, \varphi_x^{(3)}, \tag{5.1.1a}$$

$$u_y(x, z) = z^0\, \varphi_y^{(0)} + z^1\, \varphi_y^{(1)} + z^2\, \varphi_y^{(2)} + z^3\, \varphi_y^{(3)}, \tag{5.1.1b}$$

$$u_z(x, z) = z^0\, \varphi_z^{(0)} + z^1\, \varphi_z^{(1)} + z^2\, \varphi_z^{(2)} + z^3\, \varphi_z^{(3)}, \tag{5.1.1c}$$

where (u_x, u_y, u_z) are the total displacements components of the displacement vector $\mathbf{u}$ and $(\varphi_x^{(i)}, \varphi_y^{(i)}, \varphi_z^{(i)})$ for $i = 1, 2, 3$ are linearly independent functions of x only. The displacement fields of the CBT and the TBT are special cases of Eqs. (5.1.1a)–(5.1.1c):

1. **CBT:**

$$\varphi_x^{(0)} = u(x), \quad \varphi_x^{(1)} = -\frac{\partial w}{\partial x}, \quad \varphi_x^{(2)} = 0, \quad \varphi_x^{(3)} = 0;$$

$$\varphi_y^{(0)} = 0, \quad \varphi_y^{(1)} = 0, \quad \varphi_y^{(2)} = 0, \quad \varphi_y^{(3)} = 0;$$

$$\varphi_z^{(0)} = w(x), \quad \varphi_z^{(1)} = 0, \quad \varphi_z^{(2)} = 0, \quad \varphi_z^{(3)} = 0.$$

2. **TBT:**

$$\varphi_x^{(0)} = u(x), \quad \varphi_x^{(1)} = \phi_x(x), \quad \varphi_x^{(2)} = 0, \quad \varphi_x^{(3)} = 0;$$

$$\varphi_y^{(0)} = 0, \quad \varphi_y^{(1)} = 0, \quad \varphi_y^{(2)} = 0, \quad \varphi_y^{(3)} = 0;$$

$$\varphi_z^{(0)} = w(x), \quad \varphi_z^{(1)} = 0, \quad \varphi_z^{(2)} = 0, \quad \varphi_z^{(3)} = 0.$$

From the discussions presented in Chapters 3 and 4, it is clear that the transverse shear stress distribution through the beam height, computed using

DOI: 10.1201/9781003240846-5

the stress–strain relation $\sigma_{xz} = 2G\varepsilon_{xz}$, is either zero (in the CBT) or constant (in the TBT), although the shear force N_{xz} cannot be zero in any beam. The actual variation of $\sigma_{xz}(x, z)$ with z can be qualitatively determined using the 3-D equations of equilibrium of linearized elasticity [see Eq. (1.4.11) with $f_1 = f_x$, $f_2 = f_y$, and $f_3 = f_y$ and set the inertia term to zero] in Cartesian rectangular coordinate system:

$$\frac{\partial \sigma_{xx}}{\partial x} + \frac{\partial \sigma_{yx}}{\partial y} + \frac{\partial \sigma_{zx}}{\partial z} + f_x = 0, \tag{5.1.2a}$$

$$\frac{\partial \sigma_{xy}}{\partial x} + \frac{\partial \sigma_{yy}}{\partial y} + \frac{\partial \sigma_{zy}}{\partial z} + f_y = 0, \tag{5.1.2b}$$

$$\frac{\partial \sigma_{xz}}{\partial x} + \frac{\partial \sigma_{yz}}{\partial y} + \frac{\partial \sigma_{zz}}{\partial z} + f_z = 0, \tag{5.1.2c}$$

where σ_{ij} are the stress components[1] and (f_x, f_y, f_z) are the components of the body force vector (measured per unit volume). Based on the kinematic assumptions of the CBT and TBT, we know that σ_{xx} is a linear function of z (when E is not a function of z). This implies that σ_{xz} is a quadratic function of z because:

$$\frac{\partial \sigma_{xz}}{\partial z} \sim -\frac{\partial \sigma_{xx}}{\partial x} \quad \rightarrow \quad \sigma_{xz} \sim -\int \frac{\partial \sigma_{xx}}{\partial x} \, dz + \text{ constant}.$$

In order for σ_{xz} to be a quadratic function, it is necessary that $\gamma_{xz} = 2\varepsilon_{xz}$,

$$\gamma_{xz} = \frac{\partial u_x}{\partial z} + \frac{\partial u_z}{\partial x}$$

be a quadratic function of z. This, in turn, implies that $u_z = w$ be a *quadratic* function of z and u_x to be a *cubic* function of z. *Thus, a displacement expansion in which u_x is cubic and u_z is quadratic in z is needed to represent a quadratic variation of the transverse shear stress through the beam thickness.*

5.1.2   PRESENT STUDY

The displacement field in Eqs. (5.1.1a)–(5.1.1c) can be used to generate several third-order theories. First, we present a general third-order beam theory which accounts for the von Kármán geometric nonlinearity, thickness stretch, through thickness variation of the material, and modified couple stress effect. Following this introduction, the kinematics of the third-order theory, governing equations of motion will be presented. Many of the steps are the same as those used in the development of the CBT and TBT, except that the algebraic complexity increases, and we are required to introduce "higher-order" stress

---

[1] No distinction between the second Piola–Kirchhoff stress components and Cauchy stress components is made in the linearized elasticity.

resultants that are not easy to interpret physically as those already intro-duced in CBT and FST. The results will be specialized to couple of different third-order theories, including the Levinson beam theory (LBT) for rectan-gular cross-section beams [124, 125], the Bickford beam theory [126], and the Reddy beam theory (RBT) [6, 10]. These theories account for the vanishing of transverse shear stress on the bottom and top surfaces of the beam. The analytical solutions of the linearized theories for the equilibrium case will also be presented.

5.2 A GENERAL THIRD-ORDER THEORY

5.2.1 KINEMATICS

We begin with the displacement field of the following form for a straight FGM beam bent by forces in the xz plane (i.e., bending about the y-axis):

$$
\begin{aligned}
u_x(x, z, t) &= u(x, t) + z\phi_x(x, t) + z^2\theta_x(x, t) + z^3\psi_x(x, t), \\
u_y(x, z, t) &= 0, \\
u_z(x, z, t) &= w(x, t) + z\phi_z(x, t) + z^2\theta_z(x, t),
\end{aligned}
\tag{5.2.1}
$$

where (u, w) are midplane displacements along the x and z directions, respec-tively, and $(\phi_x, \theta_x, \psi_x, \phi_z, \theta_z)$ have the meaning

$$
\phi_x = \left.\frac{\partial u_x}{\partial z}\right|_{z=0}, \quad 2\theta_x = \left.\frac{\partial^2 u_x}{\partial z^2}\right|_{z=0}, \quad 6\psi_x = \left.\frac{\partial^3 u_x}{\partial z^3}\right|_{z=0},
$$

$$
\phi_z = \left.\frac{\partial u_z}{\partial z}\right|_{z=0}, \quad 2\theta_z = \left.\frac{\partial^2 u_z}{\partial z^2}\right|_{z=0}.
\tag{5.2.2}
$$

The nonzero von Kármán type nonlinear strains associated with the dis-placement field in Eq. (5.2.1) are obtained by retaining only small strain and moderate rotation terms (i.e., squares and products of ϕ_x, θ_x, ψ_x, ϕ_z, and θ_z and their derivatives are omitted):

$$
\begin{aligned}
\varepsilon_{xx} &= \left[\frac{\partial u}{\partial x} + \frac{1}{2}\left(\frac{\partial w}{\partial x}\right)^2\right] + z\frac{\partial \phi_x}{\partial x} + z^2\frac{\partial \theta_x}{\partial x} + z^3\frac{\partial \psi_x}{\partial x} \\
&= \varepsilon_{xx}^{(0)} + z\varepsilon_{xx}^{(1)} + z^2\varepsilon_{xx}^{(2)} + z^3\varepsilon_{xx}^{(3)},
\end{aligned}
\tag{5.2.3a}
$$

$$
\begin{aligned}
\gamma_{xz} &= \phi_x + \frac{\partial w}{\partial x} + z\left(2\theta_x + \frac{\partial \phi_z}{\partial x}\right) + z^2\left(3\psi_x + \frac{\partial \theta_z}{\partial x}\right) \\
&= \gamma_{xz}^{(0)} + z\gamma_{xz}^{(1)} + z^2\gamma_{xz}^{(2)},
\end{aligned}
\tag{5.2.3b}
$$

$$
\varepsilon_{zz} = \phi_z + 2z\theta_z = \varepsilon_{zz}^{(0)} + z\varepsilon_{zz}^{(1)},
\tag{5.2.3c}
$$

where the higher-order (nonlinear) terms in the thickness strain ε_{zz} are neglected, and

$$\varepsilon_{xx}^{(0)} = \frac{\partial u}{\partial x} + \frac{1}{2}\left(\frac{\partial w}{\partial x}\right)^2, \quad \varepsilon_{xx}^{(1)} = \frac{\partial \phi_x}{\partial x}, \quad \varepsilon_{xx}^{(2)} = \frac{\partial \theta_x}{\partial x}, \quad \varepsilon_{xx}^{(3)} = \frac{\partial \psi_x}{\partial x}, \quad (5.2.4a)$$

$$\varepsilon_{zz}^{(0)} = \phi_z, \quad \varepsilon_{zz}^{(1)} = 2\theta_z, \quad (5.2.4b)$$

$$\gamma_{xz}^{(0)} = \phi_x + \frac{\partial w}{\partial x}, \quad \gamma_{xz}^{(1)} = \left(2\theta_x + \frac{\partial \phi_z}{\partial x}\right), \quad \gamma_{xz}^{(2)} = \left(3\psi_x + \frac{\partial \theta_z}{\partial x}\right). \quad (5.2.4c)$$

The only nonzero components of the rotation vector $\boldsymbol{\omega}$ and the curvature tensor $\boldsymbol{\chi}$ associated with the displacement field in Eq. (5.2.1) are presented in Eqs. (5.2.5a)–(5.2.5d). These are ω_y, χ_{xy}, and χ_{yz}, which can be expressed in terms of the generalized displacements $(u, w, \phi_x, \theta_x, \psi_x, \phi_z, \theta_z)$ as

$$\omega_y = \frac{1}{2}\left(\frac{\partial u_x}{\partial z} - \frac{\partial u_z}{\partial x}\right) = \frac{1}{2}\left(\phi_x + 2z\theta_x + 3z^2\psi_x - \frac{\partial w}{\partial x} - z\frac{\partial \phi_z}{\partial x} - z^2\frac{\partial \theta_z}{\partial x}\right)$$

$$= \frac{1}{2}\left[\left(\phi_x - \frac{\partial w}{\partial x}\right) + z\left(2\theta_x - \frac{\partial \phi_z}{\partial x}\right) + z^2\left(3\psi_x - \frac{\partial \theta_z}{\partial x}\right)\right], \quad (5.2.5a)$$

$$\chi_{xy} = \frac{1}{2}\frac{\partial \omega_y}{\partial x} = \frac{1}{4}\left[\left(\frac{\partial \phi_x}{\partial x} - \frac{\partial^2 w}{\partial x^2}\right) + z\left(2\frac{\partial \theta_x}{\partial x} - \frac{\partial^2 \phi_z}{\partial x^2}\right)\right.$$

$$\left. + z^2\left(3\frac{\partial \psi_x}{\partial x} - \frac{\partial^2 \theta_z}{\partial x^2}\right)\right]$$

$$\equiv \chi_{xy}^{(0)} + z\chi_{xy}^{(1)} + z^2\chi_{xy}^{(2)}, \quad (5.2.5b)$$

$$\chi_{yz} = \frac{1}{2}\frac{\partial \omega_y}{\partial z} = \frac{1}{4}\left[\left(2\theta_x - \frac{\partial \phi_z}{\partial x}\right) + 2z\left(3\psi_x - \frac{\partial \theta_z}{\partial x}\right)\right]$$

$$\equiv \chi_{yz}^{(0)} + z\chi_{yz}^{(1)}, \quad (5.2.5c)$$

where

$$\chi_{xy}^{(0)} = \frac{1}{4}\left(\frac{\partial \phi_x}{\partial x} - \frac{\partial^2 w}{\partial x^2}\right), \quad \chi_{xy}^{(1)} = \frac{1}{4}\left(2\frac{\partial \theta_x}{\partial x} - \frac{\partial^2 \phi_z}{\partial x^2}\right),$$

$$\chi_{xy}^{(2)} = \frac{1}{4}\left(3\frac{\partial \psi_x}{\partial x} - \frac{\partial^2 \theta_z}{\partial x^2}\right), \quad (5.2.5d)$$

$$\chi_{yz}^{(0)} = \frac{1}{4}\left(2\theta_x - \frac{\partial \phi_z}{\partial x}\right), \quad \chi_{yz}^{(1)} = \frac{1}{4}\left(6\psi_x - 2\frac{\partial \theta_z}{\partial x}\right).$$

It is clear that the third-order theory brings many more terms than the TBT when couple stress effect is included. Next, we derive the governing equations of motion using Hamilton's principle.

5.2.2   EQUATIONS OF MOTION

Hamilton's principle is used to derive the equations of motion

$$\int_{t_2}^{t_1} (-\delta K + \delta U + \delta V_E)\, dt = 0, \tag{5.2.6}$$

where δK is the virtual kinetic energy, δU is the virtual strain energy, and δV_E is the virtual work done by external forces. The virtual kinetic energy expression is

$$
\begin{aligned}
\delta K =\ & \int_0^L \int_A \rho\, \dot u_i\, \delta \dot u_i\, dA dx \\
=\ & \int_0^L \Big[\Big(m_0 \dot u + m_1 \dot\phi_x + m_2 \dot\theta_x + m_3 \dot\psi_x\Big)\delta \dot u + \Big(m_1 \dot u + m_2 \dot\phi_x + m_3 \dot\theta_x \\
& + m_4 \dot\psi_x\Big)\delta \dot\phi_x + \Big(m_2 \dot u + m_3 \dot\phi_x + m_4 \dot\theta_x + m_5 \dot\psi_x\Big)\delta \dot\theta_x \\
& + \Big(m_3 \dot u + m_4 \dot\phi_x + m_5 \dot\theta_x + m_6 \dot\psi_x\Big)\delta \dot\psi_x \\
& + \Big(m_0 \dot w + m_1 \dot\phi_z + m_2 \dot\theta_z\Big)\delta w + \Big(m_1 \dot w + m_2 \dot\phi_z \\
& + m_3 \dot\theta_z\Big)\delta \dot\phi_z + \Big(m_2 \dot w + m_3 \dot\phi_z + m_4 \dot\theta_z\Big)\delta \dot\theta_z\Big] dx, \tag{5.2.7}
\end{aligned}
$$

and

$$(m_0, m_1, m_2, m_3, m_4, m_5, m_6) = \int_A \rho(1, z, z^2, z^3, z^4, z^5, z^6)\, dA. \tag{5.2.8}$$

The expression for the virtual strain energy δU, accounting for the modified couple stress, is

$$
\begin{aligned}
\delta U =\ & \int_0^L \int_A \big(\sigma_{xx}\, \delta\varepsilon_{xx} + \sigma_{zz}\, \delta\varepsilon_{zz} + \sigma_{xz}\delta\gamma_{xz} + 2\mathcal{M}_{xy}\, \delta\chi_{xy} + 2\mathcal{M}_{yz}\, \delta\chi_{yz}\big)\, dA dx \\
=\ & \int_0^L \Big[M_{xx}^{(0)}\Big(\frac{\partial \delta u}{\partial x} + \frac{\partial w}{\partial x}\frac{\partial \delta w}{\partial x}\Big) + M_{xx}^{(1)}\frac{\partial \delta\phi_x}{\partial x} + M_{xx}^{(2)}\frac{\partial \delta\theta_x}{\partial x} + M_{xx}^{(3)}\frac{\partial \delta\psi_x}{\partial x} \\
& + M_{zz}^{(0)}\delta\phi_z + 2M_{zz}^{(1)}\delta\theta_z + M_{xz}^{(0)}\Big(\delta\phi_x + \frac{\partial \delta w}{\partial x}\Big) + M_{xz}^{(1)}\Big(2\delta\theta_x + \frac{\partial \delta\phi_z}{\partial x}\Big) \\
& + M_{xz}^{(2)}\Big(3\delta\psi_x + \frac{\partial \delta\theta_z}{\partial x}\Big) + \tfrac{1}{2}P_{xy}^{(0)}\Big(\frac{\partial \delta\phi_x}{\partial x} - \frac{\partial^2 \delta w}{\partial x^2}\Big) \\
& + \tfrac{1}{2}P_{xy}^{(1)}\Big(2\frac{\partial \delta\theta_x}{\partial x} - \frac{\partial^2 \delta\phi_z}{\partial x^2}\Big) + \tfrac{1}{2}P_{xy}^{(2)}\Big(3\frac{\partial \delta\psi_x}{\partial x} - \frac{\partial^2 \delta\theta_z}{\partial x^2}\Big) \\
& + \tfrac{1}{2}Q_{yz}^{(0)}\Big(2\delta\theta_x - \frac{\partial \delta\phi_z}{\partial x}\Big) + \tfrac{1}{2}Q_{yz}^{(1)}\Big(6\delta\psi_x - 2\frac{\partial \delta\theta_z}{\partial x}\Big)\Big] dx. \tag{5.2.9}
\end{aligned}
$$

Various stress resultants used in Eq. (5.2.9) are defined as

$$M_{xx}^{(i)} = \int_A (z)^i \sigma_{xx}\, dA, \quad M_{zz}^{(i)} = \int_A (z)^i \sigma_{zz}\, dA,$$

$$M_{xz}^{(i)} = \int_A (z)^i \sigma_{xz}\, dA, \quad P_{xy}^{(0)} = \int_A \mathcal{M}_{xy}\, dA,$$

$$P_{xy}^{(1)} = \int_A z \mathcal{M}_{xy}\, dA, \quad P_{xy}^{(2)} = \int_A z^2 \mathcal{M}_{xy}\, dA, \tag{5.2.10}$$

$$Q_{yz}^{(0)} = \int_A \mathcal{M}_{yz}\, dA, \quad Q_{yz}^{(1)} = \int_A z \mathcal{M}_{yz}\, dA.$$

The virtual work done by the external distributed transverse loads $q^t(x)$ at the top and $q^b(x)$ at the bottom (assuming that there is no axial distributed load, $f = 0$) is

$$\delta V_E = -\int_0^L \left[q^t \delta u_z(x, \tfrac{h}{2}) + q^b \delta u_z(x, -\tfrac{h}{2}) \right] dx$$

$$= -\int_0^L \left[(q^t + q^b)\left(\delta w + \tfrac{h^2}{4}\delta\theta_z \right) + \tfrac{h}{2}(q^t - q^b)\,\delta\phi_z \right] dx. \tag{5.2.11}$$

Substituting the expressions for δU, δV_E, and δK into the Hamilton's principle (5.2.6), performing integration-by-parts with respect t as well as x to relieve the generalized virtual displacements δu, $\delta\phi_x$, $\delta\theta_x$, $\delta\psi_x$, δw, $\delta\phi_z$, and $\delta\theta_z$ of any differentiation, and using the fundamental lemma of calculus of variations, we obtain the following equations of motion as the Euler–Lagrange equations of motion:

$$\delta u: \quad m_0 \frac{\partial^2 u}{\partial t^2} + m_1 \frac{\partial^2 \phi_x}{\partial t^2} + m_2 \frac{\partial^2 \theta_x}{\partial t^2} + m_3 \frac{\partial^2 \psi_x}{\partial t^2} - \frac{\partial M_{xx}^{(0)}}{\partial x} = 0, \tag{5.2.12}$$

$$\delta w: \quad m_0 \frac{\partial^2 w}{\partial t^2} + m_1 \frac{\partial^2 \phi_z}{\partial t^2} + m_2 \frac{\partial^2 \theta_z}{\partial t^2}$$
$$- \frac{1}{2}\frac{\partial^2 P_{xy}^{(0)}}{\partial x^2} - \frac{\partial}{\partial x}\left(M_{xx}^{(0)} \frac{\partial w}{\partial x} \right) - \frac{\partial M_{xz}^{(0)}}{\partial x} - q_1 = 0, \tag{5.2.13}$$

$$\delta\phi_x: \quad m_1 \frac{\partial^2 u}{\partial t^2} + m_2 \frac{\partial^2 \phi_x}{\partial t^2} + m_3 \frac{\partial^2 \theta_x}{\partial t^2} + m_4 \frac{\partial^2 \psi_x}{\partial t^2}$$
$$- \frac{\partial M_{xx}^{(1)}}{\partial x} + M_{xz}^{(0)} - \frac{1}{2}\frac{\partial P_{xy}^{(0)}}{\partial x} = 0, \tag{5.2.14}$$

$$\delta\theta_x: \quad m_2 \frac{\partial^2 u}{\partial t^2} + m_3 \frac{\partial^2 \phi_x}{\partial t^2} + m_4 \frac{\partial^2 \theta_x}{\partial t^2} + m_5 \frac{\partial^2 \psi_x}{\partial t^2}$$
$$- \frac{\partial M_{xx}^{(2)}}{\partial x} - \frac{\partial P_{xy}^{(1)}}{\partial x} + Q_{yz}^{(0)} + 2M_{xz}^{(1)} = 0, \tag{5.2.15}$$

$$\delta\psi_x: \quad m_3\frac{\partial^2 u}{\partial t^2} + m_4\frac{\partial^2 \phi_x}{\partial t^2} + m_5\frac{\partial^2 \theta_x}{\partial t^2} + m_6\frac{\partial^2 \psi_x}{\partial t^2}$$

$$-\frac{3}{2}\frac{\partial P_{xy}^{(2)}}{\partial x} + 3Q_{yz}^{(1)} - \frac{\partial M_{xx}^{(3)}}{\partial x} + 3M_{xz}^{(2)} = 0, \quad (5.2.16)$$

$$\delta\phi_z: \quad m_1\frac{\partial^2 w}{\partial t^2} + m_2\frac{\partial^2 \phi_z}{\partial t^2} + m_3\frac{\partial^2 \theta_z}{\partial t^2} - \frac{\partial M_{xz}^{(1)}}{\partial x}$$

$$-\frac{1}{2}\frac{\partial^2 P_{xy}^{(1)}}{\partial x^2} + \frac{1}{2}\frac{\partial Q_{yz}^{(0)}}{\partial x} + M_{zz}^{(0)} - q_2 = 0, \quad (5.2.17)$$

$$\delta\theta_z: \quad m_2\frac{\partial^2 w}{\partial t^2} + m_3\frac{\partial^2 \phi_z}{\partial t^2} + m_4\frac{\partial^2 \theta_z}{\partial t^2} - \frac{\partial M_{xz}^{(2)}}{\partial x}$$

$$-\frac{1}{2}\frac{\partial^2 P_{xy}^{(2)}}{\partial x^2} + \frac{\partial Q_{yz}^{(2)}}{\partial x} + 2M_{zz}^{(1)} - q_3 = 0, \quad (5.2.18)$$

where

$$q_1 = q^t + q^b, \quad q_2 = \frac{h}{2}\left(q^t - q^b\right), \quad q_3 = \frac{h^2}{4}\left(q^t + q^b\right). \quad (5.2.19)$$

The natural boundary conditions resulting from Hamilton's principle involve specifying one element of each of the following ten duality pairs:

$$(u, M_{xx}^{(0)}), \quad \left(w, M_{xz}^{(0)} + M_{xx}^{(0)}\frac{\partial w}{\partial x} + \frac{1}{2}\frac{\partial P_{xy}^{(0)}}{\partial x}\right),$$

$$\left(-\frac{\partial w}{\partial x}, \frac{1}{2}P_{xy}^{(0)}\right), \quad \left(\phi_x, M_{xx}^{(1)} + \frac{1}{2}P_{xy}^{(0)}\right),$$

$$\left(\theta_x, M_{xx}^{(2)} + P_{xy}^{(1)}\right), \quad \left(\psi_x, M_{xx}^{(3)} + \frac{3}{2}P_{xy}^{(2)}\right), \quad (5.2.20)$$

$$\left(\phi_z, M_{xz}^{(1)} - \frac{1}{2}Q_{yz}^{(0)} + \frac{1}{2}\frac{\partial P_{xy}^{(1)}}{\partial x}\right), \quad \left(-\frac{\partial \phi_z}{\partial x}, \frac{1}{2}P_{xy}^{(1)}\right),$$

$$\left(\theta_z, M_{zz}^{(2)} - Q_{yz}^{(1)} + \frac{1}{2}\frac{\partial P_{xy}^{(2)}}{\partial x}\right), \quad \left(-\frac{\partial \theta_z}{\partial x}, \frac{1}{2}P_{xy}^{(2)}\right).$$

The third-order theory presented here has seven dependent unknowns (u, ϕ_x, θ_x, ψ_x, ϕ_z, θ_z, ψ_z) and there are seven partial differential equations. However, there are ten duality pairs because of the higher-order (fourth-order) nature of the derivatives of (w, ϕ_z, θ_z) in the differential equations. When a "displacement" finite element model of the theory is developed, one is required to use the Hermite cubic interpolation of (w, ϕ_z, θ_z) and the Lagrange interpolation of ($u, \phi_x, \theta_x, \psi_x$). Of course, when the modified couple stress effect is not included we will have only seven unknowns and seven duality pairs, all variables requiring only the Lagrange interpolation.

The number of dependent unknowns can be reduced either by invoking inextensibility assumption or use other conditions, such as vanishing of transverse shear stresses on the top and bottom surfaces of the beam, to express some of the variables in terms of the others. This aspect will be discussed in Section 5.3.

5.2.3   EQUATIONS OF MOTION WITHOUT COUPLE STRESS EFFECTS

The equations of motion *without* the modified couple stress terms are

$$m_0\frac{\partial^2 u}{\partial t^2} + m_1\frac{\partial^2 \phi_x}{\partial t^2} + m_2\frac{\partial^2 \theta_x}{\partial t^2} + m_3\frac{\partial^2 \psi_x}{\partial t^2} - \frac{\partial M_{xx}^{(0)}}{\partial x} = 0, \quad (5.2.21)$$

$$m_0\frac{\partial^2 w}{\partial t^2} + m_1\frac{\partial^2 \phi_z}{\partial t^2} + m_2\frac{\partial^2 \theta_z}{\partial t^2}$$
$$- \frac{\partial}{\partial x}\left(M_{xx}^{(0)}\frac{\partial w}{\partial x}\right) - \frac{\partial M_{xz}^{(0)}}{\partial x} - q_1 = 0, \quad (5.2.22)$$

$$m_1\frac{\partial^2 u}{\partial t^2} + m_2\frac{\partial^2 \phi_x}{\partial t^2} + m_3\frac{\partial^2 \theta_x}{\partial t^2} + m_4\frac{\partial^2 \psi_x}{\partial t^2} - \frac{\partial M_{xx}^{(1)}}{\partial x} + M_{xz}^{(0)} = 0, \quad (5.2.23)$$

$$m_2\frac{\partial^2 u}{\partial t^2} + m_3\frac{\partial^2 \phi_x}{\partial t^2} + m_4\frac{\partial^2 \theta_x}{\partial t^2} + m_5\frac{\partial^2 \psi_x}{\partial t^2} - \frac{\partial M_{xx}^{(2)}}{\partial x} + 2M_{xz}^{(1)} = 0, \quad (5.2.24)$$

$$m_3\frac{\partial^2 u}{\partial t^2} + m_4\frac{\partial^2 \phi_x}{\partial t^2} + m_5\frac{\partial^2 \theta_x}{\partial t^2} + m_6\frac{\partial^2 \psi_x}{\partial t^2} - \frac{\partial M_{xx}^{(3)}}{\partial x} + 3M_{xz}^{(2)} = 0, \quad (5.2.25)$$

$$m_1\frac{\partial^2 w}{\partial t^2} + m_2\frac{\partial^2 \phi_z}{\partial t^2} + m_3\frac{\partial^2 \theta_z}{\partial t^2} - \frac{\partial M_{xz}^{(1)}}{\partial x} + M_{zz}^{(0)} - q_2 = 0, \quad (5.2.26)$$

$$m_2\frac{\partial^2 w}{\partial t^2} + m_3\frac{\partial^2 \phi_z}{\partial t^2} + m_4\frac{\partial^2 \theta_z}{\partial t^2} - \frac{\partial M_{xz}^{(2)}}{\partial x} + 2M_{zz}^{(1)} - q_3 = 0. \quad (5.2.27)$$

The natural boundary conditions involve specifying one element of each of the following duality pairs:

$$(u, M_{xx}^{(0)}),\quad \left(w, M_{xz}^{(0)} + M_{xx}^{(0)}\frac{\partial w}{\partial x}\right),\quad (\phi_x, M_{xx}^{(1)}),$$
$$(\theta_x, M_{xx}^{(2)}),\quad (\psi_x, M_{xx}^{(3)}),\quad (\phi_z, M_{xz}^{(1)}),\quad (\theta_z, M_{xz}^{(2)}). \quad (5.2.28)$$

5.2.4   CONSTITUTIVE RELATIONS

In order to express all of the stress resultants in terms of the generalized displacements (in the interest of solving the equations), we must now introduce appropriate stress–strain relations. Due to the inclusion of thickness stretch in the expansion of $u_z(x, z, t)$, we have nonzero ε_{zz} [see Eqs. (5.2.3a)–(5.2.4c)]. Therefore, we are required to use two-dimensional stress-strain relations. Assuming isotropic material, we write the following stress–strain relations (obtained from the three-dimensional stress–strain relations,

$\sigma_{ij} = 2\mu\,\varepsilon_{ij} + \lambda\,\varepsilon_{kk}\,\delta_{ij} - K\alpha_T(T - T_0))$:

$$\sigma_{xx} = (2\mu + \lambda)\,\varepsilon_{xx} + \lambda\varepsilon_{zz} - \frac{E\alpha_T}{(1 - 2\nu)}(T - T_0)$$

$$= \frac{E}{(1 - 2\nu)}\left[\frac{(1 - \nu)}{(1 + \nu)}\varepsilon_{xx} + \frac{\nu}{(1 + \nu)}\varepsilon_{zz} - \alpha_T(T - T_0)\right], \qquad (5.2.29a)$$

$$\sigma_{zz} = (2\mu + \lambda)\,\varepsilon_{zz} + \lambda\varepsilon_{xx} - \frac{E\alpha_T}{(1 - 2\nu)}(T - T_0)$$

$$= \frac{E}{(1 - 2\nu)}\left[\frac{(1 - \nu)}{(1 + \nu)}\varepsilon_{zz} + \frac{\nu}{(1 + \nu)}\varepsilon_{xx} - \alpha_T(T - T_0)\right], \qquad (5.2.29b)$$

$$\sigma_{xz} = 2\mu\varepsilon_{xz} = G\,\gamma_{xz}, \quad \gamma_{xz} = 2\varepsilon_{xz}, \qquad (5.2.29c)$$

where $\mu = G$ and λ are Lamé's constants; E, G, and ν denote Young's modulus, the shear modulus, and Poisson's ratio, respectively; K is the isothermal bulk modulus ($K = 2\mu/3 + \lambda$); α_T is the coefficient of thermal expansion; and T is the temperature measured from a reference absolute temperature T_0. For FGM structures, E (hence G) is a function of z.

The constitutive relation for the couple stress field is given by Eq. (3.4.1):

$$\begin{aligned}
\mathcal{M}_{xy} &= 2\ell^2 G\chi_{xy} = 2\ell^2 G\left(\chi_{xy}^{(0)} + z\chi_{xy}^{(1)} + z^2\chi_{xy}^{(2)}\right), \\
\mathcal{M}_{yz} &= 2\ell^2 G\chi_{yz} = 2\ell^2 G\left(\chi_{yz}^{(0)} + z\chi_{yz}^{(1)}\right).
\end{aligned} \qquad (5.2.30)$$

Now various stress resultants defined in Eqs. (5.2.10) can be related to the generalized displacements and their derivatives as

$$\begin{Bmatrix} M_{xx}^{(i)} \\ M_{zz}^{(i)} \\ M_{xz}^{(i)} \end{Bmatrix} = \int_A \begin{Bmatrix} \sigma_{xx} \\ \sigma_{zz} \\ \sigma_{xz} \end{Bmatrix} z^i\,dA = \sum_{k=i}^{3+i} \begin{bmatrix} A_{11}^{(k)} & A_{12}^{(k)} & 0 \\ A_{12}^{(k)} & A_{11}^{(k)} & 0 \\ 0 & 0 & B_{11}^{(k)} \end{bmatrix} \begin{Bmatrix} \varepsilon_{xx}^{(k-i)} \\ \varepsilon_{zz}^{(k-i)} \\ \gamma_{xz}^{(k-i)} \end{Bmatrix} - \begin{Bmatrix} X_T^{(i)} \\ Z_T^{(i)} \\ 0 \end{Bmatrix}$$
$$(5.2.31)$$

for $i = 0, 1, 2, 3, 4$, and the couple stress resultants $(P_{xy}^{(0)}, P_{xy}^{(1)}, P_{xy}^{(2)})$ and $(Q_{yz}^{(0)}, Q_{yz}^{(1)})$ are given in terms of the curvature components by

$$\begin{Bmatrix} P_{xy}^{(0)} \\ P_{xy}^{(1)} \\ P_{xy}^{(2)} \end{Bmatrix} = \int_A \begin{Bmatrix} 1 \\ z \\ z^2 \end{Bmatrix} \mathcal{M}_{xy}\,dA = \begin{bmatrix} A_{xy}^{\ell} & B_{xy}^{\ell} & D_{xy}^{\ell} \\ B_{xy}^{\ell} & D_{xy}^{\ell} & E_{xy}^{\ell} \\ D_{xy}^{\ell} & E_{xy}^{\ell} & F_{xy}^{\ell} \end{bmatrix} \begin{Bmatrix} 2\chi_{xy}^{(0)} \\ 2\chi_{xy}^{(1)} \\ 2\chi_{xy}^{(2)} \end{Bmatrix}, \qquad (5.2.32)$$

$$\begin{Bmatrix} Q_{yz}^{(0)} \\ Q_{yz}^{(1)} \end{Bmatrix} = \int_A \begin{Bmatrix} 1 \\ z \end{Bmatrix} \mathcal{M}_{yz}\,dA = \begin{bmatrix} A_{yz}^{\ell} & B_{yz}^{\ell} \\ B_{xx}^{\ell} & D_{xx}^{\ell} \end{bmatrix} \begin{Bmatrix} 2\chi_{yz}^{(0)} \\ 2\chi_{yz}^{(1)} \end{Bmatrix}, \qquad (5.2.33)$$

where

$$A_{11}^{(k)} = \frac{(1-\nu)}{(1+\nu)(1-2\nu)} \int_A E(z)\, z^k \, dA,$$

$$A_{12}^{(k)} = \frac{\nu}{(1+\nu)(1-2\nu)} \int_A E(z)\, z^k \, dA,$$

$$B_{11}^{(k)} = \frac{1}{2(1+\nu)} \int_A E(z)\, z^k \, dA, \qquad\qquad (5.2.34)$$

$$X_T^{(k)} = Z_T^{(k)} = \frac{1}{(1-2\nu)} \int_A E(z)\, z^k\, \alpha_T(z,T)\Delta T(z)\, dA,$$

and $(A_{yz}^\ell = A_{xy}^\ell,\ B_{yz}^\ell = B_{xy}^\ell,\ \text{and}\ D_{yz}^\ell = D_{xy}^\ell)$

$$(A_{xy}^\ell, B_{xy}^\ell, D_{xy}^\ell, E_{xy}^\ell, F_{xy}^\ell) = \frac{\ell^2}{2(1+\nu)} \int_A E(z)(1, z, z^2, z^3, z^4)\, dA. \quad (5.2.35)$$

For functionally graded beams, the modulus $E(z)$ can be assumed to vary with z according to a specified distribution of different phases of materials. For a two-constituent material variation through thickness, we have adopted the following power-law model:

$$E(z) = E_2\left[(M-1)V_1(z) + 1\right], \quad V_1(z) = \left(\frac{1}{2} + \frac{z}{h}\right)^n, \quad M = \frac{E_1}{E_2}. \quad (5.2.36)$$

5.3 A THIRD-ORDER THEORY WITH VANISHING SHEAR STRESS ON THE TOP AND BOTTOM FACES

5.3.1 THE GENERAL CASE

A special case of the general third-order theory developed in Section 5.2 can be developed by assuming that the transverse shear stress $\sigma_{xz}(x, z, t)$ vanishes at the top and bottom faces, $z = \pm h/2$ (see Levinson [124, 125] and Reddy [6, 7, 8]). The details are presented next.

The transverse shear strain based on the displacement field in Eq. (5.2.1) is given by

$$\gamma_{xz} = \phi_x + \frac{\partial w}{\partial x} + z\left(2\theta_x + \frac{\partial \phi_z}{\partial x}\right) + z^2\left(3\psi_x + \frac{\partial \theta_z}{\partial x}\right).$$

Then the shear stress is $\sigma_{xz} = G\gamma_{xz}$; requiring $\sigma_{xz} = 0$ at $z = \pm h/2$ results in the relations

$$G_2\left[\phi_x + \frac{\partial w}{\partial x} - \frac{h}{2}\left(2\theta_x + \frac{\partial \phi_z}{\partial x}\right) + \frac{h^2}{4}\left(3\psi_x + \frac{\partial \theta_z}{\partial x}\right)\right] = 0, \qquad (5.3.1a)$$

$$G_1\left[\phi_x + \frac{\partial w}{\partial x} + \frac{h}{2}\left(2\theta_x + \frac{\partial \phi_z}{\partial x}\right) + \frac{h^2}{4}\left(3\psi_x + \frac{\partial \theta_z}{\partial x}\right)\right] = 0, \qquad (5.3.1b)$$

where G_1 and G_2 are the shear modulus of material 1 (at the top) and material 2 (at the bottom), respectively. These two equations require, for an arbitrary FGM beam, the following conditions to be met:

$$\phi_x + \frac{\partial w}{\partial x} + \frac{h^2}{4}\left(3\psi_x + \frac{\partial \theta_z}{\partial x}\right) = 0, \quad 2\theta_x + \frac{\partial \phi_z}{\partial x} = 0, \tag{5.3.2}$$

which can be used to express two of the variables (θ_x, ψ_x) in terms of the other four variables associated with bending $(w, \phi_x, \phi_z, \theta_z)$:

$$\psi_x = -\frac{4}{3h^2}\left(\phi_x + \frac{\partial w}{\partial x}\right) - \frac{1}{3}\frac{\partial \theta_z}{\partial x}, \quad \theta_x = -\frac{1}{2}\frac{\partial \phi_z}{\partial x}. \tag{5.3.3}$$

Then the displacement field in Eq. (5.2.1) becomes

$$u_x(x,z,t) = u(x,t) + z\phi_x(x,t) - \frac{z^2}{2}\frac{\partial \phi_z}{\partial x} - \frac{z^3}{3}\frac{\partial \theta_z}{\partial x} - \frac{4z^3}{3h^2}\left(\phi_x + \frac{\partial w}{\partial x}\right),$$

$$u_y(x,z,t) = 0, \tag{5.3.4}$$

$$u_z(x,z,t) = w(x,t) + z\theta_z(x,t) + z^2\phi_z(x,t),$$

where (u,w) are midplane displacements along the x and z directions, respectively, ϕ_x is the rotation of a transverse normal line about the y axis, and (ϕ_z, θ_z) denote the variables that account for linear and quadratic variations, respectively, in the stretching of a transverse normal line.

Interested reader can follow the procedure discussed in the previous section and derive the equations of motion associated with the displacement field in Eq. (5.3.4). In the rest of the chapter, we consider a third-order beam theory in which the inextensibility of the transverse normal is assumed, that is, $\phi_z = 0$ and $\theta_z = 0$, which is a reasonable assumption for a theory with small strains. The third-order theory based on this displacement field is known in the literature as the *Reddy third-order beam theory* (RBT)[2]. The kinematics of the various beam theories is summarized in Fig. 5.3.1. *The third-order beam theory being developed in this section is limited to beams with biaxially symmetric cross sections only* because of the way the displacement field is derived (i.e., the top and bottom surfaces must be at a distance of $z = \pm h/2$, respectively).

---

[2]In 1984, Reddy [6, 7] developed a variationally-derived (i.e., using the principle of virtual displacements or the principle of the minimum total potential energy) third-order plate theory with vanishing shear stresses on the top and bottom surfaces of laminated composite plates, which was later on used for beams by Heyliger and Reddy [10]. In 1980, Levinson [124] used the third-order plate theory kinematics but adopted the governing equations of the TBT. Similar, but independent works on the third-order theories of beams can be found in the papers of Levinson [125] and Bickford [126], the latter appeared in the proceedings of a conference, and no follow up work was reported by the author. It is almost certain that similar ideas have appeared in the literature long before Levinson, Bickford, and Reddy detailed them in their respective contexts, and many more works have appeared since then, which are not reviewed here.

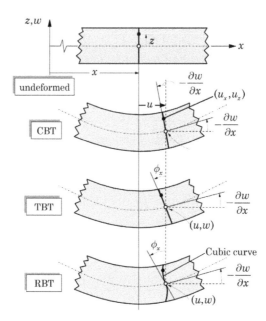

Figure 5.3.1: Kinematics of deformation of the classical beam theory (CBT), the Timoshenko beam theory (TBT), and the Reddy beam theory (RBT).

5.3.2   THE REDDY THIRD-ORDER BEAM THEORY (RBT)

5.3.2.1   Kinematics

The displacement field of the RBT is obtained from the displacement field (5.3.3) by invoking inextensibility condition: $\theta_z = \phi_z = 0$:

$$\mathbf{u}(x, z, t) = \left[u(x,t) + z\phi_x(x,t) - \alpha z^3 \left(\phi_x + \frac{\partial w}{\partial x} \right) \right] \hat{\mathbf{e}}_x + w(x,t)\hat{\mathbf{e}}_z, \quad (5.3.5)$$

which is a special case of the displacement field in Eq. (5.2.1) with the following correspondence:

$$\theta_x = 0, \quad \psi_x = -\alpha\left(\phi_x + \frac{\partial w}{\partial x}\right), \quad \phi_z = 0, \quad \theta_z = 0, \quad \alpha = \frac{4}{3h^2}. \quad (5.3.6)$$

The nonzero strain and curvature components are

$$\varepsilon_{xx} = \left[\frac{\partial u}{\partial x} + \frac{1}{2}\left(\frac{\partial w}{\partial x}\right)^2 \right] + z\frac{\partial \phi_x}{\partial x} - z^3\,\alpha\left(\frac{\partial \phi_x}{\partial x} + \frac{\partial^2 w}{\partial x^2}\right)$$

$$\equiv \varepsilon_{xx}^{(0)} + z\varepsilon_{xx}^{(1)} + z^3\varepsilon_{xx}^{(3)}, \quad (5.3.7a)$$

$$\gamma_{xz} = \phi_x + \frac{\partial w}{\partial x} - 3\alpha z^2\left(\phi_x + \frac{\partial w}{\partial x}\right) \equiv \gamma_{xz}^{(0)} + z^2\gamma_{xz}^{(2)},$$

$$\omega_y = \frac{1}{2}\left[\left(\phi_x - \frac{\partial w}{\partial x}\right) - 3\alpha z^2\left(\phi_x + \frac{\partial w}{\partial x}\right)\right],$$

$$\chi_{xy} = \frac{1}{4}\left(\frac{\partial \phi_x}{\partial x} - \frac{\partial^2 w}{\partial x^2}\right) - \frac{1}{4}\beta z^2\left(\frac{\partial \phi_x}{\partial x} + \frac{\partial^2 w}{\partial x^2}\right) \equiv \chi_{xy}^{(0)} + z^2\chi_{xy}^{(2)}, \quad (5.3.7\text{b})$$

$$\chi_{yz} = -z\frac{1}{2}\beta\left(\phi_x + \frac{\partial w}{\partial x}\right) \equiv z\chi_{yz}^{(1)},$$

where (omitting the higher-order terms in the thickness strain)

$$\varepsilon_{xx}^{(0)} = \frac{\partial u}{\partial x} + \frac{1}{2}\left(\frac{\partial w}{\partial x}\right)^2, \quad \varepsilon_{xx}^{(1)} = \frac{\partial \phi_x}{\partial x}, \quad \varepsilon_{xx}^{(3)} = -\alpha\left(\frac{\partial \phi_x}{\partial x} + \frac{\partial^2 w}{\partial x^2}\right),$$

$$\gamma_{xz}^{(0)} = \phi_x + \frac{\partial w}{\partial x}, \quad \gamma_{xz}^{(2)} = -\beta\left(\phi_x + \frac{\partial w}{\partial x}\right), \quad \beta = 3\alpha = \frac{4}{h^2}, \quad (5.3.7\text{c})$$

$$\chi_{xy}^{(0)} = \frac{1}{4}\left(\frac{\partial \phi_x}{\partial x} - \frac{\partial^2 w}{\partial x^2}\right), \quad \chi_{xy}^{(2)} = -\frac{1}{4}\beta\left(\frac{\partial \phi_x}{\partial x} + \frac{\partial^2 w}{\partial x^2}\right),$$

$$\chi_{yz}^{(1)} = -\frac{1}{2}\beta\left(\phi_x + \frac{\partial w}{\partial x}\right).$$

5.3.2.2 Equations of motion using Hamilton's principle

The equations of motion of the RBT are derived using Hamilton's principle:

$$0 = \int_{t_1}^{t_2}\int_0^L\int_A\left\{\sigma_{xx}\left(\delta\varepsilon_{xx}^{(0)} + z\delta\varepsilon_{xx}^{(1)} + z^3\delta\varepsilon_{xx}^{(3)}\right) + \sigma_{xz}\left(\delta\gamma_{xz}^{(0)} + z^2\delta\gamma_{xz}^{(2)}\right)\right.$$

$$+ 2\mathcal{M}_{xy}\left(\delta\chi_{xy}^{(0)} + z^2\delta\chi_{xy}^{(2)}\right) + 2\mathcal{M}_{yz}\,z\delta\chi_{yz}^{(1)}$$

$$- \rho\left[\dot{u} + z\dot{\phi}_x - \alpha z^3\left(\dot{\phi}_x + \frac{\partial\dot{w}}{\partial x}\right)\right]\left[\delta\dot{u} + z\delta\dot{\phi}_x - \alpha z^3\left(\delta\dot{\phi}_x + \frac{\partial\delta\dot{w}}{\partial x}\right)\right]$$

$$\left. - \rho\dot{w}\,\delta\dot{w}\right\}dAdxdt - \int_{t_1}^{t_2}\int_0^L\left(-c_f w + f\,\delta u + q\,\delta w\right)dx\,dt$$

$$= \int_{t_1}^{t_2}\int_0^L\left\{N_{xx}\delta\varepsilon_{xx}^{(0)} + M_{xx}\delta\varepsilon_{xx}^{(1)} + P_{xx}\delta\varepsilon_{xx}^{(3)} + N_{xz}\delta\gamma_{xz}^{(0)} + P_{xz}\delta\gamma_{xz}^{(2)}\right.$$

$$+ 2P_{xy}\delta\chi_{xy}^{(0)} + 2R_{xy}\delta\chi_{xy}^{(2)} + 2Q_{yz}\delta\chi_{yz}^{(1)} - \left(m_0\dot{u} + \bar{m}_1\dot{\phi}_x - \alpha m_3\frac{\partial\dot{w}}{\partial x}\right)\delta\dot{u}$$

$$- \left(\bar{m}_1\dot{u} + \hat{m}_2\dot{\phi}_x - \alpha\bar{m}_4\frac{\partial\dot{w}}{\partial x}\right)\delta\dot{\phi}_x + c_f w - m_0\dot{w}\,\delta\dot{w}$$

$$\left. + \alpha\left(m_3\dot{u} + \bar{m}_4\dot{\phi}_x - \alpha m_6\frac{\partial\dot{w}}{\partial x}\right)\frac{\partial\delta\dot{w}}{\partial x} - f\,\delta u - q\,\delta w\right\}dxdt, \quad (5.3.8)$$

where c_f is the linear elastic foundation modulus (if the beam is resting on an elastic foundation), $(N_{\alpha\beta}, M_{\alpha\beta}, P_{\alpha\beta})$ are the stress resultants, and m_i are mass inertia coefficients:

$$m_i = \int_A \rho(z)\,(z)^i\,dz \quad (i=0,1,2,\ldots,6), \tag{5.3.9a}$$

$$\bar{m}_i = m_i - \alpha\,m_{i+2}, \quad \hat{m}_i = \bar{m}_i - \alpha\,\bar{m}_{i+2}, \tag{5.3.9b}$$

$$(N_{xx}, M_{xx}, P_{xx}) = \int_A (1,z,z^3)\sigma_{xx}\,dA, \tag{5.3.9c}$$

$$(N_{xz}, P_{xz}) = \int_A (1,z^2)\sigma_{xz}\,dA, \tag{5.3.9d}$$

$$(P_{xy}, R_{xy}) = \int_A (1,z^2)\mathcal{M}_{xy}\,dA, \tag{5.3.9e}$$

$$Q_{yz} = \int_A z\mathcal{M}_{yz}\,dA. \tag{5.3.9f}$$

Next, we use the usual procedure of (1) integrating by parts with respect to both x and t to relieve all varied quantities of any derivatives, (2) using the fundamental lemma of the calculus of variations, and (3) knowing that (δu, δw, $\delta\phi_x$) are arbitrary and linearly independent of each other to obtain the following equations of motion from Eq. (5.3.8):

$$m_0\frac{\partial^2 u}{\partial t^2} - \alpha m_3 \frac{\partial^3 w}{\partial x \partial t^2} + \bar{m}_1 \frac{\partial^2 \phi_x}{\partial t^2} - \frac{\partial N_{xx}}{\partial x} = f, \tag{5.3.10}$$

$$m_0\frac{\partial^2 w}{\partial t^2} + \alpha\frac{\partial}{\partial x}\left(m_3\frac{\partial^2 u}{\partial t^2} - \alpha\, m_6 \frac{\partial^3 w}{\partial x \partial t^2} + \bar{m}_4 \frac{\partial^2 \phi_x}{\partial t^2}\right) + c_f\,w$$
$$-\frac{\partial}{\partial x}\left(N_{xx}\frac{\partial w}{\partial x}\right) - \frac{\partial \bar{N}_{xz}}{\partial x} - \alpha\frac{\partial^2 P_{xx}}{\partial x^2} - \frac{1}{2}\frac{\partial^2 \tilde{P}_{xy}}{\partial x^2} = q, \tag{5.3.11}$$

$$\bar{m}_1\frac{\partial^2 u}{\partial t^2} - \alpha\,\bar{m}_4 \frac{\partial^3 w}{\partial x \partial t^2} + \hat{m}_2 \frac{\partial^2 \phi_x}{\partial t^2} - \frac{\partial \bar{M}_{xx}}{\partial x} + \bar{N}_{xz} - \frac{1}{2}\frac{\partial \bar{P}_{xy}}{\partial x} = 0, \tag{5.3.12}$$

where [see Eqs. (5.3.20a)–(5.3.20j)]

$$\bar{M}_{xx} = M_{xx} - \alpha\,P_{xx}, \quad \bar{N}_{xz} = N_{xz} - \beta P_{xz} - \beta Q_{yz},$$
$$\bar{P}_{xy} = P_{xy} - \beta R_{xy}, \quad \tilde{P}_{xy} = P_{xy} + \beta R_{xy}, \tag{5.3.13}$$

where α and β are defined by [see Eqs. (5.3.6) and (5.3.7c)]

$$\alpha = \frac{4}{3h^2} \quad \beta = 3\alpha = \frac{4}{h^2}\,,$$

with h being the height of the beam.

The duality pairs (the first element of each pair denotes a generalized displacement while the second element denotes a generalized force):

$$(u, N_{xx}),\quad (w, V_{\text{eff}}),\quad (\phi_x, M_{\text{eff}}),\quad \left(-\frac{\partial w}{\partial x}, \tilde{M}_{xx}\right), \tag{5.3.14a}$$

where

$$M_{\text{eff}} = \bar{M}_{xx} + \tfrac{1}{2}\bar{P}_{xy}, \quad \tilde{M}_{xx} = \alpha\, P_{xx} + \tfrac{1}{2}\tilde{P}_{xy}, \tag{5.3.14b}$$

$$V_{\text{eff}} = \bar{N}_{xz} + N_{xx}\frac{\partial w}{\partial x} + \alpha\,\frac{\partial P_{xx}}{\partial x} + \frac{1}{2}\frac{\partial \tilde{P}_{xy}}{\partial x}$$

$$- \alpha\left(m_3\frac{\partial^2 u}{\partial t^2} - \alpha\, m_6\frac{\partial^3 w}{\partial x \partial t^2} + \bar{m}_4\frac{\partial^2 \phi_x}{\partial t^2}\right). \tag{5.3.14c}$$

Thus, there are four duality pairs, indicating that there are four boundary conditions at each boundary point. Requiring $\partial w/\partial x$ as well as ϕ_x to vanish at a support necessarily implies the shear stress computed using the constitutive relation $\sigma_{xz} = G\gamma_{xz}$ is zero while $V_{\text{eff}} \neq 0$.

The equations of equilibrium in terms of the stress resultants of the RBT are obtained from Eqs. (5.3.10)–(5.3.12) by omitting the time derivative terms:

$$-\frac{dN_{xx}}{dx} = f, \tag{5.3.15}$$

$$-\alpha\frac{d^2 P_{xx}}{dx^2} - \frac{d}{dx}\left(N_{xx}\frac{dw}{dx}\right) - \frac{d\bar{N}_{xz}}{dx} + c_f\, w - \frac{1}{2}\frac{d^2 \tilde{P}_{xy}}{dx^2} = q, \tag{5.3.16}$$

$$-\frac{d\bar{M}_{xx}}{dx} + \bar{N}_{xz} - \frac{1}{2}\frac{d\bar{P}_{xy}}{dx} = 0, \tag{5.3.17}$$

where $\bar{M}_{xx}$, $\bar{N}_{xz}$, $\bar{P}_{xy}$, and $\tilde{P}_{xy}$ are defined in Eq. (5.3.13).

One of the challenges of higher-order theories is the specification of boundary conditions on higher-order stress resultants P_{xx}, P_{xz}, P_{xy}, R_{xy}, and Q_{yz} introduced in Eqs. (5.3.9c)–(5.3.9f). In most cases, one does not know what are the known values of these higher-order stress resultants. Therefore, whenever the lower-order stress resultant is specified, we assume that the corresponding higher-order stress resultant is known to be zero. For example, if M_{xx} is specified at a point, we assume that $\bar{M}_{xx} = M_{xx}$ (implying that $P_{xx} = 0$ there). This problem exists in all strain- or stress-gradient theories of mechanics.

5.3.2.3 Constitutive relations

In the RBT, as in in the case of CBT and TBT, we invoked the inextensibility of the transverse normal lines, which amounts to setting $\varepsilon_{zz} = 0$. Therefore, we can use one-dimensional constitutive relations. In particular, the one-dimensional constitutive relations are

$$\sigma_{xx}(x,z) = E(z)\left(\varepsilon_{xx} - \alpha_T\,\Delta T\right), \quad \sigma_{xz}(x,z) = G(z)\,\gamma_{xz}, \tag{5.3.18}$$

$$\mathcal{M}_{xy}(x,z) = 2\ell^2 G(z)\,\chi_{xy}, \qquad \mathcal{M}_{yz}(x,z) = 2\ell^2 G(z)\,\chi_{yz}, \tag{5.3.19}$$

where $G(z) = E(z)/[2(1+\nu)]$, α_T is the coefficient of thermal expansion, and ℓ material length scale; ε_{xx}, γ_{xz}, χ_{xy}, and χ_{yz} are defined in Eqs. (5.3.7a)–(5.3.7c).

Substituting the constitutive relations in Eqs. (5.3.18) and (5.3.19) into the definition of the stress resultants in Eqs. (5.3.9b) and (5.3.9c), we obtain

$$
\begin{aligned}
N_{xx} &= A_{xx}\,\varepsilon_{xx}^{(0)} + B_{xx}\,\varepsilon_{xx}^{(1)} + E_{xx}\,\varepsilon_{xx}^{(3)} - N_{xx}^{T} \\[4pt]
&= A_{xx}\left[\frac{\partial u}{\partial x} + \frac{1}{2}\left(\frac{\partial w}{\partial x}\right)^{2}\right] + B_{xx}\frac{\partial \phi_x}{\partial x} - \alpha\,E_{xx}\left(\frac{\partial \phi_x}{\partial x} + \frac{\partial^2 w}{\partial x^2}\right) - N_{xx}^{T} \\[4pt]
&= A_{xx}\left[\frac{\partial u}{\partial x} + \frac{1}{2}\left(\frac{\partial w}{\partial x}\right)^{2}\right] + \bar{B}_{xx}\frac{\partial \phi_x}{\partial x} - \alpha\,E_{xx}\frac{\partial^2 w}{\partial x^2} - N_{xx}^{T},
\end{aligned}\tag{5.3.20a}
$$

$$
M_{xx} = B_{xx}\left[\frac{\partial u}{\partial x} + \frac{1}{2}\left(\frac{\partial w}{\partial x}\right)^{2}\right] + \bar{D}_{xx}\frac{\partial \phi_x}{\partial x} - \alpha\,F_{xx}\frac{\partial^2 w}{\partial x^2} - M_{xx}^{T},\tag{5.3.20b}
$$

$$
P_{xx} = E_{xx}\left[\frac{\partial u}{\partial x} + \frac{1}{2}\left(\frac{\partial w}{\partial x}\right)^{2}\right] + \bar{F}_{xx}\frac{\partial \phi_x}{\partial x} - \alpha\,H_{xx}\frac{\partial^2 w}{\partial x^2} - P_{xx}^{T},\tag{5.3.20c}
$$

$$
N_{xz} = A_{xz}\left(\phi_x + \frac{\partial w}{\partial x}\right) - \beta\,D_{xz}\left(\phi_x + \frac{\partial w}{\partial x}\right) = \bar{A}_{xz}\left(\phi_x + \frac{\partial w}{\partial x}\right),\tag{5.3.20d}
$$

$$
P_{xz} = D_{xz}\left(\phi_x + \frac{\partial w}{\partial x}\right) - \beta\,F_{xz}\left(\phi_x + \frac{\partial w}{\partial x}\right) = \bar{D}_{xz}\left(\phi_x + \frac{\partial w}{\partial x}\right),\tag{5.3.20e}
$$

$$
\begin{aligned}
P_{xy} &= \frac{1}{2}A_{xy}\left(\frac{\partial \phi_x}{\partial x} - \frac{\partial^2 w}{\partial x^2}\right) - \frac{\beta}{2}D_{xy}\left(\frac{\partial \phi_x}{\partial x} + \frac{\partial^2 w}{\partial x^2}\right) \\[4pt]
&= \frac{1}{2}\bar{A}_{xy}\frac{\partial \phi_x}{\partial x} - \frac{1}{2}\tilde{A}_{xy}\frac{\partial^2 w}{\partial x^2},
\end{aligned}\tag{5.3.20f}
$$

$$
R_{xy} = \frac{1}{2}\bar{D}_{xy}\frac{\partial \phi_x}{\partial x} - \frac{1}{2}\tilde{D}_{xy}\frac{\partial^2 w}{\partial x^2},\tag{5.3.20g}
$$

$$
Q_{yz} = -\beta\,D_{xy}\left(\phi_x + \frac{\partial w}{\partial x}\right),\tag{5.3.20h}
$$

$$
\bar{P}_{xy} = P_{xy} - \beta\,R_{xy} = \frac{1}{2}\hat{A}_{xy}\frac{\partial \phi_x}{\partial x} - \frac{1}{2}\bar{\bar{A}}_{xy}\frac{\partial^2 w}{\partial x^2},\tag{5.3.20i}
$$

$$
\tilde{P}_{xy} = P_{xy} + \beta\,R_{xy} = \frac{1}{2}\tilde{A}_{xy}\frac{\partial \phi_x}{\partial x} - \frac{1}{2}\bar{\bar{A}}_{xy}\frac{\partial^2 w}{\partial x^2}.\tag{5.3.20j}
$$

The various stiffness coefficients are defined by [see Eq. (3.4.8)]

$$
(A_{xx}, B_{xx}, D_{xx}, E_{xx}, F_{xx}, H_{xx}) = \int_{A}(1, z, z^2, z^3, z^4, z^6)E(z)\,dA,\tag{5.3.21a}
$$

$$
(A_{xz}, D_{xz}, F_{xz}) = \frac{1}{2(1+\nu)}\int_{A}(1, z^2, z^4)E(z)\,dA,\tag{5.3.21b}
$$

$$
(A_{xy}, D_{xy}, F_{xy}) = \frac{\ell^2}{2(1+\nu)}\int_{A}(1, z^2, z^4)E(z)\,dA,\tag{5.3.21c}
$$

$$
(N_{xx}^{T}, M_{xx}^{T}, P_{xx}^{T}) = \int_{A}(1, z, z^3)E(z)\,\alpha_T(z)\,\Delta T(z)\,dA.\tag{5.3.21d}
$$

and $[\alpha = 4/3h^2$ and $\beta = 4/h^2 = 3\alpha]$

$$\bar{B}_{xx} = B_{xx} - \alpha E_{xx}, \quad \bar{D}_{xx} = D_{xx} - \alpha F_{xx}, \quad \bar{F}_{xx} = F_{xx} - \alpha H_{xx},$$
$$\bar{A}_{xz} = A_{xz} - \beta D_{xz}, \quad \bar{D}_{xz} = D_{xz} - \beta F_{xz}, \quad \hat{A}_{xz} = \bar{A}_{xz} - \beta \bar{D}_{xz},$$
$$\bar{A}_{xy} = A_{xy} - \beta D_{xy}, \quad \bar{D}_{xy} = D_{xy} - \beta F_{xy}, \quad \hat{A}_{xy} = \bar{A}_{xy} - \beta \bar{D}_{xy},$$
$$\tilde{A}_{xy} = A_{xy} + \beta D_{xy}, \quad \tilde{D}_{xy} = D_{xy} + \beta F_{xy}, \quad \hat{D}_{xx} = \bar{D}_{xx} - \alpha \bar{F}_{xx},$$
$$\tilde{\bar{A}}_{xx} = \hat{\tilde{A}}_{xx} = A_{xy} - \beta^2 F_{xy},$$
$$\tilde{A}_{xy} = \tilde{A}_{xy} + \beta \tilde{D}_{xy} = A_{xy} + 2\beta D_{xy} + \beta^2 F_{xy}. \tag{5.3.22}$$

Beam stiffness coefficients for homogeneous beams. For a homogeneous beam (i.e., $E_1 = E_2$ and $E = E(x)$ is not a function of z) of rectangular cross section, $b \times h$ (b is the width and h is the height of the beam), we have $B_{xx} = E_{xx} = \bar{B} = 0$ and

$$A_{xx} = Ebh, \quad D_{xx} = E\frac{bh^3}{12}, \quad F_{xx} = E\frac{bh^5}{80} = \frac{3h^2}{20}D_{xx},$$
$$H_{xx} = E\frac{bh^7}{448} = \frac{3h^4}{112}D_{xx}, \quad A_{xy} = \frac{Ebh\ell^2}{2(1+\nu)}, \quad D_{xy} = \frac{Ebh^3\ell^2}{24(1+\nu)},$$
$$F_{xy} = \frac{Ebh^5\ell^2}{160(1+\nu)} = \frac{3h^2}{20}D_{xy}, \quad A_{xz} = \frac{Ebh}{2(1+\nu)}, \tag{5.3.23}$$
$$D_{xz} = \frac{Ebh^3}{24(1+\nu)} = \frac{h^2}{12}A_{xz}, \quad F_{xz} = \frac{Ebh^5}{160(1+\nu)} = \frac{h^4}{80}A_{xz}.$$

Beam stiffness coefficients for FGM beams. For the FGM beams, the integrals in Eqs. (5.3.21a)–(5.3.21d) can be evaluated using the formulas presented in **Chapter 3 Appendix**. We have

$$E(z) = E_2\left[(M-1)V_1(z) + 1\right], \quad V_1(z) = \left(\frac{1}{2} + \frac{z}{h}\right)^n, \quad M = \frac{E_1}{E_2}, \tag{5.3.24a}$$

$$m_0 = \rho_2 bh\frac{R+n}{1+n}, \quad m_1 = \rho_2\frac{bh^2}{2}\frac{n(R-1)}{(1+n)(2+n)},$$
$$m_2 = \rho_2\frac{bh^3}{12}\left[\frac{(6+3n+3n^2)R + (8n+3n^2+n^3)}{(1+n)(2+n)(3+n)}\right],$$
$$m_3 = \rho_2\frac{bh^4}{8}(M-1)\left[\frac{n(8+3n+n^2)}{(1+n)(2+n)(3+n)(4+n)}\right], \tag{5.3.24b}$$
$$m_4 = \rho_2\frac{bh^5}{80}f(n), \quad m_6 = \rho_2\frac{bh^7}{448}g(n), \quad R = \frac{\rho_1}{\rho_2},$$

$$A_{xx} = E_2 bh \frac{M+n}{1+n}, \quad B_{xx} = E_2 \frac{bh^2}{2} \frac{n(M-1)}{(1+n)(2+n)},$$

$$D_{xx} = E_2 \frac{bh^3}{12} \left[\frac{(6+3n+3n^2)M + (8n+3n^2+n^3)}{(1+n)(2+n)(3+n)} \right],$$

$$E_{xx} = E_2 \frac{bh^4}{8}(M-1) \left[\frac{n(8+3n+n^2)}{(1+n)(2+n)(3+n)(4+n)} \right],$$

$$F_{xx} = E_2 \frac{bh^5}{80} f(n), \quad H_{xx} = E_2 \frac{bh^7}{448} g(n),$$

$$A_{xy} = E_2 \ell^2 \frac{bh}{2(1+\nu)} \frac{M+n}{1+n}, \quad F_{xy} = \frac{E_2 \ell^2 bh^5}{120(1+\nu)} f(n),$$

$$D_{xy} = E_2 \ell^2 \frac{bh^3}{24(1+\nu)} \left[\frac{(6+3n+3n^2)M + (8n+3n^2+n^3)}{(1+n)(2+n)(3+n)} \right],$$

(5.3.24c)

where

$$f(n) = \frac{5f_1 M + n f_2}{f_3}, \quad g(n) = \frac{7g_1 M + n g_2}{g_3}, \quad M = \frac{E_1}{E_2},$$

$$f_1 = (24 + 18n + 23n^2 + 6n^3 + n^4),$$

$$f_2 = (184 + 110n + 55n^2 + 10n^3 + n^4),$$

$$f_3 = (1+n)(2+n)(3+n)(4+n)(5+n),$$

$$g_1 = (720 + 660n + 964n^2 + 405n^3 + 115n^4 + 15n^5 + n^6),$$

$$g_2 = (8448 + 6384n + 3934n^2 + 1155n^3 + 217n^4 + 21n^5 + n^6),$$

$$g_3 = (1+n)(2+n)(3+n)(4+n)(5+n)(6+n)(7+n).$$

(5.3.24d)

If the higher-order terms are neglected in the governing equations of motion but not in the constitutive relations, we obtain the third-order theory developed by Levinson [125]. If one neglects the higher-order terms selectively in the constitutive relations, one obtains the so-called simplified Reddy beam theory. These ideas will be discussed shortly.

5.3.2.4   Equations of motion in terms of the generalized displacements: the general case

With the help of Eqs. (5.3.20a)–(5.3.20j), the equations of motion, Eqs. (5.3.10)–(5.3.12), can be expressed in terms of the generalized displacements (u, w, ϕ_x). In view of the many stress resultants, the resulting equations contains many terms. Some of the terms contain higher-order stiffness coefficients defined in Eqs. (5.3.21a)–(5.3.21d). It is possible to neglect some of these terms on the basis of their (small) magnitude. However, for the sake of completeness, we include all of the terms in writing the equations of motion in terms of the generalized displacements.

The equations of motion, Eqs. (5.3.10)–(5.3.12), expressed in terms of the generalized displacements (u, w, ϕ_x) take the form

$$-\frac{\partial}{\partial x}\left\{A_{xx}\left[\frac{\partial u}{\partial x} + \frac{1}{2}\left(\frac{\partial w}{\partial x}\right)^2\right] + \bar{B}_{xx}\frac{\partial \phi_x}{\partial x} - \alpha\, E_{xx}\frac{\partial^2 w}{\partial x^2} - N_{xx}^T\right\}$$

$$+ m_0\frac{\partial^2 u}{\partial t^2} - \alpha\, m_3\frac{\partial^3 w}{\partial t^2 \partial x} + \bar{m}_1\frac{\partial^2 \phi_x}{\partial t^2} = f,$$

$$(5.3.25)$$

$$-\frac{\partial}{\partial x}\left[\frac{\partial w}{\partial x}\left\{A_{xx}\left[\frac{\partial u}{\partial x} + \frac{1}{2}\left(\frac{\partial w}{\partial x}\right)^2\right] + \bar{B}_{xx}\frac{\partial \phi_x}{\partial x} - \alpha\, E_{xx}\frac{\partial^2 w}{\partial x^2} - N_{xx}^T\right\}\right]$$

$$-\frac{1}{4}\frac{\partial^2}{\partial x^2}\left(\tilde{\bar{A}}_{xy}\frac{\partial \phi_x}{\partial x} - \tilde{\bar{A}}_{xy}\frac{\partial^2 w}{\partial x^2}\right) - \left(\hat{A}_{xz} + \beta^2 D_{xy}\right)\frac{\partial}{\partial x}\left(\phi_x + \frac{\partial w}{\partial x}\right)$$

$$-\alpha\frac{\partial^2}{\partial x^2}\left\{E_{xx}\left[\frac{\partial u}{\partial x} + \frac{1}{2}\left(\frac{\partial w}{\partial x}\right)^2\right] + \bar{F}_{xx}\frac{\partial \phi_x}{\partial x} - \alpha\, H_{xx}\frac{\partial^2 w}{\partial x^2} - P_{xx}^T\right\}$$

$$+ c_f\, w + m_0\frac{\partial^2 w}{\partial t^2} + \alpha\frac{\partial}{\partial x}\left(m_3\frac{\partial^2 u}{\partial t^2} - \alpha\, m_6\frac{\partial^3 w}{\partial x \partial t^2} + \bar{m}_4\frac{\partial^2 \phi_x}{\partial t^2}\right) = q,$$

$$(5.3.26)$$

$$-\frac{\partial}{\partial x}\left\{\bar{B}_{xx}\left[\frac{\partial u}{\partial x} + \frac{1}{2}\left(\frac{\partial w}{\partial x}\right)^2\right] + \hat{D}_{xx}\frac{\partial \phi_x}{\partial x} - \alpha\, \bar{F}_{xx}\frac{\partial^2 w}{\partial x^2} - \bar{M}_{xx}^T\right\}$$

$$-\frac{1}{4}\frac{\partial}{\partial x}\left(\hat{A}_{xy}\frac{\partial \phi_x}{\partial x} - \tilde{\bar{A}}_{xy}\frac{\partial^2 w}{\partial x^2}\right) + \left(\hat{A}_{xz} + \beta^2 D_{xy}\right)\left(\phi_x + \frac{\partial w}{\partial x}\right)$$

$$+ \bar{m}_1\frac{\partial^2 u}{\partial t^2} - \alpha\, \bar{m}_4\frac{\partial^3 w}{\partial x \partial t^2} + \hat{m}_2\frac{\partial^2 \phi_x}{\partial t^2} = 0,$$

$$(5.3.27)$$

where $\bar{M}_{xx}^T = M_{xx}^T - \alpha\, P_{xx}^T$. From the above equations it is clear that the length scale parameter (through D_{xy}) has the effect of increasing the shear stiffness $\hat{A}_{xz}$ to $\hat{A}_{xz} + \beta^2 D_{xy}$ and adding to other stiffness coefficients, making the RBT stiffer than the TBT, while both theories predict smaller deflections due to the presence of the couple stress terms.

5.3.2.5 Equations of motion in terms of the generalized displacements: the linear case

The equations of motion are nonlinear only because of the inclusion of the von Kármán strain. When expressed in terms of the displacements, the nonlinearity is cubic in terms of $\partial w/\partial x$. Equations of motion in terms of the displacements for the linear case are obtained by omitting the nonlinear terms in Eqs. (5.3.25)–(5.3.27). These equations are valid for FGM beams with the couple stress effect. We have

$$-\frac{\partial}{\partial x}\left(A_{xx}\frac{\partial u}{\partial x} + \bar{B}_{xx}\frac{\partial \phi_x}{\partial x} - \alpha E_{xx}\frac{\partial^2 w}{\partial x^2} - N_{xx}^T\right)$$

$$+m_0\frac{\partial^2 u}{\partial t^2} - \alpha m_3\frac{\partial^3 w}{\partial t^2 \partial x} + \bar{m}_1\frac{\partial^2 \phi_x}{\partial t^2} = f, \quad (5.3.28)$$

$$-\frac{1}{4}\frac{\partial^2}{\partial x^2}\left(\tilde{\bar{A}}_{xy}\frac{\partial \phi_x}{\partial x} - \tilde{\bar{A}}_{xy}\frac{\partial^2 w}{\partial x^2}\right) - \left(\hat{A}_{xz} + \beta^2 D_{xy}\right)\frac{\partial}{\partial x}\left(\phi_x + \frac{\partial w}{\partial x}\right)$$

$$-\alpha\frac{\partial^2}{\partial x^2}\left(E_{xx}\frac{\partial u}{\partial x} + \bar{F}_{xx}\frac{\partial \phi_x}{\partial x} - \alpha H_{xx}\frac{\partial^2 w}{\partial x^2} - P_{xx}^T\right)$$

$$+c_f w + m_0\frac{\partial^2 w}{\partial t^2} + \alpha\frac{\partial}{\partial x}\left(m_3\frac{\partial^2 u}{\partial t^2} - \alpha m_6\frac{\partial^3 w}{\partial x\partial t^2} + \bar{m}_4\frac{\partial^2 \phi_x}{\partial t^2}\right) = q, \quad (5.3.29)$$

$$-\frac{\partial}{\partial x}\left(\bar{B}_{xx}\frac{\partial u}{\partial x} + \hat{D}_{xx}\frac{\partial \phi_x}{\partial x} - \alpha\bar{F}_{xx}\frac{\partial^2 w}{\partial x^2} - \bar{M}_{xx}^T\right)$$

$$-\frac{1}{4}\frac{\partial}{\partial x}\left(\hat{A}_{xy}\frac{\partial \phi_x}{\partial x} - \tilde{\bar{A}}_{xy}\frac{\partial^2 w}{\partial x^2}\right) + \left(\hat{A}_{xz} + \beta^2 D_{xy}\right)\left(\phi_x + \frac{\partial w}{\partial x}\right)$$

$$+\bar{m}_1\frac{\partial^2 u}{\partial t^2} - \alpha\bar{m}_4\frac{\partial^3 w}{\partial x\partial t^2} + \hat{m}_2\frac{\partial^2 \phi_x}{\partial t^2} = 0. \quad (5.3.30)$$

When the modified couple stress effect is not considered (i.e., set $A_{xy} = D_{xy} = F_{xy} = 0$), the above *equations of motion* (for the linear case) become

$$-\frac{\partial}{\partial x}\left(A_{xx}\frac{\partial u}{\partial x} + \bar{B}_{xx}\frac{\partial \phi_x}{\partial x} - \alpha E_{xx}\frac{\partial^2 w}{\partial x^2} - N_{xx}^T\right)$$

$$+m_0\frac{\partial^2 u}{\partial t^2} - \alpha m_3\frac{\partial^3 w}{\partial t^2 \partial x} + \bar{m}_1\frac{\partial^2 \phi_x}{\partial t^2} = f, \quad (5.3.31)$$

$$-\alpha\frac{\partial^2}{\partial x^2}\left(E_{xx}\frac{\partial u}{\partial x} + \bar{F}_{xx}\frac{\partial \phi_x}{\partial x} - \alpha H_{xx}\frac{\partial^2 w}{\partial x^2} - P_{xx}^T\right)$$

$$-\hat{A}_{xz}\frac{\partial}{\partial x}\left(\phi_x + \frac{\partial w}{\partial x}\right) + c_f w + m_0\frac{\partial^2 w}{\partial t^2}$$

$$+\alpha\frac{\partial}{\partial x}\left(m_3\frac{\partial^2 u}{\partial t^2} - \alpha m_6\frac{\partial^3 w}{\partial x\partial t^2} + \bar{m}_4\frac{\partial^2 \phi_x}{\partial t^2}\right) = q, \quad (5.3.32)$$

$$-\frac{\partial}{\partial x}\left(\bar{B}_{xx}\frac{\partial u}{\partial x} + \hat{D}_{xx}\frac{\partial \phi_x}{\partial x} - \alpha\bar{F}_{xx}\frac{\partial^2 w}{\partial x^2} - \bar{M}_{xx}^T\right)$$

$$+\hat{A}_{xz}\left(\phi_x + \frac{\partial w}{\partial x}\right) + \bar{m}_1\frac{\partial^2 u}{\partial t^2} - \alpha\bar{m}_4\frac{\partial^3 w}{\partial x\partial t^2} + \hat{m}_2\frac{\partial^2 \phi_x}{\partial t^2} = 0. \quad (5.3.33)$$

The *equations of equilibrium* for the linear case without the modified couple stress effect are

$$-\frac{d}{dx}\left(A_{xx}\frac{du}{dx} + \bar{B}_{xx}\frac{d\phi_x}{dx} - \alpha E_{xx}\frac{d^2w}{dx^2} - N_{xx}^T\right) = f, \qquad (5.3.34)$$

$$-\alpha\frac{d^2}{dx^2}\left(E_{xx}\frac{du}{dx} + \bar{F}_{xx}\frac{d\phi_x}{dx} - \alpha H_{xx}\frac{d^2w}{dx^2} - P_{xx}^T\right)$$

$$-\hat{A}_{xz}\frac{d}{dx}\left(\phi_x + \frac{dw}{dx}\right) + c_f\,w = q, \qquad (5.3.35)$$

$$-\frac{d}{dx}\left(\bar{B}_{xx}\frac{du}{dx} + \hat{D}_{xx}\frac{d\phi_x}{dx} - \alpha\bar{F}_{xx}\frac{d^2w}{dx^2} - \bar{M}_{xx}^T\right)$$

$$+\hat{A}_{xz}\left(\phi_x + \frac{dw}{dx}\right) = 0. \qquad (5.3.36)$$

We note that $B_{xx} = E_{xx} = 0$ for beams for which E is independent of z.

5.3.3 LEVINSON'S THIRD-ORDER BEAM THEORY (LBT)

5.3.3.1 Equations of motion

The plate and beam theories advanced by Levinson [124, 125] used the same kinematics as discussed in Eqs. (5.3.5)–(5.3.7c); Levinson considered only pure bending of isotropic beams and plates, and used the equilibrium equations of the TBT while accounting for the higher-order terms in the strain–displacement relations. From energy considerations, such an approach is said to be variationally inconsistent. The LBT is extended in this book to account for the inertia terms, the von Kármán nonlinear strain, material variation through the beam thickness, and the couples stress effect.

The equations of motion of FGM beams based on the TBT are [see Eqs. (4.3.18)–(4.3.20)] which are the same for the LBT, except that the stress resultants are based on the higher-order strains. Thus, the equations of motion of the TBT as well as the LBT in terms of the stress resultants are

$$-\frac{\partial N_{xx}}{\partial x} + m_0\frac{\partial^2 u}{\partial t^2} + m_1\frac{\partial^2 \phi_x}{\partial t^2} - f = 0, \qquad (5.3.37)$$

$$-\frac{\partial N_{xz}}{\partial x} - \frac{1}{2}\frac{\partial^2 P_{xy}}{\partial x^2} - \frac{\partial}{\partial x}\left(N_{xx}\frac{\partial w}{\partial x}\right) + c_f\,w + m_0\frac{\partial^2 w}{\partial t^2} - q = 0, \qquad (5.3.38)$$

$$-\frac{\partial M_{xx}}{\partial x} - \frac{1}{2}\frac{\partial P_{xy}}{\partial x} + N_{xz} + m_2\frac{\partial^2 \phi_x}{\partial t^2} + m_1\frac{\partial^2 u}{\partial t^2} = 0, \qquad (5.3.39)$$

where the effect of the couple stress through P_{xy} as well as the elastic foundation term $c_f w$ are included.

The stress resultants N_{xx}, N_{xz}, and M_{xx} appearing in Eqs. (5.3.37)–(5.3.39) for the LBT are defined [see Eqs. (5.3.20a)–(5.3.20j)] by

$$N_{xx} = A_{xx} \left[\frac{\partial u}{\partial x} + \frac{1}{2} \left(\frac{\partial w}{\partial x} \right)^2 \right] + \bar{B}_{xx} \frac{\partial \phi_x}{\partial x} - \alpha E_{xx} \frac{\partial^2 w}{\partial x^2} - N_{xx}^T, \qquad (5.3.40a)$$

$$M_{xx} = B_{xx} \left[\frac{\partial u}{\partial x} + \frac{1}{2} \left(\frac{\partial w}{\partial x} \right)^2 \right] + \bar{D}_{xx} \frac{\partial \phi_x}{\partial x} - \alpha F_{xx} \frac{\partial^2 w}{\partial x^2} - M_{xx}^T, \qquad (5.3.40b)$$

$$N_{xz} = \bar{A}_{xz} \left(\phi_x + \frac{\partial w}{\partial x} \right), \qquad (5.3.40c)$$

$$P_{xy} = \int_A m_{xy} \, dA = \frac{\ell^2}{2(1+\nu)} \int_A 2E(z) \chi_{xy} \, dA$$
$$= \frac{1}{2} A_{xy} \left(\frac{\partial \phi_x}{\partial x} - \frac{\partial^2 w}{\partial x^2} \right) - \frac{1}{2} \beta D_{xy} \left(\frac{\partial \phi_x}{\partial x} + \frac{\partial^2 w}{\partial x^2} \right). \qquad (5.3.40d)$$

The duality pairs for this theory are the same as those for the TBT (because LBT is not derived using energy considerations):

$$(u, N_{xx}), \quad (\phi_x, M_{xx}), \quad \left(w, N_{xz} + N_{xx} \frac{\partial w}{\partial x} \right). \qquad (5.3.40e)$$

The beam stiffness coefficients A_{xx}, B_{xx}, D_{xx}, E_{xx}, F_{xx}, $\bar{A}_{xz}$, and A_{xy} are the same as those defined in Eqs. (5.3.21a)–(5.3.21d) and (5.3.22). We note that *the equations of motion of the LBT and their special cases do not have variational principles from which they can be derived.* Whether such a variational consistency is required to obtain a reasonable representation of the deformation and stress state in a beam structure is a separate matter.

5.3.3.2   Equations of motion in terms of the displacements

The equations of motion of the generalized LBT (valid for FGM beams with the von Kármán nonlinearity and couples stress effect) are given by

$$-\frac{\partial}{\partial x} \left\{ A_{xx} \left[\frac{\partial u}{\partial x} + \frac{1}{2} \left(\frac{\partial w}{\partial x} \right)^2 \right] + \bar{B}_{xx} \frac{\partial \phi_x}{\partial x} - \alpha E_{xx} \frac{\partial^2 w}{\partial x^2} - N_{xx}^T \right\}$$
$$+ m_0 \frac{\partial^2 u}{\partial t^2} + m_1 \frac{\partial^2 \phi_x}{\partial t^2} - f = 0, \qquad (5.3.41)$$

$$-\frac{1}{4} A_{xy} \frac{\partial^2}{\partial x^2} \left(\frac{\partial \phi_x}{\partial x} - \frac{\partial^2 w}{\partial x^2} \right) + \frac{1}{4} \beta D_{xy} \frac{\partial^2}{\partial x^2} \left(\frac{\partial \phi_x}{\partial x} + \frac{\partial^2 w}{\partial x^2} \right)$$
$$-\frac{\partial}{\partial x} \left[\frac{\partial w}{\partial x} \left\{ A_{xx} \left[\frac{\partial u}{\partial x} + \frac{1}{2} \left(\frac{\partial w}{\partial x} \right)^2 \right] + \bar{B}_{xx} \frac{\partial \phi_x}{\partial x} - \alpha E_{xx} \frac{\partial^2 w}{\partial x^2} - N_{xx}^T \right\} \right]$$
$$-\bar{A}_{xz} \frac{\partial}{\partial x} \left(\phi_x + \frac{\partial w}{\partial x} \right) + c_f w + m_0 \frac{\partial^2 w}{\partial t^2} - q = 0, \qquad (5.3.42)$$

$$-\frac{\partial}{\partial x}\left\{B_{xx}\left[\frac{\partial u}{\partial x}+\frac{1}{2}\left(\frac{\partial w}{\partial x}\right)^2\right]+\bar{D}_{xx}\frac{\partial \phi_x}{\partial x}-\alpha F_{xx}\frac{\partial^2 w}{\partial x^2}-M_{xx}^T\right\}$$

$$-\frac{1}{4}A_{xy}\frac{\partial}{\partial x}\left(\frac{\partial \phi_x}{\partial x}-\frac{\partial^2 w}{\partial x^2}\right)+\frac{1}{4}\beta D_{xy}\frac{\partial}{\partial x}\left(\frac{\partial \phi_x}{\partial x}+\frac{\partial^2 w}{\partial x^2}\right)$$

$$+\bar{A}_{xz}\left(\phi_x+\frac{\partial w}{\partial x}\right)+m_1\frac{\partial^2 u}{\partial t^2}+m_2\frac{\partial^2 \phi_x}{\partial t^2}=0. \qquad (5.3.43)$$

We note that these equations *cannot* be derived using the principle of minimum total potential energy because the equations of motion of the TBT in terms of the new stress resultants were adopted in the LBT.

5.3.3.3 Equations of motion for the linear case

The linearized equations of motion of the LBT (valid for FGM beams with the couple stress effect) are obtained from Eqs. (5.3.41)–(5.3.43) by setting the nonlinear terms to zero:

$$-\frac{\partial}{\partial x}\left(A_{xx}\frac{\partial u}{\partial x}+\bar{B}_{xx}\frac{\partial \phi_x}{\partial x}-\alpha E_{xx}\frac{\partial^2 w}{\partial x^2}-N_{xx}^T\right)$$

$$+m_0\frac{\partial^2 u}{\partial t^2}+m_1\frac{\partial^2 \phi_x}{\partial t^2}=f, \qquad (5.3.44)$$

$$-\frac{1}{4}A_{xy}\frac{\partial^2}{\partial x^2}\left(\frac{\partial \phi_x}{\partial x}-\frac{\partial^2 w}{\partial x^2}\right)+\frac{1}{4}\beta D_{xy}\frac{\partial^2}{\partial x^2}\left(\frac{\partial \phi_x}{\partial x}+\frac{\partial^2 w}{\partial x^2}\right)$$

$$-\bar{A}_{xz}\frac{\partial}{\partial x}\left(\phi_x+\frac{\partial w}{\partial x}\right)+c_f w+m_0\frac{\partial^2 w}{\partial t^2}=q, \qquad (5.3.45)$$

$$-\frac{\partial}{\partial x}\left(B_{xx}\frac{\partial u}{\partial x}+\bar{D}_{xx}\frac{\partial \phi_x}{\partial x}-\alpha F_{xx}\frac{\partial^2 w}{\partial x^2}-M_{xx}^T\right)$$

$$-\frac{1}{4}A_{xy}\frac{\partial}{\partial x}\left(\frac{\partial \phi_x}{\partial x}-\frac{\partial^2 w}{\partial x^2}\right)+\frac{1}{4}\beta D_{xy}\frac{\partial}{\partial x}\left(\frac{\partial \phi_x}{\partial x}+\frac{\partial^2 w}{\partial x^2}\right)$$

$$+\bar{A}_{xz}\left(\phi_x+\frac{\partial w}{\partial x}\right)+m_1\frac{\partial^2 u}{\partial t^2}+m_2\frac{\partial^2 \phi_x}{\partial t^2}=0. \qquad (5.3.46)$$

5.3.3.4 Equations of equilibrium for the linear case

The linearized equations of equilibrium for the LBT that are valid for FGM beams with the couple stress effect are obtained from Eqs. (5.3.44)–(5.3.46) by setting the time derivative terms to zero:

$$-\frac{d}{dx}\left(A_{xx}\frac{du}{dx}+\bar{B}_{xx}\frac{d\phi_x}{dx}-\alpha E_{xx}\frac{d^2 w}{dx^2}-N_{xx}^T\right)-f=0, \qquad (5.3.47)$$

$$-\tfrac{1}{4}A_{xy}\frac{d^2}{dx^2}\left(\frac{d\phi_x}{dx}-\frac{d^2w}{dx^2}\right)+\tfrac{1}{4}\beta D_{xy}\frac{d}{dx}\left(\frac{d\phi_x}{dx}+\frac{d^2w}{dx^2}\right)$$

$$-\bar{A}_{xz}\frac{d}{dx}\left(\phi_x+\frac{dw}{dx}\right)+c_f\,w-q=0, \qquad (5.3.48)$$

$$-\frac{d}{dx}\left(B_{xx}\frac{du}{dx}+\bar{D}_{xx}\frac{d\phi_x}{dx}-\alpha F_{xx}\frac{d^2w}{dx^2}-M_{xx}^T\right)$$

$$-\tfrac{1}{4}A_{xy}\frac{d}{dx}\left(\frac{d\phi_x}{dx}-\frac{d^2w}{dx^2}\right)+\tfrac{1}{4}\beta D_{xy}\frac{d^2}{dx^2}\left(\frac{d\phi_x}{dx}+\frac{d^2w}{dx^2}\right)$$

$$+\bar{A}_{xz}\left(\phi_x+\frac{dw}{dx}\right)=0. \qquad (5.3.49)$$

We note that the equations of equilibrium contain the second-order derivatives of u, the fourth-order derivatives of w, and the third-order derivatives of ϕ_x.

5.3.3.5 Linearized equations without the couple stress effect

The linearized equations of equilibrium without the couple stress effect are

$$-\frac{d}{dx}\left(A_{xx}\frac{du}{dx}+\bar{B}_{xx}\frac{d\phi_x}{dx}-\alpha E_{xx}\frac{d^2w}{dx^2}-N_{xx}^T\right)-f=0, \qquad (5.3.50)$$

$$-\bar{A}_{xz}\frac{d}{dx}\left(\phi_x+\frac{dw}{dx}\right)+c_f\,w-q=0, \qquad (5.3.51)$$

$$-\frac{d}{dx}\left(B_{xx}\frac{du}{dx}+\bar{D}_{xx}\frac{d\phi_x}{dx}-\alpha F_{xx}\frac{d^2w}{dx^2}-M_{xx}^T\right)$$

$$+\bar{A}_{xz}\left(\phi_x+\frac{dw}{dx}\right)=0. \qquad (5.3.52)$$

In this case, the equations of equilibrium contain the second-order derivatives of u, the third-order derivatives of w, and the second-order derivatives of ϕ_x.

5.4 EXACT SOLUTIONS FOR BENDING

5.4.1 THE REDDY BEAM THEORY

In this section, we (try to) develop exact solutions to the *linear* equations of equilibrium of the RBT for FGM beams without the effect of the modified couple stress, namely, Eqs. (5.3.30)–(5.3.33). As we shall see shortly, it is not possible to determine analytical solutions of the RBT without having to solve additional second-order differential equation in terms of the transverse deflection. Due to this complication, we shall look at a simplified RBT, which admits exact solution of the governing equations. The simplified RBT, as described in the sequel, is a third-order beam theory in which certain higher-order terms are omitted.

First, the equations of equilibrium in terms of the stress resultants can be obtained from Eqs. (5.3.15)–(5.3.17) by omitting the nonlinear and couple stress terms and setting $c_f = 0$:

$$-\frac{dN_{xx}}{dx} = 0, \tag{5.4.1}$$

$$-\frac{d\bar{N}_{xz}}{dx} - \alpha\frac{d^2 P_{xx}}{dx^2} = q, \tag{5.4.2}$$

$$-\frac{d\bar{M}_{xx}}{dx} + \bar{N}_{xz} = 0. \tag{5.4.3}$$

Integrating the first two equations with respect to x, we obtain

$$N_{xx} = K_1, \tag{5.4.4}$$

$$\bar{N}_{xz} + \alpha\frac{dP_{xx}}{dx} = -\int q(x)\, dx + K_2, \tag{5.4.5}$$

where K_1 and K_2 are constants of integration. Using Eq. (5.4.5) in Eq. (5.4.3) and integrating, we arrive at

$$M_{xx} = -\int^x \int^\xi q(\eta)\, d\eta\, d\xi + K_2 x + K_3 \equiv F(x). \tag{5.4.6}$$

The dummy variables of integration, ξ and η, are introduced only to indicate the sequence of integration; otherwise, one may simply use x. For example, when we have $q(x) = q_0(x/L)$, the expression for $F(x)$ becomes

$$F(x) = -\int^x \int^\xi q(\eta)\, d\eta\, d\xi + K_2 x + K_3 = -\int^x \int^\xi \frac{q_0}{L}\eta\, d\eta\, d\xi + K_2 x + K_3$$

$$= -\int^x \frac{q_0}{2L}\xi^2\, d\xi + K_2 x + K_3 = -\frac{q_0 x^3}{6L} + K_2 x + K_3.$$

Now expressing N_{xx} and M_{xx} in Eqs. (5.4.4) and (5.4.6) in terms of the generalized displacements [see Eqs. (5.3.20a) and (5.3.20b)], we have

$$A_{xx}\frac{du}{dx} + \bar{B}_{xx}\frac{d\phi_x}{dx} = \alpha E_{xx}\frac{d^2 w}{dx^2} + K_1 + N_{xx}^T, \tag{5.4.7}$$

$$\bar{B}_{xx}\frac{du}{dx} + \bar{D}_{xx}\frac{d\phi_x}{dx} = \alpha F_{xx}\frac{d^2 w}{dx^2} + F(x) + M_{xx}^T. \tag{5.4.8}$$

Solving for du/dx and $d\phi_x/dx$ in terms of $d^2 w/dx^2$, we obtain

$$\frac{du}{dx} = \frac{P_1}{\bar{D}_{xx}^*} + \frac{\bar{D}_{xx}}{\bar{D}_{xx}^*}K_1 - \frac{\bar{B}_{xx}}{\bar{D}_{xx}^*}F(x) + \frac{J_1}{\bar{D}_{xx}^*}\frac{d^2 w}{dx^2}, \tag{5.4.9a}$$

$$\frac{d\phi_x}{dx} = \frac{P_2}{\bar{D}_{xx}^*} - \frac{\bar{B}_{xx}}{\bar{D}_{xx}^*}K_1 + \frac{A_{xx}}{\bar{D}_{xx}^*}F(x) + \frac{J_2}{\bar{D}_{xx}^*}\frac{d^2 w}{dx^2}, \tag{5.4.9b}$$

where [see Eq. (5.3.22)]

$$
\begin{aligned}
&P_1 = \bar{D}_{xx} N_{xx}^T - \bar{B}_{xx} M_{xx}^T, \quad P_2 = A_{xx} M_{xx}^T - B_{xx} N_{xx}^T, \\
&J_1 = \alpha \left(\bar{D}_{xx} E_{xx} - \bar{B}_{xx} F_{xx} \right) = \alpha \left(D_{xx} E_{xx} - B_{xx} F_{xx} \right), \\
&J_2 = \alpha \left(A_{xx} F_{xx} - B_{xx} E_{xx} \right), \\
&\bar{D}_{xx}^* = A_{xx} \bar{D}_{xx} - B_{xx} \bar{B}_{xx} \\
&\qquad = D_{xx}^* - \alpha \left(A_{xx} F_{xx} - B_{xx} E_{xx} \right) = D_{xx}^* - J_2 \, \bar{D}_{xx}^*, \\
&D_{xx}^* = A_{xx} D_{xx} - B_{xx} B_{xx}.
\end{aligned}
\tag{5.4.9c}
$$

Integrating the two equations in (5.4.9a) and (5.4.9b), we obtain

$$
\bar{D}_{xx}^* \, u(x) = -\bar{B}_{xx} \left(-\int^x \int^\xi \int^\eta q(\zeta)\, d\zeta\, d\eta\, d\xi + K_2 \frac{x^2}{2} + K_3 x + K_4 \right)
$$

$$
+ J_1 \frac{dw}{dx} + \left(P_1 + \bar{D}_{xx} K_1 \right) x, \tag{5.4.10}
$$

$$
\bar{D}_{xx}^* \, \phi_x(x) = A_{xx} \left(-\int^x \int^\xi \int^\eta q(\zeta)\, d\zeta\, d\eta\, d\xi + K_2 \frac{x^2}{2} + K_3 x + K_5 \right)
$$

$$
+ J_2 \frac{dw}{dx} + \left(P_2 - B_{xx} K_1 \right) x, \tag{5.4.11}
$$

where K_4 and K_5 are the constants of integration.

Now we return to Eq. (5.4.5) and write it in terms of the generalized displacements [the differential of the constant part involving P_1, P_2, and K_1 is set to zero; P_{xx} is defined in terms of the displacements in Eq. (5.3.20c)]:

$$
0 = -\hat{A}_{xz} \left(\phi_x + \frac{dw}{dx} \right) + \left(-\int q(x)\, dx + K_2 \right)
$$

$$
- \alpha \frac{d}{dx} \left(E_{xx} \frac{du}{dx} + \bar{F}_{xx} \frac{d\phi_x}{dx} - \alpha H_{xx} \frac{d^2 w}{dx^2} \right)
$$

$$
= -\frac{\hat{A}_{xz}}{\bar{D}_{xx}^*} \left[A_{xx} \left(-\int^x \int^\xi \int^\eta q(\zeta)\, d\zeta\, d\eta\, d\xi + K_2 \frac{x^2}{2} + K_3 x + K_5 \right) \right.
$$

$$
\left. + J_2 \frac{dw}{dx} + \left(P_2 - B_{xx} K_1 \right) x + \bar{D}_{xx}^* \frac{dw}{dx} \right]
$$

$$
- \frac{1}{\bar{D}_{xx}^*} \frac{d}{dx} \left[P_3 F(x) + \alpha \left(E_{xx} J_1 + \bar{F}_{xx} J_2 - \alpha H_{xx} \bar{D}_{xx}^* \right) \frac{d^2 w}{dx^2} \right]
$$

$$
+ \left(-\int q(x)\, dx + K_2 \right), \tag{5.4.12a}
$$

where

$$
P_3 = \alpha \left(A_{xx} \bar{F}_{xx} - \bar{B}_{xx} E_{xx} \right), \quad \hat{D}_{xx}^* = \bar{D}_{xx}^* - P_3 = A_{xx} \hat{D}_{xx} - \bar{B}_{xx} \bar{B}_{xx}.
\tag{5.4.12b}
$$

Integrating Eq. (5.4.12a) once and collecting the like terms together, we obtain a second-order differential equation for $w(x)$:

$$
\begin{aligned}
0 = & -\frac{\hat{A}_{xz}}{\bar{D}^*_{xx}} \left(\bar{D}^*_{xx} + J_2\right) w + \frac{\alpha}{\bar{D}^*_{xx}} \left(\alpha \bar{D}^*_{xx} H_{xx} - E_{xx} J_1 - \bar{F}_{xx} J_2\right) \frac{d^2 w}{dx^2} \\
& -\frac{\hat{A}_{xz}}{\bar{D}^*_{xx}} \left[A_{xx} \left(-\int^x \int^\xi \int^\eta \int^\zeta q(\mu)\, d\mu\, d\zeta\, d\eta\, d\xi \right. \right. \\
& \left. \left. + K_2 \frac{x^3}{6} + K_3 \frac{x^2}{2} + K_5\, x + K_6 \right) + (P_2 - B_{xx} K_1) \frac{x^2}{2} \right] \\
& +\frac{\hat{D}^*_{xx}}{\bar{D}^*_{xx}} \left(-\int^x \int^\xi q(\eta)\, d\eta\, d\xi + K_2 x \right) - \frac{P_3}{\bar{D}^*_{xx}} K_3 \\
= & -c_1\, w + c_2 \frac{d^2 w}{dx^2} + g(x),
\end{aligned}
\tag{5.4.13}
$$

where

$$
\begin{aligned}
c_1 &= \frac{\hat{A}_{xz}}{\bar{D}^*_{xx}} (\bar{D}^*_{xx} + J_2) = \frac{\hat{A}_{xz} D^*_{xx}}{\bar{D}^*_{xx}}, \\
c_2 &= \frac{\alpha}{\bar{D}^*_{xx}} \left(\alpha \bar{D}^*_{xx} H_{xx} - E_{xx} J_1 - \bar{F}_{xx} J_2\right),
\end{aligned}
\tag{5.4.14a}
$$

and

$$
\begin{aligned}
g(x) = & -\frac{\hat{A}_{xz}}{\bar{D}^*_{xx}} \left[A_{xx} \left(-\int^x \int^\xi \int^\eta \int^\zeta q(\mu)\, d\mu\, d\zeta\, d\eta\, d\xi + K_2 \frac{x^3}{6} + K_3 \frac{x^2}{2} \right. \right. \\
& \left. \left. + K_5\, x + K_6 \right) + (P_2 - B_{xx} K_1) \frac{x^2}{2} \right] \\
& +\frac{\hat{D}^*_{xx}}{\bar{D}^*_{xx}} \left(-\int^x \int^\xi q(\eta)\, d\eta\, d\xi + K_2 x \right) - \frac{P_3}{\bar{D}^*_{xx}} K_3.
\end{aligned}
\tag{5.4.14b}
$$

It is clear from Eq. (5.4.13) that the analytical solution to the RBT is not algebraic but hyperbolic (because $\alpha, \beta > 0$). The homogeneous solution of Eq. (5.4.13) is

$$
w_h(x) = K_7 \cosh \mu x + K_8 \sinh \mu x, \quad \mu = \sqrt{\frac{c_1}{c_2}}.
\tag{5.4.15}
$$

The total solution $w(x)$ is obtained by adding the particular solution, $w_p(x)$ due to $g(x)$:

$$
w(x) = w_h(x) + w_p(x).
$$

Also, there are eight constants of integration, including the two constants of integration introduced in Eq. (5.4.15). In the RBT, one is required specify $-dw/dx$ in addition to ϕ_x (or their dual variables, P_{xx} and $\bar{M}_{xx}$, respectively), providing the required eight boundary conditions. In the next section, we consider a simplified Reddy beam theory that avoids the solution of a second-order differential equation.

5.4.2   THE SIMPLIFIED RBT

5.4.2.1   FGM beams

The second-order derivative term appearing in Eq. (5.4.13) comes from P_{xx} of Eq. (5.4.2). If we neglect the second-order derivative of w in Eq. (5.4.13) (i.e., set $c_2 = 0$) by reasoning that it is very small compared to the other terms in the equation, we obtain the governing equations of the simplified RBT. This simplification is not necessary if one is to obtain the solution by a numerical method (e.g., the finite element method). It is only in the rest of finding an analytical solution, we simplify Eq. (5.4.13).

Setting $c_2 = 0$ and solving Eq. (5.4.13) for w, we obtain

$$\frac{\hat{A}_{xz}D^*_{xx}}{\bar{D}^*_{xx}} w(x) = -\frac{\hat{A}_{xz}}{\bar{D}^*_{xx}}\left[A_{xx}\left(-\int^x\int^\xi\int^\eta\int^\zeta q(\mu)\,d\mu\,d\zeta\,d\eta\,d\xi\right.\right.$$

$$\left.+ K_2\frac{x^3}{6} + K_3\frac{x^2}{2} + K_5\,x + K_6\right)$$

$$\left.+ \left(A_{xx}M^T_{xx} - B_{xx}N^T_{xx} - B_{xx}K_1\right)\frac{x^2}{2}\right]$$

$$+ \frac{\hat{D}^*_{xx}}{\bar{D}^*_{xx}}\left(-\int^x\int^\xi q(\eta)\,d\eta\,d\xi + K_2 x\right) - \frac{P_3}{\bar{D}^*_{xx}}K_3. \qquad (5.4.16)$$

Upon simplifying Eq. (5.4.16), we arrive at the following expression for the displacement $w(x)$:

$$w(x) = -\frac{A_{xx}}{D^*_{xx}}\left(-\int^x\int^\xi\int^\eta\int^\zeta q(\mu)\,d\mu\,d\zeta\,d\eta\,d\xi\right.$$

$$\left.+ K_2\frac{x^3}{6} + K_3\frac{x^2}{2} + K_5\,x + K_6\right)$$

$$- \frac{1}{D^*_{xx}}\left(A_{xx}M^T_{xx} - B_{xx}N^T_{xx} - B_{xx}K_1\right)\frac{x^2}{2} - \frac{P_3}{\hat{A}_{xz}D^*_{xx}}K_3$$

$$+ \frac{\hat{D}^*_{xx}}{\hat{A}_{xz}D^*_{xx}}\left(-\int^x\int^\xi q(\eta)\,d\eta\,d\xi + K_2\,x\right). \qquad (5.4.17)$$

There are six constants, K_1 through K_6 [K_4 appears in the expression for $u(x)$], which are to be determined using the boundary conditions of the beam problem being studied.

Toward computing $u(x)$ and $\phi_x(x)$, we need dw/dx [see Eqs. (5.4.10) and (5.4.11)]:

$$\frac{dw}{dx} = -\frac{A_{xx}}{D_{xx}^*}\left(-\int^x\int^\xi\int^\eta q(\zeta)\,d\zeta\,d\eta\,d\xi + K_2\frac{x^2}{2} + K_3\,x + K_5\right)$$
$$-\frac{1}{D_{xx}^*}\left(A_{xx}M_{xx}^T - B_{xx}N_{xx}^T - B_{xx}K_1\right)x$$
$$+\frac{\hat{D}_{xx}^*}{\hat{A}_{xz}D_{xx}^*}\left(-\int^x q(\xi)\,d\xi + K_2\right),\tag{5.4.18}$$

Substituting this expression into Eqs. (5.4.10) and (5.4.11), we obtain

$$u(x) = -\frac{B_{xx}}{D_{xx}^*}\left(-\int^x\int^\xi\int^\eta q(\zeta)\,d\zeta\,d\xi d\eta + K_2\frac{x^2}{2} + K_3 x\right)$$
$$+\frac{1}{D_{xx}^*}\left(D_{xx}N_{xx}^T - B_{xx}M_{xx}^T + D_{xx}K_1\right)x$$
$$+\frac{J_1}{\bar{D}_{xx}^*}\frac{\hat{D}_{xx}^*}{\hat{A}_{xz}D_{xx}^*}\left(-\int^x q(\xi)\,d\xi + K_2\right)$$
$$-\frac{\bar{B}_{xx}}{\bar{D}_{xx}^*}K_4 - \frac{J_1}{\bar{D}_{xx}^*}\frac{A_{xx}}{D_{xx}^*}K_5,\tag{5.4.19}$$

$$\phi_x(x) = \frac{A_{xx}}{D_{xx}^*}\left(-\int^x\int^\xi\int^\eta q(\zeta)\,d\zeta\,d\xi d\eta + K_2\frac{x^2}{2} + K_3x + K_5\right)$$
$$+\frac{1}{D_{xx}^*}\left(A_{xx}M_{xx}^T - B_{xx}N_{xx}^T - B_{xx}K_1\right)x - \frac{B_{xx}}{D_{xx}^*}K_1x$$
$$+\frac{J_2\hat{D}_{xx}^*}{D_{xx}^*\hat{A}_{xz}\bar{D}_{xx}^*}\left(-\int^x q(\xi)\,d\xi + K_2\right),\tag{5.4.20}$$

where we have used the following identities:

(a) $\bar{B}_{xx} + J_1\dfrac{A_{xx}}{D_{xx}^*} = \dfrac{B_{xx}\bar{D}_{xx}^*}{D_{xx}^*},$

(b) $\bar{D}_{xx} + J_1\dfrac{B_{xx}}{D_{xx}^*} = \dfrac{D_{xx}\bar{D}_{xx}^*}{D_{xx}^*},$

(c) $P_1 - \dfrac{J_1}{D_{xx}^*}P_2 = \dfrac{1}{D_{xx}^*}\left(D_{xx}\bar{D}_{xx}^*M_T^{(0)} - B_{xx}D_{xx}^*M_T^{(1)}\right),$ (5.4.21)

(d) $1 + J_2\dfrac{A_{xx}}{\bar{D}_{xx}^*} = D_{xx}^*,$

(e) $\bar{D}_{xx}^* + J_2 = D_{xx}^*.$

Various stiffness quantities appearing in Eqs. (5.4.17)–(5.4.21) are defined in Eqs. (5.3.22) and (5.4.9c).

5.4.2.2 Homogeneous beams

For an homogeneous isotropic beam ($B_{xx} = E_{xx} = \bar{B} = 0$) of rectangular cross-section, we have [see Eq. (5.3.22); $\alpha = 4/(3h^2)$ and $\beta = 4/h^2$]

$$\bar{F}_{xx} = \frac{4h^2}{35} D_{xx}, \quad \bar{D}_{xx} = \frac{4}{5} D_{xx}, \quad \bar{A}_{xz} = \frac{2}{3} A_{xz}, \quad \hat{A}_{xz} = \frac{8}{15} A_{xz},$$

$$\hat{D}_{xx} = \frac{68}{105} D_{xx}, \quad D^*_{xx} = A_{xx} D_{xx}, \quad \bar{D}^*_{xx} = \frac{4}{5} A_{xx} D_{xx}, \quad \hat{D}^*_{xx} = \frac{68}{105} A_{xx} D_{xx},$$

$$P_3 = \alpha A_{xx} \bar{F}_{xx} = \frac{16}{105} A_{xx} D_{xx}, \quad \bar{D}^*_{xx} - P_3 = \frac{17}{315} E^2 b^2 h^4,$$

$$c_1 = \frac{\hat{A}_{xz} D_{xx}}{\bar{D}_{xx}} = \frac{2}{3} A_{xz}, \quad c_2 = \alpha^2 \left(D_{xx} H_{xx} - F_{xx} F_{xx} \right) = \frac{1}{18900} E^2 b^2 h^8.$$

$$(5.4.22)$$

It is clear that c_2 is considerably smaller than c_1, and one may justifiably neglect it in Eq. (5.4.13).

For the present case, the solutions become ($P_3 = \alpha A_{xx} \bar{F}_{xx} = A_{xx}(4Ebh^3/315)$, $D^*_{xx} = A_{xx} D_{xx}$, $\bar{D}^*_{xx} = A_{xx} \bar{D}_{xx} = A_{xx}(Ebh^3/15)$, $\hat{D}^*_{xx} = A_{xx} \hat{D}_{xx} = 17E^2 b^2 h^4/315$, and $\hat{A}_{xz} = 8A_{xz}/15$)

$$u(x) = \frac{1}{A_{xx}} \left(N^T_{xx} + K_1 \right) x, \tag{5.4.23}$$

$$\phi_x(x) = \frac{1}{D_{xx}} \left(-\int^x \int^\xi \int^\eta q(\zeta)\, d\zeta\, d\eta\, d\xi + K_2 \frac{x^2}{2} + K_3 x \right)$$

$$+ \frac{17}{56} \frac{1}{A_{xz}} \left(-\int^x q(\xi)\, d\xi + K_2 \right) + \frac{1}{D_{xx}} M^T_{xx} x + \frac{1}{D_{xx}} K_5, \tag{5.4.24}$$

$$w(x) = -\frac{1}{D_{xx}} \left(-\int^x \int^\xi \int^\eta \int^\zeta q(\mu)\, d\mu\, d\zeta\, d\eta\, d\xi + K_2 \frac{x^3}{6} + K_3 \frac{x^2}{2} \right.$$

$$\left. + K_5 x \right) - \frac{1}{D_{xx}} M^T_{xx} \frac{x^2}{2} - \frac{1}{D_{xx}} K_6 - \frac{2}{7} \frac{1}{A_{xz}} K_3$$

$$+ \frac{17}{14} \frac{1}{A_{xz}} \left(-\int^x \int^\xi q(\eta)\, d\eta\, d\xi + K_2 x \right), \tag{5.4.25}$$

$$-\frac{dw}{dx} = \frac{1}{D_{xx}} \left(-\int^x \int^\xi \int^\eta q(\zeta)\, d\zeta\, d\eta\, d\xi + K_2 \frac{x^2}{2} + K_3 x \right)$$

$$- \frac{17}{14} \frac{1}{A_{xz}} \left(-\int^x q(\xi)\, d\xi + K_2 \right) + \frac{1}{D_{xx}} M^T_{xx} x + \frac{1}{D_{xx}} K_5. \tag{5.4.26}$$

The corresponding solutions obtained using the TBT are [see Eqs. (4.6.12)–(4.6.16); the constants c_i are replaced with K_i for direct comparison with the RBT solutions presented above; $K_s A_{xz}$ from the TBT is replaced with A_{xz}

in the RBT]

$$u(x) = \frac{1}{A_{xx}} \left(N_{xx}^T + K_1 \right) x, \tag{5.4.27}$$

$$\phi_x(x) = \frac{1}{D_{xx}} \left(-\int^x \int^\xi \int^\eta q(\zeta)\, d\zeta\, d\eta d\xi + K_2 \frac{x^2}{2} + K_3\, x \right)$$
$$+ \frac{1}{D_{xx}} M_{xx}^T x + \frac{1}{D_{xx}} K_5, \tag{5.4.28}$$

$$w(x) = -\frac{1}{D_{xx}} \left(-\int^x \int^\xi \int^\eta \int^\zeta q(\mu)\, d\mu\, d\zeta\, d\eta d\xi + K_2 \frac{x^3}{6} \right.$$
$$\left. + K_3 \frac{x^2}{2} + K_5\, x \right) - \frac{1}{D_{xx}} M_{xx}^T \frac{x^2}{2} - \frac{1}{D_{xx}} K_6$$
$$+ \frac{1}{A_{xz}} \left(-\int^x \int^\xi q(\eta)\, d\eta\, d\xi + K_2\, x \right), \tag{5.4.29}$$

$$-\frac{dw}{dx} = \frac{1}{D_{xx}} \left(-\int^x \int^\xi \int^\eta q(\zeta)\, d\zeta\, d\eta d\xi + K_2 \frac{x^2}{2} + K_3\, x \right)$$
$$- \frac{1}{A_{xz}} \left(-\int^x q(\xi)\, d\xi + K_2 \right) + \frac{1}{D_{xx}} M_{xx}^T x + \frac{1}{D_{xx}} K_5. \tag{5.4.30}$$

A comparison of Eq. (5.4.29) with Eq. (5.4.25) shows that $A_{xz} = K_s A_{xz}$ the shear correction coefficient predicted by the "simplified" RBT is $K_s = 14/17$, which is slightly smaller than the value suggested for rectangular cross-section beams, which is $K_s = 5/6$. Of course, the TBT solution is different from the RBT solution, especially the expression for ϕ_x in the RBT contains a term due to transverse shear coefficient.

5.4.3   THE LEVINSON BEAM THEORY

5.4.3.1   FGM beams

Here, we develop the exact solutions of the LBT, without accounting for the couple stress effect. Much of the development from the RBT presented in the preceding section is valid here with some modifications. The equations of equilibrium of the LBT in terms of the stress resultants N_{xx}, N_{xz}, and M_{xx} are:

$$-\frac{dN_{xx}}{dx} = 0, \quad -\frac{dN_{xz}}{dx} = q, \quad -\frac{dM_{xx}}{dx} + N_{xz} = 0. \tag{5.4.31}$$

Integrating the first two equations in Eq. (5.4.31) with respect to x, we obtain

$$N_{xx} = K_1, \quad N_{xz} = -\int q(x)\, dx + K_2, \tag{5.4.32}$$

where K_1 and K_2 are constants of integration. Using the second equation in the third equation of Eq. (5.4.31) and integrating, we arrive at

$$M_{xx} = -\int^x \int^\xi q(\eta)\, d\xi d\eta + K_2 x + K_3 \equiv F(x). \tag{5.4.33}$$

Equations (5.4.7)–(5.4.11) are valid here. In particular, $u(x)$ and $\phi_x(x)$ are given by

$$\bar{D}^*_{xx}\, u(x) = -\bar{B}_{xx}\left(-\int^x \int^\xi \int^\eta q(\zeta)\, d\zeta\, d\xi d\eta + K_2\frac{x^2}{2} + K_3\, x + K_4\right)$$

$$+ J_1\frac{dw}{dx} + P_1\, x, +\bar{D}_{xx}K_1\, x, \tag{5.4.34}$$

$$\bar{D}^*_{xx}\, \phi_x(x) = \frac{A_{xx}}{\bar{D}^*_{xx}}\left(-\int^x \int^\xi \int^\eta q(\zeta)\, d\zeta\, d\xi d\eta + K_2\frac{x^2}{2} + K_3\, x + K_5\right)$$

$$+ J_2\frac{dw}{dx} + P_2\, x - B_{xx}K_1\, x, \tag{5.4.35}$$

where the expressions for P_1, P_2, J_1, and J_2 are given in Eq. (5.4.9c). Writing the second equation in Eq. (5.4.32) in terms of the generalized displacements, we obtain

$$0 = -\bar{A}_{xz}\left(\phi_x + \frac{dw}{dx}\right) + \left(-\int q(x)\, dx + K_2\right)$$

$$= -\frac{\bar{A}_{xz}}{\bar{D}^*_{xx}}\left[A_{xx}\left(-\int^x \int^\xi \int^\eta q(\zeta)\, d\zeta\, d\eta\, d\xi + K_2\frac{x^2}{2} + K_3x + K_5\right)\right.$$

$$\left. + J_2\frac{dw}{dx} + P_2\, x + \bar{D}^*_{xx}\frac{dw}{dx}\right] + \left(-\int q(x)\, dx + K_2\right). \tag{5.4.36}$$

It is clear from Eq. (5.4.36) that the analytical solution to the LBT does not require solving a second-order differential equation to obtain $w(x)$. The six constants of integration require the conventional six boundary conditions, as in the case of CBT and TBT.

Solving Eq. (5.4.36) for dw/dx

$$\frac{dw}{dx} = -\frac{A_{xx}}{D^*_{xx}}\left(-\int^x \int^\xi \int^\eta q(\zeta)\, d\zeta\, d\eta\, d\xi + K_2\frac{x^2}{2} + K_3\, x + K_5\right)$$

$$- \frac{1}{D^*_{xx}}\left(A_{xx}\, M^T_{xx} - B_{xx}N^T_{xx} - B_{xx}K_1\right)x$$

$$+ \frac{\bar{D}^*_{xx}}{D^*_{xx}\, \bar{A}_{xz}}\left(-\int q(x)\, dx + K_2\right), \tag{5.4.37}$$

and integrating once and simplifying the expressions, we obtain

$$
\begin{aligned}
w(x) = -\frac{A_{xx}}{D_{xx}^*}\Bigg(& -\int^x\int^\xi\int^\eta\int^\zeta q(\mu)\,d\mu\,d\zeta\,d\eta\,d\xi \\
& + K_2\frac{x^3}{6} + K_3\frac{x^2}{2} + K_5 x + K_6 \Bigg) \\
& -\frac{1}{D_{xx}^*}\left(A_{xx}M_{xx}^T - B_{xx}N_{xx}^T - B_{xx}K_1 \right)\frac{x^2}{2} \\
& +\frac{\bar{D}_{xx}^*}{D_{xx}^*\,\bar{A}_{xz}}\left(-\int^x\int^\xi q(\eta)\,d\eta\,d\xi + K_2\,x \right).
\end{aligned}
\tag{5.4.38}
$$

Substituting Eq. (5.4.38) into Eqs. (5.4.34) and (5.4.35), we obtain

$$
\begin{aligned}
u(x) = -\frac{B_{xx}}{D_{xx}^*}\Bigg(& -\int^x\int^\xi\int^\eta q(\zeta)\,d\zeta\,d\xi d\eta + K_2\frac{x^2}{2} + K_3 x \Bigg) \\
& +\frac{1}{D_{xx}^*}\left(D_{xx}N_{xx}^T - B_{xx}M_{xx}^T + D_{xx}K_1 \right)x - \frac{\bar{B}_{xx}}{D_{xx}^*}K_4 \\
& - J_1\frac{A_{xx}}{D_{xx}^*}K_5 + J_1\frac{1}{D_{xx}^*\,\bar{A}_{xz}}\left(-\int^x q(\xi)\,d\xi + K_2 \right),
\end{aligned}
\tag{5.4.39}
$$

$$
\begin{aligned}
\phi_x(x) = \frac{A_{xx}}{D_{xx}^*}\Bigg(& -\int^x\int^\xi\int^\eta q(\zeta)\,d\zeta\,d\eta d\xi + K_2\frac{x^2}{2} + K_3 x + K_5 \Bigg) \\
& +\frac{1}{D_{xx}^*}\left(A_{xx}M_{xx}^T - B_{xx}N_{xx}^T - B_{xx}K_1 \right)x \\
& +\frac{J_2}{D_{xx}^*\,\bar{A}_{xz}}\left(-\int^x q(\xi)\,d\xi + K_2 \right).
\end{aligned}
\tag{5.4.40}
$$

In Eq. (5.4.40), ζ, η, and ξ are dummy coordinates to indicate the sequence of integration. Also, various stiffness quantities (e.g., $\hat{D}_{xx}^*$, $\bar{D}_{xx}^*$, etc.) are defined in Eqs. (5.3.22) and (5.4.9c).

5.4.3.2   Homogeneous beams

For homogeneous beams, Eqs. (5.4.37)–(5.4.40) take the following simpler form:

$$
u(x) = \frac{1}{A_{xx}}\left(N_{xx}^T + K_1 \right)x,
\tag{5.4.41}
$$

$$
\begin{aligned}
\phi_x(x) = \frac{1}{D_{xx}}\Bigg(& -\int^x\int^\xi\int^\eta q(\zeta)\,d\zeta\,d\eta\,d\xi + K_2\frac{x^2}{2} + K_3 x \Bigg) \\
& +\frac{3}{10}\frac{1}{A_{xz}}\left(-\int^x q(\xi)\,d\xi + K_2 \right) + \frac{1}{D_{xx}}M_{xx}^T + \frac{1}{D_{xx}}K_5,
\end{aligned}
\tag{5.4.42}
$$

$$
w(x) = -\frac{1}{D_{xx}} \left(-\int^x \int^\xi \int^\eta \int^\zeta q(\mu)\, d\mu\, d\zeta\, d\eta\, d\xi \right.
$$

$$
\left. + K_2 \frac{x^3}{6} + K_3 \frac{x^2}{2} + K_5 x \right) - \frac{1}{D_{xx}} M_{xx}^T \frac{x^2}{2} - \frac{1}{D_{xx}} K_6
$$

$$
+ \frac{6}{5} \frac{1}{A_{xz}} \left(-\int^x \int^\xi q(\eta)\, d\eta\, d\xi + K_2\, x \right), \tag{5.4.43}
$$

$$
-\frac{dw}{dx} = \frac{1}{D_{xx}} \left(-\int^x \int^\xi \int^\eta q(\zeta)\, d\zeta\, d\eta\, d\xi + K_2 \frac{x^2}{2} + K_3\, x \right)
$$

$$
-\frac{6}{5} \frac{1}{A_{xz}} \left(-\int^x q(\xi)\, d\xi + K_2 \right) + \frac{1}{D_{xx}} M_{xx}^T\, x + \frac{1}{D_{xx}} K_5. \tag{5.4.44}
$$

From Eq. (5.4.43) it is clear that the LBT estimates the shear correction coefficient to be $K_s = 5/6$, which is the same as that often used in the TBT.

Next, we consider an example to illustrate the determination of the constants of integration using the boundary conditions.

Example 5.4.1 ———————————————————

Determine the exact solutions using the simplified RBT and LBT for a functionally graded beam with left end clamped, the right end subjected to an upward point load F_0 (i.e., a cantilevered beam with a point load at the free end). Take the origin of the x-coordinate at the left end of the beam.

Solution: We consider the simplified RBT and LBT, one at a time. In each case, we list the boundary conditions.

Simplified Reddy beam theory. For this case, the boundary conditions are [see the duality pairs in Eq. (5.3.14a); see, also, Eq. (5.4.5) with the P_{xx} term set to zero]:

$$
u(0) = w(0) = \phi_x(0) = 0; \quad N_{xx}(L) = \bar{M}_{xx}(L) = 0, \quad N_{xz}(L) = F_0, \tag{5.4.45}
$$

where $\alpha = 4/3h^2$, h being the beam height. The condition $N_{xx}(L) = 0$ results, using Eq. (5.4.4), in $K_1 = 0$. The shear force boundary condition at $x = L$ gives $K_2 = F_0$. The vanishing of the bending moment at $x = L$, namely, $\bar{M}_{xx}(L) = 0$, which is equivalent to $M_{xx}(L) = 0$ for the simplified RBT, gives [see Eq. (5.4.6)] $K_2 L + K_3 = 0$. Thus, we have

$$
K_1 = 0, \quad K_2 = F_0, \quad K_3 = -F_0 L. \tag{5.4.46}
$$

Next, using the boundary conditions on u, w, and ϕ_x at $x = 0$ in Eqs. (5.4.19), (5.4.17), and (5.4.20), respectively, we obtain the following relations among K_4, K_5, and K_6 (in terms of K_2 and K_3):

$$
u(0) = -\frac{\bar{B}_{xx}}{\bar{D}_{xx}^*} K_4 - \frac{J_1 A_{xx}}{D_{xx}^* \bar{D}_{xx}^*} K_5 + \frac{J_1}{\bar{D}_{xx}^*} \frac{\hat{D}_{xx}^*}{\hat{A}_{xz} D_{xx}^*} K_2 = 0,
$$

$$
\phi_x(0) = -\frac{A_{xx}}{D_{xx}^*} K_6 - \frac{P_3}{\hat{A}_{xz} D_{xx}^*} K_3 = 0, \tag{5.4.47}
$$

$$
w(0) = \frac{A_{xx}}{D_{xx}^*} K_5 + \frac{J_2 \hat{D}_{xx}^*}{D_{xx}^* \hat{A}_{xz} \bar{D}_{xx}^*} K_2 = 0.
$$

Then the solutions in Eqs. (5.4.19), (5.4.20), and (5.4.17) become (after some algebraic simplifications):

$$u(x) = \frac{\bar{B}_{xx}}{D_{xx}^*} \frac{F_0 L^2}{2} \frac{x}{L}\left(2 - \frac{x}{L}\right) + \left(\frac{D_{xx}}{D_{xx}^*}N_{xx}^T - \frac{B_{xx}}{D_{xx}^*}M_{xx}^T\right)x, \tag{5.4.48}$$

$$\phi_x(x) = -\frac{A_{xx}}{D_{xx}^*} \frac{F_0 L^2}{2} \frac{x}{L}\left(2 - \frac{x}{L}\right) - \left(\frac{B_{xx}}{D_{xx}^*}N_{xx}^T - \frac{A_{xx}}{D_{xx}^*}M_{xx}^T\right)x, \tag{5.4.49}$$

$$w(x) = \frac{A_{xx}}{D_{xx}^*} \frac{F_0 L^3}{6} \frac{x^2}{L^2}\left(3 - \frac{x}{L}\right) + \left(\frac{B_{xx}}{D_{xx}^*}N_{xx}^T - \frac{A_{xx}}{D_{xx}^*}M_{xx}^T\right)\frac{x^2}{2}$$
$$+ \frac{\hat{D}_{xx}^*}{\hat{A}_{xz}\bar{D}_{xx}^*}F_0 x, \tag{5.4.50}$$

$$-\frac{dw}{dx} = -\frac{A_{xx}}{D_{xx}^*} \frac{F_0 L^2}{2} \frac{x}{L}\left(2 - \frac{x}{L}\right) - \left(\frac{B_{xx}}{D_{xx}^*}N_{xx}^T - \frac{A_{xx}}{D_{xx}^*}M_{xx}^T\right)x$$
$$- \frac{\hat{D}_{xx}^*}{\hat{A}_{xz}\bar{D}_{xx}^*}F_0. \tag{5.4.51}$$

The expressions presented in Eqs. (5.4.48)–(5.4.51) match with those of TBT presented in Eqs. (4.7.58)–(4.7.60), except that the simplified RBT has extra term in the expression for $w(x)$, which is clearly a higher-order term. We also note that the slope $-dw/dx$ in the simplified RBT is not zero at $x = 0$, as in the TBT, although $\phi_x(0) = 0$.

Levinson beam theory. In this case, the boundary conditions are:

$$u(0) = w(0) = \phi_x(0) = 0; \quad N_{xx}(L) = M_{xx}(L) = 0, \quad N_{xz}(L) = F_0, \tag{5.4.52}$$

which result in the same values for K_1, K_2, and K_3 as listed in Eq. (5.4.48), and

$$u(0) = -\frac{\bar{B}_{xx}}{\bar{D}_{xx}^*}K_4 - \frac{J_1 A_{xx}}{D_{xx}^*}K_5 + \frac{J_1}{D_{xx}^*\bar{A}_{xz}}K_2 = 0,$$

$$\phi_x(0) = \frac{A_{xx}}{D_{xx}^*}K_5 + \frac{J-2}{D_{xx}^*\bar{A}_{xz}}K_2 = 0, \tag{5.4.53}$$

$$w(0) = -\frac{A_{xx}}{D_{xx}^*}K_6 = 0.$$

Then the solution according to the LBT becomes:

$$u(x) = \frac{B_{xx}}{D_{xx}^*} \frac{F_0 L^2}{2} \frac{x}{L}\left(2 - \frac{x}{L}\right) + \left(\frac{D_{xx}}{D_{xx}^*}N_{xx}^T - \frac{B_{xx}}{D_{xx}^*}M_{xx}^T\right)x, \tag{5.4.54}$$

$$\phi_x(x) = -\frac{A_{xx}}{D_{xx}^*} \frac{F_0 L^2}{2} \frac{x}{L}\left(2 - \frac{x}{L}\right) - \left(\frac{B_{xx}}{D_{xx}^*}N_{xx}^T - \frac{A_{xx}}{D_{xx}^*}M_{xx}^T\right)x, \tag{5.4.55}$$

$$w(x) = \frac{A_{xx}}{D_{xx}^*} \frac{F_0 L^3}{6} \frac{x^2}{L^2}\left(3 - \frac{x}{L}\right) + \left(\frac{B_{xx}}{D_{xx}^*}N_{xx}^T - \frac{A_{xx}}{D_{xx}^*}M_{xx}^T\right)\frac{x^2}{2} + \frac{\bar{D}_{xx}^*F_0}{\bar{A}_{xz}D_{xx}^*}x, \tag{5.4.56}$$

$$-\frac{dw}{dx} = -\frac{A_{xx}}{D_{xx}^*} \frac{F_0 L^2}{2} \frac{x}{L}\left(2 - \frac{x}{L}\right) - \left(\frac{B_{xx}}{D_{xx}^*}N_{xx}^T - \frac{A_{xx}}{D_{xx}^*}M_{xx}^T\right)x - \frac{\bar{D}_{xx}^*F_0}{\bar{A}_{xz}D_{xx}^*}. \tag{5.4.57}$$

The solutions presented here can be specialized to isotropic beams by setting $B_{xx} = E_{xx} = 0$, $\bar{D}_{xx}^* = A_{xx}\bar{D}_{xx} = A_{xx}(D_{xx} - \alpha F_{xx})$, $D_{xx}^* = A_{xx}D_{xx}$, $\hat{D}_{xx}^* = A_{xx}(D_{xx} - 2\alpha F_{xx} + \alpha^2 H_{xx})$, and $\hat{A}_{xz}^* = A_{xz} - 2\beta D_{xz} + \beta^2 F_{xz}$. We have

$$u(x) = \frac{1}{A_{xx}} N_{xx}^T x, \tag{5.4.58}$$

$$\phi_x(x) = -\frac{1}{D_{xx}} \frac{F_0 L^2}{2} \frac{x}{L} \left(2 - \frac{x}{L}\right) + \frac{1}{D_{xx}} M_{xx}^T x, \tag{5.4.59}$$

$$w(x) = \frac{1}{D_{xx}} \frac{F_0 L^3}{6} \frac{x^2}{L^2} \left(3 - \frac{x}{L}\right) - \frac{1}{D_{xx}} M_{xx}^T \frac{x^2}{2} + \frac{\hat{D}_{xx}}{\hat{A}_{xz}\bar{D}_{xx}} F_0 x, \tag{5.4.60}$$

for the RBT and

$$u(x) = \frac{1}{A_{xx}} N_{xx}^T x, \tag{5.4.61}$$

$$\phi_x(x) = -\frac{1}{D_{xx}} \frac{F_0 L^2}{2} \frac{x}{L} \left(2 - \frac{x}{L}\right) + \frac{1}{D_{xx}} M_{xx}^T x, \tag{5.4.62}$$

$$w(x) = \frac{1}{D_{xx}} \frac{F_0 L^3}{6} \frac{x^2}{L^2} \left(3 - \frac{x}{L}\right) - \frac{1}{D_{xx}} M_{xx}^T \frac{x^2}{2} + \frac{\bar{D}_{xx}F_0}{\bar{A}_{xz}D_{xx}} x, \tag{5.4.63}$$

for the LBT.

Both the RBT and LBT solutions reduce to the TBT solutions when we set $\alpha = \beta = 0$ (i.e., $\hat{D}_{xx}/\bar{D}_{xx} = 1$, $\bar{D}_{xx}/D_{xx} = 1$, and $\hat{A}_{xz} = A_{xz}$). The CBT solution is obtained by taking $1/A_{xz} = 0$.

Example 5.4.2

Determine the exact solutions using the simplified RBT and LBT of a functionally graded beam with both ends *pinned* and subjected to a uniformly distributed load of magnitude q_0. Take the origin of the x-coordinate at the left end of the beam (i.e., $0 \le x \le L$).

Solution:

Simplified RBT. For this case, the boundary conditions at the both ends, $x = 0$ and $x = L$, are: $u = w = 0$, $\bar{M}_{xx} = 0 \rightarrow M_{xx} = 0$. The condition $M_{xx}(0) = 0$ results in $K_3 = 0$ and condition $M_{xx}(L) = 0$ yields $K_2 = q_0 L/2$. Using the boundary conditions on the displacements at $x = 0$ in Eqs. (5.4.19) and (5.4.17), we obtain

$$u(0) = \frac{J_1 \hat{D}_{xx}^*}{\bar{D}_{xx}^* D_{xx}^* \hat{A}_{xz}} K_2 - \frac{\bar{B}_{xx}}{\bar{D}_{xx}^*} K_4 - \frac{J_1 A_{xx}}{D_{xx}^* \bar{D}_{xx}^*} K_5 = 0, \tag{5.4.64a}$$

$$w(0) = -\frac{A_{xx}}{D_{xx}^*} K_6 - \frac{P_3}{\hat{A}_{xz} D_{xx}^*} K_3 = 0, \tag{5.4.64b}$$

where J_1 is defined in Eq. (5.4.9c) and P_3 in Eq. (5.4.12b). Thus, since $K_3 = 0$, we have $K_6 = 0$.

Using the relation from Eq. (5.4.64a), $K_3 = K_6 = 0$, and $K_2 = q_0 L/2$, we have

$$u(L) = -\frac{B_{xx}}{D_{xx}^*}\frac{q_0 L^3}{12} + \frac{1}{D_{xx}^*}\left(D_{xx}N_{xx}^T - B_{xx}M_{xx}^T\right)L$$

$$+ \frac{D_{xx}}{D_{xx}^*}K_1 L - \frac{J_1 \hat{D}_{xx}^*}{D_{xx}^* \hat{A}_{xz}\bar{D}_{xx}^*}q_0 L = 0, \qquad (5.4.65a)$$

$$w(L) = -\frac{A_{xx}}{D_{xx}^*}\left(\frac{q_0 L^4}{24} + K_5 L\right) - \frac{L^2}{2D_{xx}^*}\left(A_{xx}M_{xx}^T - B_{xx}N_{xx}^T\right)$$

$$+ \frac{B_{xx}}{D_{xx}^*}\frac{L^2}{2}K_1 = 0. \qquad (5.4.65b)$$

Solving the above two equations for K_1 and K_5 (after some algebraic simplifications), we obtain

$$K_1 = \frac{B_{xx}}{D_{xx}}\frac{q_0 L^2}{12} + \frac{1}{D_{xx}}\left(B_{xx}M_{xx}^T - D_{xx}N_{xx}^T\right) + \frac{J_1 \hat{D}_{xx}^*}{D_{xx}\hat{A}_{xz}\bar{D}_{xx}^*}q_0 L, \qquad (5.4.66a)$$

$$K_5 = -\frac{D_{xx}^*}{A_{xx}D_{xx}}\frac{q_0 L^3}{24} + \frac{D_{xx}^* L}{2A_{xx}D_{xx}}M_{xx}^T - \frac{J_1 B_{xx}\hat{D}_{xx}^*}{\hat{A}_{xz}D_{xx}A_{xx}\bar{D}_{xx}^*}\frac{q_0 L^2}{2}. \qquad (5.4.66b)$$

Due to the relation in Eq. (5.4.64a), all constant parts of $u(x)$ disappear. The solution for the generalized displacements become [J_1 is defined in Eq. (5.4.9c)]:

$$u(x) = \frac{B_{xx}}{D_{xx}^*}\frac{q_0 L^3}{12}\frac{x}{L}\left(1 - 3\frac{x}{L} + 2\frac{x^2}{L^2}\right) + \frac{J_1 \hat{D}_{xx}^*}{D_{xx}^* \hat{A}_{xz}\bar{D}_{xx}^*}q_0 L\left(1 - \frac{x}{L}\right), \qquad (5.4.67)$$

$$\phi_x(x) = -\frac{A_{xx}}{D_{xx}^*}\frac{q_0 L^3}{12}\frac{x}{L}\left(1 - 3\frac{x}{L} + 2\frac{x^2}{L^2}\right) - \frac{1}{D_{xx}}\frac{q_0 L^3}{24}\left(1 - 2\frac{x}{L}\right)$$

$$- \frac{M_{xx}^T L}{2D_{xx}}\left(1 - 2\frac{x}{L}\right) + \frac{J_1}{D_{xx}^* \hat{A}_{xz}}\frac{\hat{D}_{xx}^*}{\bar{D}_{xx}^*}\frac{B_{xx}}{D_{xx}}\frac{q_0 L^2}{2}\left(1 - 2\frac{x}{L}\right), \qquad (5.4.68)$$

$$w(x) = \frac{A_{xx}}{D_{xx}^*}\frac{q_0 L^4}{24}\frac{x}{L}\left(1 - 2\frac{x^2}{L^2} + \frac{x^3}{L^3}\right) - \frac{q_0 L^4}{24D_{xx}}\frac{x}{L}\left(1 - \frac{x}{L}\right)$$

$$+ \frac{\hat{D}_{xx}^*}{D_{xx}^*}\frac{q_0 L^2}{24\hat{A}_{xz}}\frac{x}{L}\left(1 - \frac{x}{L}\right) + \frac{M_{xx}^T L^2}{2D_{xx}}\frac{x}{L}\left(1 - \frac{x}{L}\right)$$

$$- \frac{J_1 B_{xx}}{\hat{A}_{xz}A_{xx}}\frac{\hat{D}_{xx}^*}{\bar{D}_{xx}^*}\frac{q_0 L^3}{2D_{xx}}\frac{x}{L}\left(1 - \frac{x}{L}\right). \qquad (5.4.69)$$

LBT solution. As before, we have $K_2 = q_0 L/2$ and $K_3 = 0$. The displacement boundary conditions at $x = 0$ give [from Eqs. (5.4.39) and (5.4.37)]:

$$u(0) = 0 = -\frac{\bar{B}_{xx}}{\bar{D}_{xx}^*}K_4 - \frac{J_1 A_{xx}}{D_{xx}^*}K_5 - \frac{J_1}{D_{xx}^* \hat{A}_{xz}}K_2, \qquad (5.4.70a)$$

$$w(0) = 0 = -\frac{A_{xx}}{D_{xx}^*}K_6 \quad \rightarrow \quad K_6 = 0. \qquad (5.4.70b)$$

The displacement boundary conditions at $x = L$ result [with $K_2 = -q_0 L/2$, $K_3 = 0$, and using the conditions in Eqs. (5.4.70a) and (5.4.70b)]

$$u(L) = 0 = -\frac{B_{xx}}{D_{xx}^*}\frac{q_0 L^3}{12} + \frac{1}{D_{xx}^*}\left(D_{xx}N_{xx}^T - B_{xx}M_{xx}^T\right)L$$

$$-J_1\frac{\bar{D}_{xx}^*}{D_{xx}^*\bar{A}_{xz}}q_0 L + \frac{D_{xx}}{D_{xx}^*}K_1 L, \qquad (5.4.70c)$$

$$w(L) = 0 = -\frac{A_{xx}}{D_{xx}^*}\left(\frac{q_0 L^4}{24} + K_5 L\right) - \frac{L^2}{2D_{xx}^*}\left(A_{xx}M_{xx}^T - B_{xx}N_{xx}^T\right)$$

$$+ \frac{B_{xx}}{D_{xx}^*}\frac{L^2}{2}K_1. \qquad (5.4.70d)$$

Solving the above two equations, we obtain

$$K_1 = \frac{B_{xx}}{D_{xx}}\frac{q_0 L^2}{12} - \frac{1}{D_{xx}}\left(D_{xx}N_{xx}^T - B_{xx}M_{xx}^T\right) + \frac{J_1\bar{D}_{xx}^*}{\bar{A}_{xz}D_{xx}}q_0, \qquad (5.4.71a)$$

$$K_5 = \frac{D_{xx}^*}{A_{xx}D_{xx}}\left(-\frac{q_0 L^3}{24} - M_{xx}^T\frac{L}{2}\right) + \frac{J_1 B_{xx}\bar{D}_{xx}^*}{A_{xx}D_{xx}\bar{A}_{xz}}\frac{q_0 L}{2}. \qquad (5.4.71b)$$

The solution for the generalized displacements according to the LBT is

$$u(x) = \frac{B_{xx}}{D_{xx}^*}\frac{q_0 L^3}{12}\frac{x}{L}\left(1 - 3\frac{x}{L} + 2\frac{x^2}{L^2}\right) + \frac{J_1 q_0 L}{\bar{A}_{xz}D_{xx}^*}\left(1 - \frac{x}{L}\right), \qquad (5.4.72)$$

$$\phi_x(x) = -\frac{A_{xx}}{D_{xx}^*}\frac{q_0 L^3}{12}\frac{x}{L}\left(1 - 3\frac{x}{L} + 2\frac{x^2}{L^2}\right) - \frac{1}{D_{xx}}\frac{q_0 L^3}{24}\left(1 - 2\frac{x}{L}\right)$$

$$-\frac{M_{xx}^T L}{2D_{xx}}\left(1 - 2\frac{x}{L}\right) + \frac{q_0 L}{2D_{xx}^*\bar{A}_{xz}}\left(J_2 + J_1\frac{B_{xx}\bar{D}_{xx}^*}{D_{xx}}\right)\left(1 - 2\frac{x}{L}\right), \qquad (5.4.73)$$

$$w(x) = \frac{A_{xx}}{D_{xx}^*}\frac{q_0 L^4}{24}\frac{x}{L}\left(\frac{x}{L} - 2\frac{x^2}{L^2} + \frac{x^3}{L^3}\right) + \frac{\bar{D}_{xx}^*}{\bar{A}_{xz}D_{xx}^*}\frac{q_0 L^2}{2}\frac{x}{L}\left(1 - \frac{x}{L}\right)$$

$$+ \frac{q_0 L^4}{24D_{xx}}\frac{x}{L}\left(1 - \frac{x}{L}\right) + \frac{M_{xx}^T L^2}{2D_{xx}}\frac{x}{L}\left(1 - \frac{x}{L}\right)$$

$$-\frac{\bar{D}_{xx}^*}{D_{xx}^*}\frac{J_1 B_{xx}}{D_{xx}}\frac{q_0 L^2}{2\bar{A}_{xz}}\frac{x}{L}\left(1 - \frac{x}{L}\right). \qquad (5.4.74)$$

The RBT solution for homogeneous beams can be obtained from Eqs. (5.4.67)–(5.4.69) by setting B_{xx}, E_{xx} and J_1 to zero:

$$u(x) = 0, \qquad (5.4.75)$$

$$\phi_x(x) = -\frac{1}{D_{xx}}\frac{q_0 L^3}{12}\frac{x}{L}\left(1 - 3\frac{x}{L} + 2\frac{x^2}{L^2}\right) - \frac{1}{D_{xx}}\frac{q_0 L^3}{24}\left(1 - 2\frac{x}{L}\right)$$

$$-\frac{M_{xx}^T L}{2D_{xx}}\left(1 - 2\frac{x}{L}\right), \qquad (5.4.76)$$

$$w(x) = \frac{1}{D_{xx}}\frac{q_0 L^4}{24}\frac{x}{L}\left(1 - 2\frac{x^2}{L^2} + \frac{x^3}{L^3}\right) - \frac{q_0 L^4}{24D_{xx}}\frac{x}{L}\left(1 - \frac{x}{L}\right)$$

$$+ \frac{\bar{D}_{xx}}{D_{xx}}\frac{q_0 L^2}{24\bar{A}_{xz}}\frac{x}{L}\left(1 - \frac{x}{L}\right) + \frac{M_{xx}^T L^2}{2D_{xx}}\frac{x}{L}\left(1 - \frac{x}{L}\right). \qquad (5.4.77)$$

5.5 BENDING RELATIONSHIPS FOR THE RBT

5.5.1 PRELIMINARY COMMENTS

The relationships between CBT and the first-order shear deformation (or Timoshenko) beam theory (TBT) were established by Wang [116], and they were extended to include functionally graded beams in Section 4.7. The objective of this section is to establish (see Wang, Reddy, and Lee [120]) bending (not including the axial displacement) relationships between the bending solutions (i.e., deflection, rotation, bending moment, and shear force) of the Reddy beam theory (RBT) in terms of the corresponding quantities of the classical (Euler–Bernoulli) beam theory (CBT). The discussion presented here is limited to homogeneous beams without thermal and couple stress effects for the moment (one may extend them to include these effects as well as the material grading through the beam height).

In the case of pure bending without couple stress effect, we note that both CBT and TBT are fourth-order theories whereas the RBT is a sixth-order theory. The "order" referred to here is the *total* order of the differential equations expressed in terms of the generalized displacements (the order goes up by 2 when the axial displacement u is also included). In the CBT, the bending deflection is described by a fourth-order differential equation, while in the TBT the bending is described by a pair of coupled differential equations among w and ϕ_x. The RBT is governed by a fourth-order equation in w and a second-order equation in ϕ_x. Therefore, the relationships between the solutions of two different order theories can only be established by solving an additional second-order equation, as will be shown shortly.

5.5.2 SUMMARY OF EQUATIONS

The linearized bending equations of the RBT are given by Eqs. (5.4.1)–(5.4.3):

$$-\frac{d\bar{N}_{xz}^R}{dx} - \alpha\frac{d^2 P_{xx}^R}{dx^2} = q, \quad -\frac{d\bar{M}_{xx}^R}{dx} + \bar{N}_{xz}^R = 0, \qquad (5.5.1a)$$

where

$$\bar{N}_{xz}^R = N_{xz}^R - \beta P_{xz}^R, \quad \bar{M}_{xx}^R = M_{xx}^R - \alpha P_{xx}^R. \qquad (5.5.1b)$$

The above two equations together yield

$$-\frac{d^2 M_{xx}^R}{dx^2} = q, \qquad (5.5.2)$$

which is exactly in the same form as the governing equation for the CBT [see Eq. (3.7.3b)]

$$-\frac{d^2 M_{xx}^C}{dx^2} = q \quad \text{and} \quad \frac{dM_{xx}^C}{dx} = N_{xz}^C. \qquad (5.5.3)$$

Equating Eqs. (5.5.2) and (5.5.3) and integrating twice, we obtain

$$\frac{dM_{xx}^R}{dx} = \frac{dM_{xx}^C}{dx} + c_1 \quad \rightarrow \quad M_{xx}^R = M_{xx}^C + c_1 x + c_2. \tag{5.5.4}$$

The stress resultant–displacement relationships in Eqs. (5.3.20a) and (5.3.20b) in the present case (i.e., homogeneous beams without thermal effects) take the form

$$M_{xx}^R = \bar{D}_{xx}\frac{d\phi_x^R}{dx} - \alpha F_{xx}\frac{d^2 w^R}{dx^2}, \tag{5.5.5a}$$

$$P_{xx}^R = \bar{F}_{xx}\frac{d\phi_x^R}{dx} - \alpha H_{xx}\frac{d^2 w^R}{dx^2}, \tag{5.5.5b}$$

$$N_{xz}^R = \bar{A}_{xz}\left(\phi_x^R + \frac{dw^R}{dx}\right), \tag{5.5.5c}$$

$$P_{xz}^R = \bar{D}_{xz}\left(\phi_x^R + \frac{dw^R}{dx}\right). \tag{5.5.5d}$$

From Eq. (5.5.5c), we have

$$\phi_x^R = \frac{1}{\bar{A}_{xz}}N_{xz}^R - \frac{dw^R}{dx}. \tag{5.5.6}$$

Then Eqs. (5.5.5a)–(5.5.5d) can be expressed as

$$M_{xx}^R = \frac{\bar{D}_{xx}}{\bar{A}_{xz}}\frac{dN_{xz}^R}{dx} - D_{xx}\frac{d^2 w^R}{dx^2}, \tag{5.5.7a}$$

$$P_{xx}^R = \frac{\bar{F}_{xx}}{\bar{A}_{xz}}\frac{dN_{xz}^R}{dx} - F_{xx}\frac{d^2 w^R}{dx^2},$$

$$= \left(\frac{\bar{F}_{xx}}{\bar{A}_{xz}} - \frac{F_{xx}\bar{D}_{xx}}{D_{xx}\bar{A}_{xz}}\right)\frac{dN_{xz}^R}{dx} + \frac{F_{xx}}{D_{xx}}M_{xx}^R, \tag{5.5.7b}$$

$$P_{xz}^R = \frac{\bar{D}_{zz}}{\bar{A}_{xz}}N_{xz}^R. \tag{5.5.7c}$$

5.5.3 GENERAL RELATIONSHIPS

The second equation of Eq. (5.5.1a) can be expressed in terms of M_{xx}^R, P_{xx}^R, N_{xz}^R, and P_{xz}^R as

$$-\frac{dM_{xx}^R}{dx} + \alpha\frac{dP_{xx}^R}{dx} + N_{xz}^R - \beta P_{xz}^R = 0. \tag{5.5.8}$$

Using the expressions for P_{xx}^R and P_{xz}^R from Eqs. (5.5.7b) and (5.5.7c), respectively, in Eq. (5.5.8), we obtain

$$-\frac{\bar{D}_{xx}}{D_{xx}}\frac{dM_{xx}^R}{dx} = \frac{\hat{A}_{xz}}{\bar{A}_{xz}}N_{xz}^R - \frac{\alpha}{\bar{A}_{xz}}\left(\frac{F_{xx}\bar{D}_{xx} - D_{xx}\bar{F}_{xx}}{D_{xx}}\right)\frac{d^2 N_{xz}^R}{dx^2}. \tag{5.5.9}$$

Simplifying the coefficients and making use of Eqs. (5.5.3) and (5.5.4), we arrive at

$$\frac{\alpha}{\bar{A}_{xz}}\left(\frac{F_{xx}\bar{D}_{xx} - D_{xx}\bar{F}_{xx}}{D_{xx}}\right)\frac{d^2 N_{xz}^R}{dx^2} - \frac{\hat{A}_{xz}}{\bar{A}_{xz}}N_{xz}^R - \left(\frac{\bar{D}_{xx}}{D_{xx}}\right)\left(N_{xz}^C + c_1\right) = 0.$$
(5.5.10)

Thus, a second-order differential equation must be solved to determine the shear force N_{xz}^R in the RBT in terms of the shear force N_{xz}^C in the CBT. Once N_{xz}^R is known, we can determine M_{xx}^R, ϕ_x^R, and w^R, as will be illustrated through examples.

The effective shear force V_{eff}^R in the RBT can be computed from

$$V_{\text{eff}}^R(x) = N_{xz}^R - \beta P_{xz}^R + \alpha\frac{dP_{xx}^R}{dx} = \frac{dM_{xx}^R}{dx}$$

$$= N_{xz}^C(x) + c_1,$$
(5.5.11)

where the second equation in (5.5.1a) and Eqs. (5.5.3) and (5.5.4) are used to derive Eq. (5.5.11).

To determine ϕ_x^R, we use Eq. (5.5.5a):

$$D_{xx}\frac{d\phi_x^R}{dx} = M_{xx}^R + \alpha F_{xx}\left(\frac{d\phi_x^R}{dx} + \frac{d^2 w^R}{dx^2}\right)$$

$$= M_{xx}^C + c_1 x + c_2 + \frac{\alpha F_{xx}}{\bar{A}_{xz}}\frac{dN_{xz}^R}{dx}$$

$$= -D_{xx}\frac{d^2 w^C}{dx^2} + c_1 x + c_2 + \frac{\alpha F_{xx}}{\bar{A}_{xz}}\frac{dN_{xz}^R}{dx},$$
(5.5.12)

or

$$D_{xx}\phi_x^R(x) = -D_{xx}\frac{dw^C}{dx} + \frac{\alpha F_{xx}}{\bar{A}_{xz}}N_{xz}^R + c_1\frac{x^2}{2} + c_2 x + c_3,$$
(5.5.13)

where we have used the relation

$$M_{xx}^C = -D_{xx}\frac{d^2 w^C}{dx^2},$$
(5.5.14)

and Eq. (5.5.4) in arriving at the last equation.

Lastly, we derive the relation between $w^R(x)$ and $w^C(x)$. Using Eqs. (5.5.5c) and (5.5.13), we can write

$$D_{xx}\frac{dw^R}{dx} = -D_{xx}\phi_x^R(x) + \frac{D_{xx}}{\bar{A}_{xz}}N_{xz}^R$$

$$= D_{xx}\frac{dw^C}{dx} + \frac{\bar{D}_{xx}}{\bar{A}_{xz}}N_{xz}^R - c_1\frac{x^2}{2} - c_2 x - c_3.$$
(5.5.15)

Integrating with respect to x once, we obtain

$$D_{xx}w^R(x) = D_{xx}w^C(x) + \frac{\bar{D}_{xx}}{\bar{A}_{xz}}\left(\int^x N_{xz}^R(\xi)d\xi\right) - c_1\frac{x^3}{6} - c_2\frac{x^2}{2} - c_3x - c_4.$$

$$(5.5.16)$$

In summary, the relationships between the solutions of the CBT and the RBT are given by Eqs. (5.5.11), (5.5.4), (5.5.13), and (5.5.16):

$$V_{\text{eff}}^R(x) = N_{xz}^C + c_1, \qquad (5.5.17)$$

$$M_{xx}^R(x) = M_{xx}^C + c_1x + c_2, \qquad (5.5.18)$$

$$w^R(x) = w^C(x) + \frac{\bar{D}_{xx}}{\bar{A}_{xz}D_{xx}}\left(\int^x N_{xz}^R(\xi)d\xi\right)$$

$$- \frac{1}{D_{xx}}\left(c_1\frac{x^3}{6} + c_2\frac{x^2}{2} + c_3x + c_4\right), \qquad (5.5.19)$$

$$\bar{D}_{xx}\phi_x^R(x) = -D_{xx}\frac{dw^C}{dx} + \alpha F_{xx}\frac{dw^R}{dx} + c_1\frac{x^2}{2} + c_2x + c_3. \qquad (5.5.20)$$

The constants of integration, c_i ($i = 1, 2, 3, 4$) appearing in the expressions for $w^R(x)$, ϕ_x^R, M_{xx}^R, and N_{xz}^R are determined using the boundary conditions. Since there are six boundary conditions in the RBT, the remaining two boundary conditions are used in the solving of the second-order differential equation, Eq. (5.5.10), for N_{xz}^R. The boundary conditions for various types of supports (F - free, P - pinned, and C - clamped) are defined below, consistent with the duality of RBT [see Eq. (5.3.14a)]:

$$\mathbf{F}: \ V_{\text{eff}}^R \equiv N_{xz}^R - \beta P_{xz}^R + \alpha\frac{dP_{xx}^R}{dx} = 0,$$

$$M_{xx}^R - \alpha P_{xx}^R = 0, \quad P_{xx}^R = 0, \qquad (5.5.21a)$$

$$\mathbf{P}: \ w^R = 0, \quad M_{xx}^R - \alpha P_{xx}^R = 0, \quad P_{xx}^R = 0, \qquad (5.5.21b)$$

$$\mathbf{C}: \ w^R = 0, \quad \phi_x^R = 0, \quad \frac{dw^R}{dx} = 0. \qquad (5.5.21c)$$

Since the second-order equation (5.5.10) requires boundary conditions on the shear force N_{xz}^R, we reduce the force boundary conditions in Eqs. (5.5.21a)–(5.5.21c) to one in terms of N_{xz}^R:

1. **Free (F):** Equations (5.5.21a) and (5.5.7b) imply

$$\frac{dN_{xz}^R}{dx} = 0. \qquad (5.5.22a)$$

2. **Pinned or hinged (P):** Equations (5.5.21b) and (5.5.7b) imply

$$\frac{dN_{xz}^R}{dx} = 0. \qquad (5.5.22b)$$

3. **Clamped (C):** Equation (5.5.21c) implies

$$N_{xz}^R = 0. \tag{5.5.22c}$$

As discussed previously, requiring dw^R/dx as well as ϕ_x^R to vanish at a support necessarily implies that the shear force, when computed using the constitutive relation (5.5.5c), is zero at the support, but the effective shear force is not zero there.

5.5.4 BENDING RELATIONSHIPS FOR THE SIMPLIFIED RBT

As stated earlier, the RBT requires, unlike in the TBT, the solution of an additional second-order equation to establish the relationships. The reason is that both the CBT and the TBT are fourth-order theories, whereas the RBT is a sixth-order beam theory. In this section we develop relationships between a simplified RBT and the CBT. The phrase *simplified RBT* refers to the fourth-order RBT obtained by dropping the second-derivative term in the additional differential equation for w^R (see Section 5.4.2 for the details). While this is an approximation of the original RBT, it is as simple and as accurate as the TBT while not requiring a shear correction coefficient.

For the simplified RBT, we first derive the second-order differential equation in terms of w^R. Expressing Eq. (5.5.4) in terms of the respective generalized displacements of the CBT and RBT, we obtain

$$\bar{D}_{xx}\frac{d\phi_x^R}{dx} - \alpha F_{xx}\frac{d^2 w^R}{dx^2} = -D_{xx}\frac{d^2 w^C}{dx^2} + c_1 x + c_2. \tag{5.5.23}$$

Integrating the above equation once, we obtain

$$\bar{D}_{xx}\phi_x^R - \alpha F_{xx}\frac{dw^R}{dx} = -D_{xx}\frac{dw^C}{dx} + c_1\frac{x^2}{2} + c_2 x + c_3. \tag{5.5.24}$$

From the second equation in (5.5.1a), we have

$$\frac{d\bar{M}_{xx}^R}{dx} = \bar{N}_{xz}^R.$$

Expressing $\bar{N}_{xz}^R$ in terms of the displacements, we have

$$\frac{d\bar{M}_{xx}^R}{dx} = \hat{A}_{xz}\left(\phi_x^R + \frac{dw^R}{dx}\right). \tag{5.5.25}$$

Solving the above equation for ϕ_x^R,

$$\phi_x^R = \frac{1}{\hat{A}_{xz}}\frac{d\bar{M}_{xx}^R}{dx} - \frac{dw^R}{dx}. \tag{5.5.26}$$

Substituting Eq. (5.5.26) into Eq. (5.5.24), we obtain

$$\left(\frac{\bar{D}_{xx}}{\hat{A}_{xz}}\right)\frac{d\bar{M}_{xx}^R}{dx} - D_{xx}\frac{dw^R}{dx} = -D_{xx}\frac{dw^C}{dx} + c_1\frac{x^2}{2} + c_2 x + c_3, \qquad (5.5.27)$$

which on integration yields

$$D_{xx}w^R(x) = \left(\frac{\bar{D}_{xx}}{\hat{A}_{xz}}\right)\bar{M}_{xx}^R + D_{xx}w^C(x) - c_1\frac{x^3}{6} - c_2\frac{x^2}{2} - c_3 x - c_4. \quad (5.5.28)$$

From Eqs. (5.5.5a) and (5.5.5b), we have

$$M_{xx}^R = \bar{D}_{xx}\frac{d\phi^R}{dx} - \alpha F_{xx}\frac{d^2 w^R}{dx^2}, \qquad (5.5.29)$$

$$\bar{M}_{xx}^R = M_{xx}^R - \alpha P_{xx}^R = \hat{D}_{xx}\frac{d\phi_x^R}{dx} - \alpha\bar{F}_{xx}\frac{d^2 w^R}{dx^2}. \qquad (5.5.30)$$

Solving the above equations for $d\phi_x^R/dx$ and $d^2 w^R/dx^2$, we obtain

$$\frac{d\phi_x^R}{dx} = \frac{M_{xx}^R \bar{F}_{xx} - \bar{M}_{xx}^R F_{xx}}{\bar{F}_{xx}\bar{D}_{xx} - \hat{D}_{xx}F_{xx}}, \qquad (5.5.31)$$

$$\frac{d^2 w^R}{dx^2} = \frac{\hat{D}_{xx}M_{xx}^R - \bar{D}_{xx}\bar{M}_{xx}^R}{\alpha(\bar{F}_{xx}\bar{D}_{xx} - \hat{D}_{xx}F_{xx})}. \qquad (5.5.32)$$

Rewriting Eq. (5.5.32) as

$$\hat{D}_{xx}M_{xx}^R - \bar{D}_{xx}\bar{M}_{xx}^R = \alpha\left(\bar{F}_{xx}\bar{D}_{xx} - F_{xx}\hat{D}_{xx}\right)\frac{d^2 w^R}{dx^2}, \qquad (5.5.33)$$

and using Eq. (5.5.4), we can write Eq. (5.5.33) as

$$\bar{M}_{xx}^R = \frac{\hat{D}_{xx}}{\bar{D}_{xx}}\left(M_{xx}^C + c_1 x + c_2\right) - \alpha\left(\frac{\bar{F}_{xx}\bar{D}_{xx} - F_{xx}\hat{D}_{xx}}{\bar{D}_{xx}}\right)\frac{d^2 w^R}{dx^2}. \qquad (5.5.34)$$

Finally, substituting for $\bar{M}_{xx}^R$ from Eq. (5.5.34) into Eq. (5.5.28), we obtain

$$\begin{aligned}
D_{xx}w^R(x) &- \frac{\alpha}{\hat{A}_{xz}}\left(F_{xx}\hat{D}_{xx} - \bar{F}_{xx}\bar{D}_{xx}\right)\frac{d^2 w^R}{dx^2} \\
&= D_{xx}w^C(x) + \left(\frac{\bar{D}_{xx}}{\hat{A}_{xz}}\right)M_{xx}^C - c_1\left[\frac{x^3}{6} - \left(\frac{\bar{D}_{xx}}{\hat{A}_{xz}}\right)x\right] \\
&\quad - c_2\left[\frac{x^2}{2} - \left(\frac{\bar{D}_{xx}}{\hat{A}_{xz}}\right)\right] - c_3 x - c_4.
\end{aligned} \qquad (5.5.35)$$

The above equation is equivalent to Eq. (5.4.13).

The simplified RBT is one in which we neglect the second-order derivative term in Eq. (5.5.35). This amounts to reducing the order of the theory from six to four. We obtain

$$
D_{xx}w^S(x) = D_{xx}w^C(x) + \left(\frac{\bar{D}_{xx}}{\hat{A}_{xz}}\right)M_{xx}^C - c_1\left[\frac{x^3}{6} - \left(\frac{\bar{D}_{xx}}{\hat{A}_{xz}}\right)x\right]
$$
$$
- c_2\left[\frac{x^2}{2} - \left(\frac{\bar{D}_{xx}}{\hat{A}_{xz}}\right)\right] - c_3 x - c_4, \tag{5.5.36}
$$

which replaces Eq. (5.5.16). Here the superscript "S" denotes the quantities in the simplified RBT.

Thus, in summary, we have the following relations between the simplified RBT and the CBT (we note that the relationships for the shear force and bending moment remain unchanged between the original RBT and the simplified RBT):

$$
V_{\text{eff}}^S = N_{xz}^C + c_1, \tag{5.5.37}
$$

$$
M_{xx}^S = M_{xx}^C + c_1 x + c_2, \tag{5.5.38}
$$

$$
\phi_x^S(x) = -\frac{dw^C}{dx} + \frac{1}{D_{xx}}\left(c_1\frac{x^2}{2} + c_2 x + c_3\right), \tag{5.5.39}
$$

$$
w^S(x) = w^C(x) + \left(\frac{\bar{D}_{xx}}{\hat{A}_{xz}D_{xx}}\right)M_{xx}^C - \frac{1}{D_{xx}}\left\{c_1\left[\frac{x^3}{6} + \left(\frac{\bar{D}_{xx}}{\hat{A}_{xz}}\right)x\right]\right.
$$
$$
\left. + c_2\left[\frac{x^2}{2} + \left(\frac{\bar{D}_{xx}}{\hat{A}_{xz}}\right)\right] + c_3 x + c_4\right\}, \tag{5.5.40}
$$

where we have introduced the following equivalent slope [Eq. (5.5.24)]:

$$
\phi_x^S(x) = \frac{\bar{D}_{xx}}{D_{xx}}\phi^R - \alpha\frac{F_{xx}}{D_{xx}}\frac{dw^R}{dx}. \tag{5.5.41}
$$

The constants of integration, c_1, c_2, c_3, and c_4 are determined using the following types of boundary conditions:

$$
\textbf{F:} \quad M_{xx}^C = M_{xx}^S = N_{xz}^C = V_{\text{eff}}^S = 0, \tag{5.5.42a}
$$

$$
\textbf{S:} \quad w^C = w^S = M_{xx}^C = M_{xx}^S = 0, \tag{5.5.42b}
$$

$$
\textbf{C:} \quad w^C = w^S = \frac{dw^C}{dx} = \phi_x^S = 0. \tag{5.5.42c}
$$

5.5.5   RELATIONSHIPS BETWEEN THE LBT AND THE CBT

The objective of this section is to establish relationships between the bending solutions (i.e., deflection, rotation, bending moment, and shear force) of the LBT in terms of the corresponding quantities of the CBT. Since, the LBT is a fourth-order theory, like the TBT is, the relationships between LBT and CBT are similar, as will be shown shortly, to those between TBT and CBT (see Reddy et al. [121]). We note that the equations of equilibrium of the LBT are

$$-\frac{dM_{xx}^L}{dx} + N_{xz}^L = 0, \tag{5.5.43}$$

$$-\frac{dN_{xz}^L}{dx} = q, \tag{5.5.44}$$

and the constitutive relations are

$$M_{xx}^L = \bar{D}_{xx}\frac{d\phi_x^L}{dx} - \alpha F_{xx}\frac{d^2 w^L}{dx^2}, \tag{5.5.45}$$

$$N_{xz}^L = \bar{A}_{xz}\left(\phi_x^L + \frac{dw^L}{dx}\right). \tag{5.5.46}$$

Here the superscript "L" denotes the quantities in the LBT.

As in the case of TBT, RBT, and simplified RBT, we have

$$N_{xz}^L = N_{xz}^C + c_1, \tag{5.5.47}$$

$$M_{xx}^L = M_{xx}^C + c_1 x + c_2. \tag{5.5.48}$$

Using the stress resultant–displacement relationships in Eqs. (5.5.45) and (5.5.46) and making use of the relation in Eq. (5.5.14), we obtain

$$\bar{A}_{xz}\left(\phi_x^L + \frac{dw^L}{dx}\right) = N_{xz}^C + c_1, \tag{5.5.49}$$

$$\bar{D}_{xx}\frac{d\phi_x^L}{dx} - \alpha F_{xx}\frac{d^2 w^L}{dx^2} = -D_{xx}\frac{d^2 w^C}{dx^2} + c_1 x + c_2. \tag{5.5.50}$$

Integrating Eq. (5.5.50) with respect to x, we arrive at

$$\bar{D}_{xx}\phi_x^L - \alpha F_{xx}\frac{dw^L}{dx} = -D_{xx}\frac{dw^C}{dx} + c_1\frac{x^2}{2} + c_2 x + c_3. \tag{5.5.51}$$

Solving Eq. (5.5.51) for dw^L/dx,

$$\frac{dw^L}{dx} = \frac{1}{\alpha F_{xx}}\left(\bar{D}_{xx}\phi_x^L + D_{xx}\frac{dw^C}{dx} - c_1\frac{x^2}{2} - c_2 x - c_3\right), \tag{5.5.52}$$

and substituting it into Eq. (5.5.49), and solving for ϕ_x^L, we obtain

$$\phi_x^L = -\frac{dw^C}{dx} + \alpha\frac{F_{xx}}{\bar{A}_{xz}D_{xx}}\left(N_{xz}^C + c_1\right) + \frac{1}{D_{xx}}\left(c_1\frac{x^2}{2} + c_2 x + c_3\right). \tag{5.5.53}$$

Returning to Eq. (5.5.46), using Eq. (5.5.47) and solving for dw^L/dx, we obtain

$$
\begin{aligned}
\frac{dw^L}{dx} &= \frac{1}{\bar{A}_{xz}}\left(N_{xz}^C + c_1\right) - \phi_x^L \\
&= \frac{1}{\bar{A}_{xz}}\left(1 - \alpha\frac{F_{xx}}{D_{xx}}\right)\left(N_{xz}^C + c_1\right) + \frac{dw^C}{dx} - \frac{1}{D_{xx}}\left(c_1\frac{x^2}{2} + c_2 x + c_3\right) \\
&= \frac{dw^C}{dx} + \frac{\bar{D}_{xx}}{\bar{A}_{xz}D_{xx}}\left(N_{xz}^C + c_1\right) - \frac{1}{D_{xx}}\left(c_1\frac{x^2}{2} + c_2 x + c_3\right).
\end{aligned}
\tag{5.5.54}
$$

Upon integration, we arrive at the deflection relationship

$$
\begin{aligned}
w^L(x) = w^C(x) &+ \frac{\bar{D}_{xx}}{\bar{A}_{xz}D_{xx}}\left(M_{xx}^C + c_1 x\right) \\
&- \frac{1}{D_{xx}}\left(c_1\frac{x^3}{6} + c_2\frac{x^2}{2} + c_3 x + c_4\right).
\end{aligned}
\tag{5.5.55}
$$

Thus, Eqs. (5.5.47), (5.5.48), (5.5.53), and (5.5.55) provide the relationships between the LBT and CBT.

The constants of integration, c_1, c_2, c_3, and c_4 are determined using the following types of boundary conditions:

$$
\mathbf{F}:\ M_{xx}^C = M_{xx}^L = N_{xz}^C = N_{xz}^L = 0, \tag{5.5.56a}
$$

$$
\mathbf{S}:\ w^C = w^L = M_{xx}^C = M_{xx}^L = 0, \tag{5.5.56b}
$$

$$
\mathbf{C}:\ w^C = w^L = \frac{dw^C}{dx} = \phi_x^L = 0. \tag{5.5.56c}
$$

5.5.6 NUMERICAL EXAMPLES

In this section, we consider numerical examples to illustrate the use of the developed bending relationships between various third-order theories and the CBT. Note that the constants of integration, c_i, introduced in developing the relationships are different from the constants of integration used in developing the exact solutions in Section 5.4.

Example 5.5.1 ————————————————————————————————

Determine the exact (a) RBT , (b) simplified RBT, and (c) LBT bending solutions (i.e., not account for the axial displacement) of a beam with both ends hinged (for pure bending case, there is no difference between the hinged and pinned boundary condition as far as the bending variables are concerned) and subjected to a uniformly distributed load of magnitude q_0.

Solution: We take the origin of the x-coordinate at the left end of the beam (i.e., $0 \leq x \leq L$). We begin with the CBT solution first.

The CBT Solution: For this case, the stress-resultants and the deflection are:

$$N_{xz}^C(x) = \frac{q_0 L}{2}\left(1 - 2\frac{x}{L}\right),$$ (5.5.57)

$$M_{xx}^c(x) = \frac{q_0 L^2}{2}\frac{x}{L}\left(1 - \frac{x}{L}\right),$$ (5.5.58)

$$w^c(x) = \frac{q_0 L^4}{24 D_{xx}}\frac{x}{L}\left(1 - 2\frac{x^2}{L^2} + \frac{x^3}{L^3}\right),$$ (5.5.59)

$$-\frac{dw^c}{dx} = -\frac{q_0 L^3}{24 D_{xx}}\left(1 - 6\frac{x^2}{L^2} + 4\frac{x^3}{L^3}\right).$$ (5.5.60)

The TBT solution using relationships: Using the relationships for simply supported (or hinged–hinged) beams in Table 4.7.1, the corresponding bending solutions for the TBT are:

$$N_{xz}^F(x) = N_{xz}^C(x) = \frac{q_0 L}{2}\left(1 - 2\frac{x}{L}\right),$$ (5.5.61)

$$M_{xx}^F(x) = M_{xx}^C(x) = \frac{q_0 L^2}{2}\frac{x}{L}\left(1 - \frac{x}{L}\right),$$ (5.5.62)

$$\phi_x^F(x) = -\frac{dw^C}{dx} = -\frac{q_0 L^3}{24 D_{xx}}\left(1 - 6\frac{x^2}{L^2} + 4\frac{x^3}{L^3}\right),$$ (5.5.63)

$$w^F(x) = w^C(x) + \frac{1}{K_s A_{xz}}M_{xx}^C(x)$$
$$= \frac{q_0 L^4}{24 D_{xx}}\frac{x}{L}\left(1 - 2\frac{x^2}{L^2} + \frac{x^3}{L^3}\right) + \frac{1}{K_s A_{xz}}\frac{q_0 L^2}{2}\frac{x}{L}\left(1 - \frac{x}{L}\right).$$ (5.5.64)

(a) *The RBT solution.* In the case of the RBT, we must first solve the second-order differential for the transverse shear force, N_{xz}^R. From Eqs. (5.5.10) and (5.5.60), we have

$$\frac{d^2 N_{xz}^R}{dx^2} - \lambda^2 N_{xz}^R = -\mu\left[\frac{q_0 L}{2}\left(1 - 2\frac{x}{L}\right) + c_1\right],$$ (5.5.65a)

where

$$\lambda^2 = \frac{\hat{A}_{xz}D_{xx}}{\alpha(F_{xx}\bar{D}_{xx} - \bar{F}_{xx}D_{xx})}, \quad \mu = \frac{\bar{A}_{xz}\bar{D}_{xz}}{\alpha(F_{xx}\bar{D}_{xx} - \bar{F}_{xx}D_{xx})}.$$ (5.5.65b)

The solution to this differential equation is

$$N_{xz}^R(x) = c_5 \sinh\lambda x + c_6 \cosh\lambda x + \frac{\mu}{\lambda^2}\left[\frac{q_0 L}{2}\left(1 - 2\frac{x}{L}\right) + c_1\right],$$ (5.5.66)

where c_5 and c_6 are constants of integration to be determined, along with the constants c_1 through c_4, using the boundary conditions.

For a rectangular cross-section beam, we have [see Eq. (5.4.22)]

$$\frac{\bar{D}_{xx}\bar{D}_{xx}}{\hat{A}_{xz}D_{xx}D_{xx}} = \frac{6}{5}\frac{1}{A_{xz}}, \quad \frac{\bar{D}_{xx}}{\bar{A}_{xz}D_{xx}} = \frac{6}{5}\frac{1}{A_{xz}}, \quad \frac{\mu}{\lambda^2} = 1.$$ (5.5.67)

The boundary conditions for the hinged–hinged beam are

$$w^C = w^R = M_{xx}^C = M_{xx}^R = P_{xx}^R = 0 \ \text{ at } \ x = 0, L. \tag{5.5.68a}$$

We note from Eq. (5.5.7b) that $M_{xx}^R = P_{xx}^R = 0$ implies

$$\frac{dN_{xz}^R}{dx} = 0. \tag{5.5.68b}$$

Using the boundary conditions, we find that

$$c_1 = c_2 = c_3 = 0, \quad c_4 = \frac{\bar{D}_{xx}}{\bar{A}_{xz}} \frac{\mu}{\lambda^4} q_0, \quad c_5 = \frac{\mu}{\lambda^3} q_0, \quad c_6 = -c_5 \tanh \frac{\lambda L}{2}. \tag{5.5.69}$$

With all the constants of integration determined, the solution for hinged–hinged beam according to the RBT can be expressed as follows (noting that $\mu/\lambda^2 = 1$):

$$N_{xz}^R(x) = \frac{q_0}{\lambda}\left(\sinh \lambda x - \tanh \frac{\lambda L}{2} \cosh \lambda x\right) + \frac{q_0 L}{2}\left(1 - 2\frac{x}{L}\right)$$

$$= \frac{q_0}{\lambda} \frac{\sinh \lambda(x - 0.5L)}{\cosh \frac{\lambda L}{2}} + \frac{q_0 L}{2}\left(1 - 2\frac{x}{L}\right), \tag{5.5.70}$$

$$M_{xx}^R(x) = M_{xx}^C(x) = \frac{q_0 L^2}{2} \frac{x}{L}\left(1 - \frac{x}{L}\right), \tag{5.5.71}$$

$$w^R(x) = w^C(x) + \frac{\bar{D}_{xx}}{\bar{A}_{xz} D_{xx}} \frac{q_0 L^2}{2} \frac{x}{L}\left(1 - \frac{x}{L}\right)$$

$$+ \frac{q_0}{\lambda^2} \frac{\bar{D}_{xx}}{\bar{A}_{xz} D_{xx}}\left[\frac{\cosh \lambda(x - 0.5L)}{\cosh \frac{\lambda L}{2}} - 1\right]$$

$$= \frac{q_0 L^4}{24 D_{xx}} \frac{x}{L}\left(1 - 2\frac{x^2}{L^2} + \frac{x^3}{L^3}\right) + \frac{6}{5\bar{A}_{xz}} \frac{q_0 L^2}{2} \frac{x}{L}\left(1 - \frac{x}{L}\right)$$

$$+ \frac{q_0}{\lambda^2} \frac{\bar{D}_{xx}}{\bar{A}_{xz} D_{xx}}\left[\frac{\cosh \lambda(x - 0.5L)}{\cosh \frac{\lambda L}{2}} - 1\right], \tag{5.5.72}$$

$$\phi_x^R(x) = -\frac{D_{xx}}{\bar{D}_{xx}} \frac{dw^c}{dx} + \alpha \frac{F_{xx}}{\bar{D}_{xx}} \frac{dw^R}{dx} + \frac{1}{\bar{D}_{xx}}\left(c_1 \frac{x^2}{2} + c_2 x + c_3\right)$$

$$= -\frac{q_0 L^3}{24 D_{xx}}\left(1 - 6\frac{x^2}{L^2} + 4\frac{x^3}{L^3}\right) + \frac{3}{10} \frac{1}{\bar{A}_{xz}} \frac{q_0 L^2}{2}\left(1 - 2\frac{x}{L}\right)$$

$$+ \frac{3}{10} \frac{1}{\bar{A}_{xz}} \frac{q_0}{\lambda} \frac{\sinh \lambda(x - 0.5L)}{\cosh \frac{\lambda L}{2}}. \tag{5.5.73}$$

By comparing the RBT solution for $w(x)$ with that of the TBT solution, we identify the shear correction coefficient K_s to be $K_s = 5/6$ [see Eq. (5.5.67)]. Of course, the RBT does not require a shear correction coefficient but provides one for the TBT. We also note that the rotation predicted by the RBT has a contribution due to transverse shear strain.

(b) *The simplified RBT solution using the relationships.* Using the boundary conditions $w^S = M_{xx}^S = 0$ at $x = 0$ and $x = L$ in Eqs. (5.5.38) and (5.5.40), we obtain

$c_1 = c_2 = c_3 = c_4 = 0$. The solution according to the simplified RBT becomes

$$V_{\text{eff}}^S = N_{xz}^C = \frac{q_0 L}{2}\left(1 - 2\frac{x}{L}\right),\tag{5.5.74}$$

$$M_{xx} = M_{xx}^C = \frac{q_0 L^2}{2}\frac{x}{L}\left(1 - \frac{x}{L}\right),\tag{5.5.75}$$

$$\phi_x^S(x) = -\frac{dw^C}{dx} = -\frac{q_0 L^3}{24 D_{xx}}\left(1 - 6\frac{x^2}{L^2} + 4\frac{x^3}{L^3}\right),\tag{5.5.76}$$

$$w^S(x) = w^C(x) + \left(\frac{\bar{D}_{xx}}{\hat{A}_{xz}D_{xx}}\right)M_{xx}^C$$

$$= \frac{q_0 L^4}{24 D_{xx}}\frac{x}{L}\left(1 - 2\frac{x^2}{L^2} + \frac{x^3}{L^3}\right) + \frac{3}{2}\frac{1}{A_{xz}}\frac{q_0 L^2}{2}\frac{x}{L}\left(1 - \frac{x}{L}\right).\tag{5.5.77}$$

It is clear that the shear correction coefficient predicted by the simplified RBT is $K_s = 2/3$.

(c) *The LBT solution using the relationships.* Using the boundary conditions $w^L = M_{xx}^L = 0$ at $x = 0$ and $x = L$ in Eqs. (5.5.48) and (5.5.49), we obtain $c_1 = c_2 = c_3 = c_4 = 0$. The solution according to the LBT becomes [see Eqs. (5.7.5), (5.5.48), (5.5.53), and (5.5.55)]

$$N_{xz}^L = N_{xz}^C = \frac{q_0 L}{2}\left(1 - 2\frac{x}{L}\right),\tag{5.5.78}$$

$$M_{xx}^L = M_{xx}^C = \frac{q_0 L^2}{2}\frac{x}{L}\left(1 - \frac{x}{L}\right),\tag{5.5.79}$$

$$\phi_x^L = -\frac{dw^C}{dx} + \alpha\frac{F_{xx}}{D_{xx}A_{xz}}N_{xz}^C$$

$$= -\frac{q_0 L^3}{24 D_{xx}}\left(1 - 6\frac{x^2}{L^2} + 4\frac{x^3}{L^3}\right) + \frac{1}{5A_{xz}}\frac{q_0 L}{2}\left(1 - 2\frac{x}{L}\right),\tag{5.5.80}$$

$$w^L(x) = w^C(x) + \frac{\bar{D}_{xx}}{A_{xz}D_{xx}}M_{xx}^C$$

$$= \frac{q_0 L^4}{24 D_{xx}}\frac{x}{L}\left(1 - 2\frac{x^2}{L^2} + \frac{x^3}{L^3}\right) + \frac{6}{5}\frac{1}{A_{xz}}\frac{q_0 L^2}{2}\frac{x}{L}\left(1 - \frac{x}{L}\right).\tag{5.5.81}$$

The shear correction coefficient predicted by the LBT is $K_s = 5/6$. We also note that the rotation ϕ_x predicted by the LBT has a contribution due to transverse shear strain, which is different that is predicted by the RBT. The rotation estimated by the simplified RBT does not contains any transverse shear contribution.

Example 5.5.2

We wish to determine the RBT exact bending solutions of a cantilevered beam (the end $x = 0$ is clamped and end $x = L$ is free) subjected to a uniformly distributed load of magnitude q_0.

Solution:

The CBT solution: For a cantilever beam under uniformly distributed load of intensity q_0, the stress–resultants and the deflection of the CBT are:

$$N_{xz}^C(x) = q_0 L \left(1 - \frac{x}{L}\right), \tag{5.5.82}$$

$$M_{xx}^C(x) = -\frac{q_0 L^2}{2} \left(1 - \frac{x}{L}\right)^2, \tag{5.5.83}$$

$$w^C(x) = \frac{q_0 L^4}{24 D_{xx}} \frac{x^2}{L^2} \left(6 - 4\frac{x}{L} + \frac{x^2}{L^2}\right), \tag{5.5.84}$$

$$-\frac{dw^C}{dx} = -\frac{q_0 L^3}{6 D_{xx}} \frac{x}{L} \left(3 - 3\frac{x}{L} + \frac{x^2}{L^2}\right). \tag{5.5.85}$$

The TBT solution using the relationships: Using the relationship for clamped-free (CF) beams from Table 4.7.1, the bending solutions for the TBT are:

$$N_{xz}^F(x) = N_{xz}^C(x) = q_0 L \left(1 - \frac{x}{L}\right), \tag{5.5.86}$$

$$M_{xx}^F(x) = M_{xx}^C(x) = -\frac{q_0 L^2}{2} \left(1 - \frac{x}{L}\right)^2, \tag{5.5.87}$$

$$\begin{aligned}
w^F(x) &= w^C(x) + \frac{1}{K_s A_{xz}} \left[M_{xx}^C(x) - M_{xx}^C(0)\right] \\
&= \frac{q_0 L^4}{24 D_{xx}} \frac{x^2}{L^2} \left(6 - 4\frac{x}{L} + \frac{x^2}{L^2}\right) + \frac{1}{K_s A_{xz}} \frac{q_0 L^2}{2} \frac{x}{L} \left(2 - \frac{x}{L}\right),
\end{aligned} \tag{5.5.88}$$

$$\phi_x^F(x) = -\frac{dw^C}{dx} = -\frac{q_0 L^3}{6 D_{xx}} \frac{x}{L} \left(3 - 3\frac{x}{L} + \frac{x^2}{L^2}\right). \tag{5.5.89}$$

The RBT solution using the relationships: In the case of the RBT, we need to first solve the second-order differential equation for the transverse shear force N_{xz}^R:

$$\frac{d^2 N_{xz}^R}{dx^2} - \lambda^2 N_{xz}^R = -\mu \left[q_0 L \left(1 - \frac{x}{L}\right) + c_1\right], \tag{5.5.90}$$

where λ and μ are defined by Eq. (5.5.65b). The general solution of Eq. (5.5.90) is

$$N_{xz}^R(x) = c_5 \sinh \lambda x + c_6 \cosh \lambda x + \frac{\mu}{\lambda^2} \left[q_0 L \left(1 - \frac{x}{L}\right) + c_1\right]. \tag{5.5.91}$$

The boundary conditions for the cantilever beam are

$$w^C = w^R = \frac{dw^C}{dx} = \frac{dw^R}{dx} = \phi_x^R = 0 \quad \text{at} \quad x = 0,$$

$$N_{xz}^C = V_{\text{eff}}^R = M_{xx}^C = M_{xx}^R = P_{xx}^R = 0 \quad \text{at} \quad x = L. \tag{5.5.92a}$$

Again, $M_{xx}^R(L) = P_{xx}^R(L) = 0$ implies that

$$\frac{dN_{xz}^R}{dx} = 0 \quad \text{at} \quad x = L \tag{5.5.92b}$$

and the boundary conditions at $x = 0$, in view of Eq. (5.5.5c), imply that $N_{xz}^R(0) = 0$. Although $N_{xz}^R(0)$ obtained from the constitutive relation (5.5.5c) is zero at the clamped edge, the effective shear force of the theory V_{eff}^R at $x = 0$ is indeed not zero. It is given by Eq. (5.5.11).

Using the boundary conditions, we find that

$$c_1 = c_2 = c_3 = 0, \quad c_4 = \frac{\mu}{\lambda^4} q_0 \left(\frac{1 + \lambda L \sinh \lambda L}{\cosh \lambda L} \right) \left(\frac{\bar{D}_{xx}}{\hat{A}_{xz}} \right),$$

$$c_5 = \frac{\mu}{\lambda^3} q_0 \left(\frac{1 + \lambda L \sinh \lambda L}{\cosh \lambda L} \right), \quad c_6 = -\frac{\mu}{\lambda^3} q_0 L. \tag{5.5.93}$$

The RBT solution becomes (noting that $\mu / \lambda^2 = 1$)

$$N_{xz}^R(x) = \frac{q_0}{\lambda} \left(\frac{\sinh \lambda x - \lambda L \cosh \lambda (L - x)}{\cosh \lambda L} \right) + q_0 L \left(1 - \frac{x}{L} \right), \tag{5.5.94}$$

$$M_{xx}^R(x) = M_{xx}^E(x) = -\frac{q_0 L^2}{2} \left(1 - \frac{x}{L} \right)^2, \tag{5.5.95}$$

$$w^R(x) = w^C(x) + \frac{\bar{D}_{xx}}{\bar{A}_{xz} D_{xx}} \frac{q_0 L^2}{2} \frac{x}{L} \left(2 - \frac{x}{L} \right)$$
$$+ \frac{\bar{D}_{xx}}{\bar{A}_{xz} D_{xx}} \frac{q_0}{\lambda^2} \left[\frac{\cosh \lambda x + \lambda L \sinh \lambda (L - x)}{\cosh \lambda L} \right]$$
$$- \frac{\bar{D}_{xx}}{\bar{A}_{xz} D_{xx}} \frac{q_0}{\lambda^2} \left(\frac{1 + \lambda L \sinh \lambda L}{\cosh \lambda L} \right)$$
$$= \frac{q_0 L^4}{24 D_{xx}} \frac{x^2}{L^2} \left(6 - 4\frac{x}{L} + \frac{x^2}{L^2} \right) + \frac{6}{5} \frac{1}{A_{xz}} \frac{q_0 L^2}{2} \frac{x}{L} \left(2 - \frac{x}{L} \right)$$
$$+ \frac{6}{5} \frac{1}{A_{xz}} \frac{q_0}{\lambda^2} \left[\frac{\cosh \lambda x + \lambda L \sinh \lambda (L - x)}{\cosh \lambda L} \right]$$
$$- \frac{6}{5} \frac{1}{A_{xz}} \frac{q_0}{\lambda^2} \left(\frac{1 + \lambda L \sinh \lambda L}{\cosh \lambda L} \right), \tag{5.5.96}$$

$$\phi_x^R(x) = -\frac{D_{xx}}{\bar{D}_{xx}} \frac{dw^c}{dx} + \alpha \frac{F_{xx}}{\bar{D}_{xx}} \frac{dw^R}{dx} + \frac{1}{\bar{D}_{xx}} \left(c_1 \frac{x^2}{2} + c_2 x + c_3 \right)$$
$$= -\frac{q_0 L^3}{6 D_{xx}} \frac{x}{L} \left(3 - 3\frac{x}{L} + \frac{x^2}{L^2} \right) + \frac{3}{10} \frac{1}{A_{xz}} q_0 L \left(1 - \frac{x}{L} \right)$$
$$+ \frac{3}{10} \frac{1}{A_{xz}} \frac{q_0}{\lambda} \left[\frac{\sinh \lambda x - \lambda L \cosh \lambda (L - x)}{\cosh \lambda L} \right]. \tag{5.5.97}$$

Again, it can be seen that the effective shear correction factor of the RBT is $K_s = 5/6$ and that the rotation is influenced by the transverse shear effect.

The simplified RBT solution using the relationships: Using the boundary conditions $V_{\text{eff}}^S = N_{xz}^S = M_{xx}^S = M_{xx}^C = 0$ at $x = L$ give $c_1 = c_2 = 0$, and the boundary conditions $w^S = w^C = \phi_x^S = dw^C/dx = 0$ at $x = 0$ yield $c_3 = c_4 = 0$. Hence the solution as per the simplified RBT is

$$V_{\text{eff}}^S = N_{xz}^C = q_0 L\left(1 - \frac{x}{L}\right), \tag{5.5.98}$$

$$M_{xx}^S = M_{xx}^C - \frac{q_0 L^2}{2}\left(1 - \frac{x}{L}\right)^2, \tag{5.5.99}$$

$$\phi_x^S(x) = -\frac{dw^C}{dx} = -\frac{q_0 L^3}{6D_{xx}}\frac{x}{L}\left(3 - 3\frac{x}{L} + \frac{x^2}{L^2}\right), \tag{5.5.100}$$

$$w^S(x) = w^C(x) + \left(\frac{\bar{D}_{xx}}{\hat{A}_{xz}D_{xx}}\right)M_{xx}^C$$

$$= \frac{q_0 L^4}{24D_{xx}}\frac{x^2}{L^2}\left(6 - 4\frac{x}{L} + \frac{x^2}{L^2}\right) - \frac{3}{2}\frac{1}{A_{xz}}\frac{q_0 L^2}{2}\left(1 - \frac{x}{L}\right)^2. \tag{5.5.101}$$

The LBT solution using the relationships: Using the boundary conditions $N_{xz}^L = N_{xz}^C = M_{xx}^L = M_{xx}^C = 0$ at $x = L$ give $c_1 = c_2 = 0$, and the boundary conditions $w^L = w^C = \phi_x^L = dw^C/dx = 0$ at $x = 0$ yield $c_3 = c_4 = 0$. Hence the solution as per the LBT is [see Eqs. (5.5.47), (5.5.48), (5.5.53), and (5.5.55)]

$$N_{xz}^L(x) = N_{xz}^C = q_0 L\left(1 - \frac{x}{L}\right), \tag{5.5.102}$$

$$M_{xx}^L(x) = M_{xx}^C = -\frac{q_0 L^2}{2}\left(1 - \frac{x}{L}\right)^2, \tag{5.5.103}$$

$$\phi_x^L(x) = -\frac{q_0 L^3}{6D_{xx}}\frac{x}{L}\left(3 - 3\frac{x}{L} + \frac{x^2}{L^2}\right) + \frac{3}{10}\frac{1}{A_{xz}}q_0 L\left(1 - \frac{x}{L}\right), \tag{5.5.104}$$

$$w^L(x) = \frac{q_0 L^4}{24D_{xx}}\frac{x^2}{L^2}\left(6 - 4\frac{x}{L} + \frac{x^2}{L^2}\right) - \frac{6}{5}\frac{1}{A_{xz}}\frac{q_0 L^2}{2}\left(1 - \frac{x}{L}\right)^2. \tag{5.5.105}$$

5.5.7 BUCKLING RELATIONSHIPS

5.5.7.1 Summary of equations of the CBT

From Eqs. (4.7.141) we have the following equations, expressed in terms of the stress resultants, governing buckling of homogeneous beams (i.e., $B_{xx} = 0$) according to the CBT:

$$-\frac{dM_{xx}^C}{dx} + N_{xz}^C = 0, \quad -\frac{dN_{xz}^C}{dx} + N_{xx}^C\frac{d^2w^C}{dx^2} = 0, \tag{5.5.106}$$

where the superscript C signifies that these equations are valid in the CBT. Solving the first equation for N_{xz}^C and substituting into the second equation of (5.5.106), we obtain

$$-\frac{d^2 M_{xx}^C}{dx^2} + N_{xx}^C\frac{d^2w^C}{dx^2} = 0.$$

5.5.7.2 Summary of equations of the RBT

The equations governing buckling of beams according to the RBT are [see Eqs. (5.3.11)–(5.3.14c); omit all time-derivative and couple stress terms, $c_f = q = 0$, and assume that $N_{xx} = -N_{xx}^R$ is a constant]:

$$\frac{d^2 M_{xx}^R}{dx^2} - N_{xx}^R \frac{d^2 w^R}{dx^2} = 0, \tag{5.5.107}$$

$$\alpha \frac{d^2 P_{xx}^R}{dx^2} + \frac{d\bar{N}_{xz}^R}{dx} - N_{xx}^R \frac{d^2 w^R}{dx^2} = 0. \tag{5.5.108}$$

The boundary conditions are of the form

$$\left\{ \begin{array}{c} w^R \\ \dfrac{dw^R}{dx} \\ \phi_x^R \end{array} \right\} \quad \text{or} \quad \left\{ \begin{array}{c} \dfrac{dM_{xx}^R}{dx} - N_{xx}^R \dfrac{dw^R}{dx} \\ \alpha\, P_{xx} \\ \bar{M}_{xx}^R \end{array} \right\}. \tag{5.5.109}$$

In order to obtain differential equations for buckling in terms of a single variable, it is useful to express P_{xx}^R, dN_{xz}^R/dx, and dP_{xz}^R/dx in terms of M_{xx}^R and $d^2 w^R/dx^2$ by using Eqs. (5.3.20b) and (5.3.20c) with $B_{xx} = 0$ and $M_{xx}^T = P_{xx}^T = 0$:

$$M_{xx}^R = \bar{D}_{xx} \frac{d\phi_x^R}{dx} - \alpha F_{xx} \frac{d^2 w^R}{dx^2}, \tag{5.5.110a}$$

$$P_{xx}^R = \bar{F}_{xx} \frac{d\phi_x^R}{dx} - \alpha H_{xx} \frac{d^2 w^R}{dx^2}, \tag{5.5.110b}$$

$$N_{xz}^R = \bar{A}_{xz} \left(\phi_x^R + \frac{dw^R}{dx} \right), \tag{5.5.110c}$$

$$P_{xz}^R = \bar{D}_{xz} \left(\phi_x^R + \frac{dw^R}{dx} \right), \tag{5.5.110d}$$

where $[\alpha = 4/(3h^2)$ and $\beta = 4/h^2]$

$$\bar{D}_{xx} = D_{xx} - \alpha F_{xx}, \quad \bar{F}_{xx} = F_{xx} - \alpha H_{xx}, \\ \bar{A}_{xz} = A_{xz} - \beta D_{xz}, \quad \bar{D}_{xz} = D_{xz} - \beta F_{xz}. \tag{5.5.111}$$

First, from Eq. (5.5.110a), it is seen that

$$\frac{d\phi_x^R}{dx} = \frac{1}{\bar{D}_{xx}} \left(M_{xx}^R + \alpha F_{xx} \frac{d^2 w^R}{dx^2} \right), \tag{5.5.112}$$

which, on substitution into Eqs. (5.5.110b)–(5.5.110d) yields

$$P_{xx}^R = \frac{1}{\bar{D}_{xx}} \left[\bar{F}_{xx} M_{xx}^R - \alpha \left(D_{xx} H_{xx} - F_{xx}^2 \right) \frac{d^2 w^R}{dx^2} \right], \tag{5.5.113a}$$

$$\frac{dN_{xz}^R}{dx} = \frac{\bar{A}_{xz}}{\bar{D}_{xz}} \left(M_{xx}^R + \alpha F_{xx} \frac{d^2 w^R}{dx^2} \right) + \bar{A}_{xz} \frac{d^2 w^R}{dx^2}, \tag{5.5.113b}$$

$$\frac{dP_{xz}^R}{dx} = \frac{\bar{D}_{xz}}{\bar{D}_{xx}}\left(M_{xx}^R + \alpha F_{xx}\frac{d^2 w^R}{dx^2}\right) + \bar{D}_{xz}\frac{d^2 w^R}{dx^2}. \qquad (5.5.113c)$$

By using the foregoing expressions for P_{xx}^R, dN_{xz}^R/dx and dP_{xz}^R/dx in Eqs. (5.5.107) and (5.5.108), the following buckling equation in terms of M_{xx}^R can be derived:

$$\frac{d^4 M_{xx}^R}{dx^4} + \left[\frac{\hat{D}_{xx}N_{xx}^R - D_{xx}\bar{A}_{xz}}{\alpha^2 \tilde{D}_{xx}}\right]\frac{d^2 M_{xx}^R}{dx^2} - \left[\frac{\bar{A}_{xz}N_{xx}^R}{\alpha^2 \tilde{D}_{xx}}\right]M_{xx}^R = 0, \qquad (5.5.114)$$

where

$$\hat{D}_{xx} = \bar{D}_{xx} - \alpha\bar{F}_{xx}, \quad \hat{A}_{xz} = \bar{A}_{xz} - \beta\bar{D}_{xz}, \quad \tilde{D}_{xx} = D_{xx}H_{xx} - F_{xx}^2. \quad (5.5.115)$$

Buckling equations in terms of ϕ_x^R and w^R can also be derived in a similar manner. These are

$$\frac{d^5 \phi_x^R}{dx^5} + \left[\frac{\hat{D}_{xx}N_{xx}^R - D_{xx}\hat{A}_{xz}}{\alpha^2 \tilde{D}_{xx}}\right]\frac{d^3 \phi_x^R}{dx^3} - \left[\frac{\hat{A}_{xz}N_{xz}^R}{\alpha^2 \tilde{D}_{xx}}\right]\frac{d\phi_x^R}{dx} = 0, \qquad (5.5.116)$$

$$\frac{d^6 w^R}{dx^6} + \left[\frac{\hat{D}_{xx}N_{xz}^R - D_{xx}\hat{A}_{xz}}{\alpha^2 \tilde{D}_{xx}}\right]\frac{d^4 w^R}{dx^4} - \left[\frac{\hat{A}_{xz}N_{xz}^R}{\alpha^2 \tilde{D}_{xx}}\right]\frac{d^2 w^R}{dx^2} = 0. \qquad (5.5.117)$$

Equation (5.5.117) shows that the RBT is a sixth-order theory in comparison to the CBT and the TBT.

Equation (5.5.114) may be factored into

$$\left(\frac{d^2}{dx^2} + \lambda_1^R\right)\left(\frac{d^2}{dx^2} + \lambda_2^R\right)M_{xx}^R = 0, \qquad (5.5.118)$$

where

$$\lambda_j^R = (-1)^j \sqrt{\left[\frac{\bar{D}_{xx}N_{xx}^R - D_{xx}\hat{A}_{xz}}{2\alpha^2 \tilde{D}_{xx}}\right]^2 + \frac{\hat{A}_{xz}N_{xx}^R}{\alpha^2 \tilde{D}_{xx}}}$$

$$+ \left[\frac{\hat{D}_{xx}N_{xx}^R - D_{xx}\bar{A}_{xz}}{2\alpha^2 \tilde{D}_{xx}}\right], \quad j = 1, 2. \qquad (5.5.119)$$

Let

$$\mu_i = \left(\frac{d^2}{dx^2} + \lambda_i^R\right)M_{xx}^R. \qquad (5.5.120)$$

Then Eq. (5.5.118) can be written as

$$\left(\frac{d^2}{dx^2} + \lambda_j^R\right)\mu_i = 0, \quad j = 1 \text{ or } 2, \quad j \neq i. \qquad (5.5.121)$$

In the case of the buckling of beams using the CBT, it can be shown that Eq. (4.7.143),

$$D_{xx}\frac{d^4 w^C}{dx^4} + \frac{N_{xx}^C}{D_{xx}}\frac{d^2 w^C}{dx^2} = 0 \tag{5.5.122}$$

can be written as (because $M_{xx}^C = D_{xx}d^2 w^C/dx^2$)

$$\left(\frac{d^2}{dx^2} + \lambda^C\right)M_{xx}^C = 0, \tag{5.5.123}$$

where M_{xx}^C is the bending moment in the CBT and

$$\lambda^C = \frac{N_{xx}^C}{D_{xx}}. \tag{5.5.124}$$

It is clear that Eqs. (5.5.121) and (5.5.123) are similar in form. Therefore, one may relate the buckling loads of the RBT to the buckling loads of CBT:

$$\lambda_j^R = \lambda^C. \tag{5.5.125}$$

The substitution for λ^C and λ_j^R from Eqs. (5.5.119) and (5.5.124) into Eq. (5.5.125) provides the buckling load relationship:

$$N_{xx}^R = \frac{N_{xx}^C\left[1 + \frac{N_{xx}^C \alpha^2 \tilde{D}_{xx}}{\hat{A}_{xz}D_{xx}^2}\right]}{1 + \frac{N_{xx}^c \hat{D}_{xx}}{D_{xx}\hat{A}_{xz}}}. \tag{5.5.126}$$

The relationship in Eq. (5.5.126) is valid provided the two boundary conditions required for solving the second-order differential equations (5.5.121) or (5.5.123) for both models (i.e., CBT and RBT) of beams must also have the same form. This is the case, as shown in the example to follow, for (see Fig. 3.7.8) (a) clamped–clamped (C–C), (b) clamped–free (C–F), and (c) pinned–pinned beams with rotational springs of equal stiffness added to their ends.

Example 5.5.3

Determine the buckling load of a beam–column based on the RBT, when its both ends are pinned (P–P); see Fig. 3.7.8(a).

Solution: The two boundary conditions for the CBT, to be solved with Eq. (5.5.123), are

$$M_{xx}^C = 0 \quad \text{at} \quad x = 0, L. \tag{5.5.127}$$

Thus, it must be shown in the RBT that $\mu = 0$ at $x = 0$ and $x = L$. Now the boundary conditions in the RBT are

$$M_{xx}^R - \alpha P_{xx}^R = 0 \quad \text{and} \quad \alpha P_{xx}^R = 0 \quad \text{at} \quad x = 0, L,$$
$$\text{which implies that} \quad M_{xx}^R = 0 \quad \text{at} \quad x = 0, L. \tag{5.5.128}$$

In view of Eqs. (5.5.110a), (5.5.110b), and (5.5.128), it may be deduced that

$$\frac{d\phi^R}{dx} = \frac{d^2 w^R}{dx} = 0 \quad \text{at} \quad x = 0, L. \tag{5.5.129}$$

It follows from Eq. (5.5.107) that

$$\frac{d^2 M_{xx}^R}{dx^2} = 0 \quad \text{at} \quad x = 0, L. \tag{5.5.130}$$

It is clear from Eqs. (5.5.120), (5.5.128), and (5.5.130) that

$$\mu_i = 0 \quad \text{at} \quad x = 0, L. \tag{5.5.131}$$

Equations (5.5.127) and (5.5.131) show an exact matching of the form of the boundary conditions and the thus the buckling load relationship is valid for the pinned–pinned case.

Table 5.5.1 contains a comparison of the buckling load parameters $\lambda = N_{xx}^0 L^2 / D_{xx}$, where L is the length of the beam, between the TBT and RBT beam-columns for P–P boundary conditions and parameter $\Omega = D_{xx}/(A_{xz}L^2)$. The TBT buckling loads are computed using Eq. (4.7.150) with $K_s = 5/6$ for a square cross section and $K_s = 9/10$ for a circular cross section.

Table 5.5.1: Comparison of buckling load parameters of pinned-pinned (P-P) beam-columns between the TBT and RBT.

P-P	Beam of square cross section		Beam of circular cross section	
Ω	λ^F	λ^R	λ^F	λ^R
0.0	9.8696	9.8696	9.8696	9.8696
0.1	4.5183	4.5526	4.7074	4.7342
0.2	2.9298	2.9874	3.0908	3.1370

Example 5.5.4

Determine the buckling loads of a beam–column based on the RBT, when its both ends are clamped (C–C); see Fig. 3.7.8(c).

Solution: For a clamped-clamped beam-column, the lateral shearing force, N_{xz}, is zero along the column length. This means that the two boundary conditions for the CBT to be solved with Eq. (5.5.123) are

$$\frac{d M_{xx}^C}{dx} = 0 \quad \text{at} \quad x = 0, L. \tag{5.5.132}$$

This means that one has to prove that $d\mu/dx = 0$ at $x = 0, L$ in the RBT for the clamped-clamped beam-columns. To achieve this, Eq. (5.5.107) is integrated with respect to x to give

$$\frac{dM_{xx}^R}{dx} - N_{xx}^R \frac{dw^R}{dx} = c, \tag{5.5.133}$$

where c is a constant of integration. Since the effective shear force is zero for the RBT clamped–clamped beam–column, we have $c = 0$. Moreover, by using the fact that $dw^R/dx = 0$ for a fixed end, it can be deduced from Eq. (5.5.133), with $c = 0$, that

$$\frac{dM_{xx}^R}{dx} = 0 \quad \text{at} \quad x = 0, L. \tag{5.5.134}$$

The substitution of Eqs. (5.5.110c) and (5.5.110d) into Eq. (5.5.108) leads to

$$\alpha \frac{dP_{xx}^R}{dx} + \hat{A}_{xz}\left(\phi_x^R + \frac{dw^R}{dx}\right) = 0 \quad \text{at} \quad x = 0, L. \tag{5.5.135}$$

Using the fact that $\phi_x^R = 0$ and $dw^R/dx = 0$ at a fixed end, Eq. (5.5.135) reduces to

$$\frac{dP_{xx}^R}{dx} = 0 \quad \text{at} \quad x = 0, L. \tag{5.5.136}$$

In view of Eqs. (5.5.134) and (5.5.136), Eqs. (5.5.110a) and (5.5.110b) provide the result

$$\frac{d^2\phi_x^R}{dx^2} = \frac{d^3 w^R}{dx^3} = 0 \quad \text{at} \quad x = 0, L, \tag{5.5.137}$$

and together with Eq. (5.5.107), we have

$$\frac{d^3 M_{xx}^R}{dx^3} = 0 \quad \text{at} \quad x = 0, L. \tag{5.5.138}$$

Thus, in view of Eqs. (5.5.134) and (5.5.138), we have

$$\frac{d\mu_i}{dx} = \frac{d^3 M_{xx}^R}{dx^3} + \lambda_i^R \frac{dM_{xx}^R}{dx} = 0 \quad \text{at} \quad x = 0, L. \tag{5.5.139}$$

As in the case of pinned-pinned beams, we have a matching of the form of the boundary conditions [c.f., Eqs. (5.5.132) and (5.5.139)] and therefore the buckling load relationship holds for the clamped-clamped beam-columns.

A comparison of the buckling load parameters $\lambda = N_{xx}^0 L^2/D_{xx}$ between the TBT and the RBT beam–columns for C–C boundary conditions and parameter $\Omega = D_{xx}/(A_{xz}L^2)$ are presented in Table 5.5.2 contains. The TBT buckling loads are computed using Eq. (4.7.150) with $K_s = 5/6$ for a square cross section and $K_s = 9/10$ for a circular cross section.

Table 5.5.2: Comparison of buckling load parameters of clamped–clamped (C–C) beam–columns between the TBT and the RBT.

C–C	Beam of square cross section		Beam of circular cross section	
Ω	λ^F	λ^R	λ^F	λ^R
0.0	39.4784	39.4784	39.4784	39.4784
0.1	6.8809	7.1982	7.3292	7.5888
0.2	3.7689	4.1498	4.0395	4.3549

Example 5.5.5 ————————————————————————

Determine the buckling loads of a beam column, as predicted by the RBT, with one end fixed and the other end free (C–F); see Fig. 3.7.8(d).

Solution: For the fixed end at $x = 0$, the boundary conditions of the CBT and RBT have already been shown to match in form. Now for the free end at $x = L$, the boundary condition of the CBT, to be solved with Eq. (5.5.123), is

$$M_{xx}^C(L) = 0. \qquad (5.5.140)$$

It is thus necessary to show that $\mu = 0$ at $x = L$ in the RBT. First, the free end boundary conditions are

$$M_{xx}^R - \alpha P_{xx}^R = 0 \;\; \text{and} \;\; \alpha P_{xx} = 0 \;\; \text{at} \;\; x = L \;\; \rightarrow \;\; M_{xx}^R(L) = 0. \qquad (5.5.141)$$

Next, from Eqs. (5.5.110a), (5.5.110b), and (5.5.141), it can be shown that

$$\frac{d\phi_x^R}{dx} = \frac{d^2 w^R}{dx^2} = 0 \;\; \text{at} \;\; x = L. \qquad (5.5.142)$$

It follows from Eq. (5.5.107) that

$$\frac{d^2 M_{xx}^R}{dx^2} = 0 \;\; \text{at} \;\; x = L. \qquad (5.5.143)$$

Thus, in view of Eqs. (5.5.120), (5.5.141), and (5.5.143), one obtains

$$\mu = \frac{d^2 M_{xx}^R}{dx^2} + \lambda_i^R = 0 \;\; \text{at} \;\; x = L. \qquad (5.5.144)$$

As before, there is a matching in the form of the boundary conditions for the clamped–free beam–columns and thus the buckling load relationship holds.

A comparison of the buckling load parameters $\lambda = N_{xx}^0 L^2 / D_{xx}$ between the TBT and RBT beam-columns for C–F boundary conditions is presented in Table 5.5.3. It is clear from the results that the buckling loads predicted by the RBT are slightly larger than those predicted by the TBT.

Table 5.5.3: Comparison of buckling load parameters of clamped-free (C–F) beam–columns between the TBT and RBT; $\Omega = D_{xx}/(A_{xz}L^2)$.

C-F	Beam of square cross section		Beam of circular cross section	
Ω	λ^F	λ^R	λ^F	λ^R
0.0	2.4674	2.4674	2.4674	2.4674
0.1	1.9037	1.9053	1.9365	1.9376
0.2	1.5497	1.5537	1.5936	1.5967

In view of the fact that the buckling load relationship is valid for the pinned–pinned columns and the clamped–clamped beam columns, it can be readily proved that it also holds for the case of pinned–pinned columns with ends having additional elastic rotational springs of equal stiffness.

We can simplify the form of the buckling load relationship for columns with square or circular cross-section. Noting the definition of the stiffness coefficients in Eqs. (5.3.27) and (5.5.111), it can be readily shown that the buckling load relationships between the RBT and CBT reduce to

$$N_{xx}^R = \frac{N_{xx}^C \left(1 + \frac{N_{xx}^C}{70 A_{xz}}\right)}{1 + \frac{17 N_{xx}^E}{14 A_{xz}}} \quad \text{for square cross-section,} \tag{5.5.145}$$

$$N_{xx}^R = \frac{N_{xx}^C \left(1 + \frac{N_{xx}^C}{90 A_{xz}}\right)}{1 + \frac{101 N^E}{90 A_{xz}}} \quad \text{for circular cross-section.} \tag{5.5.146}$$

Recall the following buckling load relationships between the TBT and the CBT:

$$N_{xx}^C = \frac{N_{xx}^T}{1 - \frac{N_{xx}^F}{S_{xz}}} \quad \text{or} \quad N_{xx}^F = \frac{N_{xx}^C}{1 + \frac{N_{xx}^C}{S_{xz}}}, \tag{5.5.147}$$

where $S_{xz} = K_s A_{xz}$, K_s being the shear correction coefficient. We can estimate the shear correction coefficient required in the TBT from Eqs. (5.5.145) and (5.5.146). We have $K_s = 14/17$ for the square cross section and $K_s = 90/101$ for the circular cross section.

The buckling loads predicted by the RBT are higher than their TBT counterparts due to the factor found in the numerator of the buckling load relationship. The buckling load parameters predicted by the RBT and the TBT are in close agreement but except for the case of clamped–clamped columns; the CBT predicts higher buckling loads, especially for the clamped–clamped case. We also note that the difference between the buckling loads predicted by the RBT and the TBT increases with higher modes of buckling. These higher modes of buckling become important when dealing with beam–columns with internal restraints. As shown by Rozvany and Mröz [127] and Olhoff and Akesson [128], the optimal locations of internal supports for maximizing the buckling load of a column are found at the nodal points of an appropriate higher-order buckling mode. The advantage of the RBT over the TBT is that it does not require shear correction factors.

5.6 NAVIER SOLUTIONS

5.6.1 THE REDDY BEAM THEORY (RBT)

In this section, we present the Navier solutions of the linearized equations of motion (with constant coefficients) of the RBT [see Eqs. (5.3.28)–(5.3.30), and add the elastic foundation term] for simply-supported boundary conditions.

The equations of motion of the RBT are

$$-A_{xx}\frac{\partial^2 u}{\partial x^2} - \bar{B}_{xx}\frac{\partial^2 \phi_x}{\partial x^2} + \alpha\,E_{xx}\frac{\partial^3 w}{\partial x^3} + \frac{\partial N_{xx}^T}{\partial x}$$
$$+m_0\frac{\partial^2 u}{\partial t^2} - \alpha\,m_3\frac{\partial^3 w}{\partial t^2 \partial x} + \bar{m}_1\frac{\partial^2 \phi_x}{\partial t^2} = f, \quad (5.6.1)$$

$$-\alpha\,E_{xx}\frac{\partial^3 u}{\partial x^3} - \left(\hat{A}_{xz} + \beta^2 D_{xy}\right)\frac{\partial^2 w}{\partial x^2} + \left(\alpha^2 H_{xx} + \tfrac{1}{4}\tilde{\tilde{A}}_{xy}\right)\frac{\partial^4 w}{\partial x^4}$$
$$-\left(\hat{A}_{xz} + \beta^2 D_{xy}\right)\frac{\partial \phi_x}{\partial x} - \left(\tfrac{1}{4}\tilde{\tilde{A}}_{xy} + \alpha\bar{F}_{xx}\right)\frac{\partial^3 \phi_x}{\partial x^3} - \alpha\frac{\partial^2 P_{xx}^T}{\partial x^2}$$
$$+\alpha m_3\frac{\partial^3 u}{\partial x \partial t^2} + m_0\frac{\partial^2 w}{\partial t^2} - \alpha^2 m_6\frac{\partial^4 w}{\partial x^2 \partial t^2}$$
$$+\alpha\bar{m}_4\frac{\partial^3 \phi_x}{\partial x \partial t^2} + c_f\,w = q, \quad (5.6.2)$$

$$-\bar{B}_{xx}\frac{\partial^2 u}{\partial x^2} + \left(\alpha\bar{F}_{xx} + \tfrac{1}{4}\tilde{\tilde{A}}_{xy}\right)\frac{\partial^3 w}{\partial x^3} + \left(\hat{A}_{xz} + \beta^2 D_{xy}\right)\frac{\partial w}{\partial x}$$
$$-\left(\hat{D}_{xx} + \tfrac{1}{4}\hat{A}_{xy}\right)\frac{\partial^2 \phi_x}{\partial x^2} + \left(\hat{A}_{xz} + \beta^2 D_{xy}\right)\phi_x + \frac{\partial \bar{M}_{xx}^T}{\partial x}$$
$$+\bar{m}_1\frac{\partial^2 u}{\partial t^2} - \alpha\,\bar{m}_4\frac{\partial^3 w}{\partial x \partial t^2} + \hat{m}_2\frac{\partial^2 \phi_x}{\partial t^2} = 0, \quad (5.6.3)$$

where $(N_{xx}^T, M_{xx}^T, P_{xx}^T)$ are the thermal force resultants, which are assumed to be known (through a known temperature distribution), and the stiffness and mass coefficients appearing in Eqs. (5.6.1)–(5.6.3) are defined as [see Eq. (5.3.22)]

$$\begin{aligned}
\bar{B}_{xx} &= B_{xx} - \alpha\,E_{xx}, \quad \bar{D}_{xx} = D_{xx} - \alpha\,F_{xx}, \quad \bar{F}_{xx} = F_{xx} - \alpha\,H_{xx},\\
\bar{A}_{xz} &= A_{xz} - \beta\,D_{xz}, \quad \bar{D}_{xz} = D_{xz} - \beta\,F_{xz}, \quad \bar{A}_{xy} = A_{xy} - \beta\,D_{xy},\\
\bar{D}_{xy} &= D_{xy} - \beta\,F_{xy}, \quad \tilde{A}_{xy} = A_{xy} + \beta\,D_{xy}, \quad \tilde{D}_{xy} = D_{xy} + \beta\,F_{xy},\\
\hat{D}_{xx} &= \bar{D}_{xx} - \alpha\,\bar{F}_{xx}, \quad \hat{A}_{xz} = \bar{A}_{xz} - \beta\,\bar{D}_{xz}, \quad \bar{m}_i = m_i - \alpha m_{i+2},\\
\hat{m}_i &= \bar{m}_i - \alpha\bar{m}_{i+2}, \quad \tilde{\tilde{A}}_{xx} = \tilde{\tilde{A}}_{xx} = A_{xy} - \beta^2 F_{xy},\\
\tilde{\tilde{A}}_{xy} &= \tilde{A}_{xy} + \beta\tilde{D}_{xy} = A_{xy} + 2\beta D_{xy} + \beta^2 F_{xy},\\
&\qquad\qquad \alpha = \frac{4}{3h^2}, \quad \beta = 3\alpha.
\end{aligned} \quad (5.6.4)$$

The simply supported boundary conditions at $x = 0$ and $x = L$ for the RBT are [see the duality pairs in Eq. (5.3.14a)]:

$$N_{xx} = 0, \quad w = 0, \quad \tilde{M}_{xx} \equiv M_{xx} - \alpha P_{xx} + \tfrac{1}{2}P_{xy} = 0,$$
$$M_{\text{eff}} \equiv \tfrac{1}{2}P_{xy} + \alpha P_{xx} + \alpha^2 m_6\frac{\partial^2 w}{\partial t^2} = 0. \quad (5.6.5)$$

Recall that N_{xx}, M_{xx}, P_{xx}, and P_{xy} are related to the generalized displacements by [see Eqs. (5.3.20a)–(5.3.20h)]:

$$N_{xx} = A_{xx}\frac{\partial u}{\partial x} + \bar{B}_{xx}\frac{\partial \phi_x}{\partial x} - \alpha E_{xx}\frac{\partial^2 w}{\partial x^2} - N_{xx}^T, \tag{5.6.6a}$$

$$M_{xx} = B_{xx}\frac{\partial u}{\partial x} + \bar{D}_{xx}\frac{\partial \phi_x}{\partial x} - \alpha F_{xx}\frac{\partial^2 w}{\partial x^2} - M_{xx}^T, \tag{5.6.6b}$$

$$P_{xx} = E_{xx}\frac{\partial u}{\partial x} + \bar{F}_{xx}\frac{\partial \phi_x}{\partial x} - \alpha H_{xx}\frac{\partial^2 w}{\partial x^2} - P_{xx}^T, \tag{5.6.6c}$$

$$P_{xy} = \frac{1}{2}A_{xy}\left(\frac{\partial \phi_x}{\partial x} - \frac{\partial^2 w}{\partial x^2}\right) - \frac{1}{2}\beta D_{xy}\left(\frac{\partial \phi_x}{\partial x} + \frac{\partial^2 w}{\partial x^2}\right), \tag{5.6.6d}$$

$$R_{xy} = \frac{1}{2}\beta D_{xy}\left(\frac{\partial \phi_x}{\partial x} + \frac{\partial^2 w}{\partial x^2}\right) - \frac{1}{2}\beta^2 F_{xy}\left(\frac{\partial \phi_x}{\partial x} + \frac{\partial^2 w}{\partial x^2}\right). \tag{5.6.6e}$$

It is clear from the expressions in Eqs. (5.6.6a)–(5.6.6e) that the expansions used for u, w, and ϕ_x should be such that the following quantities vanish at $x = 0$ and $x = L$ so that the simply-supported boundary conditions in Eq. (5.6.5) are satisfied:

$$\frac{\partial u}{\partial x}, \quad w, \quad \frac{\partial^2 w}{\partial x^2}, \quad \frac{\partial \phi_x}{\partial x}.$$

As discussed in Sections 3.8 and 4.8, the solution of the following form meets the simply-supported boundary conditions:

$$u(x,t) = \sum_{m=1}^{\infty} U_m(t)\cos\alpha_m x,$$

$$w(x,t) = \sum_{m=1}^{\infty} W_m(t)\sin\alpha_m x, \quad \alpha_m = \frac{m\pi}{L}, \tag{5.6.7}$$

$$\phi_x(x,t) = \sum_{m=1}^{\infty} S_m(t)\cos\alpha_m x,$$

where $U_m(t)$, $W_m(t)$, and S_m are the coefficients to be determined such that Eqs. (5.6.1)–(5.6.3) are satisfied everywhere in the domain $(0, L)$ for any arbitrary time t. The generalized forces are expanded in the cosine and sine series as

$$f(x,t) = \sum_{m=1}^{\infty} F_m(t)\cos\alpha_m x, \quad q(x,t) = \sum_{m=1}^{\infty} Q_m(t)\sin\alpha_m x,$$

$$N_{xx}^T(x,t) = \sum_{m=1}^{\infty} N_m^T(t)\sin\alpha_m x, \quad M_{xx}^T(x,t) = \sum_{m=1}^{\infty} M_m^T(t)\sin\alpha_m x, \tag{5.6.8}$$

$$P_{xx}^T(x,t) = \sum_{m=1}^{\infty} P_m^T(t)\sin\alpha_m x,$$

where the coefficients F_m, Q_m, N_m^T, M_m^T, and P_m^T are determined from

$$Q_m(t) = \frac{2}{L} \int_0^L q(x,t)\, \sin \alpha_m x\, dx,$$

$$N_m^T(t) = \frac{2}{L} \int_0^L N_{xx}^T(x,t)\, \sin \alpha_m x\, dx,$$

$$M_m^T(t) = \frac{2}{L} \int_0^L M_{xx}^T(x,t)\, \sin \alpha_m x\, dx, \qquad (5.6.9)$$

$$F_m(t) = \frac{2}{L} \int_0^L f(x,t)\, \cos \alpha_m x\, dx.$$

Substituting the expansions from Eqs. (5.6.7) and (5.6.8) into Eqs. (5.6.1)–(5.6.3), we obtain ($\alpha = 4/3h^2$ and $\beta = 4/h^2$; do not confuse between α and α_m as they are different)

$$0 = \sum_{m=1}^{\infty} \left(A_{xx} \alpha_m^2 U_m - \alpha\, E_{xx} \alpha_m^3 W_m + \bar{B}_{xx} \alpha_m^2 S_m + m_0 \frac{d^2 U_m}{dt^2} - \alpha m_3 \alpha_m \frac{d^2 W_m}{dt^2} \right.$$

$$\left. + \bar{m}_1 \frac{d^2 S_m}{dt^2} + \alpha_m N_m^T - F_m \right) \cos \alpha_m x = 0, \qquad (5.6.10a)$$

$$0 = \sum_{m=1}^{\infty} \left\{ -\alpha E_{xx} \alpha_m^3\, U_m + \left[\left(\hat{A}_{xz} + \beta^2 D_{xy} \right) \alpha_m^2 + \left(\alpha^2 H_{xx} + \tfrac{1}{4} \tilde{\tilde{A}}_{xx} \right) \alpha_m^4 \right] W_m \right.$$

$$+ c_f W_m + \left[\left(\hat{A}_{xz} + \beta^2 D_{xy} \right) \alpha_m - \left(\tfrac{1}{4} \tilde{\tilde{A}}_{xy} + \alpha \bar{F}_{xx} \right) \alpha_m^3 \right] S_m$$

$$+ \alpha m_3 \alpha_m \frac{d^2 U_m}{dt^2} + \left(m_0 + \alpha^2 m_6 \alpha_m^2 \right) \frac{d^2 W_m}{dt^2} - \alpha \bar{m}_4 \alpha_m \frac{d^2 S_m}{dt^2}$$

$$\left. - \alpha\, \alpha_m^2 P_{xx}^T - Q_m \right\} \sin \alpha_m x, \qquad (5.6.10b)$$

$$0 = \sum_{m=1}^{\infty} \left\{ \bar{B}_{xx} \alpha_m^2 U_m + \left[\left(\hat{A}_{xz} + \beta^2 D_{xy} \right) \alpha_m - \left(\alpha \bar{F}_{xx} + \tfrac{1}{4} \tilde{\tilde{A}}_{xy} \right) \alpha_m^3 \right] W_m \right.$$

$$+ \left[\left(\hat{D}_{xx} + \hat{A}_{xy} \right) \alpha_m^2 + \hat{A}_{xz} + \beta^2 D_{xy} \right] S_m$$

$$\left. + \bar{m}_1 \frac{d^2 U_m}{dt^2} - \alpha \bar{m}_4 \alpha_m \frac{d^2 W_m}{dt^2} + \hat{m}_2 \frac{d^2 S_m}{dt^2} + \alpha_m \bar{M}_m^T \right\} \cos \alpha_m x, \qquad (5.6.10c)$$

where

$$\bar{m}_1 = m_1 - \alpha m_3, \quad \bar{m}_2 = m_2 - \alpha m_4,$$

$$\bar{m}_4 = m_4 - \alpha m_6, \quad \hat{m}_2 = \bar{m}_2 - \alpha \bar{m}_4. \qquad (5.6.11)$$

Since the above equations must hold for any x and m, we obtain the following ordinary differential equations in time:

$$\begin{bmatrix} R_{11} & R_{12} & R_{13} \\ R_{21} & R_{22} & R_{23} \\ R_{31} & R_{32} & R_{33} \end{bmatrix} \begin{Bmatrix} U_m \\ W_m \\ S_m \end{Bmatrix} + \begin{bmatrix} M_{11} & M_{12} & M_{13} \\ M_{31} & M_{22} & M_{23} \\ M_{31} & M_{32} & M_{33} \end{bmatrix} \begin{Bmatrix} \ddot{U}_m \\ \ddot{W}_m \\ \ddot{S}_m \end{Bmatrix} = \begin{Bmatrix} F_1 \\ F_2 \\ F_3 \end{Bmatrix}, \quad (5.6.12)$$

where $R_{ij} = R_{ji}$ and $M_{ij} = M_{ji}$ (i.e., the matrices $\mathbf{R}$ and $\mathbf{M}$ are symmetric)

$$R_{11} = A_{xx}\alpha_m^2, \quad R_{12} = -\alpha\, E_{xx}\alpha_m^3, \quad R_{13} = \bar{B}_{xx}\alpha_m^2,$$

$$R_{22} = \left[\alpha^2\, H_{xx} + \frac{1}{4}\left(A_{xy} + 2\beta\, D_{xy} + \beta^2\, F_{xy}\right)\right]\alpha_m^4$$

$$+ \left(\hat{A}_{xz} + \beta^2\, D_{xy}\right)\alpha_m^2 + c_f, \quad (5.6.13a)$$

$$R_{23} = \left(\hat{A}_{xz} + \beta^2 D_{xy}\right)\alpha_m - \left[\alpha\bar{F}_{xx} + \frac{1}{4}\left(A_{xy} - \beta^2 F_{xy}\right)\right]\alpha_m^3,$$

$$R_{33} = \hat{A}_{xz} + \beta^2 D_{xy} + \left[\hat{D}_{xx} + \frac{1}{4}\left(A_{xy} + 2\beta D_{xy} + \beta^2 F_{xy}\right)\right]\alpha_m^2,$$

$$M_{11} = m_0, \quad M_{12} = -\alpha\, m_3\,\alpha_m, \quad M_{13} = \bar{m}_1,$$

$$M_{22} = m_0 + \alpha^2\, m_6\alpha^2, \quad M_{23} = -\alpha\,\bar{m}_4\alpha_m,$$

$$M_{33} = \hat{m}_2, \quad F_1 = F_m - \alpha_m N_m^T, \quad (5.6.13b)$$

$$F_2 = Q_m + \alpha\alpha_m^2\, P_m^T, \quad F_3 = -\alpha_m \bar{M}_m^T.$$

5.6.1.1   Bending analysis

For bending analysis, we set the time derivative terms to zero and setup the equations to solve, for each m, for the coefficients (U_m, W_m, S_m):

$$\begin{bmatrix} R_{11} & R_{12} & R_{13} \\ R_{21} & R_{22} & R_{23} \\ R_{31} & R_{32} & R_{33} \end{bmatrix} \begin{Bmatrix} U_m \\ W_m \\ S_m \end{Bmatrix} = \begin{Bmatrix} F_1 \\ F_2 \\ F_3 \end{Bmatrix}, \quad (5.6.14a)$$

and the solution is given by Cramer's rule as ($\Delta_m = \det|\mathbf{R}|$)

$$U_m = \frac{1}{\Delta_m}\begin{vmatrix} F_1 & R_{12} & R_{13} \\ F_2 & R_{22} & R_{23} \\ F_3 & R_{32} & R_{33} \end{vmatrix}, \quad W_m = \frac{1}{\Delta_m}\begin{vmatrix} R_{11} & F_1 & R_{13} \\ R_{21} & F_2 & R_{23} \\ R_{31} & F_3 & R_{33} \end{vmatrix}, \quad S_m = \frac{1}{\Delta_m}\begin{vmatrix} R_{11} & R_{12} & F_1 \\ R_{21} & R_{22} & F_2 \\ R_{31} & R_{32} & F_3 \end{vmatrix}.$$

$$(5.6.14b)$$

With (U_m, W_m, S_m) known, Eq. (5.6.7) gives the complete solution.

5.6.1.2 Natural vibration

For natural (or free) vibration analysis, we set all force coefficients to zero and assume a periodic solution in the form (see Sections 3.8 and 4.8 for details)

$$
\begin{aligned}
u(x,t) &= U_m^0(x)\,\cos\omega_m t, \\
w(x,t) &= w_m^0(x)\,\cos\omega_m t, \\
\phi_x(x,t) &= S_m^0(x)\,\cos\omega_m t,
\end{aligned}
\tag{5.6.15a}
$$

where ω is the natural frequency of vibration and (u_0, w_0, S_0) define the mode shape. We obtain the following eigenvalue equation:

$$
\left(
\begin{bmatrix}
R_{11} & R_{12} & R_{13} \\
R_{21} & R_{22} & R_{23} \\
R_{31} & R_{32} & R_{33}
\end{bmatrix}
- \omega_m^2
\begin{bmatrix}
M_{11} & M_{12} & M_{13} \\
M_{31} & M_{22} & M_{23} \\
M_{31} & M_{32} & M33
\end{bmatrix}
\right)
\begin{Bmatrix}
U_m^0 \\
W_m^0 \\
S_m^0
\end{Bmatrix}
=
\begin{Bmatrix}
0 \\
0 \\
0
\end{Bmatrix},
\tag{5.6.15b}
$$

where the coefficients R_{ij} and M_{ij} are defined in Eqs. (5.6.13a) and (5.6.13b). Equation (5.6.15b) is solved for each m to find ω_m and amplitudes (U_m^0, W_m^0, S_m^0).

5.6.2 THE LEVINSON BEAM THEORY (LBT)

In this section, we present the Navier solutions of the following linearized equations of motion of the LBT [the LBT is generalized in this book to account for the couple stress effect; see Eqs. (5.3.41)–(5.3.43)] for simply-supported boundary conditions:

$$
\begin{aligned}
-A_{xx}\frac{\partial^2 u}{\partial x^2} + \alpha E_{xx}\frac{\partial^3 w}{\partial x^3} - \bar{B}_{xx}\frac{\partial^2 \phi_x}{\partial x^2} + \frac{\partial N_{xx}^T}{\partial x} \\
+ m_0 \frac{\partial^2 u}{\partial t^2} + m_1 \frac{\partial^2 \phi_x}{\partial t^2} = f,
\end{aligned}
\tag{5.6.16}
$$

$$
\begin{aligned}
-\bar{A}_{xz}\frac{\partial^2 w}{\partial x^2} + \tfrac{1}{4}\left(A_{xy} + \beta D_{xy}\right)\frac{\partial^4 w}{\partial x^4} + c_f w \\
-\bar{A}_{xz}\frac{\partial \phi_x}{\partial x} - \tfrac{1}{4}\left(A_{xy} - \beta D_{xy}\right)\frac{\partial^3 \phi_x}{\partial x^3} + m_0 \frac{\partial^2 w}{\partial t^2} = q,
\end{aligned}
\tag{5.6.17}
$$

$$
\begin{aligned}
-B_{xx}\frac{\partial^2 u}{\partial x^2} + \left[\alpha F_{xx} + \tfrac{1}{4}\left(A_{xy} + \beta D_{xy}\right)\right]\frac{\partial^3 w}{\partial x^3} + \bar{A}_{xz}\frac{\partial w}{\partial x} \\
- \left[\bar{D}_{xx} + \tfrac{1}{4}\left(A_{xy} - \beta D_{xy}\right)\right]\frac{\partial^2 \phi_x}{\partial x^2} + \bar{A}_{xz}\phi_x \\
+ \frac{\partial M_{xx}^T}{\partial x} + m_2 \frac{\partial^2 \phi_x}{\partial t^2} + m_1 \frac{\partial^2 u}{\partial t^2} = 0.
\end{aligned}
\tag{5.6.18}
$$

Substituting the expansions in Eqs. (5.6.7) and (5.6.8) into Eqs. (5.6.16)–(5.6.18), we obtain

$$
\begin{bmatrix} L_{11} & L_{12} & L_{13} \\ L_{21} & L_{22} & L_{23} \\ L_{31} & L_{32} & L_{33} \end{bmatrix} \begin{Bmatrix} U_m \\ W_m \\ S_m \end{Bmatrix} + \begin{bmatrix} M_{11} & M_{12} & M_{13} \\ M_{31} & M_{22} & M_{23} \\ M_{31} & M_{32} & M33 \end{bmatrix} \begin{Bmatrix} \ddot{U}_m \\ \ddot{W}_m \\ \ddot{S}_m \end{Bmatrix} = \begin{Bmatrix} F_1 \\ F_2 \\ F_3 \end{Bmatrix}, \quad (5.6.19)
$$

where (we note that the matrices **L** and **M** are *not* symmetric in this case)

$$
\begin{aligned}
& L_{11} = A_{xx}\alpha_m^2, \quad L_{12} = -\alpha\,E_{xx}\alpha_m^3, \\
& L_{13} = \bar{B}_{xx}\alpha_m^2, \quad L_{21} = 0, \\
& L_{22} = \bar{A}_{xz}\alpha_m^2 + \tfrac{1}{4}\left(A_{xy} + \beta\,D_{xy}\right)\alpha_m^4 + c_f, \\
& L_{23} = \bar{A}_{xz}\alpha_m - \tfrac{1}{4}\left(A_{xy} - \beta\,D_{xy}\right)\alpha_m^3, \quad L_{31} = B_{xx}\alpha_m^2, \\
& L_{32} = \bar{A}_{xz}\alpha_m - \left[\alpha\,F_{xx} + \tfrac{1}{4}\left(A_{xy} + \beta\,F_{xy}\right)\right]\alpha_m^3, \quad (5.6.20) \\
& L_{33} = \bar{A}_{xz} + \left[\bar{D}_{xx} + \tfrac{1}{4}\left(A_{xy} - \beta\,D_{xy}\right)\right]\alpha_m^2, \\
& M_{11} = m_0, \quad M_{12} = 0, \quad M_{13} = m_1, \quad M_{21} = 0, \\
& M_{22} = m_0, \quad M_{23} = 0, \quad M_{31} = m_1, \quad M_{32} = 0, \quad M_{33} = m_2, \\
& F_1 = F_m - \alpha_m N_m^T, \quad F_2 = Q_m, \quad F_3 = -\alpha_m M_m^T.
\end{aligned}
$$

5.6.3 NUMERICAL RESULTS

Example 5.6.1 ──────────────────────────────────

Determine the Navier solutions based on CBT, TBT, RBT, and LBT of a simply supported micro-beam subjected to three types of loads: (a) sinusoidal load $q(x) = q_0 \sin\frac{\pi x}{L}$, (b) uniformly distributed load, $q(x) = q_0$, and (c) point load at the center, $q(x) = Q_0\delta(x - x_0)$ with $x_0 = 0.5L$. Use the following material and geometric parameters of micro-beams (with $c_f = 0$):

$$
E_1 = 14.4 \text{ GPa}, \quad E_2 = E = 1.44 \text{ GPa}, \quad \nu = 0.38, \quad K_s = \frac{5(1+\nu)}{(6+5\nu)}, \quad (5.6.21)
$$

$$
h = 5 \times 17.6 \times 10^{-6} \text{ m}, \quad b = 2h, \quad L = 20h, \quad q_0 = 1.0 \text{ N/m}.
$$

Solution: For bending analysis, we use, for example, Eq. (5.6.14b) to determine (U_m, W_m, S_m) for each m and then use Eq. (5.6.7), which requires the sum over the number of terms (i.e., $m = 1, 2, \ldots, N$, N being the number of terms in the series). The value of N is taken to be 100 for the uniform and point loads. Also, note that the results obtained for the TBT depend on K_s.

Numerical results of the normalized center deflection $\bar{w} = w(0.5L)(E_2 I/q_0 L^4) \times 10^3$ of simply supported FGM beams under various types of loads and for the three different beam theories, CBT, TBT, and RBT, are presented in Tables 5.6.1–5.6.3 for the sinusoidal load, uniform load, and point load at the center, respectively, as a function of the volume fraction index n, for four different ratios of the length scale to the beam height, $\mu = \ell/h$. All four theories give almost the same results when the couple stress effect is not included. The RBT model exhibits more stiffening effect due to the couple stress terms. Plots of the center deflection, axial deflection, and rotation as functions of x/L were presented in Chapters 3, 4, or 5, and they are not shown again because the difference between the solutions cannot be seen in the plots.

Table 5.6.1: Center deflections $\bar{w}$ of simply supported FGM beams under sinusoidal load q_0 and for different values of the index n and the ratio $\mu \equiv \ell/h$ $[\bar{w} = w(0.5L)(E_2I/q_0L^4) \times 10^3]$.

	$\mu = 0$				$\mu = 0.2$			
n	CBT	TBT	RBT	LBT	CBT	TBT	RBT	LBT
0.0	1.0266	1.0333	1.0336	1.0853	0.8745	0.8802	0.8388	0.9176
1.0	2.4027	2.4148	2.4154	2.5731	1.9632	1.9732	1.8600	2.0775
5.0	4.6794	4.7061	4.7162	5.2067	3.9054	3.9279	3.6462	4.2707
10.0	5.1454	5.1820	5.1988	5.6942	4.4415	4.4733	4.1915	4.8502
20.0	5.8546	5.9012	5.9156	6.4179	5.1280	5.1691	4.8827	5.5614
	$\mu = 0.4$				$\mu = 1.0$			
n	CBT	TBT	RBT	LBT	CBT	TBT	RBT	LBT
0.0	0.6054	0.6096	0.5360	0.6272	0.1920	0.1943	0.1520	0.1959
1.0	1.2676	1.2747	1.1009	1.3171	0.3642	0.3682	0.2855	0.3713
5.0	2.6102	2.6264	2.1695	2.7754	0.7859	0.7949	0.5657	0.8076
10.0	3.1491	3.1729	2.6511	3.3583	1.0369	1.0502	0.7423	1.0693
20.0	3.7368	3.7681	3.2046	3.9726	1.2889	1.3063	0.9411	1.3294

Table 5.6.2: Center deflections $\bar{w}$ of simply supported FGM beams under uniform load q_0 and for different values of the index n and the ratio $\mu \equiv \ell/h$ $[\bar{w} = w(0.5L)(E_2I/q_0L^4) \times 10^3]$.

	$\mu = 0$				$\mu = 0.2$			
n	CBT	TBT	RBT	LBT	CBT	TBT	RBT	LBT
0.0	1.3021	1.3103	1.3107	1.3762	1.1092	1.1162	1.0638	1.1637
1.0	3.0474	3.0624	3.0631	3.2631	2.4900	2.5023	2.3588	2.6346
5.0	5.9351	5.9680	5.9805	6.6026	4.9534	4.9811	4.6238	5.4157
10.0	6.5261	6.5714	6.5920	7.2204	5.6333	5.6726	5.3149	6.1502
20.0	7.4256	7.4832	7.5009	8.1378	6.5041	6.5547	6.1914	7.0519
	$\mu = 0.4$				$\mu = 1.0$			
n	CBT	TBT	RBT	LBT	CBT	TBT	RBT	LBT
0.0	0.7679	0.7731	0.6796	0.7954	0.2435	0.2464	0.1927	0.2484
1.0	1.6077	1.6165	1.3961	1.6703	0.4620	0.4669	0.3621	0.4709
5.0	3.3107	3.3306	2.7513	3.5195	0.9967	1.0080	0.7174	1.0240
10.0	3.9941	4.0235	3.3618	4.2585	1.3152	1.3315	0.9413	1.3558
20.0	4.7395	4.7781	4.0637	5.0372	1.6348	1.6562	1.1935	1.6856

Table 5.6.3: Center deflections $\bar{w}$ of simply supported FGM beams under central point load Q_0 and for different values of n and μ $[\bar{w} = w(0.5L)(E_2 I/Q_0 L^3) \times 10^4]$.

n	$\mu = 0$				$\mu = 0.2$			
	CBT	TBT	RBT	LBT	CBT	TBT	RBT	LBT
0.0	0.0367	0.0370	0.0370	0.0388	0.0312	0.0315	0.0300	0.0328
0.5	0.0607	0.0611	0.0611	0.0644	0.0505	0.0509	0.0483	0.0531
1.0	0.0858	0.0863	0.0864	0.0920	0.0701	0.0705	0.0665	0.0743
2.0	0.1259	0.1266	0.1267	0.1380	0.1016	0.1022	0.0952	0.1095
5.0	0.1671	0.1683	0.1687	0.1862	0.1395	0.1404	0.1304	0.1527
10.0	0.1838	0.1854	0.1861	0.2038	0.1586	0.1600	0.1500	0.1735
20.0	0.2091	0.2111	0.2117	0.2297	0.1832	0.1849	0.1747	0.1990
100.0	0.2967	0.2993	0.2995	0.3187	0.2572	0.2595	0.2469	0.2740

n	$\mu = 0.4$				$\mu = 1.0$			
	CBT	TBT	RBT	LBT	CBT	TBT	RBT	LBT
0.0	0.0216	0.0218	0.0218	0.0192	0.0069	0.0070	0.0054	0.0070
0.5	0.0336	0.0339	0.0336	0.0296	0.0101	0.0102	0.0080	0.0103
1.0	0.0453	0.0456	0.0453	0.0394	0.0130	0.0132	0.0102	0.0133
2.0	0.0644	0.0648	0.0644	0.0546	0.0181	0.0183	0.0137	0.0185
5.0	0.0932	0.0939	0.0932	0.0776	0.0281	0.0285	0.0202	0.0289
10.0	0.1125	0.1135	0.0948	0.1201	0.0370	0.0376	0.0265	0.0383
20.0	0.1335	0.1348	0.1146	0.1421	0.0460	0.0468	0.0336	0.0476
100.0	0.1839	0.1856	0.1839	0.1617	0.0614	0.0624	0.0474	0.0631

The center deflection (indistinguishable between TBT and LBT) versus n for sinusoidal distributed load for different ratios of ℓ/h are shown in Fig. 5.6.1.

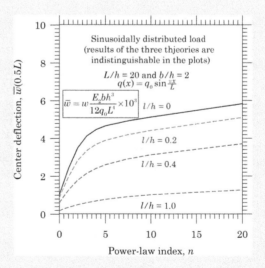

Figure 5.6.1: Normalized center deflection, $\bar{w} = w(0.5L)(E_2 I_2/q_0 L^4)10^3$, versus the volume fraction index n for four different values of the ratio, ℓ/h, for simply-supported beams under sinusoidal distributed load, $q(x) = q_0 \sin(\pi x/L)$.

Example 5.6.2 ———————————————————————————

Determine the first three natural frequencies of a simply supported micro-beam according to the four theories. Use the same material and geometric properties as before [see **Examples 3.8.2** and **4.8.2**]:

$$h = 17.6 \times 10^{-6} \text{ m}, \quad \rho_1 = \rho_2 = 1.22 \times 10^3 \text{ kg/m},$$

$$E_1 = 14.4 \text{ GPa}, \quad E_2 = 1.44 \text{ GPa}, \quad \rho_1 = 12.2 \times 10^3 \text{ kg/m}, \quad (5.6.22)$$

$$\rho_2 = 1.22 \times 10^3 \text{ kg/m}, \quad h = 17.6 \times 10^{-6} \text{ m}.$$

Solution: In the case of natural vibration, we solve Eq. (5.6.14b) for each m. For the present case, we take $m = 1, 2, 3$ to obtain the first three natural frequencies. Most eigenvalue solvers also print out the amplitudes to determine the mode shapes [see Eq. (5.6.7)]: $U_m^0 \cos \alpha_m x$, $W_m^0 \sin \alpha_m x$, and $S_m^0 \cos \alpha_m x$.

Table 5.6.4 contains a comparison of the first three normalized natural frequencies, $\bar{\omega}_m = \omega_m L^2 / \sqrt{\rho_2 A_0 / E_2}$, obtained with the CBT, TBT, and RBT for isotropic ($n = 0$) and functionally graded ($n \neq 0$) beams. The rotary inertia, m_2, is included in all cases (as well as the higher-order inertias in RBT). The results for various cases of including or not including the length scale effect in homogeneous ($n = 0$) and functionally graded ($n \neq 0$) beams are presented. Figures 5.6.3 and 5.6.4 contain plots of dimensionless fundamental frequencies versus the power-law index n for $\ell = 0$ and $\ell = h = 17.6 \times 10^{-6}$ m, respectively. The effect of shear deformation is more significant when the micro-structure dependent constitutive model is considered, which increases the magnitude of the frequency due to the stiffening effect.

Table 5.6.4: First three natural frequencies $\bar{\omega}_m$ of functionally graded simply supported beams.

		$\bar{\omega}_1$			$\bar{\omega}_2$			$\bar{\omega}_3$		
n	ℓ/h	CBT	TBT	RBT	CBT	TBT	RBT	CBT	TBT	RBT
0	0.0	9.86	9.83	9.83	39.32	38.82	38.85	88.02	85.63	85.75
	0.2	10.68	10.65	10.65	42.60	42.06	42.10	95.36	92.78	92.97
	0.4	12.84	12.80	12.80	51.20	50.52	50.60	114.61	111.34	111.77
	0.6	15.79	15.73	15.74	62.97	62.01	62.18	140.97	136.39	137.21
	0.8	19.18	19.08	19.10	76.47	75.05	75.35	171.18	164.51	165.91
	1.0	22.80	22.66	22.69	90.92	88.84	89.32	203.54	193.82	196.06
1	0.0	8.69	8.67	8.67	34.69	34.29	34.32	77.74	75.79	75.97
	0.2	9.62	9.59	9.61	38.37	37.93	38.36	86.00	83.84	85.95
	0.4	11.97	11.93	11.97	47.75	47.16	47.74	107.72	104.15	106.97
	0.6	15.09	15.04	15.09	60.23	59.35	60.21	134.97	130.77	134.90
	0.8	18.61	18.52	18.60	74.25	72.91	74.23	166.39	160.69	166.30
	1.0	22.41	22.28	22.32	89.43	87.42	89.06	200.42	190.99	199.53
10	0.0	10.33	10.28	10.21	41.17	40.47	39.47	92.12	88.80	84.77
	0.2	11.12	11.07	11.04	44.31	43.56	43.14	99.15	95.58	93.58
	0.4	13.20	13.14	13.11	52.63	51.70	51.23	117.75	113.38	111.14
	0.6	16.09	16.00	15.98	64.13	62.88	62.43	143.49	137.66	135.43
	0.8	19.42	19.30	19.29	77.42	75.67	75.37	173.22	165.14	163.48
	1.0	23.09	22.92	22.85	92.04	89.57	89.28	205.95	194.63	193.65

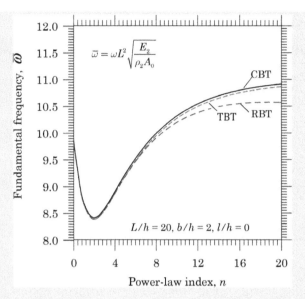

Figure 5.6.2: The effect of the the index on the fundamental natural frequencies ω_1 of a simply-supported FGM beam for $\ell/h = 0$, as predicted by the three beam theories.

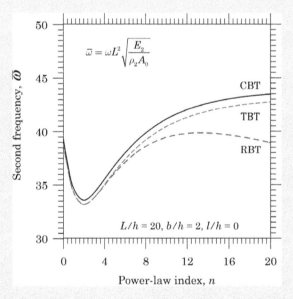

Figure 5.6.3: The effect of the the index on the second natural frequency ω_2 of a simply-supported FGM beam for $\ell/h = 0$, as predicted by the three beam theories.

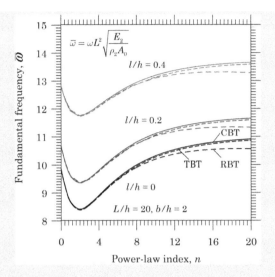

Figure 5.6.4: The effect of the index n and the length scale parameter ℓ/h on the fundamental frequencies of a simply supported FGM beam for $\ell/h = 0$, 0.2, 0.4.

In summary, the bending solutions obtained with the TBT, RBT, and LBT are close to each other when the couple stress effect is not included. When the couple stress effect is included, the RBT models the beams much stiffer than its counterparts. On the other hand, the natural frequencies predicted by the LBT are slightly different from those predicted by the TBT and RBT.

Table 5.6.5: First three natural frequencies $\bar{\omega}_m$ $(m = 1, 2, 3)$ of homogeneous and functionally graded simply supported beams predicted by various beam theories.

m	n	ℓ/h	CBT	TBT	RBT	LBT
1	0	0.0	9.859	9.828	9.829	8.789
2			39.317	38.824	38.847	34.704
3			88.016	85.633	85.746	76.494
1		0.2	10.682	10.648	10.650	9.553
2			42.599	42.062	42.103	37.731
3			95.363	92.777	92.972	83.212
1		0.4	12.839	12.795	12.800	11.543
2			51.198	50.515	50.604	45.602
3			114.612	111.344	111.772	100.617
1		0.8	19.176	19.083	19.102	17.342
2			76.467	75.050	75.348	68.372
3			171.181	164.506	165.915	150.460
1		1.0	22.800	22.663	22.695	20.637
2			90.922	88.840	89.317	81.190
3			203.540	193.820	196.058	178.154

Table 5.6.5 is continued on the next page.

(Table 5.6.5 continued from the previous page)

m	n	ℓ/h	CBT	TBT	RBT	LBT
1	2	0.0	8.415	8.388	8.391	6.843
2			33.587	33.167	33.199	27.001
3			75.290	73.254	73.413	59.438
1		0.2	9.367	9.336	9.364	7.773
2			37.383	36.913	37.334	30.692
3			83.801	81.528	83.553	67.645
1		0.4	11.768	11.727	11.764	10.059
2			46.966	46.334	46.904	39.736
3			105.283	102.250	104.968	87.677
1		0.8	18.479	18.387	18.473	16.240
2			73.750	72.353	73.651	64.039
3			165.324	158.743	164.813	140.989
1		1.0	22.219	22.083	22.212	19.626
2			88.679	86.607	88.559	77.238
3			198.790	189.123	198.160	169.599
1	5	0.0	9.237	9.201	9.190	7.285
2			36.850	36.286	36.139	28.596
3			82.543	79.834	79.227	62.459
1		0.2	10.111	10.071	10.099	8.115
2			40.336	39.717	40.141	31.892
3			90.353	87.387	89.382	69.795
1		0.4	12.368	12.316	12.353	10.208
2			49.339	48.544	49.100	40.171
3			110.519	106.740	109.329	88.138
1		0.8	18.865	18.759	18.842	16.037
2			75.257	73.654	74.890	63.081
3			168.575	161.097	166.740	138.411
1		1.0	22.541	22.388	22.513	19.276
2			89.920	87.617	89.480	75.706
3			201.420	190.791	199.208	165.826

5.7 SOLUTIONS BY VARIATIONAL METHODS

In this section, we consider approximate solutions of beams using Ritz and Galerkin methods. Other weighted–residual methods can also be used but the example presented herein is limited to these two methods. As noted in Section 3.9, the Ritz and Galerkin methods differ in the selection of the approximation functions, ϕ_i and ϕ_0 (see Section 3.9.2.3). In the Ritz method, the approximation functions are required only to satisfy the homogeneous form of the specified essential boundary conditions (because the variational statement used accounts for the actual specified natural boundary conditions), whereas in the Galerkin method they are required to satisfy the homogeneous form of all specified boundary conditions, both essential and natural. In the example presented next, we shall consider both algebraic and trigonometric approximation functions.

Example 5.7.1

Consider a beam of length L and constant bending stiffness EI, loaded with uniformly distributed transverse load of intensity q_0. Develop the N-parameter Ritz solution using the RBT for simply supported boundary conditions, and then compute the one-, two-, and three-parameter solutions and compare with the Navier solution of the problem. Use (a) algebraic and (b) trigonometric approximation functions.

Solution: The total potential energy for this pure bending problem (since the axial displacement u is decoupled from the bending variables w and ϕ_x and it is zero because there is not force to cause it, the strain energy due to the pure extensional strain du/dx is zero) as per the RBT (not including the effect of the couple stress) is given by

$$
\begin{aligned}
\Pi(w,\phi_x) &= \frac{1}{2}\int_0^L \int_A (\sigma_{xx}\varepsilon_{xx} + 2\sigma_{xz}\varepsilon_{xz})\, dA\, dx - \int_0^L q\,w\, dx \\
&= \frac{1}{2}\int_0^L \left[\bar{M}_{xx}\frac{d\phi_x}{dx} - \alpha P_{xx}\frac{d^2 w}{dx^2} + \bar{N}_{xz}\left(\phi_x + \frac{dw}{dx}\right)\right] dx - \int_0^L q\,w\, dx \\
&= \frac{1}{2}\int_0^L \left[\hat{D}_{xx}\left(\frac{d\phi_x}{dx}\right)^2 + \alpha^2 H_{xx}\left(\frac{d^2 w}{dx^2}\right)^2 - 2\alpha \bar{F}_{xx}\frac{d\phi_x}{dx}\frac{d^2 w}{dx^2} \right.\\
&\quad \left. + \alpha^2 H_{xx}\frac{d\phi_x}{dx}\frac{d^2 w}{dx^2} + \hat{A}_{xz}\left(\phi_x + \frac{dw}{dx}\right)^2 \right] dx - \int_0^L q\,w\, dx, \qquad (5.7.1)
\end{aligned}
$$

where the following strain–displacement relations (5.3.7a) and beam constitutive relations (5.3.9b) ($\sigma_{xx} = E\varepsilon_{xx}$ and $\sigma_{xz} = 2G\varepsilon_{xz}$) are used in arriving at Eq. (5.7.1):

$$
\varepsilon_{xx} = z\frac{d\phi_x}{dx} - \alpha z^3\left(\frac{d\phi_x}{dx} + \frac{d^2 w}{dx^2}\right),
$$

$$
2\varepsilon_{xz} = (1 - \beta z^2)\left(\phi_x + \frac{dw}{dx}\right),
$$

$$
M_{xx} = \int_A \sigma_{xx}\, z\, dA = \bar{D}_{xx}\frac{d\phi_x}{dx} - \alpha F_{xx}\frac{d^2 w}{dx^2},
$$

$$
\tag{5.7.2a}
P_{xx} = \int_A z^3 \sigma_{xx}\, dA = \bar{F}_{xx}\frac{d\phi_x}{dx} - \alpha H_{xx}\frac{d^2 w}{dx^2},
$$

$$
N_{xz} = \int_A \sigma_{xz}\, dA = \bar{A}_{xz}\left(\phi_x + \frac{dw}{dx}\right),
$$

$$
R_{xz} = \int_A z^2 \sigma_{xz}\, dA = \bar{D}_{xz}\left(\phi_x + \frac{dw}{dx}\right),
$$

and [$D_{xx} = EI = Ebh^3/12$ and $A_{xz} = GA = Gbh$; see Eq. (5.3.22) for other stiffness coefficients]

$$
\begin{aligned}
&\bar{M}_{xx} = M_{xx} - \alpha P_{xx}, \quad \bar{N}_{xz} = N_{xz} - \beta R_{xz}, \quad \beta = 3\alpha = \frac{4}{h^2}, \\
&\bar{D}_{xx} = D_{xx} - \alpha F_{xx}, \quad \bar{F}_{xx} = F_{xx} - \alpha H_{xx}, \quad \hat{D}_{xx} = \bar{D}_{xx} - \alpha \bar{F}_{xx}, \qquad (5.7.2b)\\
&\bar{A}_{xz} = A_{xz} - \beta D_{xz}, \quad \bar{D}_{xz} = D_{xz} - \beta F_{xz}, \quad \hat{A}_{xz} = \bar{A}_{xz} - \beta \bar{D}_{xz}.
\end{aligned}
$$

The principle of minimum total potential energy, $\delta\Pi(w,\phi_x) = 0$, for the RBT is equivalent to the following two integral statements:

$$0 = \int_0^L \left[\hat{A}_{xz}\frac{d\delta w}{dx}\left(\phi_x + \frac{dw}{dx}\right) + \alpha^2 H_{xx}\frac{d^2\delta w}{dx^2}\frac{d^2w}{dx^2} - \alpha\bar{F}_{xx}\frac{d^2\delta w}{dx^2}\frac{d\phi_x}{dx} - \delta w\,q \right] dx, \tag{5.7.3a}$$

$$0 = \int_0^L \left[\hat{D}_{xx}\frac{d\delta\phi_x}{dx}\frac{d\phi_x}{dx} + \hat{A}_{xz}\,\delta\phi_x\left(\phi_x + \frac{dw}{dx}\right) - \alpha\bar{F}_{xx}\frac{d\delta\phi_x}{dx}\frac{d^2w}{dx^2} \right] dx, \tag{5.7.3b}$$

We assume the Ritz approximations of w and ϕ_x are in the form

$$w(x) \approx W_M = \sum_{j=1}^{M} b_j\psi_j(x) + \psi_0(x), \quad \phi_x(x) \approx \Phi_N = \sum_{j=1}^{N} c_j\theta_j(x) + \theta_0(x), \tag{5.7.4}$$

where the approximation functions $\psi_j(x)$ and $\theta_j(x)$ are independent of each other. Substituting the Ritz approximations in Eq. (5.7.4) for w and ϕ_x and $\delta w = \psi_i$ and $\delta\phi_x = \theta_i$ into the variational statements in Eq. (5.7.3a) and (5.7.3b), we obtain

$$0 = \int_0^L \left[\hat{A}_{xz}\frac{d\psi_i}{dx}\left(\sum_{j=1}^{N} c_j\theta_j + \sum_{j=1}^{M} \frac{d\psi_j}{dx}b_j\right) + \alpha^2 H_{xx}\frac{d^2\psi_i}{dx^2}\left(\sum_{j=1}^{M} b_j\frac{d^2\psi_j}{dx^2}\right) \right.$$
$$\left. - \alpha\bar{F}_{xx}\frac{d^2\psi_i}{dx^2}\left(\sum_{j=1}^{N} c_j\frac{d\theta_j}{dx}\right) - \psi_i q \right] dx$$
$$= \sum_{j=1}^{M} A_{ij}b_j + \sum_{j=1}^{N} B_{ij}c_j - F_i^1, \tag{5.7.5}$$

$$0 = \int_0^L \left[\hat{D}_{xx}\frac{d\theta_i}{dx}\left(\sum_{j=1}^{N} \frac{d\theta_j}{dx}c_j\right) + \hat{A}_{xz}\theta_i\left(\sum_{j=1}^{N} \theta_j c_j + \sum_{j=1}^{M} \frac{d\psi_j}{dx}b_j\right) \right.$$
$$\left. - \alpha\bar{F}_{xx}\frac{d\theta_i}{dx}\left(\sum_{j=1}^{M} b_j\frac{d^2\psi_j}{dx^2}\right) \right] dx$$
$$= \sum_{j=1}^{M} C_{ij}b_j + \sum_{j=1}^{N} D_{ij}c_j - F_i^2, \tag{5.7.6}$$

or in matrix form

$$\begin{bmatrix} \mathbf{A} & \mathbf{B} \\ \mathbf{C} & \mathbf{D} \end{bmatrix}\begin{Bmatrix} \mathbf{b} \\ \mathbf{c} \end{Bmatrix} = \begin{Bmatrix} \mathbf{F}^1 \\ \mathbf{F}^2 \end{Bmatrix}. \tag{5.7.7}$$

The coefficients A_{ij}, B_{ij}, and so on are given by ($F_i^2 = 0$)

$$A_{ij} = \int_0^L \left(\hat{A}_{xz}\frac{d\psi_i}{dx}\frac{d\psi_j}{dx} + \alpha^2 H_{xx}\frac{d^2\psi_i}{dx^2}\frac{d^2\psi_j}{dx^2} \right) dx,$$
$$B_{ij} = \int_0^L \left(\hat{A}_{xz}\frac{d\psi_i}{dx}\theta_j - \alpha\bar{F}_{xx}\frac{d^2\psi_i}{dx^2}\frac{d\theta_j}{dx} \right) dx,$$
$$C_{ij} = \int_0^L \left(\hat{A}_{xz}\theta_i\frac{d\psi_j}{dx} - \alpha\bar{F}_{xx}\frac{d\theta_i}{dx}\frac{d^2\psi_j}{dx^2} \right) dx, \tag{5.7.8}$$
$$D_{ij} = \int_0^L \left(\hat{D}_{xx}\frac{d\theta_i}{dx}\frac{d\theta_j}{dx} + \hat{A}_{xz}\theta_i\,\theta_j \right) dx, \quad F_i^1 = \int_0^L \psi_i q\,dx.$$

The simply-supported boundary conditions for RBT are:

$$w = 0, \quad \bar{M}_{xx} = M_{xx} - \alpha P_{xx} = 0, \quad P_{xx} = 0 \quad \text{at } x = 0, L. \qquad (5.7.9)$$

Of the six boundary conditions, only the first two ($w(0) = w(L) = 0$) are the specified geometric boundary conditions and they are homogeneous. Hence, we have $\psi_0 = \theta_0 = 0$. The ψ_i must satisfy the homogeneous geometric boundary conditions: $\psi_i(0) = 0$ and $\psi_i(L) = 0$.

(a) *Algebraic functions.* As discussed in **Example 4.9.1**, the approximations of w and ϕ_x can be of the form ($M \geq 1$ and $N \geq 2$):

$$W_M(x) = b_1\psi_1(x) + b_2\psi_2(x) + \cdots + b_M\psi_M(x), \quad \psi_i = x^i(L - x), \qquad (5.7.10a)$$

$$\Phi_N(x) = c_1\theta_1(x) + c_2\theta_2(x) + \cdots + c_N\theta_N(x), \quad \theta_i = x^{i-1}. \qquad (5.7.10b)$$

Substituting the Ritz approximation functions into Eq. (5.7.8), we obtain

$$A_{ij} = \hat{A}_{xz}\left[\frac{ij}{i+j-1} - \frac{i+j+2ij}{i+j} + \frac{(i+1)(j+1)}{i+j+1}\right]L^{i+j+1}$$

$$+ \alpha^2 H_{xx}\, ij\left[\frac{(i-1)(j-1)}{i+j-3} - \frac{(i+1)(j-1) - (i-1)(j+1)}{i+j-2}\right.$$

$$\left. + \frac{(i+1)(j+1)}{i+j-1}\right]L^{i+j-1}, \quad (i,j = 1,2,\ldots,M), \qquad (5.7.11a)$$

$$B_{ij} = \hat{A}_{xz}\left[\frac{i}{i+j-1} - \frac{i+1}{i+j}\right]L^{i+j} - \alpha \bar{F}_{xx}\, i(j-1)\left[\frac{i-1}{i+j-3} - \frac{i+1}{i+j-2}\right]L^{i+j-2},$$

$$F_i^1 = q_0 L^{i+2}\frac{1}{(i+1)(i+2)}, \quad (i = 1,2,\ldots,M; j = 1,2,\ldots,N), \qquad (5.7.11b)$$

$$D_{ij} = \left[\hat{D}_{xx}\frac{(i-1)(j-1)}{i+j-3} + \hat{A}_{xz}L^2\frac{1}{i+j-1}\right]L^{i+j-3} \quad (i,j = 1,2,\ldots,N), \quad F_i^2 = 0. \qquad (5.7.11c)$$

Note that when the denominator of the expressions for A_{ij}, $B_{ij} = C_{ji}$, and D_{ij} is less than equal to zero for fixed values of i and j, then that expression is not to be included in the evaluation.

(b) *Trigonometric functions.* The following approximations of w and ϕ_x satisfy the geometric boundary conditions (there are no geometric boundary conditions on ϕ_x):

$$w(x) \approx W_M(x) = \sum_{i=1}^{N} b_i\psi_i(x), \quad \psi_i(x) = \sin \alpha_i x, \quad \alpha_i = \frac{i\pi}{L}, \qquad (5.7.12a)$$

$$\phi_x \approx \Phi_N(x) = \sum_{i=1}^{N} c_i\theta_i(x), \quad \theta_i(x) = \cos \alpha_i x. \qquad (5.7.12b)$$

Substituting for ψ and θ_i into the expressions for A_{ij}, $B_{ij} = C_{ji}$, and D_{ij} and carrying out the indicated integration, we find that

$$A_{ij} = B_{ij} = C_{ij} = D_{ij} = 0 \quad \text{for } i \neq j, \quad \text{and} \quad F_i^1 = 0 \text{ when } i \text{ is even.} \qquad (5.7.13)$$

This result is due to the orthogonality of the trigonometric functions:

$$\int_0^L \sin\frac{m\pi x}{L}\sin\frac{n\pi x}{L}\,dx = 0, \quad \int_0^L \cos\frac{m\pi x}{L}\cos\frac{n\pi x}{L}\,dx = 0, \text{ for } m \neq n,$$
(5.7.14a)

$$\int_0^L \sin\frac{m\pi x}{L}\sin\frac{n\pi x}{L}\,dx = \frac{L}{2}, \quad \int_0^L \cos\frac{m\pi x}{L}\cos\frac{n\pi x}{L}\,dx = \frac{L}{2}, \text{ for } m = n,$$

$$F_i^1 = \int_0^L \sin\frac{i\pi x}{L}\,dx = -\frac{1}{\alpha_i}\left[(-1)^i - 1\right].$$
(5.7.14b)

Thus, all of the matrices $\mathbf{A}$, $\mathbf{C} = \mathbf{B}^{\mathrm{T}}$, and $\mathbf{D}$ are diagonal with the coefficients (no sum on i):

$$A_{ii} = \left(\hat{A}_{xz}\alpha_i\alpha_i + \alpha^2 H_{xx}\alpha_i^2\alpha_i^2\right)\frac{L}{2}, \quad B_{ii} = \left(\hat{A}_{xz} - \alpha\hat{F}_{xx}\alpha_i\alpha_i\right)\alpha_i\frac{L}{2},$$

$$C_{ii} = \left(\hat{A}_{xz} - \alpha\hat{F}_{xx}\alpha_i\alpha_i\right)\alpha_i\frac{L}{2}, \quad D_{ii} = \left(\hat{A}_{xz} + \hat{D}_{xx}\alpha_i\alpha_i\right)\frac{L}{2}.$$
(5.7.15)

Table 5.7.6 contains the numerical values for various number of terms used in the series for the algebraic and trigonometric functions: (a) trigonometric functions: Case 1 ($M = N = 1$), Case 2 ($M = N = 3$), Case 3 ($M = N = 5$), and Case 4 ($M = N = 9$); and (b) algebraic functions: Case 1 ($M = 1, N = 2$) and Case 2 ($M = 3, N = 4$). Figure 5.7.5 shows a comparison of the Ritz solutions obtained with the algebraic functions and the trigonometric functions (the same as the Navier solution).

Table 5.7.6: Comparison of the Ritz solutions obtained using algebraic and trigonometric functions ($\bar{w} = w(0.5L)(E_2 I/q_0 L^4)10^3$) for simply supported beam.

	Algebraic				Trigonometric	
x/L	Case 1	Case 2	Case 3	Case 4	Case 1	Case 2
0.05	0.6818	0.6899	0.6909	0.6912	0.6599	0.6913
0.10	1.3467	1.3613	1.3627	1.3629	1.2504	1.3629
0.15	1.9786	1.9963	1.9973	1.9972	1.7714	1.9972
0.20	2.5616	2.5787	2.5787	2.5784	2.2229	2.5785
0.25	3.0817	3.0944	3.0934	3.0933	2.6050	3.0933
0.30	3.5258	3.5314	3.5300	3.5301	2.9176	3.5301
0.35	3.8831	3.8803	3.8793	3.8795	3.1607	3.8795
0.40	4.1448	4.1343	4.1343	4.1344	3.3344	4.1344
0.45	4.3045	4.2885	4.2895	4.2894	3.4386	4.2894
0.50	4.3581	4.3402	4.3416	4.3414	3.4733	4.3414
0.55	4.3045	4.2885	4.2895	4.2894	3.4386	4.2894
0.60	4.1448	4.1343	4.1343	4.1344	3.3344	4.1344
0.65	3.8831	3.8803	3.8793	3.8795	3.1607	3.8795
0.70	3.5258	3.5314	3.5300	3.5301	2.9176	3.5301
0.75	3.0817	3.0944	3.0934	3.0933	2.6050	3.0933
0.80	2.5616	2.5787	2.5787	2.5784	2.2229	2.5785
0.85	1.9786	1.9963	1.9973	1.9972	1.7714	1.9972
0.90	1.3467	1.3613	1.3627	1.3629	1.2504	1.3629
0.95	0.6818	0.6899	0.6909	0.6912	0.6599	0.6913

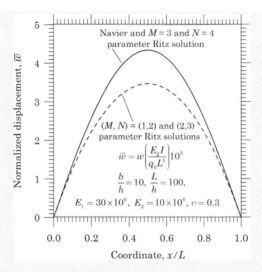

Figure 5.7.5: A comparison of the Ritz solutions $\bar{w}$ with the Navier (exact) solution for the simply supported beam under uniform load of intensity q_0.

We note that the RBT and the TBT are very similar, and the generalized displacements from both theories are close, except for the fact that the RBT accounts for quadratic variation of the transverse shear strain and stress and does not require a shear correction coefficient. The shear stress in the TBT is a constant, which can be interpreted as the value of shear stress at $z = 0$. Since the TBT is algebraically simple and permits easy calculations, one may adopt the shear strain expression from the RBT [see Eq. (5.3.7a); $\beta = 4/h^2$]:

$$\gamma_{xz} = \left(1 - \beta z^2\right) \left(\phi_x^T + \frac{\partial w^T}{\partial x}\right), \tag{5.7.16}$$

with ϕ_x^T and w^T are the TBT solutions. Then the shear stress in the TBT can be computed from

$$\sigma_{xz}^T(x, z) = G(z) \left(1 - \beta z^2\right) \left(\phi_x^T + \frac{\partial w^T}{\partial x}\right). \tag{5.7.17}$$

This way, a parabolic description of the shear stress in the TBT is possible. These expressions hold in all cases (linear or nonlinear, FGM or not, and static or dynamic analyses).

As an example, consider the static bending of a simply supported beam with uniformly distributed load of intensity q_0. The Navier solution for the TBT and the RBT are known from Eqs. (4.8.15) and (5.6.14b), respectively. Using the data in Eq. (5.6.16), we solve the problem using the EBT, TBT, RBT, and LBT. The shear stress distributions at $x = 0$ $\bar{\sigma}_{xz} = \sigma_{xz}(0, z) \times (A/q_0 L)$ through the beam thickness, as predicted by various theories, are shown in Fig. 5.7.6 for homogeneous $(n = 0)$ and FGM (with $n = 1$) beams. It is interesting to note that the shear stress distribution predicted by the CBT using $\sigma_{xz} = VQ/Ib$ (from equilibrium equations) match those predicted (using constitutive relations) by the RBT and the LBT for homogeneous beams (i.e., $n = 0$). The CBT and the TBT distributions for $n \neq 0$ are different from those predicted by the RBT and LBT models.

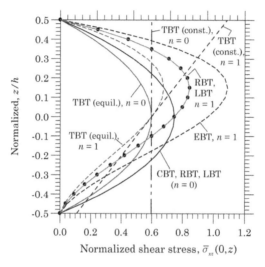

Figure 5.7.6: Normalized transverse shear stress, $\bar{\sigma}_{xz} = \sigma_{xz}(0, z) \times (A/q_0 L)$, distributions through the beam height predicted by various theories.

5.8 CHAPTER SUMMARY

In this chapter, a general third-order theory is introduced first. Then the third-order beam theory with vanishing shear stress at the top and bottom of the beam, referred by many researchers (see, e.g., [129]–[134], which have applications of the RBT to a number of problems) as *the Reddy beam theory* (RBT), is developed accounting for material variation through the thickness, the von Kármán nonlinearity, and the modified couple stress effect. The theory represents a quadratic variation of transverse shear strains and shear stresses through the beam thickness, and thus does not require a shear correction coefficient. In fact, the theory can be used to deduce the shear correction factor needed in the TBT. Exact solutions (by direct integration), the Navier solution, and variational solution of the theory for the linear case are presented. Algebraic relations between the RBT, the CBT, and the TBT are also included.

The LBT, which is based on the same third-order kinematics as the RBT but adopts the TBT equations of motion, is also discussed, and its exact and Navier solutions are presented. The LBT also does not require a shear correction coefficient, and no higher-order stress resultants are introduced. However, the governing equations are not consistent with the principle of virtual displacements.

SUGGESTED EXERCISES

5.1 Derive the equations of motion associated with the displacement field in Eq. (5.3.4).

5.2 Verify the equations of motion presented in Eqs. (5.3.10)–(5.3.12).

5.3 Suppose that the displacements (u, v, w) along the three coordinate axes (x, y, z) in a beam can be expressed as

$$u_1(x, z) = u(x) + z\left[c_0\frac{dw}{dx} + c_1\phi_x(x)\right] + c_2 z^2 \psi_x(x) + c_3\left(\frac{z}{h}\right)^3\left(\phi_x + \frac{dw}{dx}\right),$$

$$u_2(x, z) = 0,$$

$$u_3(x, z) = w(x), \tag{1}$$

where (u, w) denote the displacements of a point $(x, y, 0)$ along the x and z directions, respectively, ϕ_x denotes the rotation of a transverse normal about the y-axis. Show that the nonzero linear strains are given by

$$\varepsilon_{xx} = \varepsilon_{xx}^{(0)} + z\varepsilon_{xx}^{(1)} + z^2\varepsilon_{xx}^{(2)} + z^3\varepsilon_{xx}^{(3)},$$

$$\gamma_{xz} = \gamma_{xz}^{(0)} + z\gamma_{xz}^{(1)} + z^2\gamma_{xz}^{(2)}, \tag{2a}$$

where

$$\varepsilon_{xx}^{(0)} = \frac{du}{dx}, \quad \varepsilon_{xx}^{(1)} = c_0\frac{d^2w}{dx^2} + c_1\frac{d\phi_x}{dx}, \quad \varepsilon_{xx}^{(2)} = c_2\frac{d\psi}{dx}, \quad \varepsilon_{xx}^{(3)} = \frac{c_3}{h^3}\left(\frac{d\phi_x}{dx} + \frac{d^2w}{dx^2}\right),$$

$$\gamma_{xz}^{(0)} = (1 + c_0)\frac{dw}{dx} + c_1\phi_x, \quad \gamma_{xz}^{(1)} = 2c_2\psi_x, \quad \gamma_{xz}^{(2)} = \frac{3c_3}{h^3}\left(\phi_x + \frac{dw}{dx}\right). \tag{2b}$$

5.4 (*Continuation of* **Problem 5.3**). Use the principle of virtual displacements to derive the equations of equilibrium and the natural and essential boundary conditions associated with the displacement field of **Problem 5.3**. In particular show that

$$\delta u: \quad \frac{dN_{xx}}{dx} - f = 0,$$

$$\delta\phi_x: \quad \frac{d}{dx}(c_1 M_{xx}) + \frac{d}{dx}\left(\frac{c_3}{h^3}P_{xx}\right) - c_1 N_{xz} - \frac{3c_3}{h^3}S_{xz} = 0,$$

$$\delta\psi_x: \quad \frac{d}{dx}(c_2 L_{xx}) - 2c_2 R_{xz} = 0, \tag{1}$$

$$\delta w: \quad \frac{d^2}{dx^2}\left(c_0 M_{xx} + \frac{c_3}{h^3}P_{xx}\right) - (1 + c_0)\frac{dN_{xz}}{dx} - \frac{3c_3}{h^3}\frac{dS_{xz}}{dx} - q = 0,$$

and the boundary conditions involve specifying

$$N_{xx} \quad \text{or} \quad u,$$

$$c_1 M_{xx} + \frac{c_3}{h^3}P_{xx} \quad \text{or} \quad \phi_x,$$

$$c_2 L_{xx} \quad \text{or} \quad \psi_x,$$

$$-\frac{d}{dx}\left(c_0 M_{xx} + \frac{c_3}{h^3}P_{xx}\right) + (1 + c_0)N_{xz} + \frac{3c_3}{h^3}S_{xz} \quad \text{or} \quad w,$$

$$c_0 M_{xx} + \frac{c_3}{h^3}P_{xx} \quad \text{or} \quad \frac{dw}{dx}, \tag{2}$$

where

$$(N_{xx}, M_{xx}, L_{xx}, P_{xx}) = \int_A (1, z, z^2, z^3)\sigma_{xx}dA,$$

$$(N_{xz}, R_{xz}, S_{xz}) = \int_A (1, z, z^2)\sigma_{xz}dA. \tag{3}$$

Note that the displacement field of **Problem 5.2**, hence the equations of equilibrium, contain those of the CBT ($c_0 = -1$, $c_1 = 0$, $c_2 = 0$, $c_3 = 0$), the TBT ($c_0 = 0$, $c_1 = 1$, $c_2 = 0$, $c_3 = 0$), and the RBT ($c_0 = 0$, $c_1 = 1$, $c_2 = 0$, $c_3 = -4h/3$).

5.5 (*Continuation of* **Problems 5.3** and **5.4**) Assume linear elastic constitutive behavior and show that the beam's constitutive equations are given by

$$\begin{Bmatrix} N_{xx} \\ M_{xx} \\ L_{xx} \\ P_{xx} \end{Bmatrix} = \begin{bmatrix} A_{xx} & B_{xx} & D_{xx} & E_{xx} \\ B_{xx} & D_{xx} & E_{xx} & F_{xx} \\ D_{xx} & E_{xx} & F_{xx} & G_{xx} \\ E_{xx} & F_{xx} & G_{xx} & H_{xx} \end{bmatrix} \begin{Bmatrix} \varepsilon_{xx}^{(0)} \\ \varepsilon_{xx}^{(1)} \\ \varepsilon_{xx}^{(2)} \\ \varepsilon_{xx}^{(3)} \end{Bmatrix}, \tag{1}$$

$$\begin{Bmatrix} N_{xz} \\ R_{xz} \\ S_{xz} \end{Bmatrix} = \begin{bmatrix} A_{xz} & B_{xz} & D_{xz} \\ B_{xz} & D_{xz} & E_{xz} \\ D_{xz} & E_{xz} & F_{xz} \end{bmatrix} \begin{Bmatrix} \gamma_{xz}^{(0)} \\ \gamma_{xz}^{(1)} \\ \gamma_{xz}^{(2)} \end{Bmatrix}, \tag{2}$$

where

$$(A_{xx}, B_{xx}, D_{xx}, F_{xx}, G_{xx}, H_{xx}) = \int_A E(1, z, z^2, z^4, z^5, z^6)\,dA,$$

$$(A_{xz}, B_{xz}, D_{xz}, E_{xz}, F_{xz}) = \int_A G(1, z, z^2, z^3, z^4)\,dA. \tag{3}$$

5.6 Determine the generalized displacements and generalized forces of a simply supported (i.e., pinned–hinged) isotropic beam subjected to uniformly distributed load of intensity q_0 using the relationships of Section 5.5.3. The solutions for the same problem by the CBT and TBT are as follows:

$$Q_x^C(x) = \frac{q_0 L}{2}\left(1 - 2\frac{x}{L}\right), \tag{1}$$

$$M_{xx}^C(x) = \frac{q_0 L^2}{2}\frac{x}{L}\left(1 - \frac{x}{L}\right), \tag{2}$$

$$w^C(x) = \frac{q_0 L^4}{24 D_{xx}}\left(\frac{x}{L} - \frac{2x^3}{L^3} + \frac{x^4}{L^4}\right), \tag{3}$$

$$Q_x^F(x) = Q_x^C(x) = \frac{q_0 L}{2}\left(1 - 2\frac{x}{L}\right), \tag{4}$$

$$M_{xx}^F(x) = M_{xx}^C(x) = \frac{q_0 L^2}{2}\frac{x}{L}\left(1 - \frac{x}{L}\right), \tag{5}$$

$$w^F(x) = w^C(x) + \frac{1}{K_s A_{xz}}M_{xx}^C(x)$$
$$= \frac{q_0 L^4}{24 D_{xx}}\left(\frac{x}{L} - 2\frac{x^3}{L^3} + \frac{x^4}{L^4}\right) + \frac{q_0 L^2}{2K_s A_{xz}}\left(\frac{x}{L} - \frac{x^2}{L^2}\right). \tag{6}$$

Also, note that $\phi_x^F = -dw/dx$. In particular, show that

$$Q_x^R(x) = \left(\frac{q_0 \mu}{\lambda^3}\right)\left[\sinh \lambda x - \tanh\left(\frac{\lambda L}{2}\right)\cosh \lambda x + \frac{\lambda L}{2}\left(1 - 2\frac{x}{L}\right)\right], \tag{7}$$

$$M_{xx}^R(x) = M_{xx}^C(x) = \frac{q_0 L^2}{2}\frac{x}{L}\left(1 - \frac{x}{L}\right), \tag{8}$$

$$w^R(x) = w^C(x) + \left(\frac{q_0\mu}{\lambda^4}\right)\left(\frac{\bar{D}_{xx}}{\bar{A}_{xz}D_{xx}}\right)\left[-\tanh\left(\frac{\lambda L}{2}\right)\sinh\lambda x\right.$$
$$\left. + \cosh\lambda x + \frac{\lambda^2 L^2}{2}\frac{x}{L}\left(1 - \frac{x}{L}\right) - 1\right], \tag{9}$$

$$\phi^R(x) = -\frac{dw^C}{dx} + \frac{\alpha F_{xx}}{D_{xx}\hat{A}_{xz}}Q_x^R, \tag{10}$$

where

$$\lambda^2 = \frac{\hat{A}_{xz}D_{xx}}{\alpha(F_{xx}\bar{D}_{xx} - \bar{F}_{xx}D_{xx})}, \quad \mu = \frac{\bar{A}_{xz}\bar{D}_{xz}}{\alpha(F_{xx}\bar{D}_{xx} - \bar{F}_{xx}D_{xx})}. \tag{11}$$

and

$$\bar{D}_{xx} = D_{xx} - \alpha F_{xx}, \quad \bar{F}_{xx} = F_{xx} - \alpha H_{xx}, \tag{12a}$$
$$\bar{A}_{xz} = A_{xz} - \beta D_{xz}, \quad \bar{D}_{xz} = D_{xz} - \beta F_{xz}, \tag{12b}$$
$$\hat{A}_{xz} = \bar{A}_{xz} - \beta\bar{D}_{xz}, \quad \hat{D}_{xx} = \bar{D}_{xx} - \alpha\bar{F}_{xx}, \tag{12c}$$
$$\hat{F}_{xx} = \bar{F}_{xx} - \alpha\bar{H}_{xx}. \tag{12d}$$

For a rectangular cross-section beam, it can be shown that

$$\frac{\bar{D}_{xx}\bar{D}_{xx}}{\hat{A}_{xz}D_{xx}D_{xx}} = \frac{\bar{D}_{xx}}{\bar{A}_{xz}D_{xx}} = \frac{6}{5}\frac{1}{A_{xz}}. \tag{12}$$

A close examination of Eq. (10) shows that the RBT solution has an effective shear coefficient, based on the coefficient in the expression for $w^R(x)$, of $K_s = 5/6$. Of course, RBT does not require a shear correction coefficient. Also the shear correction coefficient for the Timoshenko beam theory can be obtained, for example, by comparing the maximum deflections predicted by the TBT with those of the RBT.

5.7 Determine the generalized displacements and generalized forces of a cantilevered (i.e., fixed–free) isotropic beam subjected to uniformly distributed load of intensity q_0 using the relationships of Section 5.5.3. In particular, establish the following relations:

$$Q_x^R(x) = \left(\frac{q_0\mu}{\lambda^3\cosh\lambda L}\right)[\sinh\lambda x - \lambda L\cosh\lambda(L-x)] + \frac{q_0 L\mu}{\lambda^2}\left(1 - \frac{x}{L}\right), \tag{1}$$

$$M_{xx}^R(x) = M_{xx}^C(x) = \frac{q_0 L^2}{2}\left(2\frac{x}{L} - \frac{x^2}{L^2} - 1\right), \tag{2}$$

$$w^R(x) = w^C(x) + \left(\frac{q_0 L^2\mu}{2\lambda^2}\right)\left(\frac{\bar{D}_{xx}}{\bar{A}_{xz}D_{xx}}\right)\left(2\frac{x}{L} - \frac{x^2}{L^2}\right)$$
$$+ \left(\frac{q_0\mu}{\lambda^4\cosh\lambda L}\right)\left(\frac{\bar{D}_{xx}}{\bar{A}_{xz}D_{xx}}\right)[\cosh\lambda x + \lambda L\sinh\lambda(L-x)]$$
$$- \left(\frac{q_0\mu}{\lambda^4}\right)\left(\frac{\bar{D}_{xx}}{\bar{A}_{xz}D_{xx}}\right)\left(\frac{1 + \lambda L\sinh\lambda L}{\cosh\lambda L}\right). \tag{3}$$

The stiffness coefficients are the same as those defined in **Problem 5.6**. From Eqs. (7)–(9), one can be shown that the effective shear correction factor of the RBT is $K_s = 5/6$.

5.8 Determine the generalized displacements and generalized forces of a propped cantilever beam subjected to uniformly distributed load using the relationships in Eqs. (4.7.16)–(4.7.21). The solutions for the same problem by the CBT are available from Eqs. (3.7.42)–(3.7.47).

5.9 Establish the buckling relationship in Eq. (5.5.126).

Science can purify religion from error and superstition. Religion can purify science from idolatry and false absolutes.

Pope John Paul II

6 Classical Theory of Circular Plates

Science and technology are going to be the basis for many of the solutions to social problems.
Frances Arnold

6.1 GENERAL RELATIONS

6.1.1 PRELIMINARY COMMENTS

Geometrically, circular plates are thin (i.e., thickness h is very small compared to the outer radius a of the plate) discs subjected to transverse loads which tend to bend and stretch them. For the convenience of analysis, we use the cylindrical coordinate system (r, θ, z) to describe the deformation and stress state in circular plates. The word "axisymmetry" refers to the case in which the solution (i.e., displacements as well as stresses) is independent of the angular coordinate θ (see Fig. 6.1.1). This is possible if and only if the geometry, material properties, loads, and boundary conditions are also independent of θ. We assume such is the case in this chapter and in the coming chapters dealing with circular plates.

In this section, we present some general equations that are common to all theories of axisymmetric circular plates to be covered in this book. The remaining sections of this chapter are devoted to the development of the classical plate theory (CPT), which is a counter part of the classical beam theory (CBT) (i.e., the transverse shear strain is neglected).

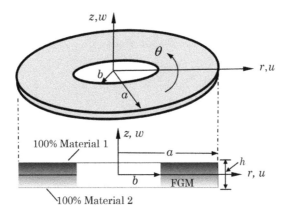

Figure 6.1.1: Geometry and coordinate system used for annular/circular plates.

DOI: 10.1201/9781003240846-6

6.1.2 KINEMATIC RELATIONS

Let $\mathbf{u}$ denote the displacement vector with components (u_r, u_θ, u_z) in the (r, θ, z) coordinate directions, respectively. Due to the assumed axisymmetry, we have $u_\theta = 0$, and u_r and u_z are independent of θ. In addition, if we *assume* the inextensibility of the transverse normal lines, then u_z is only a function of the radial coordinate r. When the inside radius $b = 0$, we have solid circular plates of radius a.

6.1.2.1 Modified Green–Lagrange strains

The modified Green–Lagrange strain tensor that accounts for small strains but moderate rotations of normal lines perpendicular to the plane of the plate is given by (see Reddy[9]–[11])

$$\mathbf{E} = \frac{1}{2} \left[\nabla\mathbf{u} + (\nabla\mathbf{u})^{\mathrm{T}} + \nabla\mathbf{u} \cdot (\nabla\mathbf{u})^{\mathrm{T}} \right]$$

$$\approx \frac{1}{2} \left[\nabla\mathbf{u} + (\nabla\mathbf{u})^{\mathrm{T}} + \frac{\partial u_z}{\partial r} \frac{\partial u_z}{\partial r} \hat{\mathbf{e}}_r \hat{\mathbf{e}}_r \right] \equiv \boldsymbol{\varepsilon}, \qquad (6.1.1)$$

where $(\hat{\mathbf{e}}_r, \hat{\mathbf{e}}_\theta, \hat{\mathbf{e}}_z)$ are the basis vectors in the cylindrical coordinate system. Thus, the nonzero strain components in the cylindrical coordinate system for the axisymmetric case are:

$$\varepsilon_{rr} = \frac{\partial u_r}{\partial r} + \frac{1}{2} \left(\frac{\partial u_z}{\partial r} \right)^2, \quad \varepsilon_{rz} = \frac{1}{2} \left(\frac{\partial u_r}{\partial z} + \frac{\partial u_z}{\partial r} \right), \quad \varepsilon_{\theta\theta} = \frac{u_r}{r}. \qquad (6.1.2)$$

6.1.2.2 Curvature tensor

The couple stress theory is based on the hypothesis that *the rate of change* of macro-rotations cause additional stresses, called *couple stresses*, in the continuum. Toward including the couple stress effect (as we did in the case of beams), we define the curvature tensor $\boldsymbol{\chi}$ as

$$\boldsymbol{\chi} = \frac{1}{2} \left[\nabla\boldsymbol{\omega} + (\nabla\boldsymbol{\omega})^{\mathrm{T}} \right], \qquad (6.1.3)$$

where $\boldsymbol{\omega}$ is the macro-rotation vector

$$\boldsymbol{\omega} = \frac{1}{2} \nabla \times \mathbf{u}. \qquad (6.1.4)$$

The only nonzero component of the rotation vector $\boldsymbol{\omega}$ for axisymmetric deformation is

$$\omega_\theta = \frac{1}{2} \left(\frac{\partial u_r}{\partial z} - \frac{\partial u_z}{\partial r} \right). \qquad (6.1.5)$$

Then the nonzero components of χ for the axisymmetric case are

$$
\begin{aligned}
\chi_{r\theta} &= \frac{1}{2}\left(\frac{\partial\omega_\theta}{\partial r} - \frac{\omega_\theta}{r}\right) \\
&= \frac{1}{4}\left(\frac{\partial^2 u_r}{\partial z\partial r} - \frac{\partial^2 u_z}{\partial r^2}\right) - \frac{1}{4r}\left(\frac{\partial u_r}{\partial z} - \frac{\partial u_z}{\partial r}\right),
\end{aligned}
\tag{6.1.6a}
$$

$$
\chi_{z\theta} = \frac{1}{2}\frac{\partial\omega_\theta}{\partial z} = \frac{1}{4}\left(\frac{\partial^2 u_r}{\partial z^2} - \frac{\partial^2 u_z}{\partial z\partial r}\right).
\tag{6.1.6b}
$$

6.1.3 STRESS–STRAIN RELATIONS

For a two-constituent functionally graded linear elastic material, the plane stress–strain equations relating the nonzero stresses $(\sigma_{rr}, \sigma_{\theta\theta}, \sigma_{rz})$ to the nonzero strains $(\varepsilon_{rr}, \varepsilon_{\theta\theta}, \varepsilon_{rz})$ of the axisymmetric case are

$$
\left\{\begin{array}{c}\sigma_{rr}\\\sigma_{\theta\theta}\\\sigma_{rz}\end{array}\right\} = \frac{E(z)}{1-\nu^2}\begin{bmatrix}1 & \nu & 0\\\nu & 1 & 0\\0 & 0 & \frac{1-\nu}{2}\end{bmatrix}\left\{\begin{array}{c}\varepsilon_{rr}\\\varepsilon_{\theta\theta}\\2\varepsilon_{rz}\end{array}\right\},
\tag{6.1.7}
$$

where Young's modulus E varies with z according to Eq. (3.4.5a)

$$
E(z) = (E_1 - E_2)\,V_1(z) + E_2, \quad V_1(z) = \left(\frac{1}{2}+\frac{z}{h}\right)^n,
\tag{6.1.8}
$$

and ν is a constant. The modified couple stress constitutive relation is [55]

$$
\mathcal{M}_{r\theta} = 2G\ell^2\,\chi_{r\theta}, \quad G(z) = \frac{E(z)}{2(1+\nu)},
\tag{6.1.9}
$$

where $\mathcal{M}_{r\theta}$ is the nonzero component of the couple stress tensor $\mathbf{m}$.

6.1.4 STRAIN ENERGY FUNCTIONAL

According to the modified couple stress theory, the strain energy potential for linear elastic case can be expressed as

$$
\begin{aligned}
U &= \frac{1}{2}\int_b^a\int_{-\frac{h}{2}}^{\frac{h}{2}}(\boldsymbol{\sigma}:\boldsymbol{\varepsilon}+\mathbf{m}:\boldsymbol{\chi})\,dz\,rdr + \frac{1}{2}\int_b^a c_f w^2\,rdr
\tag{6.1.10a}\\
&= \frac{1}{2}\int_b^a\int_{-\frac{h}{2}}^{\frac{h}{2}}(\sigma_{rr}\,\varepsilon_{rr} + \sigma_{\theta\theta}\,\varepsilon_{\theta\theta} + 2\sigma_{rz}\,\varepsilon_{rz} + 2\mathcal{M}_{r\theta}\,\chi_{r\theta})\,dz\,rdr\\
&\quad + \frac{1}{2}\int_b^a c_f w^2\,rdr,
\tag{6.1.10b}
\end{aligned}
$$

where a and b are the outside and inside radii, respectively, and h is the thickness of the plate, $\boldsymbol{\sigma}$ is the Cauchy stress tensor, and $\boldsymbol{\varepsilon}$ is the modified Green–Lagrange strain tensor defined in Eq. (6.1.1).

6.2 GOVERNING EQUATIONS OF THE CPT

6.2.1 DISPLACEMENTS AND STRAINS

The total displacements (u_r, u_z) along the coordinate directions (r, z), as implied by the Kirchhoff hypothesis [141] for axisymmetric bending of circular plates, which is the same as the Euler–Bernoulli hypothesis for beams, are assumed in the form

$$\mathbf{u} = u_r\,\hat{\mathbf{e}}_r + u_z\,\hat{\mathbf{e}}_z,$$

$$u_r(r, z, t) = u(r, t) - z\frac{\partial w}{\partial r}, \quad u_\theta = 0, \quad u_z(r, z, t) = w(r, t), \tag{6.2.1}$$

where u is the radial displacement and w is the transverse deflection of a point on the midplane of the plate at time t. The Kirchhoff hypothesis is that straight lines normal to the midplane before deformation remain: (1) inextensible, (2) straight, and (3) normal to the midsurface after deformation. The hypothesis amounts to neglecting both transverse shear and transverse normal strains, $\varepsilon_{rz} = \varepsilon_{zz} = 0$.

The von Kármán strains in (6.1.2) for the CPT take the form

$$\varepsilon_{rr} = \varepsilon_{rr}^{(0)} + z\varepsilon_{rr}^{(1)}, \quad \varepsilon_{\theta\theta} = \varepsilon_{\theta\theta}^{(0)} + z\varepsilon_{\theta\theta}^{(1)}, \tag{6.2.2}$$

where

$$\varepsilon_{rr}^{(0)} = \frac{\partial u}{\partial r} + \frac{1}{2}\left(\frac{\partial w}{\partial r}\right)^2, \quad \varepsilon_{rr}^{(1)} = -\frac{\partial^2 w}{\partial r^2},$$

$$\varepsilon_{\theta\theta}^{(0)} = \frac{u}{r}, \quad \varepsilon_{\theta\theta}^{(1)} = -\frac{1}{r}\frac{\partial w}{\partial r}. \tag{6.2.3}$$

The rotation and curvature components are

$$\omega_\theta = \frac{1}{2}\left(\frac{\partial u_r}{\partial z} - \frac{\partial u_z}{\partial r}\right) = -\frac{\partial w}{\partial r}, \quad \chi_{z\theta} = 0,$$

$$\chi_{r\theta} = \frac{1}{2}\left(\frac{\partial \omega_\theta}{\partial r} - \frac{\omega_\theta}{r}\right) = \frac{1}{2}\left(-\frac{\partial^2 w}{\partial r^2} + \frac{1}{r}\frac{\partial w}{\partial r}\right). \tag{6.2.4}$$

6.2.2 EQUATIONS OF MOTION

One can obtain the equations of motion of axisymmetric annular/circular plates using either the vector approach or the energy approach. It is much simpler to derive them using Hamilton's principle and obtain the associated duality pairs to specify boundary conditions, especially when we account for the von Kármán nonlinearity and couple stress effect. Therefore, here we consider the energy approach and derive the equations of motion of circular plates loaded with transverse distributed load $q(r)$ and resting on elastic foundation with modulus c_f.

The statement of Hamilton's principle is $(\delta L = \delta K - \delta V - \delta U)$,

$$
0 = \int_0^T \int_b^a \left[\int_{-\frac{h}{2}}^{\frac{h}{2}} \rho \left(\dot{u}_r \, \delta \dot{u}_r + \dot{u}_z \, \delta \dot{u}_z \right) dz + (q - c_f u_z) \delta u_z(r,t) \right] r \, dr \, dt
$$

$$
- \int_0^T \int_b^a \int_{-\frac{h}{2}}^{\frac{h}{2}} \left(\sigma_{rr} \, \delta \varepsilon_{rr} + \sigma_{\theta\theta} \, \delta \varepsilon_{\theta\theta} + 2 m_{r\theta} \delta \chi_{r\theta} \right) r \, dr \, dz \, dt
$$

$$
= \int_0^T \int_b^a \left[m_0 \left(\dot{u} \, \delta \dot{u} + \dot{w} \, \delta \dot{w} \right) - m_1 \left(\frac{\partial \dot{w}}{\partial r} \, \delta \dot{u} + \dot{u} \, \frac{\partial \delta \dot{w}}{\partial r} \right) + m_2 \frac{\partial \dot{w}}{\partial r} \frac{\partial \delta \dot{w}}{\partial r} \right.
$$

$$
- N_{rr} \left(\frac{\partial \delta u}{\partial r} + \frac{\partial w}{\partial r} \frac{\partial \delta w}{\partial r} \right) - N_{\theta\theta} \left(\frac{\delta u}{r} \right) + M_{rr} \frac{\partial^2 \delta w}{\partial r^2} - c_f w \, \delta w
$$

$$
\left. + \frac{1}{r} M_{\theta\theta} \frac{\partial \delta w}{\partial r} - P_{r\theta} \left(-\frac{\partial^2 \delta w}{\partial r^2} + \frac{1}{r} \frac{\partial \delta w}{\partial r} \right) + q \, \delta w \right] r \, dr \, dt, \quad (6.2.5)
$$

where the stress resultants $(N_{rr}, N_{\theta\theta}, M_{rr}, M_{\theta\theta}, P_{r\theta})$ and the mass inertia m_i are defined by (see Fig. 6.2.1):

$$
N_{rr}(r,t) = \int_{-\frac{h}{2}}^{\frac{h}{2}} \sigma_{rr} \, dz, \qquad N_{\theta\theta}(r,t) = \int_{-\frac{h}{2}}^{\frac{h}{2}} \sigma_{\theta\theta} \, dz,
$$

$$
M_{rr}(r,t) = \int_{-\frac{h}{2}}^{\frac{h}{2}} \sigma_{rr} z \, dz, \qquad M_{\theta\theta}(r,t) = \int_{-\frac{h}{2}}^{\frac{h}{2}} \sigma_{\theta\theta} z \, dz, \qquad (6.2.6)
$$

$$
P_{r\theta}(r,t) = \int_{-\frac{h}{2}}^{\frac{h}{2}} m_{r\theta} \, dz, \qquad m_i = \int_{-\frac{h}{2}}^{\frac{h}{2}} \rho(z)(z)^i \, dz, \quad (i = 0, 1, 2).
$$

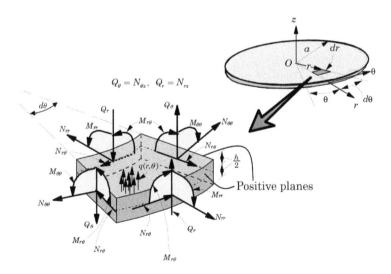

Figure 6.2.1: Stress resultants at a point of a circular plate.

Here the superposed dot denotes the time derivative, and ρ denotes the mass density. The modulus is assumed to vary according to the power law in Eq. (6.1.8). We also assume that the density ρ varies according to the same power-law:

$$\rho(z) = (\rho_1 - \rho_2) V_1(z) + \rho_2, \quad V_1(z) = \left(\frac{1}{2} + \frac{z}{h}\right)^n, \tag{6.2.7}$$

where ρ_1 and ρ_2 are the mass densities of materials 1 and 2, respectively.

The Euler–Lagrange equations obtained from Eq. (6.2.5) are the equations of motion of the CPT, which account for the von Kármán nonlinearity and the couple stress effect:

$$\frac{1}{r}\left[\frac{\partial}{\partial r}(rN_{rr}) - N_{\theta\theta}\right] = m_0\frac{\partial^2 u}{\partial t^2} - m_1\frac{\partial^3 w}{\partial r\partial t^2}, \tag{6.2.8}$$

$$\frac{1}{r}\frac{\partial}{\partial r}(rV_r) + \frac{1}{r}\frac{\partial}{\partial r}\left[\frac{\partial}{\partial r}(rP_{r\theta}) + P_{r\theta}\right] - c_f w + q$$

$$= m_0\frac{\partial^2 w}{\partial t^2} + \frac{1}{r}\frac{\partial}{\partial r}\left[r\left(m_1\frac{\partial^2 u}{\partial t^2} - m_2\frac{\partial^3 w}{\partial r\partial t^2}\right)\right], \tag{6.2.9}$$

where V_r denotes the *effective* shear force acting on the rz-plane and N_{rz} is the transverse shear force,

$$V_r = N_{rz} + N_{rr}\frac{\partial w}{\partial r},$$

$$N_{rz} = \frac{1}{r}\left[\frac{\partial}{\partial r}(rM_{rr}) - M_{\theta\theta}\right] - m_1\frac{\partial^2 u}{\partial t^2} + m_2\frac{\partial^3 w}{\partial r\partial t^2}. \tag{6.2.10}$$

When all the stress resultants in Eqs. (6.2.8) and (6.2.9) are expressed in terms of the displacements (u, w), we will have two partial differential equations in two displacements (u, w).

Using the second relation in Eq. (6.2.10) as an additional equation, we have a set of three equations:

$$\frac{1}{r}\left[\frac{\partial}{\partial r}(rN_{rr}) - N_{\theta\theta}\right] = m_0\frac{\partial^2 u}{\partial t^2} - m_1\frac{\partial^3 w}{\partial r\partial t^2}, \tag{6.2.11}$$

$$\frac{1}{r}\frac{\partial}{\partial r}(rN_{rz}) + \frac{1}{r}\frac{\partial}{\partial r}\left[\frac{\partial}{\partial r}(rP_{r\theta}) + P_{r\theta}\right]$$

$$+ \frac{1}{r}\frac{\partial}{\partial r}\left(rN_{rr}\frac{\partial w}{\partial r}\right) - c_f w + q = m_0\frac{\partial^2 w}{\partial t^2}, \tag{6.2.12}$$

$$-\frac{1}{r}\left[\frac{\partial}{\partial r}(rM_{rr}) - M_{\theta\theta}\right] + N_{rz} = m_1\frac{\partial^2 u}{\partial t^2} - m_2\frac{\partial^3 w}{\partial r\partial t^2}. \tag{6.2.13}$$

The boundary conditions involve specifying one element of each of the following duality pairs:

$$u \quad \text{or} \quad rN_{rr},$$

$$w \quad \text{or} \quad r\left[V_r + \frac{\partial}{\partial r}(rP_{r\theta}) + P_{r\theta} - m_1\frac{\partial^2 u}{\partial t^2} + m_2\frac{\partial^3 w}{\partial r \partial t^2}\right] \equiv r\hat{V}_r,$$

$$\tag{6.2.14}$$

$$-\frac{\partial w}{\partial r} \quad \text{or} \quad rM_{rr} + rP_{r\theta} \equiv r\hat{M}_{rr}.$$

The equations of motion of the CPT *without the couple stress effect* are obtained by setting $P_{r\theta}$ to zero in Eqs. (6.2.8) and (6.2.9):

$$\frac{1}{r}\left[\frac{\partial}{\partial r}(rN_{rr}) - N_{\theta\theta}\right] = m_0\frac{\partial^2 u}{\partial t^2} - m_1\frac{\partial^3 w}{\partial r \partial t^2}, \tag{6.2.15}$$

$$\frac{1}{r}\frac{\partial}{\partial r}\left[\frac{\partial}{\partial r}(rM_{rr}) - M_{\theta\theta} + rN_{rr}\frac{\partial w}{\partial r}\right] - c_f w + q$$

$$= m_0\frac{\partial^2 w}{\partial t^2} + \frac{1}{r}\frac{\partial}{\partial r}\left(rm_1\frac{\partial^2 u}{\partial t^2} - rm_2\frac{\partial^3 w}{\partial r \partial t^2}\right). \tag{6.2.16}$$

The equations of equilibrium (or motion) of the CPT in terms of the stress resultants can also be obtained using the vector approach by summing forces in the radial and transverse directions and summing the moments about the angular direction. However, it is limited to the case where we do not account for the couple stress effect, unless we know how to include the forces associated with the couple stress in the free-body-diagram. To use the vector approach, one needs to identify the forces and moments on a typical element of the plate, one similar to that is shown in Fig. 6.2.1, which shows the forces and moments only at a point (i.e., the element has no dimensions). In order to derive the equations by vector approach, the element in Fig. 6.2.1 must be modified to be an "infinitesimally small plate element," which has finite dimensions of $d\theta$ angularly, dr radially, and h in the transverse direction. On this element, the force on the positive planes must be incremented. For example, N_{rr} acting on the plane at a distance of $r + dr$ should be $N_{rr} + dN_{rr}$ or $N_{rr} + (\partial N_{rr}/\partial r)dr$. Similar increments should be used for all forces acting on this plane. For the axisymmetric case, the forces in the angular (θ) direction are not incremented. The resulting equations of equilibrium are

$$-\frac{1}{r}\left[\frac{d}{dr}(rN_{rr}) - N_{\theta\theta}\right] = 0, \tag{6.2.17}$$

$$-\frac{1}{r}\frac{d}{dr}(rN_{rz}) + c_f w - q = 0, \tag{6.2.18}$$

$$-\frac{1}{r}\left[\frac{d}{dr}(rM_{rr}) - M_{\theta\theta}\right] + N_{rz} = 0. \tag{6.2.19}$$

These equations are the same as those obtained from Eqs. (6.2.11)–(6.2.13), when the couple stress and the nonlinear terms are set to zero.

When we combine Eqs. (6.2.18) and (6.2.19) (because we have only two unknowns, namely, u and w), we obtain

$$-\frac{1}{r}\frac{d}{dr}\left[\frac{d}{dr}\left(rM_{rr}\right) - M_{\theta\theta}\right] + c_f w - q = 0. \tag{6.2.20}$$

Equations (6.2.17) and (6.2.20) are the equilibrium equations for the axisymmetric bending of circular plates, and Eq. (6.2.19) can be viewed as a relation that allows us to compute the transverse shear force N_{rz} from the bending moments M_{rr} and $M_{\theta\theta}$ (all these arguments should be familiar to the reader from the discussion presented for the CBT in Chapter 3).

6.2.3 ISOTROPIC CONSTITUTIVE RELATIONS

The stress resultants N_{rr}, $N_{\theta\theta}$, M_{rr}, $M_{\theta\theta}$, $P_{r\theta}$ of the CPT are related to the displacements (u, w) according to the following equations:

$$N_{rr} = A_{rr}\left[\frac{\partial u}{\partial r} + \frac{1}{2}\left(\frac{\partial w}{\partial r}\right)^2 + \nu\frac{u}{r}\right] - B_{rr}\left(\frac{\partial^2 w}{\partial r^2} + \frac{\nu}{r}\frac{\partial w}{\partial r}\right), \tag{6.2.21a}$$

$$N_{\theta\theta} = A_{rr}\left[\frac{u}{r} + \nu\frac{\partial u}{\partial r} + \frac{\nu}{2}\left(\frac{\partial w}{\partial r}\right)^2\right] - B_{rr}\left(\nu\frac{\partial^2 w}{\partial r^2} + \frac{1}{r}\frac{\partial w}{\partial r}\right), \tag{6.2.21b}$$

$$M_{rr} = B_{rr}\left[\frac{\partial u}{\partial r} + \frac{1}{2}\left(\frac{\partial w}{dr}\right)^2 + \nu\frac{u}{r}\right] - D_{rr}\left(\frac{\partial^2 w}{\partial r^2} + \frac{\nu}{r}\frac{\partial w}{\partial r}\right), \tag{6.2.21c}$$

$$M_{\theta\theta} = B_{rr}\left[\frac{u}{r} + \nu\frac{\partial u}{\partial r} + \frac{\nu}{2}\left(\frac{\partial w}{\partial r}\right)^2\right] - D_{rr}\left(\nu\frac{\partial^2 w}{\partial r^2} + \frac{1}{r}\frac{\partial w}{\partial r}\right), \tag{6.2.21d}$$

$$P_{r\theta} = S_{r\theta}\left(-\frac{\partial^2 w}{\partial r^2} + \frac{1}{r}\frac{\partial w}{\partial r}\right), \tag{6.2.21e}$$

$$N_{rz} = \frac{1}{r}\left[\frac{d}{dr}\left(rM_{rr}\right) - M_{\theta\theta}\right], \tag{6.2.21f}$$

where A_{rr}, B_{rr}, D_{rr}, and $S_{r\theta}$ are the extensional, extensional-bending, bending, and shear stiffnesses, respectively:

$$A_{rr} = \frac{1}{(1-\nu^2)}\int_{-\frac{h}{2}}^{\frac{h}{2}} E(z)\, dz, \qquad B_{rr} = \frac{1}{(1-\nu^2)}\int_{-\frac{h}{2}}^{\frac{h}{2}} E(z)z\, dz,$$

$$\tag{6.2.22}$$

$$D_{rr} = \frac{1}{(1-\nu^2)}\int_{-\frac{h}{2}}^{\frac{h}{2}} E(z)z^2\, dz, \qquad S_{r\theta} = \frac{\ell^2}{2(1+\nu)}\int_{-\frac{h}{2}}^{\frac{h}{2}} E(z)\, dz.$$

For the material distribution through the thickness according to Eq. (6.1.8), the following integrals are useful in computing A_{rr}, B_{rr}, and so on:

$$\int_{-\frac{h}{2}}^{\frac{h}{2}} V_1(z)\,dz = \frac{h}{n+1},$$

$$\int_{-\frac{h}{2}}^{\frac{h}{2}} V_1(z)\,z\,dz = \frac{nh^2}{2(n+1)(n+2)}, \qquad (6.2.23)$$

$$\int_{-\frac{h}{2}}^{\frac{h}{2}} V_1(z)\,z^2\,dz = \frac{(2+n+n^2)h^3(r)}{4(n+1)(n+2)(n+3)}.$$

6.2.4 DISPLACEMENT FORMULATION OF THE CPT

We now can write the governing equations of the CPT solely in terms of u and w with the help of the plate constitutive equations. The resulting differential equations would be second order in u and fourth order in w, the total order being six. Here we present such equations for the case in which the couple stress effect is omitted.

The *equilibrium equations* of the CPT without the couple stress effect are obtained by setting the time derivative terms and $P_{r\theta}$ to zero:

$$-\frac{1}{r}\left[\frac{d}{dr}(rN_{rr}) - N_{\theta\theta}\right] = 0, \qquad (6.2.24)$$

$$-\frac{1}{r}\frac{d}{dr}\left[\frac{d}{dr}(rM_{rr}) - M_{\theta\theta}\right] - \frac{1}{r}\frac{d}{dr}\left[r\left(N_{rr}\frac{dw}{dr}\right)\right] + c_f w - q = 0. \qquad (6.2.25)$$

Substituting for N_{rr}, $N_{\theta\theta}$, M_{rr}, and $M_{\theta\theta}$ from Eqs. (6.2.21a)–(6.2.21d) into the equations of equilibrium, Eqs. (6.2.24) and (6.2.25), we obtain

$$-\frac{1}{r}\frac{d}{dr}\left\{ rA_{rr}\left[\frac{du}{dr} + \frac{1}{2}\left(\frac{dw}{dr}\right)^2 + \nu\frac{u}{r}\right] - rB_{rr}\left(\frac{d^2 w}{dr^2} + \frac{\nu}{r}\frac{dw}{dr}\right)\right\}$$
$$+ A_{rr}\frac{1}{r}\left[\frac{u}{r} + \nu\frac{du}{dr} + \frac{\nu}{2}\left(\frac{dw}{dr}\right)^2\right] - B_{rr}\frac{1}{r}\left(\nu\frac{d^2 w}{dr^2} + \frac{1}{r}\frac{dw}{dr}\right) = 0, \qquad (6.2.26)$$

$$-\frac{1}{r}\frac{d}{dr}\left[\frac{d}{dr}\left\{ rB_{rr}\left[\frac{du}{dr} + \frac{1}{2}\left(\frac{dw}{dr}\right)^2 + \nu\frac{u}{r}\right] - rD_{rr}\left(\frac{d^2 w}{dr^2} + \frac{\nu}{r}\frac{dw}{dr}\right)\right\}\right.$$
$$\left. - B_{rr}\left[\frac{u}{r} + \nu\frac{du}{dr} + \frac{\nu}{2}\left(\frac{dw}{dr}\right)^2\right] + D_{rr}\left(\nu\frac{d^2 w}{dr^2} + \frac{1}{r}\frac{dw}{dr}\right)\right]$$
$$- \frac{1}{r}\frac{d}{dr}\left\{ A_{rr}r\frac{dw}{dr}\left[\frac{du}{dr} + \frac{1}{2}\left(\frac{dw}{dr}\right)^2 + \nu\frac{u}{r}\right]\right.$$
$$\left. - B_{rr}r\frac{dw}{dr}\left(\frac{d^2 w}{dr^2} + \frac{\nu}{r}\frac{dw}{dr}\right)\right\} + c_f w - q = 0. \qquad (6.2.27)$$

The nonlinear equations of equilibrium for *homogeneous* plates (i.e., without FGM) can be obtained from Eqs. (6.2.26) and (6.2.27) by setting $B_{rr} = 0$:

$$-\frac{1}{r}\frac{d}{dr}\left\{rA_{rr}\left[\frac{du}{dr} + \frac{1}{2}\left(\frac{dw}{dr}\right)^2 + \nu\frac{u}{r}\right]\right\}$$

$$+ A_{rr}\frac{1}{r}\left[\frac{u}{r} + \nu\frac{du}{dr} + \frac{\nu}{2}\left(\frac{dw}{dr}\right)^2\right] = 0, \qquad (6.2.28)$$

$$-\frac{1}{r}\frac{d}{dr}\left\{-\frac{d}{dr}\left[rD_{rr}\left(\frac{d^2w}{dr^2} + \frac{\nu}{r}\frac{dw}{dr}\right)\right] + D_{rr}\left(\nu\frac{d^2w}{dr^2} + \frac{1}{r}\frac{dw}{dr}\right)\right\}$$

$$-\frac{1}{r}\frac{d}{dr}\left\{A_{rr}r\frac{dw}{dr}\left[\frac{du}{dr} + \frac{1}{2}\left(\frac{dw}{dr}\right)^2 + \nu\frac{u}{r}\right]\right\} + c_f w - q = 0. \qquad (6.2.29)$$

The *linearized* governing equations of equilibrium for homogeneous plates (i.e., $B_{rr} = 0$) can be obtained from Eqs. (6.2.28) and (6.2.29) by setting the nonlinear terms to zero:

$$-\frac{1}{r}\frac{d}{dr}\left[rA_{rr}\left(\frac{du}{dr} + \nu\frac{u}{r}\right)\right] + A_{rr}\frac{1}{r}\left(\frac{u}{r} + \nu\frac{du}{dr}\right) = 0, \qquad (6.2.30)$$

$$-\frac{1}{r}\frac{d}{dr}\left\{-\frac{d}{dr}\left[rD_{rr}\left(\frac{d^2w}{dr^2} + \frac{\nu}{r}\frac{dw}{dr}\right)\right] + D_{rr}\left(\nu\frac{d^2w}{dr^2} + \frac{1}{r}\frac{dw}{dr}\right)\right\}$$

$$+ c_f w - q = 0. \qquad (6.2.31)$$

The above two equations are decoupled between u and w.

6.2.5   MIXED FORMULATION OF THE CPT

To reduce the order of the differential equations expressed in terms of the generalized displacements in CPT, from the fourth-order to the second-order differential equations, here we reformulate the governing equations as a set of second-order differential equations in terms of the displacements (u, w) and the bending moment (M_{rr}). Such formulations are known as *mixed* formulations (i.e., mixing displacement variables with force variables). We omit, for simplicity, the thermal as well as the couple stress effects in this part and limit the discussion to static equilibrium problems, although the development can be extended to time-dependent problems (one needs to add the mass inertia terms to the equations and replace ordinary derivatives with partial derivatives).

Consider the equilibrium equations in Eqs. (6.2.24) and (6.2.25), with

$$N_{rr} = A_{rr}\,\hat{\varepsilon}_{rr}^{(0)} + B_{rr}\,\hat{\varepsilon}_{rr}^{(1)}, \quad N_{\theta\theta} = A_{rr}\,\hat{\varepsilon}_{\theta\theta}^{(0)} + B_{rr}\,\hat{\varepsilon}_{\theta\theta}^{(1)},$$

$$M_{rr} = B_{rr}\,\hat{\varepsilon}_{rr}^{(0)} + D_{rr}\,\hat{\varepsilon}_{rr}^{(1)}, \quad M_{\theta\theta} = B_{rr}\,\hat{\varepsilon}_{\theta\theta}^{(0)} + D_{rr}\,\hat{\varepsilon}_{\theta\theta}^{(1)}, \qquad (6.2.32)$$

where $\hat{\varepsilon}$ are the effective strains

$$\hat{\varepsilon}_{rr}^{(0)} = \left[\frac{du}{dr} + \frac{1}{2}\left(\frac{dw}{dr}\right)^2 + \nu\frac{u}{r}\right], \quad \hat{\varepsilon}_{rr}^{(1)} = -\left(\frac{d^2w}{dr^2} + \frac{\nu}{r}\frac{dw}{dr}\right),$$

$$\hat{\varepsilon}_{\theta\theta}^{(0)} = \left[\frac{u}{r} + \nu\frac{du}{dr} + \frac{\nu}{2}\left(\frac{dw}{dr}\right)^2\right], \quad \hat{\varepsilon}_{\theta\theta}^{(1)} = -\left(\nu\frac{d^2w}{dr^2} + \frac{1}{r}\frac{dw}{dr}\right).$$

$$(6.2.33)$$

Rewriting Eq. (6.2.33) for the effective strains $\hat{\varepsilon}_{rr}^{(0)}$ and $\hat{\varepsilon}_{rr}^{(1)}$ in terms of the stress resultants N_{rr} and M_{rr}, we have

$$\hat{\varepsilon}_{rr}^{(0)} = \frac{1}{D_{rr}^*}\left(D_{rr}\,N_{rr} - B_{rr}\,M_{rr}\right), \quad \hat{\varepsilon}_{rr}^{(1)} = \frac{1}{D_{rr}^*}\left(-B_{rr}\,N_{rr} + A_{rr}\,M_{rr}\right),$$

$$(6.2.34a)$$

$$\hat{\varepsilon}_{\theta\theta}^{(0)} = \frac{1}{D_{rr}^*}\left(D_{rr}\,N_{\theta\theta} - B_{rr}\,M_{\theta\theta}\right), \quad \hat{\varepsilon}_{\theta\theta}^{(1)} = \frac{1}{D_{rr}^*}\left(-B_{rr}\,N_{\theta\theta} + A_{rr}\,M_{\theta\theta}\right),$$

$$(6.2.34b)$$

where

$$D_{rr}^* = D_{rr}\,A_{rr} - B_{rr}\,B_{rr}. \quad (6.2.34c)$$

Now, we solve Eq. (6.2.34a) for N_{rr} in terms of bending moment M_{rr} and Eq. (6.2.34b) for $N_{\theta\theta}$ in terms of bending moment $M_{\theta\theta}$ (and displacements) to obtain

$$N_{rr} = \bar{A}_{rr}\,\hat{\varepsilon}_{rr}^{(0)} + \bar{B}_{rr}\,M_{rr}, \quad N_{\theta\theta} = \bar{A}_{rr}\,\hat{\varepsilon}_{\theta\theta}^{(0)} + \bar{B}_{rr}\,M_{\theta\theta}, \quad (6.2.35)$$

where

$$\bar{A}_{rr} = \frac{D_{rr}^*}{D_{rr}}, \quad \bar{B}_{rr} = \frac{B_{rr}}{D_{rr}}. \quad (6.2.36)$$

Using Eq. (6.2.35), we rewrite the second equation in Eq. (6.2.34a) and obtain

$$\hat{\varepsilon}_{rr}^{(1)} = -\bar{B}_{rr}\hat{\varepsilon}_{rr}^{(0)} + \frac{1}{D_{rr}}M_{rr}. \quad (6.2.37)$$

Next, we write all stress resultants N_{rr}, $N_{\theta\theta}$, and $M_{\theta\theta}$ in terms of (u, w, M_{rr}). We already have N_{rr} in terms of (u, w, M_{rr}) through Eq. (6.2.35). First, we use the last two equations of Eq. (6.2.32) to write $M_{\theta\theta}$ in terms of (u, w, M_{rr}) as

$$M_{\theta\theta} = \nu M_{rr} + (1 - \nu^2)\frac{1}{r}\left(B_{rr}u - D_{rr}\frac{\partial w}{\partial r}\right). \quad (6.2.38)$$

Then, we use the first two equations of Eq. (6.2.32) to write $N_{\theta\theta}$ in terms of

(u, w, N_{rr}) and finally in terms of (u, w, M_{rr}) (note that $A_{rr} = \bar{A}_{rr} + \bar{B}_{rr} B_{rr}$):

$$N_{\theta\theta} = \nu N_{rr} + (1 - \nu^2)\frac{1}{r}\left(A_{rr} u - B_{rr}\frac{dw}{dr}\right)$$

$$= \nu\bar{A}_{rr}\left[\frac{du}{dr} + \left(\frac{dw}{dr}\right)^2 + \nu\frac{u}{r}\right] + \frac{(1-\nu^2)}{r}\left(A_{rr} u - B_{rr}\frac{dw}{dr}\right) + \nu\bar{B}_{rr} M_{rr}$$

$$= \bar{A}_{rr}\left[\nu\frac{du}{dr} + \nu\left(\frac{dw}{dr}\right)^2 + \frac{u}{r}\right] + \frac{(1-\nu^2)B_{rr}}{r}\left(\bar{B}_{rr} u - \frac{dw}{dr}\right) + \nu\bar{B}_{rr} M_{rr}.$$

$$\text{(6.2.39)}$$

The two equilibrium equations, Eq. (6.2.24) and (6.2.25), when N_{rr}, $N_{\theta\theta}$, and $M_{\theta\theta}$ are expressed in terms of u, w, and M_{rr}, and Eq. (6.2.37), with $\hat{\varepsilon}_{rr}^{(0)}$ and $\hat{\varepsilon}_{rr}^{(1)}$ written in terms of u and w, provide the three required equations of the mixed formulation (additional algebraic identities are used in arriving at these equations):

$$-\frac{1}{r}\frac{d}{dr}\left\{r\bar{A}_{rr}\left[\frac{du}{dr} + \frac{1}{2}\left(\frac{dw}{dr}\right)^2 + \nu\frac{u}{r}\right] + r\bar{B}_{rr} M_{rr}\right\}$$

$$+\frac{1}{r}\bar{A}_{rr}\left[\nu\frac{du}{dr} + \frac{\nu}{2}\left(\frac{dw}{dr}\right)^2 + \frac{u}{r}\right] + \frac{1}{r}\nu\bar{B}_{rr} M_{rr}$$

$$+\frac{1}{r}(1-\nu^2)B_{rr}\left(\bar{B}_{rr}\frac{u}{r} - \frac{1}{r}\frac{dw}{dr}\right) = 0, \quad \text{(6.2.40)}$$

$$-\frac{1}{r}\frac{d}{dr}\left[r\frac{dM_{rr}}{dr} + (1-\nu)M_{rr} - (1-\nu^2)D_{rr}\left(\bar{B}_{rr}\frac{u}{r} - \frac{1}{r}\frac{dw}{dr}\right)\right] + c_f w - q$$

$$-\frac{1}{r}\frac{d}{dr}\left\{r\bar{A}_{rr}\left[\frac{dw}{dr}\frac{du}{dr} + \frac{1}{2}\left(\frac{dw}{dr}\right)^3 + \nu\frac{dw}{dr}\frac{u}{r}\right] + r\bar{B}_{rr}\frac{dw}{dr}M_{rr}\right\} = 0, \quad \text{(6.2.41)}$$

$$-\frac{1}{r}\frac{d}{dr}\left(r\frac{dw}{dr}\right) + (1-\nu)\frac{1}{r}\frac{dw}{dr} + \bar{B}_{rr}\left[\frac{du}{dr} + \frac{1}{2}\left(\frac{dw}{dr}\right)^2 + \nu\frac{u}{r}\right]$$

$$-\frac{1}{D_{rr}} M_{rr} = 0. \quad \text{(6.2.42)}$$

The governing equations of equilibrium for homogeneous plates can be obtained from Eqs. (6.2.40)–(6.2.42) by setting $B_{rr} = 0$ and $\bar{A}_{rr} = A_{rr}$.

6.3   SOLUTIONS FOR HOMOGENEOUS PLATES IN BENDING

6.3.1   GOVERNING EQUATIONS

In this section, we develop exact solutions of homogeneous (i.e., $B_{rr} = 0$) circular plates for axisymmetric boundary conditions and loads (see Szilard [142] and Reddy [9, 113] for additional details). For homogeneous circular plates under axisymmetric assumptions concerning the loads and boundary conditions, Eq. (6.2.31) simplifies to

$$\frac{D_{rr}}{r}\frac{d}{dr}\left\{r\frac{d}{dr}\left[\frac{1}{r}\frac{d}{dr}\left(r\frac{dw}{dr}\right)\right]\right\} + c_f w = q, \quad D_{rr} = \frac{Eh^3}{12(1-\nu^2)}. \qquad (6.3.1)$$

Equation (6.3.1) is a fourth-order equation with the duality pairs (w, rN_{rz}) and $(\frac{dw}{dr}, rM_{rr})$. The bending moments M_{rr} and $M_{\theta\theta}$ and the shear force N_{rz} are known in terms of the displacement w (using the identity in **Problem 6.2**) as

$$M_{rr} = -D_{rr}\left(\frac{d^2 w}{dr^2} + \nu\frac{1}{r}\frac{dw}{dr}\right), \quad M_{\theta\theta} = -D_{rr}\left(\nu\frac{d^2 w}{dr^2} + \frac{1}{r}\frac{dw}{dr}\right), \qquad (6.3.2)$$

and

$$rN_{rz} = \frac{d}{dr}\left(rM_{rr}\right) - M_{\theta\theta} = -D_{rr}r\frac{d}{dr}\left[\frac{1}{r}\frac{d}{dr}\left(r\frac{dw}{dr}\right)\right]. \qquad (6.3.3)$$

6.3.2 EXACT SOLUTIONS

The general solution of Eq. (6.3.1), for the case where $c_f = 0$, can be obtained by successive integrations:

$$D_{rr}\frac{d}{dr}\left[\frac{1}{r}\frac{d}{dr}\left(r\frac{dw}{dr}\right)\right] = \frac{1}{r}\int rq(r)\,dr + \frac{c_1}{r} \qquad (=-N_{rz}), \qquad (6.3.4)$$

$$D_{rr}\frac{d}{dr}\left(r\frac{dw}{dr}\right) = r\int\frac{1}{r}\left(\int^r \xi q(\xi)\,d\xi\right)dr + c_1 r\log r + c_2 r, \qquad (6.3.5)$$

$$D_{rr}\frac{dw}{dr} = F(r) + c_1\frac{r}{4}(2\log r - 1) + c_2\frac{r}{2} + c_3\frac{1}{r}, \qquad (6.3.6)$$

$$D_{rr}w = G(r) + c_1\frac{r^2}{4}(\log r - 1) + c_2\frac{r^2}{4} + c_3\log r + c_4, \qquad (6.3.7)$$

where $F(r)$ and $G(r)$ are integrals of the distributed load $q(r)$:

$$F(r) = \frac{1}{r}\int^r \xi\left[\int^\xi\frac{1}{\eta}\left(\int^\eta \mu q(\mu)\,d\mu\right)d\eta\right]d\xi, \qquad (6.3.8a)$$

$$G(r) = \int F(r)\,dr, \quad \int r\log r\,dr = \frac{r^2}{4}(2\log r - 1), \qquad (6.3.8b)$$

and c_i ($i = 1, 2, 3, 4$) are constants of integration that are to be evaluated using the boundary conditions. For solid circular plates the requirement that the slope be zero due to symmetry at $r = 0$ requires $c_3 = 0$. For annular plates with internal radius b, the constant c_3 is not zero but is determined using the boundary conditions at edge $r = b$.

Once the displacement w is determined, one can calculate the bending moments and stresses using the equations

$$M_{rr}(r) = -\left[\frac{dF}{dr} + \frac{c_1}{4}(2\log r + 1) + \frac{c_2}{2} - \frac{c_3}{r^2}\right]$$
$$- \frac{\nu}{r}\left[F + \frac{c_1 r}{4}(2\log r - 1) + \frac{c_2 r}{2} + \frac{c_3}{r}\right], \tag{6.3.9}$$

$$M_{\theta\theta}(r) = -\left[\nu\frac{dF}{dr} + \nu\frac{c_1}{4}(2\log r + 1) + \nu\frac{c_2}{2} - \nu\frac{c_3}{r^2}\right]$$
$$- \frac{1}{r}\left[F + \frac{c_1 r}{4}(2\log r - 1) + \frac{c_2 r}{2} + \frac{c_3}{r}\right], \tag{6.3.10}$$

$$\sigma_{rr}(r,z) = \frac{12z}{h^3}M_{rr}(r) = -\frac{Ez}{(1-\nu^2)}\left(\frac{d^2w}{dr^2} + \frac{\nu}{r}\frac{dw}{dr}\right), \tag{6.3.11}$$

$$\sigma_{\theta\theta}(r,z) = \frac{12z}{h^3}M_{\theta\theta}(r) = -\frac{Ez}{(1-\nu^2)}\left(\nu\frac{d^2w}{dr^2} + \frac{1}{r}\frac{dw}{dr}\right). \tag{6.3.12}$$

In the case of plates subjected to point loads, the maximum values of stresses can occur near to the point of application of the point load. According to Nadai [135] and Westergaard [136], the maximum bending moments and stresses within a small radius $r_c \ll a$ can be computed using the equivalent radius r_e:

$$r_e = \sqrt{1.6r_c^2 + h^2} - 0.675h \text{ for } r_c < 0.5h;$$
$$r_e = r_c \text{ for } r_c \geq 0.5h, \tag{6.3.13}$$

where h is the plate thickness.

6.3.3 NUMERICAL EXAMPLES

In this section, we consider some examples of the exact solutions of circular plates with different boundary conditions and loads. We note that one element of each of the pairs (w, N_{rz}) and $(dw/dr, M_{rr})$ must be specified (or a relation between the two elements of each of the pair must be known).

Example 6.3.1 ————————————————————————————

Determine the exact solutions for deflection, moments, and stresses in a circular plate with clamped edge, $r = a$. Consider two different load cases: (a) uniformly distributed load of intensity q_0 [see Fig. 6.3.1(a)] and (b) point load Q_0 at the center, $r = 0$ [see Fig. 6.3.1(b)].

Solution: (a) *Uniform load.* The boundary conditions at $r = 0$ for circular plates without a point load at the center $r = 0$ are (one element of each of the duality pairs must be known)

$$rN_{rz} = 0, \quad \frac{dw}{dr} = 0 \text{ at } r = 0. \tag{6.3.14}$$

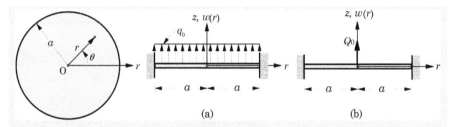

Figure 6.3.1: A clamped circular plate under uniformly distributed load.

The first boundary condition in Eq. (6.3.14), in view of Eq. (6.3.4), gives $c_1 = 0$. The second condition, in light of Eq. (6.3.6), yields $c_3 = 0$, which holds for any solid circular plate, independent of the boundary conditions at $r = a$. Thus, we have

$$c_1 = c_3 = 0. \tag{6.3.15}$$

The boundary conditions associated with the clamped edge at $r = a$ are

$$w = 0, \quad \frac{dw}{dr} = 0. \tag{6.3.16}$$

These boundary conditions yield

$$c_2 = -\frac{2F(a)}{a}, \quad c_4 = -G(a) - c_2\frac{a^2}{4} = -G(a) + \frac{a}{2}F(a). \tag{6.3.17}$$

For uniformly distributed load of intensity q_0, the functions $F(r)$ and $G(r)$ have the form

$$F(r) = \frac{q_0 r^3}{16}, \quad G(r) = \frac{q_0 r^4}{64}. \tag{6.3.18}$$

Therefore, we have

$$c_2 = -\frac{q_0 a^2}{8}, \quad c_4 = \frac{q_0 a^4}{64}. \tag{6.3.19}$$

Hence, the deflection of a clamped circular plate under uniform transverse load is

$$w(r) = \frac{q_0 a^4}{64 D_{rr}}\left[1 - \left(\frac{r}{a}\right)^2\right]^2. \tag{6.3.20}$$

Expressions for the bending moments and stresses from Eqs. (6.3.9)–(6.3.12) become

$$M_{rr}(r) = \frac{q_0 a^2}{16}\left[(1+\nu) - (3+\nu)\left(\frac{r}{a}\right)^2\right], \tag{6.3.21}$$

$$M_{\theta\theta}(r) = \frac{q_0 a^2}{16}\left[(1+\nu) - (1+3\nu)\left(\frac{r}{a}\right)^2\right], \tag{6.3.22}$$

$$\sigma_{rr}(r, z) = \frac{12z}{h^3}M_{rr} = \frac{3q_0 a^2 z}{4h^3}\left[(1+\nu) - (3+\nu)\left(\frac{r}{a}\right)^2\right], \tag{6.3.23}$$

$$\sigma_{\theta\theta}(r, z) = \frac{12z}{h^3}M_{\theta\theta} = \frac{3q_0 a^2 z}{4h^3}\left[(1+\nu) - (1+3\nu)\left(\frac{r}{a}\right)^2\right]. \tag{6.3.24}$$

The maximum deflection occurs at the center of the plate $(r = 0)$:

$$w_{max} = \frac{q_0 a^4}{64 D_{rr}}, \qquad (6.3.25a)$$

while the maximum (in magnitude) values of the bending moments are found to be at the fixed edge $r = a$:

$$M_{rr}(a) = -\frac{q_0 a^2}{8}, \quad M_{\theta\theta}(a) = -\frac{\nu q_0 a^2}{8}. \qquad (6.3.25b)$$

Therefore, the maximum stresses also occur at the fixed edge and they are given by

$$\sigma_{rr}^{max}\left(a, -\frac{h}{2}\right) = -\frac{6 M_{rr}(a)}{h^2} = \frac{3 q_0}{4}\left(\frac{a}{h}\right)^2, \quad \sigma_{\theta\theta}^{max}\left(a, -\frac{h}{2}\right) = -\frac{6 M_{\theta\theta}(a)}{h^2} = \frac{3 \nu q_0}{4}\left(\frac{a}{h}\right)^2.$$
$$(6.3.25c)$$

(b) *Point load.* For a point load Q_0 applied upward at the center, the boundary conditions are [we note that the shear force N_{rz} at $r = 0$, by our notation, acts downward; see Fig. 6.3.1(b)]

$$2\pi\left(r N_{rz}\right)_{r=0} = -Q_0, \quad \left.\frac{dw}{dr}\right|_{r=0} = 0, \quad w(a) = 0, \quad \left.\frac{dw}{dr}\right|_{r=a} = 0. \qquad (6.3.26)$$

Since $q = 0$, we have $F(r) = G(r) = 0$, and from Eq. (6.3.4), we have $2\pi c_1 = Q_0$. The second boundary condition gives, as before, $c_3 = 0$. Thus, the first two boundary conditions in Eq. (6.3.26) give

$$c_1 = \frac{Q_0}{2\pi}, \quad c_3 = 0. \qquad (6.3.27a)$$

The remaining two boundary conditions in Eq. (6.3.26), which correspond to the clamped edge, yield

$$\frac{Q_0 a}{8\pi}\left(2\log a - 1\right) + \frac{a}{2}c_2 = 0, \quad \frac{Q_0 a^2}{8\pi}\left(\log a - 1\right) + \frac{a^2}{4}c_2 + c_4 = 0.$$

Solving for c_2 and c_4, we obtain

$$c_2 = -\frac{Q_0}{4\pi}\left(2\log a - 1\right), \quad c_4 = \frac{Q_0 a^2}{16\pi}. \qquad (6.3.27b)$$

Hence, the solution for a clamped circular plate with a point load Q_0 at the center becomes

$$w(r) = \frac{Q_0 a^2}{16\pi D_{rr}}\left[1 - \left(\frac{r}{a}\right)^2 + 2\left(\frac{r}{a}\right)^2 \log\left(\frac{r}{a}\right)\right], \qquad (6.3.28)$$

$$M_{rr}(r) = -\frac{Q_0}{4\pi}\left[1 + (1+\nu)\log\left(\frac{r}{a}\right)\right], \qquad (6.3.29)$$

$$M_{\theta\theta}(r) = -\frac{Q_0}{4\pi}\left[\nu + (1+\nu)\log\left(\frac{r}{a}\right)\right], \qquad (6.3.30)$$

$$\sigma_{rr}(r, z) = -\frac{3 Q_0 z}{\pi h^3}\left[1 + (1+\nu)\log\left(\frac{r}{a}\right)\right], \qquad (6.3.31)$$

$$\sigma_{\theta\theta}(r, z) = -\frac{3 Q_0 z}{\pi h^3}\left[\nu + (1+\nu)\log\left(\frac{r}{a}\right)\right]. \qquad (6.3.32)$$

The maximum deflection occurs at $r = 0$ [the last term in Eq. (6.3.28) goes to zero as $r \to 0$] is given by

$$w_{\max} = \frac{Q_0 a^2}{16 \pi D_{rr}}.$$

We note that M_{rr} and $M_{\theta\theta}$ (and the corresponding stresses) are not defined at $r = 0$; the maximum values occur at $r = r_e$, where r_e is given by Eq. (6.3.13).

Example 6.3.2

Consider an annular plate with inner radius b and outer radius a, and subjected to uniformly distributed load of intensity q_0, as shown in Fig. 6.3.2. Suppose that the edge $r = a$ is simply supported. Determine the expressions for the transverse deflection, bending moments, and stresses in the plate. Specialize the results for the case of simply supported solid circular plates (i.e., $b = 0$) for two different load cases: (a) uniformly distributed load and (b) point load at the center.

Solution: (a) *Annular plate.* The boundary conditions for this case (i.e., simply supported at $r = a$ and free at $r = b$) are

$$\text{At } r = b: \quad M_{rr} = 0, \quad (rN_{rz}) = 0, \tag{6.3.33a}$$

$$\text{At } r = a: \quad w = 0, \quad M_{rr} = 0. \tag{6.3.33b}$$

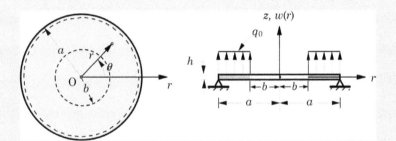

Figure 6.3.2: A simply supported annular plate under uniformly distributed load.

The functions $F(r)$ and $G(r)$ are as given in Eq. (6.3.18). The second condition in Eq. (6.3.33a) gives $c_1 = -q_0 b^2/2$; the substitution of the remaining boundary conditions into Eqs. (6.3.7) and (6.3.9) yield, after some algebraic manipulations, the following relations among c_2, c_3, and c_4:

$$-\left(\frac{3+\nu}{16}\right) q_0 b^2 + \left[\frac{1+\nu}{4} \log b + \frac{1-\nu}{8}\right] q_0 b^2 - \frac{1+\nu}{2} c_2 + \frac{1-\nu}{b^2} c_3 = 0,$$

$$-\left(\frac{3+\nu}{16}\right) q_0 a^2 + \left[\frac{1+\nu}{4} \log a + \frac{1-\nu}{8}\right] q_0 b^2 - \frac{1+\nu}{2} c_2 + \frac{1-\nu}{a^2} c_3 = 0,$$

$$\frac{q_0 a^4}{64} - \frac{a^2}{8} (\log a - 1) q_0 b^2 + \frac{a^2}{4} c_2 + c_3 \log a + c_4 = 0.$$

Solving for the constants c_2, c_3, and c_4, we obtain

$$c_2 = -\left(\frac{1+3\nu}{1+\nu}\right)\frac{q_0 b^2}{8} - \left(\frac{3+\nu}{1+\nu}\right)\frac{q_0 a^2}{8} + \frac{q_0 b^2}{2}\log a - \frac{q_0 b^4}{2(a^2-b^2)}\log\beta, \quad (6.3.34a)$$

$$c_3 = -\left(\frac{3+\nu}{1-\nu}\right)\frac{q_0 b^2 a^2}{16} - \left(\frac{1+\nu}{1-\nu}\right)\frac{q_0 a^2 b^4}{4(a^2-b^2)}\log\beta, \quad (6.3.34b)$$

$$c_4 = -\frac{q_0 a^4}{64} + \left(\frac{3+\nu}{1+\nu}\right)\frac{q_0 a^2(a^2-b^2)}{32} + \left(\frac{3+\nu}{1-\nu}\right)\frac{q_0 b^2 a^2}{16}\log a$$

$$+ \frac{q_0 b^4 a^2}{8(a^2-b^2)}\log\beta + \left(\frac{1+\nu}{1-\nu}\right)\frac{q_0 b^4 a^2}{4(a^2-b^2)}\log a\log\beta. \quad (6.3.34c)$$

The deflection, bending moments, and stresses in the simply supported annular plate under uniformly distributed transverse load are

$$w(r) = \frac{q_0 a^4}{64 D_{rr}}\left\{-1 + \left(\frac{r}{a}\right)^4 + \frac{2\alpha_1}{1+\nu}\left[1 - \left(\frac{r}{a}\right)^2\right] - \frac{4\alpha_2\beta^2}{1-\nu}\log\left(\frac{r}{a}\right)\right\}, \quad (6.3.35)$$

$$M_{rr}(r) = \frac{q_0 a^2}{16}\left\{(3+\nu)\left[1 - \left(\frac{r}{a}\right)^2\right] + \beta^2\left[3+\nu + 4(1+\nu)\kappa\right.\right.$$

$$\left.\left. -\alpha_2\left(\frac{a}{r}\right)^2 + 4(1+\nu)\log\left(\frac{r}{a}\right)\right]\right\}, \quad (6.3.36)$$

$$M_{\theta\theta}(r) = \frac{q_0 a^2}{16}\left\{(3+\nu) - (1+3\nu)\left(\frac{r}{a}\right)^2 + \beta^2\left[(5\nu-1) + 4(1+\nu)\kappa\right.\right.$$

$$\left.\left. +\alpha_2\left(\frac{a}{r}\right)^2 + 4(1+\nu)\log\left(\frac{r}{a}\right)\right]\right\}, \quad (6.3.37)$$

$$\sigma_{rr}(r,z) = \frac{12z}{h^3}M_{rr}(r), \quad \sigma_{\theta\theta}(r,z) = \frac{12z}{h^3}M_{\theta\theta}(r), \quad (6.3.38)$$

where

$$\alpha_1 = (3+\nu)(1-\beta^2) - 4(1+\nu)\beta^2\kappa, \quad \alpha_2 = (3+\nu) + 4(1+\nu)\kappa,$$

$$\kappa = \frac{\beta^2}{1-\beta^2}\log\beta, \quad \beta = \frac{b}{a}.$$

(b) *Solid circular plate.* For a simply supported solid circular plate under uniform load, we set $b=0$ (i.e., $\beta=0$, $\kappa=0$, and $\alpha_1=\alpha_2=3+\nu$) in the previous equations and obtain

$$w(r) = \frac{q_0 a^4}{64 D_{rr}}\left[\left(\frac{r}{a}\right)^4 - 2\left(\frac{3+\nu}{1+\nu}\right)\left(\frac{r}{a}\right)^2 + \frac{5+\nu}{1+\nu}\right], \quad (6.3.39)$$

$$M_{rr}(r) = \frac{q_0 a^2}{16}(3+\nu)\left[1 - \left(\frac{r}{a}\right)^2\right], \quad (6.3.40)$$

$$M_{\theta\theta}(r) = \frac{q_0 a^2}{16}\left[(3+\nu) - (1+3\nu)\left(\frac{r}{a}\right)^2\right]. \quad (6.3.41)$$

The maximum deflection, bending moments, and stresses occur at $r=0$, and they are

$$w_{\max} = \left(\frac{5+\nu}{1+\nu}\right)\frac{q_0 a^4}{64 D_{rr}}, \quad M_{\max} = (3+\nu)\frac{q_0 a^2}{16}, \quad \sigma_{\max} = 3(3+\nu)\frac{q_0 a^2}{8h^2}. \quad (6.3.42)$$

Finally, for a simply supported solid circular plate with a central point load Q_0, Eqs. (6.3.4) and (6.3.6) result in the values $c_1 = Q_0/2\pi$ and $c_3 = 0$. Using the boundary conditions $w(a) = 0$ and $M_{rr}(a) = 0$ in Eqs. (6.3.7) and (6.3.9), we obtain

$$\frac{Q_0 a^2}{8\pi}(\log a - 1) + \frac{a^2}{4}c_2 + c_4 = 0,$$

$$-\frac{Q_0}{2\pi}\left[\frac{1+\nu}{2}\log a + \frac{1-\nu}{4}\right] - \frac{1+\nu}{2}c_2 = 0.$$

Solving for the constants c_2 and c_4, we obtain

$$c_2 = -\frac{Q_0}{4\pi}\left[2\log a + \left(\frac{1-\nu}{1+\nu}\right)\right], \quad c_4 = \left(\frac{3+\nu}{1+\nu}\right)\frac{Q_0 a^2}{16\pi}. \tag{6.3.43}$$

Thus, the solution for any $r \neq 0$ is given by

$$w(r) = \frac{Q_0 a^2}{16\pi D_{rr}}\left[\left(\frac{3+\nu}{1+\nu}\right)\left(1 - \frac{r^2}{a^2}\right) + 2\left(\frac{r}{a}\right)^2\log\left(\frac{r}{a}\right)\right], \tag{6.3.44}$$

$$M_{rr}(r) = -\frac{Q_0(1+\nu)}{4\pi}\log\left(\frac{r}{a}\right), \tag{6.3.45}$$

$$M_{\theta\theta}(r) = -\frac{Q_0}{4\pi}\left[(1+\nu)\log\left(\frac{r}{a}\right) - (1-\nu)\right], \tag{6.3.46}$$

$$\sigma_{rr}(r) = -\frac{3zQ_0(1+\nu)}{h^3\pi}\log\left(\frac{r}{a}\right), \tag{6.3.47}$$

$$\sigma_{\theta\theta}(r) = -\frac{3zQ_0}{h^3\pi}\left[(1+\nu)\log\left(\frac{r}{a}\right) - (1-\nu)\right]. \tag{6.3.48}$$

The maximum deflection is given by

$$w_{max} = w(0) = \frac{Q_0 a^2}{16\pi D_{rr}}\left(\frac{3+\nu}{1+\nu}\right). \tag{6.3.49}$$

The stresses and bending moments cannot be calculated at $r = 0$ due to the logarithmic singularity. The maximum finite stresses produced by load Q_0 on a very small circular area of radius r_c can be calculated using the equivalent radius r_e in Eqs. (6.3.45)–(6.3.48) (see Roark and Young [137]):

$$r_e = \sqrt{1.6r_c^2 + h^2} - 0.675h \quad \text{when } r_c < 1.7h,$$
$$r_e = r_c \quad \text{when } r_c \geq 1.7h, \tag{6.3.50}$$

where h is the plate thickness.

Example 6.3.3

Obtain the exact solution for the deflection $w(r)$ of a simply supported circular plate of radius a subjected to bending moment M_a at $r = a$.

Solution: The boundary conditions in this case are

$$\text{At } r = 0: \quad rN_{rz} = 0, \quad \frac{dw}{dr} = 0; \quad \text{At } r = a: \quad w(a) = 0, \quad M_{rr}(a) = M_a. \quad (6.3.51)$$

In the present problem we have $q = 0$ (hence $F(r) = G(r) = 0$). From Example 6.3.2, for a solid circular plate, we have $c_1 = c_3 = 0$. From Eq. (6.3.6), we have

$$D_{rr} \frac{1}{r} \frac{dw}{dr} = \frac{1}{2} c_2 \quad \rightarrow \quad D_{rr} \frac{d^2 w}{dr^2} = \frac{1}{2} c_2.$$

Substituting these results into Eq. (6.2.21c) (with $B_{rr} = 0$) and equating the result to M_a, we obtain

$$M_{rr}(a) = -\frac{1+\nu}{2} c_2 \quad \rightarrow \quad c_2 = -\frac{2 M_a}{1+\nu}.$$

Then from Eq. (6.3.7), setting $w(a) = 0$ gives

$$\frac{a^2}{4} c_2 + c_4 = 0 \quad \rightarrow \quad c_4 = \frac{M_a a^2}{2(1+\nu)}.$$

Hence, the deflection due to the applied edge moment M_a at $r = a$ is $[M_{rr}(r) = M_a$ everywhere$]$

$$w(r) = \frac{M_a a^2}{2(1+\nu) D_{rr}} \left(1 - \frac{r^2}{a^2} \right). \quad (6.3.52)$$

Example 6.3.4

Consider a simply supported circular plate of radius a and subjected to uniformly distributed load of intensity q_0, where the edge $r = a$ is hinged and elastically restrained from rotation. Determine the analytical solution of the problem.

Solution: The boundary conditions for this case are:

$$\text{At } r = 0: \quad rN_{rz} = 0, \quad \frac{dw}{dr} = 0,$$

$$\text{At } r = a: \quad w(a) = 0, \quad M_{rr}(a) = -\beta \left(-\frac{dw}{dr} \right). \quad (6.3.53)$$

For this problem, without a point load at the center, we again have $c_1 = c_3 = 0$. Then Eqs. (6.3.7) and (6.3.6), respectively, become

$$D_{rr} w(r) = \frac{q_0 r^4}{64} + \frac{r^2}{4} c_2 + c_4,$$

$$D_{rr} \frac{dw}{dr} = \frac{q_0 r^3}{16} + \frac{r}{2} c_2.$$

Using the above relations in Eq. (6.2.21c), we obtain

$$M_{rr} = -\frac{3+\nu}{16} q_0 r^2 - \frac{1+\nu}{2} c_2.$$

Using the boundary conditions,

$$w = 0, \quad M_{rr} = \beta \frac{dw}{dr} \quad \text{at} \quad r = a,$$

we obtain

$$w(a) = 0: \quad \frac{q_0 a^4}{64} + \frac{a^2}{4} c_2 + c_4 = 0,$$

$$M_{rr}(a) = \beta \frac{dw}{dr}(a): \quad \frac{\beta}{D_{rr}} \left(q_0 a^3 + 8ac_2 \right) + (3 + \nu) q_0 a^2 + 8(1 + \nu) c_2 = 0.$$

Solving for c_2 and c_4, we obtain

$$c_2 = -\frac{q_0 a^2}{8} \left[\frac{\beta a + D_{rr}(3 + \nu)}{\beta a + (1 + \nu) D_{rr}} \right], \quad c_4 = \frac{q_0 a^4}{64} \left[\frac{\beta a + D_{rr}(5 + \nu)}{\beta a + (1 + \nu) D_{rr}} \right].$$

Substituting c_2 and c_4 into Eqs. (6.3.7), (6.2.21c), and (6.2.21d), we obtain the expressions:

$$w(r) = \frac{q_0 a^4}{64 D_{rr}} \left\{ \frac{(5 + \nu) D_{rr} + \beta a}{(1 + \nu) D_{rr} + \beta a} - 2 \left[\frac{(3 + \nu) D_{rr} + \beta a}{(1 + \nu) D_{rr} + \beta a} \right] \left(\frac{r}{a} \right)^2 + \left(\frac{r}{a} \right)^4 \right\},$$
$$\tag{6.3.54}$$

$$M_{rr}(r) = \frac{q_0 a^2}{16} \left[(1 + \nu) \frac{(3 + \nu) D_{rr} + \beta a}{(1 + \nu) D_{rr} + \beta a} - (3 + \nu) \left(\frac{r}{a} \right)^2 \right], \tag{6.3.55}$$

$$M_{\theta\theta}(r) = \frac{q_0 a^2}{16} \left[(1 + \nu) \frac{(3 + \nu) D_{rr} + \beta a}{(1 + \nu) D_{rr} + \beta a} - (1 + 3\nu) \left(\frac{r}{a} \right)^2 \right]. \tag{6.3.56}$$

6.4   BENDING SOLUTIONS FOR FGM PLATES

6.4.1   GOVERNING EQUATIONS

In this section, we develop the exact solutions of the equilibrium equations governing functionally graded material (FGM) circular plates. Of course, the homogeneous plate solutions developed in the preceding section can be deduced from the results to be derived here.

First, we summarize the relevant equations for the purpose of this section. The equations of equilibrium in terms of the stress resultants are (without the couple stress effect and with $c_f = 0$):

$$-\frac{1}{r} \left[\frac{d}{dr} (r N_{rr}) - N_{\theta\theta} \right] = 0, \tag{6.4.1}$$

$$-\frac{1}{r} \frac{d}{dr} \left[\frac{d}{dr} (r M_{rr}) - M_{\theta\theta} \right] - q = 0. \tag{6.4.2}$$

The bending equation (6.4.2) can be cast as a pair of equations [the vector approach gives these equations, which, when combined, results in Eq. (6.4.2); see the discussion of the vector approach used to derive the CBT equations in Chapter 3]:

$$-\frac{1}{r}\frac{d}{dr}(rN_{rz}) - q = 0, \tag{6.4.3}$$

$$-\frac{d}{dr}(rM_{rr}) + M_{\theta\theta} + rN_{rz} = 0. \tag{6.4.4}$$

Equation (6.4.4) defines the bending moment–shear force relationship:

$$rN_{rz} = \frac{d}{dr}(rM_{rr}) - M_{\theta\theta}.$$

In the linearized theory, the stress resultants N_{rr}, $N_{\theta\theta}$, M_{rr}, and $M_{\theta\theta}$ are related to the displacements by

$$N_{rr} = A_{rr}\left(\frac{du}{dr} + \nu\frac{u}{r}\right) - B_{rr}\left(\frac{d^2w}{dr^2} + \frac{\nu}{r}\frac{dw}{dr}\right), \tag{6.4.5}$$

$$N_{\theta\theta} = A_{rr}\left(\frac{u}{r} + \nu\frac{du}{dr}\right) - B_{rr}\left(\nu\frac{d^2w}{dr^2} + \frac{1}{r}\frac{dw}{dr}\right), \tag{6.4.6}$$

$$M_{rr} = B_{rr}\left(\frac{du}{dr} + \nu\frac{u}{r}\right) - D_{rr}\left(\frac{d^2w}{dr^2} + \frac{\nu}{r}\frac{dw}{dr}\right), \tag{6.4.7}$$

$$M_{\theta\theta} = B_{rr}\left(\frac{u}{r} + \nu\frac{du}{dr}\right) - D_{rr}\left(\nu\frac{d^2w}{dr^2} + \frac{1}{r}\frac{dw}{dr}\right), \tag{6.4.8}$$

$$\sigma_{rr} = \frac{E}{(1-\nu^2)}\left[\left(\frac{du}{dr} + \nu\frac{u}{r}\right) - z\left(\frac{d^2w}{dr^2} + \frac{\nu}{r}\frac{dw}{dr}\right)\right], \tag{6.4.9}$$

$$\sigma_{\theta\theta} = \frac{E}{(1-\nu^2)}\left[\left(\nu\frac{du}{dr} + \frac{u}{r}\right) - z\left(\nu\frac{d^2w}{dr^2} + \frac{1}{r}\frac{dw}{dr}\right)\right], \tag{6.4.10}$$

where the expressions for σ_{rr} and $\sigma_{\theta\theta}$ are obtained using Eqs. (6.1.7), (6.2.2), and (6.2.3) (omitting the nonlinear contribution).

6.4.2 EXACT SOLUTIONS

From Eq. (6.4.3), we obtain

$$rN_{rz} = -\int^r \xi q(\xi)\,d\xi + c_1. \tag{6.4.11}$$

Substituting Eqs. (6.4.7), (6.4.8), and (6.4.11) in Eq. (6.4.4), we obtain

$$B_{rr}\left[\frac{d}{dr}\left(r\frac{du}{dr}\right) - \frac{u}{r}\right] - D_{rr}\left[\frac{d}{dr}\left(r\frac{d^2w}{dr^2}\right) - \frac{1}{r}\frac{dw}{dr}\right] = -\int^r \xi q(\xi)\,d\xi + c_1 \tag{6.4.12}$$

We can use the following identities to further simplify Eq. (6.4.12):

$$\frac{d}{dr}\left(r\frac{du}{dr}\right) - \frac{u}{r} = r\frac{d}{dr}\left[\frac{1}{r}\frac{d}{dr}(ru)\right],$$

$$\frac{d}{dr}\left(r\frac{d^2w}{dr^2}\right) - \frac{1}{r}\frac{dw}{dr} = r\frac{d}{dr}\left[\frac{1}{r}\frac{d}{dr}\left(r\frac{dw}{dr}\right)\right]. \tag{6.4.13}$$

Then Eq. (6.4.12) becomes

$$B_{rr}r\frac{d}{dr}\left[\frac{1}{r}\frac{d}{dr}(ru)\right] - D_{rr}r\frac{d}{dr}\left[\frac{1}{r}\frac{d}{dr}\left(r\frac{dw}{dr}\right)\right] = -\int^r \xi q(\xi)\, d\xi + c_1.$$

Integrating once, we obtain

$$B_{rr}\left[\frac{1}{r}\frac{d}{dr}(ru)\right] - D_{rr}\left[\frac{1}{r}\frac{d}{dr}\left(r\frac{dw}{dr}\right)\right]$$

$$= -\int^r \left[\frac{1}{\xi}\int^\xi \eta q(\eta)\, d\eta\right] d\xi + c_1 \log r + c_2.$$

One more integration yields

$$B_{rr}\,ru - D_{rr}\,r\frac{dw}{dr} = -\int^r \left\{\xi \int^\xi \left[\frac{1}{\eta}\int^\eta \mu q(\mu)\, d\mu\right] d\eta\right\} d\xi$$

$$+ c_1\frac{r^2}{4}(2\log r - 1) + c_2\frac{r^2}{2} + c_3. \tag{6.4.14}$$

Following the same procedure with Eq. (6.4.1) as we did with Eq. (6.4.4), we obtain (the first step in the sequence)

$$A_{rr}\left[\frac{d}{dr}\left(r\frac{du}{dr}\right) - \frac{u}{r}\right] - B_{rr}\left[\frac{d}{dr}\left(r\frac{d^2w}{dr^2}\right) - \frac{1}{r}\frac{dw}{dr}\right] = 0, \tag{6.4.15}$$

and (the final step)

$$A_{rr}\,ru - B_{rr}\,r\frac{dw}{dr} = c_4\frac{r^2}{2} + c_5. \tag{6.4.16}$$

From Eqs. (6.4.14) and (6.4.16), we can solve for ru and $r(dw/dr)$ as

$$ru(r) = \bar{D}_{rr}^*\left(c_4\frac{r^2}{2} + c_5\right) - \bar{B}_{rr}^*\left(F(r) + c_2\frac{r^2}{2} + c_3\right), \tag{6.4.17}$$

$$r\frac{dw}{dr} = \bar{B}_{rr}^*\left(c_4\frac{r^2}{2} + c_5\right) - \bar{A}_{rr}^*\left(F(r) + c_2\frac{r^2}{2} + c_3\right), \tag{6.4.18}$$

where

$$\bar{A}_{rr}^* = \frac{A_{rr}}{D_{rr}^*}, \quad \bar{B}_{rr}^* = \frac{B_{rr}}{D_{rr}^*}, \quad \bar{D}_{rr}^* = \frac{D_{rr}}{D_{rr}^*}, \quad D_{rr}^* = A_{rr}D_{rr} - B_{rr}B_{rr},$$

(6.4.19)

$$F(r) = -\int^r \left\{ \xi \int^\xi \left[\frac{1}{\eta} \int^\eta \mu q(\mu)\, d\mu \right] d\eta \right\} d\xi + c_1 \frac{r^2}{4} (2\log r - 1). \quad (6.4.20)$$

Integrating Eq. (6.4.18) once, we arrive at the expression for $w(r)$

$$w(r) = \bar{B}_{rr}^* \left(c_4 \frac{r^2}{4} + c_5 \log r \right) - \bar{A}_{rr}^* \left(\int^r \frac{1}{\xi} F(\xi)\, d\xi + c_2 \frac{r^2}{4} + c_3 \log r + c_6 \right).$$

(6.4.21)

The six constants of integration will be determined using six boundary conditions, three at $r = b$ and three at $r = a$ from the duality pairs:

$$(u, rN_{rr}), \quad (w, rN_{rz}), \quad \left(\frac{dw}{dr}, rM_{rr} \right). \quad (6.4.22)$$

To facilitate the determination of the constants of integration using the boundary conditions, we write N_{rr} and M_{rr} in terms of the displacements u and w. First, we compute du/dr and d^2w/dr^2:

$$\frac{du}{dr} = \bar{D}_{rr}^* \left(\frac{1}{2} c_4 - \frac{c_5}{r^2} \right) - \bar{B}_{rr}^* \left[\frac{d}{dr} \left(\frac{F}{r} \right) + \frac{1}{2} c_2 - \frac{c_3}{r^2} \right], \quad (6.4.23)$$

$$\frac{d^2w}{dr^2} = \bar{B}_{rr}^* \left(\frac{1}{2} c_4 - \frac{c_5}{r^2} \right) - \bar{A}_{rr}^* \left[\frac{d}{dr} \left(\frac{F}{r} \right) + \frac{1}{2} c_2 - \frac{c_3}{r^2} \right]. \quad (6.4.24)$$

Next, we write N_{rr} and M_{rr} in terms of the constants of integration (after some simplifications) as:

$$N_{rr} = \frac{1+\nu}{2} c_4 - \frac{1-\nu}{r^2} c_5, \quad (6.4.25)$$

$$M_{rr} = \frac{c_2}{2}(1+\nu) - \frac{c_3}{r^2}(1-\nu) + \frac{d}{dr}\left(\frac{F}{r} \right) + \nu \frac{F}{r^2}. \quad (6.4.26)$$

The function $F(r)$ depends on $q(r)$ and the constant of integration c_1. Two cases that are of interest are when $q(r) = 0$ and $q(r) = q_0$, a constant. In these two cases, we have

$$F(r) = c_1 \frac{r^2}{4} (2\log r - 1), \quad \text{for } q = 0, \quad (6.4.27)$$

$$F(r) = -\frac{q_0 r^4}{16} + c_1 \frac{r^2}{4} (2\log r - 1), \quad \text{for } q = q_0. \quad (6.4.28)$$

Example 6.4.1

Determine the exact solutions for deflection, moments, and stresses in an FGM circular plate with clamped edge, $r = a$. Assume uniform distributed load of intensity q_0.

Solution: The boundary conditions are

$$u = 0, \quad rN_{rz} = 0, \quad \frac{dw}{dr} = 0 \text{ at } r = 0, \tag{6.4.29a}$$

$$u = 0, \quad w = 0, \quad \frac{dw}{dr} = 0, \text{ at } r = a. \tag{6.4.29b}$$

The first boundary condition in Eq. (6.4.29a), in view of Eq. (6.4.17), gives

$$-D_{rr} c_5 + B_{rr} c_3 = 0. \tag{6.4.30a}$$

Using the second boundary condition in Eq. (6.4.29a) in Eq. (6.4.9), we find that $c_1 = 0$. The third boundary condition in Eq. (6.4.29a), in light of Eq. (6.4.18), yields

$$-B_{rr} c_5 + A_{rr} c_3 = 0, \tag{6.4.30b}$$

which holds for any solid circular plate, independent of the boundary conditions at $r = a$. Equations (6.4.30a) and (6.4.30b) together yield $c_3 = c_5 = 0$. The three boundary conditions in Eq. (6.4.29b) yield

$$c_4 = 0, \quad c_2 = \frac{q_0 a^2}{8}, \quad c_6 = -\frac{q_0 a^4}{64}. \tag{6.4.31}$$

Hence, the displacements of a clamped FGM circular plate under uniform transverse load are [the coefficients $\bar{A}_{rr}^*$ and $\bar{B}_{rr}^*$ are defined in Eq. (6.3.19)]

$$u(r) = -\bar{B}_{rr}^* \frac{q_0 a^3}{16} \frac{r}{a} \left(1 - \frac{r^2}{a^2}\right), \tag{6.4.32}$$

$$w(r) = \bar{A}_{rr}^* \frac{q_0 a^4}{64} \left[1 - \left(\frac{r}{a}\right)^2\right]^2. \tag{6.4.33}$$

Expressions for the stress resultants from Eqs. (6.4.5)–(6.4.8) become ($N_{rr} = N_{\theta\theta} = 0$)

$$M_{rr}(r) = \frac{q_0 a^2}{16} \left[(1 + \nu) - (3 + \nu)\left(\frac{r}{a}\right)^2\right], \tag{6.4.34}$$

$$M_{\theta\theta}(r) = \frac{q_0 a^2}{16} \left[(1 + \nu) - (1 + 3\nu)\left(\frac{r}{a}\right)^2\right], \tag{6.4.35}$$

$$\sigma_{rr}(r, z) = \frac{q_0 a^2}{16} \frac{E(z)}{1 - \nu^2} (\bar{B}_{rr}^* + z\bar{A}_{rr}^*)\left[(1 + \nu) - (3 + \nu)\left(\frac{r}{a}\right)^2\right], \tag{6.4.36}$$

$$\sigma_{\theta\theta}(r, z) = \frac{q_0 a^2}{16} \frac{E(z)}{1 - \nu^2} (\bar{B}_{rr}^* + z\bar{A}_{rr}^*)\left[(1 + \nu) - (1 + 3\nu)\left(\frac{r}{a}\right)^2\right]. \tag{6.4.37}$$

The maximum deflection and bending moments are:

$$w(0) = \frac{A_{rr} q_0 a^4}{64 D_{rr}^*}, \quad M_{rr}(a) = -\frac{q_0 a^2}{8}, \quad M_{\theta\theta}(a) = -\frac{\nu q_0 a^2}{8}. \tag{6.4.38}$$

All of the results presented in this example coincide with those derived in **Example 6.3.1** for a homogeneous plate ($B_{rr} = \bar{B}_{rr}^* = 0$), when we set $\bar{A}_{rr}^* = 1/D_{rr}$ and $\bar{D}_{rr}^* = 1/A_{rr}$.

To generate numerical results, we consider circular plates of radius $a = 10$ in., thickness $h = 0.1$ in., and modulus ratio $E_1/E_2 = 10$ with $E_2 = 30 \times 10^6$ psi and $\nu = 0.3$. Figure 6.4.1 shows plots of the transverse deflection $w(r)$ as a function of the normalized radial distance r/a for various values of the volume fraction index n ($n = 0$ corresponds to the homogeneous plate). The deflections are normalized with respect to the load q_0. Figure 6.4.2 shows plots of the bending moment $M_{rr}(r)$ as a function of r/a (independent of n).

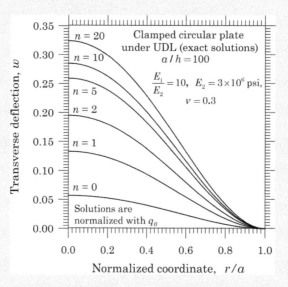

Figure 6.4.1: Variation of the transverse deflection $w(r)/q_0$ with r/a for various values of the volume fraction index n; clamped FGM circular plate under uniformly distributed load ($a/h = 100$).

Example 6.4.2

Consider a FGM circular plate with pinned edge at $r = a$ and subjected to uniformly distributed load of intensity q_0. Determine the expressions for the transverse deflection, bending moments, and stresses in the plate. The plate is loaded with uniformly distributed transverse load and an applied bending moment M_a at $r = a$.

Solution: The function $F(r)$ is given by Eq. (6.4.28). The boundary conditions for the case at hand (i.e., pinned at $r = a$) are

$$\text{At } r = b: \qquad u = 0, \qquad \frac{dw}{dr} = 0, \qquad (rN_{rz}) = 0, \qquad (6.4.39a)$$

$$\text{At } r = a: \qquad u = 0, \qquad w = 0, \qquad M_{rr} = M_a. \qquad (6.4.39b)$$

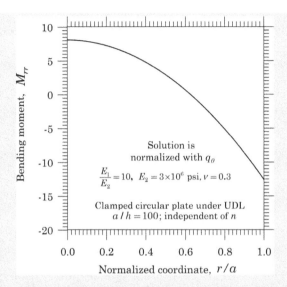

Figure 6.4.2: Variation of the bending moment $M_{rr}(r)/q_0$ with r/a for various values of the n; clamped FGM circular plate under uniformly distributed load ($a/h = 100$); in this case, the bending moment is found to be independent of n.

As in **Example 6.4.1**, the boundary conditions in Eq. (6.4.39a) give: $c_1 = c_3 = c_5 = 0$. The use of the boundary conditions from Eq. (6.4.39b) in Eqs. (6.4.26), (6.4.17), and (6.4.21) yield the following values for c_2, c_4, and c_6:

$$c_2 = \left(\frac{3+\nu}{1+\nu}\right)\frac{q_0 a^2}{8} + \frac{2M_a}{(1+\nu)},$$

$$c_4 = \frac{\bar{B}_{rr}^*}{\bar{D}_{rr}^*(1+\nu)}\left(\frac{q_0 a^2}{4} + 2M_a\right),$$

(6.4.40)

$$\bar{A}_{rr}^* c_6 = \frac{\bar{B}_{rr}^*\bar{B}_{rr}^*}{\bar{D}_{rr}^*(1+\nu)}\frac{q_0 a^4}{16} - \bar{A}_{rr}^*\frac{5+\nu}{1+\nu}\frac{q_0 a^4}{64} - \frac{M_a a^2}{2D_{rr}(1+\nu)}.$$

Then the displacements and bending moments are given by

$$u(r) = -\bar{B}_{rr}^*\frac{q_0 a^3}{16}\frac{r}{a}\left(1 - \frac{r^2}{a^2}\right),$$

(6.4.41)

$$-\frac{dw}{dr} = \bar{A}_{rr}^*\frac{q_0 a^3}{16}\frac{r}{a}\left[\left(\frac{3+\nu}{1+\nu}\right) - \frac{r^2}{a^2}\right] - \frac{\bar{B}_{rr}^*\bar{B}_{rr}^*}{(1+\nu)\bar{D}_{rr}^*}\frac{q_0 a^3}{8}\frac{r}{a}$$

$$+ \frac{M_a a}{2(1+\nu)D_{rr}}\frac{r}{a},$$

(6.4.42)

$$w(r) = \bar{A}_{rr}^*\frac{q_0 a^4}{64}\left[\left(\frac{5+\nu}{1+\nu}\right) - 2\left(\frac{3+\nu}{1+\nu}\right)\frac{r^2}{a^2} + \frac{r^4}{a^4}\right]$$

$$- \frac{\bar{B}_{rr}^*\bar{B}_{rr}^*}{(1+\nu)\bar{D}_{rr}^*}\frac{q_0 a^4}{16}\left(1 - \frac{r^2}{a^2}\right) + \frac{M_a a^2}{2(1+\nu)D_{rr}}\left(1 - \frac{r^2}{a^2}\right),$$

(6.4.43)

$$M_{rr}(r) = (3+\nu)\frac{q_0 a^2}{16}\left(1 - \frac{r^2}{a^2}\right) + M_a, \qquad (6.4.44)$$

where the coefficients $\bar{A}_{rr}^*$, $\bar{B}_{rr}^*$, and $\bar{D}_{rr}^*$ are defined in Eq. (6.3.19). We note that M_a does not contribute to $u(r)$.

The maximum deflection and bending moment occur at $r = 0$, and they are

$$w_{\max} = \bar{A}_{rr}^*\left(\frac{5+\nu}{1+\nu}\right)\frac{q_0 a^4}{64} + \frac{M_a a^2}{2(1+\nu)D_{rr}}, \quad M_{\max} = (3+\nu)\frac{q_0 a^2}{16} + M_a. \quad (6.4.45)$$

The results presented here can be specialized to isotropic plates, and Eqs. (6.4.43) and (6.4.44) coincide with those in Eqs. (6.3.39) and (6.3.40), respectively (note that $B_{rr} = \bar{B}_{rr}^* = 0$, $\bar{A}_{rr}^* = 1/D_{rr}$, and $D_{rr}^* = D_{rr}A_{rr}$ for homogeneous plates).

Numerical results are obtained for the case in which $M_a = 0$ using the following data:

$$a = 10 \text{ in.}, \quad h = 0.1 \text{ in.}, \quad \frac{E_1}{E_2} = 10, \quad E_2 = 30 \times 10^6 \text{ psi}, \quad \nu = 0.3. \qquad (6.4.46)$$

All results are normalized by the load q_0. Figure 6.4.3 contains plots of the deflections $w(r)$ predicted for the pinned FGM plates as a function of the normalized radial coordinate, r/a, for various values of the power-law index n; Figure 6.4.4 contains plots of the variation of the bending moment M_{rr} as function of the normalized radial coordinate, r/a; the results are independent of n.

Figure 6.4.5 shows the center deflection $w(0)$ as a function of the power-law index n for the pinned and clamped circular plates. We note that the rate of increase of the deflection has two different regions; the first region has a rapid increase of the

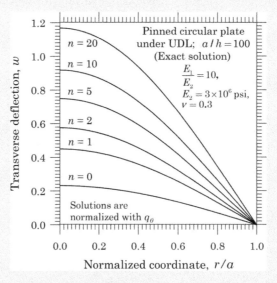

Figure 6.4.3: Plots of the center deflection $w(r)$ of pinned circular plates as functions of the normalized radial coordinate, r/a for various value of n ($a/h = 100$) (uniformly distributed load is used).

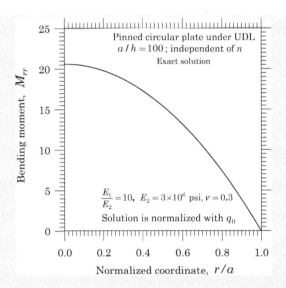

Figure 6.4.4: Plots of the bending moment M_{rr} of pinned circular plates as a function of the normalized radial coordinate, r/a ($a/h = 100$). The results are independent of the n (uniformly distributed load is used).

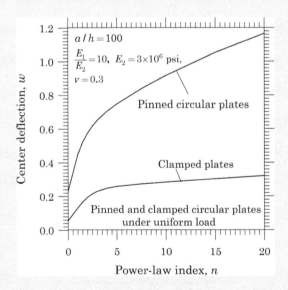

Figure 6.4.5: Plots of the center deflection $w(0)$ of pinned/hinged and clamped circular plates as a function of the power-law index, n ($a/h = 100$; uniformly distributed load is used).

deflection while the second region is marked with a relatively slow increase. This is primarily because of the fact that the coupling coefficient B_{rr} varies with n rapidly for the smaller values of n followed by a slow decay after $n > 3$ [see Fig. 3.4.1(b)]. The rate of increase in the deflection or slope in the second part is less for clamped plates than for the pinned plates. The reason is the fact that the clamped plate is relatively stiffer than the pinned plate.

6.5 BUCKLING AND NATURAL VIBRATION

6.5.1 BUCKLING SOLUTIONS

Here we consider buckling of circular plates under in-plane radial compressive load $N_{rr} = -N_0$ per unit length (see Fig. 6.5.1). The governing equation can be deduced from Eq. (6.2.12):

$$-\frac{1}{r}\frac{d}{dr}\left(rN_{rz}\right) + N_0\frac{1}{r}\frac{d}{dr}\left(r\frac{dw}{dr}\right) = 0. \tag{6.5.1}$$

Integrating the above equation once with respect to r, and obtain

$$rN_{rz} - rN_0\frac{dw}{dr} = c_1. \tag{6.5.2}$$

For a circular plate, the condition that $rN_{rz} = 0$ at $r = 0$ gives $c_1 = 0$, and we obtain

$$N_{rz} = N_0\frac{dw}{dr} = N_0\psi, \quad \psi \equiv \frac{dw}{dr}, \tag{6.5.3}$$

where ψ represents the angle between the central axis of the plate and the normal to the deflected surface at any point along the coordinate r.

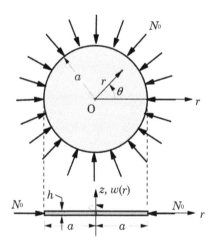

Figure 6.5.1: Buckling of a circular plate under uniform radial compressive load.

From Eq. (6.3.3), we recognize that

$$rN_{rz} = \frac{d}{dr}(rM_{rr}) - M_{\theta\theta} = -D_{rr}\, r \frac{d}{dr}\left[\frac{1}{r}\frac{d}{dr}\left(r\frac{dw}{dr}\right)\right]$$

$$= -D_{rr}\, r\left[\frac{1}{r}\frac{d}{dr}\left(r\frac{d^2w}{dr^2}\right) - \frac{1}{r^2}\frac{dw}{dr}\right]$$

$$= -D_{rr}\, r\left[\frac{1}{r}\frac{d}{dr}\left(r\frac{d\psi}{dr}\right) - \frac{\psi}{r^2}\right]. \tag{6.5.4}$$

Then Eq. (6.5.3) becomes

$$r\frac{d}{dr}\left(r\frac{d\psi}{dr}\right) + \left(\frac{N_0}{D_{rr}}r^2 - 1\right)\psi = 0. \tag{6.5.5}$$

Equation (6.5.5) can be recast in an alternative form by invoking the transformation

$$\bar{r} = r\alpha, \quad \alpha^2 = \frac{N_0}{D_{rr}}, \tag{6.5.6}$$

so that we obtain the *Bessel differential equation*

$$\bar{r}\frac{d}{d\bar{r}}\left(\bar{r}\frac{d\psi}{d\bar{r}}\right) + \left(\bar{r}^2 - n^2\right)\psi = 0, \tag{6.5.7}$$

with $n = 1$, which corresponds to the critical buckling load.
  The general solution of Eq. (6.5.7) is

$$\psi(\bar{r}) = K_1\, J_n(\bar{r}) + K_2\, Y_n(\bar{r}), \tag{6.5.8}$$

where J_n is the Bessel function of the first kind of order n, Y_n is the Bessel function of the second kind of order n, and K_1 and K_2 are constants to be determined using the boundary conditions. In the case of a buckling problem, we do not actually find these constants but determine the stability condition that gives the possible values of N_0. We illustrate the use of the developed equation through examples.

Example 6.5.1

We wish to determine the critical buckling loads for solid circular plates under in-plane compressive load when the outer edge is (a) clamped, (b) simply supported, and (c) simply supported with rotational restraint at the outer edge (i.e., three different problems).

Solution: (a) *Clamped plate.* For a clamped plate, the boundary conditions require that $dw/dr = \psi$ is zero at $r = 0, a$:

$$\psi(0) = 0, \quad \psi(a) = 0, \tag{6.5.9}$$

where a is the radius of the plate. Using the general solution in Eq. (6.5.8) for a homogeneous plate, we obtain

$$K_1 J_1(0) + K_2 Y_1(0) = 0, \quad K_1 J_1(\alpha a) + K_2 Y_1(\alpha a) = 0, \quad \alpha^2 = \frac{N_0}{D_{rr}}. \tag{6.5.10}$$

Since $J_1(0) = 0$ and $Y_1(0)$ is unbounded, we must have $K_2 = 0$, which reduces the second equation (since K_1 cannot be zero for a non-trivial solution) to the condition

$$J_1(\alpha a) = 0, \tag{6.5.11}$$

which is the stability criterion. The smallest root of the condition in Eq. (6.5.11) is $\alpha a = 3.8317$. Thus we have the following *critical buckling load* for a solid clamped circular plate of radius a:

$$N_{cr} = 14.682 \frac{D_{rr}}{a^2}. \tag{6.5.12}$$

(b) *Simply supported plate.* For a simply supported plate, the boundary conditions require that $dw/dr = \psi$ is zero at $r = 0$ and $M_{rr} = 0$ at $r = a$)

$$\psi(0) = 0, \quad \left[\frac{d\psi}{dr} + \nu \frac{1}{r}\psi\right]_{r=a} = 0. \tag{6.5.13}$$

The second boundary condition can be written in terms of $\bar{r}$ as

$$\left[\frac{d\psi}{d\bar{r}} + \frac{\nu}{\bar{r}}\psi\right]_{\bar{r}=\alpha a} = 0. \tag{6.5.14}$$

Using the general solution in Eq. (6.5.8) and the boundary conditions, we obtain

$$K_1 J_1(0) + K_2 Y_1(0) = 0, \quad K_1 J_1'(\alpha a) + K_2 Y_1'(\alpha a) + \frac{\nu}{\alpha a}[K_1 J_1(\alpha a) + K_2 Y_1(\alpha a)] = 0. \tag{6.5.15}$$

The first equation in Eq. (6.5.15) gives $K_2 = 0$, and the second equation, in view of $K_2 = 0$ and the identity

$$\frac{dJ_n}{d\bar{r}} = J_{n-1}(\bar{r}) - \frac{1}{\bar{r}}J_n(\bar{r}), \tag{6.5.16}$$

gives

$$\alpha a J_0'(\alpha a) - (1 - \nu)J_1(\alpha a) = 0, \tag{6.5.17}$$

which is the stability condition for the simply supported plate. For $\nu = 0.3$, the smallest root of the transcendental equation in Eq. (6.5.17) is $\alpha a = 2.05$. Hence, the buckling load for simply supported solid circular plate becomes

$$N_{cr} = 4.198 \frac{D_{rr}}{a^2}. \tag{6.5.18}$$

(c) *Simply supported plate with rotational restraint.* For a simply supported plate with rotational restraint (see Fig. 6.5.2), the boundary conditions are

$$\psi(0) = 0, \quad D_{rr}\left[r\frac{d\psi}{dr} + \nu\psi\right]_{r=a} + a k_R \psi(a) = 0, \tag{6.5.19}$$

where k_R denotes the rotational spring constant. Using the boundary conditions, we obtain the following stability condition:

$$\alpha a\, J_0'(\alpha a) - (1 - \nu - \beta) J_1(\alpha a) = 0, \quad \beta = \frac{a k_R}{D_{rr}}. \tag{6.5.20}$$

When $\beta = 0$, we obtain Eq. (6.5.17) and, when $\beta = \infty$, we obtain Eq. (6.5.11) as special cases. The numerical values of the buckling load factor $\bar{N} = N_{cr}(a^2/D_{rr})$ for various values of the parameter β (for $\nu = 0.3$) are:

$\beta \rightarrow$	0	0.1	0.5	1	5	10	100	∞
$\bar{N} \rightarrow$	4.198	4.449	5.369	6.353	10.462	12.173	14.392	14.682

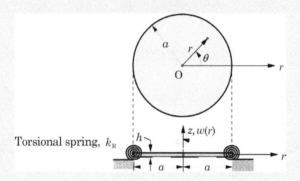

Figure 6.5.2: Buckling of a rotationally-restrained circular plate under uniform radial compressive load.

6.5.2 NATURAL FREQUENCIES

The equation of motion of a homogeneous plate in invariant form (i.e., does not depend on a coordinate system) is given by (see Reddy [113])

$$D_{rr}\nabla^2\nabla^2 w + c_f w + N_0\nabla^2 w + m_0\frac{\partial^2 w}{\partial t^2} - m_2\frac{\partial^2}{\partial t^2}\left(\nabla^2 w\right) = q, \tag{6.5.21}$$

where $m_0 = \rho h$ and $m_2 = \rho h^3/12$ are the principal and rotatory inertias, respectively, N_0 is the compressive biaxial compressive force, c_f is the modulus of the elastic foundation (if any), and ∇^2 is the Laplace operator. In the polar coordinate system (r, θ), ∇^2 can be expressed as

$$\nabla^2 = \frac{1}{r}\frac{\partial}{\partial r}\left(r\frac{\partial}{\partial r}\right) + \frac{1}{r^2}\frac{\partial^2}{\partial\theta^2}. \tag{6.5.22}$$

For free harmonic motion (i.e., natural vibration), the deflection $w(r, \theta)$ can be expressed as

$$w(r, \theta, t) = W(r, \theta)\cos\omega t, \tag{6.5.23}$$

where ω is the circular frequency of vibration (radians per unit time) and $W(r,\theta)$ (mode shape) is a function of only r and θ. Substituting Eq. (6.5.23) into Eq. (6.5.21) (set $q = 0$ for natural vibration), we obtain

$$D_{rr}\nabla^2\nabla^2 W + c_f W + N_0\nabla^2 W - m_0\omega^2 W + m_2\omega^2\nabla^2 W = 0. \qquad (6.5.24)$$

The presence of the rotatory inertia m_2 presents difficulties in obtaining analytical solutions while it contributes little to the frequencies, especially the fundamental frequency. Hence, we neglect the rotatory inertia term in this part of the discussion. For natural vibration without the in-plane force N_0, Eq. (6.5.24) becomes

$$\left(\nabla^4 - \beta^4\right) W = 0, \qquad (6.5.25a)$$

where

$$\beta^4 = \frac{m_0\omega^2 - c_f}{D_{rr}}. \qquad (6.5.25b)$$

Equation (6.5.25a) can be factored into

$$\left(\nabla^2 + \beta^2\right)\left(\nabla^2 - \beta^2\right) W = 0, \qquad (6.5.26)$$

so that the complete solution to Eq. (6.5.26) can be obtained by superimposing the solutions of the equations

$$\nabla^2 W_1 + \beta^2 W_1 = 0, \quad \nabla^2 W_2 - \beta^2 W_2 = 0. \qquad (6.5.27)$$

We assume solution to Eq. (6.5.25a) in the form of the general Fourier series:

$$W(r,\theta) = \sum_{n=0}^{\infty} W_n(r)\cos n\theta + \sum_{n=1}^{\infty} W_n^*(r)\sin n\theta. \qquad (6.5.28)$$

Substitution of Eq. (6.5.28)) into Eq. (6.5.27) yields

$$r^2\frac{d^2 W_{n1}}{dr^2} + r\frac{dW_{n1}}{dr} + \left(\beta^2 r^2 - n^2\right) W_{n1} = 0,$$

$$r^2\frac{d^2 W_{n2}}{dr^2} + r\frac{dW_{n2}}{dr} - \left(\beta^2 r^2 + n^2\right) W_{n2} = 0, \qquad (6.5.29)$$

and two identical equations for W_n^* (W_{n1}^* and W_{n2}^*). Equations (6.5.29) are a form of Bessel's equations, which have the solutions

$$W_{n1} = A_n J_n(\beta r) + B_n Y_n(\beta r), \quad W_{n2} = C_n I_n(\beta r) + D_n K_n(\beta r), \qquad (6.5.30)$$

respectively, where J_n and Y_n are the *Bessel functions of first and second kind*, respectively, and I_n and K_n are the *modified Bessel functions of the first and second kind*, respectively [138]. The coefficients A_n, B_n, C_n, and D_n, which determine the mode shapes, are solved using the boundary conditions. Thus, the general solution of Eq. (6.5.25a) is

$$
W(r, \theta) = \sum_{n=0}^{\infty} [A_n \, J_n(\beta r) + B_n \, Y_n(\beta r) + C_n \, I_n(\beta r) + D_n \, K_n(\beta r)] \cos n\theta
$$

$$
+ \sum_{n=1}^{\infty} [A_n^* \, J_n(\beta r) + B_n^* \, Y_n(\beta r) + C_n^* \, I_n(\beta r) + D_n^* \, K_n(\beta r)] \sin n\theta.
$$

$$(6.5.31)$$

For solid circular plates (i.e., when the inner radius $b = 0$), the terms involving Y_n and K_n in the solution in Eq. (6.5.31) must be discarded in order to avoid singularity of deflections and stresses (i.e., avoid infinite values) at $r = 0$. In addition, if the boundary conditions are symmetrically applied about a diameter of the plate, then the second expression containing $\sin n\theta$ is not needed to represent the solution. Then, the nth term of Eq. (6.5.31) becomes

$$
W_n(r, \theta) = [A_n \, J_n(\beta r) + C_n \, I_n(\beta r)] \cos n\theta, \tag{6.5.32}
$$

for $n = 0, 1, \cdots, \infty$. A *nodal line* is one that has zero deflection (i.e., $W_n = 0$). For circular plates, nodal lines are either concentric circles or diameters. The nodal diameters are determined by $n\theta = \pi/2, 3\pi/2, \ldots$.

When the boundary conditions (which are homogeneous) are applied, a pair of algebraic equations among A_n and C_n for any n are obtained. For a non-trivial solution of the pair, the determinant of the matrix representing the two equations is set to zero. The resulting equation is known as the characteristic polynomial. In the following examples, we illustrate the procedure.

Example 6.5.2

Determined the natural frequencies of a homogeneous clamped circular plate.

Solution: The boundary conditions for the clamped circular plate of radius a are

$$
W_n = 0 \quad \text{and} \quad \frac{\partial W_n}{\partial r} = 0 \quad \text{at } r = a \text{ for any } \theta. \tag{6.5.33}
$$

Using Eq. (6.5.32) in Eq. (6.5.33), we obtain

$$
\begin{bmatrix} J_n(\lambda) & I_n(\lambda) \\ J_n'(\lambda) & I_n'(\lambda) \end{bmatrix} \begin{Bmatrix} A_n \\ C_n \end{Bmatrix} = \begin{Bmatrix} 0 \\ 0 \end{Bmatrix}, \tag{6.5.34}
$$

where $\lambda = \beta a$ and, the prime denotes differentiation with respect to the argument, βr. For nontrivial solution, we set the determinant of the coefficient matrix in Eq. (6.5.34) to zero

$$
\begin{vmatrix} J_n(\lambda) & I_n(\lambda) \\ J_n'(\lambda) & I_n'(\lambda) \end{vmatrix} = 0. \tag{6.5.35}
$$

Expanding the determinant and using the recursion relations

$$\lambda J_n'(\lambda) = nJ_n(\lambda) - \lambda J_{n+1}(\lambda), \quad \lambda I_n'(\lambda) = nI_n(\lambda) + \lambda I_{n+1}(\lambda), \quad (6.5.36)$$

we find

$$J_n(\lambda)I_{n+1}(\lambda) + I_n(\lambda)J_{n+1}(\lambda) = 0, \quad (6.5.37a)$$

or

$$\frac{J_{n+1}(\lambda)}{J_n(\lambda)} + \frac{I_{n+1}(\lambda)}{I_n(\lambda)} = 0, \quad (6.5.37b)$$

which is called the *frequency equation*. The roots λ of Eq. (6.5.37b) are the the eigenvalues, which are used to determine the frequencies ω [see Eq. (6.5.25a)]

$$\omega^2 = \frac{D_{rr}\beta^4 + k}{I_0} = \frac{D_{rr}\lambda^4 + ka^4}{a^4 I_0}. \quad (6.5.38)$$

When $k = 0$, Eq. (6.5.38) reduces to

$$\omega^2 = \frac{D_{rr}\lambda^4}{a^4 I_0}, \quad (6.5.39a)$$

$$\lambda^2 = \omega a^2 \sqrt{I_0/D_{rr}}. \quad (6.5.39b)$$

Note that the frequencies of a clamped circular plate do not depend on Poisson's ratio.

There are an infinite number of roots λ of Eq. (6.5.37b) for each value of n, which represents the number of nodal diameters. For instance, when $n = 0$ (i.e., when the only nodal diameter is the boundary circle) the roots in order of magnitude correspond with $1, 2, \cdots, m$ nodal circles. The mode shape associated with λ is determined using Eq. (6.5.35)

$$\frac{A_n}{C_n} = -\frac{I_n(\lambda)}{J_n(\lambda)}, \quad (6.5.40)$$

where λ is the solution (i.e., root) of Eq. (6.5.37b). The radii of nodal circles $\xi = r/a$ are determined from Eqs. (6.5.40) and (6.5.32):

$$\frac{J_n(\lambda\xi)}{J_n(\lambda)} = \frac{I_n(\lambda\xi)}{I_n(\lambda)}. \quad (6.5.41)$$

Values of λ^2 (see McLachlan [138], Kantham [139], and Leissa [140]) are presented in Table 6.5.1, where n denotes the number of nodal diameters and m is the number of nodal circles, not including the circle $r = a$. Figure 6.5.3 shows typical nodal patterns for a clamped circular plate.

Table 6.5.1: Values of $\lambda^2 = \omega a^2 \sqrt{I_0/D_{rr}}$ for a clamped circular plate.

m	$n = 0$	$n = 1$	$n = 2$	$n = 3$	$n = 4$	$n = 5$
0	10.216	21.26	34.88	51.04	69.6659	90.7390
1	39.771	60.82	84.58	111.01	140.1079	171.8029
2	89.104	120.08	153.81	190.30	229.5185	271.4283
3	158.183	199.06	242.71	289.17	338.4113	390.3896

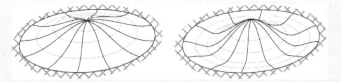

Figure 6.5.3: Typical modes of free vibration of a clamped circular plate, showing the nodal diameters and nodal circles.

Example 6.5.3

Determine the natural frequencies of a simply supported circular plate of radius a.

Solution: The boundary conditions of the simply supported plate are

$$W_n = 0 \quad \text{and} \quad M_{rr} = 0 \quad \text{at } r = a \text{ for any } \theta. \tag{6.5.42}$$

In addition, we note that $\partial^2 W/\partial\theta^2 = 0$ on the boundary. Use of Eq. (6.5.32) in Eq. (6.5.42) results in the following two equations (with $\lambda = \beta a$):

$$A_n J_n(\lambda) + C_n I_n(\lambda) = 0,$$
$$A_n \left[J_n''(\lambda) + \frac{\nu}{\lambda} J_n'(\lambda) \right] + C_n \left[I_n''(\lambda) + \frac{\nu}{\lambda} I_n'(\lambda) \right] = 0. \tag{6.5.43}$$

These equations lead to the frequency equation

$$\frac{J_{n+1}(\lambda)}{J_n(\lambda)} + \frac{I_{n+1}(\lambda)}{I_n(\lambda)} = \frac{2\lambda}{1 - \nu}. \tag{6.5.44}$$

The mode shape is determined using Eq. (6.5.43):

$$\frac{A_n}{C_n} = -\frac{I_n(\lambda)}{J_n(\lambda)}, \tag{6.5.45}$$

where λ is a solution of Eq. (6.5.44). Table 6.5.2 contains values of λ^2 for various values of n and m and Poisson's ratio $\nu = 0.3$.

Table 6.5.2: Values of $\lambda^2 = \omega a^2 \sqrt{I_0/D_{rr}}$ for a simply supported circular plate ($\nu = 0.3$).

m	$n = 0$	$n = 1$	$n = 2$
0	4.98	13.94	25.65
1	29.76	48.51	70.14
2	74.20	102.80	134.33
3	138.34	176.84	218.24

6.6 VARIATIONAL SOLUTIONS

6.6.1 INTRODUCTORY COMMENTS

In this section, we consider the Ritz and Galerkin solutions for linear axisymmetric bending of circular plates. As discussed in the previous chapters, the Ritz method requires a variational statement of the problem, while the Galerkin method only requires the governing equation. We first identify the variational statement for circular plates based on the CPT.

6.6.2 VARIATIONAL STATEMENT

From the statement of Hamilton's principle in Eq. (6.2.5), omitting terms involving time derivatives and couple stress, we have $(N_{rr} = -N_0)$

$$
0 = - \int_b^a \left(M_{rr} \frac{d^2 \delta w}{dr^2} + M_{\theta\theta} \frac{1}{r} \frac{d\delta w}{dr} - c_f w \, \delta w + q \delta w + N_0 \frac{dw}{dr} \frac{d\delta w}{dr} \right) r \, dr
$$

$$
+ \left[a \bar{Q}_a \delta w(a) - b \bar{Q}_b \delta w(b) - a M_a \left(\frac{d\delta w}{dr} \right)_a + b M_b \left(\frac{d\delta w}{dr} \right)_b \right]
$$

$$
= \int_b^a \left\{ D_{rr} \left[\left(\frac{d^2 w}{dr^2} + \frac{\nu}{r} \frac{dw}{dr} \right) \frac{d^2 \delta w}{dr^2} + \left(\nu \frac{d^2 w}{dr^2} + \frac{1}{r} \frac{dw}{dr} \right) \frac{1}{r} \frac{d\delta w}{dr} \right] \right.
$$

$$
\left. + c_f w \, \delta w - q \delta w - N_0 \frac{dw}{dr} \frac{d\delta w}{dr} \right\} r \, dr
$$

$$
+ \left[a \bar{Q}_a \, \delta w(a) - b \bar{Q}_b \, \delta w(b) \right]
$$

$$
+ \left[-a M_a \left(\frac{d\delta w}{dr} \right)_a + b M_b \left(\frac{d\delta w}{dr} \right)_b \right], \tag{6.6.1}
$$

where a and b denote the outer and inner radii of an annular plate, c_f is the modulus of the linear elastic foundation, D_{rr} is the bending stiffness, q is the distributed transverse load, $\bar{Q}_a$ and $\bar{Q}_b$ are the intensities of effective line loads (include in-plane compressive force) at the outer and inner edges, respectively,

$$
\bar{Q}_a = \left[Q_r - N_0 \frac{dw}{dr} \right]_{r=a}, \quad \bar{Q}_b = \left[Q_r - N_0 \frac{dw}{dr} \right]_{r=b}, \tag{6.6.2}
$$

and M_a and M_b the distributed edge moments at the outer and inner edges, respectively. When $b = 0$ (for solid circular plate), we have $M_b = 0$ and $2\pi b \bar{Q}_b = Q_0$, Q_0 being the applied point load at the center of the plate. Equation (6.6.1) is the weak form used in the Ritz method.

The variational statement in Eq. (6.6.1) should be modified to include any nonzero natural boundary conditions. For example, if the edge $r = a$ of the plate is elastically restrained (see Fig. 6.6.1):

$$\left(r\bar{Q}_r\right)_{r=a} + k_E w(a) = 0,$$

$$(-rM_{rr})_{r=a} + k_R \left(\frac{dw}{dr}\right)_{r=a} = 0,$$ (6.6.3)

where k_E and k_R are spring constants associated with extensional and rotational springs, respectively. One can simulate the simply supported boundary condition ($k_E = \infty$ and $k_R = 0$), the clamped boundary condition ($k_E = \infty$ and $k_R = \infty$), and free edge condition ($k_E = 0$ and $k_R = 0$). Similar expressions can be written for the edge at $r = b$.

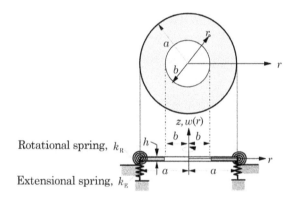

Figure 6.6.1: An elastically restrained annular plate.

The variational statement in Eq. (6.6.1) takes the form

$$0 = \int_b^a \left\{ D_{rr} \left[\left(\frac{d^2w}{dr^2} + \frac{\nu}{r}\frac{dw}{dr} \right) \frac{d^2\delta w}{dr^2} + \left(\nu\frac{d^2w}{dr^2} + \frac{1}{r}\frac{dw}{dr} \right) \frac{1}{r}\frac{d\delta w}{dr} \right] \right.$$

$$\left. + c_f w\delta w - q\delta w - N_0 \frac{dw}{dr}\frac{d\delta w}{dr} \right\} r dr$$

$$+ \left[-k_E \delta w(a)w(a) - b\bar{Q}_b\delta w(b) \right]$$

$$+ \left[-k_R \left(\frac{d\delta w}{dr}\frac{dw}{dr} \right)_a + bM_b \left(\frac{d\delta w}{dr} \right)_b \right]. \tag{6.6.4}$$

6.6.3   THE RITZ METHOD

We presume that all specified geometric boundary conditions are homogeneous so that we can take $\phi_0(r) = 0$. Then the N-parameter Ritz solution is assumed to be of the form

$$w(r) \approx W_N(r) = \sum_{j=1}^{N} c_j \phi_j(r). \qquad (6.6.5)$$

The approximation functions $\phi_j(r)$ must satisfy three conditions: (1) ϕ_j must be continuous as required by the variational statement in Eq. (6.6.1), (2) the set $\{\phi_j\}_i^N$ must be linearly independent, and (3) each ϕ_j must satisfy the homogeneous form of any specified essential boundary conditions. The selection of $\phi_j(r)$, $j = 1, 2, \ldots, N$ subjected to these guidelines is the main task in setting up the Ritz solutions of any circular plate problem. Any specified nonzero natural (i.e., force) boundary conditions enter the variational statement in Eq. (6.6.1)

Substituting Eq. (6.6.5) into Eq. (6.6.1), we obtain a set of algebraic equations in terms of the undetermined parameters $(c_1, c_2, \ldots, c_N)$

$$\sum_{j=1}^{N} A_{ij} c_j - b_i = 0 \quad i = 1, 2, \ldots, N; \quad \text{or} \quad \mathbf{Ac} = \mathbf{b}, \qquad (6.6.6a)$$

where

$$A_{ij} = D_{rr} \int_b^a \left[\frac{d^2\phi_i}{dr^2} \frac{d^2\phi_j}{dr^2} + \frac{\nu}{r} \left(\frac{d\phi_i}{dr} \frac{d^2\phi_j}{dr^2} + \frac{d^2\phi_i}{dr^2} \frac{d\phi_j}{dr} \right) + \frac{1}{r^2} \frac{d\phi_i}{dr} \frac{d\phi_j}{dr} \right] r \, dr$$

$$+ c_f \int_b^a \phi_i \phi_j \, r \, dr - \int_b^a N_0 \frac{d\phi_i}{dr} \frac{d\phi_j}{dr} \, r \, dr, \qquad (6.6.6b)$$

$$b_i = \int_b^a q \phi_i \, r \, dr - \left[a\bar{Q}_a \phi_i(a) - b\bar{Q}_b \phi_i(b) - aM_a \frac{d\phi_i}{dr} \Big|_a + bM_b \frac{d\phi_i}{dr} \Big|_b \right]. \qquad (6.6.6c)$$

Once the approximation functions ϕ_j are selected, the coefficients A_{ij} and b_i can be computed by evaluating the integrals in Eqs. (6.6.6b) and (6.6.6c), and Eq. (6.6.6a) can be solved for the parameters c_i $(i = 1, 2, \ldots, N)$. Then, the N-parameter solution $W_N(r)$ is given by Eq. (6.6.5).

Example 6.6.1

Consider a simply supported circular plate with uniformly distributed transverse load of intensity q_0, as shown in Fig. 6.6.2. Formulate the N-parameter Ritz solution and determine the solution for $N = 1$, 2, and 3. Note that $c_f = 0$ and $N_0 = 0$ for this problem.

Solution: The specified essential (or geometric) boundary conditions are

$$\frac{dw}{dr}(0) = 0, \quad w(a) = 0,$$

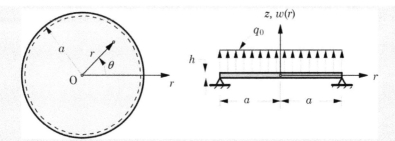

Figure 6.6.2: A simply supported solid circular plate under uniformly distributed load.

and the specified natural boundary conditions are homogeneous:

$$M_{rr} = 0 \text{ at } r = a, \quad N_{rz} = \frac{1}{r}\left[\frac{d}{dr}(rM_{rr}) - M_{\theta\theta}\right] = 0 \text{ at } r = 0.$$

Consequently, $\phi_0(r) = 0$ and each ϕ_j must satisfy the homogeneous form of the geometric boundary conditions

$$\frac{d\phi_j}{dr}(0) = 0, \quad \phi_j(a) = 0.$$

For the choice of algebraic polynomials, we assume

$$\phi_1(r) = \alpha_1 + \alpha_2 r + \alpha_3 r^2,$$

and determine that $\alpha_2 = 0$ and $\alpha_1 + \alpha_3 a^2 = 0$. Thus, we have (for the choice of $\alpha_1 = 1$)

$$\phi_1 = 1 - \frac{r^2}{a^2}.$$

This procedure can be used to obtain a linearly independent and complete set of functions

$$\phi_1 = 1 - \frac{r^2}{a^2}, \quad \phi_2 = 1 - \frac{r^3}{a^3}, \quad \ldots, \quad \phi_j = 1 - \left(\frac{r}{a}\right)^{j+1}. \tag{6.6.7}$$

Since the specified bending moment and shear force are zero, the variational statement in Eq. (6.6.1) is valid. Substituting Eq. (6.6.7) into Eqs. (6.6.6b) and (6.6.6c) and noting that $c_f = 0$ and $N_0 = 0$ in the present case, we obtain

$$A_{ij} = \frac{D_{rr}}{a^2}\left(\frac{ij+1}{i+j} + \nu\right)(i+1)(j+1),$$

$$b_i = q_0 a^2 \frac{(i+1)}{2(i+3)}. \tag{6.6.8}$$

In particular, for $N = 3$ we have the algebraic equations

$$\frac{D_{rr}}{a^2}\begin{bmatrix} 4(1+\nu) & 6(1+\nu) & 8(1+\nu) \\ 6(1+\nu) & 9(1.25+\nu) & 12(1.4+\nu) \\ 8(1+\nu) & 12(1.4+\nu) & 16(\frac{5}{3}+\nu) \end{bmatrix}\begin{Bmatrix} c_1 \\ c_2 \\ c_3 \end{Bmatrix} = \frac{q_0 a^2}{60}\begin{Bmatrix} 15 \\ 18 \\ 20 \end{Bmatrix}. \tag{6.6.9}$$

The one-, two- and three-parameter Ritz solutions are

$$W_1(r) = \frac{q_0 a^4}{16 D_{rr}(1+\nu)} \left(1 - \frac{r^2}{a^2}\right),$$

$$\tag{6.6.10}$$

$$W_2(r) = \frac{q_0 a^4}{80 D_{rr}} \left(\frac{9+4\nu}{1+\nu}\right)\left(1 - \frac{r^2}{a^2}\right) - \frac{q_0 a^4}{30 D_{rr}}\left(1 - \frac{r^3}{a^3}\right),$$

$$\tag{6.6.11}$$

$$W_3(r) = \frac{q_0 a^4}{64 D_{rr}} \left(\frac{6+2\nu}{1+\nu}\right)\left(1 - \frac{r^2}{a^2}\right) - \frac{q_0 a^4}{64 D_{rr}}\left(1 - \frac{r^4}{a^4}\right)$$

$$= \frac{q_0 a^4}{64 D_{rr}}\left[\frac{5+\nu}{1+\nu} - 2\left(\frac{3+\nu}{1+\nu}\right)\left(\frac{r}{a}\right)^2 + \left(\frac{r}{a}\right)^4\right].$$

$$\tag{6.6.12}$$

The three-parameter solution in Eq. (6.6.12) coincides with the exact solution in Eq. (6.3.39). If one tries to use $N \geq 4$, the associated coefficients c_4, c_5, and so on will be computed as zeros because the solution for $N = 3$ is already exact. One can use $W_N(r)$ to compute the stresses, bending moments, and shear force using the relations in Eqs. (6.3.9)–(6.3.12).

6.6.4   THE GALERKIN METHOD

For the weighted-residual family of methods, including the Galerkin method, we use the weighted-integral statement of Eq. (6.3.1):

$$0 = \int_b^a \psi_i(r)\left[\frac{D_{rr}}{r}\frac{d}{dr}\left\{r\frac{d}{dr}\left[\frac{1}{r}\frac{d}{dr}\left(r\frac{dw}{dr}\right)\right]\right\} + c_f w + \frac{1}{r}\frac{d}{dr}\left(rN_0\frac{dw}{dr}\right) - q\right]rdr,$$

$$\tag{6.6.13}$$

where $\psi_i(r)$ is the weight function that takes different forms depending on the method used; for the Galerkin method, we have $\psi_i = \phi_i$. The approximation functions ϕ_i must be (1) continuous as required by the weighted-integral statement in Eq. (6.6.13) and (2) linearly independent and (3) satisfy the homogeneous form of *all* specified boundary conditions, whereas ϕ_0 must satisfy *all* specified boundary conditions. Of course, $\phi_0 = 0$ when all specified boundary conditions are homogeneous.

Substituting the N-parameter approximation of the form

$$w(r) \approx W_N(r) = \sum_{j=1}^{N} c_j \phi_j(r) + \phi_0(r) \tag{6.6.14}$$

into Eq. (6.6.13), we obtain

$$\sum_{j=1}^{N} A_{ij}c_j - b_i = 0 \ (i = 1, 2, \dots, N); \quad \text{or} \quad \mathbf{Ac} = \mathbf{b}, \tag{6.6.15a}$$

where

$$A_{ij} = \int_b^a \psi_i \left[\frac{D_{rr}}{r} \frac{d}{dr} \left\{ r \frac{d}{dr} \left[\frac{1}{r} \frac{d}{dr} \left(r \frac{d\phi_j}{dr} \right) \right] \right\} + c_f \phi_j \right.$$

$$\left. + \frac{1}{r} \frac{d}{dr} \left(r N_0 \frac{d\phi_j}{dr} \right) \right] r \, dr, \qquad (6.6.15b)$$

$$b_i = \int_b^a q \psi_i \, r \, dr - \int_b^a \psi_i \left[\frac{D_{rr}}{r} \frac{d}{dr} \left\{ r \frac{d}{dr} \left[\frac{1}{r} \frac{d}{dr} \left(r \frac{d\phi_0}{dr} \right) \right] \right\} \right.$$

$$\left. + \frac{1}{r} \frac{d}{dr} \left(r N_0 \frac{d\phi_0}{dr} \right) + c_f \phi_0 \right] r \, dr. \qquad (6.6.15c)$$

Example 6.6.2

Consider the bending of a clamped (at $r = a$) solid circular plate under uniformly distributed transverse load of intensity q_0. Determine the two-parameter Ritz and Galerkin solutions.

Solution: The boundary conditions of the problem are

Geometric (essential) : $\quad w(a) = 0, \quad \dfrac{dw}{dr} = 0 \quad$ at $r = 0, a,$ $\qquad (6.6.16a)$

Force (natural) : $\quad Q_r(0) \equiv -D \left\{ \dfrac{d}{dr} \left[\dfrac{1}{r} \dfrac{d}{dr} \left(r \dfrac{dw}{dr} \right) \right] \right\}_{r=0} = 0.$ $\qquad (6.6.16b)$

The three boundary conditions in Eq. (6.6.16a) are of the essential type, while that in Eq. (6.6.16b) is of the natural type.

Obviously, ϕ_0 is zero in both methods since all specified boundary conditions are homogeneous. The choice of ϕ_i depends on the method we use, the Ritz method or the Galerkin method. We derive them for both methods.

The Ritz method. In the Ritz method the approximation functions are required to satisfy the homogeneous form of only the geometric boundary conditions. Since there are *three* essential boundary conditions, we begin with a four-parameter polynomial,

$$\phi_1(r) = c_0 + c_1 r + c_2 r^2 + c_3 r^3,$$

and determine the constants (one of them is arbitrary but nonzero) such that the three conditions

$$\phi_i'(0) = 0, \quad \phi_i(a) = 0, \quad \phi_i'(a) = 0$$

are satisfied. We obtain ($c_0 = a^3 c_3/2, c_1 = 0, c_2 = -3ac_3/2$)

$$\phi_1 = 1 - 3 \left(\frac{r}{a} \right)^2 + 2 \left(\frac{r}{a} \right)^3.$$

Similarly, we pick the five-term polynomial for ϕ_2 and determine the constants (two of them are arbitrary but the coefficient of the highest-order term, c_4, should never be taken as zero)

$$\phi_2(r) = c_0 + c_1 r + c_2 r^2 + c_3 r^3 + c_4 r^4.$$

We obtain

$$c_0 = \frac{3}{2}a^3 c_3 + c_4 a^4, \quad c_1 = 0, \quad c_2 = -\frac{3}{2}ac_3 - 2a^2 c_4.$$

For simplicity, we take $c_3 = 0$ and obtain

$$\phi_2 = \left[1 - \left(\frac{r}{a}\right)^2\right]^2.$$

For the two-parameter Ritz approximation,

$$W_2(r) = c_1\phi_1(r) + c_2\phi_2(r)$$
$$= c_1\left[1 - 3\left(\frac{r}{a}\right)^2 + 2\left(\frac{r}{a}\right)^3\right] + c_2\left[1 - \left(\frac{r}{a}\right)^2\right]^2. \tag{6.6.17}$$

Using the approximation functions ϕ_1 and ϕ_2, compute can A_{ij} and b_i from Eqs. (6.6.6b) and (6.6.6c), noting that k, $\bar{Q}_a$, $\bar{Q}_b$, M_a, and M_b are zero for the present problem. We have $(c_f = 0)$:

$$\frac{1}{r}\frac{d\phi_1}{dr} = -\frac{6}{a^2}\left(1 - \frac{r}{a}\right), \qquad \frac{d^2\phi_1}{dr^2} = -\frac{6}{a^2}\left[1 - 2\left(\frac{r}{a}\right)\right],$$

$$\frac{1}{r}\frac{d\phi_2}{dr} = -\frac{4}{a^2}\left[1 - \left(\frac{r}{a}\right)^2\right], \qquad \frac{d^2\phi_2}{dr^2} = -\frac{4}{a^2}\left[1 - 3\left(\frac{r}{a}\right)^2\right],$$

$$a_{11} = \frac{18\pi D_{rr}}{3a^2}, \quad a_{12} = a_{21} = \frac{96\pi D_{rr}}{5a^2}, \quad a_{22} = \frac{64\pi D_{rr}}{3a^2}, \tag{6.6.18}$$

$$b_1 = \frac{3\pi q_0 a^2}{10}, \quad b_2 = \frac{\pi q_0 a^2}{3}, \quad c_1 = 0, \quad c_2 = \frac{q_0 a^4}{64 D_{rr}}.$$

The one-parameter solution is given by $(c_1 = b_1/a_{11})$

$$W_1(r) = c_1\phi_1(r) = \frac{9q_0 a^4}{640 D_{rr}}\left[1 - 3\left(\frac{r}{a}\right)^2 + 2\left(\frac{r}{a}\right)^3\right]. \tag{6.6.19}$$

The maximum deflection obtained with the one-parameter Ritz method is in 10% error.

The two-parameter Ritz solution coincides with the exact solution in Eq. (6.3.20):

$$w(r) = \frac{q_0 a^4}{64 D_{rr}}\left[1 - \left(\frac{r}{a}\right)^2\right]^2. \tag{6.6.20}$$

The Galerkin method. In the weighted-residual method, the approximation functions are required to satisfy the homogeneous form of all specified boundary conditions. Since there are *four* specified boundary conditions, we begin with a five-term complete polynomial,

$$\phi_1(r) = c_0 + c_1 r + c_2 r^2 + c_3 r^3 + c_4 r^4,$$

and determine the constants such that the four conditions

$$\phi_i'(0) = 0, \quad \phi_i(a) = 0, \quad \phi_i'(a) = 0, \quad \left\{\frac{d}{dr}\left[\frac{1}{r}\frac{d}{dr}\left(r\frac{d\phi_i}{dr}\right)\right]\right\}_{r=0} = 0$$

are satisfied. We obtain $(c_0 = a^4 c_4, c_1 = 0, c_2 = -2a^2 c_4, c_3 = 0)$

$$\phi_1(r) = \left[1 - \left(\frac{r}{a}\right)^2\right]^2.$$

Next, we pick the six-term complete polynomial for ϕ_2 and determine the constants (two of them are arbitrary but the coefficient of the highest-order term, c_5, should not be taken as zero)

$$\phi_2(r) = c_0 + c_1 r + c_2 r^2 + c_3 r^3 + c_4 r^4 + c_5 r^5.$$

We obtain

$$c_0 = a^4 c_4 + \frac{3}{2} a^5 c_5, \quad c_1 = 0, \quad c_2 = -2a^2 c_4 - \frac{5}{2} a^3 c_5, \quad c_3 = 0,$$

and

$$\phi_2(r) = \left[1 - \left(\frac{r}{a}\right)^2\right]^2 a^4 c_4 + \frac{1}{2} a^5 \left[3 - 5\left(\frac{r}{a}\right)^2 + \left(\frac{r}{a}\right)^5\right] c_5.$$

Since the first part is already represented by ϕ_1, we set $c_4 = 0$ and obtain

$$\phi_2(r) = 3 - 5\left(\frac{r}{a}\right)^2 + \left(\frac{r}{a}\right)^5.$$

Thus, two-parameter Galerkin (or any weighted-residual) method is of the form

$$W_2(r) = c_1 \phi_1(r) + c_2 \phi_2(r)$$

$$= c_1 \left[1 - \left(\frac{r}{a}\right)^2\right]^2 + c_2 \left[3 - 5\left(\frac{r}{a}\right)^2 + \left(\frac{r}{a}\right)^5\right]. \tag{6.6.21}$$

The residual using the approximation in Eq. (6.6.21) is computed as follows:

$$\frac{dW_2}{dr} = c_1 \frac{d\phi_1}{dr} + c_2 \frac{d\phi_2}{dr} = \frac{4}{a}\left(\frac{r}{a}\right)\left[1 - \left(\frac{r}{a}\right)^2\right] c_1 + \frac{5}{a}\left[-2\left(\frac{r}{a}\right) + \left(\frac{r}{a}\right)^4\right] c_2,$$

$$\frac{d^2 W_2}{dr^2} = \frac{4}{a^2}\left[1 - 3\left(\frac{r}{a}\right)^2\right] c_1 + \frac{10}{a^2}\left[-1 + 2\left(\frac{r}{a}\right)^3\right] c_2,$$

$$\nabla^2 W_2 \equiv \frac{d^2 W_2}{dr^2} + \frac{1}{r}\frac{dW_2}{dr} = \frac{8}{a^2}\left[1 - 2\left(\frac{r}{a}\right)^2\right] c_1 + \frac{5}{a^2}\left[-4 + 5\left(\frac{r}{a}\right)^3\right] c_2,$$

$$\nabla^4 W_2 \equiv \frac{d^2}{dr^2}\left(\nabla^2 W_2\right) + \frac{1}{r}\frac{d}{dr}\left(\nabla^2 W_2\right) = -\frac{64}{a^4} c_1 + \frac{225}{a^4}\left(\frac{r}{a}\right) c_2.$$

$$\tag{6.6.22}$$

Also, note that

$$N_{rz}(r) \equiv -D_{rr}\frac{d}{dr}\left[\frac{1}{r}\frac{d}{dr}\left(r\frac{dW_2}{dr}\right)\right]$$

$$= D_{rr}\left[-\frac{32}{a^3}\left(\frac{r}{a}\right) c_1 + \frac{75}{a^3}\left(\frac{r}{a}\right)^2 c_2\right], \tag{6.6.23}$$

$$\frac{d}{dr}(rN_{rz}) = -D_{rr}\frac{d}{dr}\left\{r\frac{d}{dr}\left[\frac{1}{r}\frac{d}{dr}\left(r\frac{dW_2}{dr}\right)\right]\right\}$$

$$= D_{rr}\left[-\frac{64}{a^3}\left(\frac{r}{a}\right) c_1 + \frac{225}{a^3}\left(\frac{r}{a}\right)^2 c_2\right]. \tag{6.6.24}$$

Hence, the coefficients A_{ij} and b_i of Eqs. (6.6.15b) and (6.6.15c) are

$$a_{11} = -\frac{64\pi D_{rr}}{3a^2}, \quad a_{12} = \frac{240\pi D_{rr}}{7a^2}, \quad a_{21} = -\frac{352\pi D_{rr}}{7a^2},$$

$$a_{22} = \frac{225\pi D_{rr}}{4a^2}, \quad b_1 = -\frac{\pi q_0 a^2}{3}, \quad b_2 = -\frac{11\pi q_0 a^2}{14}.$$

(6.6.25)

The solution of the Galerkin equations yields

$$c_1 = \frac{q_0 a^4}{64D}, \quad c_2 = 0.$$

(6.6.26)

The fact that $c_2 = 0$ indicates that the two-parameter Galerkin solution is the same as the one-parameter Galerkin solution. The one-parameter Galerkin solution already coincides with the exact solution:

$$W_1(r) = \frac{q_0 a^4}{64D}\left[1 - \left(\frac{r}{a}\right)^2\right]^2.$$

(6.6.27)

6.6.5 NATURAL FREQUENCIES AND BUCKLING LOADS

As seen from **Examples 6.5.1–6.5.3**, the exact determination of buckling loads of circular plates with various boundary conditions results in solutions involving Bessel functions. This is also true when we try to determine the natural frequencies. Here, we illustrate the use of the Ritz method to determine the natural frequencies and buckling loads of circular plates. First, we present the governing equation for free vibration of circular plates.

6.6.5.1 Variational statement

The governing equation for natural vibration was derived in Eqs. (6.5.21)–(6.5.24) (in invariant form). The variational statement of Eq. (6.5.24), with W replaced by w_0 and ∇^2 expressed in terms of the polar coordinates r and θ (see the appendix at the end of the chapter), is given by ($D = D_{rr}$)

$$
\begin{aligned}
0 = \int_\Omega \Bigg\{ & D_{rr}\frac{\partial^2 w_0}{\partial r^2}\frac{\partial^2 \delta w_0}{\partial r^2} + D_{rr}\frac{\nu}{r}\left(\frac{\partial w_0}{\partial r}\frac{\partial^2 \delta w_0}{\partial r^2} + \frac{\partial \delta w_0}{\partial r}\frac{\partial^2 w_0}{\partial r^2}\right. \\
& \left. + \frac{1}{r}\frac{\partial^2 w_0}{\partial \theta^2}\frac{\partial^2 \delta w_0}{\partial r^2} + \frac{1}{r}\frac{\partial^2 \delta w_0}{\partial \theta^2}\frac{\partial^2 w_0}{\partial r^2}\right) + c_f w_0 \delta w_0 \\
& + D_{rr}\left(\frac{1}{r}\frac{\partial w_0}{\partial r} + \frac{1}{r^2}\frac{\partial^2 w_0}{\partial \theta^2}\right)\left(\frac{1}{r}\frac{\partial \delta w_0}{\partial r} + \frac{1}{r^2}\frac{\partial^2 \delta w_0}{\partial \theta^2}\right) \\
& + 2(1-\nu)D_{rr}\left(\frac{1}{r}\frac{\partial^2 w_0}{\partial r \partial \theta} - \frac{1}{r^2}\frac{\partial w_0}{\partial \theta}\right)\left(\frac{1}{r}\frac{\partial^2 \delta w_0}{\partial r \partial \theta} - \frac{1}{r^2}\frac{\partial \delta w_0}{\partial \theta}\right)
\end{aligned}
$$

$$- N_0 \left(\frac{\partial \delta w_0}{\partial r} \frac{\partial w_0}{\partial r} + \frac{1}{r^2} \frac{\partial \delta w_0}{\partial \theta} \frac{\partial w_0}{\partial \theta} \right)$$

$$- \omega^2 \left[m_0 w_0 \delta w_0 + m_2 \left(\frac{\partial w_0}{\partial r} \frac{\partial \delta w_0}{\partial r} + \frac{1}{r^2} \frac{\partial w_0}{\partial \theta} \frac{\partial \delta w_0}{\partial \theta} \right) \right] \bigg\} r dr d\theta.$$

$$(6.6.28)$$

Assume an N-parameter Ritz approximation of the form

$$w_0(r, \theta) \approx W_N(r, \theta) = \sum_{j=1}^{N} c_j \phi_j(r, \theta).$$

$$(6.6.29)$$

Substituting Eq. (6.6.29) into Eq. (6.6.28), we obtain

$$0 = \sum_{j=1}^{N} \left(A_{ij}^{(1)} + A_{ij}^{(2)} - \omega^2 M_{ij} - N_0 G_{ij} \right) c_j,$$

$$(6.6.30)$$

where M_{ij}, G_{ij}, $A_{ij}^{(1)}$, and $A_{ij}^{(2)}$ are defined as

$$M_{ij} = \int_{\Omega} \left[m_0 \phi_i \phi_j + m_2 \left(\frac{\partial \phi_i}{\partial r} \frac{\partial \phi_j}{\partial r} + \frac{1}{r^2} \frac{\partial \phi_i}{\partial \theta} \frac{\partial \phi_j}{\partial \theta} \right) \right] r dr d\theta,$$

$$(6.6.31)$$

$$G_{ij} = \int_0^{2\pi} \int_0^a \left(\frac{\partial \phi_i}{\partial r} \frac{\partial \phi_j}{\partial r} + \frac{1}{r^2} \frac{\partial \phi_i}{\partial \theta} \frac{\partial \phi_j}{\partial \theta} \right) r dr d\theta,$$

$$(6.6.32)$$

$$A_{ij}^{(1)} = D_{rr} \int_{\Omega} \left[\frac{\partial^2 \phi_i}{\partial r^2} \frac{\partial^2 \phi_j}{\partial r^2} + \left(\frac{1}{r} \frac{\partial \phi_i}{\partial r} + \frac{1}{r^2} \frac{\partial^2 \phi_i}{\partial \theta^2} \right) \left(\frac{1}{r} \frac{\partial \phi_j}{\partial r} + \frac{1}{r^2} \frac{\partial^2 \phi_j}{\partial \theta^2} \right) \right.$$

$$+ \frac{\nu}{r} \left(\frac{\partial \phi_i}{\partial r} \frac{\partial^2 \phi_j}{\partial r^2} + \frac{\partial \phi_j}{\partial r} \frac{\partial^2 \phi_i}{\partial r^2} + \frac{1}{r} \frac{\partial^2 \phi_i}{\partial \theta^2} \frac{\partial^2 \phi_j}{\partial r^2} + \frac{1}{r} \frac{\partial^2 \phi_j}{\partial \theta^2} \frac{\partial^2 \phi_i}{\partial r^2} \right)$$

$$\left. + 2(1 - \nu) \left(\frac{1}{r} \frac{\partial^2 \phi_i}{\partial r \partial \theta} - \frac{1}{r^2} \frac{\partial \phi_i}{\partial \theta} \right) \left(\frac{1}{r} \frac{\partial^2 \phi_j}{\partial r \partial \theta} - \frac{1}{r^2} \frac{\partial \phi_j}{\partial \theta} \right) \right] r dr d\theta,$$

$$(6.6.33)$$

$$A_{ij}^{(2)} = c_f \int_{\Omega} \phi_i \phi_j \, r dr d\theta.$$

$$(6.6.34)$$

In matrix notation, Eq. (6.6.30) has the form of an eigenvalue problem:

$$\left(\mathbf{A}^{(1)} + \mathbf{A}^{(2)} - N_0 \mathbf{G} - \omega^2 \mathbf{M} \right) \mathbf{c} = \mathbf{0}.$$

$$(6.6.35)$$

For a non-trivial solution, the coefficient matrix in Eq. (6.6.35) should be singular, that is,

$$|\mathbf{A}^{(1)} + \mathbf{A}^{(2)} - N_0 \mathbf{G} - \omega^2 \mathbf{M}| = 0.$$

$$(6.6.36)$$

The aforementioned formulation is valid also for annular plates. Next, we consider couple of examples.

Example 6.6.3

Determine the fundamental (i.e., the lowest) frequency of a clamped solid circular plate on elastic foundation (i.e., $c_f \neq 0$) using a one-parameter Ritz approximation.

Solution: For free vibration, we set $N_0 = 0$ in Eq. (6.6.35). For a one-parameter $(N = 1)$ Ritz solution, let

$$\phi_1(r, \theta) = f_1(r) \cos n\theta \tag{6.6.37}$$

and compute $A_{11}^{(1)}$, $A_{11}^{(2)}$, and M_{11} as

$$A_{11}^{(1)} = D_{rr} \int_0^{2\pi} \int_0^a \left\{ \left[f_1'' f_1'' + \left(\frac{1}{r} f_1' - \frac{n^2}{r^2} f_1 \right)^2 + \frac{2\nu}{r} \left(f_1' f_1'' - \frac{n^2}{r} f_1 f_1'' \right) \right] \cos^2 n\theta \right.$$

$$\left. + 2(1-\nu) \left(\frac{n}{r} f_1' - \frac{n}{r^2} f_1 \right)^2 \sin^2 n\theta \right\} r\, dr\, d\theta, \tag{6.6.38}$$

$$A_{11}^{(2)} = c_f \int_0^{2\pi} \int_0^a f_1 f_1 \cos^2 n\theta \, r\, dr\, d\theta, \tag{6.6.39}$$

$$M_{11} = \int_0^{2\pi} \int_0^a \left[(m_0 f_1 f_1 + m_2 f_1' f_1') \cos^2 n\theta + m_2 \frac{n^2}{r^2} f_1 f_1 \sin^2 n\theta \right] r\, dr\, d\theta. \tag{6.6.40}$$

The conditions on the approximation functions are

$$\phi_i(r, \theta) = \frac{\partial \phi_i}{\partial r} = 0 \text{ at } r = a, \text{ and } \frac{\partial \phi_i}{\partial r} = 0 \text{ at } r = 0, \tag{6.6.41}$$

which translate into the conditions $f_1(a) = f_1'(a) = f_1'(0) = 0$. Clearly, the choice

$$f_1(r) = 1 - 3 \left(\frac{r}{a} \right)^2 + 2 \left(\frac{r}{a} \right)^3 \tag{6.6.42}$$

satisfies the conditions. Since

$$f_1' = -\frac{6r}{a^2} + \frac{6r^2}{a^3}, \quad f_1'' = -\frac{6}{a^2} + \frac{12r}{a^3},$$

it is clear that the integrals (for $n > 0$)

$$\int_0^a \frac{1}{r} f_1 f_1'' \, dr, \quad \int_0^a \frac{1}{r^3} f_1 f_1 \, dr$$

required in $A_{11}^{(1)}$ of Eq. (6.6.38) do not exist because of the logarithmic singularity. Thus $f_1(r)$ defined in Eq. (6.6.42) is not admissible for $n > 0$. The next function that satisfies the boundary conditions $f_1(a) = f_1'(a) = f_1'(0) = 0$ is

$$f_1(r) = \left[1 - \left(\frac{r}{a} \right)^2 \right]^2 = 1 - 2 \left(\frac{r}{a} \right)^2 + \left(\frac{r}{a} \right)^4, \tag{6.6.43}$$

which is also not admissible for $n > 0$. The next admissible function is

$$f_1(r) = \frac{r}{a} \left[1 - \left(\frac{r}{a} \right)^2 \right]^2 = \frac{r}{a} - 2 \left(\frac{r}{a} \right)^3 + \left(\frac{r}{a} \right)^5, \tag{6.6.44}$$

which will not present any problem in evaluating the integrals for $n > 0$.

The fundamental frequency corresponding to the axisymmetric mode, $n = 0$. For this case, we use the function in Eq. (6.6.42) and obtain

$$A_{11}^{(1)} = 2\pi D_{rr} \int_0^a \left(f_1'' f_1'' + \frac{1}{r^2} f_1' f_1' + 2\nu \frac{1}{r} f_1' f_1'' \right) r\,dr = 2\pi \frac{9D_{rr}}{a^2},$$

$$A_{11}^{(2)} = 2\pi c_f \int_0^a f_1 f_1\, r\,dr = 2\pi \frac{3c_f a^2}{35},$$

$$M_{11} = 2\pi \int_0^a \left(m_0 f_1 f_1 + m_2 f_1' f_1' \right) r\,dr = 2\pi \left(\frac{3a^2}{35} m_0 + \frac{3}{5} m_2 \right).$$

Hence,

$$\omega^2 = \left(\frac{9D_{rr}}{a^2} + \frac{3c_f a^2}{35} \right) \left[\frac{3a^2}{35} m_0 + \frac{3}{5} m_2 \right]^{-1} = \frac{D_{rr}}{a^4 m_0} \left[\frac{105 + (c_f a^4/D_{rr})}{1 + 7(m_2/m_0 a^2)} \right]. \tag{6.6.45}$$

Clearly, rotatory inertia has the effect of reducing the frequency of vibration while the elastic foundation modulus increases it.

For a clamped plate without elastic foundation (i.e., $c_f = 0$), the frequency parameter becomes ($m_2 = m_0 h^2/12$)

$$\lambda^2 \equiv \omega a^2 \sqrt{\frac{m_0}{D_{rr}}} = 10.247 \left[\frac{1}{1 + 0.583 \frac{h^2}{a^2}} \right]^{\frac{1}{2}}. \tag{6.6.46}$$

For very thin plates, say, $h/a = 0.01$, the effect of rotatory inertia is negligible. Even for $h/a = 0.1$, the effect is less than 1%. Note that the one-parameter Ritz solution differs from the exact solution $\omega = 10.216$ in Table 5.5.1 of Reddy [113]) by less than half a percent!

For $n = 1$ and $c_f \neq 0$, the function in Eq. (6.6.44) may be used. We obtain

$$a_{11}^{(1)} = \frac{\pi D_{rr}}{a^2} \left[6 + \frac{2}{3} + 2 \left(\frac{1}{2} - \frac{7}{6} \right) \nu + \frac{4}{3}(1 - \nu) \right],$$

$$a_{11}^{(2)} = \frac{\pi a^2}{60} c_f, \quad m_{11} = \pi \left(\frac{a^2}{60} m_0 + \frac{1}{6} m_2 \right).$$

The frequency parameter is given by

$$\lambda^2 = \omega a^2 \sqrt{\frac{m_0}{D_{rr}}} = \left[\frac{480 + \frac{a^4 c_f}{60 D_{rr}}}{1 + \frac{10 m_2}{a^2 m_0}} \right]^{\frac{1}{2}}. \tag{6.6.47}$$

For a clamped plate without elastic foundation (i.e., $c_f = 0$), the frequency for the case $m = 0$, $n = 1$ becomes ($m_2 = m_0 h^2/12$)

$$\lambda^2 = \omega a^2 \sqrt{\frac{m_0}{D_{rr}}} = 21.909 \left[\frac{1}{1 + \frac{5}{6} \frac{h^2}{a^2}} \right]^{\frac{1}{2}}. \tag{6.6.48}$$

When rotatory inertia is neglected, the frequency predicted by Eq. (6.6.48) differs from the analytical solution $\omega = 21.26$ listed in Table 6.5.1 for $m = 0, n = 1$ (see also Reddy [113]) only by 3%.

Example 6.6.4

Determine the critical buckling load of a solid circular plate under uniform compression force $N_{rr} = -N_0$ in the middle plane of the plate (see Fig. 6.5.1). First formulate the N-parameter Ritz approximation and then determine the critical buckling load.

Solution: For this case, we set $\omega^2 = 0$ in Eq. (6.6.35) and obtain

$$(\mathbf{A} - N_0 \mathbf{G})\mathbf{c} = \mathbf{0}, \qquad (6.6.49)$$

where A_{ij} is the same as $A_{ij}^{(1)}$ in Eq. (6.6.33), and G_{ij} is given by Eq. (6.6.32).

Since we are interested in the minimum buckling load, which occurs in the axisymmetric mode $(n = 0)$, we can use the one-parameter approximation $W_1(r) = c_1 f_1(r)$, where $f_1(r)$ is defined in Eq. (6.6.42). We obtain

$$9 \frac{D_{rr}}{a^2} - \frac{3}{5} N_0 = 0,$$

or

$$N_0 = 15 \frac{D_{rr}}{a^2}, \qquad (6.6.50)$$

which differs from the exact solution in **Example 6.5.1** (for $\beta = \infty$), namely, $14.682(D_{rr}/a^2)$, by 2.18%.

6.7 CHAPTER SUMMARY

In this chapter, the CPT for axisymmetric analysis of circular plates is developed accounting for material variation through the thickness, the von Kármán nonlinearity, and the effect of the modified couple stress. Exact solutions (by direct integration) and variational solution of the theory for the linear case are presented. The nonlinear solutions by the finite element method are presented in Chapter 10.

SUGGESTED EXERCISES

6.1 Derive the equations of motion in (6.2.8) and (6.2.9) using Hamilton's principle, and verify the duality pairs listed in Eq. (6.2.11).

6.2 Verify equations (6.2.35) and (6.2.36).

6.3 Show that Eq. (6.2.25) can be reduced to Eq. (6.3.1) for linear homogeneous plates without the couple stress effect. *Hint:* Prove the following identity first:

$$\frac{d}{dr} \left(r \frac{d^2 w}{dr^2} \right) - \frac{1}{r} \frac{dw}{dr} = r \frac{d}{dr} \left[\frac{1}{r} \frac{d}{dr} \left(r \frac{dw}{dr} \right) \right].$$

6.4 Show that the maximum deflection and bending moment of a simply supported homogeneous circular plate under linearly varying load $q(r) = q_0(1 - r/a)$ are

$$w_{\max} = \frac{q_0 a^4}{4800 D_{rr}} \left(\frac{183 + 43\nu}{1 + \nu} \right), \qquad M_{\max} = q_0 a^2 \left(\frac{71 + 29\nu}{720} \right).$$

6.5 Show that the maximum deflection and bending moment of a simply supported homogeneous circular plate under linearly varying load $q(r) = q_1(r/a)$ are

$$w_{max} = \frac{q_1 a^4}{150 D_{rr}}\left(\frac{6+\nu}{1+\nu}\right), \quad M_{max} = q_1 a^2\left(\frac{4+\nu}{45}\right).$$

6.6 Show that the deflection of a clamped homogeneous circular plate under the load $q(r) = q_0(r^2/a^2)$ is given by

$$w(r) = \frac{q_0 a^4}{576 D_{rr}}\left[2 - 3\left(\frac{r}{a}\right)^2 + \left(\frac{r}{a}\right)^6\right].$$

6.7 Show that the expression for the deflection of a clamped homogeneous circular plate under linearly varying load, $q(r) = q_0(1 - r/a)$ is

$$w(r) = \frac{q_0 a^4}{14400 D_{rr}}\left(129 - 290\frac{r^2}{a^2} + 225\frac{r^4}{a^4} - 64\frac{r^5}{a^5}\right).$$

6.8 Show that the expressions for the deflection and bending moments of a clamped homogeneous circular plate under linearly varying load $q(r) = q_1(r/a)$ are

$$w(r) = \frac{q_1 a^4}{450 D_{rr}}\left(3 - 5\frac{r^2}{a^2} + 2\frac{r^5}{a^5}\right),$$

$$M_{rr}(r) = \frac{q_1 a^2}{45}\left[(1+\nu) - (4+\nu)\frac{r^3}{a^3}\right],$$

$$M_{\theta\theta}(r) = \frac{q_1 a^2}{45}\left[(1+\nu) - (1+4\nu)\frac{r^3}{a^3}\right].$$

6.9 Determine the one-parameter Ritz solution for the deflection under linearly varying load $q = q_1(r/a)$ for a homogeneous circular plate when the edge $r = a$ is (a) simply supported plate and (b) clamped.

6.10 Determine the fundamental natural frequency of a homogeneous circular plate using a one-parameter Ritz approximation when the plate is (a) simply supported and (b) plate is clamped. Use algebraic polynomials.

APPENDIX

The divergence, curl, and gradient of a vector **u**, and the Laplacian in the cylindrical coordinate system are given by (set $z = 0$ for the polar coordinates system):

$$\nabla \cdot \mathbf{u} = \frac{1}{r}\left[\frac{\partial(ru_r)}{\partial r} + \frac{\partial u_\theta}{\partial \theta} + r\frac{\partial u_z}{\partial z}\right], \tag{A.1}$$

$$\nabla \times \mathbf{u} = \left(\frac{1}{r}\frac{\partial u_z}{\partial \theta} - \frac{\partial u_\theta}{\partial z}\right)\hat{\mathbf{e}}_r + \left(\frac{\partial u_r}{\partial z} - \frac{\partial u_z}{\partial r}\right)\hat{\mathbf{e}}_\theta + \frac{1}{r}\left[\frac{\partial(ru_\theta)}{\partial r} - \frac{\partial u_r}{\partial \theta}\right]\hat{\mathbf{e}}_z, \tag{A.2}$$

$$\nabla \mathbf{u} = \frac{\partial u_r}{\partial r}\hat{\mathbf{e}}_r\hat{\mathbf{e}}_r + \frac{\partial u_\theta}{\partial r}\hat{\mathbf{e}}_r\hat{\mathbf{e}}_\theta + \frac{1}{r}\left(\frac{\partial u_r}{\partial \theta} - u_\theta\right)\hat{\mathbf{e}}_\theta\hat{\mathbf{e}}_r + \frac{\partial u_z}{\partial r}\hat{\mathbf{e}}_r\hat{\mathbf{e}}_z + \frac{\partial u_r}{\partial z}\hat{\mathbf{e}}_z\hat{\mathbf{e}}_r$$

$$+ \frac{1}{r}\left(u_r + \frac{\partial u_\theta}{\partial \theta}\right)\hat{\mathbf{e}}_\theta\hat{\mathbf{e}}_\theta + \frac{1}{r}\frac{\partial u_z}{\partial \theta}\hat{\mathbf{e}}_\theta\hat{\mathbf{e}}_z + \frac{\partial u_\theta}{\partial z}\hat{\mathbf{e}}_z\hat{\mathbf{e}}_\theta + \frac{\partial u_z}{\partial z}\hat{\mathbf{e}}_z\hat{\mathbf{e}}_z, \tag{A.3}$$

$$\nabla^2 = \frac{1}{r}\left[\frac{\partial}{\partial r}\left(r\frac{\partial}{\partial r}\right) + \frac{1}{r}\frac{\partial^2}{\partial \theta^2} + r\frac{\partial^2}{\partial z^2}\right]. \tag{A.4}$$

If you can't fly, then run. If you can't run, then walk. If you can't walk, then crawl, but by all means, keep moving.

Martin Luther King, Jr.

7 First-Order Theory of Circular Plates

Science does not know its debt to imagination.   Ralph Waldo Emerson

7.1 GOVERNING EQUATIONS

7.1.1 DISPLACEMENTS AND STRAINS

The *first-order shear deformation plate theory* (FST) is the simplest theory that accounts for nonzero transverse shear strain [4]. Thus, FST is based on the same kinematics as the TBT for beams in Chapter 4. The assumed displacement is

$$\mathbf{u} = u_r\,\hat{\mathbf{e}}_r + u_z\,\hat{\mathbf{e}}_z, \quad u_r(r,z,t) = u(r,t) + z\phi_r(r,t), \quad u_z(r,z,t) = w(r,t), \tag{7.1.1}$$

where ϕ_r denotes the rotation of a transverse normal in the plane $\theta = \text{constant}$. The FST includes a constant state of transverse shear strain with respect to the thickness coordinate, and hence, requires the use of a shear correction coefficient, which depends, in general, not only on the material and geometric parameters but also on the loading and boundary conditions.

The nonzero von Kármán strains of the theory are

$$\varepsilon_{rr} = \varepsilon_{rr}^{(0)} + z\varepsilon_{rr}^{(1)}, \quad \varepsilon_{\theta\theta} = \varepsilon_{\theta\theta}^{(0)} + z\varepsilon_{\theta\theta}^{(1)}, \quad \varepsilon_{rz} = \varepsilon_{rz}^{(0)}, \tag{7.1.2}$$

where

$$\varepsilon_{rr}^{(0)} = \frac{\partial u}{\partial r} + \frac{1}{2}\left(\frac{\partial w}{\partial r}\right)^2, \quad \varepsilon_{rr}^{(1)} = \frac{\partial \phi_r}{\partial r},$$

$$\varepsilon_{\theta\theta}^{(0)} = \frac{u}{r}, \quad \varepsilon_{\theta\theta}^{(1)} = \frac{\phi_r}{r}, \quad 2\varepsilon_{rz}^{(0)} = \phi_r + \frac{\partial w}{\partial r}. \tag{7.1.3}$$

The rotation and curvature components are [cf. Eqs. (6.1.5) and (6.1.6a)]

$$\omega_\theta = \frac{1}{2}\left(\frac{\partial u_r}{\partial z} - \frac{\partial u_z}{\partial r}\right) = \frac{1}{2}\left(\phi_r - \frac{\partial w}{\partial r}\right), \quad \chi_{z\theta} = 0,$$

$$\chi_{r\theta} = \frac{1}{2}\left(\frac{\partial \omega_\theta}{\partial r} - \frac{\omega_\theta}{r}\right) = \frac{1}{4}\left[\frac{\partial \phi_r}{\partial r} - \frac{1}{r}\phi_r - \left(\frac{\partial^2 w}{\partial r^2} - \frac{1}{r}\frac{\partial w}{\partial r}\right)\right]. \tag{7.1.4}$$

DOI: 10.1201/9781003240846-7

7.1.2 EQUATIONS OF MOTION

Hamilton's principle for the FST takes the form

$$
\begin{aligned}
0 &= \int_0^T \left(\delta K - \delta U - \delta V\right) dt \\
&= \int_0^T \int_0^a \int_{-\frac{h}{2}}^{\frac{h}{2}} \rho \left(\dot{u}_r\, \delta \dot{u}_r + \dot{u}_z\, \delta \dot{u}_z\right) r\, dr\, dz\, dt + \int_0^T \int_0^a q\, \delta w\, r\, dr\, dt \\
&\quad - \int_0^T \int_0^a \int_{-\frac{h}{2}}^{\frac{h}{2}} \left(\sigma_{rr}\, \delta \varepsilon_{rr} + \sigma_{\theta\theta}\, \delta \varepsilon_{\theta\theta} + 2 K_s \sigma_{rz}\, \delta \varepsilon_{rz} + 2 m_{r\theta}\, \delta \chi_{r\theta}\right) r\, dr\, dz\, dt \\
&= \int_0^T \int_0^a \left\{ m_0 \left(\dot{u}\, \delta \dot{u} + \dot{w}\, \delta \dot{w}\right) + m_1 \left(\dot{\phi}_r\, \delta \dot{u} + \dot{u}\, \delta \dot{\phi}_r\right) + m_2 \dot{\phi}_r\, \delta \dot{\phi}_r + q \delta w \right. \\
&\qquad - N_{rr} \left(\frac{\partial \delta u}{\partial r} + \frac{\partial w}{\partial r} \frac{\partial \delta w}{\partial r}\right) - N_{\theta\theta} \left(\frac{\delta u}{r}\right) - M_{rr} \frac{\partial \delta \phi_r}{\partial r} \\
&\qquad - \frac{1}{r} M_{\theta\theta}\, \delta \phi_r - N_{rz} \left(\delta \phi_r + \frac{\partial \delta w}{\partial r}\right) \\
&\qquad \left. - \frac{1}{2} P_{r\theta} \left[\frac{\partial \delta \phi_r}{\partial r} - \frac{1}{r} \delta \phi_r - \left(\frac{\partial^2 \delta w}{\partial r^2} - \frac{1}{r} \frac{\partial \delta w}{\partial r}\right)\right] \right\} r\, dr\, dt, \quad (7.1.5)
\end{aligned}
$$

where the various stress resultants and mass inertia coefficients m_i are defined by

$$
(N_{rr}, M_{rr}) = \int_{-\frac{h}{2}}^{\frac{h}{2}} (1, z)\sigma_{rr}\, dz, \quad N_{rz} = K_s \int_{-\frac{h}{2}}^{\frac{h}{2}} \sigma_{rz}\, dz
$$

$$
(N_{\theta\theta}, M_{\theta\theta}) = \int_{-\frac{h}{2}}^{\frac{h}{2}} (1, z)\sigma_{\theta\theta}\, dz, \quad P_{r\theta} = \int_{-\frac{h}{2}}^{\frac{h}{2}} m_{r\theta}\, dz, \quad (7.1.6)
$$

$$
(m_0, m_1, m_2) = \int_{-\frac{h}{2}}^{\frac{h}{2}} (1, z, z^2)\, \rho(z)\, dz,
$$

and K_s denotes the shear correction coefficient.

The Euler–Lagrange equations obtained from Hamilton's principle in Eq. (7.1.5) are the governing equations of motion of the FST, and they are:

$$
\frac{1}{r} \left[\frac{\partial}{\partial r} (r N_{rr}) - N_{\theta\theta}\right] = m_0 \frac{\partial^2 u}{\partial t^2} + m_1 \frac{\partial^2 \phi_r}{\partial t^2}, \quad (7.1.7)
$$

$$
\frac{1}{r} \frac{\partial}{\partial r} (r V_r) + \frac{1}{2r} \frac{\partial}{\partial r} \left[\frac{\partial}{\partial r} (r P_{r\theta}) + P_{r\theta}\right] + q = m_0 \frac{\partial^2 w}{\partial t^2}, \quad (7.1.8)
$$

$$
\frac{1}{r} \left[\frac{\partial}{\partial r} (r M_{rr}) - M_{\theta\theta} + \frac{1}{2} \frac{\partial}{\partial r} (r P_{r\theta}) + \frac{1}{2} P_{r\theta}\right] - N_{rz}
$$

$$
= m_1 \frac{\partial^2 u}{\partial t^2} + m_2 \frac{\partial^2 \phi_r}{\partial t^2}, \quad (7.1.9)
$$

where

$$V_r = N_{rz} + N_{rr}\frac{\partial w}{\partial r}. \tag{7.1.10}$$

When the time derivative, nonlinear, and the couple stress terms are omitted, Eqs. (7.1.7)–(7.1.9) reduce to [the same as those in Eqs. (6.2.17)–(6.2.19)]:

$$-\frac{1}{r}\left[\frac{d}{dr}(rN_{rr}) - N_{\theta\theta}\right] = 0, \tag{7.1.11}$$

$$-\frac{1}{r}\frac{d}{dr}(rN_{rz}) - q = 0, \tag{7.1.12}$$

$$-\frac{1}{r}\left[\frac{d}{dr}(rM_{rr}) - M_{\theta\theta}\right] + N_{rz} = 0. \tag{7.1.13}$$

The boundary conditions obtained from Hamilton's principle involve specifying one element of each of the following pairs:

$$u \ \text{or} \ rN_{rr}; \quad w \ \text{or} \ rV_r + \frac{r}{2}\left[\frac{d}{dr}(rP_{r\theta}) + P_{r\theta}\right] \equiv r\bar{V}_r,$$

$$\tag{7.1.14}$$

$$-\frac{dw}{dr} \ \text{or} \ -\tfrac{1}{2}rP_{r\theta}; \quad \phi_r \ \text{or} \ rM_{rr} + \tfrac{1}{2}rP_{r\theta} \equiv r\bar{M}_{rr}.$$

7.1.3 PLATE CONSTITUTIVE RELATIONS

The stress resultants appearing in Eqs. (7.1.7)–(7.1.9) can be expressed in terms of the generalized displacements (u, w, ϕ_r) as (thermal effects are not included)

$$N_{rr} = A_{rr}\left[\frac{\partial u}{\partial r} + \frac{1}{2}\left(\frac{\partial w}{\partial r}\right)^2 + \nu\frac{u}{r}\right] + B_{rr}\left(\frac{\partial \phi_r}{\partial r} + \frac{\nu}{r}\phi_r\right),$$

$$N_{\theta\theta} = A_{rr}\left[\frac{u}{r} + \nu\frac{\partial u}{\partial r} + \frac{\nu}{2}\left(\frac{\partial w}{\partial r}\right)^2\right] + B_{rr}\left(\nu\frac{\partial \phi_r}{\partial r} + \frac{1}{r}\phi_r\right),$$

$$M_{rr} = B_{rr}\left[\frac{\partial u}{\partial r} + \frac{1}{2}\left(\frac{\partial w}{dr}\right)^2 + \nu\frac{u}{r}\right] + D_{rr}\left(\frac{\partial \phi_r}{\partial r} + \frac{\nu}{r}\phi_r\right), \tag{7.1.15}$$

$$M_{\theta\theta} = B_{rr}\left[\frac{u}{r} + \nu\frac{\partial u}{\partial r} + \frac{\nu}{2}\left(\frac{\partial w}{\partial r}\right)^2\right] + D_{rr}\left(\nu\frac{\partial \phi_r}{\partial r} + \frac{1}{r}\phi_r\right),$$

$$N_{rz} = S_{rz}\left(\phi_r + \frac{\partial w}{\partial r}\right),$$

$$P_{r\theta} = \frac{1}{2}S_{r\theta}\left[\frac{\partial \phi_r}{\partial r} - \frac{\partial^2 w}{\partial r^2} - \frac{1}{r}\left(\phi_r - \frac{\partial w}{\partial r}\right)\right],$$

where A_{rr}, B_{rr}, D_{rr}, $S_{rz} = K_s A_{rz}$, and $S_{r\theta}$ are the extensional, extensional-bending, bending, shear, and couple stress stiffness coefficients, respectively:

$$A_{rr} = \frac{1}{(1-\nu^2)} \int_{-\frac{h}{2}}^{\frac{h}{2}} E(z)\,dz, \qquad B_{rr} = \frac{1}{(1-\nu^2)} \int_{-\frac{h}{2}}^{\frac{h}{2}} E(z)z\,dz,$$

$$D_{rr} = \frac{1}{(1-\nu^2)} \int_{-\frac{h}{2}}^{\frac{h}{2}} E(z)z^2\,dz, \quad A_{rz} = \frac{1}{2(1+\nu)} \int_{-\frac{h}{2}}^{\frac{h}{2}} E(z)\,dz, \qquad (7.1.16)$$

$$S_{r\theta} = \frac{\ell^2}{2(1+\nu)} \int_{-\frac{h}{2}}^{\frac{h}{2}} E(z)\,dz.$$

For FGM plates, the power-law distribution given in Eq. (6.1.8) is used:

$$E(z) = (E_1 - E_2) V_1(z) + E_2, \quad V_1(z) = \left(\frac{1}{2} + \frac{z}{h}\right)^n, \qquad (7.1.17)$$

with a similar expression for the mass density $\rho(z)$ [see Eq. (6.2.7)]; ν is assumed to be a constant. The constitutive law for modified couple stress is given by Eq. (6.1.9).

7.1.4 EQUATIONS OF MOTION IN TERMS OF THE DISPLACEMENTS

7.1.4.1 The general case

The equations of motion in Eqs. (7.1.7)–(7.1.9) can be expressed in terms of the generalized displacements (u, w, ϕ_r) by invoking Eqs. (7.1.15):

$$-\frac{1}{r}\frac{\partial}{\partial r}\left\{ rA_{rr}\left[\frac{\partial u}{\partial r} + \frac{1}{2}\left(\frac{\partial w}{\partial r}\right)^2 + \nu\frac{u}{r}\right] + rB_{rr}\left(\frac{\partial\phi_r}{\partial r} + \frac{\nu}{r}\phi_r\right)\right\}$$

$$+\frac{1}{r}\left\{ A_{rr}\left[\frac{u}{r} + \nu\frac{\partial u}{\partial r} + \frac{\nu}{2}\left(\frac{\partial w}{\partial r}\right)^2\right] + B_{rr}\left(\nu\frac{\partial\phi_r}{\partial r} + \frac{1}{r}\phi_r\right)\right\}$$

$$+ m_0\frac{\partial^2 u}{\partial t^2} + m_1\frac{\partial^2\phi_r}{\partial t^2} = 0, \qquad (7.1.18)$$

$$-\frac{1}{r}\frac{\partial}{\partial r}\left\{ rS_{rz}\left(\phi_r + \frac{\partial w}{\partial r}\right) + rA_{rr}\frac{\partial w}{\partial r}\left[\frac{\partial u}{\partial r} + \frac{1}{2}\left(\frac{\partial w}{\partial r}\right)^2 + \nu\frac{u}{r}\right]\right.$$

$$\left. + rB_{rr}\frac{\partial w}{\partial r}\left(\frac{\partial\phi_r}{\partial r} + \frac{\nu}{r}\phi_r\right)\right\}$$

$$-\frac{1}{4r}\frac{\partial^2}{\partial r^2}\left\{ rS_{r\theta}\left[\frac{\partial\phi_r}{\partial r} - \frac{\partial^2 w}{\partial r^2} - \frac{1}{r}\left(\phi_r - \frac{\partial w}{\partial r}\right)\right]\right\}$$

$$-\frac{1}{4r}\frac{\partial}{\partial r}\left\{ S_{r\theta}\left[\frac{\partial\phi_r}{\partial r} - \frac{\partial^2 w}{\partial r^2} - \frac{1}{r}\left(\phi_r - \frac{\partial w}{\partial r}\right)\right]\right\}$$

$$- q + m_0\frac{\partial^2 w}{\partial t^2} = 0, \qquad (7.1.19)$$

$$-\frac{1}{r}\frac{\partial}{\partial r}\left\{rB_{rr}\left[\frac{\partial u}{\partial r}+\frac{1}{2}\left(\frac{\partial w}{\partial r}\right)^2+\nu\frac{u}{r}\right]+rD_{rr}\left(\frac{\partial \phi_r}{\partial r}+\frac{\nu}{r}\phi_r\right)\right\}$$

$$+\frac{1}{r}\left\{B_{rr}\left[\frac{u}{r}+\nu\frac{\partial u}{\partial r}+\frac{\nu}{2}\left(\frac{\partial w}{\partial r}\right)^2\right]+D_{rr}\left(\nu\frac{\partial \phi_r}{\partial r}+\frac{1}{r}\phi_r\right)\right\}$$

$$-\frac{1}{4r}\frac{\partial}{\partial r}\left\{rS_{r\theta}\left[\frac{\partial \phi_r}{\partial r}-\frac{\partial^2 w}{\partial r^2}-\frac{1}{r}\left(\phi_r-\frac{\partial w}{\partial r}\right)\right]\right\}$$

$$-\frac{1}{4r}S_{r\theta}\left[\frac{\partial \phi_r}{\partial r}-\frac{\partial^2 w}{\partial r^2}-\frac{1}{r}\left(\phi_r-\frac{\partial w}{\partial r}\right)\right]+S_{rz}\left(\phi_r+\frac{\partial w}{\partial r}\right)$$

$$+m_1\frac{\partial^2 u}{\partial t^2}+m_2\frac{\partial^2 \phi_r}{\partial t^2}=0. \tag{7.1.20}$$

Several special cases of these equations are presented next.

7.1.4.2 Nonlinear equations of equilibrium

$$-\frac{1}{r}\frac{d}{dr}\left\{rA_{rr}\left[\frac{du}{dr}+\frac{1}{2}\left(\frac{dw}{dr}\right)^2+\nu\frac{u}{r}\right]+rB_{rr}\left(\frac{d\phi_r}{dr}+\frac{\nu}{r}\phi_r\right)\right\}$$

$$+\frac{1}{r}\left\{A_{rr}\left[\frac{u}{r}+\nu\frac{du}{dr}+\frac{\nu}{2}\left(\frac{dw}{dr}\right)^2\right]+B_{rr}\left(\nu\frac{d\phi_r}{dr}+\frac{1}{r}\phi_r\right)\right\}=0, \tag{7.1.21}$$

$$-\frac{1}{r}\frac{d}{dr}\left\{rS_{rz}\left(\phi_r+\frac{dw}{dr}\right)+rA_{rr}\frac{dw}{dr}\left[\frac{du}{dr}+\frac{1}{2}\left(\frac{dw}{dr}\right)^2+\nu\frac{u}{r}\right]\right.$$

$$\left.+rB_{rr}\frac{dw}{dr}\left(\frac{d\phi_r}{dr}+\frac{\nu}{r}\phi_r\right)\right\}$$

$$-\frac{1}{4r}\frac{d^2}{dr^2}\left\{rS_{r\theta}\left[\frac{d\phi_r}{dr}-\frac{d^2 w}{dr^2}-\frac{1}{r}\left(\phi_r-\frac{dw}{dr}\right)\right]\right\}$$

$$-\frac{1}{4r}\frac{d}{dr}\left\{S_{r\theta}\left[\frac{d\phi_r}{dr}-\frac{d^2 w}{dr^2}-\frac{1}{r}\left(\phi_r-\frac{dw}{dr}\right)\right]\right\}-q=0, \tag{7.1.22}$$

$$-\frac{1}{r}\frac{d}{dr}\left\{rB_{rr}\left[\frac{du}{dr}+\frac{1}{2}\left(\frac{dw}{dr}\right)^2+\nu\frac{u}{r}\right]+rD_{rr}\left(\frac{d\phi_r}{dr}+\frac{\nu}{r}\phi_r\right)\right\}$$

$$+\frac{1}{r}\left\{B_{rr}\left[\frac{u}{r}+\nu\frac{du}{dr}+\frac{\nu}{2}\left(\frac{dw}{dr}\right)^2\right]+D_{rr}\left(\nu\frac{d\phi_r}{dr}+\frac{1}{r}\phi_r\right)\right\}$$

$$-\frac{1}{4r}\frac{d}{dr}\left\{rS_{r\theta}\left[\frac{d\phi_r}{dr}-\frac{d^2 w}{dr^2}-\frac{1}{r}\left(\phi_r-\frac{dw}{dr}\right)\right]\right\}$$

$$-\frac{1}{4r}S_{r\theta}\left[\frac{d\phi_r}{dr}-\frac{d^2 w}{dr^2}-\frac{1}{r}\left(\phi_r-\frac{dw}{dr}\right)\right]+S_{rz}\left(\phi_r+\frac{dw}{dr}\right)=0. \tag{7.1.23}$$

7.1.4.3 Linear equations of equilibrium without couple stress

By setting the nonlinear terms and terms involving $S_{r\theta}$ in Eqs. (7.1.21)–(7.1.23), we obtain (accounts for FGM)

$$-\frac{1}{r}\frac{d}{dr}\left[rA_{rr}\left(\frac{du}{dr}+\nu\frac{u}{r}\right)+rB_{rr}\left(\frac{d\phi_r}{dr}+\frac{\nu}{r}\phi_r\right)\right]$$

$$+\frac{1}{r}\left[A_{rr}\left(\frac{u}{r}+\nu\frac{du}{dr}\right)+B_{rr}\left(\nu\frac{d\phi_r}{dr}+\frac{1}{r}\phi_r\right)\right]=0, \qquad (7.1.24)$$

$$-\frac{1}{r}\frac{d}{dr}\left[rS_{rz}\left(\phi_r+\frac{dw}{dr}\right)\right]-q=0, \qquad (7.1.25)$$

$$-\frac{1}{r}\frac{d}{dr}\left[rB_{rr}\left(\frac{du}{dr}+\nu\frac{u}{r}\right)+rD_{rr}\left(\frac{d\phi_r}{dr}+\frac{\nu}{r}\phi_r\right)\right]$$

$$+\frac{1}{r}\left[B_{rr}\left(\frac{u}{r}+\nu\frac{du}{dr}\right)+D_{rr}\left(\nu\frac{d\phi_r}{dr}+\frac{1}{r}\phi_r\right)\right]$$

$$+S_{rz}\left(\phi_r+\frac{dw}{dr}\right)=0. \qquad (7.1.26)$$

7.1.4.4 Linear equations of equilibrium without couple stress and FGM

By setting the terms involving B_{rr} in Eqs. (7.1.24)–(7.1.26), we obtain

$$-\frac{1}{r}\frac{d}{dr}\left[rA_{rr}\left(\frac{du}{dr}+\nu\frac{u}{r}\right)\right]+\frac{1}{r}\left[A_{rr}\left(\frac{u}{r}+\nu\frac{du}{dr}\right)\right]=0, \qquad (7.1.27)$$

$$-\frac{1}{r}\frac{d}{dr}\left[rS_{rz}\left(\phi_r+\frac{dw}{dr}\right)\right]-q=0, \qquad (7.1.28)$$

$$-\frac{1}{r}\frac{d}{dr}\left[rD_{rr}\left(\frac{d\phi_r}{dr}+\frac{\nu}{r}\phi_r\right)\right]+\frac{1}{r}\left[D_{rr}\left(\nu\frac{d\phi_r}{dr}+\frac{1}{r}\phi_r\right)\right]$$

$$+S_{rz}\left(\phi_r+\frac{dw}{dr}\right)=0. \qquad (7.1.29)$$

7.2 EXACT SOLUTIONS OF ISOTROPIC CIRCULAR PLATES

For axisymmetric bending, the equilibrium equations of isotropic plates in Eq. (7.1.27) is decoupled from Eqs. (7.1.28) and (7.1.29). Returning to the stress-resultant form of the bending equations for this case, we have

$$-\frac{1}{r}\left[\frac{d}{dr}\left(rM_{rr}\right)-M_{\theta\theta}\right]+N_{rz}=0, \qquad (7.2.1)$$

$$-\frac{1}{r}\frac{d}{dr}\left(rN_{rz}\right)-q=0, \qquad (7.2.2)$$

where

$$M_{rr} = D_{rr} \left(\frac{d\phi_r}{dr} + \nu \frac{\phi_r}{r} \right),$$ (7.2.3)

$$M_{\theta\theta} = D_{rr} \left(\nu \frac{d\phi_r}{dr} + \frac{\phi_r}{r} \right),$$ (7.2.4)

$$N_{rz} = S_{rz} \left(\phi_r + \frac{dw}{dr} \right),$$ (7.2.5)

and $D_{rr} = Eh^3/[12(1 - \nu^2)]$ and $S_{rz} = K_s Gh$.
Integration of Eq. (7.2.2) gives

$$rN_{rz} = - \int rq(r) \, dr + c_1.$$ (7.2.6)

Use of Eqs. (7.2.3), (7.2.4), and (7.2.6) in Eq. (7.2.1), using the identity in
Problem 7.1, and integrating the result twice, we obtain

$$D_{rr}\phi_r(r) = -\frac{dF}{dr} + \frac{r}{4}(2\log r - 1)c_1 + \frac{r}{2}c_2 + \frac{1}{r}c_3,$$ (7.2.7)

where

$$F(r) = \int^r \frac{1}{\xi} \int^\xi \eta \int^\eta \frac{1}{\zeta} \int^\zeta \mu q(\mu) \, d\mu \, d\zeta \, d\eta \, d\xi,$$ (7.2.8)

where μ, ζ, η, and ξ are dummy coordinates to indicate the sequence of integration (ultimately, the resulting expression will be in terms of r only).
Finally, from Eqs. (7.2.5)–(7.2.7), we arrive at ($S_{rz} = K_s Gh$)

$$S_{rz} \frac{dw}{dr} = -\frac{S_{rz}}{D_{rr}} \left[-\frac{dF}{dr} + \frac{r}{4}(2\log r - 1)c_1 + \frac{r}{2}c_2 + \frac{1}{r}c_3 \right]$$
$$- \frac{1}{r} \int^r \xi q(\xi) \, d\xi + \frac{c_1}{r},$$ (7.2.9)

$$S_{rz} w(r) = -\frac{S_{rz}}{D_{rr}} \left[-F(r) + \frac{r^2}{4}(\log r - 1)c_1 + \frac{r^2}{4}c_2 + c_3 \log r \right]$$
$$- \int^r \frac{1}{\xi} \int^\xi \eta q(\eta) \, d\eta \, d\xi + c_1 \log r + c_4.$$ (7.2.10)

The constants of integration, c_1, c_2, c_3, and c_4 are determined using the boundary conditions. For the axisymmetric bending problem, the duality pairs are
(i.e., one must know one element of each of the following pairs, or a relation
between the two elements of each pair must be known in solving a boundary-
value problem):

$$(w, rN_{rz}) \text{ and } (\phi_r, rM_{rr}).$$

To compute the bending moments, we require the derivative of ϕ_r,

$$D_{rr}\frac{d\phi_r}{dr} = -\frac{d^2 F}{dr^2} + \frac{1}{4}(2\log r + 1)c_1 + \frac{1}{2}c_2 - \frac{1}{r^2}c_3.$$

Then the bending moments are

$$
\begin{aligned}
M_{rr}(r) = {}& -\frac{d^2 F}{dr^2} + \frac{1}{4}(2\log r + 1)c_1 + \frac{1}{2}c_2 - \frac{1}{r^2}c_3 \\
& + \frac{\nu}{r}\left[-\frac{dF}{dr} + \frac{r}{4}(2\log r - 1)c_1 + \frac{r}{2}c_2 + \frac{1}{r}c_3\right],
\end{aligned}
\qquad (7.2.11)
$$

$$
\begin{aligned}
M_{\theta\theta}(r) = {}& \nu\left(-\frac{d^2 F}{dr^2} + \frac{1}{4}(2\log r + 1)c_1 + \frac{1}{2}c_2 - \frac{1}{r^2}c_3\right) \\
& + \frac{1}{r}\left[-\frac{dF}{dr} + \frac{r}{4}(2\log r - 1)c_1 + \frac{r}{2}c_2 + \frac{1}{r}c_3\right].
\end{aligned}
\qquad (7.2.12)
$$

For solid circular plates, the condition that the rotation ϕ_r be finite at $r = 0$ requires [from Eq. (7.2.7)] that $c_3 = 0$. In addition, if the plate is not subjected to a point load at $r = 0$, the shear force must be zero there. This implies that [from Eq. (7.2.6)] $c_1 = 0$. Thus, for solid circular plates without a point load at the center, we must have

$$c_1 = 0, \quad c_3 = 0. \qquad (7.2.13)$$

If a solid circular plate is subjected to a point load Q_0 at the center, we have

$$2\pi(rN_{rz}) = -Q_0 \text{ at } r = 0 \quad \rightarrow \quad c_1 = -\frac{Q_0}{2\pi}. \qquad (7.2.14)$$

Obviously, $c_3 \neq 0$ for annular plates.

Example 7.2.1

Determine the exact solutions of an isotropic clamped circular plate of radius a under two different loads: (a) uniformly distributed load of intensity q_0 (see Fig. 7.2.1) and (b) point load Q_0 at the center of the plate.

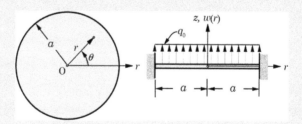

Figure 7.2.1: A clamped circular plate under uniformly distributed load.

Solution: (a) *Uniformly distributed load.* For this case, we have $c_3 = 0$. The boundary conditions associated with the clamped edge $r = a$ are

$$w(a) = 0, \quad \phi_r(a) = 0. \tag{7.2.15}$$

For uniform load of intensity q_0, we have $c_1 = 0$ and $(S_{rz} = K_s Gh)$

$$c_2 = \frac{q_0 a^2}{8}, \quad c_4 = \frac{q_0 a^2}{4} + \frac{S_{rz}}{D} \frac{q_0 a^4}{64}.$$

Hence, the deflection and rotation becomes

$$w(r) = \frac{q_0 a^4}{64 D_{rr}} \left(1 - \frac{r^2}{a^2}\right)^2 + \frac{q_0 a^2}{4 S_{rz}} \left(1 - \frac{r^2}{a^2}\right), \tag{7.2.16}$$

$$\phi_r(r) = \frac{q_0 a^3}{16 D_{rr}} \frac{r}{a} \left(1 - \frac{r^2}{a^2}\right). \tag{7.2.17}$$

Note that the deflection in Eq. (7.2.15) has two parts, $w(r) = w^b(r) + w^s(r)$, one due to bending $w^b(r)$ and the other due to shear $w^s(r)$:

$$w^b(r) = \frac{q_0 a^4}{64 D_{rr}} \left(1 - \frac{r^2}{a^2}\right)^2, \quad w^s(r) = \frac{q_0 a^2}{4 S_{rz}} \left(1 - \frac{r^2}{a^2}\right). \tag{7.2.18}$$

The bending deflection $w^b(r)$ is the same as that predicted by the classical plate theory (CPT) [see Eq. (6.3.20) of **Example 6.3.1**]. Thus, the effect of including transverse shear strain in the formulation is to increase the deflection by $w^s(r)$. In other words, the CPT underpredicts deflection compared with the shear deformation plate theory. We also note that the rotation ϕ_r, in the present case, is equal to $\phi_r(r) = -dw^b/dr$. That is, the rotation of a transverse normal is not affected by the shear deformation. Also, we note that (dw^s/dr) is not zero at $r = a$.

The maximum deflection of the plate is

$$w_{max} = \frac{q_0 a^4}{64 D_{rr}} + \frac{q_0 a^2}{4 S_{rz}} = \frac{q_0 a^4}{64 D_{rr}} \left(1 + \frac{8}{3(1-\nu) K_s} \frac{h^2}{a^2}\right). \tag{7.2.19}$$

For $\nu = 0.3, K_s = 5/6$, and $a/h = 100$ (a thin plate), the difference between the maximum deflection predicted by the CPT and the FST is 4.57×10^{-4}, which is negligible. For $a/h = 10$ (a moderately thick plate), it is 4.57×10^{-2}, or 4.57%. Thus, the effect of shear deformation on the deflection is more for thick plates. Of course, this too may be negligible in some applications but it is important to know the difference.

The expressions for the bending moments and stresses are the same as those in Eqs. (6.3.21)–(6.3.24) of **Example 6.3.1**. This is because the bending moment does not have any contribution from transverse shear strain. The shear force N_{rz} and shear stress σ_{rz} are given by (through constitutive relation; $S_{rz} = K_s Gh$)

$$N_{rz}(r) = S_{rz} \left(\phi_r + \frac{dw}{dr}\right) = S_{rz} \frac{dw^s}{dr} = -\frac{q_0 r}{2}, \tag{7.2.20}$$

$$\sigma_{rz}(r) = G \left(\phi_r + \frac{dw}{dr}\right) = -\frac{K_s q_0 r}{2h}. \tag{7.2.21}$$

(b) *Point load Q_0 at the center.* For this case, from Eq. (7.2.14) we have

$$c_1 = -\frac{Q_0}{2\pi}, \quad c_3 = 0.$$

The boundary conditions of the clamped edge give

$$c_2 = \frac{Q_0}{4\pi}(2\log a - 1), \quad c_4 = \frac{S_{rz}}{D_{rr}}\frac{Q_0 a^2}{16\pi} + \frac{Q_0}{2\pi}\log a.$$

Hence, the solution becomes

$$w(r) = \frac{Q_0 a^2}{16\pi D_{rr}}\left[1 - \frac{r^2}{a^2} + 2\frac{r^2}{a^2}\log\left(\frac{r}{a}\right)\right] - \frac{Q_0}{2\pi K_s Gh}\log\frac{r}{a}, \qquad (7.2.22)$$

$$\phi_r(r) = -\frac{Q_0 r}{4\pi D_{rr}}\log\frac{r}{a}. \qquad (7.2.23)$$

As before, the deflection predicted by the FST has two parts: one due to bending and the other due to shear. The bending part is the same as that predicted by the CPT, and $\phi_r(r) = -(dw^b/dr)$. However, the shear part $w^s(r)$ is singular at $r = 0$, making it difficult to determine the maximum deflection (see the comments in **Example 6.3.1**).

Example 7.2.2

Determine the exact solution of a simply supported isotropic circular plate subjected to uniformly distributed load of intensity q_0.

Solution: The boundary conditions are

$$\text{At } r = 0: \ (rN_{rz}) = 0, \quad \text{At } r = a: \ w = 0, \ M_{rr} = 0. \qquad (7.2.24)$$

The constants of integration are determined to be ($c_1 = c_3 = 0$; and $S_{rz} = K_s Gh$ and $D_{rr} = Eh^3/[12(1-\nu^2)]$)

$$c_2 = \frac{q_0 a^2}{8}\frac{(3+\nu)}{(1+\nu)}, \quad c_4 = \frac{q_0 a^2}{4} + \frac{S_{rz}}{D_{rr}}\frac{q_0 a^4}{64}\frac{(5+\nu)}{(1+\nu)}.$$

The deflection and rotation become

$$w(r) = \frac{q_0 a^4}{64 D_{rr}}\left(\frac{r^4}{a^4} - 2\frac{3+\nu}{1+\nu}\frac{r^2}{a^2} + \frac{5+\nu}{1+\nu}\right) + \frac{q_0 a^2}{4 S_{rz}}\left(1 - \frac{r^2}{a^2}\right), \qquad (7.2.25)$$

$$\phi_r(r) = \frac{q_0 a^3}{16 D_{rr}}\frac{r}{a}\left[\frac{(3+\nu)}{(1+\nu)} - \frac{r^2}{a^2}\right]. \qquad (7.2.26)$$

Note that the deflection is affected by shear deformation [the second term in Eq. (7.2.25)]; however, for this problem also, the rotation is not affected by transverse shear deformation.

The maximum deflection and rotation are

$$w_{\max} = w(0) = \left(\frac{5+\nu}{1+\nu}\right)\frac{q_0 a^4}{64 D_{rr}} + \frac{q_0 a^2}{4 S_{rz}}, \quad \phi_{\max} = \phi_r(a) = \frac{q_0 a^3}{8 D_{rr}(1+\nu)}. \qquad (7.2.27)$$

7.3   EXACT SOLUTIONS FOR FGM CIRCULAR PLATES

7.3.1   GOVERNING EQUATIONS

In this section, we develop the exact solutions of functionally graded material (FGM) plates using the FST. The couples stress effect is not included here, although it is possible to include but algebraically a bit more complicated. The developments to be presented are similar to those presented in Section 6.4.

In general, we consider an annular plate of inner radius b and outer radius a, with the origin of the cylindrical coordinate system (r, θ, z) being at the center of the plate; that is, the inner edge is at $r = b$ and the outer edge is at $r = a$, as shown in Fig. 7.3.2.

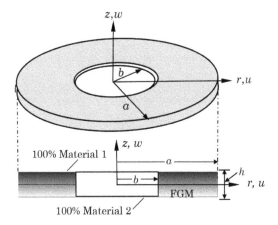

Figure 7.3.2: The geometry and coordinate system of a two-constituent functionally graded annular plate.

The equations of equilibrium of the FST in terms of the stress resultants are (without the couple stress effect):

$$-\frac{1}{r}\left[\frac{d}{dr}(rN_{rr}) - N_{\theta\theta}\right] = 0, \qquad (7.3.1)$$

$$-\frac{1}{r}\frac{d}{dr}(rN_{rz}) - q = 0, \qquad (7.3.2)$$

$$-\frac{d}{dr}(rM_{rr}) + M_{\theta\theta} + rN_{rz} = 0. \qquad (7.3.3)$$

The stress resultants $(N_{rr}, N_{\theta\theta}, M_{rr}, M_{\theta\theta}, N_{rz})$ and stresses $(\sigma_{rr}, \sigma_{\theta\theta}, \sigma_{rz})$

of the linearized FST are related to the displacements by

$$N_{rr} = A_{rr}\left(\frac{du}{dr} + \nu\frac{u}{r}\right) + B_{rr}\left(\frac{d\phi_r}{dr} + \frac{\nu}{r}\phi_r\right), \tag{7.3.4}$$

$$N_{\theta\theta} = A_{rr}\left(\frac{u}{r} + \nu\frac{du}{dr}\right) + B_{rr}\left(\nu\frac{d\phi_r}{dr} + \frac{1}{r}\phi_r\right), \tag{7.3.5}$$

$$M_{rr} = B_{rr}\left(\frac{du}{dr} + \nu\frac{u}{r}\right) + D_{rr}\left(\frac{d\phi_r}{dr} + \frac{\nu}{r}\phi_r\right), \tag{7.3.6}$$

$$M_{\theta\theta} = B_{rr}\left(\frac{u}{r} + \nu\frac{du}{dr}\right) + D_{rr}\left(\nu\frac{d\phi_r}{dr} + \frac{1}{r}\phi_r\right), \tag{7.3.7}$$

$$\sigma_{rr} = \frac{E}{(1-\nu^2)}\left[\left(\frac{du}{dr} + \nu\frac{u}{r}\right) + z\left(\frac{d\phi_r}{dr} + \frac{\nu}{r}\phi_r\right)\right], \tag{7.3.8}$$

$$\sigma_{\theta\theta} = \frac{E}{(1-\nu^2)}\left[\left(\nu\frac{du}{dr} + \frac{u}{r}\right) + z\left(\nu\frac{d\phi_r}{dr} + \frac{1}{r}\phi_r\right)\right], \tag{7.3.9}$$

$$N_{rz} = S_{rz}\left(\phi_r + \frac{dw}{dr}\right), \quad \sigma_{rz} = \frac{E}{2(1+\nu)}\left(\phi_r + \frac{dw}{dr}\right). \tag{7.3.10}$$

7.3.2 EXACT SOLUTIONS

From Eq. (7.3.2), we obtain

$$rN_{rz} = -\int^r \xi q(\xi)\, d\xi + c_1. \tag{7.3.11}$$

Substituting Eqs. (7.3.6), (7.3.7), and (7.3.11) in Eq. (7.3.3), we obtain

$$B_{rr}\left[\frac{d}{dr}\left(r\frac{du}{dr}\right) - \frac{u}{r}\right] + D_{rr}\left[\frac{d}{dr}\left(r\frac{d\phi_r}{dr}\right) - \frac{1}{r}\phi_r\right] = -\int^r \xi q(\xi)\, d\xi + c_1. \tag{7.3.12}$$

We can use the following identities to further simplify Eq. (7.3.12):

$$\frac{d}{dr}\left(r\frac{du}{dr}\right) - \frac{u}{r} = r\frac{d}{dr}\left[\frac{1}{r}\frac{d}{dr}(ru)\right],$$
$$\frac{d}{dr}\left(r\frac{d\phi_r}{dr}\right) - \frac{\phi_r}{r} = r\frac{d}{dr}\left[\frac{1}{r}\frac{d}{dr}(r\phi_r)\right]. \tag{7.3.13}$$

Then Eq. (7.3.12) becomes

$$B_{rr}r\frac{d}{dr}\left[\frac{1}{r}\frac{d}{dr}(ru)\right] + D_{rr}r\frac{d}{dr}\left[\frac{1}{r}\frac{d}{dr}(r\phi_r)\right] = -\int^r \xi q(\xi)\, d\xi + c_1.$$

Integrating once, we obtain

$$B_{rr}\left[\frac{1}{r}\frac{d}{dr}(ru)\right] + D_{rr}\left[\frac{1}{r}\frac{d}{dr}(r\phi_r)\right]$$

$$= -\int^r\left[\frac{1}{\xi}\int^\xi \eta q(\eta)\,d\eta\right]d\xi + c_1\log r + c_2.$$

One more integration yields

$$B_{rr}\,ru + D_{rr}\,r\phi_r = -\int^r\left\{\xi\int^\xi\left[\frac{1}{\eta}\int^\eta \mu q(\mu)\,d\mu\right]d\eta\right\}d\xi$$

$$+ c_1\frac{r^2}{4}(2\log r - 1) + c_2\frac{r^2}{2} + c_3. \tag{7.3.14}$$

Following the same procedure with Eq. (7.3.1) as we did with Eq. (7.3.3), we obtain

$$A_{rr}\left[\frac{d}{dr}\left(r\frac{du}{dr}\right) - \frac{u}{r}\right] + B_{rr}\left[\frac{d}{dr}\left(r\frac{d\phi_r}{dr}\right) - \frac{1}{r}\phi_r\right] = 0, \tag{7.3.15}$$

and

$$A_{rr}\,ru + B_{rr}\,r\phi_r = c_4\frac{r^2}{2} + c_5. \tag{7.3.16}$$

Solving Eqs. (7.3.14) and (7.3.16) for ru and $r\phi_r$ as

$$u(r) = \bar{D}_{rr}^*\left(c_4\frac{r}{2} + \frac{c_5}{r}\right) - \bar{B}_{rr}^*\left(\frac{1}{r}F(r) + c_2\frac{r}{2} + \frac{c_3}{r}\right), \tag{7.3.17}$$

$$\phi_r(r) = -\bar{B}_{rr}^*\left(c_4\frac{r}{2} + \frac{c_5}{r}\right) + \bar{A}_{rr}^*\left(\frac{1}{r}F(r) + c_2\frac{r}{2} + \frac{c_3}{r}\right), \tag{7.3.18}$$

where

$$\bar{A}_{rr}^* = \frac{A_{rr}}{D_{rr}^*}, \quad \bar{B}_{rr}^* = \frac{B_{rr}}{D_{rr}^*}, \quad \bar{D}_{rr}^* = \frac{D_{rr}}{D_{rr}^*}, \quad D_{rr}^* = A_{rr}D_{rr} - B_{rr}B_{rr}, \tag{7.3.19}$$

$$F(r) = -\int^r\left\{\xi\int^\xi\left[\frac{1}{\eta}\int^\eta \mu q(\mu)\,d\mu\right]d\eta\right\}d\xi + c_1\frac{r^2}{4}(2\log r - 1). \tag{7.3.20}$$

Substituting for N_{rz} from Eq. (7.3.10) into Eq. (7.3.11), we obtain

$$-\int^r \xi q(\xi)\,d\xi + c_1 = rN_{rz} = rS_{rz}\left(\phi_r + \frac{dw}{dr}\right)$$

$$= rS_{rz}\frac{dw}{dr} + S_{rz}\left[-\bar{B}_{rr}^*\left(c_4\frac{r^2}{2} + c_5\right) + \bar{A}_{rr}^*\left(F(r) + c_2\frac{r^2}{2} + c_3\right)\right],$$

or

$$\frac{dw}{dr} = \frac{1}{r}\left[\bar{B}^*_{rr}\left(c_4\frac{r^2}{2} + c_5\right) - \bar{A}^*_{rr}\left(F(r) + c_2\frac{r^2}{2} + c_3\right)\right]$$
$$- \frac{1}{rS_{rz}}\left(\int^r \xi q(\xi)\,d\xi + c_1\right). \tag{7.3.21}$$

Integrating the above expression, we arrive at

$$w(r) = \bar{B}^*_{rr}\left(c_4\frac{r^2}{4} + c_5\log r\right) - \bar{A}^*_{rr}\left(\int^r \frac{1}{\xi}F(\xi)\,d\xi + c_2\frac{r^2}{4} + c_3\log r + c_6\right)$$
$$- \frac{1}{S_{rz}}\left(\int^r \frac{1}{\xi}\int^\xi \eta q(\eta)\,d\eta\,d\xi + c_1\log r\right). \tag{7.3.22}$$

The constant of integration, c_6, can be included in any part of the expression (and its value will depend on where it is included). Thus, the solution for the generalized displacements is provided by Eqs. (7.3.17), (7.3.18), and (7.3.22). The constants of integration, c_i, will be determined in a specific problem using the boundary conditions arising from the specification of one element of each of the following three duality pairs:

$$(u, rN_{rr}), \quad (w, rN_{rz}), \quad (\phi_r, rM_{rr}). \tag{7.3.23}$$

7.3.3 EXAMPLES

Here we consider couple of examples to illustrate the use of the boundary conditions to determine the exact solutions. Expressions for the function $F(r)$ [see Eq. (7.3.20)] and the integrals involving it are needed in the examples to be discussed. Two cases that are of interest are when $q(r) = 0$ and $q(r) = q_0$, a constant. In these two cases, we have

$$F(r) = c_1\frac{r^2}{4}\left(2\log r - 1\right), \quad \text{for } q = 0, \tag{7.3.24}$$

$$F(r) = -\frac{q_0 r^4}{16} + c_1\frac{r^2}{4}\left(2\log r - 1\right), \quad \text{for } q(r) = q_0, \tag{7.3.25}$$

$$\int \frac{1}{r}F(r)\,dr = -\frac{q_0 r^4}{64} + c_1\frac{r^2}{4}\left(2\log r - 1\right), \quad \text{for } q(r) = q_0, \tag{7.3.26}$$

where we have used the following identities:

$$\int r\log r\,dr = \int \frac{r}{4}\left(2\log r - 1\right)\,dr = \frac{r^2}{4}\left(2\log r - 1\right). \tag{7.3.27}$$

We also need the following integral when $q(r) = q_0$:

$$\int^r \frac{1}{\xi}\int^\xi \eta q(\eta)\,d\eta\,d\xi = \frac{q_0 r^2}{4}. \tag{7.3.28}$$

Example 7.3.1 ———————————————————————

Determine the exact solutions for deflection, moments, and stresses in an FGM circular plate with clamped edge, $r = a$. Assume uniformly distributed load of intensity q_0.

Solution: The boundary conditions of the FST for this case are

$$u = 0, \quad rN_{rz} = 0, \quad \phi_r = 0 \quad \text{at} \quad r = 0, \tag{7.3.29a}$$
$$u = 0, \quad w = 0, \quad \phi_r = 0, \quad \text{at} \quad r = a. \tag{7.3.29b}$$

The first boundary condition in Eq. (7.3.29a), in view of Eq. (7.3.17), gives (for finite value of u at $r = 0$)

$$-\bar{D}_{rr}^* \, c_5 + \bar{B}_{rr}^* \, c_3 = 0. \tag{7.3.30a}$$

Using the second boundary condition in Eq. (7.3.29a) in Eq. (7.3.11) we find that $c_1 = 0$. The third boundary condition in Eq. (7.3.29a), in light of Eq. (7.3.18), yields

$$-\bar{B}_{rr}^* \, c_5 + \bar{A}_{rr}^* \, c_3 = 0. \tag{7.3.30b}$$

Equations (7.3.30a) and (7.3.30b) together yield $c_3 = c_5 = 0$. Then $u(r)$, $\phi_r(r)$, and $w(r)$ become

$$u(r) = \bar{D}_{rr}^* c_4 \frac{r}{2} - \bar{B}_{rr}^* \left(-\frac{q_0 r^3}{16} + c_2 \frac{r}{2}\right),$$

$$\phi_r(r) = -\bar{B}_{rr}^* c_4 \frac{r}{2} + \bar{A}_{rr}^* \left(-\frac{q_0 r^3}{16} + c_2 \frac{r}{2}\right),$$

$$w(r) = \left[\bar{B}_{rr}^* c_4 \frac{r^2}{4} - \bar{A}_{rr}^* \left(-\frac{q_0 r^4}{64} + c_2 \frac{r^2}{4} + c_6\right)\right] - \frac{1}{S_{rz}} \frac{q_0 r^2}{4}.$$

Use of the three boundary conditions from Eq. (7.3.29b) in the above three expressions for $u(a)$, $\phi_r(a)$, and $w(a)$ yield

$$c_4 = 0, \quad c_2 = \frac{q_0 a^2}{8}, \quad c_6 = -\frac{q_0 a^4}{64} - \frac{1}{S_{rz} \bar{A}_{rr}^*} \frac{q_0 a^2}{4}.$$

It is clear that the solutions predicted by FST will differ from those by CPT only in $w(r)$ (because c_6 appears only in the expression for w).

Hence, the displacements of a clamped FGM circular plate under uniform transverse load are [the coefficients $\bar{A}_{rr}^*$ and $\bar{B}_{rr}^*$ are defined in Eq. (7.3.19)]

$$u(r) = -\bar{B}_{rr}^* \frac{q_0 a^3}{16} \frac{r}{a} \left(1 - \frac{r^2}{a^2}\right), \tag{7.3.31}$$

$$\phi_r(r) = \bar{A}_{rr}^* \frac{q_0 a^3}{16} \frac{r}{a} \left(1 - \frac{r^2}{a^2}\right), \tag{7.3.32}$$

$$w(r) = \bar{A}_{rr}^* \frac{q_0 a^4}{64} \left[1 - \left(\frac{r}{a}\right)^2\right]^2 + \frac{1}{S_{rz}} \frac{q_0 a^2}{4} \left(1 - \frac{r^2}{a^2}\right). \tag{7.3.33}$$

Expressions for the stress resultants from Eqs. (7.3.5)–(7.3.8) become (they are the same as in the CPT)

$$M_{rr}(r) = \frac{q_0 a^2}{16}\left[(1+\nu) - (3+\nu)\left(\frac{r}{a}\right)^2\right], \tag{7.3.34}$$

$$M_{\theta\theta}(r) = \frac{q_0 a^2}{16}\left[(1+\nu) - (1+3\nu)\left(\frac{r}{a}\right)^2\right], \tag{7.3.35}$$

$$\sigma_{rr}(r,z) = \frac{q_0 a^2}{16}\frac{E(z)}{1-\nu^2}(\bar{B}_{rr}^* + z\bar{A}_{rr}^*)\left[(1+\nu) - (3+\nu)\left(\frac{r}{a}\right)^2\right], \tag{7.3.36}$$

$$\sigma_{\theta\theta}(r,z) = \frac{q_0 a^2}{16}\frac{E(z)}{1-\nu^2}(\bar{B}_{rr}^* + z\bar{A}_{rr}^*)\left[(1+\nu) - (1+3\nu)\left(\frac{r}{a}\right)^2\right]. \tag{7.3.37}$$

The maximum deflection and bending moments are:

$$w(0) = \bar{A}_{rr}^*\frac{q_0 a^4}{64} + \frac{1}{S_{xz}}\frac{q_0 a^2}{4}, \quad M_{rr}(a) = -\frac{q_0 a^2}{8}, \quad M_{\theta\theta}(a) = -\frac{\nu q_0 a^2}{8}. \tag{7.3.38}$$

All of the solutions presented in this example coincide with the solutions developed in **Example 7.2.1** for a clamped plate, when we set $B_{rr} = \bar{B}_{rr}^* = 0$, $\bar{A}_{rr}^* = 1/D_{rr}$, and $\bar{D}_{rr}^* = 1/A_{rr}$.

The numerical results generated with the following data:

$$a = 10 \text{ in.,} \quad h = 0.1 \text{ in.,} \quad E_1/E_2 = 10, \quad E_2 = 30 \times 10^6 \text{ psi,} \quad \nu = 0.3$$

coincide with the plots presented in Figs. 6.4.1 and 6.4.2, indicating that the effect of shear deformation is negligible for this thin plate ($a/h = 100$). Figure 7.3.3 shows $\bar{w} = w(0)h^3 \times 10^3$ versus the power-law exponent n for two different ratios $a/h = 10$ (thick) and $a/h = 100$ (thin), showing the effect of shear deformation. Figure 7.3.3 also contains results for pinned circular plates presented in **Example 7.3.2** next.

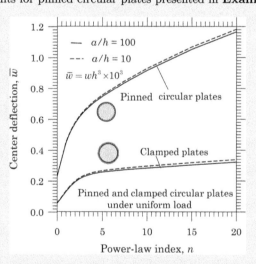

Figure 7.3.3: Variation of the transverse maximum deflection $\bar{w}$ versus the power-law index n for clamped and pinned circular plates for two different radius-to-thickness ratios, $a/h = 10, 100$.

Example 7.3.2

Consider a FGM circular plate with pinned edge at $r = a$ and subjected to uniformly distributed load of intensity q_0. Determine the expressions for the transverse deflection and bending moments in the plate. The plate is loaded with uniformly distributed transverse load and an applied bending moment M_a at $r = a$.

Solution: The boundary conditions for the case at hand are

$$\text{At } r = b: \qquad u = 0, \qquad \phi_r = 0, \quad (rN_{rz}) = 0, \qquad (7.3.39a)$$
$$\text{At } r = a: \qquad u = 0, \qquad w = 0, \qquad M_{rr} = M_a. \qquad (7.3.39b)$$

As in **Example 7.3.1**, the boundary conditions in Eq. (7.3.39a) give: $c_1 = c_3 = c_5 = 0$. The bending moment M_{rr} in this case (with $c_1 = c_3 = c_5 = 0$) becomes

$$M_{rr} = -(3 + \nu)\frac{q_0 r^2}{16} + (1 + \nu)\frac{c_2}{2}.$$

The use of the boundary conditions from Eq. (7.3.39b) in Eqs. (7.3.17), (7.3.22), and M_{rr} above yields the following values for c_2, c_4, and c_6:

$$c_2 = \left(\frac{3 + \nu}{1 + \nu}\right)\frac{q_0 a^2}{8} + \frac{2M_a}{(1 + \nu)},$$

$$c_4 = \frac{2\bar{B}_{rr}^*}{\bar{D}_{rr}^*(1 + \nu)}\left(\frac{q_0 a^2}{8} + M_a\right), \qquad (7.3.40)$$

$$\bar{A}_{rr}^* c_6 = \frac{\bar{B}_{rr}^*\bar{B}_{rr}^*}{\bar{D}_{rr}^*(1 + \nu)}\frac{q_0 a^4}{16} - \bar{A}_{rr}^*\frac{5 + \nu}{1 + \nu}\frac{q_0 a^4}{64} - \frac{M_a a^2}{2D_{rr}(1 + \nu)} + \frac{q_0 a^2}{4S_{rz}}.$$

Then the displacements and bending moments are given by

$$u(r) = -\bar{B}_{rr}^*\frac{q_0 a^3}{16}\frac{r}{a}\left(1 - \frac{r^2}{a^2}\right), \qquad (7.3.41)$$

$$\phi_r(r) = \frac{q_0 a^3}{16}\frac{r}{a}\left[\left(\frac{3 + \nu}{1 + \nu}\right)\bar{A}_{rr}^* - 2\frac{\bar{B}_{rr}^*\bar{B}_{rr}^*}{(1 + \nu)\bar{D}_{rr}^*} - \bar{A}_{rr}^*\frac{r^2}{a^2}\right], \qquad (7.3.42)$$

$$w(r) = \bar{A}_{rr}^*\frac{q_0 a^4}{64}\left[\left(\frac{5 + \nu}{1 + \nu}\right) - 2\left(\frac{3 + \nu}{1 + \nu}\right)\frac{r^2}{a^2} + \frac{r^4}{a^4}\right] - \frac{\bar{B}_{rr}^*\bar{B}_{rr}^*}{(1 + \nu)\bar{D}_{rr}^*}\frac{q_0 a^4}{16}\left(1 - \frac{r^2}{a^2}\right)$$

$$+ \frac{M_a a^2}{2(1 + \nu)D_{rr}}\left(1 - \frac{r^2}{a^2}\right) + \frac{q_0 a^2}{4S_{rz}}\left(1 - \frac{r^2}{a^2}\right), \qquad (7.3.43)$$

$$M_{rr}(r) = (3 + \nu)\frac{q_0 a^2}{16}\left(1 - \frac{r^2}{a^2}\right) + M_a, \qquad (7.3.44)$$

where the coefficients $\bar{A}_{rr}^*$, $\bar{B}_{rr}^*$, and $\bar{D}_{rr}^*$ are defined in Eq. (6.4.19). We note that M_a does not contribute to $u(r)$.

The maximum deflection and bending moment occur at $r = 0$, and they are

$$w_{max} = \bar{A}_{rr}^*\left(\frac{5 + \nu}{1 + \nu}\right)\frac{q_0 a^4}{64} + \frac{M_a a^2}{2(1 + \nu)D_{rr}} + \frac{q_0 a^2}{4S_{rz}},$$

$$(7.3.45)$$

$$M_{max} = (3 + \nu)\frac{q_0 a^2}{16} + M_a.$$

7.4   BENDING RELATIONSHIPS BETWEEN CPT AND FST

7.4.1   SUMMARY OF THE GOVERNING EQUATIONS

The governing equations of the classical and first-order shear deformation plate theories are summarized in the following for isotropic (E and ν) plates. The superscripts on variables refer to the theory used: C for CPT and F for FST (see [143]–[147] for some of the details).

Classical Plate Theory (CPT):

$$\frac{1}{r}\frac{d}{dr}\left(rN_{rz}^{C}\right) = -q, \tag{7.4.1}$$

where

$$rN_{rz}^{C} \equiv \frac{d}{dr}\left(rM_{rr}^{C}\right) - M_{\theta\theta}^{C}, \tag{7.4.2}$$

$$M_{rr}^{C} = -D_{rr}\left(\frac{d^{2}w^{C}}{dr^{2}} + \frac{\nu}{r}\frac{dw^{C}}{dr}\right), \tag{7.4.3}$$

$$M_{\theta\theta}^{C} = -D_{rr}\left(\nu\frac{d^{2}w^{C}}{dr^{2}} + \frac{1}{r}\frac{dw^{C}}{dr}\right). \tag{7.4.4}$$

First-order Shear deformation Plate Theory (FST):

$$\frac{1}{r}\frac{d}{dr}\left(rN_{rz}^{F}\right) = -q, \tag{7.4.5}$$

$$rN_{rz}^{F} = \frac{d}{dr}\left(rM_{rr}^{F}\right) - M_{\theta\theta}^{F}, \tag{7.4.6}$$

where

$$M_{rr}^{F} = D_{rr}\left(\frac{d\phi_{r}^{F}}{dr} + \frac{\nu}{r}\phi_{r}^{F}\right), \tag{7.4.7}$$

$$M_{\theta\theta}^{F} = D_{rr}\left(\nu\frac{d\phi_{r}^{F}}{dr} + \frac{1}{r}\phi_{r}^{F}\right), \tag{7.4.8}$$

$$N_{rz}^{F} = S_{rz}\left(\phi_{r}^{F} + \frac{dw^{F}}{dr}\right). \tag{7.4.9}$$

7.4.2   RELATIONSHIPS

First, we introduce the *moment sum* [148] for the convenience of developing the relationships:

$$\mathcal{M} = \frac{M_{rr} + M_{\theta\theta}}{1 + \nu}. \tag{7.4.10}$$

Using Eqs. (7.4.1) to (7.4.9), we can show that

$$\mathcal{M}^C = -D_{rr}\left(\frac{d^2 w^C}{dr^2} + \frac{1}{r}\frac{dw^C}{dr}\right) = -D_{rr}\frac{1}{r}\frac{d}{dr}\left(r\frac{dw^C}{dr}\right), \tag{7.4.11}$$

$$\mathcal{M}^F = D_{rr}\left(\frac{d\phi_r^F}{dr} + \frac{1}{r}\phi_r^F\right) = D_{rr}\frac{1}{r}\frac{d}{dr}\left(r\phi_r^F\right), \tag{7.4.12}$$

$$\frac{1}{r}\frac{d}{dr}\left(r\frac{d\mathcal{M}^C}{dr}\right) = -q, \quad \frac{1}{r}\frac{d}{dr}\left(r\frac{d\mathcal{M}^F}{dr}\right) = -q, \tag{7.4.13}$$

$$r\frac{d\mathcal{M}^C}{dr} = \frac{d}{dr}\left(r M_{rr}^C\right) - M_{\theta\theta}^C = r N_{rz}^C, \tag{7.4.14}$$

$$r\frac{d\mathcal{M}^F}{dr} = \frac{d}{dr}\left(r M_{rr}^F\right) - M_{\theta\theta}^F = r N_{rz}^F. \tag{7.4.15}$$

We are now ready to establish relationships between the solutions of the two theories. From Eqs. (7.4.1) and (7.4.5), because the load q being the same, it follows that

$$r N_{rz}^F = r N_{rz}^C + c_1, \tag{7.4.16}$$

where c_1 is a constant of integration. From Eqs. (7.4.13)–(7.4.16), we have

$$r\frac{d\mathcal{M}^F}{dr} = r\frac{d\mathcal{M}^C}{dr} + c_1 \;\Rightarrow\; \mathcal{M}^F(r) = \mathcal{M}^C + c_1 \log r + c_2, \tag{7.4.17}$$

where c_1 and c_2 are constants of integration. Next, from Eqs. (7.4.11), (7.4.12), and (7.4.17), we have

$$\phi_r^F(r) = -\frac{dw^C}{dr} + \frac{c_1 r}{4 D_{rr}}(2\log r - 1) + \frac{c_2 r}{2 D_{rr}} + \frac{c_3}{r D_{rr}}. \tag{7.4.18}$$

Finally, from Eqs. (7.4.9) and (7.4.16), we obtain

$$\frac{dw^F}{dr} = -\phi_r^F + \frac{1}{S_{rz}}\left(N_{rz}^C + \frac{c_1}{r}\right), \tag{7.4.19}$$

and noting that $N_{rz}^C = d\mathcal{M}^C/dr$, we have

$$w^F(r) = w^C(r) + \frac{\mathcal{M}^C}{K_s G h} + \frac{c_1 r^2}{4 D_{rr}}(1 - \log r) + \frac{c_1}{K_s G h}\log r$$

$$- \frac{c_2 r^2}{4 D_{rr}} - \frac{c_3 \log r}{D_{rr}} + \frac{c_4}{D_{rr}}. \tag{7.4.20}$$

The four constants of integration, c_1, c_2, c_3, and c_4, are determined using the boundary conditions.

The boundary conditions for various cases are given as follows.

Free edge:

$$r N_{rz}^F = r N_{rz}^C = 0, \quad r M_{rr}^F = r M_{rr}^C = 0. \tag{7.4.21}$$

Pinned or hinged:

$$w^F = w^C = 0, \quad rM^F_{rr} = rM^C_{rr} = 0. \tag{7.4.22}$$

Clamped edge:

$$w^F = w^C = 0, \quad \phi^F_r = \frac{dw^C}{dr} = 0. \tag{7.4.23}$$

Solid circular plate at $r = 0$ (i.e., at the plate center):

$$(rN^F_{rz}) = (rN^C_{rz}) = \phi^F_r = \frac{dw^C}{dr} = 0 \;\rightarrow\; c_1 = c_3 = 0. \tag{7.4.24}$$

7.4.3 EXAMPLES

Here we consider couple of examples to illustrate the use of the relationships developed in this section. In particular, solutions for pinned and clamped circular plates are presented, as these are the problems for which solutions are needed to compare with solutions generated by numerical methods.

Example 7.4.1 ──

Consider a circular plate of radius a and subjected to a uniformly distributed load of intensity q_0. Determine the relationships between the solutions of the CPT and FST for (a) clamped and (b) pinned boundary conditions at $r = a$.

Solution: We first recall from Eq. (7.4.24) that for solid circular plates under uniform load, we have $c_1 = c_3 = 0$.

(a) *Clamped plates.* Substitution of the conditions $\phi^F = dw^C/dr = 0$ at $r = a$ in Eq. (7.4.18) gives $c_2 = 0$. Using the condition $w^F(a) = w^C(a) = 0$ in Eq. (7.4.20) yields

$$c_4 = -\frac{D_{rr}\mathcal{M}^C_a}{K_s G h}, \tag{7.4.25}$$

where $\mathcal{M}^C_a$ is the moment sum at the edge ($r = a$) of the CPT. Hence, the relationships in Eqs (7.4.18) and (7.4.20) become

$$\phi^F_r(r) = -\frac{dw^C}{dr}, \quad w^F(r) = w^C + \frac{\mathcal{M}^C(r) - \mathcal{M}^C_a}{K_s G h}. \tag{7.4.26}$$

The CPT solution for $w^C(r)$ and $\mathcal{M}^C(r)$ for the clamped plate are [see Eqs. (6.3.20)–(6.3.22)]

$$w^C(r) = \frac{q_0 a^4}{64 D_{rr}}\left[1 - \left(\frac{r}{a}\right)^2\right]^2,$$

$$M^C_{rr}(r) = \frac{q_0 a^2}{16}\left[(1+\nu) - (3+\nu)\left(\frac{r}{a}\right)^2\right],$$

$$M^C_{\theta\theta}(r) = \frac{q_0 a^2}{16}\left[(1+\nu) - (1+3\nu)\left(\frac{r}{a}\right)^2\right], \tag{7.4.27}$$

$$\mathcal{M}^C(r) = \frac{M^C_{rr} + M^C_{\theta\theta}}{(1+\nu)} = \frac{q_0 a^2}{8}\left(1 - 2\frac{r^2}{a^2}\right),$$

and
$$\mathcal{M}^C(r) - \mathcal{M}^C(a) = \frac{q_0 a^2}{4}\left(1 - \frac{r^2}{a^2}\right). \tag{7.4.28}$$

Substitution of Eq. (7.4.28) into Eq. (7.4.26) yields the rotation and deflection of the clamped circular plate according to the FST:

$$\phi_r^F(r) = \frac{q_0 a^3}{16 D_{rr}}\frac{r}{a}\left[1 - \left(\frac{r}{a}\right)^2\right], \tag{7.4.29}$$

$$w^F(r) = \frac{q_0 a^4}{64 D_{rr}}\left[1 - \left(\frac{r}{a}\right)^2\right]^2 + \frac{1}{S_{rz}}\frac{q_0 a^2}{4}\left(1 - \frac{r^2}{a^2}\right). \tag{7.4.30}$$

The maximum deflection (occurs at $r = 0$) is

$$w_{max}^F = \frac{1}{64}\frac{q_0 a^4}{D_{rr}} + \frac{1}{4}\frac{q_0 a^2}{S_{rz}}. \tag{7.4.31}$$

(b) *Pinned plates.* For this case, Substitution of the conditions $M_{rr}^C = M_{rr}^F = 0$ at $r = a$ in Eq. (7.4.18) gives, in view of the relation $M_{rr}^F(r) = M_{rr}^C + c_2$, $c_2 = 0$. Using the condition $w^F(a) = w^C(a) = 0$ in Eq. (7.4.20) gives the value of c_4 to be the same as in Eq. (7.4.25).

The CPT solution for $w^C(r)$ and $\mathcal{M}^C(r)$ for pinned circular plates is [see Eqs. (6.3.39)–(6.3.41)]

$$w^C(r) = \frac{q_0 a^4}{64 D_{rr}}\left[\left(\frac{r}{a}\right)^4 - 2\left(\frac{3+\nu}{1+\nu}\right)\left(\frac{r}{a}\right)^2 + \frac{5+\nu}{1+\nu}\right], \tag{7.4.32a}$$

$$M_{rr}^C(r) = \frac{q_0 a^2}{16}(3+\nu)\left[1 - \left(\frac{r}{a}\right)^2\right], \tag{7.4.32b}$$

$$M_{\theta\theta}^C(r) = \frac{q_0 a^2}{16}\left[(3+\nu) - (1+3\nu)\left(\frac{r}{a}\right)^2\right], \tag{7.4.32c}$$

$$\mathcal{M}^C(r) = \frac{q_0 a^2}{16(1+\nu)}\left[(3+\nu) - 4(1+\nu)\left(\frac{r}{a}\right)^2\right], \tag{7.4.32d}$$

and
$$\mathcal{M}^C(r) - \mathcal{M}^C(a) = \frac{q_0 a^2}{4}\left[1 - \left(\frac{r}{a}\right)^2\right]. \tag{7.4.33}$$

Substitution of Eqs. (7.4.33) into Eq. (7.4.26) yields the rotation and deflection of the simply supported plate according to the FST:

$$\phi_r^F(r) = \frac{q_0 a^3}{16 D_{rr}}\frac{r}{a}\left(\frac{3+\nu}{1+\nu} - \frac{r^2}{a^2}\right), \tag{7.4.34}$$

$$w^F(r) = \frac{q_0 a^4}{64 D_{rr}}\left[\frac{5+\nu}{1+\nu} - 2\left(\frac{3+\nu}{1+\nu}\right)\left(\frac{r}{a}\right)^2 + \left(\frac{r}{a}\right)^4\right] + \frac{q_0 a^2}{4 S_{rz}}\left(1 - \frac{r^2}{a^2}\right). \tag{7.4.35}$$

The maximum deflection (occurs at $r = 0$) is

$$w_{max}^F = \frac{5+\nu}{1+\nu}\frac{q_0 a^4}{64 D_{rr}} + \frac{1}{4}\frac{q_0 a^2}{S_{rz}}. \tag{7.4.36}$$

Example 7.4.2

Consider a circular plate of radius a and subjected to a linearly varying axisymmetric (or pyramid) load $q = q_0(1 - r/a)$ (see Fig. 7.4.1). Determine the relationships between the solutions of the CPT and FST for (a) simply supported and (b) clamped plates.

Solution: As discussed in the previous example, for simply supported as well as clamped (at $r = a$) circular plates, the boundary conditions in Eqs. (7.4.22) and (7.4.23) give $c_1 = c_2 = c_3 = 0$ and $c_4 = -D_{rr}\mathcal{M}_a^C/K_sGh$.

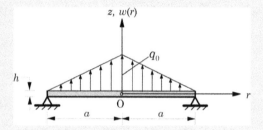

Figure 7.4.1: Simply supported circular plate under linearly varying axisymmetric load.

Hence, the relationships in Eqs. (7.4.18) and (7.4.20) become

$$\phi_r^F(r) = -\frac{dw^C}{dr}, \quad w^F(r) = w^C + \frac{\mathcal{M}^C(r) - \mathcal{M}_a^C}{K_sGh}. \tag{7.4.37}$$

(a) Simply supported plates. The CPT solution $w^C(r)$ and its derivatives for this case are

$$w^C(r) = \frac{q_0a^4}{14400D_{rr}}\left[\frac{3(183+43\nu)}{1+\nu} - \frac{10(71+29\nu)}{1+\nu}\left(\frac{r}{a}\right)^2 + 225\left(\frac{r}{a}\right)^4 - 64\left(\frac{r}{a}\right)^5\right],$$

$$\frac{1}{r}\frac{dw^C}{dr} = \frac{q_0a^2}{14400D_{rr}}\left[-\frac{20(71+29\nu)}{1+\nu} + 4\times225\left(\frac{r}{a}\right)^2 - 5\times64\left(\frac{r}{a}\right)^3\right],$$

$$\frac{d^2w^C}{dr^2} = \frac{q_0a^2}{14400D_{rr}}\left[-\frac{20(71+29\nu)}{1+\nu} + 12\times225\left(\frac{r}{a}\right)^2 - 20\times64\left(\frac{r}{a}\right)^3\right].$$

Substituting into Eq. (7.4.11), we obtain

$$\mathcal{M}^C(r) = -D_{rr}\left(\frac{d^2w^C}{dr^2} + \frac{1}{r}\frac{dw^C}{dr}\right)$$

$$= \frac{q_0a^2}{14400}\left[-\frac{40(71+29\nu)}{1+\nu} + 16\times225\left(\frac{r}{a}\right)^2 - 25\times64\left(\frac{r}{a}\right)^3\right],$$

$$\mathcal{M}^C(a) = \frac{q_0a^2}{14400}\left[-\frac{40(71+29\nu)}{1+\nu} + 16\times225 - 25\times64\right],$$

$$\mathcal{M}^C(r) - \mathcal{M}^C(a) = \frac{q_0a^2}{36}\left[-5 + 9\left(\frac{r}{a}\right)^2 - 4\left(\frac{r}{a}\right)^3\right]. \tag{7.4.38}$$

Substitution of Eqs. (7.4.38) into Eq. (7.4.26) yields the deflection of the simply supported plate according to the FST:

$$\phi_r^F(r) = \frac{q_0 a^2}{720 D_{rr}} \frac{r}{a} \left[\frac{(71 + 29\nu)}{1 + \nu} - 45 \left(\frac{r}{a} \right)^2 + 16 \left(\frac{r}{a} \right)^3 \right], \qquad (7.4.39)$$

$$w^F(r) = w^C(r) + \frac{q_0 a^2}{36 K_s G h} \left[5 - 9 \left(\frac{r}{a} \right)^2 + 4 \left(\frac{r}{a} \right)^3 \right]. \qquad (7.4.40)$$

The maximum deflection (occurs at $r = 0$) is

$$w_{max}^F = w_{max}^C + \frac{5 q_0 a^4}{36 S_{rz}} = \frac{(183 + 43\nu)}{4800} \frac{q_0 a^4}{D_{rr}} + \frac{5}{36} \frac{q_0 a^2}{S_{rz}}. \qquad (7.4.41)$$

Clamped plates (CPT). The CPT solution $w^C(r)$ and the moment sum for the clamped plate are

$$w^C(r) = \frac{q_0 a^4}{14400 D_{rr}} \left[129 - 290 \left(\frac{r}{a} \right)^2 + 225 \left(\frac{r}{a} \right)^4 - 64 \left(\frac{r}{a} \right)^5 \right],$$

$$\mathcal{M}^C(r) = \frac{q_0 a^2}{14400} \left[-4 \times 290 + 12 \times 225 \left(\frac{r}{a} \right)^2 - 25 \times 64 \left(\frac{r}{a} \right)^3 \right].$$

$$\mathcal{M}^C(a) = \frac{q_0 a^2}{14400} \left(-4 \times 290 + 12 \times 225 - 25 \times 64 \right),$$

$$\mathcal{M}^C(r) - \mathcal{M}^C(a) = \frac{q_0 a^2}{144} \left[-11 + 27 \left(\frac{r}{a} \right)^2 - 16 \left(\frac{r}{a} \right)^3 \right]. \qquad (7.4.42)$$

Substitution of Eqs. (7.4.37) into Eq. (7.4.26) yields the deflection of the clamped circular plate according to the FST:

$$\phi_r^F(r) = \frac{q_0 a^2}{720 D_{rr}} \left[29 - 45 \left(\frac{r}{a} \right)^2 + 16 \left(\frac{r}{a} \right)^3 \right], \qquad (7.4.43)$$

$$w^F(r) = w^C(r) + \frac{q_0 a^2}{144 D_{rr}} \left[11 - 27 \left(\frac{r}{a} \right)^2 + 16 \left(\frac{r}{a} \right)^3 \right]. \qquad (7.4.44)$$

The maximum deflection (occurs at $r = 0$) is

$$w_{max}^F = w_{max}^C + \frac{11 q_0 a^2}{144 S_{rz}} = \frac{183 + 43\nu}{4800} \frac{q_0 a^4}{D_{rr}} + \frac{11}{144} \frac{q_0 a^2}{S_{rz}}. \qquad (7.4.45)$$

Example 7.4.3

Consider a circular plate of radius a and subjected to a uniformly distributed load of intensity q_0 over the inner portion of the plate $r \leq \alpha a$ $(0 < \alpha \leq 1)$ as shown in Fig.7.4.2. Determine the FST solution using the developed relationships for (a) simply supported and (b) clamped outer edge.

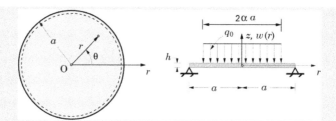

Figure 7.4.2: Circular plate under partial uniformly distributed load.

Solution: The CPT solutions for the simply supported and clamped plates are outlined next (see Szilard [142] and Reddy [113]):

(a) *Simply supported plate.* For $0 \le r \le \alpha a$: For $0 \le r \le a$:

$$w^C(r) = \frac{q_0 a^4}{64 D_{rr}} \left\{ \left(\frac{r}{a}\right)^4 + \alpha^2 \left[4 - 5\alpha^2 + 4\left(2+\alpha^2\right)\log\alpha\right] \right.$$
$$\left. + 2\frac{\alpha^2}{1+\nu}\left[1 - \left(\frac{r}{a}\right)^2\right]\left[4 - (1-\nu)\alpha^2 - 4(1+\nu)\log\alpha\right] \right\}, \quad (7.4.46a)$$

and for $\alpha a \le r \le a$:

$$w_0^C = \frac{q_0 \alpha^2 a^4}{32 D_{rr}} \left\{ 2\left[\alpha^2 + 2\left(\frac{r}{a}\right)^2\right]\log\frac{r}{a} \right.$$
$$\left. + \frac{2(3+\nu) - \alpha^2(1-\nu)}{1+\nu}\left[1 - \left(\frac{r}{a}\right)^2\right] \right\}. \quad (7.4.46b)$$

The case of uniformly distributed load on the entire plate is obtained by setting $\alpha = 1$ in Eq. (7.4.46a) and Eq. (7.4.46b).

(b) *Clamped plate.* For this case, we have for $0 \le r \le \alpha a$

$$w^C(r) = \frac{q_0 a^4}{64 D_{rr}} \left[\left(\frac{r}{a}\right)^4 + \alpha^2\left(4 - 3\alpha^2 + 4\alpha^2\log\alpha\right) \right.$$
$$\left. - 2\alpha^2\left(\frac{r}{R}\right)^2\left(\alpha^2 - 4\log\alpha\right) \right], \quad (7.4.47a)$$

and for $\alpha a \le r \le a$:

$$w^C(r) = \frac{q_0 \alpha^2 a^4}{32 D_{rr}} \left\{ 2\left[\alpha^2 + 2\left(\frac{r}{a}\right)^2\right]\log\left(\frac{r}{a}\right) + (2+\alpha^2)\left[1 - \left(\frac{r}{a}\right)^2\right] \right\}. \quad (7.4.47b)$$

The case of uniformly distributed load on the entire plate is obtained by setting $\alpha = 1$ in Eq. (7.4.47a) and Eq. (7.4.47b).

We note that the deflection component due to the transverse shear deformation is the same regardless of the support being simply supported or clamped. Substituting

Eqs. (7.4.46a), (7.4.46b), (7.4.47a), and (7.4.47b) into Eqs. (7.4.11) and (7.4.20) yield the corresponding FST deflections ($S_{rz} = K_s Gh$ and $D_{rr} = Eh^3/[12(1 - \nu^2)]$:

$$w^F(r) = w^C(r) + \begin{cases} \frac{q_0 a^2}{4 S_{xz}} \left[\alpha^2 \left(1 - 2\log\alpha\right) - \left(\frac{r}{a}\right)^2 \right], & 0 \le r \le \alpha a, \\ \frac{q_0 a^2 \alpha^2}{2 S_{xz}} \log\frac{a}{r}, & \alpha a \le r \le a. \end{cases} \tag{7.4.48}$$

7.5 BENDING RELATIONSHIPS FOR FUNCTIONALLY GRADED CIRCULAR PLATES

7.5.1 INTRODUCTION

In this section, bending relationships between the CPT and FST for axisymmetric bending of through-the-thickness functionally graded circular plates are presented (see [143]–[149] and references therein). Due to the general material variation through the thickness of the plate, the bending-stretching coupling exists. General solution of the FST problem for arbitrary variation of the constituents is derived in terms of the isotropic CPT solution. Particular solution are developed for a number of boundary conditions. The effect of material distribution through the thickness and boundary conditions on deflections and stresses are presented (see [120] for additional details).

7.5.2 SUMMARY OF EQUATIONS

Consider a functionally graded circular plate of radius a and total thickness h, and subjected to axisymmetric transverse load q. We presume that the grading of the material through the thickness, applied loads, and boundary conditions are axisymmetric so that the displacement u_θ is identically zero and (u_r, u_z) are only functions of r and z. At the moment, we assume that $E = E(z)$ and $\nu = \nu(z)$, and their specific variation will be discussed in the sequel.

The displacement field of the CPT is

$$u_r(r, z) = u(r) - z\frac{dw}{dr}, \quad u_z(r, z) = w(r), \tag{7.5.1}$$

and that of the FST is

$$u_r(r, z) = u(r) + z\phi_r(r), \quad u_z(r, z) = w(r). \tag{7.5.2}$$

Because of the bending-stretching coupling present in functionally graded plates, we must include the in-plane displacement u along with the bending variables (w, ϕ_r). First, we summarize the governing equations of the CPT and the FST plate theories.

Both of the theories are governed by the following equations:

$$-\frac{d}{dr}(rN_{rr}) - N_{\theta\theta} = 0, \tag{7.5.3}$$

$$-\frac{d}{dr}(rN_{rz}) - rq = 0, \tag{7.5.4}$$

$$\frac{d}{dr}(rM_{rr}) - M_{\theta\theta} - rN_{rz} = 0. \tag{7.5.5}$$

where N_{rr} and $N_{\theta\theta}$ are the radial and circumferential in-plane forces, N_{rz} is the transverse shear force, and M_{rr} and $M_{\theta\theta}$ are the radial and circumferential moments defined as follows:

$$(N_{rr}, N_{\theta\theta}, N_{rz}) = \int_{-\frac{h}{2}}^{\frac{h}{2}} (\sigma_{rr}, \sigma_{\theta\theta}, \sigma_{rz})\, dz, \tag{7.5.6}$$

$$(M_{rr}, M_{\theta\theta}) = \int_{-\frac{h}{2}}^{\frac{h}{2}} (\sigma_{rr}, \sigma_{\theta\theta})z\, dz. \tag{7.5.7}$$

The plate constitutive equations of the two theories are as follows.

CPT for isotropic plates:

$$N_{rr}^C = A_{rr}\left(\frac{du^C}{dr} + \nu\frac{u^C}{r}\right), \tag{7.5.8}$$

$$N_{\theta\theta}^C = A_{rr}\left(\nu\frac{du^C}{dr} + \frac{u^C}{r}\right), \tag{7.5.9}$$

$$M_{rr}^C = -D_{rr}\left(\frac{d^2w^C}{dr^2} + \nu\frac{1}{r}\frac{dw^C}{dr}\right), \tag{7.5.10}$$

$$M_{\theta\theta}^C = -D_{rr}\left(\nu\frac{d^2w^C}{dr^2} + \frac{1}{r}\frac{dw^C}{dr}\right), \tag{7.5.11}$$

$$N_{rz}^C = \frac{1}{r}\left[\frac{d}{dr}(rM_{rr}^C) - M_{\theta\theta}^C\right]. \tag{7.5.12}$$

FST for functionally graded plates:

$$N_{rr}^F = A_{rr}\left(\frac{du^F}{dr} + \nu\frac{u^F}{r}\right) + B_{rr}\left(\frac{d\phi_r^F}{dr} + \nu\frac{\phi_r^F}{r}\right), \tag{7.5.13}$$

$$N_{\theta\theta}^F = A_{rr}\left(r\frac{du^F}{dr} + \frac{u^F}{r}\right) + B_{rr}\left(\nu\frac{d\phi_r^F}{dr} + \frac{\phi_r^F}{r}\right), \tag{7.5.14}$$

$$M_{rr}^F = B_{rr}\left(\frac{du^F}{dr} + \nu\frac{u^F}{r}\right) + D_{rr}\left(\frac{d\phi_r^F}{dr} + \nu\frac{\phi_r^F}{r}\right), \tag{7.5.15}$$

$$M_{\theta\theta}^F = B_{rr}\left(\nu\frac{du^F}{dr} + \frac{u^F}{r}\right) + D_{rr}\left(\nu\frac{d\phi_r^F}{dr} + \frac{\phi_r^F}{r}\right), \tag{7.5.16}$$

$$N_{rz}^F = S_{rz}\left(\phi_r^F + \frac{dw^F}{dr}\right), \tag{7.5.17}$$

where, as indicated in Chapter 4, the superscript C on stress resultants denotes that they belong to the CPT and F denotes the same in the FST (we note that ϕ_r^F appears only in the FST). The plate stiffness coefficients A_{rr}, B_{rr}, and D_{rr} (common to both theories) are defined by

$$(A_{rr}, B_{rr}, D_{rr}) = \int_{-\frac{h}{2}}^{\frac{h}{2}} Q_{ij}(1, z, z^2)\, dz \quad (i, j = 1, 2), \tag{7.5.18a}$$

$$S_{rz} = K_s \int_{-\frac{h}{2}}^{\frac{h}{2}} \frac{E}{2(1+\nu)}\, dz, \tag{7.5.18b}$$

$$Q_{11} = Q_{22} = \frac{E}{(1-\nu^2)}, \quad Q_{12} = \nu Q_{11}, \tag{7.5.18c}$$

where E is the modulus of elasticity, ν is the Poisson ratio, and K_s is the shear correction coefficient.

The plan here is to develop relations for the deflections, forces, and moments of functionally graded plates based on the FST in terms of the associated quantities of isotropic plates based on the CPT. Then the relations developed are specialized for plates with various boundary conditions.

7.5.3 RELATIONSHIPS BETWEEN THE CPT AND FST

From Eqs. (7.5.3), (7.5.13), and (7.5.14), when used for the FST, we obtain

$$
\begin{aligned}
0 &= -\frac{d}{dr}\left(rN_{rr}^F\right) - N_{\theta\theta}^F \\
&= A_{rr}\left[\frac{d}{dr}\left(r\frac{du^F}{dr}\right) - \frac{u^F}{r}\right] + B_{rr}\left[\frac{d}{dr}\left(r\frac{d\phi_r^F}{dr}\right) - \frac{\phi_r^F}{r}\right] \\
&= A_{rr}\left[r\frac{d}{dr}\left(\frac{1}{r}\frac{d}{dr}(ru^F)\right)\right] + B_{rr}\left[r\frac{d}{dr}\left(\frac{1}{r}\frac{d}{dr}(r\phi_r^F)\right)\right],
\end{aligned}
$$

or

$$\left[r\frac{d}{dr}\left(\frac{1}{r}\frac{d}{dr}(ru^F)\right)\right] = -\frac{B_{rr}}{A_{rr}}\left[r\frac{d}{dr}\left(\frac{1}{r}\frac{d}{dr}(r\phi_r^F)\right)\right].$$

Upon integration, we obtain

$$\frac{d}{dr}\left(ru^F\right) = -\frac{B_{rr}}{A_{rr}}\frac{d}{dr}\left(r\phi_r^F\right) + K_1 r, \tag{7.5.19a}$$

$$u = -\frac{B_{rr}}{A_{rr}}\phi_r^F + K_1\frac{r}{2} + \frac{K_2}{r}, \tag{7.5.19b}$$

from which we can compute

$$\frac{du^F}{dr} = -\frac{B_{rr}}{A_{rr}}\frac{d\phi_r^F}{dr} + \frac{K_1}{2} - \frac{K_2}{r^2}, \tag{7.5.20}$$

where K_1 and K_2 are integration constants. Using Eqs. (7.5.19a), (7.5.19b), and (7.5.20), the forces and moments of Eqs. (7.5.13)–(7.5.16) can be expressed in terms of ϕ_r as

$$M_{rr}^F = \Omega_1 \frac{d\phi_r^F}{dr} + \Omega_2 \frac{\phi_r^F}{r} + \frac{1}{2}\Omega_3 K_1 + \frac{1}{r^2}\Omega_4 K_2, \tag{7.5.21}$$

$$M_{\theta\theta}^F = \Omega_2 \frac{d\phi_r^F}{dr} + \Omega_1 \frac{\phi_r^F}{r} + \frac{1}{2}\Omega_3 K_1 - \frac{1}{r^2}\Omega_4 K_2, \tag{7.5.22}$$

$$N_{rr}^F = \Omega_5 \frac{\phi_r^F}{r} + \frac{1}{2}\Omega_6 K_1 + \frac{1}{r^2}\Omega_7 K_2, \tag{7.5.23}$$

$$N_{\theta\theta}^F = \Omega_5 \frac{d\phi_r^F}{dr} + \frac{1}{2}\Omega_6 K_1 - \frac{1}{r^2}\Omega_7 K_2, \tag{7.5.24}$$

where

$$\Omega_1 = D_{rr} - \frac{B_{rr}^2}{A_{rr}}, \quad \Omega_2 = \nu\Omega_1, \quad \Omega_3 = (1+\nu)B_{rr},$$

$$\Omega_4 = -(1-\nu)B_{rr}, \quad \Omega_5 = \nu\left(B_{rr} - \frac{A_{rr}B_{rr}}{A_{rr}}\right) = 0, \tag{7.5.25}$$

$$\Omega_6 = (1+\nu)A_{rr}, \quad \Omega_7 = -(1-\nu)A_{rr}.$$

Based on load equivalence, we have

$$\frac{d}{dr}\left(rN_{rz}^F\right) = \frac{d}{dr}\left(rN_{rz}^C\right), \tag{7.5.26}$$

and after integration yields

$$rN_{rz}^F = rN_{rz}^C + c_1, \tag{7.5.27}$$

where c_1 is a constant of integration. But from Eqs. (7.5.5), (7.5.21), and (7.5.22), we have

$$rN_{rz}^F = \frac{d}{dr}\left(rM_{rr}^F\right) - M_{\theta\theta}^F = \Omega_1\left[r\frac{d}{dr}\left(\frac{1}{r}\frac{d}{dr}(r\phi_r^F)\right)\right]. \tag{7.5.28}$$

Similarly,

$$rN_{rz}^C = \frac{d}{dr}\left(rM_{rr}^C\right) - M_{\theta\theta}^C = -D_{rr}\left[r\frac{d}{dr}\left(\frac{1}{r}\frac{d}{dr}(r\frac{dw^C}{dr})\right)\right]. \tag{7.5.29}$$

Using relations (7.5.28) and (7.5.29) in (7.5.27), we obtain

$$\Omega_1\left[r\frac{d}{dr}\left(\frac{1}{r}\frac{d}{dr}(r\phi_r^F)\right)\right] = -D_{rr}\left[r\frac{d}{dr}\left(\frac{1}{r}\frac{d}{dr}(r\frac{dw^C}{dr})\right)\right] + c_1. \tag{7.5.30}$$

Upon integration, we have [integral of $r \log r$ is $r^2(2 \log r - 1)/4$]

$$\Omega_1 \frac{d}{dr}\left(r\phi_r^F\right) = -D_{rr}\frac{d}{dr}\left(r\frac{dw^C}{dr}\right) + c_1 r \log r + c_2 r, \tag{7.5.31a}$$

$$\Omega_1 \phi_r^F = -D_{rr}\frac{dw^C}{dr} + \frac{1}{4}c_1 r\left(2 \log r - 1\right) + \frac{1}{2}c_2 r + \frac{1}{r}c_3. \tag{7.5.31b}$$

Next from Eqs. (7.5.17) and (7.5.28), we have

$$S_{rz}\left(\phi_r^F + \frac{dw^F}{dr}\right) = \frac{d\mathcal{M}^C}{dr} + \frac{1}{r}c_1, \tag{7.5.32}$$

where $\mathcal{M}^C$ is the *moment sum* [148, 142]

$$\mathcal{M}^C = \frac{M_{rr}^C + M_{\theta\theta}^C}{(1+\nu)}. \tag{7.5.33}$$

Substituting for ϕ_r^F from Eq. (7.5.31b) into Eq. (7.5.32), we obtain

$$S_{rz}\left[-\frac{D_{rr}}{\Omega_1}\frac{dw^C}{dr} + \frac{c_1}{4\Omega_1}r\left(2 \log r - 1\right) + \frac{c_2}{2\Omega_1}r + \frac{c_3}{\Omega_1 r} + \frac{dw^F}{dr}\right]$$
$$= \frac{d\mathcal{M}^C}{dr} + \frac{1}{r}c_1. \tag{7.5.34}$$

Integrating the above expression, we obtain

$$w^F(r) = \frac{D_{rr}}{\Omega_1}w^C(r) + \frac{c_1}{\Omega_1}\left[\frac{r^2}{4}\left(1 - \log r\right) + \frac{\Omega_1}{S_{rz}}\log r\right]$$
$$- \frac{c_2}{4\Omega_1}r^2 - \frac{c_3}{\Omega_1}\log r - c_4 + \frac{\mathcal{M}^C}{S_{rz}}. \tag{7.5.35}$$

From Eq. (7.5.31b), we have

$$\Omega_1 \frac{d\phi_r^F}{dr} = -D_{rr}\frac{d^2w^C}{dr^2} + \frac{c_1}{4}\left(2 \log r + 1\right) + \frac{c_2}{2} - \frac{c_3}{r^2}. \tag{7.5.36}$$

Substituting Eqs. (7.5.31b) and (7.5.36) into Eq. (7.5.21), we obtain

$$M_{rr}^F = -D_{rr}\frac{d^2w^C}{dr^2} + \frac{c_1}{4}\left(2 \log r + 1\right) + \frac{c_2}{2} - \frac{c_3}{r^2}$$
$$+ \frac{\Omega_2}{r}\left[-\hat{D}_{rr}\frac{dw^C}{dr} + \hat{c}_1\frac{r}{4}\left(2 \log r - 1\right) + \hat{c}_2\frac{r}{2} + \hat{c}_3\frac{1}{r}\right] + \Omega_3\frac{K_1}{2} + \Omega_4\frac{K_2}{r^2},$$

or

$$M_{rr}^F = M_{rr}^C + \frac{D_{rr}}{r}\frac{dw^C}{dr}\left(\nu - \hat{\Omega}_2\right) + \frac{c_1}{2}\left[\frac{1}{2}\left(1 - \hat{\Omega}_2\right) + \left(1 + \hat{\Omega}_2\right)\log r\right]$$
$$+ \frac{c_2}{2}\left(1 + \hat{\Omega}_2\right) - \frac{c_3}{r^2}\left(1 - \hat{\Omega}_2\right) + \Omega_3\frac{K_1}{2} + \Omega_4\frac{K_2}{r^2}, \tag{7.5.37}$$

where $\hat{D}_{rr} = \frac{D_{rr}}{\Omega_1}, \hat{c}_i = \frac{c_i}{\Omega_1}$ and so on. Similarly, we can write

$$
\begin{aligned}
M_{\theta\theta}^F = M_{\theta\theta}^C &+ D_{rr}\frac{d^2 w^C}{dr^2}\left(\nu - \hat{\Omega}_2\right) + \frac{c_1}{2}\left[-\frac{1}{2}\left(1 - \hat{\Omega}_2\right) + \left(1 + \hat{\Omega}_2\right)\log r\right] \\
&+ \frac{c_2}{2}\left(1 + \hat{\Omega}_2\right) + \frac{c_3}{r^2}\left(1 - \hat{\Omega}_2\right) + \Omega_3\frac{K_1}{2} - \Omega_4\frac{K_2}{r^2}.
\end{aligned}
\tag{7.5.38}
$$

Next we substitute (7.5.31b) into (7.5.23) and obtain

$$
N_{rr}^F = \Omega_6\frac{K_1}{2} + \Omega_7\frac{K_2}{r^2}.
\tag{7.5.39}
$$

Substituting Eq. (7.5.36) into Eq. (7.5.24), we obtain

$$
N_{\theta\theta}^F = \Omega_6\frac{K_1}{2} - \Omega_7\frac{K_2}{r^2},
\tag{7.5.40}
$$

where

$$
\hat{D}_{rr} = \frac{D_{rr}}{\Omega_1}, \quad \hat{c}_i = \frac{c_i}{\Omega_1}.
$$

Define

$$
\mathcal{M}^{\mathcal{F}} \equiv \frac{M_{rr}^F + M_{\theta\theta}^F}{(1 + \hat{\Omega}_2)}, \quad \hat{\Omega}_2 = \frac{\Omega_2}{\Omega_1},
\tag{7.5.41}
$$

$$
\mathcal{N}^{\mathcal{F}} \equiv \left(N_{rr}^F + N_{\theta\theta}^F\right)\Omega_1.
\tag{7.5.42}
$$

Then we have

$$
\mathcal{M}^{\mathcal{F}} = \mathcal{M}^C + c_1\log r + c_2 + \frac{\Omega_3}{(1 + \hat{\Omega}_2)}K_1,
\tag{7.5.43}
$$

and

$$
\mathcal{N}^{\mathcal{F}} = \Omega_1\,\Omega_6\,K_1.
\tag{7.5.44}
$$

This completes the development of equations for the deflections, forces, and moments of functionally graded plates based on the FST in terms of the associated quantities of isotropic plates based on the CPT. Now it remains that we develop relationships for axisymmetric bending of plates with different boundary conditions.

Example 7.5.1

Consider a solid circular plate with a *roller support* at $r = a$, a being the radius of the plate. Determine the bending solutions of a functionally graded plate using the relationships developed.

Solution: The boundary conditions are

$$\text{At } r = 0: \quad u = 0, \quad \phi_r^F = 0, \quad \frac{dw^C}{dr} = 0, \quad N_{rz}^C = N_{rz}^F = 0, \quad (7.5.45a)$$

$$\text{At } r = a: \quad w^C = w^F = 0, \quad N_{rr}^C = N_{rr}^F = 0, \quad M_{rr}^C = M_{rr}^F = 0. \quad (7.5.45b)$$

The above boundary conditions give

$$K_1 = 0, \quad K_2 = 0, \quad (7.5.46a)$$

$$c_2 = -\frac{2D_{rr}}{a} \left(\frac{\nu - \hat{\Omega}_2 + \Omega_8}{1 + \hat{\Omega}_2 - \Omega_8} \right) \frac{dw^C(a)}{dr}, \quad (7.5.46b)$$

$$c_4 = \frac{\mathcal{M}^C(a)}{S_{rz}} - \frac{\hat{c}_2 a^2}{4}, \quad c_1 = c_3 = 0, \quad (7.5.46c)$$

where $\Omega_8 = \hat{\Omega}_3 \Omega_5 / \Omega_6 = 0$. Hence, we have the following relations between the deflections, forces, and moments of the two theories:

$$w^F(r) = \hat{D}_{rr} w^C(r) + \frac{\mathcal{M}^C(r) - \mathcal{M}^C(a)}{S_{rz}} + \frac{1}{4}\hat{c}_2(a^2 - r^2), \quad (7.5.47)$$

$$N_{rz}^F(r) = N_{rz}^C(r), \quad (7.5.48)$$

$$N_{rr}^F(r) = 0, \quad (7.5.49)$$

$$M_{rr}^F(r) = M_{rr}^C(r) + D_{rr}\frac{1}{r}\frac{dw^C}{dr}\left(\nu - \hat{\Omega}_2\right) + \frac{1}{2}c_2\left(1 + \hat{\Omega}_2\right) + \frac{1}{2}K_1\Omega_3, \quad (7.5.50)$$

$$M_{\theta\theta}^F(r) = M_{\theta\theta}^C(r) + D_{rr}\frac{d^2w^C}{dr^2}\left(\nu - \hat{\Omega}_2\right) + \frac{1}{2}c_2\left(1 + \hat{\Omega}_2\right) + \frac{1}{2}K_1\Omega_3. \quad (7.5.51)$$

To be specific, consider the case of isotropic circular plates under uniformly distributed transverse load of intensity q_0. The CPT solution for the deflection $[D_{rr} = Eh^3/12(1 - \nu^2)]$ is (the CPT solution for an isotropic plate is independent of the roller support or pinned support at $r = a$)

$$w^C(r) = \frac{q_0 a^4}{64 D_{rr}} \left[\left(\frac{r}{a}\right)^4 - 2 \left(\frac{3 + \nu}{1 + \nu}\right) \left(\frac{r}{a}\right)^2 + \frac{5 + \nu}{1 + \nu} \right]. \quad (7.5.52)$$

For a functionally graded plate, we assume that the modulus $E(z)$ is assumed to vary according to Eq. (6.1.8) (with $E_1 = E_c$ and $E_2 = E_m$)

$$E(z) = E_m\left[(M - 1)V_1(z) + 1\right], \quad V_1(z) = \left(\frac{1}{2} + \frac{z}{h}\right)^n, \quad M = \frac{E_c}{E_m}, \quad (7.5.53)$$

and ν is independent of z. Here E_c and E_m denote the modulus of two different material constituents, namely, ceramic (c) and metal (m), n is the power-law exponent, which is related to the volume fraction of ceramic and metal. Then the plate stiffness coefficients are given in terms of the constituent material properties

$Q_{11}^c = E_c/(1-\nu^2)$ and $Q_{11}^m = E_m/(1-\nu^2)$ as

$$A_{rr} = (Q_{11}^c - Q_{11}^m)h\left(\frac{1}{1+n}\right) + Q_{11}^m h,$$

$$B_{rr} = (Q_{11}^c - Q_{11}^m)\frac{h^2}{2}\left(\frac{n}{(1+n)(2+n)}\right), \tag{7.5.54}$$

$$D_{rr} = (Q_{11}^c - Q_{11}^m)\frac{h^3}{12}\left(\frac{3(2+n+n^2)}{(1+n)(2+n)(3+n)}\right) + Q_{11}^m\frac{h^3}{24},$$

or, in terms of the constituent moduli E_c and E_m and power-law index n as

$$A_{rr} = \frac{h(E_m + nE_c)}{(1+n)(1-\nu^2)},$$

$$S_{rz} = \frac{hK_s(E_m + nE_c)}{2(1+n)(1+\nu)},$$

$$B_{rr} = \frac{nh^2(E_c - E_m)}{2(1+n)(2+n)(1-\nu^2)}, \tag{7.5.55}$$

$$D_{rr} = \frac{h^3\left[n(n^2 + 3n + 8)E_c + 3(n^2 + n + 2)E_m\right]}{12(1+n)(2+n)(3+n)(1-\nu^2)},$$

where K_s denotes the shear correction coefficient.
If we define

$$D_c = \frac{E_c h^3}{12(1-\nu^2)}, \quad G_c = \frac{E_c}{2(1+\nu)}, \quad M = \frac{E_m}{E_c}, \tag{7.5.56}$$

then the expressions in Eq. (7.5.55) can be expressed as

$$A_{rr} = \frac{12(n+E_r)D_c}{h^2(1+n)}, \quad S_{rz} = \frac{K_s G h(n+E_r)}{(1+n)},$$

$$B_{rr} = \frac{6n(1-E_r)D_c}{h(1+n)(2+n)} \tag{7.5.57}$$

$$D_{rr} = \frac{\left[n(n^2 + 3n + 8) + 3(n^2 + n + 2)E_r\right]D_c}{(1+n)(2+n)(3+n)}.$$

The constants Ω_i have the following values (note that $\Omega_5 = \Omega_8 = 0$):

$$\Omega_1 = D_{rr} - \frac{B_{rr}^2}{A_{rr}},$$

$$= \frac{D_c[(n^4 + 4n^3 + 7n^2) + 4E_r(n^3 + 4n^2 + 7n + 3E_r)]}{(n+E_r)(3+n)(2+n)^2},$$

$$\Omega_2 = \nu\Omega_1,$$

$$\Omega_3 = (1+\nu)B_{rr} = \frac{6n(1-E_r)(1+\nu)D_c}{h(1+n)(2+n)},$$

$$\Omega_4 = -(1-\nu)B_{rr} = -\frac{6n(1-E_r)(1-\nu)D_c}{h(1+n)(2+n)}, \tag{7.5.58}$$

$$\Omega_6 = \frac{12(n + E_r)(1 + \nu)D_c}{h^2(1 + n)},$$

$$\Omega_7 = -\frac{12(n + E_r)(1 - \nu)D_c}{h^2(1 + n)},$$

$$\Omega_9 = \frac{3(1 - E_r)^2 n^2 (3 + n)(1 + \nu)}{(1 + n)[7n^2 + 4n^3 + n^4 + 4E_r(3E_r + 7n + 4n^2 + n^3)]}.$$

By substituting the CPT solution given by Eq. (7.5.52) into (7.5.47), (7.5.10), (7.5.11), and (7.5.33), one obtains the deflection of the FGM plate as

$$\bar{w}^F = \frac{64D_c}{q_0 a^4} w^F$$

$$= \frac{D_c}{\Omega_1} \left[\left(\frac{r}{a}\right)^4 - 2\left(\frac{3 + \nu}{1 + \nu}\right)\left(\frac{r}{a}\right)^2 + \frac{5 + \nu}{1 + \nu} \right]$$

$$+ \frac{8}{3K_s(1 - \nu)} \left(\frac{h}{a}\right)^2 \left[1 - \left(\frac{r}{a}\right)^2\right]\left(\frac{1 + n}{E_r + n}\right). \qquad (7.5.59)$$

From Eqs. (7.5.48)–(7.5.52) and (7.5.53), the stress resultants for the FGM plates are (since $\Omega_5 = \Omega_8 = K_1 = K_2 = 0$)

$$N_{rr}^F = 0, \quad N_{\theta\theta}^F = 0, \quad M_{rr}^F = M_{rr}^C, \quad M_{\theta\theta}^F = M_{\theta\theta}^C. \qquad (7.5.60)$$

For presenting numerical results, we consider Titanium-Zirconica FGM plate with $\nu = 0.288$ and modulus ratio $M = E_m/E_c = 0.396$, and take $K_s = 5/6$. The maximum dimensionless deflections are tabulated in Table 7.5.1 for various values of n and h/a ratio. The deflection parameter increases with increasing h/a ratios ($h/a = 0$ corresponds to the CPT solution) but decreases with increasing values of n.

Table 7.5.1: Maximum deflection $\bar{w}^F$ of functionally graded *roller-supported* circular plates ($\nu = 0.288$, $M = E_m/E_c = 0.396$, and $K_s = 5/6$).

n	Thickness-to-radius ratio, h/a				
	0.0	0.05	0.10	0.15	0.20
0	10.368	10.396	10.481	10.623	10.822
2	5.700	5.714	5.756	5.826	5.925
4	5.210	5.223	5.261	5.325	5.414
6	4.958	4.970	5.007	5.069	5.155
8	4.800	4.812	4.848	4.909	4.993
10	4.692	4.704	4.739	4.799	4.882
20	4.434	4.446	4.480	4.538	4.619
30	4.334	4.345	4.379	4.437	4.517
40	4.280	4.291	4.326	4.383	4.463
50	4.247	4.258	4.293	4.349	4.429
10^2	4.178	4.189	4.223	4.280	4.359
10^3	4.113	4.124	4.158	4.214	4.293
10^4	4.106	4.118	4.151	4.208	4.286
10^5	4.106	4.117	4.151	4.207	4.285

Example 7.5.2

Consider a solid circular plate with a *pin supported* (or simply supported) at $r = a$ and loaded with uniform distributed load of intensity q_0. Determine the bending solutions of a functionally graded plate using the relationships developed.

Solution: The boundary conditions for this case are slightly different from those of the roller-supported plate. They are

$$\text{At } r = 0: \quad u = 0, \quad \phi^F = 0, \quad \frac{dw^C}{dr} = 0, \quad N_{rz}^F = N_{rz}^C = 0, \tag{7.5.61a}$$

$$\text{At } r = a: \quad u^C = u^F = 0, \quad w^C = w^F = 0, \quad M_{rr}^C = M_{rr}^F = 0. \tag{7.5.61b}$$

The boundary conditions give ($c_1 = c_3 = 0$ and $K_2 = 0$)

$$K_1 = \frac{2B_{rr}}{aA_{rr}}\left(-\hat{D}_{rr}\frac{dw^C(a)}{dr} + \frac{\hat{c}_2 a}{2}\right), \tag{7.5.62a}$$

$$c_2 = -\frac{2D_{rr}}{a}\left(\frac{\nu - \hat{\Omega}_2 - \Omega_9}{1 + \hat{\Omega}_2 + \Omega_9}\right)\frac{dw^C(a)}{dr}, \tag{7.5.62b}$$

$$c_4 = \frac{\mathcal{M}^C(a)}{S_{rz}} - \frac{\hat{c}_2 a^2}{4}, \tag{7.5.62c}$$

where $\Omega_9 = \hat{\Omega}_3 B_{rr}/A_{rr}$ and $\hat{c}_i = c_i/\Omega_1$. The relations follow the same form as those given in Eqs. (7.5.47)–(7.5.51) but K_1, c_2, and c_4 take the expressions given in Eqs. (7.5.62a)–(7.5.62c).

Using the CPT solution in **Example 7.5.1**, we obtain the following expressions for the transverse deflection of an FGM plate:

$$\begin{aligned}
\bar{w}^F &= \frac{64D_c}{q_0 a^4}w^F \\
&= \frac{D_c}{\Omega_1}\left[\left(\frac{r}{a}\right)^4 - 2\left(\frac{3+\nu}{1+\nu}\right)\left(\frac{r}{a}\right)^2 + \frac{5+\nu}{1+\nu}\right] \\
&\quad + \frac{8}{3K_s(1-\nu)}\left(\frac{h}{a}\right)^2\left[1 - \left(\frac{r}{a}\right)^2\right]\left(\frac{1+n}{M+n}\right) \\
&\quad - \frac{4\eta D_c}{(1+\nu)\Omega_1}\left(\frac{\Omega_9}{1+\nu+\Omega_9}\right)\left[1 - \left(\frac{r}{a}\right)^2\right].
\end{aligned} \tag{7.5.63}$$

We consider Titanium-Zirconica FGM plate with $\nu = 0.288$ and modulus ratio $M = E_m/E_c = 0.396$, and shear correction factor $K_s = 5/6$ to obtain numerical results. The maximum dimensionless deflections are tabulated in Table 7.5.2 for various values of n and h/a ratio.

Example 7.5.3

Consider a solid circular plate with a *clamped support* at $r = a$ and loaded with uniform distributed load of intensity q_0. Determine the bending solutions of a functionally graded plate using the relationships developed.

Table 7.5.2: Maximum deflection $\bar{w}^F$ of functionally graded *simply-supported* circular plates ($\nu = 0.288$, $M = E_m/E_c = 0.396$, and $K_s = 5/6$).

		Thickness-to-radius ratio, h/a			
n	0.0	0.05	0.10	0.15	0.20
0	10.368	10.396	10.481	10.623	10.822
2	5.483	5.497	5.539	5.610	5.708
4	5.102	5.115	5.153	5.217	5.307
6	4.897	4.909	4.946	5.007	5.094
8	4.761	4.773	4.810	4.870	4.954
10	4.665	4.677	4.712	4.772	4.855
20	4.426	4.438	4.473	4.531	4.612
30	4.330	4.342	4.376	4.433	4.513
40	4.278	4.289	4.324	4.381	4.461
50	4.246	4.257	4.291	4.348	4.428
10^2	4.178	4.189	4.223	4.280	4.359
10^3	4.113	4.124	4.158	4.214	4.293
10^4	4.106	4.118	4.151	4.208	4.286
10^5	4.106	4.117	4.151	4.207	4.285

Solution: The boundary conditions are

$$\text{At } r = 0: \quad u^F = u^C = 0, \quad \phi^F = 0, \quad \frac{dw^C}{dr} = 0, \quad N_{rz}^F = N_{rz}^C = 0, \qquad (7.5.64a)$$

$$\text{At } r = a: \quad u^F = u^C = 0, \quad w^F = w^C = 0, \quad \phi^F = 0, \quad \frac{dw^C}{dr} = 0. \qquad (7.5.64b)$$

The boundary conditions yield

$$K_1 = K_2 = 0, \quad c_1 = c_2 = c_3 = 0, \quad c_4 = \frac{\mathcal{M}^C(a)}{S_{rz}}. \qquad (7.5.65)$$

The CPT solution for a clamped plate is given by

$$w^C(r) = \frac{q_0 a^4}{64 D_{rr}} \left[1 - \left(\frac{r}{a}\right)^2 \right]^2. \qquad (7.5.66)$$

Hence, we have the following relations between the deflections, forces, and moments of the two theories:

$$w^F(r) = \hat{D}_{rr} w^C(r) + \frac{\mathcal{M}^C(r) - \mathcal{M}^C(a)}{S_{rz}}, \qquad (7.5.67)$$

$$N_{rz}^F(r) = N_{rz}^C(r), \qquad (7.5.68)$$

$$N_{rr}^F(r) = -\Omega_5 \hat{D}_{rr} \frac{1}{r} \frac{dw^C}{dr} = 0, \qquad (7.5.69)$$

$$N_{\theta\theta}^F(r) = -\Omega_5 \hat{D}_{rr} \frac{d^2 w^C}{dr^2} = 0, \qquad (7.5.70)$$

$$M_{rr}^F(r) = M_{rr}^C(r) + D_{rr} \frac{1}{r} \frac{dw^C}{dr} \left(\nu - \hat{\Omega}_2 \right), \qquad (7.5.71)$$

$$M_{\theta\theta}^F(r) = M_{\theta\theta}^C(r) + D_{rr} \frac{d^2 w^C}{dr^2} \left(\nu - \hat{\Omega}_2 \right). \qquad (7.5.72)$$

As before, we consider Titanium-Zirconica FGM plate with $\nu = 0.288$ and modulus ratio $M = E_m/E_c = 0.396$, and shear correction factor $K_s = 5/6$ to obtain numerical results. The maximum dimensionless deflections are tabulated in Table 7.5.3 for various values of n and h/a ratio.

Table 7.5.3: Maximum deflection $\bar{w}^F$ of functionally graded *clamped* circular plates ($\nu = 0.288$, $M = E_m/E_c = 0.396$, and $K_s = 5/6$).

	Thickness-to-radius ratio, h/a				
n	0.0	0.05	0.10	0.15	0.20
0	2.525	2.554	2.639	2.781	2.979
2	1.388	1.402	1.444	1.515	1.613
4	1.269	1.282	1.320	1.384	1.473
6	1.208	1.220	1.257	1.318	1.404
8	1.169	1.181	1.217	1.278	1.362
10	1.143	1.155	1.190	1.250	1.333
20	1.080	1.092	1.126	1.184	1.265
30	1.056	1.067	1.101	1.159	1.239
40	1.043	1.054	1.088	1.145	1.225
50	1.034	1.046	1.080	1.137	1.216
10^2	1.018	1.029	1.063	1.119	1.199
10^3	1.002	1.013	1.047	1.103	1.182
10^4	1.000	1.011	1.045	1.101	1.180

Example 7.5.4

Consider an annular plate with a clamped support at the inner edge $r = b$ and free at the outer edge $r = a$. Determine the bending solutions of a functionally graded plate with the same boundary conditions using the relationships developed.

Solution: The boundary conditions are

$$\text{At } r = b: \quad u^F = u^C = 0, \quad w^F = w^C = 0, \quad \phi^F = 0, \quad \frac{dw^C}{dr} = 0, \qquad (7.5.73a)$$

$$\text{At } r = a: \quad N_{rr}^F = N_{rr}^C = 0, \quad M_{rr}^F = M_{rr}^C = 0, \quad N_{rz}^F = N_{rz}^C = 0. \qquad (7.5.73b)$$

The boundary conditions give

$$K_1 = \Omega_5 \left(\frac{-2a\hat{D}_{rr} \frac{dw^C(a)}{dr} + \hat{c}_2 a^2 + 2\hat{c}_3}{b^2 \Omega_7 - a^2 \Omega_6} \right) = 0, \quad K_2 = -\frac{b^2}{2} K_1 = 0, \qquad (7.5.74a)$$

$$c_1 = 0, \quad c_2 = -2ad_1 D_{rr} \frac{dw^C(a)}{dr}, \quad c_3 = -\frac{b^2}{2} c_2, \qquad (7.5.74b)$$

$$c_4 = \frac{M^C(b)}{S_{rz}} - \frac{b^2}{4} \hat{c}_2 (1 - 2\log b), \quad d_2 = \left(\frac{a^2 \Omega_3 - b^2 \Omega_4}{b^2 \Omega_7 - a^2 \Omega_6} \right), \qquad (7.5.74c)$$

$$d_1 = \frac{\nu - \hat{\Omega}_2}{a^2 \left(1 + \hat{\Omega}_2\right) + b^2 \left(1 - \hat{\Omega}_2\right)}. \tag{7.5.74d}$$

$$d_2 = \frac{a^2 \Omega_3 - b^2 \Omega_4}{b^2 \Omega_7 - a^2 \Omega_6}. \tag{7.5.74e}$$

Example 7.5.5

Consider an annular plate with clamped inner edge $r = b$ and outer edge $r = a$. Determine the bending solutions of a functionally graded annular plate with the same boundary conditions using the relationships developed.

Solution: The boundary conditions are

$$\text{At } r = b: \quad u^F = u^C = 0, \quad w^F = w^C = 0, \quad \phi_r^F = 0, \quad \frac{dw^C}{dr} = 0, \tag{7.5.75a}$$

$$\text{At } r = a: \quad u^F = u^C = 0, \quad w^F = w^C = 0, \quad \phi_r^F = 0, \quad \frac{dw^C}{dr} = 0. \tag{7.5.75b}$$

The boundary conditions give $K_1 = K_2 = 0$ and

$$c_1 = \frac{\Omega_1}{S_{rz} c_5} \left(\mathcal{M}^C(b) - \mathcal{M}^C(a)\right), \quad c_2 = \left(\frac{b^2 \log b - a^2 \log a}{a^2 - b^2} + \frac{1}{2}\right) c_1, \tag{7.5.76a}$$

$$c_3 = \left(\frac{a^2 b^2}{2(a^2 - b^2)} \log \frac{a}{b}\right) c_1, \tag{7.5.76b}$$

$$c_4 = \left[\frac{a^2 + b^2}{16} + \frac{\Omega_1}{2 S_{rz}} \log ab + \frac{a^2 b^2 \left(\log \frac{b}{a} - (\log ab)^2\right)}{4(b^2 - a^2)}\right] \hat{c}_1$$

$$+ \frac{\mathcal{M}^C(a) + \mathcal{M}^C(b)}{2 S_{rz}}, \tag{7.5.76c}$$

$$c_5 = \frac{a^2 - b^2}{8} + \frac{\Omega_1}{S_{rz}} \log \frac{a}{b} - \frac{a^2 b^2}{2(a^2 - b^2)} \left(\log \frac{a}{b}\right)^2. \tag{7.5.76d}$$

Example 7.5.6

Lastly, consider an annular plate with clamped inner edge $r = b$ and roller supported at the outer edge $r = a$. Determine the bending solutions of a functionally graded annular plate using the relationships developed.

Solution: The boundary conditions are

$$\text{At } r = b: \quad u^F = u^C = 0, \quad w^F = w^C = 0, \quad \phi_r^F = 0, \quad \frac{dw^C}{dr} = 0, \tag{7.5.77a}$$

$$\text{At } r = a: \quad w^F = w^C = 0, \quad N_{rr}^F = N_{rr}^C = 0, \quad M_{rr}^F = M_{rr}^C = 0. \tag{7.5.77b}$$

The boundary conditions yield

$$K_1 = \Omega_5 \left(\frac{-2a\hat{D}\frac{dw^C(a)}{dr} + \hat{c}_1\frac{a^2}{2}(2\ln a - 1) + \hat{c}_2 a^2 + 2\hat{c}_3}{b^2\Omega_7 - a^2\Omega_6} \right) = 0, \qquad (7.5.78a)$$

$$K_2 = -\frac{b^2}{2}K_1 = 0, \quad c_1 = \frac{f_1 e_4 - f_2 e_2}{e_1 e_4 - e_2 e_3}, \quad c_2 = \frac{f_1 - C_1 e_1}{e_2}, \qquad (7.5.78b)$$

$$c_3 = -\frac{c_1 b^2}{4}(2\ln b - 1) - \frac{c_2 b^2}{2}, \qquad (7.5.78c)$$

$$c_4 = \frac{\mathcal{M}^C(b) + \mathcal{M}^C(a)}{2S_{rz}} + \frac{\hat{c}_1}{2}\left[\frac{b^2}{4}(1 - \ln b) + \frac{a^2}{4}(1 - \ln a) + \frac{\Omega_1}{S_{rz}}\ln ab \right]$$
$$- \hat{c}_2 \frac{(a^2 + b^2)}{8} - \frac{1}{2}\hat{c}_3 \ln ab, \qquad (7.5.78d)$$

$$e_1 = \frac{b^2}{4}(1 - \ln b) - \frac{a^2}{4}(1 - \ln a) + \frac{b^2}{4}(2\ln b - 1)\ln\frac{b}{a} + \frac{\Omega_1}{S_{rz}}\ln\frac{b}{a}, \qquad (7.5.79a)$$

$$e_2 = \frac{b^2}{2}\ln\frac{b}{a} - \frac{(b^2 - a^2)}{4}, \qquad (7.5.79b)$$

$$e_3 = \frac{1}{2}\left(1 + \hat{\Omega}_2\right)\ln a + \frac{1}{4}\left(1 - \hat{\Omega}_2\right) + \frac{b^2}{4a^2}\left(1 - \hat{\Omega}_2\right), \qquad (7.5.79c)$$

$$e_4 = \frac{1}{2}\left[\left(1 + \hat{\Omega}_2\right) + \frac{b^2}{a^2}\left(1 - \hat{\Omega}_2\right)\right], \qquad (7.5.79d)$$

$$f_1 = \frac{\mathcal{M}^C(a) - \mathcal{M}^C(b)}{S_{rz}}\Omega_1, \quad f_2 = \frac{D_{rr}}{a}\left(\hat{\Omega}_2 - \nu\right)\frac{dw^C(a)}{dr}. \qquad (7.5.80a)$$

7.6   CHAPTER SUMMARY

The first-order shear deformation plate theory (FST) of axisymmetric circular plates, which is an extension of the Timoshenko beam theory, is developed accounting for material variation through the thickness, the von Kármán nonlinearity and the effect of the modified couple stress. Exact solutions (by direct integration) and variational solution of the theory for the linear case are presented. Bending relationships between the classical and first-order plate theories are also presented. The nonlinear solutions by the finite element method are presented in Chapter 10.

SUGGESTED EXERCISES

7.1 Show that substitution of Eqs. (7.2.3) and (7.2.4) for M_{rr} and $M_{\theta\theta}$ into Eq. (7.2.1) yields the result

$$rN_{rz} = D_{rr}r\frac{d}{dr}\left[\frac{1}{r}\frac{d}{dr}(r\phi_r)\right].$$

7.2 For axisymmetric bending of isotropic circular plates according to the CPT, Eq. (6.4.3) reduces to

$$-\frac{1}{r}\frac{d}{dr}\left(rN_{rz}^C\right) = q; \quad rN_{rz}^C \equiv \frac{d}{dr}\left(rM_{rr}^C\right) - M_{\theta\theta}^C, \tag{1}$$

where

$$M_{rr}^C = -D_{rr}\left(\frac{d^2w^C}{dr^2} + \nu\frac{1}{r}\frac{dw^C}{dr}\right), \quad M_{\theta\theta}^C = -D_{rr}\left(\nu\frac{d^2w^C}{dr^2} + \frac{1}{r}\frac{dw^C}{dr}\right), \tag{2}$$

where $D_{rr} = Eh^3/12(1-\nu^2)$ and $S_{rz} = K_sGh$, h being the thickness of the plate. The corresponding equations of the first-order theory are

$$-\frac{1}{r}\frac{d}{dr}\left(rN_{rz}^F\right) = q, \quad -\frac{d}{dr}\left(rM_{rr}^F\right) + M_{\theta\theta}^F + rN_{rz}^F = 0, \tag{3}$$

with

$$M_{rr}^F = D_{rr}\left(\frac{d\phi_r^F}{dr} + \nu\frac{1}{r}\phi_r^F\right), \quad M_{\theta\theta}^F = D_{rr}\left(\nu\frac{d\phi_r^F}{dr} + \frac{1}{r}\phi_r^F\right),$$

$$N_{rz}^F = S_{rz}\left(\frac{dw^F}{dr} + \phi_r^F\right). \tag{4}$$

Introduce the moment sum

$$\mathcal{M} = \frac{M_{rr} + M_{\theta\theta}}{1+\nu} \tag{5}$$

and show that

$$\mathcal{M}^C = -D_{rr}\frac{1}{r}\frac{d}{dr}\left(r\frac{dw^C}{dr}\right), \tag{6}$$

$$-\frac{1}{r}\frac{d}{dr}\left(r\frac{d\mathcal{M}^C}{dr}\right) = q, \tag{7}$$

$$r\frac{d\mathcal{M}^C}{dr} = \frac{d}{dr}\left(rM_{rr}^C\right) - M_{\theta\theta}^C = rN_{rz}^C, \tag{8}$$

$$\mathcal{M}^F = D_{rr}\frac{1}{r}\frac{d}{dr}\left(r\phi_r^F\right), \tag{9}$$

$$-\frac{1}{r}\frac{d}{dr}\left(r\frac{d\mathcal{M}^F}{dr}\right) = q, \tag{10}$$

$$r\frac{d\mathcal{M}^F}{dr} = \frac{d}{dr}\left(rM_{rr}^F\right) - M_{\theta\theta}^F = rN_{rz}^F. \tag{11}$$

7.3 Using Eqs (1)–(11) of **Problem 7.2**, establish the following relationships:

$$rN_{rz}^F = rN_{rz}^C + c_1, \tag{1}$$

$$\mathcal{M}^F = \mathcal{M}^C + c_1 \log r + c_2, \tag{2}$$

$$\phi_r^F = -\frac{dw^C}{dr} + \frac{c_1}{4D_{rr}}r\left(2\log r - 1\right) + \frac{c_2}{2D_{rr}}r + \frac{c_3}{D_{rr}}\frac{1}{r}, \tag{3}$$

$$w^F = w^C + \frac{c_1}{4D_{rr}}r^2\left(1 - \log r\right) + \frac{c_1}{S_{rz}}\log r - \frac{c_2}{4D_{rr}}r^2$$

$$\qquad - \frac{c_3}{D_{rr}}\log r + \frac{c_4}{D_{rr}} + \frac{\mathcal{M}^C}{S_{rz}}. \tag{4}$$

---

*To get to know, to **discover**, to **publish** – this is the **destiny** of a scientist.*

François Ara

8 Third-Order Theory of Circular Plates

Science knows no country, because knowledge belongs to humanity, and is the torch which illuminates the world. Science is the highest personification of the nation because that nation will remain the first which carries the furthest the works of thought and intelligence.

Louis Pasteur

8.1 GOVERNING EQUATIONS

8.1.1 PRELIMINARY COMMENTS

In this chapter, we develop the Reddy third-order shear deformation plate theory (TST) of the axisymmetric circular plates. As explained in Chapter 5, we use a higher-order expansion of the radial displacement u_r through the thickness of the plate, and thus, further relax the Kirchhoff hypothesis by removing the assumption of straightness of a transverse normal (in all theories, the inextensibility of a transverse normal can be removed by assuming that the transverse deflection also varies through the thickness; that is, include the thickness stretch, as explained in Section 5.2).

8.1.2 DISPLACEMENTS AND STRAINS

The third-order plate theory of Reddy [6]–[9] is based on the displacement field

$$\mathbf{u}(r, z, t) = u_r(r, z, t)\,\hat{\mathbf{e}}_r + u_z(r, z, t)\,\hat{\mathbf{e}}_z, \tag{8.1.1a}$$

$$u_r(r, z, t) = u(r, t) + z\phi_r(r, t) - \alpha z^3\left(\phi_r + \frac{\partial w}{\partial r}\right),$$

$$u_z(r, z, t) = w(r, t), \quad \alpha = \frac{4}{3h^2}, \tag{8.1.1b}$$

where (u_r, u_z) the total displacement components along the r and z coordinates, respectively, (u, w, ϕ_r) are the generalized displacements, and h is the total thickness of the plate. The displacement field accommodates quadratic variation of transverse shear strains and shear stresses and vanishing of transverse shear stress on the top $z = h/2$ and bottom $z = -h/2$ planes of a plate, and there is no need to use shear correction coefficient in the third-order theory.

DOI: 10.1201/9781003240846-8

The nonzero von Kármán nonlinear strains can be written as

$$\varepsilon_{rr} = \varepsilon_{rr}^{(0)} + z\varepsilon_{rr}^{(1)} + z^3\varepsilon_{rr}^{(3)}, \quad \varepsilon_{\theta\theta} = \varepsilon_{\theta\theta}^{(0)} + z\varepsilon_{\theta\theta}^{(1)} + z^3\varepsilon_{\theta\theta}^{(3)}, \quad \varepsilon_{rz} = \varepsilon_{rz}^{(0)} + z^2\varepsilon_{rz}^{(2)},$$
(8.1.2)

where

$$\varepsilon_{rr}^{(0)} = \frac{\partial u}{\partial r} + \frac{1}{2}\left(\frac{\partial w}{\partial r}\right)^2, \quad \varepsilon_{rr}^{(1)} = \frac{\partial \phi_r}{\partial r}, \quad \varepsilon_{rr}^{(3)} = -\alpha\left(\frac{\partial \phi_r}{\partial r} + \frac{\partial^2 w}{\partial r^2}\right),$$

$$\varepsilon_{\theta\theta}^{(0)} = \frac{u}{r}, \quad \varepsilon_{\theta\theta}^{(1)} = \frac{\phi_r}{r}, \quad \varepsilon_{\theta\theta}^{(3)} = -\alpha\frac{1}{r}\left(\phi_r + \frac{\partial w}{\partial r}\right),$$
(8.1.3)

$$2\varepsilon_{rz}^{(0)} = \phi_r + \frac{\partial w}{\partial r}, \quad 2\varepsilon_{rz}^{(2)} = -\beta\left(\phi_r + \frac{\partial w}{\partial r}\right), \quad \beta = \frac{4}{h^2}.$$

The rotation and curvature components are [cf. Eqs. (6.1.6a) and (6.1.6b)]

$$\omega_\theta = \frac{1}{2}\left(\frac{\partial u_r}{\partial z} - \frac{\partial u_z}{\partial r}\right) = \frac{1}{2}\left[\phi_r - \frac{\partial w}{\partial r} - \beta z^2\left(\phi_r + \frac{\partial w}{\partial r}\right)\right],$$

$$\chi_{r\theta} = \frac{1}{2}\left(\frac{\partial \omega_\theta}{\partial r} - \frac{\omega_\theta}{r}\right) = \frac{1}{4}\left[(1 - \beta z^2)\left(\frac{\partial \phi_r}{\partial r} - \frac{1}{r}\phi_r\right)\right.$$
$$\left. - (1 + \beta z^2)\left(\frac{\partial^2 w}{\partial r^2} - \frac{1}{r}\frac{\partial w}{\partial r}\right)\right],$$
(8.1.4)

$$\chi_{\theta z} = \frac{1}{4}\left(\frac{\partial^2 u_r}{\partial z^2} - \frac{\partial^2 u_z}{\partial z \partial r}\right) = -\frac{1}{2}\beta z\left(\phi_r + \frac{\partial w}{\partial r}\right).$$

8.1.3 EQUATIONS OF MOTION

The statement of Hamilton's principle for the third-order theory is

$$0 = \int_0^T \int_\Omega \left(\sigma_{rr}\,\delta\varepsilon_{rr} + \sigma_{\theta\theta}\,\delta\varepsilon_{\theta\theta} + 2\sigma_{rz}\,\delta\varepsilon_{rz} + 2\mathcal{M}_{r\theta}\,\delta\chi_{r\theta} + 2\mathcal{M}_{z\theta}\,\delta\chi_{\theta z}\right.$$

$$\left. - \rho\dot{\mathbf{u}}\cdot\delta\dot{\mathbf{u}}\right)dz\,r dr\,dt + \int_0^T \int_b^a (c_f w\,\delta w - q\,\delta w)\,r dr\,dt$$

$$= \int_0^T \int_b^a \left[N_{rr}\left(\frac{\partial \delta u}{\partial r} + \frac{\partial w}{\partial r}\frac{\partial \delta w}{\partial r}\right) + \frac{1}{r}N_{\theta\theta}\,\delta u + \bar{M}_{rr}\frac{\partial \delta \phi_r}{\partial r} + \frac{1}{r}\bar{M}_{\theta\theta}\,\delta\phi_r\right.$$

$$- \alpha P_{rr}\frac{\partial^2 \delta w}{\partial r^2} - \frac{\alpha}{r}P_{\theta\theta}\frac{\partial \delta w}{\partial r} + \bar{N}_{rz}\left(\delta\phi_r + \frac{\partial \delta w}{\partial r}\right) - q\,\delta w + c_f\,w\,\delta w$$

$$- m_0\left(\dot{u}\,\delta u + \dot{w}\,\delta w\right) - \hat{m}_2\dot{\phi}_r\,\delta\phi_r + \alpha\bar{m}_4\left(\delta\dot{\phi}_r\frac{\partial \dot{w}}{\partial r} + \dot{\phi}_r\frac{\partial \delta\dot{w}}{\partial r}\right)$$

$$- \alpha^2 m_6\frac{\partial \dot{w}}{\partial r}\frac{\partial \delta\dot{w}}{\partial r} - \bar{m}_1\left(\dot{\phi}_r\,\delta\dot{u} + \delta\dot{\phi}_r\,\dot{u}\right) + \alpha m_3\left(\dot{u}\frac{\partial \delta\dot{w}}{\partial r} + \delta\dot{u}\frac{\partial \dot{w}}{\partial r}\right)$$

$$+ \frac{1}{2}\bar{P}_{r\theta}\left(\frac{\partial \delta\phi_r}{\partial r} - \frac{1}{r}\delta\phi_r\right) - \frac{1}{2}\tilde{P}_{r\theta}\left(\frac{\partial^2 \delta w}{\partial r^2} - \frac{1}{r}\frac{\partial \delta w}{\partial r}\right)\right]r dr\,dt,$$
(8.1.5)

where Ω denotes the midplane of the plate (i.e., $b \leq r \leq a$ and $-h/2 \leq z \leq h/2$), T is the arbitrary final time, c_f is the modulus of the elastic foundation (if any), and the stress resultants $(N_{rr}, N_{\theta\theta}, N_{rz}, M_{rr}, M_{\theta\theta})$, higher-order stress resultants $(P_{rr}, P_{\theta\theta}, P_{rz})$, $(P_{r\theta}, R_{r\theta})$, and $Q_{\theta z}$ are defined as follows:

$$(N_{rr}, N_{\theta\theta}) = \int_{-\frac{h}{2}}^{\frac{h}{2}} (\sigma_{rr}, \sigma_{\theta\theta}) \, dz, \quad (M_{rr}, M_{\theta\theta}) = \int_{-\frac{h}{2}}^{\frac{h}{2}} (\sigma_{rr}, \sigma_{\theta\theta}) z \, dz \quad (8.1.6a)$$

$$(N_{rz}, P_{rz}) = \int_{-\frac{h}{2}}^{\frac{h}{2}} \sigma_{rz} \left(1, z^2\right) dz, \quad (P_{rr}, P_{\theta\theta}) = \int_{-\frac{h}{2}}^{\frac{h}{2}} (\sigma_{rr}, \sigma_{\theta\theta}) z^3 \, dz, \quad (8.1.6b)$$

$$(P_{r\theta}, R_{r\theta}) = \int_{-\frac{h}{2}}^{\frac{h}{2}} \mathcal{M}_{r\theta} \left(1, z^2\right) dz, \qquad Q_{\theta z} = \int_{-\frac{h}{2}}^{\frac{h}{2}} \mathcal{M}_{\theta z} \, z \, dz, \quad (8.1.6c)$$

and

$$\bar{M}_{rr} = M_{rr} - \alpha \, P_{rr}, \quad \bar{M}_{\theta\theta} = M_{\theta\theta} - \alpha \, P_{\theta\theta}, \quad (8.1.7)$$

$$\bar{N}_{rz} = N_{rz} - \beta \left(P_{rz} + Q_{\theta z}\right), \quad \bar{P}_{r\theta} = P_{r\theta} - \beta R_{r\theta}, \quad \tilde{P}_{r\theta} = P_{r\theta} + \beta R_{r\theta}.$$

The mass inertia coefficients m_i $(i = 0, 1, \ldots, 6)$ are defined by

$$m_i = \int_{-\frac{h}{2}}^{\frac{h}{2}} \rho(z)^i \, dz, \quad \bar{m}_i = m_i - \alpha m_{i+2}, \quad \hat{m}_2 = \bar{m}_2 - \alpha \bar{m}_4. \quad (8.1.8)$$

Then the Euler–Lagrange equations of motion of the TST, accounting for the von Kármán nonlinearity and the couple stress effect, are

$$-\frac{1}{r}\left[\frac{\partial}{\partial r}\left(rN_{rr}\right) - N_{\theta\theta}\right] + m_0 \frac{\partial^2 u}{\partial t^2} - \alpha m_3 \frac{\partial^3 w}{\partial r \partial t^2} + \bar{m}_1 \frac{\partial^2 \phi_r}{\partial t^2} = 0, \quad (8.1.9)$$

$$-\frac{1}{r}\frac{\partial}{\partial r}\left(r\bar{N}_{rz} + rN_{rr}\frac{\partial w}{\partial r}\right) - \alpha \frac{1}{r}\frac{\partial}{\partial r}\left[\frac{\partial}{\partial r}\left(rP_{rr}\right) - P_{\theta\theta}\right]$$

$$-\frac{1}{2r}\frac{\partial}{\partial r}\left[\tilde{P}_{r\theta} + \frac{\partial}{\partial r}\left(r\tilde{P}_{r\theta}\right)\right] + c_f w + m_0 \frac{\partial^2 w}{\partial t^2}$$

$$+\alpha \frac{1}{r}\frac{\partial}{\partial r}\left[r\left(m_3 \frac{\partial^2 u}{\partial t^2} - \alpha m_6 \frac{\partial^3 w}{\partial r \partial t^2} + \bar{m}_4 \frac{\partial^2 \phi_r}{\partial t^2}\right)\right] = q, \quad (8.1.10)$$

$$-\frac{1}{r}\left[\frac{\partial}{\partial r}\left(r\bar{M}_{rr}\right) - \bar{M}_{\theta\theta}\right] + \bar{N}_{rz} - \frac{1}{2r}\left[\bar{P}_{r\theta} + \frac{\partial}{\partial r}\left(r\bar{P}_{r\theta}\right)\right]$$

$$+\bar{m}_1 \frac{\partial^2 u}{\partial t^2} + \hat{m}_2 \frac{\partial^2 \phi_r}{\partial t^2} - \alpha \bar{m}_4 \frac{\partial^3 w}{\partial r \partial t^2} = 0. \quad (8.1.11)$$

The boundary conditions involve specifying one element of each of the following four duality pairs:

$$(u, rN_{rr}); \quad (w, rV_{\text{eff}}), \quad (\theta_r, M_{\text{eff}}), \quad (\phi_r, r\tilde{M}_{rr}), \quad (8.1.12)$$

where $\theta_r = -\partial w/\partial r$ is the slope and

$$\tilde{M}_{rr} = \bar{M}_{rr} + \tfrac{1}{2}\bar{P}_{r\theta}, \quad M_{\text{eff}} = \alpha r P_{rr} + \tfrac{1}{2}\tilde{P}_{r\theta}, \tag{8.1.13}$$

$$V_{\text{eff}} = \bar{N}_{rz} + N_{rr}\frac{\partial w}{\partial r} + \alpha\left[\frac{\partial}{\partial r}(rP_{rr}) - P_{\theta\theta}\right] + \tfrac{1}{2}\left[\frac{\partial}{\partial r}\left(r\tilde{P}_{r\theta}\right) + \tilde{P}_{r\theta}\right]$$

$$- \alpha\left(m_3\frac{\partial^2 u}{\partial t^2} - \alpha m_6\frac{\partial^3 w}{\partial r \partial t^2} + \bar{m}_4\frac{\partial^2 \phi_r}{\partial t^2}\right). \tag{8.1.14}$$

The equilibrium equations of the TST are

$$-\frac{1}{r}\left[\frac{d}{dr}(rN_{rr}) - N_{\theta\theta}\right] = 0, \tag{8.1.15}$$

$$-\frac{1}{r}\frac{d}{dr}\left(r\bar{N}_{rz} + r\frac{dw}{dr}N_{rr}\right) - \alpha\frac{1}{r}\frac{d}{dr}\left[\frac{d}{dr}(rP_{rr}) - P_{\theta\theta}\right]$$

$$-\tfrac{1}{2}\frac{1}{r}\frac{d}{dr}\left[\frac{d}{dr}\left(r\tilde{P}_{r\theta}\right) + \tilde{P}_{r\theta}\right] + c_f w = q, \tag{8.1.16}$$

$$-\frac{1}{r}\left[\frac{d}{dr}\left(r\bar{M}_{rr}\right) - \bar{M}_{\theta\theta}\right] - \tfrac{1}{2}\frac{1}{r}\left[\frac{d}{dr}\left(r\bar{P}_{r\theta}\right) + \bar{P}_{r\theta}\right] + \bar{N}_{rz} = 0. \tag{8.1.17}$$

The equilibrium equations without the couple stress effect are (now $\bar{N}_{rz} = N_{rz} - \beta P_{rz}$ because $Q_{\theta z} = 0$)

$$-\frac{1}{r}\left[\frac{d}{dr}(rN_{rr}) - N_{\theta\theta}\right] = 0, \tag{8.1.18}$$

$$-\frac{1}{r}\frac{d}{dr}\left[r\left(\bar{N}_{rz} + N_{rr}\frac{dw}{dr}\right)\right] - \frac{\alpha}{r}\frac{d}{dr}\left[\frac{d}{dr}(rP_{rr}) - P_{\theta\theta}\right] + c_f w = q, \tag{8.1.19}$$

$$-\frac{1}{r}\left[\frac{d}{dr}\left(r\bar{M}_{rr}\right) - \bar{M}_{\theta\theta}\right] + \bar{N}_{rz} = 0. \tag{8.1.20}$$

8.1.4 PLATE CONSTITUTIVE EQUATIONS

The stress relations used are the same as those in Eqs. (6.1.7) and (6.1.9), with $E(z)$ and $\rho(z)$ following the power-law variations listed in Eqs. (6.1.8) and (6.2.7). They are repeated here for a ready reference:

$$\left\{\begin{matrix}\sigma_{rr}\\\sigma_{\theta\theta}\\\sigma_{rz}\end{matrix}\right\} = \frac{E(z)}{1-\nu^2}\begin{bmatrix}1 & \nu & 0\\\nu & 1 & 0\\0 & 0 & \frac{1-\nu}{2}\end{bmatrix}\left\{\begin{matrix}\varepsilon_{rr}\\\varepsilon_{\theta\theta}\\2\varepsilon_{rz}\end{matrix}\right\}, \tag{8.1.21}$$

$$\mathcal{M}_{r\theta} = 2G\ell^2\,\chi_{r\theta}, \quad \mathcal{M}_{\theta z} = 2G\ell^2\,\chi_{\theta z},$$

and

$$E(z) = (E_1 - E_2)\,V_1(z) + E_2, \quad V_1(z) = \left(\frac{1}{2} + \frac{z}{h}\right)^n, \tag{8.1.22}$$

$$\rho(z) = (\rho_1 - \rho_2)\,V_1(z) + \rho_2. \tag{8.1.23}$$

The stress resultants appearing in Eqs. (8.1.6a)–(8.1.6c) can be expressed in terms of the generalized displacements (u, w, ϕ_r) as

$$
N_{rr} = A_{rr} \left[\frac{\partial u}{\partial r} + \frac{1}{2} \left(\frac{\partial w}{\partial r} \right)^2 + \nu \frac{u}{r} \right] + \bar{B}_{rr} \left(\frac{\partial \phi_r}{\partial r} + \frac{\nu}{r} \phi_r \right)
$$
$$
- \alpha E_{rr} \left(\frac{\nu}{r} \frac{\partial w}{\partial r} + \frac{\partial^2 w}{\partial r^2} \right), \tag{8.1.24a}
$$

$$
N_{\theta\theta} = A_{rr} \left[\frac{u}{r} + \nu \frac{\partial u}{\partial r} + \frac{\nu}{2} \left(\frac{\partial w}{\partial r} \right)^2 \right] + \bar{B}_{rr} \left(\nu \frac{\partial \phi_r}{\partial r} + \frac{1}{r} \phi_r \right)
$$
$$
- \alpha E_{rr} \left(\frac{1}{r} \frac{\partial w}{\partial r} + \nu \frac{\partial^2 w}{\partial r^2} \right), \tag{8.1.24b}
$$

$$
M_{rr} = B_{rr} \left[\frac{\partial u}{\partial r} + \frac{1}{2} \left(\frac{\partial w}{\partial r} \right)^2 + \nu \frac{u}{r} \right] + \bar{D}_{rr} \left(\frac{\partial \phi_r}{\partial r} + \frac{\nu}{r} \phi_r \right)
$$
$$
- \alpha F_{rr} \left(\frac{\nu}{r} \frac{\partial w}{\partial r} + \frac{\partial^2 w}{\partial r^2} \right), \tag{8.1.24c}
$$

$$
M_{\theta\theta} = B_{rr} \left[\frac{u}{r} + \nu \frac{\partial u}{\partial r} + \frac{\nu}{2} \left(\frac{\partial w}{\partial r} \right)^2 \right] + \bar{D}_{rr} \left(\nu \frac{\partial \phi_r}{\partial r} + \frac{1}{r} \phi_r \right)
$$
$$
- \alpha F_{rr} \left(\frac{1}{r} \frac{\partial w}{\partial r} + \nu \frac{\partial^2 w}{\partial r^2} \right), \tag{8.1.24d}
$$

$$
P_{rr} = E_{rr} \left[\frac{\partial u}{\partial r} + \frac{1}{2} \left(\frac{\partial w}{\partial r} \right)^2 + \nu \frac{u}{r} \right] + \bar{F}_{rr} \left(\frac{\partial \phi_r}{\partial r} + \frac{\nu}{r} \phi_r \right)
$$
$$
- \alpha H_{rr} \left(\frac{\nu}{r} \frac{\partial w}{\partial r} + \frac{\partial^2 w}{\partial r^2} \right), \tag{8.1.24e}
$$

$$
P_{\theta\theta} = E_{rr} \left[\frac{u}{r} + \nu \frac{\partial u}{\partial r} + \frac{\nu}{2} \left(\frac{\partial w}{\partial r} \right)^2 \right] + \bar{F}_{rr} \left(\nu \frac{\partial \phi_r}{\partial r} + \frac{1}{r} \phi_r \right)
$$
$$
- \alpha H_{rr} \left(\frac{1}{r} \frac{\partial w}{\partial r} + \nu \frac{\partial^2 w}{\partial r^2} \right), \tag{8.1.24f}
$$

$$
N_{rz} = \bar{A}_{rz} \left(\phi_r + \frac{\partial w}{\partial r} \right), \tag{8.1.24g}
$$

$$
P_{rz} = \bar{D}_{rz} \left(\phi_r + \frac{\partial w}{\partial r} \right), \tag{8.1.24h}
$$

$$
Q_{\theta z} = -\beta \, D_{z\theta} \left(\phi_r + \frac{\partial w}{\partial r} \right), \tag{8.1.24i}
$$

$$P_{r\theta} = \frac{1}{2} A_{r\theta} \left(\frac{\partial \phi_r}{\partial r} - \frac{1}{r} \phi_r - \frac{\partial^2 w}{\partial r^2} + \frac{1}{r} \frac{\partial w}{\partial r} \right)$$

$$- \frac{1}{2} \beta D_{r\theta} \left(\frac{\partial \phi_r}{\partial r} - \frac{1}{r} \phi_r + \frac{\partial^2 w}{\partial r^2} - \frac{1}{r} \frac{\partial w}{\partial r} \right)$$

$$= \frac{1}{2} \bar{A}_{r\theta} \left(\frac{\partial \phi_r}{\partial r} - \frac{1}{r} \phi_r \right) + \frac{1}{2} \tilde{A}_{r\theta} \left(-\frac{\partial^2 w}{\partial r^2} + \frac{1}{r} \frac{\partial w}{\partial r} \right), \tag{8.1.25a}$$

$$R_{r\theta} = \frac{1}{2} \bar{D}_{r\theta} \left(\frac{\partial \phi_r}{\partial r} - \frac{1}{r} \phi_r \right) + \frac{1}{2} \tilde{D}_{r\theta} \left(-\frac{\partial^2 w}{\partial r^2} + \frac{1}{r} \frac{\partial w}{\partial r} \right), \tag{8.1.25b}$$

$$\bar{P}_{r\theta} = \frac{1}{2} \hat{A}_{r\theta} \left(\frac{\partial \phi_r}{\partial r} - \frac{1}{r} \phi_r \right) + \frac{1}{2} \bar{\tilde{A}}_{r\theta} \left(-\frac{\partial^2 w}{\partial r^2} + \frac{1}{r} \frac{\partial w}{\partial r} \right), \tag{8.1.25c}$$

$$\tilde{P}_{r\theta} = \frac{1}{2} \tilde{\bar{A}}_{r\theta} \left(\frac{\partial \phi_r}{\partial r} - \frac{1}{r} \phi_r \right) + \frac{1}{2} \tilde{\tilde{A}}_{r\theta} \left(-\frac{\partial^2 w}{\partial r^2} + \frac{1}{r} \frac{\partial w}{\partial r} \right), \tag{8.1.25d}$$

where A_{rr}, B_{rr}, D_{rr}, E_{rr}, F_{rr}, H_{rr}, $A_{r\theta}$, $D_{r\theta}$, and $F_{r\theta}$ are the extensional, extensional-bending, bending, and higher-order stiffness coefficients:

$$(A_{rr}, B_{rr}, D_{rr}, E_{rr}, F_{rr}, H_{rr}) = \frac{1}{(1 - \nu^2)} \int_{-\frac{h}{2}}^{\frac{h}{2}} (1, z, z^2, z^3, z^4, z^6) E(z) \, dz,$$

$$(A_{rz}, D_{rz}, F_{rz}) = \frac{1}{2(1 + \nu)} \int_{-\frac{h}{2}}^{\frac{h}{2}} (1, z^2, z^4) E(z) \, dz, \tag{8.1.26}$$

$$(A_{r\theta}, D_{r\theta}, F_{r\theta}) = \frac{\ell^2}{2(1 + \nu)} \int_{-\frac{h}{2}}^{\frac{h}{2}} (1, z^2, z^4) E(z) \, dz,$$

and

$$\bar{B}_{rr} = B_{rr} - \alpha E_{rr}, \quad \bar{D}_{rr} = D_{rr} - \alpha F_{rr}, \quad \bar{F}_{rr} = F_{rr} - \alpha H_{rr},$$

$$\bar{A}_{rz} = A_{rz} - \beta D_{rz}, \quad \bar{D}_{rz} = D_{rz} - \beta F_{rz}, \quad \bar{A}_{r\theta} = A_{r\theta} - \beta D_{r\theta},$$

$$\bar{D}_{r\theta} = D_{r\theta} - \beta F_{r\theta}, \quad \tilde{A}_{r\theta} = A_{r\theta} + \beta D_{r\theta}, \quad \tilde{D}_{r\theta} = D_{r\theta} + \beta F_{r\theta}, \tag{8.1.27}$$

$$\hat{A}_{r\theta} = \bar{A}_{r\theta} - \beta \bar{D}_{r\theta}, \quad \bar{\tilde{A}}_{r\theta} = \tilde{\bar{A}}_{r\theta} = A_{r\theta} - \beta^2 F_{r\theta}, \quad D_{z\theta} = D_{r\theta},$$

$$\tilde{\tilde{A}}_{r\theta} = A_{r\theta} + 2\beta D_{r\theta} + \beta^2 F_{r\theta}.$$

8.2 EXACT SOLUTIONS OF THE TST

As shown in the following pages, it is not possible to determine the exact solution of the TST equations due to the presence of higher-order stress resultants P_{rr} and $P_{\theta\theta}$. We outline the steps similar to those followed for the FST in Section 7.3 to find the exact solutions of the linearized equations without the foundation modulus (i.e., $c_f = 0$) and the couple stress terms (i.e., $P_{r\theta} = R_{r\theta} = Q_{\theta z} = 0$).

We begin with the following mathematical identities:

$$\frac{d}{dr}(rN_{rr}) - N_{\theta\theta} = A_{rr}\left[\frac{d}{dr}\left(r\frac{du}{dr}\right) - \frac{u}{r}\right] + \bar{B}_{rr}\left[\frac{d}{dr}\left(r\frac{d\phi_r}{dr}\right) - \frac{\phi_r}{r}\right]$$

$$- \alpha E_{rr}\left[\frac{d}{dr}\left(r\frac{d^2w}{dr^2}\right) - \frac{1}{r}\frac{dw}{dr}\right]$$

$$= A_{rr}r\frac{d}{dr}\left[\frac{1}{r}\frac{d}{dr}(ru)\right] + \bar{B}_{rr}r\frac{d}{dr}\left[\frac{1}{r}\frac{d}{dr}(r\phi_r)\right]$$

$$- \alpha E_{rr}r\frac{d}{dr}\left[\frac{1}{r}\frac{d}{dr}\left(r\frac{dw}{dr}\right)\right], \tag{8.2.1}$$

$$\frac{d}{dr}(rM_{rr}) - M_{\theta\theta} = B_{rr}r\frac{d}{dr}\left[\frac{1}{r}\frac{d}{dr}(ru)\right] + \bar{D}_{rr}r\frac{d}{dr}\left[\frac{1}{r}\frac{d}{dr}(r\phi_r)\right]$$

$$- \alpha F_{rr}r\frac{d}{dr}\left[\frac{1}{r}\frac{d}{dr}\left(r\frac{dw}{dr}\right)\right], \tag{8.2.2}$$

$$\frac{d}{dr}(rP_{rr}) - P_{\theta\theta} = E_{rr}r\frac{d}{dr}\left[\frac{1}{r}\frac{d}{dr}(ru)\right] + \bar{F}_{rr}r\frac{d}{dr}\left[\frac{1}{r}\frac{d}{dr}(r\phi_r)\right]$$

$$- \alpha H_{rr}r\frac{d}{dr}\left[\frac{1}{r}\frac{d}{dr}\left(r\frac{dw}{dr}\right)\right]. \tag{8.2.3}$$

Then from Eq. (8.1.18) we have

$$A_{rr}\frac{d}{dr}\left[\frac{1}{r}\frac{d}{dr}(ru)\right] + \bar{B}_{rr}\frac{d}{dr}\left[\frac{1}{r}\frac{d}{dr}(r\phi_r)\right] - \alpha E_{rr}\frac{d}{dr}\left[\frac{1}{r}\frac{d}{dr}\left(r\frac{dw}{dr}\right)\right] = 0.$$
$$\tag{8.2.4}$$

Integration with respect r twice yields

$$A_{rr}u + \bar{B}_{rr}\phi_r - \alpha E_{rr}\frac{dw}{dr} = \frac{c_1 r}{2} + \frac{c_2}{r}, \tag{8.2.5}$$

where c_1 and c_2 are constants of integration.

Integrating Eq. (8.1.19) (note that $\bar{V}_r = \bar{N}_{rz}$ for the linear case and $c_f = 0$) with respect to r results in

$$r\bar{N}_{rz} + \alpha\left[\frac{d}{dr}(rP_{rr}) - P_{\theta\theta}\right] = -\int rq(r)\,dr + c_3. \tag{8.2.6}$$

Substituting for $\bar{N}_{rz}$ from Eq. (8.1.20) into Eq. (8.2.6), we obtain

$$\left[\frac{d}{dr}(rM_{rr}) - M_{\theta\theta}\right] = -\int rq(r)\,dr + c_3. \tag{8.2.7}$$

Then, in view of the identity in Eq. (8.2.2), we obtain

$$\bar{B}_{rr} r \frac{d}{dr}\left[\frac{1}{r}\frac{d}{dr}(ru)\right] + \bar{D}_{rr} r \frac{d}{dr}\left[\frac{1}{r}\frac{d}{dr}(r\phi_r)\right] - \alpha \bar{F}_{rr} r \frac{d}{dr}\left[\frac{1}{r}\frac{d}{dr}\left(r\frac{dw}{dr}\right)\right]$$

$$= -\int r\,q(r)\,dr + c_3. \tag{8.2.8}$$

Integrating the above equation twice with respect to r, we obtain:

$$\bar{B}_{rr}\,u + \bar{D}_{rr}\,\phi_r - \alpha \bar{F}_{rr}\frac{dw}{dr} = -\frac{1}{r}\int r\left[\int^r \frac{1}{\xi}\left(\int^\xi \eta\,q(\eta)\,d\eta\right)d\xi\right]dr$$

$$+ c_3 \frac{r}{4}(2\log r - 1) + \frac{c_4\,r}{2} + \frac{c_5}{r}$$

$$= -F(r) + c_3 \frac{r}{4}(2\log r - 1) + \frac{c_4\,r}{2} + \frac{c_5}{r}, \tag{8.2.9}$$

where

$$F(r) = \frac{1}{r}\int r\left[\int^r \frac{1}{\xi}\left(\int^\xi \eta\,q(\eta)\,d\eta\right)d\xi\right]dr. \tag{8.2.10}$$

Solving Eqs. (8.2.5) and (8.2.9) for u and ϕ_r in terms of dw/dr, we obtain

$$u(r) = \frac{\bar{D}_{rr}\,p(r) - \bar{B}_{rr}\,g(r)}{\bar{D}^*}, \tag{8.2.11}$$

$$\phi_r(r) = \frac{A_{rr}\,g(r) - \bar{B}_{rr}\,p(r)}{\bar{D}^*}, \tag{8.2.12}$$

where

$$p(r) = \alpha E_{rr}\frac{dw}{dr} + \frac{c_1 r}{2} + \frac{c_2}{r}, \tag{8.2.13a}$$

$$g(r) = \alpha F_{rr}\frac{dw}{dr} - F(r) + c_3\frac{r}{4}(2\log r - 1) + \frac{c_4\,r}{2} + \frac{c_5}{r}, \tag{8.2.13b}$$

$$\bar{D}^* = A_{rr}\bar{D}_{rr} - \bar{B}_{rr}\bar{B}_{rr}. \tag{8.2.13c}$$

We see that the solution for u and ϕ_r includes the unknown dw/dr. In the first-order shear deformation plate theory (FST), we have used Eq. (8.2.6) without the higher-order stress resultants. However, the presence of these higher-order terms makes the task of solving for dw/dr difficult. To see this, use Eqs. (8.2.3) and (8.2.6) and obtain

$$\bar{A}_{rz} r\left(\phi_r + \frac{dw}{dr}\right) + \alpha\left\{E_{rr} r \frac{d}{dr}\left[\frac{1}{r}\frac{d}{dr}(ru)\right] + \bar{F}_{rr} r \frac{d}{dr}\left[\frac{1}{r}\frac{d}{dr}(r\phi_r)\right]\right.$$

$$\left. - \alpha H_{rr} r \frac{d}{dr}\left[\frac{1}{r}\frac{d}{dr}\left(r\frac{dw}{dr}\right)\right]\right\} = -\int r\,q(r)\,dr + c_3. \tag{8.2.14}$$

The form of the above equation makes it very difficult (if not impossible) to obtain the exact solution. The exact solutions of the simplified Reddy beam theory (SBT) were discussed in Chapter 5. Readers may follow similar approaches to determine the exact solutions to a *simplified TST*.

8.3   RELATIONSHIPS BETWEEN CPT AND TST

8.3.1   BENDING RELATIONSHIPS

Here, we develop the relationships between the bending solutions of CPT and TST for the case in which we do not account for the couple stress effect. At the outset, we note that both the classical and first-order plate theories are fourth-order theories whereas the Reddy third-order plate theory is a sixth-order theory. The order referred to here is the *total* order of all equations of equilibrium expressed in terms of the generalized displacements. The third-order plate theory is governed by a fourth-order equation in w^R and a second-order equation in ϕ_r^R. Therefore, the relationships between the solutions of two different order theories can only be established by solving an additional second-order equation (see [120] for additional discussion).

8.3.1.1   Classical plate theory (CPT)

The governing equations of the classical plate theory (CPT) are:

$$-\frac{d}{dr}\left(rM_{rr}^C\right) + M_{\theta\theta}^C + rN_{rz}^C = 0, \tag{8.3.1}$$

$$-\frac{d}{dr}\left(rN_{rz}^C\right) = rq. \tag{8.3.2}$$

The stress resultant–displacement relations of the CPT for an isotropic plate are

$$M_{rr}^C = -D_{rr}\left(\frac{d^2w^C}{dr^2} + \frac{\nu}{r}\frac{dw^C}{dr}\right), \tag{8.3.3}$$

$$M_{\theta\theta}^C = -D_{rr}\left(\nu\frac{d^2w^C}{dr^2} + \frac{1}{r}\frac{dw^C}{dr}\right). \tag{8.3.4}$$

Introducing the moment sum [148] for the CPT,

$$\mathcal{M}^C \equiv \frac{M_{rr}^C + M_{\theta\theta}^C}{(1+\nu)}, \tag{8.3.5}$$

we can write

$$\mathcal{M}^C = -D_{rr}\left(\frac{d^2w^C}{dr^2} + \frac{1}{r}\frac{dw^C}{dr}\right) = -D_{rr}\left[\frac{1}{r}\frac{d}{dr}\left(r\frac{dw^C}{dr}\right)\right]. \tag{8.3.6}$$

Using Eqs. (8.3.3) and (8.3.4) [and in view of Eq. (8.3.6)], we obtain:

$$r\frac{d\mathcal{M}^C}{dr} = \frac{d}{dr}\left(rM_{rr}^C\right) - M_{\theta\theta}^C = rN_{rz}^C. \tag{8.3.7}$$

Then, we have

$$-\frac{1}{r}\frac{d}{dr}\left(r\frac{d\mathcal{M}^C}{dr}\right) = q, \quad \mathcal{M}^C = -D_{rr}\frac{1}{r}\frac{d}{dr}\left(r\frac{dw^C}{dr}\right). \tag{8.3.8}$$

8.3.1.2 Third-order shear deformation plate theory (TST)

The governing equations of the TST are [see Eqs. (8.1.19) and (8.1.20), where $c_f w$ term and the nonlinear term are omitted]:

$$-\frac{d}{dr}\left(r\bar{N}_{rz}^R\right) - \alpha\left[\frac{d^2}{dr^2}\left(rP_{rr}^R\right) - \frac{dP_{\theta\theta}^R}{dr}\right] = rq, \tag{8.3.9}$$

$$-\frac{d}{dr}\left(r\bar{M}_{rr}^R\right) + r\bar{N}_{rz}^R + \bar{M}_{\theta\theta}^R = 0, \tag{8.3.10}$$

where

$$\bar{M}_{rr}^R = M_{rr}^R - \alpha P_{rr}^R, \quad \bar{M}_{\theta\theta}^R = M_{\theta\theta}^R - \alpha P_{\theta\theta}^R, \quad \bar{N}_{rz}^R = N_{rz}^R - \beta P_{rz}^R. \tag{8.3.11}$$

The stress resultant–displacement relations of the TST are

$$M_{rr}^R = \bar{D}_{rr}\left(\frac{d\phi_r^R}{dr} + \frac{\nu}{r}\phi_r^R\right) - \alpha F_{rr}\left(\frac{d^2 w^R}{dr^2} + \frac{\nu}{r}\frac{dw^R}{dr}\right), \tag{8.3.12}$$

$$M_{\theta\theta}^R = \bar{D}_{rr}\left(\nu\frac{d\phi_r^R}{dr} + \frac{1}{r}\phi_r^R\right) - \alpha F_{rr}\left(\nu\frac{d^2 w^R}{dr^2} + \frac{1}{r}\frac{dw^R}{dr}\right), \tag{8.3.13}$$

$$P_{rr}^R = \bar{F}_{rr}\left(\frac{d\phi_r^R}{dr} + \frac{\nu}{r}\phi_r^R\right) - \alpha H_{rr}\left(\frac{d^2 w^R}{dr^2} + \frac{\nu}{r}\frac{dw^R}{dr}\right), \tag{8.3.14}$$

$$P_{\theta\theta}^R = \bar{F}_{rr}\left(\nu\frac{d\phi_r^R}{dr} + \frac{1}{r}\phi_r^R\right) - \alpha H_{rr}\left(\nu\frac{d^2 w^R}{dr^2} + \frac{1}{r}\frac{dw^R}{dr}\right), \tag{8.3.15}$$

$$N_{rz}^R = \bar{A}_{rz}\left(\phi_r^R + \frac{dw^R}{dr}\right), \quad P_{rz}^R = \bar{D}_{rz}\left(\phi_r^R + \frac{dw^R}{dr}\right). \tag{8.3.16}$$

First, we note that Eqs. (8.3.9) and (8.3.10) together yield

$$-\frac{d^2}{dr^2}\left(rM_{rr}^R\right) + \frac{dM_{\theta\theta}^R}{dr} = rq. \tag{8.3.17}$$

Defining the effective shear force V_r^R as

$$rV_r^R = r\bar{N}_{rz}^R + \alpha\left[\frac{d}{dr}\left(rP_{rr}^R\right) - P_{\theta\theta}^R\right], \tag{8.3.18}$$

Eq. (8.3.10) can be expressed in terms of rV_r^R as

$$-\frac{d}{dr}\left(rM_{rr}^R\right) + M_{\theta\theta}^R + rV_r^R = 0. \tag{8.3.19}$$

From Eqs. (8.3.17) and (8.3.19), we have

$$-\frac{d}{dr}\left(rV_r^R\right) = rq. \tag{8.3.20}$$

Next, we introduce the following moment sums for the TST:

$$\mathcal{M}^R = \frac{M_{rr}^R + M_{\theta\theta}^R}{(1+\nu)}, \quad \mathcal{P}^R = \frac{P_{rr}^R + P_{\theta\theta}^R}{(1+\nu)}. \tag{8.3.21}$$

Using the definitions in Eq. (8.3.21) and Eqs. (8.3.12)–(8.3.15) one can show that

$$\mathcal{M}^R = \bar{D}_{rr}\frac{1}{r}\frac{d}{dr}\left(r\phi^R\right) - \alpha F_{rr}\frac{1}{r}\frac{d}{dr}\left(r\frac{dw^R}{dr}\right), \tag{8.3.22}$$

$$\mathcal{P}^R = \bar{F}_{rr}\frac{1}{r}\frac{d}{dr}\left(r\phi^R\right) - \alpha H_{rr}\frac{1}{r}\frac{d}{dr}\left(r\frac{dw^R}{dr}\right), \tag{8.3.23}$$

$$r\frac{d\mathcal{M}^R}{dr} = \frac{d}{dr}\left(rM_{rr}^R\right) - M_{\theta\theta}^R = rV_r^R, \tag{8.3.24}$$

$$r\frac{d\mathcal{P}^R}{dr} = \frac{d}{dr}\left(rP_{rr}^R\right) - P_{\theta\theta}^R. \tag{8.3.25}$$

Substituting for $r\phi_r^R$ from Eq. (8.3.16) in terms of N_{rz}^R into Eqs. (8.3.22) and (8.3.23), we arrive at

$$r\mathcal{M}^R = \frac{\bar{D}_{rr}}{\bar{A}_{rz}}\frac{d}{dr}\left(rN_{rz}^R\right) - D_{rr}\frac{d}{dr}\left(r\frac{dw^R}{dr}\right), \tag{8.3.26}$$

$$r\mathcal{P}^R = \frac{\bar{F}_{rr}}{\bar{A}_{rz}}\frac{d}{dr}\left(rN_{rz}^R\right) - F_{rr}\frac{d}{dr}\left(r\frac{dw^R}{dr}\right). \tag{8.3.27}$$

Now solving Eq. (8.3.26) for $(d/dr)(rdw^R/dr)$, we obtain

$$\frac{d}{dr}\left(r\frac{dw^R}{dr}\right) = -\frac{1}{D_{rr}}r\mathcal{M}^R + \frac{\bar{D}_{rr}}{D_{rr}\bar{A}_{rz}}\frac{d}{dr}\left(rN_{rz}^R\right). \tag{8.3.28}$$

Substituting the above result into Eq. (8.3.27), we obtain

$$r\mathcal{P}^R = \alpha\left(\frac{F_{rr}^2 - D_{rr}H_{rr}}{D_{rr}\bar{A}_{rz}}\right)\frac{d}{dr}\left(rN_{rz}^R\right) + \frac{F_{rr}}{D_{rr}}\left(r\mathcal{M}^R\right). \tag{8.3.29}$$

From Eqs. (8.3.11) and (8.3.16) we have

$$\bar{N}_{rz}^R = \frac{\hat{A}_{rz}}{\bar{A}_{rz}}N_{rz}^R, \quad \hat{A}_{rz} = \bar{A}_{rz} - \beta\bar{D}_{rz}. \tag{8.3.30}$$

8.3.2   RELATIONSHIPS

Then it follows from Eqs. (8.3.20) and (8.3.2) (by the load equivalence) that

$$rV_r^R = rN_{rz}^C + c_1. \tag{8.3.31}$$

From Eqs. (8.3.7), (8.3.24), and (8.3.31), we obtain the result

$$\mathcal{M}^R = \mathcal{M}^C + c_1 \log r + c_2. \tag{8.3.32}$$

Next, we use Eqs. (8.3.8) and (8.3.22) in Eq. (8.3.32) to arrive at

$$\bar{D}_{rr}\phi_r^R - \alpha F_{rr} \frac{dw^R}{dr} = -D_{rr}\frac{dw^C}{dr} + \frac{c_1 r}{4}(2\log r - 1) + \frac{c_2 r}{2} + \frac{c_3}{r}, \tag{8.3.33}$$

where c_3 is a constant of integration. From Eqs. (8.3.18) and (8.3.19), we have

$$r\bar{N}_{rz}^R = \frac{d}{dr}\left(rM_{rr}^R\right) - M_{\theta\theta}^R - \alpha\left[\frac{d}{dr}(rP_{rr}) - P_{\theta\theta}\right] \tag{8.3.34a}$$

$$= r\left(\frac{d\mathcal{M}^R}{dr} - \alpha\frac{d\mathcal{P}^R}{dr}\right). \tag{8.3.34b}$$

Using Eqs. (8.3.29) and (8.3.30) in Eq. (8.3.34b), we obtain

$$\frac{\hat{A}_{rz}}{\bar{A}_{rz}}\left(rN_{rz}^R\right) = \frac{\bar{D}_{rr}}{D_{rr}}r\frac{d\mathcal{M}^R}{dr} - \alpha^2\left(\frac{F_{rr}^2 - D_{rr}H_{rr}}{D_{rr}\bar{A}_{xz}}\right)r\frac{d}{dr}\left[\frac{1}{r}\frac{d}{dr}(rN_{rz}^R)\right], \tag{8.3.35}$$

and using Eqs. (8.3.30) and (8.3.31), we arrive at the final equation for N_{xz}^R:

$$r\frac{d}{dr}\left[\frac{1}{r}\frac{d}{dr}\left(rN_{rz}^R\right)\right] + \frac{D_{rr}}{\alpha^2}\left(\frac{\hat{A}_{rz}}{F_{rr}^2 - D_{rr}H_{rr}}\right)\left(rN_{rz}^R\right)$$
$$= \left(\frac{\bar{D}_{rr}\bar{A}_{rz}}{\alpha^2(F_{rr}^2 - D_{rr}H_{rr})}\right)\left(rN_{rz}^C + c_1\right). \tag{8.3.36}$$

Thus, a second-order equation must be solved to determine the shear force, N_{rz}^R.

Next, we derive the relationships between deflections w^R and w^C and rotation ϕ^R and slope $-dw_0^C/dr$. Replacing $\mathcal{M}^R$ in terms of $\mathcal{M}^C$ by means of Eq. (8.3.32) and using Eq. (8.3.6), Eq. (8.3.28) can be written as

$$\frac{d}{dr}\left(r\frac{dw^R}{dr}\right) = \frac{d}{dr}\left(r\frac{dw^C}{dr}\right) - \frac{r}{D_{rr}}(c_1\log r + c_2) + \frac{\bar{D}_{rr}}{D_{rr}\bar{A}_{rz}}\frac{d}{dr}\left(rN_{rz}^R\right). \tag{8.3.37}$$

Integrating the above equation once and twice with respect to r, we obtain

$$\frac{dw^R}{dr} = \frac{dw^C}{dr} - \frac{1}{D_{rr}}\left[\frac{c_1 r}{4}(2\log r - 1) + \frac{c_2 r}{2} + \frac{c_3}{r}\right] + \left(\frac{\bar{D}_{rr}}{D_{rr}\bar{A}_{rz}}\right)N_{rz}^R, \tag{8.3.38}$$

$$w^R = w^C - \frac{1}{D_{rr}}\left[\frac{c_1 r^2}{4}(\log r - 1) + \frac{c_2 r^2}{4} + c_3\log r + c_4\right]$$
$$+ \left(\frac{\bar{D}_{rr}}{D_{rr}\bar{A}_{rz}}\right)\int N_{rz}^R\,dr. \tag{8.3.39}$$

Finally, using Eq. (8.3.38) in Eq. (8.3.33), we obtain

$$\phi_r^R = -\frac{dw^C}{dr} + \frac{1}{D_{rr}}\left[\frac{c_1 r}{4}(2\log r - 1) + \frac{c_2 r}{2} + \frac{c_3}{r}\right] + \left(\frac{\alpha F_{rr}}{D_{rr}\bar{A}_{rz}}\right)N_{rz}^R. \quad (8.3.40)$$

Next, we discuss various types of boundary conditions in terms of the dependent variables for the TST. Since a second–order equation for N_{rz}^R must be solved to determine solutions of the TST, it is also useful to have the boundary conditions on N_{rz}^R for various types of edge supports. These are listed below.

Clamped Edge

$$\phi^R = 0, \quad \frac{dw^R}{dr} = 0 \quad \text{which imply} \quad N_{rz}^R = 0; \quad w^R = 0. \quad (8.3.41)$$

Simply Supported Edge

$$M_{rr}^R = 0, \quad P_{rr}^R = 0 \quad \text{which imply} \quad r\frac{dN_{rz}^R}{dr} + \nu N_{rz}^R = 0; \quad w^R = 0. \quad (8.3.42)$$

Free Edge

$$M_{rr}^R = 0, \quad P_{rr}^R = 0 \quad \text{which imply} \quad r\frac{dN_{rz}^R}{dr} + \nu N_{rz}^R = 0;$$

$$rV_r^R = r\bar{N}_{rz}^R + \alpha r\frac{d\mathcal{P}^R}{dr} = 0. \quad (8.3.43)$$

For solid circular plates, we have the additional conditions at the center of the plate (i.e., at $r = 0$):

$$N_{rz}^R = 0, \quad \phi_r^R = 0, \quad \frac{dw^R}{dr} = 0, \quad rV_r^R = r\bar{N}_{rz}^R + \alpha r\frac{d\mathcal{P}^R}{dr} = 0. \quad (8.3.44)$$

For annular plates, the boundary conditions at the inner edge (i.e., $r = b$) are given by the type of edge support there.

Next, we present an example to illustrate the derivation of the solutions of the TST using the relationships developed in the present section between CPT and TST.

Example 8.3.1

Consider a solid circular plate under uniformly distributed load of intensity q_0 and clamped at the edge $r = a$. Determined the analytical solution using the relationships derived.

Solution: First note that Eq. (8.3.36) can be expressed in the alternative form

$$\frac{d^2 N_{rz}^R}{dr^2} + \frac{1}{r}\frac{dN_{rz}^R}{dr} - \left(\frac{1}{r^2} + \xi_1\right)N_{rz}^R = -\xi_2\left(N_{rz}^C + \frac{c_1}{r}\right), \quad (8.3.45)$$

where

$$\xi_1 = \frac{D_{rr}\hat{A}_{rz}}{\alpha^2(D_{rr}H_{rr} - F_{rr}^2)}, \quad \xi_2 = \frac{\bar{D}_{rr}\bar{A}_{rz}}{\alpha^2(D_{rr}H_{rr} - F_{rr}^2)}. \tag{8.3.46}$$

The solution to the homogeneous differential equation,

$$\frac{d^2 N_{rz}^R}{dr^2} + \frac{1}{r}\frac{dN_{rz}^R}{dr} - \left(\frac{1}{r^2} + \xi_1\right)N_{rz}^R = 0, \tag{8.3.47}$$

is given by

$$N_{rz}^R(r) = c_5 I_1(\sqrt{\xi_1}\,r) + c_6 K_1(\sqrt{\xi_1}\,r), \tag{8.3.48}$$

where I_1 and K_1 are the first-order modified Bessel functions of the first and second kind, respectively.

Next, we use the boundary conditions to determine the constants of integration. For the clamped circular plate at hand, the boundary conditions at $r = a$ give $N_{rz}^R(a) = 0$ and $c_2 = 0$, and those at $r = 0$ give $N_{rz}^R(0) = 0$ and $c_1 = c_3 = 0$. Then the general solution to Eq. (8.3.45) is given by ($\xi_1 = \xi_2 \equiv \xi$)

$$N_{rz}^R(r) = c_5 I_1(\sqrt{\xi}\,r) + c_6 K_1(\sqrt{\xi}\,r) - \frac{q_0 r}{2}. \tag{8.3.49}$$

Using the boundary conditions on N_{rz}^R, we obtain

$$c_6 = 0, \quad c_5 = \frac{q_0 a}{2 I_1(\sqrt{\xi}a)}.$$

Hence the solution becomes

$$N_{rz}^R(r) = \frac{q_0 a}{2}\left[\frac{I_1(\sqrt{\xi}\,r)}{I_1(\sqrt{\xi}\,a)} - \frac{r}{a}\right], \tag{8.3.50}$$

and

$$\int N_{rz}^R\,dr = \left(\frac{q_0 a^2}{4}\right)\left[\frac{2 I_0(\sqrt{\xi}\,r)}{a I_1(\sqrt{\xi}\,a)\sqrt{\xi}} - \left(\frac{r}{a}\right)^2\right]. \tag{8.3.51}$$

Then the exact deflection of the TST is given by

$$w^R(r) = w_0^C(r) + \left(\frac{\bar{D}_{rr}}{D_{rr}\bar{A}_{rz}}\right)\left(\frac{q_0 a^2}{4}\right)\left[\frac{2 I_0(\sqrt{\xi}\,r)}{a I_1(\sqrt{\xi}\,a)\sqrt{\xi}} - \left(\frac{r}{a}\right)^2\right] - \frac{c_4}{D_{rr}}, \tag{8.3.52}$$

where the constant c_4 is evaluated using the boundary conditions $w_0^R = w_0^C = 0$ at $r = a$:

$$c_4 = \left(\frac{\bar{D}_{rr}}{\bar{A}_{rz}}\right)\left(\frac{q_0 a^2}{4}\right)\left[\frac{2 I_0(\sqrt{\xi}\,a)}{a I_1(\sqrt{\xi}\,a)\sqrt{\xi}} - 1\right].$$

The complete solution is given by

$$w^R(r) = w_0^C(r) + \left(\frac{\bar{D}_{rr}}{D_{rr}\bar{A}_{rz}}\right)\left(\frac{q_0 a^2}{4}\right)\left[\frac{2 I_0(\sqrt{\xi}\,r)}{a I_1(\sqrt{\xi}\,a)\sqrt{\xi}} - \left(\frac{r}{a}\right)^2\right]$$

$$- \left(\frac{\bar{D}_{rr}}{\bar{A}_{rz}D_{rr}}\right)\left(\frac{q_0 a^2}{4}\right)\left[\frac{2 I_0(\sqrt{\xi}\,a)}{a I_1(\sqrt{\xi}\,a)\sqrt{\xi}} - 1\right]. \tag{8.3.53}$$

Note that the deflection $w_0^C(r)$ of the CPT for the problem is given by Eq. (6.3.20):

$$w^C(r) = \frac{q_0 a^4}{64 D_{rr}} \left[1 - \left(\frac{r}{a}\right)^2 \right]^2.$$

(8.3.54)

The maximum deflection as per the TST is ($I_0(0) = 1$)

$$w_{max}^R = w^R(0) = \frac{q_0 a^4}{64 D_{rr}} + \frac{6 q_0 a^2}{20 A_{rz}}.$$

(8.3.55)

Comparing w_{max}^R with w_{max}^F (due to the FST) from Eq. (7.2.19), we note that for $K_s = 5/6$ the maximum deflection predicted by the the FST, with $K_s = 5/6$, coincides with that predicted by the TST. Of course, the third-order theory does not require a shear correction coefficient. Further, the comparison of the solutions of the FST and the TST for different boundary conditions and loads may lead to different shear correction factors.

8.3.3 BUCKLING RELATIONSHIPS

8.3.3.1 Governing equations

Here we develop buckling relationships for elastic, isotropic circular plate of radius a and uniform thickness h, and subjected to a uniform radial load N_0 (see Fig. 8.3.1). The governing equations of the CPT, the FST, and the TST for axisymmetric buckling of isotropic circular plates are summarized here (see [120] for additional discussion).

CPT:

$$\frac{d}{dr}\left(r N_{rz}^C\right) = r N_{rr}^C \nabla^2 w^R, \quad r N_{rz}^C \equiv \frac{d}{dr}\left(r M_{rr}^C\right) - M_{\theta\theta}^C,$$

(8.3.56)

$$M_{rr}^C = -D_{rr}\left(\frac{d^2 w^C}{dr^2} + \nu\frac{1}{r}\frac{dw^C}{dr}\right), \quad M_{\theta\theta}^C = -D_{rr}\left(\nu\frac{d^2 w^C}{dr^2} + \frac{1}{r}\frac{dw^C}{dr}\right).$$

(8.3.57)

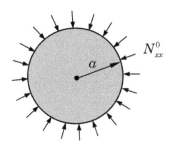

Figure 8.3.1: Circular plate under uniform compression.

FST:

$$\frac{d}{dr}\left(rN_{rz}^F\right) = rN_{rr}^F \nabla^2 w^F, \quad rN_{rz}^F = \frac{d}{dr}\left(rM_{rr}^F\right) - M_{\theta\theta}^F, \tag{8.3.58}$$

$$M_{rr}^F = D_{rr}\left(\frac{d\phi_r^F}{dr} + \nu\frac{1}{r}\phi_r^F\right), \quad M_{\theta\theta}^F = D_{rr}\left(\nu\frac{d\phi_r^F}{dr} + \frac{1}{r}\phi_r^F\right), \tag{8.3.59}$$

$$N_{rz}^F = S_{xz}\left(\phi_r^F + \frac{dw^F}{dr}\right). \tag{8.3.60}$$

TST:

$$\left(N_{rz}^R - \beta R_{rz}^R\right) = \frac{d}{dr}\left(rM_{rr}^R - \alpha r P_{rr}^R\right) - \left(M_{\theta\theta}^R - \alpha P_{\theta\theta}^R\right), \tag{8.3.61}$$

$$\frac{d}{dr}\left(rN_{rz}^R - \beta r R_{rz}^R\right) + \alpha\left[\frac{d^2}{dr^2}\left(rP_{rr}^R\right) - \frac{dP_{\theta\theta}^R}{dr}\right] = rN_{rr}^R \nabla^2 w^R, \tag{8.3.62}$$

$$M_{rr}^R = \frac{4D_{rr}}{5}\left(\frac{d\phi_r^R}{dr} + \frac{\nu}{r}\phi_r^R\right) - \frac{D_{rr}}{5}\left(\frac{d^2 w^R}{dr^2} + \frac{\nu}{r}\frac{dw^R}{dr}\right), \tag{8.3.63}$$

$$M_{\theta\theta}^R = \frac{4D_{rr}}{5}\left(\nu\frac{d\phi_r^R}{dr} + \frac{1}{r}\phi_r^R\right) - \frac{D_{rr}}{5}\left(\nu\frac{d^2 w^R}{dr^2} + \frac{1}{r}\frac{dw^R}{dr}\right), \tag{8.3.64}$$

$$P_{rr}^R = \frac{4h^2 D_{rr}}{35}\left(\frac{d\phi_r^R}{dr} + \frac{\nu}{r}\phi_r^R\right) - \frac{h^2 D_{rr}}{28}\left(\frac{d^2 w^R}{dr^2} + \frac{\nu}{r}\frac{dw^R}{dr}\right), \tag{8.3.65}$$

$$P_{\theta\theta}^R = \frac{4h^2 D_{rr}}{35}\left(\nu\frac{d\phi_r^R}{dr} + \frac{1}{r}\phi_r^R\right) - \frac{h^2 D_{rr}}{28}\left(\nu\frac{d^2 w^R}{dr^2} + \frac{1}{r}\frac{dw^R}{dr}\right), \tag{8.3.66}$$

$$N_{rz}^R = \frac{2Gh}{3}\left(\phi_r^R + \frac{dw^R}{dr}\right), \quad R_r^R = \frac{Gh^3}{30}\left(\phi_r^R + \frac{dw^R}{dr}\right), \tag{8.3.67}$$

where the Laplace operator ∇^2 in the polar coordinates is given by

$$\nabla^2 = \frac{d^2}{dr^2} + \frac{1}{r}\frac{d}{dr}. \tag{8.3.68}$$

8.3.3.2   Relationship between CPT and FST

Equations (8.3.56) and (8.3.57) of the CPT and Eqs. (8.3.58)–(8.3.60) of the FST can be reduced to

$$\frac{d^3\psi}{dr^3} + \frac{2}{r}\frac{d^2\psi}{dr^2} + \left(\lambda_0 - \frac{1}{r^2}\right)\frac{d\psi}{dr} + \frac{1}{r}\left(\lambda_0 + \frac{1}{r^2}\right)\psi = 0, \tag{8.3.69}$$

where

$$\psi = \begin{cases} -\frac{dw}{dr}, & \text{for CPT}, \\ \phi_r^F, & \text{for FST}, \end{cases} \quad \lambda_0 = \begin{cases} \frac{N_{rr}^C}{D_{rr}}, & \text{for CPT}, \\ \frac{N_{rr}^F}{1-\frac{N_{rr}^F}{S_{rz}}}, & \text{for FST}. \end{cases} \tag{8.3.70}$$

The function ψ is subject to the following boundary conditions at $r = a$:

$$\psi = 0, \quad \text{for clamped plates,} \tag{8.3.71a}$$

$$\frac{d\psi}{dr} + \frac{\nu}{r}\psi = 0, \quad \text{for simply supported plates,} \tag{8.3.71b}$$

$$\frac{d\psi}{dr} + \frac{\nu}{r}\psi = k_2\psi, \quad \text{for rotational elastic restraint,} \tag{8.3.71c}$$

and the following condition at $r = 0$:

$$\psi = 0, \quad \text{for all boundary conditions,} \tag{8.3.72}$$

where k_2 is the rotational spring constant.

In view of the similarity of the governing equations and boundary conditions, we obtain

$$N_{rr}^F = \frac{N_{rr}^C}{(1 + \frac{N_{rr}^C}{S_{rz}})}. \tag{8.3.73}$$

8.3.3.3 Relationship between CPT and TST

Introducing the higher order moment sum $\mathcal{P}^R$ as

$$\mathcal{P}^R = \frac{P_{rr}^R + P_{\theta\theta}^R}{1 + \nu} = \frac{4h^2 D_{rr}}{35} \frac{1}{r} \frac{d}{dr}(r\phi_r^R) - \frac{h^2 D_{rr}}{28} \nabla^2 w^R, \tag{8.3.74}$$

we can write Eq. (8.3.62) as

$$\frac{d}{dr}\left(rN_{rz}^R - r\frac{4}{h^2}R_r^R\right) + \frac{4}{3h^2}r\nabla^2\mathcal{P}^R = rN^R\nabla^2 w^R. \tag{8.3.75}$$

The substitution of Eq. (8.3.61) into Eq. (8.3.75) leads to

$$\mathcal{M}^R = N^R\nabla^2 w^R, \tag{8.3.76}$$

where the moment sum $\mathcal{M}^R$ is defined as

$$\mathcal{M}^R = \frac{M_{rr}^R + M_{\theta\theta}^R}{1 + \nu} = \frac{4D_{rr}}{5} \frac{1}{r} \frac{d}{dr}(r\phi_r^R) - \frac{D_{rr}}{5}\nabla^2 w^R. \tag{8.3.77}$$

Next, we substitute Eqs. (8.3.67) and (8.3.74) into Eq. (8.3.75) and obtain

$$\frac{8Gh}{15}\left[\frac{1}{r}\frac{d}{dr}(r\phi_r^R) + \nabla^2 w^R\right] + \frac{16D_{rr}}{105}\nabla^2\left[\frac{1}{r}\frac{d}{dr}(r\phi_r^R)\right] - \frac{D_{rr}}{21}\nabla^4 w^R = N_{rr}^R\nabla^2 w^R. \tag{8.3.78}$$

From Eq. (8.3.77), we have the relation

$$\nabla^2\left[\frac{1}{r}\frac{d}{dr}(r\phi_r^R)\right] = \frac{5}{4D_{rr}}\nabla^2\mathcal{M}^R + \frac{1}{4}\nabla^4 w^R. \tag{8.3.79}$$

In view of Eqs. (8.3.76), (8.3.77), and (8.3.79), we may express Eq. (8.3.78) as

$$\nabla^4 \mathcal{M}^R - \left(\frac{420(1-\nu)}{h^2} - \frac{85}{D_{rr}}N_{rr}^R\right)\nabla^2 \mathcal{M}^R - \left(\frac{420(1-\nu)}{h^2}N^R\right)\mathcal{M}^R = 0.$$

$$(8.3.80)$$

Equation (8.3.80) can be expressed in the form

$$(\nabla^2 + \lambda_1^R)(\nabla^2 + \lambda_2^R)\mathcal{M}^R = 0,$$

$$(8.3.81)$$

or

$$(\nabla^2 + \lambda_1^R)(\nabla^2 + \lambda_2^R)\nabla^2 w^R = 0,$$

$$(8.3.82)$$

where

$$\lambda_{1,2}^R = -\xi_1 \pm \sqrt{\xi_1^2 + \xi_2}\,,$$

$$(8.3.83)$$

$$\xi_1 = \frac{210(1-\nu)}{h^2} - \frac{85}{2D_{rr}}N_{rr}^R, \quad \xi_2 = \frac{410(1-\nu)}{h^2}N_{rr}^R.$$

$$(8.3.84)$$

The general solution to Eq. (8.3.83) is of the form

$$w^R(r) = c_1 + c_2 \ln r + c_3 J_0(\sqrt{\lambda_1^R}r) + c_4 K_0(\sqrt{\lambda_1^R}r)$$

$$+ c_5 J_0(\sqrt{\lambda_2^R}r) + c_6 K_0(\sqrt{\lambda_2^R}r),$$

$$(8.3.85)$$

where J_0 and K_0 are the Bessel functions, and c_i $(i = 1, 2, \ldots, 6)$ are constants to be determined using the boundary conditions. We have

At $r = a$:

$$w^R = \frac{dw^R}{dr} = \phi_r^R = 0 \quad \text{for clamped plates,}$$

$$(8.3.86a)$$

$$w^R = M_{rr}^R = P_{rr}^R = 0 \quad \text{for simply supported plates,}$$

$$(8.3.86b)$$

$$w^R = P_{rr}^R = 0, \quad M_{rr}^R = k_2\phi_r^R \quad \text{for rotational elastic restraint;}$$

$$(8.3.86c)$$

At $r = 0$:

$$\frac{dw^R}{dr} = 0 \quad \frac{d}{dr}\left(rM_{rr}^R - \alpha r P_{rr}^R\right) - \left(M_{\theta\theta}^R - \alpha P_{\theta\theta}^R\right) = 0$$

$$\text{for all boundary conditions,}$$

$$(8.3.86d)$$

where k_2 is the rotational (ie., torsional) spring constant. For example, boundary conditions for the clamped plate yield $c_2 = c_4 = c_6 = 0$ and

$$\begin{bmatrix} 1 & 1 & 1 \\ 0 & J_0'(\sqrt{\lambda_1^R}a) & J_0'(\sqrt{\lambda_2^R}a) \\ 0 & -(\lambda_1^R)^2 J_0'(\sqrt{\lambda_1^R}a) & -(\lambda_2^R)^2 J_0'(\sqrt{\lambda_2^R}a) \end{bmatrix} \begin{Bmatrix} c_1 \\ c_3 \\ c_5 \end{Bmatrix} = \begin{Bmatrix} 0 \\ 0 \\ 0 \end{Bmatrix}.$$

$$(8.3.87)$$

or

$$J_1(\sqrt{\lambda_1^R}a)\, J_1(\sqrt{\lambda_2^R}a) = 0. \qquad (8.3.88)$$

The same type of equation holds for the CPT with $\lambda_1^R = \lambda_2^R = \lambda^C$. Hence, by analogy, we have $(S_{rz} = K_s A_{rz} = K_s Gh)$

$$N_{rr}^R = \frac{N_{rr}^C \left(1 + \frac{N_{rr}^C}{70 A_{rz}}\right)}{1 + \frac{N_{rr}^C}{\frac{14}{17} A_{rz}}}. \qquad (8.3.89)$$

The relationship given in Eq. (8.3.89) is valid for circular plates with any homogeneous edge condition such as (a) simply supported edges, (b) clamped edges, (c) simply supported edges with elastic rotational restraints and (d) free edges with the center clamped. Cases (a) and (d) produce identical buckling solutions. Now, the CPT buckling solution for these plate cases may be unified and expressed as

$$\sqrt{\frac{N_{rr}^R a^2}{D_{rr}}} J_0 \left(\sqrt{\frac{N_{rr}^C a^2}{D_{rr}}}\right) + \left[\frac{k_2 a}{D_{rr}} - (1 - \nu)\right] J_1 \left(\sqrt{\frac{N_{rr}^C a^2}{D_{rr}}}\right) = 0, \qquad (8.3.90)$$

where $J_0(\cdot)$ and $J_1(\cdot)$ are Bessel functions of the first kind of order 0 and 1, respectively, and k_2 is the rotational spring stiffness with extreme values covering the two ideal edges of simply supported $(k_2 = 0)$ and clamped $(k_2 = \infty)$.

The book by Wang, Reddy, and Lee [120] contains many other results for bending, buckling, and natural frequency relationships for circular plates and sectorial plates. We close this section with a numerical example from [120], which brings out the difference between buckling loads predicted by various plate theories.

Example 8.3.2

Determine the buckling loads of circular plates as predicted by the various plate theories. Use the relationships developed in this section.

Solution: Using the relationships developed in this section, we can obtain the buckling loads. Table 8.3.1 presents the classical (CPT), first-order (FST), third-order (TST), and Ye's [149] buckling factors $\bar{N}_{rr}^0 = N_{rr}^0 a^2/D_{rr}$ for circular plates with various values of the thickness to radius ratio h/a, elastic rotational restraint parameter $k_2 a/D_{rr}$ and Poisson's ratio $\nu = 0.3$. Note that the buckling factor based on the CPT is independent of h/a due to the neglect of transverse shear deformation and the non-dimensionalization used.

Both the FST and TST results are very close to each other but are somewhat lower than the three-dimensional elasticity solution of Ye [149], who derived the buckling loads of circular plates from three-dimensional elasticity considerations. The analysis is based on a recursive formulation that results in the need to solve for only the roots of a 2×2 determinant for the buckling load.

Table 8.3.1: Comparison of buckling load factors for circular plates of radius a based on different plate theories.

$\frac{h}{a}$	$k_2 a / D_{rr}$	CPT	FST	TST	Ye [149]
0.05	0	4.1978	4.1853	4.1853	
	1	6.3532	6.3245	6.3245	
	10	12.173	12.068	12.068	
	∞	14.682	14.530	14.530	14.552
0.10	0	4.1978	4.1481	4.1481	
	1	6.3532	6.2399	6.2400	
	10	12.173	11.764	11.764	
	∞	14.682	14.091	14.091	14.177
0.20	0	4.1978	4.0056	4.0057	
	1	6.3532	5.9231	5.9235	
	10	12.173	10.686	10.688	
	∞	14.682	12.572	12.576	12.824
0.30	0	4.1978	3.7888	3.7893	
	1	6.3532	5.4610	5.4625	
	10	12.173	9.2710	9.2792	
	∞	14.682	10.658	10.671	11.024

8.4 CHAPTER SUMMARY

The Reddy third-order shear deformation plate theory (TST) of axisymmetric circular plates, which is an extension of the Reddy third-order beam theory, is developed accounting for material variation through the thickness, the von Kármán nonlinearity, and the effect of the modified couple stress. Exact solutions (by direct integration) of the theory for the linear case are presented. Bending relationships between the classical and third-order plate theories are also presented. The nonlinear solutions by the finite element method are presented in Chapter 10. For additional works on the applications of the Reddy plate theory to circular and annular plates can be found in [149]–[154].

SUGGESTED EXERCISES

8.1 Establish the relations in Eqs. (8.3.22)–(8.3.25).

8.2 Establish the relationships in Eqs. (8.3.38)–(8.3.40).

8.3 Establish the results in Eqs. (8.3.38)–(8.3.40).

When even the brightest mind in our world has been trained up from childhood in a superstition of any kind, it will never be possible for that mind, in its maturity, to examine sincerely, dispassionately, and conscientiously any evidence or any circumstance which shall seem to cast a doubt upon the validity of that superstition. I doubt if I could do it myself.   Mark Twain

9 Finite Element Analysis of Beams

To raise new questions, new possibilities, to regard old problems from a new angle, requires creative imagination and marks real advance in science.
Albert Einstein

9.1 INTRODUCTION

9.1.1 THE FINITE ELEMENT METHOD

For the sake of completeness and for the benefit of those readers who are not familiar with the finite element method (FEM), a brief introduction is presented here. The FEM is a numerical method of solving differential equations, much like the classical variational methods. The major drawback of the traditional variational method is the unique derivation of approximation functions used to represent the dependent unknown. In the FEM, the domain is represented as a collection of non-overlapping subdomains, called *finite elements*, that cover the total domain. By dividing the domain into finite elements, one can uniquely derive polynomial approximation functions for a typical finite element. The functions derived are used in the Ritz and Galerkin type methods to convert a given set of differential equations to a set of algebraic equations among the values of the duality variables at selected points of each element. Then the sets of algebraic equations derived on all finite elements are "assembled" to arrive at a system of algebraic equations over the entire domain. In one dimension, the domains and elements are lines of fixed length. The approximation functions are those which interpolate primary variable(s) of the formulation at selective points (including the end points), called *nodes*, of the element. The degree of the polynomial approximation functions depends on the number of nodes in an element.

The main steps in the finite element analysis of a problem are as follows (see Reddy [155] for further details).

1. The whole domain Ω [in the present case, the domain is a line, $\Omega = (0, L)$] is discretized into a collection, called a *mesh*, of inter-connected and non-overlapping N finite elements, $\Omega^e = (x_a^e, x_b^e)$, $e = 1, 2, \ldots, N$, where N is the number of elements in the mesh [see Fig. 9.1.1(a)]:

$$\Omega = \Omega^1 \cup \Omega^2 \cup \Omega^3 \cdots \cup \Omega^N = \cup_{e=1}^{N} \Omega^e, \quad \Omega^i \cap \Omega^j = \Gamma^{ij}, \qquad (9.1.1)$$

DOI: 10.1201/9781003240846-9

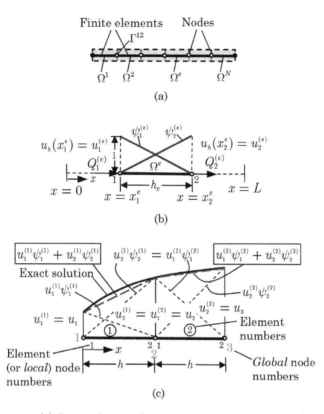

Figure 9.1.1: (a) Finite element discretization of a 1-D domain (mesh). (b) Finite element approximation of a variable u over a typical element (as an example, only linear interpolation functions are shown). (c) Assembly of two (linear) finite elements.

where $\cup$ denotes "union," $\cap$ denotes "intersection," and Γ^{ij} represents the set of points common to elements Ω^i and Ω^j. For example, when $\bar{\Omega} = [0, L]$ (including the end points, $x = 0$ and $x = L$), a typical element Ω^e is a line element of length h_e, $\bar{\Omega}^e = [x_a^e, x_b^e]$, where x_a^e is the coordinate of the left end of the line and x_b^e is the right end of the line with $h_e = x_b^e - x_a^e$. In this case, Γ^{ij} is the point common to element Ω^i and Ω^j (i.e., the right end point of element Ω^i is the same as the left end point of element Ω^j in the subdivision). The mesh shown has elements with two end nodes only but, in general, an element can have end nodes and some interior nodes for higher-order interpolation. The division of whole into parts allows the use of low degree of approximation of the field variables; otherwise, one may need higher degree of approximation if one were to seek an approximation of the variable on the whole domain, like in the Ritz and Galerkin methods. Since the domain cannot be divided into an infinite number parts, the size of the element is finite in size, hence the phrase *finite element*. The division of the whole into parts is called *mesh generation*.

2. Over each finite element Ω^e, we seek approximation of the primary variables in the form of algebraic polynomials (although other types of approximations are possible, we limit our choice here to algebraic polynomials). In order to define the geometry of the finite element (i.e., length and end points) as well as to express the unknown parameters of the polynomial approximations in terms of the values of the primary variables at selective locations of the element, we identify geometric points, called *nodes*. The resulting approximation of a typical variable $u(x)$ is of the form $u \approx u_h = \sum u_j \psi_j(x)$, where u_j is the value of u at the jth node and ψ_j are the (polynomial) approximation functions over a typical finite element Ω^e; see Fig. 9.1.1(b).

3. Over each representative finite element Ω^e, we develop integral statements equivalent to the governing differential equation(s). In structural mechanics, the integral statements, also called *variational statements*, are nothing but the statements of Hamilton's principle or the principle of virtual displacements. These statements are often called *weak forms*, because they involve lower order derivatives than the original differential equation.

4. Using the variational statement over an element Ω^e, we derive algebraic relations among the duality pairs (i.e., relations between pairs of dual variables like "displacements" and "forces") by substituting the assumed approximation for the variables (e.g., u_h for u) and obtain a set of algebraic equations, called the *finite element model*.

5. The algebraic relations of all elements are put together, called *assembly*, using continuity of the primary variables (e.g., displacements) and balance of the secondary variables (e.g., forces) of the duality pairs and obtain finite element equations of the governing equation over the whole domain (i.e., obtain the global finite element model); see Fig. 9.1.1(c).

6. Apply the known boundary conditions, which have to be converted to the known values of the duality pairs at the nodes, and solve the algebraic equations for the unknown nodal values of the primary and secondary variables.

7. Determine the values of the primary and secondary variables at points of interest using the interpolation of the primary variables and the relations between the secondary variables and the primary variables. This step is known as the *post-processing*.

9.1.2 INTERPOLATION FUNCTIONS

We recall that the weak form over an element is equivalent to the statement of the principle of virtual displacements or Hamilton's principle, which consists of the internal virtual work done and the work done by external forces at the end nodes, Q_a^e and Q_b^e, in moving through their respective virtual displacements, $\delta u(x_a^e)$ and $\delta u(x_b^e)$. Therefore, the approximate solution $u_h^e(x)$ should be selected such that the differentiability implied by the weak form is met and the end conditions on the primary variables $u(x_i^e) \equiv u_i^e$ are satisfied. For weak forms which contain only the first-order derivative of u_h^e, any function with a non-zero first derivative would be a candidate for u_h^e. Thus, the finite element approximation u_h^e of $u(x)$ can be an interpolant, that is, equal to u_a^e at x_a^e and u_b^e at x_b^e. Thus, a linear polynomial (see Fig. 9.1.2)

$$u_h^e(x) = c_1^e + c_2^e\, x \tag{9.1.2}$$

is admissible if one can select c_1^e and c_2^e such that

$$u_h^e(x_a^e) = c_1^e + c_2^e\, x_a^e = u_a^e, \qquad u_h^e(x_b^e) = c_1^e + c_2^e\, x_b^e = u_b^e, \tag{9.1.3}$$

which can be expressed in matrix form and can be solved for (u_a^e, u_b^e) in terms of (c_1^e, c_2^e) as

$$\begin{bmatrix} 1 & x_a^e \\ 1 & x_b^e \end{bmatrix} \begin{Bmatrix} c_1^e \\ c_2^e \end{Bmatrix} = \begin{Bmatrix} u_a^e \\ u_b^e \end{Bmatrix} \;\rightarrow\; c_1^e = \frac{u_a^e\, x_b^e - u_b^e\, x_a^e}{x_b^e - x_a^e}, \quad c_2^e = \frac{u_b^e - u_a^e}{x_b^e - x_a^e}. \tag{9.1.4}$$

Substitution of Eq. (9.1.4) for c_i^e into Eq. (9.1.2) yields

$$u_h^e(x) = \psi_1^e(x)\, u_1^e + \psi_2^e(x)\, u_2^e = \sum_{j=1}^{2} \psi_j^e(x)\, u_j^e, \tag{9.1.5}$$

where (noting that $h_e = x_b^e - x_a^e$)

$$\psi_1^e(x) = \frac{x_b^e - x}{h_e}, \qquad \psi_2^e(x) = \frac{x - x_a^e}{h_e}. \tag{9.1.6}$$

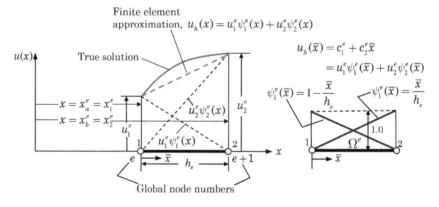

Figure 9.1.2: Linear interpolation functions for a finite element.

Here the functions (ψ_1^e, ψ_2^e) are called the *linear interpolation functions* and $u_1^e = u_a^e$ and $u_2^e = u_b^e$ are the nodal values of $u_h^e(x)$ at $x = x_a^e$ and $x = x_b^e$, respectively. An element with linear approximation is called a *linear element*.

The manner in which we derived the functions ψ_i^e $(i = 1, 2)$, they satisfy the *interpolation property*

$$\psi_i^e(x_j^e) = \delta_{ij} = \begin{cases} 1, & \text{if } i = j, \\ 0, & \text{if } i \neq j. \end{cases} \tag{9.1.7}$$

The functions that interpolate only a function (and not its derivatives) are known as the *Lagrange interpolation functions*. In addition, the Lagrange interpolation functions satisfy the property known as the *partition of unity*:

$$\sum_{j=1}^{2} \psi_j^e(x) = 1. \tag{9.1.8}$$

If we wish to approximate $u(x)$ with a quadratic polynomial, we write

$$u_h^e(x) = c_1^e + c_2^e x + c_3^e x^2, \tag{9.1.9}$$

where c_i^e $(i = 1, 2, 3)$ are constants defining a complete polynomial of degree (or order) 2. In order to express these constants in terms of the values of $u_h(x)$ at three nodes in the element, we must identify a third node in addition to the two end nodes. Identifying the third node at the center of the element, as indicated in Fig. 9.1.3(a), with nodal locations

$$x_1^e = x_a^e, \qquad x_2^e = x_a^e + \frac{h_e}{2}, \qquad x_3^e = x_a^e + h_e = x_b^e,$$

we obtain, following the same procedure as in the case of linear interpolation functions, by inverting the algebraic relations between c_i^e and u_i^e

$$u_h^e(x) = \psi_1^e(x)u_1^e + \psi_2^e(x)u_2^e + \psi_3^e(x)u_3^e = \sum_{j=1}^{3} \psi_j^e(x)u_j^e, \tag{9.1.10}$$

where $\psi_i^e(x)$ are the quadratic Lagrange interpolation functions [see Fig. 9.1.3(b)], which can be expressed in terms of the element (or local) coordinate $\bar{x} = x - x_a^e$ as

$$\psi_1^e(\bar{x}) = \left(1 - \frac{\bar{x}}{h_e}\right)\left(1 - \frac{2\bar{x}}{h_e}\right),$$

$$\psi_2^e(\bar{x}) = 4\frac{\bar{x}}{h_e}\left(1 - \frac{\bar{x}}{h_e}\right), \tag{9.1.11}$$

$$\psi_3^e(\bar{x}) = -\frac{\bar{x}}{h_e}\left(1 - \frac{2\bar{x}}{h_e}\right).$$

An element with quadratic approximation is called a *quadratic element*. The functions $\psi_i^e(\bar{x})$ can be expressed in terms of x by simply replacing $\bar{x}$ with $x - x_a^e$.

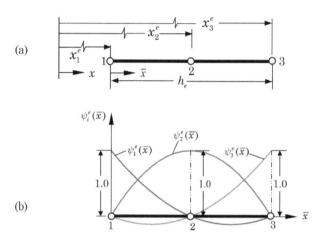

Figure 9.1.3: Quadratic (a) finite element and (b) interpolation functions.

When the weak form (or variational statement) involves second-order derivatives and the primary variable list contains the function and its derivative, as in the case of the Euler–Bernoulli beam theory, we must interpolate the field variable to have the function values and its derivatives as the nodal degrees of freedom. The resulting interpolation functions are known as the *Hermite interpolation functions*. The lowest order Hermite interpolation functions are necessarily cubic because a cubic polynomial has four parameters to replace four nodal values – two values of the function and two values of its derivative (i.e., the function and its derivative at each of the two nodes).

To fix the ideas, we begin with the cubic polynomial representation of the transverse deflection, $w_h(x)$, of an Euler–Bernoulli beam theory (for simplicity, we use the local coordinate $\bar{x}$):

$$w_h^e(\bar{x}) = c_1^e + c_2^e \bar{x} + c_3^e \bar{x}^2 + c_4^e \bar{x}^3 . \tag{9.1.12}$$

Note that the minimum continuity requirement (i.e., the existence of a nonzero second derivative of w_h^e in the element) is automatically met. In addition, the cubic approximation of w_h in the CBT allows computation of the shear force, which involves the third derivative of w_h^e. Next, we express c_i^e in terms of the primary nodal variables

$$\Delta_1^e \equiv w_h^e(0), \quad \Delta_2^e \equiv -\left.\frac{dw_h^e}{d\bar{x}}\right|_{\bar{x}=0}, \quad \Delta_3^e \equiv w_h^e(h_e), \quad \Delta_4^e \equiv -\left.\frac{dw_h^e}{d\bar{x}}\right|_{\bar{x}=h_e} . \tag{9.1.13}$$

We have

$$\Delta_1^e = w_h^e(0) \qquad = c_1^e, \quad \Delta_2^e = -\frac{dw_h}{dx}\bigg|_{x=x_a} = -\frac{dw_h^e}{d\bar{x}}\bigg|_{\bar{x}=0} = -c_2^e.$$

$$\Delta_3^e = w_h^e(h_e) = c_1^e + c_2^e h_e + c_3^e h_e^2 + c_4^e h_e^3, \tag{9.1.14}$$

$$\Delta_4^e = -\frac{dw_h^e}{dx}\bigg|_{x=x_b} = -\frac{dw_h^e}{d\bar{x}}\bigg|_{\bar{x}=h_e} = -c_2^e - 2c_3^e h_e - 3c_4^e h_e^2.$$

Solving (i.e., inverting) the above equations to express $(c_1^e, c_2^e, c_3^e, c_4^e)$ in terms of $(\Delta_1^e, \Delta_2^e, \Delta_3^e, \Delta_4^e)$ and substituting the result into Eq. (9.1.12), we obtain

$$w_h^e(\bar{x}) = \Delta_1^e \phi_1^e(\bar{x}) + \Delta_2^e \phi_2^e(\bar{x}) + \Delta_3^e \phi_3^e(\bar{x}) + \Delta_4^e \phi_4^e(\bar{x})$$

$$= \sum_{j=1}^{4} \Delta_j^e \phi_j^e(\bar{x}), \tag{9.1.15}$$

where

$$\phi_1^e(\bar{x}) = 1 - 3\left(\frac{\bar{x}}{h_e}\right)^2 + 2\left(\frac{\bar{x}}{h_e}\right)^3, \quad \phi_2^e(\bar{x}) = -\bar{x}\left(1 - \frac{\bar{x}}{h_e}\right)^2,$$

$$\phi_3^e(\bar{x}) = 3\left(\frac{\bar{x}}{h_e}\right)^2 - 2\left(\frac{\bar{x}}{h_e}\right)^3, \quad \phi_4^e(\bar{x}) = -\bar{x}\left[\left(\frac{\bar{x}}{h_e}\right)^2 - \frac{\bar{x}}{h_e}\right]. \tag{9.1.16}$$

The functions ϕ_i^e can be expressed in terms of x by simply replacing $\bar{x}$ with $x - x_a^e$. Plots of the Hermite cubic interpolation functions are shown in Fig. 9.1.4(a), and the finite element approximation of the deflection w_h using the Hermite cubic interpolation is shown in Fig. 9.1.4(b).

Recall that the Lagrange cubic interpolation functions are derived to interpolate a function, but not its derivatives, at the nodes. Hence, a Lagrange cubic element will have four nodes, with the dependent variable, not its derivative, as the nodal degree of freedom at each node. Since the derivative of w_h^e must be continuous between elements, as required by the weak form for the CBT, the Lagrange cubic approximation of w_h^e meets the continuity of w_h^e but not dw_h^e/dx, and therefore it is *not admissible* in the weak form of the CBT.

It should be noted that the order of the interpolation functions derived above is the minimum required for the weak form of the CBT. If a higher-order (i.e., higher than cubic) approximation of w_h is desired, one must either identify additional primary unknowns at each of the two nodes or add additional nodes with w_h or $(w_h, -dw_h/dx)$. For example, if we add a third node with $(w_h, -dw_h/dx)$ as the nodal degrees of freedom at each node, there will be a total of six conditions, and a fifth-order polynomial is required to interpolate the end conditions. A fourth degree polynomial with w_h and $-dw_h/dx$ at the end nodes and only w_h at the interior node may also be used.

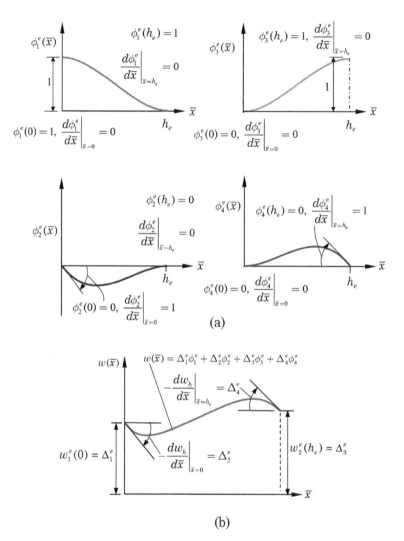

Figure 9.1.4: (a) Hermite cubic interpolations functions needed for the CBT. (b) Finite element approximation using the Hermite cubic polynomials.

9.1.3 PRESENT STUDY

In this chapter, the finite element models of the CBT, TBT, and RBT beams are presented. These are primarily displacement-based models (i.e., models based on the principle of virtual displacements) but where possible, mixed models that include the bending moment as a nodal variable along with the displacement variables are considered.

The typical steps in the development of finite element models involve: (1) setting up the variational (or weak form) statements of the governing equations, (2) assuming finite element approximation of the field variables, and (3) the derivation of the final algebraic equations among the nodal variables (called, the *finite element model*) using the Ritz method[1].

9.2 DISPLACEMENT MODEL OF THE CBT

9.2.1 GOVERNING EQUATIONS AND VARIATIONAL STATEMENTS

The conventional, displacement-based finite element model [156] of the CBT is constructed using the weak forms of Eqs. (3.4.9) and (3.4.10):

$$
-\frac{\partial}{\partial x}\left\{ A_{xx}\left[\frac{\partial u}{\partial x} + \frac{1}{2}\left(\frac{\partial w}{\partial x}\right)^2\right] - B_{xx}\frac{\partial^2 w}{\partial x^2} - N_{xx}^T \right\}
$$

$$
+ m_0\frac{\partial^2 u}{\partial t^2} - m_1\frac{\partial^3 w}{\partial t^2 \partial x} - f = 0, \tag{9.2.1}
$$

$$
-\frac{\partial^2}{\partial x^2}\left\{ B_{xx}\left[\frac{\partial u}{\partial x} + \frac{1}{2}\left(\frac{\partial w}{\partial x}\right)^2\right] - (D_{xx} + A_{xy})\frac{\partial^2 w}{\partial x^2} - M_{xx}^T \right\}
$$

$$
-\frac{\partial}{\partial x}\left\{ \frac{\partial w}{\partial x}A_{xx}\left[\frac{\partial u}{\partial x} + \frac{1}{2}\left(\frac{\partial w}{\partial x}\right)^2\right] - B_{xx}\frac{\partial w}{\partial x}\frac{\partial^2 w}{\partial x^2} - \frac{\partial w}{\partial x}N_{xx}^T \right\}
$$

$$
+ m_0\frac{\partial^2 w}{\partial t^2} + m_1\frac{\partial^3 u}{\partial t^2 \partial x} - m_2\frac{\partial^4 w}{\partial t^2 \partial x^2} + c_f w - q = 0. \tag{9.2.2}
$$

First, we divide the domain $\Omega = [0, L]$ of the beam into a set of finite elements, with a typical element being $\Omega = [x_a^e, x_b^e]$. We note that the cross-sectional geometry and material variations of the beam are accounted for in the beam stiffness coefficients A_{xx}, B_{xx}, and D_{xx} as well as the mass inertias m_0, m_1, and m_2. The variational statements or the weak forms of Eqs. (9.2.1) and (9.2.2) over Ω^e are [see the final form of the Hamilton principle in Eq. (3.3.15)]:

$$
0 = \int_{x_a^e}^{x_b^e}\left[\frac{\partial \delta u}{\partial x}\left\{ A_{xx}\left[\frac{\partial u}{\partial x} + \frac{1}{2}\left(\frac{\partial w}{\partial x}\right)^2\right] - B_{xx}\frac{\partial^2 w}{\partial x^2} - N_{xx}^T \right\} - f\,\delta u \right.
$$

$$
\left. + \delta u\left(m_0\frac{\partial^2 u}{\partial t^2} - m_1\frac{\partial^3 w}{\partial t^2 \partial x} \right)\right] dx - \delta u(x_a^e, t)Q_1^e - \delta u(x_b^e, t)Q_4^e,
$$

$$
\tag{9.2.3}
$$

---

[1]One may use other variational methods to derive the finite element equations. Thus, one can develop subdomain finite element model, least-squares finite element model, and so on. Each model, in general, will result in a different set of algebraic equations and hence the solutions.

$$0 = \int_{x_a^e}^{x_b^e} \left\{ D_{xx}^e \frac{\partial^2 \delta w}{\partial x^2} \frac{\partial^2 w}{\partial x^2} - B_{xx} \frac{\partial^2 \delta w}{\partial x^2} \left[\frac{\partial u}{\partial x} + \frac{1}{2} \left(\frac{\partial w}{\partial x} \right)^2 \right] + \frac{\partial^2 \delta w}{\partial x^2} M_{xx}^T \right. $$

$$+ A_{xx} \frac{\partial \delta w}{\partial x} \frac{\partial w}{\partial x} \left[\frac{\partial u}{\partial x} + \frac{1}{2} \left(\frac{\partial w}{\partial x} \right)^2 \right] - B_{xx} \frac{\partial \delta w}{\partial x} \frac{\partial w}{\partial x} \frac{\partial^2 w}{\partial x^2} - \frac{\partial \delta w}{\partial x} \frac{\partial w}{\partial x} N_{xx}^T $$

$$+ m_0 \delta w \frac{\partial^2 w}{\partial t^2} - m_1 \frac{\partial \delta w}{\partial x} \frac{\partial^2 u}{\partial t^2} + m_2 \frac{\partial \delta w}{\partial x} \frac{\partial^3 w}{\partial x \partial t^2} + c_f \delta w \, w - \delta w \, q \left. \right\} dx $$

$$- \delta w(x_a^e, t) Q_2 - \delta w(x_b^e, t) Q_5^e - \delta \theta_x(x_a^e, t) Q_3^e - \delta \theta_x(x_b^e, t) Q_6^e, \quad (9.2.4)$$

where (Q_1^e, Q_4^e) are the axial forces, (Q_2^e, Q_5^e) are the transverse forces, and (Q_3^e, Q_6^e) are the bending moments; the first entry in each pair is at node 1 and the second one is at node 2 of the element, as shown in Fig. 9.2.5(a):

$$Q_1^e = -N_{xx}(x_a^e, t), \quad Q_2^e = -V_x(x_a^e, t) \quad Q_3^e = -M_{xx}(x_a^e, t)$$
$$Q_4^e = N_{xx}(x_b^e, t), \quad Q_5^e = V_x(x_b^e, t), \quad Q_6^e = M_{xx}(x_b^e, t). \quad (9.2.5)$$

The associated generalized displacements are shown in Fig. 9.2.5(b).

In Eqs. (9.2.3) and (9.2.5) we have used the notation [see Eq. (3.3.19)]

$$D_{xx}^e = D_{xx} + A_{xy}, \quad V_x = \frac{\partial M_{xx}}{\partial x} + N_{xx} \frac{\partial w}{\partial x} + m_2 \frac{\partial^3 w}{\partial x \partial t^2} - m_1 \frac{\partial^2 u}{\partial t^2}. \quad (9.2.6)$$

Although the mass inertia terms appear in the effective shear force V_x, they do not present any problem. The definition in Eq. (9.2.6) is used to compute V_x in the post-computation.

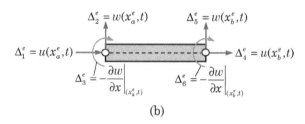

(a)

(b)

Figure 9.2.5: (a) Generalized nodal forces and (b) generalized displacements in the displacement model of the CBT.

9.2.2 FINITE ELEMENT MODEL

First, we note from the discussion of Section 3.3 and the variational statements in Eqs. (9.2.3) and (9.2.4) that: (a) the primary variables are $(u, w, -\partial w/\partial x)$ and the secondary variables are (N_{xx}, V_x, M_{xx}) and (b) the minimum continuity requirements on u is that it be differentiable once and w be differentiable twice, with continuous first derivative, $\partial w/\partial x$. Therefore, the finite element approximations must be such that the primary variables are represented as the nodal degrees of freedom so that they can be made continuous at the nodes common to elements during the assembly. This requires, at the minimum, a linear approximation of u and a quadratic approximation of w. However, a quadratic polynomial has only three parameters while there are four, two per node $(w, -\partial w/\partial x)$, degrees of freedom that need to be represented. This increases the degree of interpolation for w to be cubic, so that the four parameters of the polynomial can be expressed in terms of the four nodal degrees of freedom $(\Delta_2, \Delta_3, \Delta_5, \Delta_6)$, as already discussed previously [see Fig. 9.2.5(b)]. Thus, the linear approximation of u and Hermite cubic approximation of w are the minimum degree of approximations that are admissible.

Using the minimum degree of interpolation of u and w, we write

$$u(x,t) \approx u_h(x,t) = \sum_{j=1}^{2} \psi_j^e(x) u_j^e(t),$$

$$w(x,t) \approx w_h(x,t) = \sum_{J=1}^{4} \varphi_J^e(x) \bar{\Delta}_J^e(t),$$

$$(9.2.7)$$

where $\psi_i^e(x)$ are linear interpolation functions [see Eq. (9.1.6)], $\varphi_J^e(x)$ $(J = 1, 2, 3, 4)$ are the Hermite cubic polynomials [see Eq. (9.1.16)], u_j^e $(j = 1, 2)$ are the nodal values of u_h, and $\bar{\Delta}_J^e$ $(J = 1, 2, 3, 4)$ are the nodal values of w_h and $-(\partial w_h/\partial x)$ at the two nodes. We note that we have adopted an approximation in which the space and time are separated, with the nodal values being functions of time t only while the approximation functions being functions of x only (which is common in the finite element literature).

The substitution of Eq. (9.2.7) into the weak forms in Eqs. (9.2.3) and (9.2.4) yields the following nonlinear finite element model with the notation $\Delta_1^e = u_1^e$, $\Delta_4^e = u_2^e$, $\Delta_2^e = \bar{\Delta}_1^e$, $\Delta_3^e = \bar{\Delta}_2^e$, $\Delta_5^e = \bar{\Delta}_3^e$, and $\Delta_6^e = \bar{\Delta}_4^e$ (for brevity the element label e is omitted from the element coefficients):

$$\begin{bmatrix} \mathbf{M}^{11} & \mathbf{M}^{12} \\ \mathbf{M}^{21} & \mathbf{M}^{22} \end{bmatrix} \begin{Bmatrix} \ddot{\mathbf{u}}^e \\ \ddot{\boldsymbol{\Delta}}^e \end{Bmatrix} + \begin{bmatrix} \mathbf{K}^{11} & \mathbf{K}^{12} \\ \mathbf{K}^{21} & \mathbf{K}^{22} \end{bmatrix} \begin{Bmatrix} \mathbf{u}^e \\ \boldsymbol{\Delta}^e \end{Bmatrix} = \begin{Bmatrix} \mathbf{f}^e + \mathbf{F}^1 \\ \mathbf{q}^e + \mathbf{F}^2 \end{Bmatrix}, \qquad (9.2.8)$$

where $\mathbf{K}^{IJ}$, for example, denotes the coefficient of the Jth nodal variable in the Ith equation (there are two equations, namely, Eqs. (9.2.3) and (9.2.4), and two variables: u and w) and

$$M_{ij}^{11} = \int_{x_a^e}^{x_b^e} m_0 \psi_i \psi_j \, dx, \quad M_{iJ}^{12} = -\int_{x_a^e}^{x_b^e} m_1 \psi_i \frac{d\varphi_J}{dx} \, dx = M_{Ji}^{21},$$

$$M_{iJ}^{22} = \int_{x_a^e}^{x_b^e} \left(m_0 \varphi_I \varphi_J + m_2 \frac{d\varphi_I}{dx} \frac{d\varphi_J}{dx} \right) dx, \tag{9.2.9a}$$

$$K_{ij}^{11} = \int_{x_a^e}^{x_b^e} A_{xx} \frac{d\psi_i}{dx} \frac{d\psi_j}{dx} \, dx,$$

$$K_{iJ}^{12} = -\int_{x_a^e}^{x_b^e} B_{xx} \frac{d\psi_i}{dx} \frac{d^2\varphi_J}{dx^2} \, dx + \frac{1}{2} \int_{x_a^e}^{x_b^e} A_{xx} \frac{\partial w}{\partial x} \frac{d\psi_i}{dx} \frac{d\varphi_J}{dx} \, dx,$$

$$K_{Ij}^{21} = -\int_{x_a^e}^{x_b^e} B_{xx} \frac{d^2\varphi_I}{dx^2} \frac{d\psi_j}{dx} \, dx + \int_{x_a^e}^{x_b^e} A_{xx} \frac{\partial w}{\partial x} \frac{d\varphi_I}{dx} \frac{d\psi_j}{dx} \, dx, \tag{9.2.9b}$$

$$K_{IJ}^{22} = \int_{x_a^e}^{x_b^e} \left[D_{xx}^e \frac{d^2\varphi_I}{dx^2} \frac{d^2\varphi_J}{dx^2} + \frac{1}{2} A_{xx} \left(\frac{dw}{dx} \right)^2 \frac{d\varphi_I}{dx} \frac{d\varphi_J}{dx} \right.$$
$$\left. - B_{xx} \frac{\partial w}{\partial x} \left(\frac{1}{2} \frac{d^2\varphi_I}{dx^2} \frac{d\varphi_J}{dx} + \frac{d\varphi_I}{dx} \frac{d^2\varphi_J}{dx^2} \right) \right] dx,$$

$$f_i^e(t) = \int_{x_a^e}^{x_b^e} \left(\psi_i f(x,t) + \frac{d\psi_i}{dx} N_{xx}^T \right) dx, \quad F_i^1(t) = \psi_i(x_a^e) Q_1^e + \psi_i(x_b^e) Q_4^e,$$

$$q_i^e(t) = \int_{x_a^e}^{x_b^e} \left(\varphi_I q(x,t) - \frac{d^2\varphi_I}{dx^2} M_{xx}^T + \frac{d\varphi_I}{dx} \frac{\partial w}{\partial x} N_{xx}^T \right) dx, \tag{9.2.9c}$$

$$F_i^2(t) = \varphi_I(x_a^e) Q_2^e - \left. \frac{d\varphi_I}{dx} \right|_{x_a^e} Q_3^e + \varphi_I(x_b^e) Q_5^e - \left. \frac{d\varphi_I}{dx} \right|_{x_b^e} Q_6^e.$$

Clearly, the nonlinear stiffness matrix is not symmetric, $K_{ij}^{\alpha\beta} \neq K_{ji}^{\beta\alpha}$ while the linear stiffness matrix is symmetric. Also, we note that the coupling coefficients $\mathbf{K}^{21}$, $\mathbf{K}^{12}$, and $\mathbf{M}^{21} = \mathbf{M}^{12}$ are nonzero even when the nonlinearity is not accounted for because of the nonzero coefficients B_{xx} and m_1.

For the linear finite element model, the coefficients $K_{ij}^{\alpha\beta}$ and $M_{ij}^{\alpha\beta}$ are known explicitly, when ψ_i^e are linear functions, $i = 1, 2$; and φ_I^e are Hermite cubic functions, $I = 1, 2, 3, 4$) when the stiffness coefficients A_{xx}, B_{xx}, and D_{xx}, the mass inertia coefficients m_0, m_1, and m_2, and the load q are constant (i.e., independent of x), and the thermal effects are not included. For this case, we have

$$\mathbf{K}^{11} = \frac{A_{xx}^e}{h_e} \begin{bmatrix} 1 & -1 \\ -1 & 1 \end{bmatrix}, \qquad \mathbf{M}^{11} = \frac{m_0^e h_e}{6} \begin{bmatrix} 2 & 1 \\ 1 & 2 \end{bmatrix}, \tag{9.2.10a}$$

$$\mathbf{K}^{12} = \frac{B_{xx}^e}{h_e} \begin{bmatrix} 0 & 1 & 0 & -1 \\ 0 & -1 & 0 & 1 \end{bmatrix}, \quad \mathbf{M}^{12} = \frac{m_1^e}{12} \begin{bmatrix} 6 & h_e & -6 & h_e \\ 6 & -h_e & -6 & h_e \end{bmatrix}, \tag{9.2.10b}$$

$$
\mathbf{K}^{22} = \frac{2D^c_{xx}}{h^3_e}
\begin{bmatrix}
6 & -3h_e & -6 & -3h_e \\
-3h_e & 2h^2_e & 3h_e & h^2_e \\
-6 & 3h_e & 6 & 3h_e \\
-3h_e & h^2_e & 3h_e & 2h^2_e
\end{bmatrix}, \quad
\mathbf{q}^e = \frac{q^e_0 h_e}{12}
\begin{Bmatrix}
6 \\
-h_e \\
6 \\
h_e
\end{Bmatrix}, \quad (9.2.10\text{c})
$$

$$
\mathbf{M}^{22} = \frac{m_2 h_e}{420}
\begin{bmatrix}
156 & -22h_e & 54 & 13h_e \\
-22h_e & 4h^2_e & -13h_e & -3h^2_e \\
54 & -13h_e & 156 & 22h_e \\
13h_e & -3h^2_e & 22h_e & 4h^2_e
\end{bmatrix}. \quad (9.2.10\text{d})
$$

9.3   MIXED FINITE ELEMENT MODEL OF THE CBT

9.3.1   VARIATIONAL STATEMENTS

The mixed finite element model we consider here is one based on (u, w) and the bending moment $(\bar{M}_{xx} = M_{xx} + P_{xy})$. Equations (3.5.3)–(3.5.5) constitute the governing equations for the mixed model in terms of u, w, and $\bar{M}_{xx}$ (see Section 3.5 for details; note that the thermal resultants are not included here):

$$
-\frac{\partial}{\partial x}\left\{ \bar{A}_{xx}\left[\frac{\partial u}{\partial x} + \frac{1}{2}\left(\frac{\partial w}{\partial x}\right)^2\right] + \bar{B}_{xx}\,\bar{M}_{xx} \right\}
$$
$$
+ m_0\frac{\partial^2 u}{\partial t^2} - m_1\frac{\partial^3 w}{\partial t^2 \partial x} = f, \quad (9.3.1)
$$

$$
-\frac{\partial^2 \bar{M}_{xx}}{\partial x^2} - \frac{\partial}{\partial x}\left\{ \bar{A}_{xx}\frac{\partial w}{\partial x}\left[\frac{\partial u}{\partial x} + \frac{1}{2}\left(\frac{\partial w}{\partial x}\right)^2\right] + \bar{B}_{xx}\frac{\partial w}{\partial x}\,\bar{M}_{xx} \right\}
$$
$$
+ m_0\frac{\partial^2 w}{\partial t^2} + m_1\frac{\partial^3 u}{\partial t^2 \partial x} - m_2\frac{\partial^4 w}{\partial t^2 \partial x^2} + c_f w = q, \quad (9.3.2)
$$

$$
-\frac{\partial^2 w}{\partial x^2} - \frac{\bar{M}_{xx}}{D^e_{xx}} + \bar{B}_{xx}\left[\frac{\partial u}{\partial x} + \frac{1}{2}\left(\frac{\partial w}{\partial x}\right)^2\right] = 0. \quad (9.3.3)
$$

Here $\bar{A}_{xx}$ and $\bar{B}_{xx}$ are defined as in Eqs. (3.5.1c) and (3.5.2c) (recall that $D^e_{xx} = D_{xx} + A_{xy}$):

$$
\bar{A}_{xx} = \frac{D^{e*}_{xx}}{D^e_{xx}} = \frac{A_{xx}D^e_{xx} - B_{xx}B_{xx}}{D^e_{xx}}, \quad \bar{B}_{xx} = \frac{B_{xx}}{D^e_{xx}}. \quad (9.3.4)
$$

The weak forms of Eqs. (9.3.1)–(9.3.3) over a typical finite element $\Omega^e = [x^e_a, x^e_b]$ are:

$$
0 = \int_{x^e_a}^{x^e_b}\left\{ \bar{A}_{xx}\frac{\partial \delta u}{\partial x}\left[\frac{\partial u}{\partial x} + \frac{1}{2}\left(\frac{\partial w}{\partial x}\right)^2\right] + \bar{B}_{xx}\frac{\partial \delta u}{\partial x}\bar{M}_{xx} + m_0\delta u \ddot{u} \right.
$$
$$
\left. - m_1\delta u\frac{\partial \ddot{w}}{\partial x} - \delta u f \right\}dx - \delta u(x^e_a)Q^e_1 - \delta u(x^e_b)Q^e_4, \quad (9.3.5)
$$

$$0 = \int_{x_a^e}^{x_b^e} \left\{ \frac{\partial \delta w}{\partial x} \frac{\partial M_{xx}}{\partial x} + \bar{A}_{xx} \frac{\partial \delta w}{\partial x} \left[\frac{\partial u}{\partial x} + \frac{1}{2} \left(\frac{\partial w}{\partial x} \right)^2 \right] \frac{\partial w}{\partial x} + c_f \delta w w - \delta q \right.$$

$$+ \bar{B}_{xx} \frac{\partial w}{\partial x} \frac{\partial \delta w}{\partial x} M_{xx} + m_0 \delta w \ddot{w} - m_1 \frac{\partial \delta w}{\partial x} \ddot{u} + m_2 \frac{\partial \delta w}{\partial x} \frac{\partial \ddot{w}}{\partial x} \bigg\}$$

$$- \delta w(x_a^e) Q_2^e - \delta w(x_b^e) Q_5^e, \tag{9.3.6}$$

$$0 = \int_{x_a^e}^{x_b^e} \left\{ \frac{\partial \delta \bar{M}_{xx}}{\partial x} \frac{\partial w}{\partial x} - \frac{1}{\hat{D}_{xx}} \delta \bar{M}_{xx} M_{xx} + \bar{B}_{xx} \delta \bar{M}_{xx} \left[\frac{\partial u}{\partial x} + \frac{1}{2} \left(\frac{\partial w}{\partial x} \right)^2 \right] \right\} dx$$

$$- Q_3^e \delta \bar{M}_{xx}(x_a^e) + Q_6^e \delta \bar{M}_{xx}(x_b^e), \tag{9.3.7}$$

where double dot over the variables (e.g., $\ddot{u}$) signifies the second derivative with respect to time, and the generalized forces Q_i^e $(i = 1, 2, \cdots, 6)$ are defined by

$$Q_1^e = -N_{xx}(x_a^e, t), \quad Q_2^e = -\frac{\partial \bar{M}_{xx}}{\partial x}\bigg|_{x_a^e}, \quad Q_3^e = -\frac{\partial w}{\partial x}\bigg|_{x_a^e}, \tag{9.3.8a}$$

$$Q_4^e = N_{xx}(x_b^e, t), \quad Q_5^e = \frac{\partial \bar{M}_{xx}}{\partial x}\bigg|_{x_b^e}, \quad Q_6^e = -\frac{\partial w}{\partial x}\bigg|_{x_b^e}. \tag{9.3.8b}$$

Here, the sets (Q_1^e, Q_2^e, Q_3^e) and (Q_4^e, Q_5^e, Q_6^e) denote the axial force, transverse shear force, and rotation at nodes 1 and 2, respectively [see Fig. 9.3.1(a)].

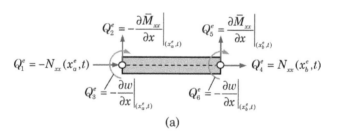

(a)

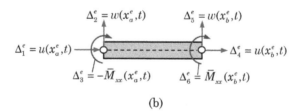

(b)

Figure 9.3.1: (a) Generalized nodal forces and (b) generalized displacements in the mixed model of the CBT.

9.3.2   FINITE ELEMENT MODEL

The finite element formulation based on the set of equations in Eqs. (9.3.1)–(9.3.2) is known as a *mixed* model because displacement variables (u, w) and a force variable $(\bar{M}_{xx})$ are used as the nodal values; the respective dual variables to $(u, w, \bar{M}_{xx})$ would be the axial force (N_{xx}), the shear force $(d\bar{M}_{xx}/dx)$, and the rotation $(\theta_x = -dw/dx)$, respectively.

The weak forms in Eqs. (9.3.5)–(9.3.7) show that Lagrange interpolations as low as linear for u, w, and $\bar{M}_{xx}$ are admissible. Let $(u, w, \bar{M}_{xx})$ be interpolated by

$$u(x) = \sum_{j=1}^{m} \psi_j^{(1)}(x) u_j^e(t), \quad w(x) = \sum_{j=1}^{n} \psi_j^{(2)}(x) w_j^e(t),$$

$$\bar{M}_{xx}(x) = \sum_{j=1}^{p} \psi_j^{(3)}(x), M_j^e(t), \tag{9.3.9}$$

where (u_j^e, w_j^e, M_j^e) are the nodal values of $(u, w, \bar{M}_{xx})$ and m is the number of nodes (or $m-1$ is the degree of the Lagrange polynomials) for the interpolation of u; similar meaning for n and p. Although unequal (i.e., $m \neq n \neq p$) interpolation is implied by Eq. (9.3.9), for practical reasons, we shall use equal interpolation (i.e., $m = n = p$). For linear interpolation of all three variables, we adopt the notation [see Fig. 9.3.1(b)]

$$\Delta_1^e = u_1^e, \quad \Delta_2^e = w_1^e, \quad \Delta_3^e = M_1^e, \quad \Delta_4^e = u_2^e, \quad \Delta_5^e = w_2^e, \quad \Delta_6^e = M_2^e. \tag{9.3.10}$$

Substitution of Eq. (9.3.9) into the weak forms in Eqs. (9.3.5)–(9.3.7), we obtain

$$\begin{bmatrix} \mathbf{M}^{11} & \mathbf{M}^{12} & \mathbf{0} \\ \mathbf{M}^{21} & \mathbf{M}^{22} & \mathbf{0} \\ \mathbf{0} & \mathbf{0} & \mathbf{0} \end{bmatrix} \begin{Bmatrix} \ddot{\mathbf{u}} \\ \ddot{\mathbf{w}} \\ \ddot{\mathbf{M}} \end{Bmatrix} + \begin{bmatrix} \mathbf{K}^{11} & \mathbf{K}^{12} & \mathbf{K}^{12} \\ \mathbf{K}^{21} & \mathbf{K}^{22} & \mathbf{K}^{23} \\ \mathbf{K}^{31} & \mathbf{K}^{32} & \mathbf{K}^{33} \end{bmatrix} \begin{Bmatrix} \mathbf{u} \\ \mathbf{w} \\ \mathbf{M} \end{Bmatrix} = \begin{Bmatrix} \mathbf{F}^1 \\ \mathbf{F}^2 \\ \mathbf{F}^3 \end{Bmatrix}, \tag{9.3.11}$$

where the superposed dot indicates the time derivative (i.e., $\dot{u} = du/dt$ and $\ddot{u} = d^2u/dt^2$) and

$$M_{ij}^{11} = \int_{x_a^e}^{x_b^e} m_0 \psi_i^{(1)} \psi_j^{(1)} \, dx, \quad M_{ij}^{12} = -\int_{x_a^e}^{x_b^e} m_1 \psi_i^{(1)} \frac{d\psi_j^{(2)}}{dx} \, dx = M_{ji}^{21},$$

$$M_{ij}^{22} = \int_{x_a^e}^{x_b^e} \left(m_0 \psi_i^{(2)} \psi_j^{(2)} + m_2 \frac{d\psi_i^{(2)}}{dx} \frac{d\psi_j^{(2)}}{dx} \right) dx,$$

$$K_{ij}^{11} = \int_{x_a^e}^{x_b^e} \bar{A}_{xx} \frac{d\psi_i^{(1)}}{dx} \frac{d\psi_j^{(1)}}{dx} \, dx, \quad K_{ij}^{12} = \frac{1}{2} \int_{x_a^e}^{x_b^e} \bar{A}_{xx} \frac{\partial w}{\partial x} \frac{d\psi_i^{(1)}}{dx} \frac{d\psi_j^{(2)}}{dx} \, dx,$$

$$K_{ij}^{13} = \int_{x_a^e}^{x_b^e} \bar{B}_{xx} \frac{d\psi_i^{(1)}}{dx} \psi_j^{(3)} \, dx, \quad K_{ij}^{21} = \int_{x_a^e}^{x_b^e} \bar{A}_{xx} \frac{\partial w}{\partial x} \frac{d\psi_i^{(2)}}{dx} \frac{d\psi_j^{(1)}}{dx} \, dx,$$

$$K_{ij}^{22} = \int_{x_a^e}^{x_b^e} \left[\frac{1}{2} \bar{A}_{xx} \left(\frac{\partial w}{\partial x} \right)^2 \frac{d\psi_i^{(2)}}{dx} \frac{d\psi_j^{(2)}}{dx} + c_f \psi_i^{(2)} \psi_j^{(2)} \right] dx,$$

$$K_{ij}^{23} = \int_{x_a^e}^{x_b^e} \left(\frac{d\psi_i^{(2)}}{dx} \frac{d\psi_j^{(3)}}{dx} + \bar{B}_{xx} \frac{\partial w}{\partial x} \frac{d\psi_i^{(2)}}{dx} \psi_j^{(3)} \right) dx,$$

$$K_{ij}^{31} = \int_{x_a^e}^{x_b^e} \bar{B}_{xx} \psi_i^{(3)} \frac{d\psi_j^{(1)}}{dx} dx, \quad K_{ij}^{33} = -\int_{x_a}^{x_b} \frac{1}{D_{xx}^e} \psi_i^{(3)} \psi_j^{(3)} dx,$$

$$K_{ij}^{32} = \int_{x_a^e}^{x_b^e} \left(\frac{d\psi_i^{(3)}}{dx} \frac{d\psi_j^{(2)}}{dx} + \frac{1}{2} \bar{B}_{xx} \frac{\partial w}{\partial x} \psi_i^{(3)} \frac{d\psi_j^{(2)}}{dx} \right) dx,$$

$$F_i^1 = \int_{x_a^e}^{x_b} \psi_i^{(1)} f(x,t)\, dx + \psi_i^{(1)}(x_a^e) Q_1^e + \psi_i^{(1)}(x_b^e) Q_4^e,$$

$$F_i^2 = \int_{x_a^e}^{x_b} \psi_i^{(2)} q(x,t)\, dx + \psi_i^{(2)}(x_a^e) Q_2^e + \psi_i^{(2)}(x_b^e) Q_4^e,$$

$$F_i^3 = \psi_i^{(3)}(x_a) Q_3^e - \psi_i^{(3)}(x_b^e) Q_6^e. \tag{9.3.12}$$

Clearly, the stiffness matrix is not symmetric, $K_{ij}^{\alpha\beta} \neq K_{ji}^{\beta\alpha}$.

9.4 DISPLACEMENT FINITE ELEMENT MODEL OF THE TBT

9.4.1 GOVERNING EQUATIONS AND VARIATIONAL STATEMENTS

The governing equations for the displacement model of the TBT, including the couple stress effect, are provided by Eqs. (4.4.3)–(4.4.5):

$$-\frac{\partial}{\partial x} \left\{ A_{xx} \left[\frac{\partial u}{\partial x} + \frac{1}{2} \left(\frac{\partial w}{\partial x} \right)^2 \right] + B_{xx} \frac{\partial \phi_x}{\partial x} - N_{xx}^T \right\}$$
$$+ m_0 \frac{\partial^2 u}{\partial t^2} + m_1 \frac{\partial^2 \phi_x}{\partial t^2} = f, \tag{9.4.1}$$

$$-\frac{\partial}{\partial x} \left[S_{xz} \left(\phi_x + \frac{\partial w}{\partial x} \right) \right] - \frac{1}{4} \frac{\partial^2}{\partial x^2} \left[A_{xy} \left(\frac{\partial \phi_x}{\partial x} - \frac{\partial^2 w}{\partial x^2} \right) \right]$$
$$-\frac{\partial}{\partial x} \left\{ A_{xx} \frac{\partial w}{\partial x} \left[\frac{\partial u}{\partial x} + \frac{1}{2} \left(\frac{\partial w}{\partial x} \right)^2 \right] + B_{xx} \frac{\partial w}{\partial x} \left(\frac{\partial \phi_x}{\partial x} \right) - \frac{\partial w}{\partial x} N_{xx}^T \right\}$$
$$-q + m_0 \frac{\partial^2 w}{\partial t^2} = 0, \tag{9.4.2}$$

$$-\frac{\partial}{\partial x} \left\{ B_{xx} \left[\frac{\partial u}{\partial x} + \frac{1}{2} \left(\frac{\partial w}{\partial x} \right)^2 \right] + D_{xx} \frac{\partial \phi_x}{\partial x} - M_{xx}^T \right\} + S_{xz} \left(\phi_x + \frac{\partial w}{\partial x} \right)$$
$$-\frac{1}{4} \frac{\partial}{\partial x} \left[A_{xy} \left(\frac{\partial \phi_x}{\partial x} - \frac{\partial^2 w}{\partial x^2} \right) \right] + m_1 \frac{\partial^2 u}{\partial t^2} + m_2 \frac{\partial^2 \phi_x}{\partial t^2} = 0. \tag{9.4.3}$$

The variational statements associated with these equations can directly be obtained from Eq. (4.3.16b), with the stress resultants $(N_{xx}, M_{xx}, N_{xz}, P_{xy})$ known in terms of the displacements (u, w, ϕ_x) through Eqs. (4.4.1a)–(4.4.1d). The statement in Eq. (4.3.16b) is equivalent to the following three variational statements when the thermal and couple stress effects are omitted:

$$0 = \int_{x_a^e}^{x_b^e} \left[\frac{\partial \delta u}{\partial x} \left\{ A_{xx} \left[\frac{\partial u}{\partial x} + \frac{1}{2} \left(\frac{\partial w}{\partial x} \right)^2 \right] + B_{xx} \frac{\partial \phi_x}{\partial x} \right\} + m_0 \delta u \frac{\partial^2 u}{\partial t^2} \right.$$

$$\left. + m_1 \delta u \frac{\partial^2 \phi_x}{\partial t^2} \right] dx - \delta u(x_a^e, t) Q_1^e - \delta u(x_b^e, t) Q_4^e, \tag{9.4.4}$$

$$0 = \int_{x_a^e}^{x_b^e} \left[m_0 \delta w \frac{\partial^2 w}{\partial t^2} + \frac{\partial \delta w}{\partial x} \frac{\partial w}{\partial x} \left\{ A_{xx} \left[\frac{\partial u}{\partial x} + \frac{1}{2} \left(\frac{\partial w}{\partial x} \right)^2 \right] + B_{xx} \frac{\partial \phi_x}{\partial x} \right\} - \delta w q \right.$$

$$\left. + S_{xz} \frac{\partial \delta w}{\partial x} \left(\phi_x + \frac{\partial w}{\partial x} \right) \right] dx - \delta w(x_a^e, t) Q_2^e - \delta w(x_b^e, t) Q_5^e, \tag{9.4.5}$$

$$0 = \int_{x_a^e}^{x_b^e} \left[\frac{\partial \delta \phi_x}{\partial x} \left\{ B_{xx} \left[\frac{\partial u}{\partial x} + \frac{1}{2} \left(\frac{\partial w}{\partial x} \right)^2 \right] + D_{xx} \frac{\partial \phi_x}{\partial x} \right\} + S_{xz} \delta \phi_x \left(\phi_x + \frac{\partial w}{\partial x} \right) \right.$$

$$\left. + m_0 \delta \phi_x \frac{\partial^2 \phi_x}{\partial t^2} + m_0 \delta \phi_x \frac{\partial^2 u}{\partial t^2} \right] dx - \delta \phi_x(x_a^e, t) Q_3 - \delta \phi_x(x_b^e, t) Q_6^e, \tag{9.4.6}$$

where Q_i^e are the generalized forces (i.e., the secondary variables) for the displacement formulation of the TBT [see Fig. 9.4.1(a)] are:

$$Q_1^e = -N_{xx}(x_a^e, t), \quad Q_2^e = \left[-Q_x - N_{xx} \frac{\partial w}{\partial x} \right]_{x_a^e}, \quad Q_3^e = -M_{xx}(x_a^e, t),$$

$$Q_4^e = N_{xx}(x_b^e, t), \quad Q_5^e = \left[Q_x + N_{xx} \frac{\partial w}{\partial x} \right]_{x_b}, \quad Q_6^e = M_{xx}(x_b^e, t). \tag{9.4.7}$$

We note that, as in the case of the CBT element, the shear forces (Q_2^e and Q_5^e) contain the component of the force $(dw/dx)N_{xx}$ due to the geometric nonlinearity. The associated generalized displacements (i.e., primary variables of the TBT) are the nodal values of (u, w, ϕ_x). In other words, the duality pairs are: (Δ_1^e, Q_1^e), (Δ_2^e, Q_2^e), (Δ_3^e, Q_3^e), (Δ_4^e, Q_4^e), (Δ_5^e, Q_5^e), and (Δ_6^e, Q_6^e).

9.4.2 THE FINITE ELEMENT MODEL

An examination of the variational statements in Eqs. (9.4.4)–(9.4.6) show that all generalized displacements (u, w, ϕ_x) admit Lagrange interpolations (because no derivatives of u, w, and ϕ_x are in the list of primary variables), with the minimum degree being linear. In general, we can seek different degree of Lagrange interpolation of u, w, and ϕ_x.

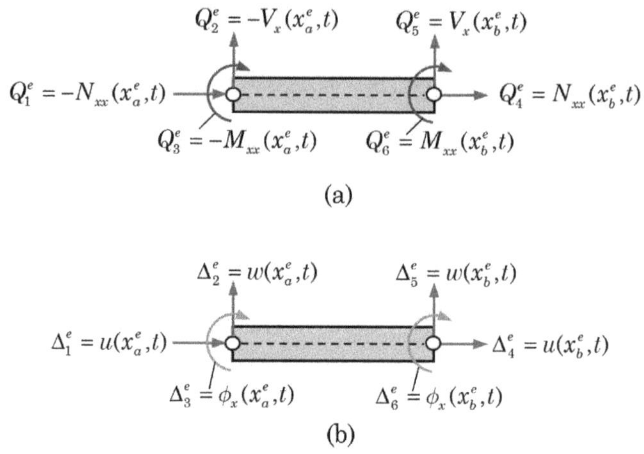

(a)

(b)

Figure 9.4.1: (a) Generalized nodal forces and (b) generalized displacements in the displacement model of the TBT.

We assume the following independent approximations (in theory) of all field variables (u, w, ϕ_x):

$$u(x) \approx \sum_{j=1}^{m} u_j(t)\psi_j^{(1)}(x), \quad w(x) \approx \sum_{j=1}^{n} w_j(t)\psi_j^{(2)}(x), \quad \phi_x \approx \sum_{j=1}^{p} S_j(t)\psi_j^{(3)}(x),$$

(9.4.8)

where $\psi_j^{(\alpha)}(x)$ $(\alpha = 1, 2, 3)$ are the Lagrange polynomials of different order used for the three variables. Substitution of Eq. (9.4.8) for (u, w, ϕ_x) and $\delta u = \psi_i^{(1)}$, $\delta w = \psi_i^{(2)}$, and $\delta \phi_x = \psi_i^{(3)}$ into the weak forms in Eqs. (9.4.4)–(9.4.6), we obtain the following finite element equations:

$$\begin{bmatrix} \mathbf{M}^{11} & \mathbf{0} & \mathbf{M}^{13} \\ \mathbf{0} & \mathbf{M}^{22} & \mathbf{0} \\ \mathbf{M}^{31} & \mathbf{0} & \mathbf{M}^{33} \end{bmatrix} \begin{Bmatrix} \ddot{\mathbf{u}} \\ \ddot{\mathbf{w}} \\ \ddot{\mathbf{s}} \end{Bmatrix} + \begin{bmatrix} \mathbf{K}^{11} & \mathbf{K}^{12} & \mathbf{K}^{13} \\ \mathbf{K}^{21} & \mathbf{K}^{22} & \mathbf{K}^{23} \\ \mathbf{K}^{31} & \mathbf{K}^{32} & \mathbf{K}^{33} \end{bmatrix} \begin{Bmatrix} \mathbf{u} \\ \bar{\mathbf{w}} \\ \mathbf{s} \end{Bmatrix} = \begin{Bmatrix} \mathbf{f}^e + \mathbf{F}^1 \\ \mathbf{q}^e + \mathbf{F}^2 \\ \mathbf{F}^3 \end{Bmatrix},$$

(9.4.9)

where

$$M_{ij}^{11} = \int_{x_a^e}^{x_b^e} m_0 \psi_i^{(1)} \psi_j^{(1)} \, dx, \quad M_{ij}^{13} = \int_{x_a^e}^{x_b^e} m_1 \psi_i^{(1)} \psi_j^{(3)} = M_{ji}^{31} \, dx,$$

$$M_{ij}^{22} = \int_{x_a^e}^{x_b^e} m_0 \psi_i^{(2)} \psi_j^{(2)} \, dx, \quad M_{ij}^{33} = \int_{x_a^e}^{x_b^e} m_2 \psi_i^{(3)} \psi_j^{(3)} \, dx,$$

$$K_{ij}^{11} = \int_{x_a^e}^{x_b^e} A_{xx} \frac{d\psi_i^{(1)}}{dx} \frac{d\psi_j^{(1)}}{dx} \, dx, \quad K_{ij}^{12} = \frac{1}{2} \int_{x_a}^{x_b} A_{xx} \frac{\partial w}{\partial x} \frac{d\psi_i^{(1)}}{dx} \frac{d\psi_j^{(2)}}{dx} \, dx,$$

$$K_{ij}^{13} = \int_{x_a^e}^{x_b^e} B_{xx} \frac{d\psi_i^{(1)}}{dx} \frac{d\psi_j^{(3)}}{dx} \, dx, \quad K_{ij}^{21} = \int_{x_a^e}^{x_b^e} A_{xx} \frac{\partial w}{\partial x} \frac{d\psi_i^{(2)}}{dx} \frac{d\psi_j^{(1)}}{dx} \, dx,$$

$$f_i^e(t) = \int_{x_a^e}^{x_b^e} f(x,t) \psi_i^{(1)} \, dx, \quad F_i^1(t) = \psi_i^{(1)}(x_a^e) Q_1^e + \psi_i^{(1)}(x_b^e) Q_4^e,$$

$$K_{ij}^{22} = \int_{x_a^e}^{x_b^e} \left[S_{xz} \frac{d\psi_i^{(2)}}{dx} \frac{d\psi_j^{(2)}}{dx} + \frac{1}{2} A_{xx} \left(\frac{\partial w}{\partial x} \right)^2 \frac{d\psi_i^{(2)}}{dx} \frac{d\psi_j^{(2)}}{dx} \right] dx,$$

$$K_{ij}^{23} = \int_{x_a^e}^{x_b^e} \left(S_{xz} \frac{d\psi^{(2)}}{dx} \psi_j^{(3)} + B_{xx} \frac{\partial w}{\partial x} \frac{d\psi_i^{(2)}}{dx} \frac{d\psi_j^{(3)}}{dx} \right) dx,$$

$$q_i^e(t) = \int_{x_a^e}^{x_b^e} q(x,t) \, \psi_i^{(2)} \, dx, \quad F_i^2(t) = \psi_i^{(2)}(x_a) Q_2^e + \psi_i^{(2)}(x_b) Q_5^e,$$

$$K_{ij}^{31} = \int_{x_a^e}^{x_b^e} B_{xx} \frac{d\psi_i^{(3)}}{dx} \frac{d\psi_j^{(1)}}{dx} \, dx,$$

$$K_{ij}^{32} = \int_{x_a^e}^{x_b^e} \left(S_{xz} \psi_i^{(3)} \frac{d\psi_j^{(2)}}{dx} + \frac{B_{xx}}{2} \frac{\partial w}{\partial x} \frac{d\psi_i^{(3)}}{dx} \frac{d\psi_j^{(2)}}{dx} \right) dx,$$

$$K_{ij}^{33} = \int_{x_a^e}^{x_b^e} \left(S_{xz} \psi_i^{(3)} \psi_j^{(3)} + D_{xx} \frac{d\psi_i^{(3)}}{dx} \frac{d\psi_j^{(3)}}{dx} \right) dx,$$

$$F_i^3(t) = \psi_i^{(3)}(x_a^e) Q_3 + \psi_i^{(3)}(x_b^e) Q_6. \tag{9.4.10}$$

9.5 MIXED FINITE ELEMENT MODEL OF THE TBT

9.5.1 GOVERNING EQUATIONS AND VARIATIONAL STATEMENTS

The governing equations for the mixed finite element model of the TBT are given by Eqs. (4.5.13)–(4.5.16) (see Section 4.5 for the details):

$$-\frac{\partial}{\partial x} \left\{ \bar{A}_{xx} \left[\frac{\partial u}{\partial x} + \frac{1}{2} \left(\frac{\partial w}{\partial x} \right)^2 \right] + \bar{B}_{xx} M_{xx} \right\}$$

$$+ m_0 \frac{\partial^2 u}{\partial t^2} + m_1 \frac{\partial^2 \phi_x}{\partial t^2} = f, \quad (9.5.1)$$

$$-\frac{\partial^2 M_{xx}}{\partial x^2} - \frac{\partial}{\partial x} \left\{ \bar{A}_{xx} \frac{\partial w}{\partial x} \left[\frac{\partial u}{\partial x} + \frac{1}{2} \left(\frac{\partial w}{\partial x} \right)^2 \right] + \bar{B}_{xx} \frac{\partial w}{\partial x} M_{xx} \right\}$$

$$+ c_f w + m_0 \frac{\partial^2 w}{\partial t^2} + m_1 \frac{\partial^3 u}{\partial t^2 \partial x} + m_2 \frac{\partial^3 \phi_x}{\partial t^2 \partial x} = q, \quad (9.5.2)$$

$$-\frac{\partial}{\partial x}\left[\frac{\partial w}{\partial x}-\frac{1}{S_{xz}}\left(\frac{\partial M_{xx}}{\partial x}-m_2\frac{\partial^2\phi_x}{\partial t^2}-m_1\frac{\partial^2 u}{\partial t^2}\right)\right]$$

$$+\bar{B}_{xx}\left[\frac{\partial u}{\partial x}+\frac{1}{2}\left(\frac{\partial w}{\partial x}\right)^2\right]-\frac{1}{D_{xx}}M_{xx}=0, \qquad (9.5.3)$$

where $\bar{A}_{xx}$ and $\bar{B}_{xx}$ are defined in (9.3.4).

The weak forms of Eqs. (9.5.1)–(9.5.3) over a typical finite element are:

$$0=\int_{x_a^e}^{x_b^e}\left\{\bar{A}_{xx}\frac{\partial\delta u}{\partial x}\left[\frac{\partial u}{\partial x}+\frac{1}{2}\left(\frac{\partial w}{\partial x}\right)^2\right]+\bar{B}_{xx}\frac{\partial\delta u}{\partial x}M_{xx}\right.$$

$$\left.+m_0\delta u\ddot{u}+m_1\delta u\,\ddot{\phi}_x-\delta u\,f(x,t)\right\}dx$$

$$-\delta u(x_a,t)\,Q_1^e-\delta u(x_b,t)\,Q_4^e, \qquad (9.5.4)$$

$$0=\int_{x_a^e}^{x_b^e}\left\{\frac{\partial\delta w}{\partial x}\frac{\partial M_{xx}}{\partial x}+\bar{A}_{xx}\frac{\partial\delta w}{\partial x}\left[\frac{\partial u}{\partial x}+\frac{1}{2}\left(\frac{\partial w}{\partial x}\right)^2\right]\frac{\partial w}{\partial x}+\bar{B}_{xx}\frac{\partial w}{\partial x}\frac{\partial\delta w}{\partial x}M_{xx}\right.$$

$$\left.+c_f\delta w\,w+m_0\delta w\ddot{w}-m_1\frac{\partial\delta w}{\partial x}\ddot{u}-m_2\frac{\partial\delta w}{\partial x}\ddot{\phi}_x-\delta w\,q(x,t)\right\}dx$$

$$-\delta w(x_a^e,t)\,Q_2^e-\delta w(x_b^e,t)\,Q_5^e, \qquad (9.5.5)$$

$$0=\int_{x_a^e}^{x_b^e}\left\{\frac{\partial\delta M_{xx}}{\partial x}\frac{\partial w}{\partial x}-\frac{1}{D_{xx}}\delta M_{xx}\,M_{xx}+\bar{B}_{xx}\,\delta M_{xx}\left[\frac{\partial u}{\partial x}+\frac{1}{2}\left(\frac{\partial w}{\partial x}\right)^2\right]\right\}dx$$

$$-\int_{x_a^e}^{x_b^e}\frac{1}{S_{xz}}\frac{\partial\delta M_{xx}}{\partial x}\left(\frac{\partial M_{xx}}{\partial x}-m_2\frac{\partial^2\phi_x}{\partial t^2}-m_1\frac{\partial^2 u}{\partial t^2}\right)dx$$

$$-Q_3^e\,\delta M_{xx}(x_a^e,t)-Q_6^e\,\delta M_{xx}(x_b^e,t). \qquad (9.5.6)$$

The secondary variables Q_i^e ($i=1,2,\ldots,6$) are defined by

$$Q_1^e=-N_{xx}(x_a^e,t),\quad Q_2^e=-\left.\frac{\partial M_{xx}}{\partial x}\right|_{x_a^e},\quad Q_3^e=\hat{\phi}_x(x_a^e), \qquad (9.5.7a)$$

$$Q_4^e=N_{xx}(x_b^e,t),\qquad Q_5^e=\left.\frac{\partial M_{xx}}{\partial x}\right|_{x_b^e},\qquad Q_6^e=\hat{\phi}_x(x_b^e), \qquad (9.5.7b)$$

where $\hat{\phi}_x$ is the effective rotation [see Eq. (4.5.16)]

$$\hat{\phi}_x=-\frac{\partial w}{\partial x}+\frac{1}{S_{xz}}\left(\frac{\partial M_{xx}}{\partial x}-m_2\frac{\partial^2\phi_x}{\partial t^2}-m_1\frac{\partial^2 u}{\partial t^2}\right). \qquad (9.5.7c)$$

The appearance of time-dependent terms in the effective rotation $\hat{\phi}_x$ presents problems in specifying boundary conditions on the slope. Furthermore, the time derivative of ϕ_x appears in the definition of the effective rotation. Therefore, the mixed model of the TBT is not recommended for time-dependent problems.

9.5.2   FINITE ELEMENT MODEL

The finite element approximations of (u, w, M_{xx}) used in the TBT are the same as those in Eqs. (9.3.9) (with $\bar{M}_{xx}$ replaced with M_{xx}). Substitution of these approximations into the weak forms in Eqs. (9.5.4)–(9.5.6), we obtain the finite element equations in the same form as in Eq. (9.3.11), and the coefficients M_{ij}^{IJ}, K_{ij}^{IJ}, and F_i^I (for $I, J = 1, 2$ and $i, j = 1, 2, \ldots, n$) also remain the same as those in Eqs. (9.3.12), except for the following stiffness coefficient (the mass coefficients will also change, and they are not listed here):

$$K_{ij}^{33} = -\int_{x_a^e}^{x_b^e} \left(\frac{1}{D_{xx}} \psi_i^{(3)} \psi_j^{(3)} + \frac{1}{S_{xz}} \frac{d\psi_i^{(3)}}{dx} \frac{d\psi_j^{(3)}}{dx} \right) dx. \tag{9.5.8}$$

9.6   DISPLACEMENT FINITE ELEMENT MODEL OF THE RBT

9.6.1   GOVERNING EQUATIONS

The governing equations of the Reddy–Bickford third-order beam theory (RBT) are [see Eqs. (5.3.25)–(5.3.27)]:

$$-\frac{\partial}{\partial x}\left\{ A_{xx}\left[\frac{\partial u}{\partial x} + \frac{1}{2}\left(\frac{\partial w}{\partial x}\right)^2\right] + \bar{B}_{xx}\frac{\partial \phi_x}{\partial x} - \alpha E_{xx}\frac{\partial^2 w}{\partial x^2} - N_{xx}^T \right\} - f$$
$$+ m_0 \frac{\partial^2 u}{\partial t^2} - \alpha\, m_3 \frac{\partial^3 w}{\partial t^2 \partial x} + \bar{m}_1 \frac{\partial^2 \phi_x}{\partial t^2} = 0, \tag{9.6.1}$$

$$-\frac{\partial}{\partial x}\left[\frac{\partial w}{\partial x}\left\{ A_{xx}\left[\frac{\partial u}{\partial x} + \frac{1}{2}\left(\frac{\partial w}{\partial x}\right)^2\right] + \bar{B}_{xx}\frac{\partial \phi_x}{\partial x} - \alpha E_{xx}\frac{\partial^2 w}{\partial x^2} - N_{xx}^T \right\}\right]$$
$$- \frac{1}{4}\frac{\partial^2}{\partial x^2}\left(\tilde{\bar{A}}_{xy}\frac{\partial \phi_x}{\partial x} - \tilde{A}_{xy}\frac{\partial^2 w}{\partial x^2} \right) - \left(\hat{A}_{xz} + \beta^2 D_{xy} \right)\frac{\partial}{\partial x}\left(\phi_x + \frac{\partial w}{\partial x} \right)$$
$$- \alpha \frac{\partial^2}{\partial x^2}\left\{ E_{xx}\left[\frac{\partial u}{\partial x} + \frac{1}{2}\left(\frac{\partial w}{\partial x}\right)^2\right] + \bar{F}_{xx}\frac{\partial \phi_x}{\partial x} - \alpha H_{xx}\frac{\partial^2 w}{\partial x^2} - P_{xx}^T \right\}$$
$$- q + c_f w + m_0 \frac{\partial^2 w}{\partial t^2} + \alpha \frac{\partial}{\partial x}\left(m_3 \frac{\partial^2 u}{\partial t^2} - \alpha\, m_6 \frac{\partial^3 w}{\partial x \partial t^2} + \bar{m}_4 \frac{\partial^2 \phi_x}{\partial t^2} \right) = 0, \tag{9.6.2}$$

$$-\frac{\partial}{\partial x}\left\{ \bar{B}_{xx}\left[\frac{\partial u}{\partial x} + \frac{1}{2}\left(\frac{\partial w}{\partial x}\right)^2\right] + \hat{D}_{xx}\frac{\partial \phi_x}{\partial x} - \alpha \bar{F}_{xx}\frac{\partial^2 w}{\partial x^2} - \bar{M}_{xx}^T \right\}$$
$$- \frac{1}{4}\frac{\partial}{\partial x}\left(\hat{A}_{xy}\frac{\partial \phi_x}{\partial x} - \tilde{\bar{A}}\frac{\partial^2 w}{\partial x^2} \right) + \left(\hat{A}_{xz} + \beta^2 D_{xy} \right)\left(\phi_x + \frac{\partial w}{\partial x} \right)$$
$$+ \bar{m}_1 \frac{\partial^2 u}{\partial t^2} - \alpha\, \bar{m}_4 \frac{\partial^3 w}{\partial x \partial t^2} + \hat{m}_2 \frac{\partial^2 \phi_x}{\partial t^2} = 0, \tag{9.6.3}$$

where $\alpha = 4/3h^2$, $\beta = 3\alpha$, and [see Eq. (5.3.22)]

$$
\begin{aligned}
&\bar{B}_{xx} = B_{xx} - \alpha\,E_{xx}, \quad \bar{D}_{xx} = D_{xx} - \alpha\,F_{xx}, \quad \bar{F}_{xx} = F_{xx} - \alpha\,H_{xx}, \\
&\bar{A}_{xz} = A_{xz} - \beta\,D_{xz}, \quad \bar{D}_{xz} = D_{xz} - \beta\,F_{xz}, \quad \bar{A}_{xy} = A_{xy} - \beta\,D_{xy}, \\
&\bar{D}_{xy} = D_{xy} - \beta\,F_{xy}, \quad \tilde{A}_{xy} = A_{xy} + \beta\,D_{xy}, \quad \tilde{D}_{xy} = D_{xy} + \beta\,F_{xy}, \\
&\hat{D}_{xx} = \bar{D}_{xx} - \alpha\,\bar{F}_{xx}, \quad \hat{A}_{xz} = \bar{A}_{xz} - \beta\,\bar{D}_{xz}, \\
&\tilde{\bar{A}}_{xx} = \tilde{A}_{xx} = A_{xy} - \beta^2 F_{xy}, \\
&\tilde{\bar{A}}_{xy} = \tilde{A}_{xy} + \beta\tilde{D}_{xy} = A_{xy} + 2\beta D_{xy} + \beta^2 F_{xy}.
\end{aligned}
\tag{9.6.4}
$$

9.6.2 WEAK FORMS

The weak forms associated with Eqs. (9.6.1)–(9.6.3) are

$$
\begin{aligned}
0 = \int_{x_a^e}^{x_b^e} \frac{\partial \delta u}{\partial x}&\left\{ A_{xx}\left[\frac{\partial u}{\partial x} + \frac{1}{2}\left(\frac{\partial w}{\partial x}\right)^2 \right] + \bar{B}_{xx}\frac{\partial \phi_x}{\partial x} - \alpha\,E_{xx}\frac{\partial^2 w}{\partial x^2} - N_{xx}^T \right\} dx \\
&+ \int_{x_a^e}^{x_b^e} \delta u \left[m_0 \frac{\partial^2 u}{\partial t^2} - \alpha\,m_3 \frac{\partial^3 w}{\partial t^2 \partial x} + \bar{m}_1 \frac{\partial^2 \phi_x}{\partial t^2} - f \right] dx \\
&- \delta u(x_a^e, t)\, Q_1^e(t) - \delta u(x_b^e, t)\, Q_5^e(t),
\end{aligned}
\tag{9.6.5}
$$

$$
\begin{aligned}
0 = \int_{x_a^e}^{x_b^e} \frac{\partial \delta w}{\partial x}\frac{\partial w}{\partial x}&\left\{ A_{xx}\left[\frac{\partial u}{\partial x} + \frac{1}{2}\left(\frac{\partial w}{\partial x}\right)^2 \right] + \bar{B}_{xx}\frac{\partial \phi_x}{\partial x} - \alpha\,E_{xx}\frac{\partial^2 w}{\partial x^2} - N_{xx}^T \right\} dx \\
&+ \int_{x_a^e}^{x_b^e} \left[-\frac{1}{4}\frac{\partial^2 \delta w}{\partial x^2}\left(\tilde{A}_{xy}\frac{\partial \phi_x}{\partial x} - \tilde{\bar{A}}_{xy}\frac{\partial^2 w}{\partial x^2} \right) + \hat{A}_{xz}\frac{\partial \delta w}{\partial x}\left(\phi_x + \frac{\partial w}{\partial x} \right) \right] dx \\
&- \alpha \int_{x_a^e}^{x_b^e} \frac{\partial^2 \delta w}{\partial x^2}\left\{ E_{xx}\left[\frac{\partial u}{\partial x} + \frac{1}{2}\left(\frac{\partial w}{\partial x}\right)^2 \right] + \bar{F}_{xx}\frac{\partial \phi_x}{\partial x} - \alpha\,H_{xx}\frac{\partial^2 w}{\partial x^2} \right\} dx \\
&+ \int_{x_a^e}^{x_b^e} \left[m_0 \delta w\, \frac{\partial^2 w}{\partial t^2} - \alpha\,\frac{\partial \delta w}{\partial x}\left(m_3 \frac{\partial^2 u}{\partial t^2} - \alpha\,m_6 \frac{\partial^3 w}{\partial x \partial t^2} + \bar{m}_4 \frac{\partial^2 \phi_x}{\partial t^2} \right) \right. \\
&\left. + c_f\,\delta w\,w - P_{xx}^T - \delta w\,q \right] dx - \delta w(x_a^e, t)\, Q_2^e(t) \\
&- \delta w(x_b^e, t)\, Q_6^e(t) - \delta \theta_x(x_a^e, t)\, Q_3^e(t) - \delta \theta_x(x_b^e, t)\, Q_7^e(t),
\end{aligned}
\tag{9.6.6}
$$

$$
\begin{aligned}
0 = \int_{x_a^e}^{x_b^e} \frac{\partial \delta \phi_x}{\partial x}&\left\{ \bar{B}_{xx}\left[\frac{\partial u}{\partial x} + \frac{1}{2}\left(\frac{\partial w}{\partial x}\right)^2 \right] + \hat{D}_{xx}\frac{\partial \phi_x}{\partial x} - \alpha\,\bar{F}_{xx}\frac{\partial^2 w}{\partial x^2} - \hat{M}_{xx}^T \right\} dx \\
&+ \int_{x_a^e}^{x_b^e} \frac{1}{4}\frac{\partial \delta \phi_x}{\partial x}\left(\hat{A}_{xy}\frac{\partial \phi_x}{\partial x} - \tilde{\bar{A}}_{xy}\frac{\partial^2 w}{\partial x^2} \right) + \hat{A}_{xz}\delta \phi_x\left(\phi_x + \frac{\partial w}{\partial x} \right) dx \\
&+ \int_{x_a^e}^{x_b^e} \delta \phi_x \left[\bar{m}_1 \frac{\partial^2 u}{\partial t^2} - \alpha\,\bar{m}_4 \frac{\partial^3 w}{\partial x \partial t^2} + \hat{m}_2 \frac{\partial^2 \phi_x}{\partial t^2} \right] dx \\
&- \delta \phi_x(x_a^e, t)\, Q_4^e(t) - \delta \phi_x(x_b^e, t)\, Q_8^e(t),
\end{aligned}
\tag{9.6.7}
$$

where $\theta_x = -\partial w/\partial x$, $\hat{A}_{xz} = \hat{A}_{xz} + \beta^2 D_{xy}$, and Q_i are the secondary variables at the nodes (four per node)

$$
\begin{aligned}
Q_1^e(t) &= -N_{xx}(x_a^e, t), & Q_5^e(t) &= N_{xx}(x_b^e, t), \\
Q_2^e(t) &= -V_{\text{eff}}(x_a^e, t), & Q_6^e(t) &= V_{\text{eff}}(x_b^e, t), \\
Q_3^e(t) &= -\tilde{M}_{xx}(x_a^e, t), & Q_7^e(t) &= \tilde{M}_{xx}(x_b^e, t), \\
Q_4^e(t) &= -M_{eff}(x_a^e, t), & Q_8^e(t) &= M_{eff}(x_b^e, t),
\end{aligned}
\tag{9.6.8a}
$$

and [see Eqs. (5.3.14a)–(5.3.14c)]

$$
\begin{aligned}
M_{\text{eff}} &= \bar{M}_{xx} + \tfrac{1}{2}\tilde{P}_{xy}, \quad \tilde{M}_{xx} = \alpha\,P_{xx} + \tfrac{1}{2}\bar{P}_{xy}, \\
V_{\text{eff}} &= \bar{N}_{xz} + N_{xx}\frac{\partial w}{\partial x} + \alpha\frac{\partial P_{xx}}{\partial x} + \tfrac{1}{2}\frac{\partial\tilde{P}_{xy}}{\partial x} \\
&\quad - \alpha\left(m_3\frac{\partial^2 u}{\partial t^2} - \alpha\, m_6\frac{\partial^3 w}{\partial x\partial t^2} + \bar{m}_4\frac{\partial^2\phi_x}{\partial t^2}\right), \\
\bar{M}_{xx} &= M_{xx} - \alpha\,P_{xx}, \quad \bar{N}_{xz} = N_{xz} - \beta P_{xz} - \beta Q_{yz}, \\
\tilde{P}_{xy} &= P_{xy} - \beta R_{xy}, \quad \bar{P}_{xy} = P_{xy} + \beta R_{xy}.
\end{aligned}
\tag{9.6.8b}
$$

9.6.3 FINITE ELEMENT MODEL

First, we note that the weak forms in Eqs. (9.6.5)–(9.6.7) have four duality pairs:

$$
(u, N_{xx}), \quad (w, V_{\text{eff}}), \quad \left(-\frac{\partial w}{\partial x}, \tilde{M}_{xx}\right), \quad (\phi_x, M_{\text{eff}}).
\tag{9.6.9}
$$

Since w and its derivative appear as primary variables, we must use Hermite family of approximation of w, while (u, ϕ_x) can be approximated using the Lagrange interpolation:

$$
u(x) \approx \sum_{j=1}^{m} u_j(t)\psi_j^{(1)}(x), \quad w(x) \approx \sum_{J=1}^{4} \bar{\Delta}_J(t)\varphi_J(x), \quad \phi_x \approx \sum_{j=1}^{n} S_j(t)\psi_j^{(2)}(x),
\tag{9.6.10}
$$

where $\psi_j^{(\alpha)}(x)$ ($\alpha = 1, 2$) are the Lagrange polynomials of different order used for (u, ϕ_x), φ_J are the Hermite cubic polynomials, and $\bar{\Delta}_J$ are the associated nodal values. Substitution of Eq. (9.6.10) for (u, w, ϕ_x) and $\delta u = \psi_i^{(1)}$, $\delta w = \varphi_I$, and $\delta\phi_x = \psi_i^{(2)}$ into the weak forms in Eqs. (9.6.5)–(9.6.7), we obtain the following finite element equations:

$$
\begin{bmatrix}
\mathbf{M}^{11} & \mathbf{M}^{12} & \mathbf{M}^{13} \\
\mathbf{M}^{21} & \mathbf{M}^{22} & \mathbf{M}^{23} \\
\mathbf{M}^{31} & \mathbf{M}^{32} & \mathbf{M}^{33}
\end{bmatrix}
\begin{Bmatrix} \ddot{\mathbf{u}} \\ \ddot{\bar{\Delta}} \\ \ddot{\mathbf{S}} \end{Bmatrix}
+
\begin{bmatrix}
\mathbf{K}^{11} & \mathbf{K}^{12} & \mathbf{K}^{13} \\
\mathbf{K}^{21} & \mathbf{K}^{22} & \mathbf{K}^{23} \\
\mathbf{K}^{31} & \mathbf{K}^{32} & \mathbf{K}^{33}
\end{bmatrix}
\begin{Bmatrix} \mathbf{u} \\ \bar{\Delta} \\ \mathbf{S} \end{Bmatrix}
=
\begin{Bmatrix} \mathbf{F}^1 \\ \mathbf{F}^2 \\ \mathbf{F}^3 \end{Bmatrix},
\tag{9.6.11}
$$

where $(M_{ij}^{IJ} = M_{ji}^{JI}$ and $\hat{\bar{A}}_{xz} = \hat{A}_{xz} + \beta^2 D_{xy})$

$$M_{ij}^{11} = \int_{x_a}^{x_b} m_0 \, \psi_i^{(1)} \psi_j^{(1)} \, dx, \quad M_{iJ}^{12} = -\alpha \int_{x_a}^{x_b} m_3 \psi_i^{(1)} \frac{d\varphi_J}{dx} \, dx,$$

$$M_{ij}^{13} = \int_{x_a}^{x_b^e} \bar{m}_1 \psi_i^{(1)} \psi_j^{(2)} \, dx, \quad M_{IJ}^{22} = \int_{x_a}^{x_b} \left(m_0 \varphi_I \varphi_J + \alpha^2 m_6 \frac{d\varphi_I}{dx} \frac{d\varphi_J}{dx} \right) dx,$$

$$M_{Ij}^{23} = -\alpha \int_{x_a}^{x_b^e} \bar{m}_4 \frac{d\varphi_I}{dx} \psi_j^{(2)} \, dx, \quad M_{ij}^{33} = \int_{x_a}^{x_b} \hat{m}_2 \psi_i^{(2)} \psi_j^{(2)} \, dx,$$

$$K_{ij}^{11} = \int_{x_a}^{x_b^e} A_{xx} \frac{d\psi_i^{(1)}}{dx} \frac{d\psi_j^{(1)}}{dx} \, dx, \quad K_{ij}^{13} = \int_{x_a}^{x_b^e} \bar{B}_{xx} \frac{d\psi_i^{(1)}}{dx} \frac{d\psi_j^{(2)}}{dx} \, dx = K_{ji}^{31},$$

$$K_{iJ}^{12} = \int_{x_a}^{x_b} \frac{d\psi_i^{(1)}}{dx} \left(\frac{1}{2} A_{xx} \frac{\partial w}{\partial x} \frac{d\varphi_J}{dx} - \alpha E_{xx} \frac{d^2 \varphi_J}{dx^2} \right) dx,$$

$$K_{Ij}^{21} = \int_{x_a}^{x_b^e} \left(A_{xx} \frac{\partial w}{\partial x} \frac{d\varphi_I}{dx} - \alpha E_{xx} \frac{d^2 \varphi_I}{dx^2} \right) \frac{d\psi_j^{(1)}}{dx} \, dx,$$

$$F_i^1 = \int_{x_a}^{x_b^e} \left(f \psi_i^{(1)} + \frac{d\psi_i^{(1)}}{dx} N_{xx}^T \right) dx + \psi_i^{(1)}(x_a^e) Q_1^e + \psi_i^{(1)}(x_b^e) Q_5^e,$$

$$K_{IJ}^{22} = \int_{x_a}^{x_b} \left[\hat{\bar{A}}_{xz} \frac{d\varphi_I}{dx} \frac{d\varphi_J}{dx} + c_f \varphi_I \varphi_J - \alpha E_{xx} \frac{\partial w}{\partial x} \left(\frac{1}{2} \frac{d^2 \varphi_I}{dx^2} \frac{d\varphi_J}{dx} + \frac{d\varphi_I}{dx} \frac{d^2 \varphi_J}{dx^2} \right) \right.$$
$$\left. + \left(\frac{1}{4} \tilde{\bar{A}}_{xy} + \alpha^2 H_{xx} \right) \frac{d^2 \varphi_I}{dx^2} \frac{d^2 \varphi_J}{dx^2} + \frac{1}{2} A_{xx} \left(\frac{\partial w}{\partial x} \right)^2 \frac{d\varphi_I}{dx} \frac{d\varphi_J}{dx} \right] dx,$$

$$K_{Ij}^{23} = \int_{x_a}^{x_b^e} \left[\hat{\bar{A}}_{xz} \frac{d\varphi_I}{dx} \psi_j^{(2)} - \left(\frac{1}{4} \tilde{\bar{A}}_{xy} + \alpha \bar{F}_{xx} \right) \frac{d^2 \varphi_I}{dx^2} \frac{d\psi_j^{(2)}}{dx} \right.$$
$$\left. + \bar{B}_{xx} \frac{\partial w}{\partial x} \frac{d\varphi_I}{dx} \frac{d\psi_j^{(2)}}{dx} \right] dx,$$

$$F_I^2 = \int_{x_a}^{x_b^e} \left(q \, \varphi_I + \frac{d\varphi_I}{dx} N_{xx}^T - \alpha P_{xx}^T \frac{d^2 \varphi_I}{dx^2} \right) dx$$
$$+ \varphi_I(x_a) Q_2^e + \varphi_I(x_b) Q_6^e - \frac{d\varphi_I}{dx}\bigg|_{x_a^e} Q_3^e - \frac{d\varphi_I}{dx}\bigg|_{x_b^e} Q_7^e,$$

$$K_{iJ}^{32} = \int_{x_a}^{x_b^e} \left[-\left(\frac{1}{4} \tilde{\bar{A}}_{xy} + \alpha \bar{F}_{xx} \right) \frac{d\psi_i^{(2)}}{dx} \frac{d^2 \varphi_J}{dx^2} + \frac{1}{2} \bar{B}_{xx} \frac{\partial w}{\partial x} \frac{d\psi_i^{(2)}}{dx} \frac{d\varphi_J}{dx} \right.$$
$$\left. + \left(\hat{A}_{xz} + \beta^2 D_{xy} \right) \psi_i^{(2)} \frac{d\varphi_J}{dx} \right] dx,$$

$$K_{ij}^{33} = \int_{x_a}^{x_b^e} \left[\left(\hat{D}_{xx} + \frac{1}{4} \hat{A}_{xy} \right) \frac{d\psi_i^{(2)}}{dx} \frac{d\psi_j^{(2)}}{dx} + \left(\hat{A}_{xz} + \beta^2 D_{xy} \right) \psi_i^{(2)} \psi_j^{(2)} \right] dx,$$

$$F_i^3 = \int_{x_a}^{x_b^e} \frac{d\psi_i^{(2)}}{dx} \bar{M}_{xx}^T \, dx + \psi_i^{(2)}(x_a^e) Q_4^e + \psi_i^{(2)}(x_b^e) Q_8^e. \tag{9.6.12}$$

9.7   TIME APPROXIMATION (FULL DISCRETIZATION)

9.7.1   INTRODUCTION

The various finite element equations, $\mathbf{M}\ddot{\boldsymbol{\Delta}} + \mathbf{K}\boldsymbol{\Delta} = \mathbf{F}$, presented in the previous sections are termed as the semidiscretized finite element models because they are ordinary differential equations in time, expressed in terms of the vector of nodal values $\boldsymbol{\Delta}(t)$. These ordinary differential equations in time need to be converted to algebraic relations using a time-approximation scheme of choice. In this section, we use the Newmark scheme (see Reddy [155, 156] for additional information) to convert the semidiscrete finite element model to the fully discretized model.

The semidiscretized finite element models presented in Eqs. (9.2.8), (9.3.11), (9.4.9), and (9.6.11) can be expressed in the following general form:

$$\mathbf{M}\ddot{\boldsymbol{\Delta}} + \mathbf{K}\boldsymbol{\Delta} = \mathbf{F}, \tag{9.7.1}$$

where $\mathbf{M}$ is the mass matrix, $\mathbf{K}$ is the stiffness matrix, $\mathbf{F}$ is the force vector, and $\boldsymbol{\Delta}$ is the column vector of nodal values (which differs from model to model).

9.7.2   NEWMARK'S METHOD

There are several numerical integration methods available to integrate second-order equations in time [157]–[160] (also see references therein). Among these, the Newmark family of time integration schemes [159] is widely used in structural dynamics. Other methods, such as the Wilson method and the Houbolt method [158], can be used to develop the algebraic equations from the second-order differential equations.

In the Newmark method, $\boldsymbol{\Delta}$ and its time derivative are approximated as

$$\boldsymbol{\Delta}_{s+1} = \boldsymbol{\Delta}_s + \delta t\,\dot{\boldsymbol{\Delta}}_s + \tfrac{1}{2}(\delta t)^2 \left[(1-\gamma)\ddot{\boldsymbol{\Delta}}_s + \gamma\,\ddot{\boldsymbol{\Delta}}_{s+1}\right], \tag{9.7.2}$$

$$\dot{\boldsymbol{\Delta}}_{s+1} = \dot{\boldsymbol{\Delta}}_s + a_2\ddot{\boldsymbol{\Delta}}_s + a_1\ddot{\boldsymbol{\Delta}}_{s+1}, \tag{9.7.3}$$

where α and γ are parameters that determine the numerical stability and accuracy of the scheme, δt is the time step size, and $a_1 = \alpha\,\delta t$ and $a_2 = \delta t\,(1-\alpha)$; $\boldsymbol{\Delta}_s$ denotes the variable $\boldsymbol{\Delta}(t)$ evaluated at time $t_s = s\,\delta t$, s being the time step number. For $\alpha = 0.5$, the following values of γ define various well-known schemes as special cases of Eqs. (9.7.2) and (9.7.3):

$$\gamma = \begin{cases} \tfrac{1}{2}, & \text{the constant-average acceleration method (stable)}, \\ \tfrac{1}{3}, & \text{the linear acceleration method (conditionally stable)}, \\ 0, & \text{the central difference method (conditionally stable)}, \\ \tfrac{8}{5}, & \text{the Galerkin method (stable)}, \\ 2, & \text{the backward difference method (stable)}. \end{cases} \tag{9.7.4}$$

9.7.3 FULLY DISCRETIZED EQUATIONS

The set of ordinary differential equations in Eq. (9.7.1) can be reduced, with the help of Eqs. (9.7.2) and (9.7.3), to a set of algebraic equations relating $\mathbf{\Delta}_{s+1}$ to $\mathbf{\Delta}_s$ (i.e., a time-marching scheme). The procedure is explained next.

Solving Eq. (9.7.2) for $\ddot{\mathbf{\Delta}}_{s+1}$, we obtain

$$
\begin{aligned}
\ddot{\mathbf{\Delta}}_{s+1} &= \frac{2}{\gamma(\delta t)^2}\left[\mathbf{\Delta}_{s+1} - \mathbf{\Delta}_s - \delta t \dot{\mathbf{\Delta}}_s - \frac{(\delta t)^2}{2}(1-\gamma)\ddot{\mathbf{\Delta}}_s\right] \\
&= a_3\left(\mathbf{\Delta}_{s+1} - \mathbf{\Delta}_s\right) - a_4\dot{\mathbf{\Delta}}_s - a_5\ddot{\mathbf{\Delta}}_s,
\end{aligned}
\tag{9.7.5}
$$

where

$$
a_3 = \frac{2}{\gamma(\delta t)^2}, \quad a_4 = \delta t\, a_3, \quad a_5 = \frac{(\delta t)^2}{2}(1-\gamma)a_3 = \frac{1-\gamma}{\gamma}.
\tag{9.7.6}
$$

Substituting the result into Eq. (9.7.3), we obtain

$$
\begin{aligned}
\dot{\mathbf{\Delta}}_{s+1} &= \dot{\mathbf{\Delta}}_s + a_2\ddot{\mathbf{\Delta}}_s + a_1\left[a_3\left(\mathbf{\Delta}_{s+1} - \mathbf{\Delta}_s\right) - a_4\dot{\mathbf{\Delta}}_s - a_5\ddot{\mathbf{\Delta}}_s\right] \\
&= a_6\left(\mathbf{\Delta}_{s+1} - \mathbf{\Delta}_s\right) - a_7\dot{\mathbf{\Delta}}_s - a_8\ddot{\mathbf{\Delta}}_s,
\end{aligned}
\tag{9.7.7}
$$

where

$$
a_6 = \frac{2\alpha}{\gamma\delta t}, \quad a_7 = \frac{2\alpha}{\gamma} - 1, \quad a_8 = \left(\frac{\alpha}{\gamma} - 1\right)\delta t.
\tag{9.7.8}
$$

Premultiplying Eq. (9.7.5) with $\mathbf{M}_{s+1}$ and substituting for $\mathbf{M}_{s+1}\ddot{\mathbf{\Delta}}_{s+1}$ from Eq. (9.7.1), we obtain

$$
\mathbf{F}_{s+1} - \mathbf{K}_{s+1}\mathbf{\Delta}_{s+1} = \mathbf{M}_{s+1}\left[a_3\left(\mathbf{\Delta}_{s+1} - \mathbf{\Delta}_s\right) - a_4\dot{\mathbf{\Delta}}_s - a_5\ddot{\mathbf{\Delta}}_s\right].
\tag{9.7.9}
$$

Collecting the terms involving $\mathbf{\Delta}_{s+1}$ on one side and the remaining terms on the other side, we obtain

$$
\begin{aligned}
\left(\mathbf{K}_{s+1} + a_3\mathbf{M}_{s+1}\right)\mathbf{\Delta}_{s+1} &= \mathbf{M}_{s+1}\left(a_3\mathbf{\Delta}_s + a_4\dot{\mathbf{\Delta}}_s + a_5\ddot{\mathbf{\Delta}}_s\right) + \mathbf{F}_{s+1} \\
\hat{\mathbf{K}}_{s+1}(\mathbf{\Delta}_{s+1})\mathbf{\Delta}_{s+1} &= \hat{\mathbf{F}}_{s,s+1},
\end{aligned}
\tag{9.7.10}
$$

where

$$
\hat{\mathbf{K}}_{s+1} = \mathbf{K}_{s+1} + a_3\mathbf{M}_{s+1}, \quad \hat{\mathbf{F}}_{s,s+1} = \mathbf{F}_{s+1} + \mathbf{G}_s,
$$
$$
\mathbf{G}_s = \mathbf{M}_{s+1}\left(a_3\mathbf{\Delta}_s + a_4\dot{\mathbf{\Delta}}_s + a_5\ddot{\mathbf{\Delta}}_s\right),
\tag{9.7.11a}
$$

and a_i $(i = 1, 2, \ldots, 8)$ are defined as

$$
a_1 = \alpha\delta t, \quad a_2 = (1-\alpha)\delta t,
$$
$$
a_3 = \frac{2}{\gamma(\delta t)^2}, \quad a_4 = a_3\delta t, \quad a_5 = \frac{1}{\gamma} - 1,
\tag{9.7.11b}
$$
$$
a_6 = \frac{2\alpha}{\gamma\delta t}, \quad a_7 = \frac{2\alpha}{\gamma} - 1, \quad a_8 = \delta t\left(\frac{\alpha}{\gamma} - 1\right).
$$

Few remarks concerning the Newmark scheme are in order:

1. *Non-self-starting nature of the scheme.* The calculation of $\hat{\mathbf{K}}$ and $\hat{\mathbf{F}}$ in Newmark's scheme requires knowledge of the initial conditions $\boldsymbol{\Delta}^0$, $\dot{\boldsymbol{\Delta}}^0$, and $\ddot{\boldsymbol{\Delta}}_0$, as can be seen from the vector $\mathbf{G}_s$. However, one does not know acceleration at time $t = 0$. As an approximation, it can be calculated from the assembled system of equations associated with Eq. (9.7.1) using initial conditions on $\boldsymbol{\Delta}$, $\dot{\boldsymbol{\Delta}}$, and $\mathbf{F}$ (often $\mathbf{F}$ is assumed to be zero at $t = 0$):

$$\ddot{\boldsymbol{\Delta}}^0 = \mathbf{M}^{-1}\left(\mathbf{F}^0 - \mathbf{K}\boldsymbol{\Delta}^0\right). \tag{9.7.12}$$

2. *Updating the velocity and accelerations at each time step.* At the end of each time step, the new velocity vector $\dot{\mathbf{u}}_{s+1}$ and acceleration vector $\ddot{\mathbf{u}}_{s+1}$ are computed using Eqs. (9.7.7) and (9.7.5), respectively:

$$\ddot{\boldsymbol{\Delta}}_{s+1} = a_3\left(\boldsymbol{\Delta}_{s+1} - \boldsymbol{\Delta}_s\right) - a_4\dot{\boldsymbol{\Delta}}_s - a_5\ddot{\boldsymbol{\Delta}}_s, \tag{9.7.13a}$$

$$\dot{\boldsymbol{\Delta}}_{s+1} = \dot{\boldsymbol{\Delta}}_s + a_2\ddot{\boldsymbol{\Delta}}_s + a_1\ddot{\boldsymbol{\Delta}}_{s+1}, \tag{9.7.13b}$$

where parameters a_1 through a_5 are defined in Eq. (9.7.11b).
3. For natural vibration, the forces and the solution are assumed to be periodic ($\mathbf{F} = \mathbf{0}$ when the source term is set to zero)

$$\boldsymbol{\Delta} = \boldsymbol{\Delta}_0\, e^{i\omega t}, \quad i = \sqrt{-1}, \tag{9.7.14}$$

where $\boldsymbol{\Delta}_0$ is the vector of amplitudes (independent of time) and ω is the frequency of natural vibration of the system. Substitution of Eq. (9.7.14) into Eq. (9.7.1), assuming that it is linear, yields (after the application of the homogeneous boundary conditions)

$$\left(-\omega^2\mathbf{M} + \mathbf{K}\right)\boldsymbol{\Delta}_0 = \mathbf{0}. \tag{9.7.15}$$

4. *Stability criterion.* Equation (9.7.10) can be expressed as

$$\boldsymbol{\Delta}_{s+1} = \mathbf{A}\boldsymbol{\Delta}_s + \mathbf{B}, \tag{9.7.16a}$$

where

$$\mathbf{A} = \hat{\mathbf{K}}^{-1}\mathbf{M}_{s+1}\left(a_3\boldsymbol{\Delta}_s + a_4\dot{\boldsymbol{\Delta}}_s + a_5\ddot{\boldsymbol{\Delta}}_s\right), \quad \mathbf{B} = \hat{\mathbf{K}}^{-1}\mathbf{F}_{s+1}. \tag{9.7.16b}$$

The matrix $\mathbf{A}$ is called the *amplification* matrix, which depends on the problem parameters (e.g., geometric and material properties, time step, and characteristic mesh size), and $\boldsymbol{\Delta}_{s+1}$ is the solution vector at time t_{s+1}. After the first time step, the solution obtained at the end of each time step is an approximate solution, error introduced into the solution $\boldsymbol{\Delta}_{s+1}$; that is, the matrix $\mathbf{A}$ can "amplify" the error in $\boldsymbol{\Delta}_s$ to calculate $\boldsymbol{\Delta}_{s+1}$, and the error can grow with each time step. A time approximation scheme is said to be *stable* if the error introduced in $\boldsymbol{\Delta}_s$ does not grow unbounded as Eq. (9.7.10) is solved repeatedly for $s = 0, 1, \ldots,$. In order for the error to remain bounded, it is necessary and sufficient that the largest eigenvalue of the amplification matrix $\mathbf{A}$ be less than or equal to unity:

$$|\lambda_{\max} = \omega_{\max}^2| \leq 1, \tag{9.7.17}$$

where $\lambda_{\max}$ is the largest eigenvalue of Eq. (9.7.15). We note that $\lambda_{\max}$ depends on the characteristic length of the mesh, time step, and material parameters (such as the density and modulus). Since modulus and density are fixed for a given problem, the stability criterion essentially provides a relation between the characteristic element size (e.g., element length h) and the time step δt used. If the condition in Eq. (9.7.17) is satisfied for any value of δt, independent of the mesh size h, the scheme is said to be *unconditionally stable*, or simply *stable*. If Eq. (9.7.17) places a restriction, for a given mesh, on the size of the time step δt, the scheme is said to be *conditionally stable*. As the mesh is refined, the value of the maximum eigenvalue increases and, hence, the value of the critical time step decreases. For all Newmark schemes in which $\gamma < \alpha$ and $\alpha \geq \frac{1}{2}$, the stability requirement is (for linear problems)

$$\delta t \leq \delta t_{cr} = \left[\frac{1}{2} \omega_{\max}^2 (\alpha - \gamma) \right]^{-1/2}, \tag{9.7.18}$$

where $\omega_{\max}$ is the maximum natural frequency of the system in Eq. (9.7.15). For all nonlinear problems, one often uses the time step restrictions imposed by linear stability criteria.

9.8   SOLUTION OF NONLINEAR ALGEBRAIC EQUATIONS

9.8.1   PRELIMINARY COMMENTS

First, we discuss the solution of nonlinear equations for the equilibrium case. Suppose that we wish to solve assembled nonlinear algebraic equations of the type

$$\mathbf{K}(\mathbf{U})\mathbf{U} = \mathbf{F}. \tag{9.8.1}$$

Since $\mathbf{K}(\mathbf{U})$ cannot be evaluated until the solution vector $\mathbf{U}$ is known and the solution is not known until we solve the equations, one must resort to an iterative procedure in which we assume that solution at the $(r-1)$st iteration is known and seek solution at the rth iteration.

At the very beginning of the iteration process, we begin with a "guess vector" $\mathbf{U}^{(0)}$ to evaluate $\mathbf{K}$ and solve, after imposing the boundary conditions of the problem, for the solution vector $\mathbf{U}^{(1)}$, where the superscript on $\mathbf{U}$ denotes the iteration number:

$$\mathbf{U}^{(1)} = \mathbf{K}^{-1}(\mathbf{U}^{(0)})\mathbf{F}. \tag{9.8.2}$$

The solution obtained can be used to evaluate $\mathbf{K}$ and obtain the solution $\mathbf{U}^{(2)}$. This iteration will continue until the difference between two consecutive solution vectors $\mathbf{U}^{(r)}$ and $\mathbf{U}^{(r+1)}$ obtained at the end of the rth and $(r+1)$st iterations, measured in a suitable norm, is less than a preselected value.

The guess vector used for structural problems, which have a linear part of the stiffness matrix, should be necessarily a zero vector so that the solution obtained at the first iteration is the linear solution to the problem. This is always the best guess vector for the nonlinear iteration. There are two types of iterative procedures. These are discussed next.

9.8.2   DIRECT ITERATION PROCEDURE

Suppose that we wish to solve the assembled system, Eq. (9.8.1), which already has the boundary conditions imposed on them. Suppose that the solution $\mathbf{U}^{(r-1)}$ at the beginning of the rth iteration is known and we wish to seek solution $\mathbf{U}^{(r)}$:

$$\mathbf{K}(\mathbf{U}^{(r-1)})\mathbf{U}^{(r)} = \mathbf{F}, \quad r = 1, 2, \ldots. \tag{9.8.3}$$

This procedure of solving Eq. (9.8.3) iteratively is known as the *direct iteration*, the *Picard iteration*, or the *method of successive substitutions*.

At the beginning of the iteration process, that is, when $r = 1$, solution $\mathbf{U}^{(0)}$ must be assumed or "guessed" consistent with the problem boundary conditions. The initial *guess vector* $\mathbf{U}^0$ should be such that: (a) it satisfies the specified boundary conditions on $\mathbf{U}$ and (b) $\mathbf{K}^{(0)}$ is invertible.

The iteration is terminated once convergence is reached or the maximum number of specified iterations is exceeded. The declaration of convergence is based on one of the two criteria: (1) residual measure or (2) solution difference measure. The residual vector is defined by

$$\mathbf{R}^{(r)} \equiv \mathbf{K}^{(r)}\mathbf{U}^{(r)} - \mathbf{F}^{(r)}. \tag{9.8.4}$$

The magnitude of this residual vector will be zero (within the specified tolerance) if the solution has converged. We terminate the iteration if the magnitude of the residual vector, measured in a suitable norm, is less than some preselected tolerance ϵ. If the problem data are such that $\mathbf{KU}$ as well as $\mathbf{F}$ are very small, the norm of the residual vector may also be very small even when the solution $\mathbf{U}$ has not converged. Therefore, it is necessary to normalize the residual vector with respect to $\mathbf{F}$. Using the Euclidean norm[2], we can express the error criterion as

$$\sqrt{\frac{\mathbf{R}^{(r)} \cdot \mathbf{R}^{(r)}}{\mathbf{F}^{(r)} \cdot \mathbf{F}^{(r)}}} \leq \epsilon. \tag{9.8.5}$$

---

[2]The *Euclidean norm* of a vector $\mathbf{r} = (r_1, r_2, \ldots, r_n)$, where r_i are real numbers, is $\|\mathbf{r}\| = (r_1^2 + r_2^2 + \cdots + r_n^2)^{1/2}$, and it is a measure of the length of the vector $\mathbf{r}$.

Alternatively, one may check to see if the normalized difference between solution vectors from two consecutive iterations, measured with the Euclidean norm, is less than a preselected tolerance ϵ:

$$\sqrt{\frac{\delta\mathbf{U}\cdot\delta\mathbf{U}}{\mathbf{U}^{(r)}\cdot\mathbf{U}^{(r)}}} = \sqrt{\frac{\sum_{I=1}^{N}|U_I^{(r)} - U_I^{(r-1)}|^2}{\sum_{I=1}^{N}|U_I^{(r)}|^2}} \le \epsilon, \tag{9.8.6}$$

where $\delta\mathbf{U} = \mathbf{U}^{(r)} - \mathbf{U}^{(r-1)}$. Thus, the iteration process is continued until the error criterion in Eq. (9.8.5) or (9.8.6) is satisfied or the number of iterations exceeded a preselected maximum number of iterations (so that the process does not go into a loop of indefinite number of iterations in the case of non-convergence).

Acceleration of convergence for some types of nonlinearities may be achieved by using a weighted-average of solutions from the last two iterations rather than the solution from the last iteration to evaluate the coefficient matrix:

$$\mathbf{U}^{(r)} = [\mathbf{K}(\bar{\mathbf{U}})]^{-1}\mathbf{F}(\bar{\mathbf{U}}), \quad \bar{\mathbf{U}} \equiv \beta\mathbf{U}^{(r-2)} + (1-\beta)\mathbf{U}^{(r-1)}, \quad 0 \le \beta \le 1, \tag{9.8.7}$$

where β is called the *acceleration parameter*. The value of β depends on the nature of nonlinearity and the type of problem. The value of β is found by trial; $\beta = 0.25\text{–}0.35$ is found to yield convergent results. In some cases, even this strategy does not help to achieve convergence, forcing us to use the Newton iteration scheme, which is discussed shortly.

Figure 9.8.1 depicts the general idea of the direct iteration procedure for a single-degree-of-freedom system. Here, K denotes the slope of the line joining the origin to the point $K(U)$ on the curve $F = K(U)U \equiv f(U)$. Note that $K(U)$ is *not* the slope of the tangent to the curve at U. Unless stated otherwise, the error criterion based on the solution rather than the residual, that is, Eq. (9.8.6) is used in this book. The direct iteration converges if the nonlinearity is mild; otherwise, it diverges. When the number of specified maximum iterations (typically about 20), one may reduce the load step size; one may also investigate convergence with the acceleration parameter. The steps involved in the direct iteration solution scheme are described in Box 9.8.1.

9.8.3   NEWTON'S ITERATION PROCEDURE

In Newton's method, we expand the residual vector $\mathbf{R}^{(r)}$ of Eq. (9.8.4) in Taylor's series about the known solution $\mathbf{U}^{(r-1)}$:

$$\mathbf{R}^{(r)} = \mathbf{R}^{(r-1)} + \left(\frac{\partial\mathbf{R}}{\partial\mathbf{U}}\right)^{(r-1)}\cdot\delta\mathbf{U} + \frac{1}{2!}\left[\left(\frac{\partial^2\mathbf{R}}{\partial\mathbf{U}^2}\right)^{(r-1)}\cdot\delta\mathbf{U}\right]\cdot\Delta\mathbf{U} + \cdots, \tag{9.8.8}$$

where

$$\delta\mathbf{U} = \mathbf{U}^{(r)} - \mathbf{U}^{(r-1)} \quad\rightarrow\quad \mathbf{U}^{(r)} = \delta\mathbf{U} + \mathbf{U}^{(r-1)}. \tag{9.8.9}$$

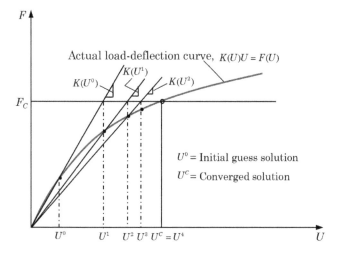

Figure 9.8.1: Convergence of the direct iteration procedure.

Box 9.8.1: Steps involved in the direct iteration scheme.

1. **Initial solution vector.** Assume an initial solution vector $\mathbf{U}^{(0)}$ such that it (a) satisfies the specified boundary conditions on $\mathbf{U}$ and (b) does not make $\mathbf{K}^e$ singular.

2. **Computation of $\mathbf{K}^e$ and $\mathbf{F}^e$.** Use the latest known vector $\mathbf{U}^{(r-1)}$ ($\mathbf{U}^{(0)}$ at the first iteration) to evaluate $\mathbf{K}^e$ and $\mathbf{F}^e$, assemble them to obtain global $\mathbf{K}$ and $\mathbf{F}$, and apply the specified boundary conditions on the assembled system.

3. **Solution of the equations.** Compute the solution vector at the rth iteration:
$$\mathbf{U}^{(r)} = [\mathbf{K}(\mathbf{U}^{(r-1)})]^{-1}\mathbf{F}^{(r-1)}.$$

4. **Convergence check.** Compute the residual vector
$$\mathbf{R}^{(r)} = \mathbf{K}(\mathbf{U}^{(r)})\mathbf{U}^{(r)} - \mathbf{F}^{(r)},$$

and check if
$$\|\mathbf{R}^{(r)}\| \leq \epsilon \|\mathbf{F}^{(r)}\| \quad \text{or} \quad \|\mathbf{U}^{(r)} - \mathbf{U}^{(r-1)}\| \leq \epsilon \|\mathbf{U}^{(r)}\|,$$

where $\|\cdot\|$ denotes the euclidean norm and ϵ is the convergence tolerance (read as an input). If the solution has converged, print the solution and move to the next "load" level or quit if it is the only or final load.

5. **Maximum iteration check.** Check if $r < itmax$, where $itmax$ is the maximum number of iterations allowed (read as an input). If *yes*, set $r \to r + 1$ and go to Step 2; if *no*, print a message that the iteration scheme did not converge and stop.

Omitting the terms of order 2 and higher in $\delta \mathbf{U}$ and requiring that $\mathbf{R}^{(r)}$ be zero in Eq. (9.8.8), we obtain

$$0 = \mathbf{R}^{(r-1)} + \left(\frac{\partial \mathbf{R}}{\partial \mathbf{U}}\right)^{(r-1)} \cdot \delta \mathbf{U},$$

or

$$\left(\frac{\partial \mathbf{R}}{\partial \mathbf{U}}\right)^{(r-1)} \delta \mathbf{U} = -\mathbf{R}^{(r-1)} \quad \text{or} \quad \mathbf{T}^{(r-1)} \, \delta \mathbf{U} = -\mathbf{R}^{(r-1)}, \tag{9.8.10}$$

where the matrix $\mathbf{T}$ is called the *tangent matrix*, and it is defined as

$$\mathbf{T}^{(r-1)} \equiv \left(\frac{\partial \mathbf{R}}{\partial \mathbf{U}}\right)^{(r-1)}. \tag{9.8.11}$$

The global component definition of the tangent matrix $\mathbf{T}$ is given by

$$\begin{aligned}
T_{IJ} &\equiv \frac{\partial R_I}{\partial U_J} = \frac{\partial}{\partial U_J}\left(\sum_{m=1}^{N} K_{Im}U_m - F_I\right) \\
&= \sum_{m=1}^{N}\left(K_{Im}\frac{\partial U_m}{\partial U_J} + \frac{\partial K_{Im}}{\partial U_J}U_m\right) - \frac{\partial F_I}{\partial U_J} \\
&= K_{IJ} + \sum_{m=1}^{N}\frac{\partial K_{Im}}{\partial U_J}U_m - \frac{\partial F_I}{\partial U_J}
\end{aligned} \tag{9.8.12}$$

for $I, J = 1, 2, \ldots, N$, where N is the total number of degrees of freedom in the mesh. If summation convention is used, the summation on m can be omitted in Eq. (9.8.12). Note that we have made use of the following identity:

$$\frac{\partial U_m}{\partial U_J} = \delta_{mJ} = \begin{cases} 0, & \text{if } m \neq J, \\ 1, & \text{if } m = J, \end{cases} \tag{9.8.13}$$

and $K_{Im}\delta_{mJ} = K_{IJ}$ (summation convention).

All of the ideas presented above also apply to the element equation $\mathbf{K}^e(\mathbf{u}^e)\mathbf{u}^e = \mathbf{F}^e$. Indeed, since $\mathbf{K}^e$ and $\mathbf{F}^e$ are computed at the element level, we apply Newton's procedure at the element level, and then assemble element equations to form the global equations. The element equation for the Newton iteration procedure is

$$\mathbf{T}^e(\mathbf{u}^{(r-1)}) \, \delta \mathbf{u}^e = -\mathbf{R}^e(\mathbf{u}^{(r-1)}), \tag{9.8.14}$$

and the elements of the tangent coefficient matrix at the element level are obtained from (when $\mathbf{F}^e$ is independent of $\mathbf{u}^e$, we have $\partial F_i^e/\partial u_j^e = 0$)

$$T_{ij}^e \equiv \frac{\partial R_i^e}{\partial u_j^e} = \frac{\partial}{\partial u_j^e} \left(\sum_{m=1}^{n} K_{im}^e u_m^e - F_i^e \right)$$

$$= K_{ij}^e + \sum_{m=1}^{n} \frac{\partial K_{im}^e}{\partial u_j^e} u_m^e - \frac{\partial F_i^e}{\partial u_j^e}. \tag{9.8.15}$$

When the element equations contain several degrees of freedom and the co-efficient matrix $\mathbf{K}^e$ is composed of several submatrices (see, e.g., Eqs. (9.2.8), (9.3.11), (9.4.9), and (9.6.11)], the procedure remains the same except that one must note that the tangent matrix $\mathbf{T}^e$ is also in the same form as $\mathbf{K}^e$. Then the residual vector must be expressed in terms of the submatrices, as will be illustrated in the sequel (see Section 9.9).

Figure 9.8.2 shows convergence of the Newton iteration procedure for a single degree-of-freedom system. Here $T(U)$ denotes the slope of the tangent to the curve $F = K(U)U$ at U. The Newton iteration converges for hardening as well as softening type nonlinearity. For the hardening type, convergence may be accelerated using the under-relaxation given in Eq. (9.8.7). The method may diverge for a saddle-point behavior. The steps involved in Newton's iteration solution approach are described in Box 9.8.2.

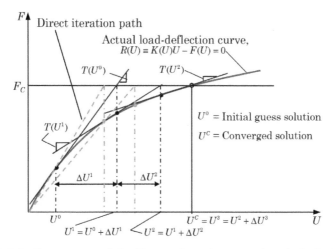

Figure 9.8.2: Convergence of the Newton iteration procedure (calculation of ΔU for a specified residual value R).

Some remarks are in order on Newton's iteration scheme.

1. In the direct iteration method, the boundary conditions are applied to the assembled system of equations, $\mathbf{KU} = \mathbf{F}$, during each iteration.

Box 9.8.2: Steps involved in the Newton iteration scheme.

---

1. **Initial solution vector.** Assume an initial solution vector $\mathbf{U}^{(0)}$ such that: (a) it satisfies the specified boundary conditions on $\mathbf{U}$ and (b) it does not make $\mathbf{T}^e$ singular.
2. **Computation of T and R.** Use the latest known vector $\mathbf{U}^{(r-1)}$ ($\mathbf{U}^{(0)}$ during the first iteration) to: (a) evaluate $\mathbf{K}^e$, $\mathbf{F}^e$, $\mathbf{T}^e$, and $-\mathbf{R}^e = \mathbf{F}^e - \mathbf{K}^e\mathbf{U}^e$, (b) assemble $\mathbf{T}^e$ and $\mathbf{R}^e$ to obtain global $\mathbf{T}$ and $\mathbf{R}$, and (c) apply the specified *homogeneous* boundary conditions (since $\mathbf{U}^{(0)}$ already satisfies the actual boundary conditions) on the assembled system, $\mathbf{TU} = -\mathbf{R}$.
3. **Computation of $\mathbf{U}^{(r)}$.** Compute the solution increment at the rth iteration and update the solution

$$\delta\mathbf{U} = -[\mathbf{T}(\mathbf{U}^{(r-1)})]^{-1}\mathbf{R}^{(r-1)}, \quad \mathbf{U}^{(r)} = \mathbf{U}^{(r-1)} + \delta\mathbf{U}.$$

4. **Convergence check.** Compute the global residual vector

$$\mathbf{R}^{(r)} = \mathbf{K}(\mathbf{U}^{(r)})\mathbf{U}^{(r)} - \mathbf{F}^{(r)},$$

and check if

$$\|\mathbf{R}^{(r)}\| \leq \epsilon\|\mathbf{F}^{(r)}\| \quad \text{or} \quad \|\delta\mathbf{U}\| \leq \epsilon\|\mathbf{U}^{(r)}\|,$$

where ϵ is the convergence tolerance (read as an input). If the solution has converged, print the solution and move to the next "load" level or stop. Otherwise, continue.
5. **Maximum iteration check.** Check if $r < itmax$, where $itmax$ is the maximum number of iterations allowed (read as an input). If *yes*, set $r \to r+1$ and go to Step 2; if *no*, print a message that the iteration scheme did not converge and quit.

In Newton's method, at the end of each iteration we only obtain the increment $\delta\mathbf{U}$ to the known solution $\mathbf{U}^{(r-1)}$, which already satisfies the specified boundary conditions, then $\delta\mathbf{U}$ must necessarily satisfy the corresponding *homogeneous* boundary conditions so that $\mathbf{U}^{(r)} = \mathbf{U}^{(r-1)} + \delta\mathbf{U}$ satisfies the specified boundary conditions. For example, if U_m is specified to be $U_m = \alpha_m$, and the initial guess is such that $U_m^{(0)} = \alpha_m$, then $\delta U_m = 0$ so that $U_m^{(1)} = U_m^{(0)} + \delta U_m = \alpha_m$. Alternatively, if the initial guess is such that $U_m^{(0)} = 0$, then $\delta U_m = \alpha_m$ for the first iteration so that $U_m^{(1)} = U_m^{(0)} + \delta U_m = \alpha_m$, and the boundary condition on δU_m during the second iteration on wards should be $\delta U_m = 0$.

2. The symmetry of the coefficient matrices $\mathbf{K}$ and $\mathbf{T}$ depends on the original differential equation as well as the weak form used to develop the finite element equations. Even when $\mathbf{K}$ is symmetric, $\mathbf{T}$ may not be symmetric, and vice versa; the symmetry of $\mathbf{T}$ depends on $\mathbf{K}$ as well as on the nature of nonlinearity in the governing differential equations.

3. The tangent matrix $\mathbf{T}$ does not have to be exact; an approximate $\mathbf{T}$ can also provide the solution but it may take more iterations. In any case, the residual vector will be computed using the definition $\mathbf{R} = \mathbf{KU} - \mathbf{F}$ and convergence is declared only when the residual vector $\mathbf{R}$ or the solution

increment $\delta \mathbf{U}$ is sufficiently small as indicated in Eqs. (9.8.5) and (9.8.6).

4. When the tangent matrix is updated only once in a certain pre-specified number of iterations (to save computational time) while updating the residual vector during each iteration, the procedure is known as the *modified Newton's method* or the *Newton–Raphson method*. Generally, the Newton–Raphson iteration solution method takes more iterations to converge than the full Newton's method, and it may even diverge for certain types of nonlinearity.

All of the discussion for the steady-state analysis can be extended to the transient analysis, where one must keep track of iteration number (superscript r) as well as time step number (subscript s). For the *direct iteration scheme*, we solve the equation,

$$\hat{\mathbf{K}}_{s+1}^{(r-1)} \boldsymbol{\Delta}_{s+1}^{(r)} = \hat{\mathbf{F}}_{s+1}. \tag{9.8.16}$$

Thus, the role of $\mathbf{K}$ in Eq. (9.8.1) is taken by $\hat{\mathbf{K}}_{s+1}$. For *Newton's iteration scheme*, we have

$$\hat{\mathbf{T}}_{s+1}^{(r-1)} \delta\boldsymbol{\Delta} = -\hat{\mathbf{R}}_{s+1}^{(r-1)} = \hat{\mathbf{F}}_{s+1} - \hat{\mathbf{K}}_{s+1}(\boldsymbol{\Delta}_{s+1}^{(r-1)})\boldsymbol{\Delta}_{s+1}^{(r-1)}, \tag{9.8.17}$$

where $\hat{\mathbf{T}}$ is the tangent matrix, and it is defined as

$$\hat{\mathbf{T}}_{s+1}^{(r-1)} \equiv \left[\frac{\partial \hat{\mathbf{R}}_{s+1}^{(r)}}{\partial \boldsymbol{\Delta}_{s+1}^{(r)}}\right]^{r-1}. \tag{9.8.18}$$

It is important to note that there are four vectors that should be kept track of in the program:

$$\boldsymbol{\Delta}_{s+1}^{(r-2)} = \text{solution vector from the previous to previous iteration at}$$
$$\text{current time } t_{s+1},$$

$$\boldsymbol{\Delta}_{s+1}^{(r-1)} = \text{solution vector from the previous iteration at current time } t_{s+1},$$

$$\boldsymbol{\Delta}_{s+1}^{(r)} = \text{solution vector from the current iteration at current time } t_{s+1},$$

$$\boldsymbol{\Delta}_{s} = \text{converged solution vector from the previous time } t_{s}.$$
$$\tag{9.8.19}$$

The vector $\boldsymbol{\Delta}_{s+1}^{(r-2)}$ is used only when the acceleration parameter β [see Eq. (9.8.7)] is used to enhance convergence.

The nonlinear convergence is checked using the following criterion:

$$\sqrt{\frac{\delta\boldsymbol{\Delta}_{s+1}^{(r)} \cdot \delta\boldsymbol{\Delta}_{s+1}^{(r)}}{\boldsymbol{\Delta}_{s+1}^{r} \cdot \boldsymbol{\Delta}_{s+1}^{r}}} \le \epsilon, \quad \delta\boldsymbol{\Delta}_{s+1}^{(r)} \equiv \boldsymbol{\Delta}_{s+1}^{r} - \boldsymbol{\Delta}_{s+1}^{r-1}, \tag{9.8.20}$$

where ϵ denotes a preselected value of the error tolerance (e.g. $\epsilon = 10^{-3}$).

9.8.4 LOAD INCREMENTS

The source of geometric nonlinearity considered in beams and plates is due to the von Kármán nonlinear strain, and it appears through the internal axial force N_{xx}. The nonlinear term, $(\partial w/\partial x)^2$ appears as a positive term in N_{xx}. As a result, the beam becomes increasingly stiff as it experiences bending deformation. Hence, for large bending loads, the nonlinearity may be too large for the numerical scheme to yield a convergent solution. Therefore, it is necessary to divide the applied total load, say F, into a number of smaller load increments $\delta F_1, \delta F_2, \ldots$, such that $F = \sum_{i=1}^{NLS} \delta F_i$, where NLS is the number of load steps. The flow chart for nonlinear bending of beams is presented in Fig. 9.8.3. Note that the outer loop is on load increments.

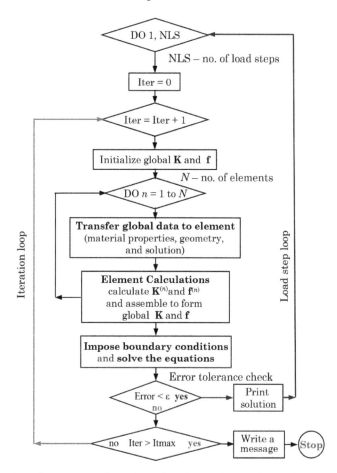

Figure 9.8.3: A computer flow chart for the nonlinear finite element analysis of structures.

For the first load step, we use the initial guess of $\mathbf{U}^{(0)} = \mathbf{0}$ so that we can obtain the linear solution of the problem for the first load, F_1, at the end of the first iteration. One of the iterative procedures outlined earlier is used to determine the solution, $\mathbf{U}^{(1)}$. If the iterative scheme does not converge within a pre-specified number of iterations (ITMAX), it may be necessary to further reduce the size of the load increment $F_1 = \delta F_1$ (and possibly invoke acceleration parameter). Once the solution Δ for the first load increment is obtained, it is used as the initial "guess vector" for the next load, $F_2 = \delta F_1 + \delta F_2$. This is continued until the total load is reached.

9.9 TANGENT STIFFNESS COEFFICIENTS

9.9.1 DEFINITION OF TANGENT STIFFNESS COEFFICIENTS

The finite element equations derived for any beam theory have the form

$$\begin{bmatrix} \mathbf{K}^{11} & \mathbf{K}^{12} & \mathbf{K}^{13} \\ \mathbf{K}^{21} & \mathbf{K}^{22} & \mathbf{K}^{23} \\ \mathbf{K}^{31} & \mathbf{K}^{32} & \mathbf{K}^{33} \end{bmatrix} \begin{Bmatrix} \Delta^{(1)} \\ \Delta^{(2)} \\ \Delta^{(3)} \end{Bmatrix} = \begin{Bmatrix} \mathbf{F}^1 \\ \mathbf{F}^2 \\ \mathbf{F}^2 \end{Bmatrix}. \tag{9.9.1}$$

Therefore, the tangent stiffness coefficients are computed in the same form. In the Newton's scheme, the equations will be of the form,

$$\begin{bmatrix} \mathbf{T}^{11} & \mathbf{T}^{12} & \mathbf{K}^{13} \\ \mathbf{T}^{21} & \mathbf{T}^{22} & \mathbf{T}^{23} \\ \mathbf{T}^{31} & \mathbf{T}^{32} & \mathbf{T}^{33} \end{bmatrix}^{(r-1)} \begin{Bmatrix} \delta\Delta^{(1)} \\ \delta\Delta^{(2)} \\ \delta\Delta^{(3)} \end{Bmatrix}^{(r)} = - \begin{Bmatrix} \mathbf{R}^1 \\ \mathbf{R}^2 \\ \mathbf{R}^3 \end{Bmatrix}^{(r-1)}, \tag{9.9.2}$$

where R_i^α are the components of the residual vector. The specific forms of $T_{ij}^{\alpha\beta}$ will be discussed for each of the finite element models developed in this chapter. We note the following identities for the CBT and TBT:

$$\text{CBT:} \quad \frac{\partial}{\partial \bar{\Delta}_P}\left(\frac{\partial w}{\partial x}\right) = \frac{d\varphi_P}{dx}; \quad \text{TBT:} \quad \frac{\partial}{\partial w_j}\left(\frac{\partial w}{\partial x}\right) = \frac{d\psi_j}{dx}. \tag{9.9.3}$$

9.9.2 THE DISPLACEMENT MODEL OF THE CBT

Identifying $\mathbf{T}^e$ to be in the same form as $\mathbf{K}^e$ in Eq. (9.2.8),

$$\begin{bmatrix} \mathbf{T}^{11} & \mathbf{T}^{12} \\ \mathbf{T}^{21} & \mathbf{T}^{22} \end{bmatrix}^{r-1} \begin{Bmatrix} \delta\Delta^{(1)} \\ \delta\Delta^{(2)} \end{Bmatrix}^{r} = - \begin{Bmatrix} \mathbf{R}^1 \\ \mathbf{R}^2 \end{Bmatrix}^{r-1}, \tag{9.9.4}$$

where symbol $\delta\Delta^{(\alpha)}$ ($\alpha = 1,2$) denotes the increment of the displacements from the $(r-1)$st iteration to the rth iteration, $\delta\Delta = \Delta^r - \Delta^{(r-1)}$. The $\Delta^{(\alpha)}$ for $\alpha = 1,2$ have the meaning $\Delta^{(1)} = \mathbf{u}$ and $\Delta^{(2)} = \bar{\Delta}$, where $\bar{\Delta}_1 = w_1$ and $\bar{\Delta}_3 = w_2$ are the transverse deflections and $\bar{\Delta}_2 = \theta_1$ and $\bar{\Delta}_4 = \theta_2$ are the rotations ($\theta = -(\partial w/\partial x)$) at the two nodes of the beam element.

The components $\mathbf{T}^{\alpha\beta}$ of the tangent stiffness matrix (evaluated at the $(r-1)$st iteration) are defined by

$$T_{ij}^{\alpha\beta} = \frac{\partial R_i^\alpha}{\partial \Delta_j^\beta}, \quad \alpha, \beta = 1, 2, \tag{9.9.5a}$$

where

$$R_i^\alpha = \sum_{\gamma=1}^{2} \sum_{p=1}^{N_\gamma} K_{ip}^{\alpha\gamma} \Delta_p^\gamma - F_i^\alpha = \sum_{p=1}^{2} K_{ip}^{\alpha 1} \Delta_p^1 + \sum_{P=1}^{4} K_{iP}^{\alpha 2} \Delta_P^2 - F_i^\alpha. \tag{9.9.5b}$$

Here N_γ $(\gamma = 1, 2)$ denotes the number of element degrees of freedom (in the present case, we have $N_1 = 2$ and $N_2 = 4$), $\Delta_p^1 = u_p$, and $\Delta_P^2 = \bar{\Delta}_P$. Using the definition in Eq. (9.9.5a), and noting that $K_{ij}^{\alpha\beta}$ depend only on $\bar{\Delta}_P$ (which represents the transverse deflections and rotations at the two nodes of the element) and not on u_j, we have $T_{ij}^{\alpha 1} = K_{ij}^{\alpha 1}$ for $\alpha = 1, 2$. Thus, we have to compute only $T_{ij}^{\alpha 2}$ for $\alpha = 1, 2$. We have

$$T_{ij}^{\alpha 2} = \frac{\partial R_i^\alpha}{\partial \bar{\Delta}_j} = K_{ij}^{\alpha 2} + \sum_{p=1}^{2} \frac{\partial}{\partial \bar{\Delta}_j} \left(K_{ip}^{\alpha 1} \right) u_p + \sum_{P=1}^{4} \frac{\partial}{\partial \bar{\Delta}_j} \left(K_{iP}^{\alpha 2} \right) \bar{\Delta}_P - \frac{\partial F_i^\alpha}{\partial \bar{\Delta}_j} \tag{9.9.6}$$

for $\alpha = 1, 2$. Then, we have

$$T_{iJ}^{12} = K_{iJ}^{12} + \sum_{p=1}^{2} \frac{\partial K_{ip}^{11}}{\partial \bar{\Delta}_J} u_p + \sum_{P=1}^{4} \frac{\partial K_{iP}^{12}}{\partial \bar{\Delta}_J} \bar{\Delta}_P = K_{iJ}^{12} + 0 + \sum_{P=1}^{4} \frac{\partial K_{iP}^{12}}{\partial \bar{\Delta}_J} \bar{\Delta}_P$$

$$= K_{iJ}^{12} + \frac{1}{2} \sum_{P=1}^{4} \bar{\Delta}_P \int_{x_a^e}^{x_b^e} A_{xx} \frac{d\varphi_J}{dx} \frac{d\psi_i}{dx} \frac{d\varphi_P}{dx} dx$$

$$= K_{iJ}^{12} + \frac{1}{2} \int_{x_a^e}^{x_b^e} A_{xx} \frac{\partial w}{\partial x} \frac{d\psi_i}{dx} \frac{d\varphi_J}{dx} dx = K_{Ji}^{21} = T_{Ji}^{21}, \tag{9.9.7a}$$

$$T_{IJ}^{22} = K_{IJ}^{22} + \sum_{p=1}^{2} \frac{\partial K_{Ip}^{21}}{\partial \bar{\Delta}_J} u_p + \sum_{P=1}^{4} \frac{\partial K_{IP}^{22}}{\partial \bar{\Delta}_J} \bar{\Delta}_P$$

$$= K_{IJ}^{22} + \int_{x_a^e}^{x_b^e} A_{xx} \frac{\partial u}{\partial x} \frac{d\varphi_I}{dx} \frac{d\varphi_J}{dx} dx + \int_{x_a^e}^{x_b^e} A_{xx} \left(\frac{\partial w}{\partial x} \right)^2 \frac{d\varphi_I}{dx} \frac{d\varphi_J}{dx} dx$$

$$- \int_{x_a^e}^{x_b^e} B_{xx} \left(\frac{1}{2} \frac{\partial w}{\partial x} \frac{d^2\varphi_I}{dx^2} + \frac{\partial^2 w}{\partial x^2} \frac{d\varphi_I}{dx} \right) \frac{d\varphi_J}{dx} dx, \tag{9.9.7b}$$

where we have used the identities [in addition to those in Eq. (9.9.3)]

$$\frac{\partial}{\partial \bar{\Delta}_P} \left(\frac{\partial w}{\partial x} \right)^2 = 2 \frac{\partial w}{\partial x} \frac{d\varphi_P}{dx}; \quad \sum_{p=1}^{2} \frac{d\psi_p}{dx} u_p = \frac{\partial u}{\partial x}; \quad \sum_{P=1}^{4} \frac{d\varphi_P}{dx} \bar{\Delta}_P = \frac{\partial w}{\partial x}.$$

$$\tag{9.9.8}$$

We note that the first expressions in both K_{Ij}^{21} and K_{Ij}^{22} are linear [see Eq. (9.2.9b)] and thus do not contribute to the additional terms to $T_{ij}^{\alpha 2}$.

9.9.3   THE MIXED MODEL OF THE CBT

The Newton's scheme for this case has the form

$$
\begin{bmatrix} \mathbf{T}^{11} & \mathbf{T}^{12} & \mathbf{T}^{13} \\ \mathbf{T}^{21} & \mathbf{T}^{22} & \mathbf{T}^{23} \\ \mathbf{T}^{31} & \mathbf{T}^{32} & \mathbf{T}^{33} \end{bmatrix}^{(r-1)} \left\{ \begin{matrix} \delta\mathbf{\Delta}^{(1)} \\ \delta\mathbf{\Delta}^{(2)} \\ \delta\mathbf{\Delta}^{(3)} \end{matrix} \right\}^{(r)} = - \left\{ \begin{matrix} \mathbf{R}^1 \\ \mathbf{R}^2 \end{matrix} \right\}^{(r-1)} \tag{9.9.9}
$$

with $\mathbf{\Delta}^{(1)} = \mathbf{u}$, $\mathbf{\Delta}^{(2)} = \mathbf{w}$, and $\mathbf{\Delta}^{(3)} = \mathbf{M}$. Thus, for this case, we have $\alpha, \beta = 1, 2, 3$ and the direct matrix coefficients $K_{ij}^{\alpha\beta}$ are at most functions of w_j; therefore, we have $T_{ij}^{\alpha 1} = K_{ij}^{\alpha 1}$ and $T_{ij}^{\alpha 3} = K_{ij}^{\alpha 3}$ $(i, j = 1, 2, \dots, n, \; n$ being the number of nodes in the element). The nodal degrees of freedom are (u_j, w_j, M_j). The remaining tangent stiffness coefficients are given by

$$
T_{ij}^{12} = K_{ij}^{12} + \frac{1}{2} \int_{x_a^e}^{x_b^e} \bar{A}_{xx} \frac{\partial w}{\partial x} \frac{d\psi_i^{(1)}}{dx} \frac{d\psi_j^{(2)}}{dx} \, dx, \tag{9.9.10a}
$$

$$
T_{ij}^{22} = K_{ij}^{22} + \int_{x_a}^{x_b} \bar{A}_{xx} \left[\frac{\partial u}{\partial x} + \left(\frac{\partial w}{\partial x} \right)^2 \right] \frac{d\psi_i^{(2)}}{dx} \frac{d\psi_j^{(2)}}{dx} \, dx
$$

$$
+ \int_{x_a^e}^{x_b^e} \bar{B}_{xx} M_{xx} \frac{d\psi_i^{(2)}}{dx} \frac{d\psi_j^{(2)}}{dx} dx, \tag{9.9.10b}
$$

$$
T_{ij}^{32} = K_{ij}^{32} + \frac{1}{2} \int_{x_a^e}^{x_b^e} \bar{B}_{xx} \frac{\partial w}{\partial x} \psi_i^{(3)} \frac{d\psi_j^{(2)}}{dx} \, dx. \tag{9.9.10c}
$$

9.9.4   THE DISPLACEMENT MODEL OF THE TBT

For this case, the Newton's scheme resembles the same as Eq. (9.9.2). The nodal degrees of freedom are for this case are (u_j, w_j, s_j), which are the nodal values of (u, w, ϕ_x). The direct stiffness matrix coefficients $K_{ij}^{\alpha\beta}$ are at most functions of the nodal values of the transverse deflection, w_j; therefore, we have $T_{ij}^{\alpha 1} = K_{ij}^{\alpha 1}$ and $T_{ij}^{\alpha 3} = K_{ij}^{\alpha 3}$ $(i, j = 1, 2, \dots, n, \; n$ being the number of nodes in the element). The remaining tangent stiffness coefficients are given by

$$
T_{ij}^{12} = K_{ij}^{12} + \frac{1}{2} \int_{x_a^e}^{x_b^e} A_{xx} \frac{\partial w}{\partial x} \frac{d\psi_i^{(1)}}{dx} \frac{d\psi_j^{(2)}}{dx} \, dx, \tag{9.9.11a}
$$

$$T_{ij}^{22} = K_{ij}^{22} + \int_{x_a^e}^{x_b^e} A_{xx} \left[\frac{\partial u}{\partial x} + \left(\frac{\partial w}{\partial x} \right)^2 \right] \frac{d\psi_i^{(2)}}{dx} \frac{d\psi_j^{(2)}}{dx} dx$$

$$+ \int_{x_a^e}^{x_b^e} B_{xx} \frac{\partial \phi_x}{\partial x} \frac{d\psi_i^{(2)}}{dx} \frac{d\psi_j^{(2)}}{dx} dx, \tag{9.9.11b}$$

$$T_{ij}^{32} = K_{ij}^{32} + \frac{1}{2} \int_{x_a^e}^{x_b^e} B_{xx} \frac{\partial w}{\partial x} \frac{d\psi_j^{(3)}}{dx} \frac{d\psi_j^{(2)}}{dx} dx. \tag{9.9.11c}$$

9.9.5   THE MIXED MODEL OF THE TBT

The mixed model of the TBT is the same as the mixed model of the EBT with only difference being the K_{ij}^{33} coefficients, as given in Eq. (9.5.6). Therefore, the tangent stiffness coefficients of the mixed model of the TBT are exactly the same as those of the mixed model of the EBT:

$$T_{ij}^{12} = K_{ij}^{12} + \frac{1}{2} \int_{x_a^e}^{x_b^e} \bar{A}_{xx} \frac{\partial w}{\partial x} \frac{d\psi_i^{(1)}}{dx} \frac{d\psi_j^{(2)}}{dx} dx, \tag{9.9.12a}$$

$$T_{ij}^{22} = K_{ij}^{22} + \int_{x_a^e}^{x_b^e} \bar{A}_{xx} \left[\frac{\partial u}{\partial x} + \left(\frac{\partial w}{\partial x} \right)^2 \right] \frac{d\psi_i^{(2)}}{dx} \frac{d\psi_j^{(2)}}{dx} dx$$

$$+ \int_{x_a^e}^{x_b^e} \bar{B}_{xx} M_{xx} \frac{d\psi_i^{(2)}}{dx} \frac{d\psi_j^{(2)}}{dx} dx, \tag{9.9.12b}$$

$$T_{ij}^{32} = K_{ij}^{32} + \frac{1}{2} \int_{x_a^e}^{x_b^e} \bar{B}_{xx} \frac{\partial w}{\partial x} \psi_i^{(3)} \frac{d\psi_j^{(2)}}{dx} dx. \tag{9.9.12c}$$

9.9.6   THE DISPLACEMENT MODEL OF THE RBT

For the displacement model of the RBT [see Eqs. (9.6.11) and (9.6.12)], the tangent stiffness coefficients that are the same as the original stiffness coefficients are

$$T_{ij}^{\alpha 1} = K_{ij}^{\alpha 1}, \quad \alpha = 1, 2, 3; \quad T_{ij}^{\alpha 3} = K_{ij}^{\alpha 3}, \quad \alpha = 1, 2, 3. \tag{9.9.13}$$

The remaining coefficients are given by

$$T_{iJ}^{12} = K_{iJ}^{12} + \frac{1}{2} \int_{x_a^e}^{x_b^e} A_{xx} \frac{\partial w}{\partial x} \frac{d\psi_i^{(1)}}{dx} \frac{d\varphi_J}{dx} dx, \tag{9.9.14a}$$

$$T_{IJ}^{22} = K_{IJ}^{22} + \frac{1}{2} \int_{x_a^e}^{x_b^e} A_{xx} \left[\frac{\partial u}{\partial x} + \left(\frac{\partial w}{\partial x} \right)^2 \right] \frac{d\varphi_I}{dx} \frac{d\varphi_J}{dx} dx$$

$$- \alpha \int_{x_a^e}^{x_b^e} E_{xx} \left(\frac{1}{2} \frac{\partial w}{\partial x} \frac{d^2\varphi_I}{dx^2} \frac{d\varphi_J}{dx} + \frac{\partial^2 w}{\partial x^2} \frac{d\varphi_I}{dx} \frac{d\varphi_J}{dx} \right) dx$$

$$+ \int_{x_a^e}^{x_b^e} \bar{B}_{xx} \frac{\partial \phi_x}{\partial x} \frac{d\varphi_I}{dx} \frac{d\varphi_J}{dx} \, dx, \tag{9.9.14b}$$

$$T_{iJ}^{32} = K_{iJ}^{32} + \frac{1}{2} \int_{x_a^e}^{x_b^e} \bar{B}_{xx} \frac{\partial w}{\partial x} \psi_i^{(2)} \frac{d\varphi_J}{dx} \, dx. \tag{9.9.14c}$$

This completes the derivation of the tangent stiffness coefficients for various finite element models of the beam theories. We note that in all finite element models, the direct stiffness matrix $\mathbf{K}$ is not symmetric while the tangent stiffness matrix $\mathbf{T}$ is symmetric.

9.10 POST-COMPUTATIONS

9.10.1 GENERAL COMMENTS

We recall that in the displacement model of the CBT, we use linear interpolation of $u(x)$ and Hermite cubic interpolation of $w(x)$. This element always gives exact linear solution at the nodes for homogeneous beams when $D_{xx} = EI$ is a constant, independent of the mesh and the load. All other models are based on Lagrange interpolation of the variables involved, and they can be linear, quadratic, or higher order. The shear locking in the displacement finite element model of the TBT is alleviated by the use of reduced integration (see Reddy [155, 156] for details). No such fix is necessary in the mixed models.

The finite element solution from various models is the vector of nodal values. The nodal values can be used to determine the solution at any desired point of the beam using the finite element approximations of the variables. The post-computation of moment and shear force in the displacement models and rotation and shear force in the mixed models can be achieved in a straightforward manner. It is well-known that the strains and stresses are closest to the actual values when they are computed at the Gauss points of the reduced integration rule [155, 156].

Here, we present the expressions for the stresses and strains in terms of the interpolation functions and nodal values for various models static case.

9.10.2 CBT FINITE ELEMENT MODELS

The membrane and bending components of strain can be computed from the known finite element solution as [see Eq. (9.2.7)]

$$\varepsilon_{xx}^{(0)}(x) = \frac{du}{dx} + \frac{1}{2}\left(\frac{dw}{dx}\right)^2 \approx \sum_{j=1}^{2} u_j^e \frac{d\psi_j^e}{dx} + \frac{1}{2}\left(\sum_{J=1}^{4} \bar{\Delta}_J^e \frac{d\varphi_J^e}{dx}\right)^2, \tag{9.10.1a}$$

$$\varepsilon_{xx}^{(1)}(x) = -\frac{d^2 w}{dx^2} \approx -\sum_{J=1}^{4} \bar{\Delta}_J^e \frac{d^2 \varphi_J^e}{dx^2}. \tag{9.10.1b}$$

The membrane and bending components of the axial stress $\sigma_{xx} = \sigma_{xx}^{(0)} + z\sigma_{xx}^{(1)}$ are given by, with strains $\varepsilon_{xx}^{(0)}$ and $\varepsilon_{xx}^{(1)}$ computed using Eqs. (9.10.1a) and (9.10.1b):

$$\sigma_{xx}^{(0)}(x, z) = E(z)\varepsilon_{xx}^{(0)}(x), \quad \sigma_{xx}^{(1)}(x, z) = E(z)\varepsilon_{xx}^{(1)}(x). \tag{9.10.2}$$

In theory, all of the expressions can be evaluated at any point in the element, including at the nodes. However, as stated earlier, they are the most accurate when evaluated at the center of the element (i.e., using one-point Gauss rule).

For the mixed model of the CBT beam element, the membrane and bending components of strain can be computed from the known finite element solution as [see Eqs. (9.3.3) and (9.3.9)]

$$\varepsilon_{xx}^{(0)} = \frac{du}{dx} + \frac{1}{2}\left(\frac{dw}{dx}\right)^2 \approx \sum_{j=1}^{m} u_j^e \frac{d\psi_j^{(1)}}{dx} + \frac{1}{2}\left(\sum_{j=1}^{n} w_j^e \frac{d\psi_j^{(2)}}{dx}\right)^2, \tag{9.10.3a}$$

$$\varepsilon_{xx}^{(1)} = -\frac{d^2 w}{dx^2} = \frac{\bar{M}_{xx}}{\hat{D}_{xx}} - \bar{B}_{xx}\varepsilon_{xx}^{(0)} \approx \frac{1}{\hat{D}_{xx}}\sum_{j=1}^{p} M_j^e \psi_j^{(3)} - \bar{B}_{xx}\varepsilon_{xx}^{(0)}. \tag{9.10.3b}$$

The membrane and bending components of the axial stress $\sigma_{xx} = \sigma_{xx}^{(0)} + z\sigma_{xx}^{(1)}$, with $\varepsilon_{xx}^{(0)}$ and $\varepsilon_{xx}^{(1)}$ given by Eqs. (9.10.3a) and (9.10.3b), are

$$\sigma_{xx}^{(0)}(x, z) = E(z)\varepsilon_{xx}^{(0)}(x), \quad \sigma_{xx}^{(1)}(x, z) = E(z)\varepsilon_{xx}^{(1)}(x). \tag{9.10.4}$$

9.10.3 TBT FINITE ELEMENT MODELS

In the TBT, the membrane, bending, and transverse shear components of strain can be computed from the known finite element solution as [see Eqs. (4.2.2a)–(4.2.2c) and (9.4.8)]

$$\varepsilon_{xx}^{(0)} = \frac{du}{dx} + \frac{1}{2}\left(\frac{dw}{dx}\right)^2 \approx \sum_{j=1}^{m} u_j^e \frac{d\psi_j^{(1)}}{dx} + \frac{1}{2}\left(\sum_{j=1}^{n} w_j^e \frac{d\psi_j^{(2)}}{dx}\right)^2, \tag{9.10.5a}$$

$$\varepsilon_{xx}^{(1)} = \frac{d\phi_x}{dx} \approx \sum_{j=1}^{p} S_j^e \frac{d\psi_j^{(3)}}{dx}, \tag{9.10.5b}$$

$$\gamma_{xz} = \phi_x + \frac{dw}{dx} \approx \sum_{j=1}^{p} S_j^e \psi_j^{(3)} + \sum_{J=1}^{n} w_j^e \frac{d\psi_j^{(2)}}{dx}. \tag{9.10.5c}$$

The membrane and bending components of the axial stress $\sigma_{xx} = \sigma_{xx}^{(0)} + z\sigma_{xx}^{(1)}$ and the transverse shear stress σ_{xz} are given by

$$\sigma_{xx}^{(0)}(x, z) = E(z)\varepsilon_{xx}^0, \quad \sigma_{xx}^{(1)}(x, z) = E(z)\varepsilon_{xx}^1,$$

$$\sigma_{xz}(x, z) = G(z)\left(\phi_x + \frac{dw}{dx}\right). \tag{9.10.6}$$

For the mixed finite element model of the TBT, the membrane strain $\varepsilon_{xx}^{(0)}$ is computed as given in Eq. (9.10.5a); the bending strain is computed from

$$\varepsilon_{xx}^{(1)} = -\frac{d^2 w}{dx^2} = \frac{\bar{M}_{xx}}{\hat{D}_{xx}} - \bar{B}_{xx}\varepsilon_{xx}^0 \approx \frac{1}{\hat{D}_{xx}}\sum_{j=1}^{p} M_j^e \psi_j^{(3)} - \bar{B}_{xx}\varepsilon_{xx}^{(0)}, \qquad (9.10.7a)$$

$$\gamma_{xz} = \phi_x + \frac{dw}{dx} = \frac{1}{S_{xz}}\frac{dM_{xx}}{dx} \approx \frac{1}{S_{xz}}\sum_{J=1}^{p} M_j^e \frac{d\psi_j^{(3)}}{dx}. \qquad (9.10.7b)$$

The membrane and bending components of the axial stress $\sigma_{xx} = \sigma_{xx}^{(0)} + z\sigma_{xx}^{(1)}$ and the shear stress are given by

$$\sigma_{xx}^{(0)}(x,z) = E(z)\varepsilon_{xx}^{(0)}(x), \quad \sigma_{xx}^{(1)}(x,z) = E(z)\varepsilon_{xx}^{(1)}(x),$$

$$\sigma_{xz}(x,z) = \frac{G(z)}{S_{xz}}\frac{dM_{xx}}{dx}. \qquad (9.10.8)$$

9.10.4 RBT DISPLACEMENT MODEL

In the case of the displacement model of the RBT, the strains $\varepsilon_{xx}^{(0)}$ and $\varepsilon_{xx}^{(1)}$ are computing using the expressions in Eqs. (9.10.1a) and (9.10.5b), respectively (with $\psi_i^{(1)} = \psi_i^{(2)}$). The higher-order strain components are computed as [see Eqs. (5.3.7a)–(5.3.7c) and (9.6.10)]

$$\varepsilon_{xx}^{(3)} = -\alpha\left(\frac{d\phi_x}{dx} + \frac{d^2 w}{dx^2}\right) \approx -\alpha\left(\sum_{j=1}^{n} S_j^e \frac{d\psi_j^{(2)}}{dx} + \sum_{J=1}^{4} \Delta_J^e \frac{d^2 \varphi_J^e}{dx^2}\right), \qquad (9.10.9a)$$

$$\gamma_{xz}^{(0)} = \phi_x + \frac{dw}{dx} \approx \sum_{j=1}^{n} S_j^e \psi_j^{(2)} + \sum_{J=1}^{n} \Delta_J^e \frac{d\varphi_J}{dx}, \qquad (9.10.9b)$$

$$\gamma_{xz}^{(2)} = -\beta\left(\phi_x + \frac{dw}{dx}\right) \approx -\beta\left(\sum_{j=1}^{n} S_j^e \psi_j^{(2)} + \sum_{J=1}^{n} \Delta_J^e \frac{d\varphi_J}{dx}\right). \qquad (9.10.9c)$$

The components of the axial stress $\sigma_{xx} = \sigma_{xx}^{(0)}(x) + z\sigma_{xx}^{(1)} + z^3\sigma_{xx}^{(3)}$ and the transverse shear stress $\sigma_{xz} = \sigma_{xz}^{(0)} + z^2\sigma_{xz}^{(2)}$ are given by

$$\sigma_{xx}^{(0)}(x,z) = E(z)\varepsilon_{xx}^{(0)}(x), \quad \sigma_{xx}^{(1)}(x,z) = E(z)\varepsilon_{xx}^{(1)}(x),$$

$$\sigma_{xx}^{(3)}(x,z) = E(z)\varepsilon_{xx}^{(3)}(x), \qquad (9.10.10)$$

$$\sigma_{xz}^{(0)}(x,z) = G(z)\gamma_{xz}^{(0)}(x), \quad \sigma_{xz}^{(2)} = G(z)\gamma_{xz}^{(2)}(x).$$

This completes the discussion of the post-computation of strains and stresses in various finite element models of beams. Although, the strains and stresses in the nonlinear case are higher-order functions of the position, in all displacement finite element models, the strains and stresses are evaluated at the reduced Gauss point locations (like in the linear analysis).

9.11   NUMERICAL RESULTS

9.11.1   GEOMETRY AND BOUNDARY CONDITIONS

Here, we consider a number of examples to evaluate the accuracy of the displacement and mixed finite element models in predicting the deflections and bending moments. All variables in the mixed models of the CBT and TBT and in the displacement model of the TBT are approximated using the same degree polynomials. In the displacement model of the CBT and RBT, u and ϕ_x are approximated using linear polynomials and w is approximated using Hermite cubic polynomials. There are five finite element models, which as designated as follows:

(1)   **CBT-D** – Displacement finite element model of the CBT
(2)   **CBT-M** – Mixed finite element model of the CBT
(3)   **TBT-D** – Displacement finite element model of the TBT
(4)   **TBT-M** – Mixed finite element model of the TBT
(5)   **RBT-D** – Displacement finite element model of the RBT

The beams considered are assumed to be straight, of length L, height h, and width b, and subjected to uniform distributed load, unless stated otherwise. The displacement boundary conditions for various types of supports are:

$$
\begin{aligned}
&\text{Hinged support (H):} && u = 0, \\
&\text{Pinned support (P):} && u = 0, \quad w = 0, \\
&\text{Clamped support (C):} && u = 0, \quad w = 0, \quad dw/dx = 0, \quad \phi_x = 0, \\
&\text{Transverse elastic support (E):} && N_{xz} + k(w - w_0) = 0,
\end{aligned}
\qquad (9.11.1)
$$

where w_0 denotes the vertical distance between the beam axis and any support. We note that the axial displacement u is decoupled from the bending displacements when the beam is isotropic and the analysis is linear. In that case, the hinged boundary condition is the same as the pinned boundary condition (when u is absent). In the nonlinear analysis, hinged boundary condition ($u \neq 0$) is not used because beam does not develop nonlinear strain.

If the beam problem being analyzed has a solution symmetry about its center, $x = L/2$, then, we use only the first half of the beam in the finite element analysis. The boundary conditions at $x = L/2$ are that the axial displacement, rotation, and shear force are all zero at $x = L/2$.

9.11.2   MATERIAL CONSTITUTION

The functionally graded beams considered here are assumed to be made of two materials (Poisson's ratio is assumed to be constant) and follow the power-law distribution given in Eq. (3.4.5a):

$$
E(z) = (E_1 - E_2)\, f(z) + E_2, \quad f(z) = \left(\frac{1}{2} + \frac{z}{h}\right)^n, \quad E_1 \geq E_2. \qquad (9.11.2)
$$

Note that $n = 0$ corresponds to homogeneous beam with $E = E_1$ (i.e., stiffer than the FGM beam). As n is increased, the beam material tends toward a homogeneous beam with modulus $E = E_2 < E_1$, and it is expected to deflect more. For the choice of $E(z)$ in Eq. (9.11.2), the beam stiffness coefficients A_{xx}, B_{xx}, D_{xx}, and so on can be evaluated analytically in terms of E_1, E_2, and n as given in Eqs. (5.3.24a)–(5.3.24d).

9.11.3 EXAMPLES

Example 9.11.1 ——

Consider a straight beam of length $L = 12$ m and rectangular cross section of width $b = 164$ mm and height $h = 400$ mm, and subjected to uniform distributed load of intensity $q_0 = 30$ kN/m, as shown in Fig. 9.11.1. Determine bending solutions (i.e., transverse deflection and bending moment) when the beam is simply supported (i.e., both ends are pinned) for the following cases: (a) linear solution for the isotropic case ($E = 200$ GPa, $\nu = 0.3$), (b) linear solution for functionally graded beams with $E_1 = 200$ GPa, $E_2 = 73.1$ GPa, and $\nu = 0.33$, and (c) nonlinear solution for isotropic and FGM beams with material properties given in case (b). Investigate the effect of mesh and finite element model on the linear solutions.

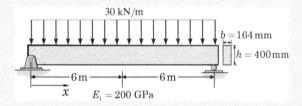

Figure 9.11.1: A simply supported beam with uniform load.

Solution: Let the origin of the coordinate system be taken at the left end, so that $x = 0$ and $x = L$ are the ends of the beam. Due to the symmetry of the solution about the center of the beam (i.e., $x = L/2$), we can use half beam as the computational domain. The essential boundary conditions (i.e., boundary conditions on the primary variables) are: $u(0) = w(0) = 0$ and $u = 0$ and $-dw/dx = 0$ (or $\phi_x = 0$) at $x = L/2$ for the displacement models; and $u(0) = w(0) = M(0) = 0$ and $-dw/dx = 0$ or $\phi_x = 0$ at $x = L/2$ for the mixed models.

(a) *Linear analysis of isotropic beams.* For linear analysis of isotropic beams, the axial displacement u is decoupled from the transverse displacement w, and since there is no axial force, we have $u(x) = 0$. For this case, there is no difference between pinned–pinned and hinged–hinged beams (and the Navier solution is valid for only hinged–hinged case) The exact solution for this problem according to the TBT is given by ($q_0 = -30$ kN/m)

$$w(\xi) = \frac{q_0 L^4}{24 D_{xx}}\left(\xi - 2\xi^3 + \xi^4\right) + \frac{q_0 L^2}{2 S_{xz}}\xi(1 - \xi), \quad M_{xx}(\xi) = \frac{q_0 L^2}{2}\xi(1 - \xi), \quad (9.11.3)$$

where $\xi = x/L$, $D_{xx} = EI = Ebh^3/12$, and $S_{xz} = Gbh\,K_s$. To obtain the CBT solution from Eq. (9.11.3), we set $1/S_{xz} = 0$.

Table 9.11.1 contains maximum deflection $w(0.5L)$ multiplied by 100, as predicted by various finite element models with different meshes. All models converge to the exact solution with mesh refinement. The effect of shear deformation is not significant for this beam, whose length-to-height ratio is $L/h = 12/0.4 = 30$. The CBT-D, and the CBT-M and TBT-M with quadratic elements (with 3 Gauss points for full and 2 Gauss points for reduced integration rules) give exact solutions for the deflection and bending moment ($M_{\max} = -540$ kN-m from all theories) for any mesh. We note that the solutions predicted by the TBT-D with linear elements converge slowly to the exact solution, although the difference is not significant enough to see in the plots presented in Figs. 9.11.2 and 9.11.3.

Table 9.11.1: Center deflections $-w(0.5L) \times 10^2$ of hinged–hinged isotropic beams under uniform distributed load as predicted by various beam finite element models.

	CBT		TBT		RBT
Mesh*	CBT-D**	CBT-M	TBT-D	TBT-M	RBT-D
4L	4.6303	4.5725	4.5274	4.5853	4.6366
8L	4.6303	4.6159	4.6142	4.6287	4.6411
16L	4.6303	4.6267	4.6359	4.6396	4.6426
32L	4.6303	4.6294	4.6414	4.6423	4.6430
2Q	– – –	4.6303	4.6432	4.6432	– – –
4Q	– – –	4.6303	4.6432	4.6432	– – –
8Q	– – –	4.6303	4.6432	4.6432	– – –
Exact	4.6303	4.6303	4.6432	4.6432	4.6432

* NL - N linear elements; NQ - N quadratic elements. ** Nodal values in the CBT-D are independent of the mesh and match with the exact CBT solution for isotropic beams; also, in the CBT-D model, the Hermite cubic finite element is used.

(b) *Linear analysis of FGM beams.* Next, we consider FGM beams with the same geometry and load as in the isotropic beam, but the material properties as listed in part (b). The exact solutions for the pinned–pinned FGM are known from Chapter 4 [see Eqs. (4.6.18)–(4.6.21), with $x/L = \xi$]:

$$u(x) = \frac{B_{xx}}{D_{xx}^*}\frac{q_0 L^3}{12}\left(\xi - 3\xi^2 + 2\xi^3\right), \tag{9.11.4a}$$

$$w(x) = \frac{A_{xx}}{D_{xx}^*}\frac{q_0 L^4}{24}\left(\xi - 2\xi^3 + \xi^4\right) + \frac{1}{S_{xz}}\frac{q_0 L^2}{2}\xi(1-\xi) - \frac{B_{xx}^2}{D_{xx}D_{xx}^*}\frac{q_0 L^4}{24}\xi(1-\xi), \tag{9.11.4b}$$

$$\phi_x(x) = -\frac{A_{xx}}{D_{xx}^*}\frac{q_0 L^3}{24}\left(1 - 6\xi^2 + 4\xi^3\right) + \frac{B_{xx}^2}{D_{xx}D_{xx}^*}\frac{q_0 L^3}{24}(1 - 2\xi), \tag{9.11.4c}$$

$$M_{xx}(x) = \frac{q_0 L^2}{2}\xi(1-\xi), \quad N_{xz}(x) = \frac{dM_{xx}}{dx} = \frac{q_0 L}{2}(1 - 2\xi), \tag{9.11.4d}$$

where the stiffness coefficients are defined in Eq. (3.4.7) with $D_{xx}^* = A_{xx}\,D_{xx} - B_{xx}^2$.

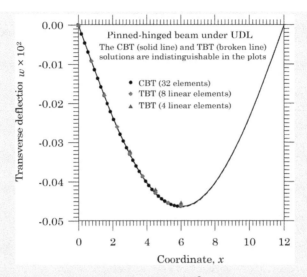

Figure 9.11.2: Transverse deflection $w(x) \times 10^2$ (m) versus the distance x (m) along the beam.

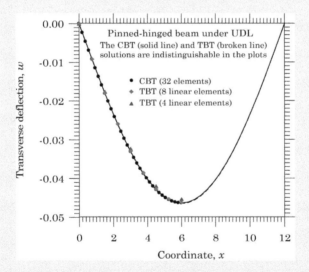

Figure 9.11.3: Bending moment $M(x)$ (N-m) versus the distance x (m) along the beam.

Table 9.11.2 contains the center transverse deflections predicted by various beam finite element models for power-law index of $n = 1.0, 2.0$ and 5. The mixed models with quadratic elements produce the exact maximum deflection. Of course, the finite element results predicted by various models agree with the exact

Table 9.11.2: Center deflections $-w(0.5L) \times 10^2$ of pinned–pinned FGM beams under uniform distributed load as predicted by various beam finite element models.

		CBT		TBT		RBT
n	Mesh	CBT-D	CBT-M	TBT-D	TBT-M	RBT-D
1.0	4L	6.8805	6.7957	6.7302	6.8150	6.8896
	8L	6.8854	6.8642	6.8623	6.8835	6.9012
	16L	6.8867	6.8814	6.8949	6.9006	6.9047
	32L	6.8870	6.8857	6.9036	6.9049	6.9056
	2Q	– – –	6.8871	6.9063	6.9063	– – –
	4Q	– – –	6.8871	6.9063	6.9063	– – –
	8Q	– – –	6.8871	6.9063	6.9063	– – –
	Exact	6.8871	6.8871	6.9063	6.9063	***
2.0	4L	7.6220	7.5285	7.4578	7.5513	7.6243
	8L	7.6292	7.6059	7.6053	7.6286	7.6465
	16L	7.6311	7.6252	7.6416	7.6480	7.6527
	32L	4.6303	0.8122	4.6414	0.5076	
	2Q	– – –	7.6317	7.6544	7.6544	– – –
	4Q	– – –	7.6317	7.6544	7.6544	– – –
	8Q	– – –	7.6317	7.6544	7.6544	– – –
	Exact	7.6317	7.6317	7.6544	7.6544	***
5.0	4L	8.5813	8.4754	8.3974	8.5033	8.5748
	8L	8.5869	8.5604	8.5619	8.5883	8.6070
	16L	8.5883	8.5817	8.6023	8.6096	8.6156
	32L	4.6303	0.8122	4.6414	0.5076	
	2Q	– – –	8.5888	8.6167	8.6167	– – –
	4Q	– – –	8.5888	8.6167	8.6167	– – –
	8Q	– – –	5.5888	8.6167	8.6167	– – –
	Exact	8.5888	8.5888	8.6167	8.6167	***

solutions (the exact RBT solution is not available) for axial displacement u, transverse deflection w, and the bending moment M. The difference between the models and theories cannot be seen in the plots, as is clear from the Figs. 9.11.4–9.11.7. Figure 9.11.7 shows a plot of the center transverse deflection versus the volume fraction index n. It is clear that the effect is more significant for smaller values than for larger values of n.

(c) *Nonlinear analysis of FGM beams.* Finally, the results of geometrically nonlinear analysis of FGM beams are presented. Fig. 9.11.8 contains plots of the normalized center deflection, $w(0.5L)/h$ versus the load parameter, $P = q_0 L^4/E_1 h^4$, for four different values of $n = 0, 1, 2$, and n. For the load range shown, the solutions predicted by the CBT and TBT are indistinguishable in the plots. Clearly, the beam experiences geometric nonlinearity as early as when w/h=0.4 for $m = 0$ (stiff beam), while it begins approximately when $w/h = 0.8$ for $n = 5$ (flexible beam).

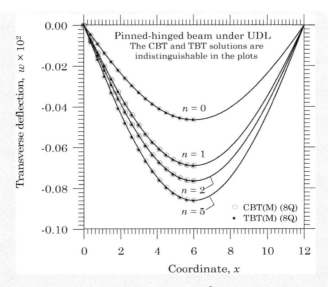

Figure 9.11.4: Transverse deflection $w(x) \times 10^2$ (m) versus the distance x (m) along the beam for $n = 0, 1, 2$, and 5.

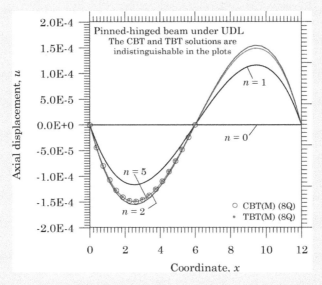

Figure 9.11.5: Axial displacement $u(x)$ (m) versus the distance x (m) along the beam for $n = 0, 1, 2$, and 5.

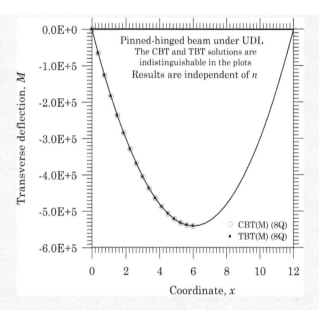

Figure 9.11.6: Bending moment $M(x)$ (N-m) versus the distance x (m) along the beam (the bending moment is independent of n).

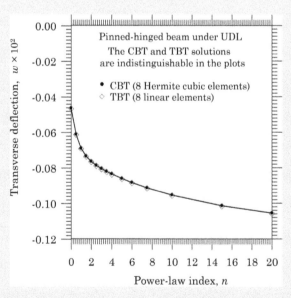

Figure 9.11.7: Maximum transverse deflection $w(0.5L) \times 10^2$ (N-m) versus n for the simply supported beam under uniform distributed load.

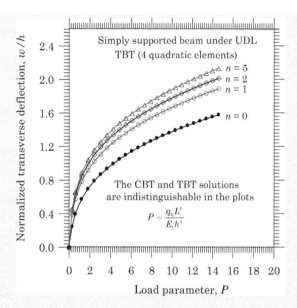

Figure 9.11.8: Maximum transverse deflection $w(0.5L)/h$ (N-m) versus $P = q_0 L^4/E_1 h^4$ for the simply supported beam under uniform distributed load of intensity q_0.

Example 9.11.2

Consider the beam in **Example 9.11.1** but with an elastic support at the center. Before the uniform distributed load is applied to the beam, there is is a small gap of 0.2 mm between the beam and the linear elastic post at point B, as shown in Fig. 9.11.9. The post at B has a diameter of $d_p = 40$ mm, length $L_p = 1$ m, and having a modulus of elasticity of $E_p = 200$ GPa. Analyze the problem using the displacement and mixed finite element models based on the CBT for isotropic and FGM beams. Assume symmetry about the centerline $x = 0.5L$, and model only the left half of the beam with suitable boundary conditions.

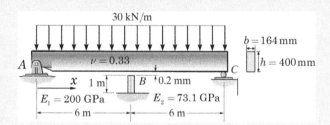

Figure 9.11.9: A simply supported straight beam with uniform load and supported at the center by a linear elastic post. Before the uniform distributed load is applied to the beam, there is a vertical gap of $w_0 = 0.2$ mm between the beam and the post.

Solution: Due to the symmetry of the solution about the center of the beam, we can use half beam as the computational domain, but with half the reaction of the post. For this problem, we must impose an additional mixed boundary condition at the center of the beam, $x = L/2$:

$$V + 0.5k(w - w_0) = 0, \quad w_0 = 0.2 \times 10^{-3} \text{ m},$$

$$0.5k = 0.5 \frac{A_p E_p}{L_p} = 0.1256637 \times 10^9, \tag{9.11.5}$$

where A_p is the cross-sectional area of the elastic post. Here, the constant k represents the axial stiffness of the post. The boundary condition in Eq. (9.10.6) is nothing but the equilibrium of transverse forces at the center point of the beam. In other words, the sum of the transverse shear force (V) at the center of the beam and the force $(0.5F_P)$ in the elastic post should be equal to zero. The force F_P is equal to the constant k times the elongation experienced by the post, and the elongation is $w(0.5L) - w_0$.

The exact CBT and TBT solutions of the linearized problem can be obtained using the method of superposition (see **Problems 3.7** and **4.4**). In the linear case, the axial displacement is decoupled from the transverse displacement and, since there is no axial force, we have $u(x) = 0$. **Problem 3.7** gives the steps to be followed to determine the force in the post. Once the forces are known, one can write shear force $V(x)$ and bending moment $M(x)$ at any point x; then use the differential relations between (V, M) and (w, ϕ_x) to obtain (by integration) the required solution for $w(x)$ and ϕ_x. The maximum deflection occurs at $\xi_0 = x/L = 0.25$, and it is equal to $w_{\max} = w(\xi_0) = -0.18963 \times 10^{-2}$ m. The maximum bending moment is $M(6) = 119.34$ kPa.

The linear analysis of the problem using 1, 2, 4, and 8 CBT elements yield the same nodal value for the transverse deflection at point B, namely, $w(0.5L) = -0.10745 \times 10^{-2}$. However, the accuracy of $w(x)$ and the post-computed bending moment $M(x)$ is a function of the mesh, as can be seen from results presented in Figs. 9.11.10 and 9.11.11. The 8-element solution for w at $x/L = 0.25$ matches with the exact solution $w_{\max}$. The bending moment computed in the 8-element mesh graphically matches with the exact solution. The mesh of 4 quadratic CBT-M elements gives solution that matches with the exact solution for both w and M at the nodes.

Figures 9.11.12 and 9.11.13 show the deflection $w(x)$ versus x/L and bending moment $M(x)$ versus x/L for FGM beams ($E_1 = 200$ GPa, $E_2 = 20$ GPa, and $\nu = 0.3$) with different values of n. The results were obtained using 8 CBT-M quadratic elements in the half beam. The bending moment does not change significantly with the volume fraction exponent n.

Next, geometrically nonlinear analysis of FGM beams (assuming pinned support: $u = w = 0$ at $x = 0$ and $u = dw/dx = 0$ at $x = L$) with the following material properties is carried out: $E_1 = 200$ GPa, $E_2 = 20$ GPa, and $\nu = 0.3$. The geometric parameters are kept the same as before. Figure 9.11.14 contains plots of the nonlinear deflection $w(x)$ versus x for various values of the load, $q = q_0$, while plots of the normalized deflection w/h versus the load parameter $P = q_0 L^4/E_1 h^4$ are shown in Fig. 9.11.15. The deflection at $x = 6$m, where an elastic post is placed, does not exhibit much nonlinearity, while the deflection at $x = 3$ m shows considerable nonlinearity, especially for larger values of n. This is due to the fact that stiffer beams require larger loads to push their response into the nonlinear range.

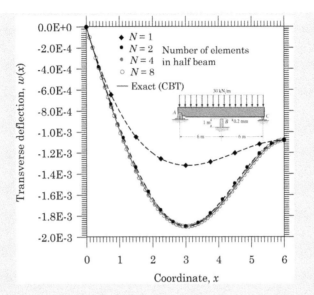

Figure 9.11.10: Transverse deflection $w(x)$ (m) versus the distance x (m) along the half beam.

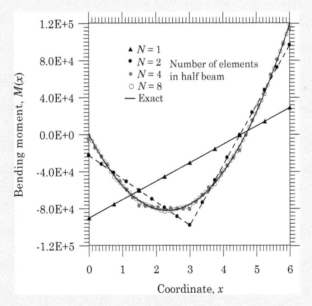

Figure 9.11.11: Bending moment $M(x)$ (N-m) versus the distance x (m) along the half beam.

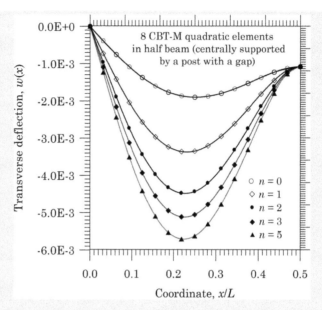

Figure 9.11.12: Transverse deflection $w(x)$ (m) versus the distance x (m) for FGM beams.

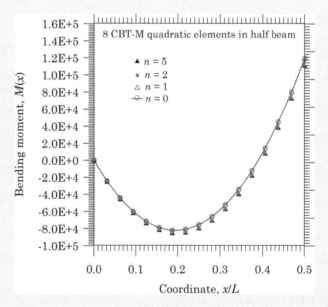

Figure 9.11.13: Bending moment $M(x)$ (N-m) versus the distance x (m) for FGM beams.

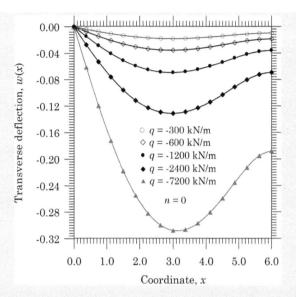

Figure 9.11.14: Transverse deflection $w(x)$ (m) versus the distance x (m) for FGM beams.

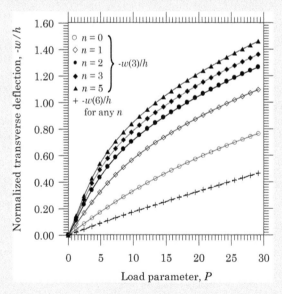

Figure 9.11.15: Normalized transverse deflection $-w/h$ versus the distance x for FGM beams.

Example 9.11.3 ————————————————————————————

Consider a beam of length $L = 100$ in., rectangular cross section with height $h = 1$ in. and width $b = 1$ in., and loaded with uniform distributed load of intensity q_0. Consider pinned and clamped boundary conditions at both ends of the beam: (a) pinned–pinned (P–P) and (b) clamped–clamped (C–C). Analyze the two types of problems for linear homogeneous ($E_1 = 30 \times 10^6$ and $\nu = 0.3$), linear FGM ($E_1 = 30 \times 10^6$, $E_1 = 3 \times 10^6$, and $\nu = 0.3$), and nonlinear FGM cases using the displacement and mixed models of the CBT and RBT.

Solution: Using the symmetry about $x = L/2$, we use half beam model for the analyses.

The exact solutions of P–P functionally graded beams, with the power-law given in Eq. (9.11.2), are given in Eqs. (9.11.4a)–(9.11.4d)]. The exact solutions for clamped-clamped beams according to the linear TBT are given by [see Eqs. (4.6.27)–(4.6.31)]

$$u(x) = \frac{B_{xx}}{D_{xx}^*}\frac{q_0L^3}{12}\left(\xi - 3\xi^2 + 2\xi^3\right),$$

$$w(x) = \frac{A_{xx}}{D_{xx}^*}\frac{q_0L^4}{24}\xi^2(1-\xi)^2 + \frac{1}{S_{xz}}\frac{q_0L^2}{2}\xi(1-\xi),$$

$$\phi_x(x) = \frac{A_{xx}}{D_{xx}^*}\frac{q_0L^3}{12}\left(-\xi + 3\xi^2 - 2\xi^3\right), \qquad (9.11.6)$$

$$M(x) = -\frac{q_0L^2}{12}\left(1 - 6\xi + 6\xi^2\right), \quad N_{xz}(x) = \frac{q_0L}{2}(1 - 2\xi),$$

$$\sigma(x, 0.5h) = -\frac{q_0L^2}{24I}\left(1 - 6\xi + 6\xi^2\right),$$

where $D_{xx}^* = A_{xx}D_{xx} - B_{xx}^2$ and $\hat{D}_{xx} = D_{xx}^*/A_{xx}$. The CBT solutions can be obtained from Eq. (9.11.7) by setting $1/S_{xz} = 0$ and replacing ϕ_x with $-dw/dx$. The RBT has no known exact solution for the clamped–clamped beams.

Linear Analysis

We note that for homogeneous beams we have $u(x) = 0$ everywhere. The bending stress, $\sigma(x, z)$, is computed at $x = L/2$ (where the bending moment is the maximum) and $z = h/2$, h being the beam height. The stress is post-computed in the displacement models at the element center using the relation

$$\sigma(x, z) = -E(z)\, z\frac{d^2w}{dx^2} \quad \text{for CBT},$$
$$\sigma(x, z) = E(z)\, z\frac{d\phi_x}{dx} \quad \text{for TBT}. \qquad (9.11.7)$$

On the other hand, the stress in the mixed models is computed using the formula

$$\sigma(x, z) = \frac{M_{xx}(x)\, z}{I}, \qquad (9.11.8)$$

where $M_{xx}(x)$ is the calculated from the finite element interpolation (i.e., $\sigma(L/2, h/2) = M(L/2)h/2I$, and $M(L/2)$ is the nodal value).

(a)*Pinned–pinned (P–P) beams.* Tables 9.11.3 and 9.11.4 contain the normalized center deflection and stress, respectively, for homogeneous ($\hat{D}_{xx} = D_{xx} = EI$ and

$\hat{B}_{xx} = B_{xx} = 0$) pinned–pinned beams for different number of elements in the half beam. The tabulated deflections and stresses are normalized as follows (with $K_s = 5/6$ and $\nu = 0.3$):

$$\bar{w} = w \, \frac{\hat{D}_{xx}}{q_0 L^4}, \quad \hat{D}_{xx} = D^*_{xx}/A_{xx}, \quad \bar{\sigma} = \sigma \, \frac{I}{h \, q_0 L^2}, \quad I = \frac{bh^3}{12}, \quad (9.11.9)$$

where I is the moment of inertia, h is height, and b is the width of the beam. From the results it is clear that the mixed models give the exact stress for any number of elements. The mixed models with 2 quadratic elements also give exact solution for the deflection (not included in the tables). We also note that for this slender beam ($L/h = 100$), the effect of shear deformation is negligible and the CBT, TBT, and RBT solutions for $\bar{w}$ are the same up to the fourth decimal point.

Table 9.11.3: The center transverse deflection $\bar{w}(L/2) \times 10$ of homogeneous P–P beams predicted by various finite element models.

Mesh	CBT-D	CBT-M	TBT-D	TBT-M	RBT-D
4	0.1302	0.1286	0.1270	0.1285	0.1301
8	0.1302	0.1298	0.1294	0.1298	0.1302
16	0.1302	0.1301	0.1300	0.1301	0.1302
32	0.1302	0.1302	0.1302	0.1302	0.1302
64	0.1302	0.1302	0.1302	0.1302	0.1302
Exact	0.1302	0.1302	0.1302	0.1302	*

Table 9.11.4: The center stress $\bar{\sigma}(L/2) \times 10$ for homogeneous P–P beams predicted by various finite element models (stresses in the displacement models are computed at the element center).

Mesh	CBT-D	CBT-M	TBT-D	TBT-M
4	0.6120	0.6250	0.6055	0.6250
8	0.6218	0.6250	0.6201	0.6250
16	0.6242	0.6250	0.6238	0.6250
32	0.6248	0.6250	0.6250	0.6250
64	0.6249	0.6250	0.6250	0.6250
Exact	0.6250	0.6250	0.6250	0.6250

Table 9.11.5 contains the normalized center deflection for functionally graded pinned–pinned beams for different values of the power-law index n. All of the results are obtained using 16 elements in the half beam. All models predict solutions that match the exact solutions up to the fourth decimal point. The stresses in the FGM and homogeneous beams are the same, because the bending moment is independent of the stiffness coefficients.

Next, we examine the effect of the power-law exponent n on the deflections. For increasing value of n, the beam becomes one made of modulus E_2, and due to the fact that $E_2 < E_1$, the deflection will increase with increasing values of n. However, if the deflection is normalized with respect to the stiffness coefficient D_{xx}, we can see the effect of the coupling stiffness B_{xx} on the deflection.

Table 9.11.5: The center transverse deflection $\bar{w}(L/2) \times 10$ of FGM P–P beams predicted by various models.

n	CBT-D	TBT-D	CBT-M	TBT-M	CBT-Exact	TBT-Exact
0.0	0.1302	0.1302	0.1302	0.1302	0.1302	0.1302
1.0	0.1069	0.1068	0.1069	0.1069	0.1070	0.1070
2.0	0.0920	0.0919	0.0919	0.0919	0.0920	0.0920
3.0	0.0879	0.0878	0.0879	0.0879	0.0880	0.0880
5.0	0.0900	0.0899	0.0899	0.0900	0.0900	0.0900
7.5	0.0959	0.0958	0.0958	0.0958	0.0959	0.0960
10.0	0.1012	0.1011	0.1012	0.1012	0.1013	0.1013
12.0	0.1048	0.1047	0.1047	0.1048	0.1048	0.1049
15.0	0.1091	0.1090	0.1090	0.1090	0.1091	0.1092
20.0	0.1142	0.1140	0.1141	0.1141	0.1142	0.1142

We note from Fig. 9.11.16 that the effect of the power-law index n on the dimensionless deflection $\bar{w} = w\,\hat{D}_{xx}/q_0 L^4$ is *not* monotonic. As n goes from zero to a value of about $n = 2$, the deflection decreases and then increases for $n > 2$. This is due to the fact that B_{xx} is not a monotonically increasing or decreasing function of n, as can be seen from Fig. 3.4.1. The finite element and analytical solutions are indistinguishable in the plots and thus the finite element results are not shown with symbols, except for $n = 1$. Figure 9.11.17 shows the variation of the normalized center deflection $\bar{w}$ with the power-law index n. It should be noted that the actual deflection w monotonically decreases with increasing value of n, as the beam goes from homogeneous beam of modulus E_1 to a homogeneous beam of modulus $E_2 < E1$.

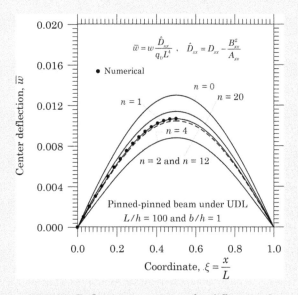

Figure 9.11.16: Deflection $\bar{w}$ versus x for different values of n.

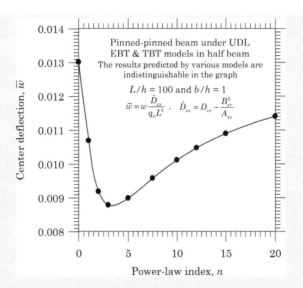

Figure 9.11.17: Variation of the normalized center deflection $\bar{w}$ with n for P–P beams.

(b) *Clamped–clamped (C–C) beams.* Tables 9.11.6 and 9.10.7 contain the normalized center deflection and stress, respectively, for the homogeneous C–C beam. The stress is computed at the center of the last element. The TBT-D model with linear elements is the only one that is slow in converging to the exact solution with mesh refinement, whereas with the mesh of 2 quadratic elements it gives the exact solution. No exact RBT solution for the clamped beam is available. The RBT mesh requires specification (to be zero) of u, w, $-dw/dx$, ϕ_x at $x = 0$ and u, $-dw/dx$, ϕ_x at $x = L/2$.

Figures 9.11.18 and 9.11.19 contains plots of the transverse deflection $w(x)$ versus x and bending moment $M(x)$ versus x, respectively, for isotropic clamped–clamped beams under uniform distributed load $q_0 = 1$ lb/in. The difference between the results predicted by various models cannot be seen in the plots.

Table 9.11.6: The center transverse deflection $\bar{w} \times 10^2$ predicted by various models for homogeneous C–C beams ($\bar{w} = w(L/2)\hat{D}_{xx}/q_0 L^4$).

Mesh	CBT-D	CBT-M	TBT-D	TBT-M	RBT-D
4	0.2604	0.2604	0.2445	0.2607	0.2571
8	0.2604	0.2604	0.2567	0.2607	0.2599
16	0.2604	0.2604	0.2597	0.2607	0.2606
32	0.2604	0.2604	0.2605	0.2607	0.2607
64	0.2604	0.2604	0.2607	0.2607	0.2607
Exact	0.2604	0.2604	0.2607	0.2607	*

Table 9.10.7: The center stress $\bar{\sigma}(0) \times 10$ predicted by various models for homogeneous C–C beams.

Mesh	CBT-D	CBT-M	TBT-D	TBT-M	RBT-D
4	0.1953	0.1953	0.1953	0.1953	0.2398
8	0.2051	0.2051	0.2051	0.2051	0.2137
16	0.2075	0.2075	0.2075	0.2075	0.2097
32	0.2081	0.2082	0.2082	0.2082	0.2088
64	0.2083	0.2083	0.2083	0.2083	0.2085
Exact	0.2083	0.2083	0.2083	0.2083	*

Table 9.11.8 contains the normalized center deflection for the functionally graded C–C beam for different values of n. All of the results were obtained with 16 elements in half beam. For the C–C beams the normalized deflections do not deviate significantly from each other (they differ only in the fourth or fifth decimal point). Figure 9.11.20 shows the variation of the normalized deflection $\bar{w}$ versus n, which has the same form as that for the P-P beams.

Table 9.11.8: The center transverse deflection $\bar{w} \times 10^2$ of FGM C–C beams predicted by various models ($\bar{w} = w(L/2)\,\hat{D}_{xx}/q_0 L^4$).

n	CBT-D	TBT-D	CBT-M	TBT-M	CBT-Exact	TBT-Exact
0.0	0.26040	0.26070	0.26040	0.26070	0.26042	0.26074
1.0	0.26019	0.25965	0.26019	0.26044	0.26042	0.26067
2.0	0.26004	0.25964	0.26004	0.26028	0.26042	0.26065
3.0	0.26000	0.25965	0.26000	0.26025	0.26042	0.26066
5.0	0.26002	0.25968	0.26002	0.26031	0.26042	0.26070
7.5	0.26008	0.25973	0.26008	0.26041	0.26040	0.26090
10.0	0.26013	0.25976	0.26013	0.26049	0.26042	0.26077
12.0	0.26017	0.25977	0.26017	0.26054	0.26042	0.26090
15.0	0.26021	0.25979	0.26021	0.26060	0.26042	0.26080
20.0	0.26026	0.25980	0.26026	0.26066	0.26042	0.26082

Nonlinear Analysis

Here we present numerical results for nonlinear analysis of the same two problems, namely, pinned-pinned (P-P) and clamped–clamped (C–C) beams. All of the results are obtained with 16 Hermite cubic elements of CBT(D) and RBT (D) models and 16 linear elements of CBT(M), TBT(D), and TBT(M) models in half beam, although 8 quadratic elements of CBT(M), TBT(D), and TBT(M) also yield almost the same results. In fact, the results predicted by all models are indistinguishable in the graphs. Hence, the symbols shown in the graphs can be interpreted as those of any of one of the finite element models.

(a) *Pinned–pinned (P–P) beams.* First, we present results for P–P beams. Figures 9.11.21 and 9.11.22 show finite element result of deflection w versus q_0 and dimensionless deflection $\bar{w}$ versus load q_0, respectively, for different values of n. The results obtained with various finite element models cannot be distinguished and thus no model label is shown in the figures. It is clear that the dimensionless deflections do not monotonically increase with n, whereas the dimensional deflections do monotonically increase with n (see Figs. 9.11.23 and 9.11.24). As discussed earlier, this is

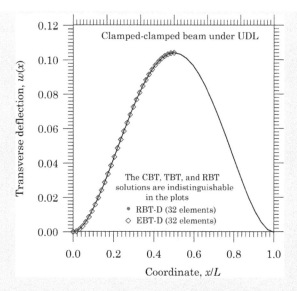

Figure 9.11.18: Deflection w versus x predicted by the CBT-D and RBT-D elements for clamped–clamped beams under uniform distributed load $q_0 = 1$ lb/in.

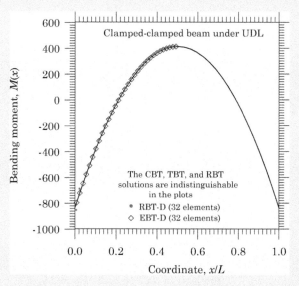

Figure 9.11.19: Bending moment M versus x predicted by the CBT-D and RBT-D elements for clamped–clamped beams under uniform distributed load $q_0 = 1$ lb/in.

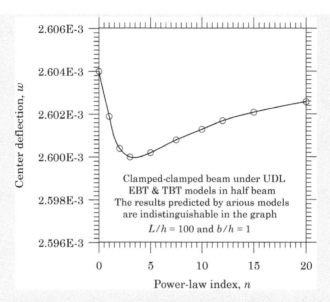

Figure 9.11.20: Variation of the normalized center deflection w with n for C–C beams.

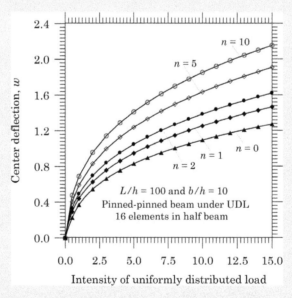

Figure 9.11.21: Variation of the center deflection $\bar{w}$ versus q_0 for P–P beams ($\bar{w} = w(L/2)\hat{D}_{xx}/q_0 L^4$).

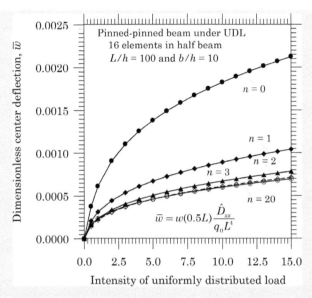

Figure 9.11.22: Variation of the dimensionless center deflection $\bar{w}$ versus q_0 for P–P beams ($\bar{w} = w(L/2)\hat{D}_{xx}/q_0 L^4$).

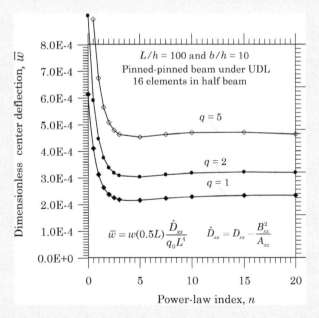

Figure 9.11.23: Variation of the center deflection $\bar{w}$ versus n for P–P beams ($\bar{w} = w(L/2)\hat{D}_{xx}/q_0 L^4$).

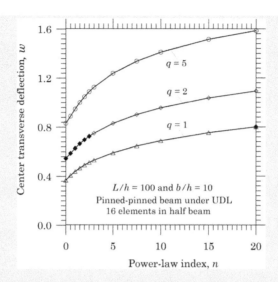

Figure 9.11.24: Variation of the center deflection w versus n for P–P beams.

due to the fact that $\bar{B}_{xx}$ does not monotonically increase/decrease with n. Figures 9.11.25 and 9.11.26 show deflection w versus axial distance x/L for $n = 0$ and $n = 5$ respectively, for different values of n.

(b) *Clamped–clamped (C-C) beams.* Next, results for C–C beams are presented. Figures 9.11.27 and 9.11.28 show finite element results for deflection w versus load q_0 and dimensionless deflection $\bar{w}$ versus load q_0, respectively, for different values of n. The results obtained with various finite models cannot be distinguished and thus no model label is indicated in the figures. Figures 9.11.29 and 9.11.30 show deflection w versus axial distance x/L for $n = 0$ and $n = 5$ respectively, for different values of n. The results for C–C beams have trends similar to those of the P–P beams.

Although the examples presented here are limited to some standard boundary conditions, loads, and geometric and material parameters, the finite element models developed in this chapter are general enough to solve a variety problems of straight beams. For example, beams with C–F (clamped–free or cantilever) and C–P (clamped–pinned) boundary conditions, over-hang beams, and beams with point loads and a variety of distributed loads (e.g., sinusoidal distributed and linear varying loads) can be analyzed. The effect of couple stress is simple to numerically investigate, as it is already included in the formulation. In addition, a variety of two constituent functionally graded beams can be analyzed for any choice of material properties. In fact, the choice of the material variation through the beam height can be changed and associated stiffness coefficients can be derived either analytically or numerically.

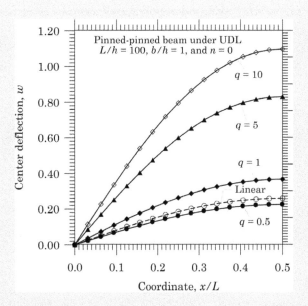

Figure 9.11.25: Variation of the center deflection w versus x/L for P–P beams $(n = 0)$.

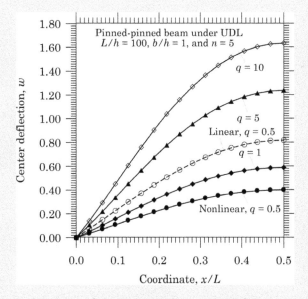

Figure 9.11.26: Variation of the dimensionless center deflection w versus x/L for P–P beams $(n = 5)$.

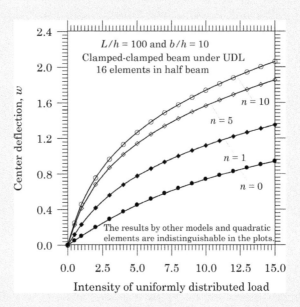

Figure 9.11.27: Variation of the center deflection w versus q_0 for C–C beams.

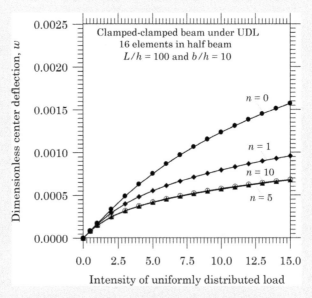

Figure 9.11.28: Variation of the dimensionless center deflection $\bar{w}$ versus q_0 for C–C beams.

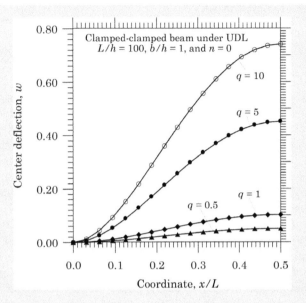

Figure 9.11.29: Center deflection w versus x/L for C–C beams $(n = 0)$.

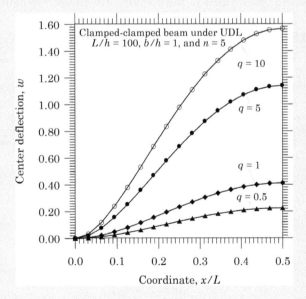

Figure 9.11.30: Center deflection w versus x/L for C–C beams $(n = 5)$.

9.12 CHAPTER SUMMARY

In this chapter, displacement and mixed finite element models of straight beams made of homogeneous as well as power-law based functionally graded (through the thickness) materials are presented. The classical (CBT), Timoshenko (TBT), and Reddy (RBT) beam theories with the von Kármán non-linearity are used. Five different finite element models are developed. The mixed models CBT-M and TBT-M have three degrees of freedom (u, w, M); the displacement models of the CBT-D and TBT-D have three degrees of freedom $(u, w, -dw/dx$ or $\phi_x)$, while that of the RBT-D have four nodal degrees of freedom $(u, w, -dw/dx, \phi_x)$. In addition, the tangent stiffness coefficients for all five models are derived. Full discretization of the structural dynamics equations using the Newmark scheme and the solution of nonlinear algebraic equations using the Picard and Newton's method are discussed.

Numerical examples of static bending are presented to illustrate the accuracy of various models and bring out certain salient features of functionally graded beams. No numerical results of the transient response are included here. The mixed models have the advantage of enabling stress calculation at the boundary nodes, instead of Gauss points in the interior of the domain. Numerical results indicate that all models yield results that are indistinguishable in graphs. The mixed models give very accurate results for the bending moments, which can be used to the compute stresses at the nodes.

SUGGESTED EXERCISES

9.1 Verify the tangent stiffness coefficients in Eqs. (9.9.7a) and (9.9.7b) for the mixed model of the CBT FGM beams.

9.2 Verify the tangent stiffness coefficients in Eqs. (9.9.14a)–(9.9.14c) for the displacement model of the RBT FGM beams.

9.3 Consider the beam of **Problem 3.6** (see Fig. P9.3). Use the connectivity of the two beams to obtain the assembled matrix and determine the finite element solution (do not include the axial degrees of freedom because they are zero for linear bending analysis). You may use the element matrices listed in Eq. (9.2.10c).

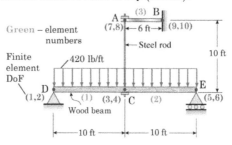

Figure P9.3

Hint: The unknown degrees of freedom (i.e., eliminate the rows and columns associated with the known degrees of freedom, Δ_1, Δ_5, Δ_9, and Δ_{10}; see Fig. P9.3) are: Δ_2, Δ_3,

Δ_4, Δ_6, Δ_7, and Δ_8. The effect of the steel rod (connecting degrees of freedom Δ_3 and Δ_7) should be included as a spring constant ($k = AE/L$).

9.4 Repeat **Problem 9.3** using the TBT element with $K_s = 5/6$ and $G = 75$ GPa. Use the following element matrices of the reduced-integration TBT element (see Reddy [155], p. 297):

$$\mathbf{K}^{22} = \frac{G_e A_e K_s}{4h_e} \begin{bmatrix} 4 & -2h_e & -4 & -2h_e \\ -2h_e & (1+4\mu_e)h_e^2 & 2h_e & (1-4\mu_e)h_e^2 \\ -4 & 2h_e & 4 & 2h_e \\ -2h_e & (1-4\mu_e)h_e^2 & 2h_e & (1+4\mu_e)h_e^2 \end{bmatrix}, \quad \mathbf{q}^e = \frac{q_0^e h_e}{2} \begin{Bmatrix} 1 \\ 0 \\ 1 \\ 0 \end{Bmatrix},$$

where $\mu_e = E_e I_e/(G_e A_e K_s h_e^2)$. With this coarse mesh, do not expect the solution to be accurate.

9.5 Consider the following equilibrium equations of the TBT for an FGM beam in the absence of distributed load $q(x)$ [see Eqs. (4.6.1)–(4.6.3), omitting the thermal resultants]:

$$-\frac{d}{dx}\left[A_{xx}\left(\frac{du}{dx}\right) + B_{xx}\frac{d\phi_x}{dx}\right] = 0, \tag{1}$$

$$-\frac{d}{dx}\left[S_{xz}\left(\phi_x + \frac{dw}{dx}\right)\right] = 0, \tag{2}$$

$$-\frac{d}{dx}\left(B_{xx}\frac{du}{dx} + D_{xx}\frac{d\phi_x}{dx}\right) + S_{xz}\left(\phi_x + \frac{dw}{dx}\right) = 0. \tag{3}$$

The exact solution of Eqs. (1)–(3) is of the form [see Eqs. (4.6.12), (4.6.13), and (4.6.16), and omitting the thermal resultants]

$$u(x) = \frac{D_{xx}}{D_{xx}^*}c_1 x - \frac{B_{xx}}{D_{xx}^*}\left(c_2\frac{x^2}{2} + c_3 x + c_4\right), \tag{4}$$

$$\phi(x) = -\frac{B_{xx}}{D_{xx}^*}c_1 x + \frac{A_{xx}}{D_{xx}^*}\left(c_2\frac{x^2}{2} + c_3 x + c_5\right), \tag{5}$$

$$w(x) = \frac{1}{S_{xz}}c_2 x + \frac{B_{xx}}{D_{xx}^*}c_1\frac{x^2}{2} - \frac{A_{xx}}{D_{xx}^*}\left(c_2\frac{x^3}{6} + c_3\frac{x^2}{2} + c_5 x + c_6\right), \tag{6}$$

where c_1 through c_6 are the constants of integration. Equations (4)–(6) suggest that one may use cubic approximation of $w(x)$ and an *interdependent* quadratic approximation of $u(x)$ and ϕ_x, because the same constants as those in the expression for $w(x)$ also appear in the expressions for $u(x)$ and $\phi_x(x)$.

For a homogeneous beam, we set $B_{xx} = 0$ and obtain the following expressions for ϕ_x and $w(x)$ (omitting the axial deformation, which is decoupled from bending deformation):

$$\phi(x) = \frac{1}{D_{xx}}\left(K_1\frac{x^2}{2} + K_2 x + K_3\right), \tag{7}$$

$$w(x) = \frac{1}{S_{xz}}K_1 x - \frac{1}{D_{xx}}\left(K_1\frac{x^3}{6} + K_2\frac{x^2}{2} + K_3 x + K_4\right), \tag{8}$$

where we have replaced cs with Ks: $K_1 = c_2$, $K_2 = c_3$, $K_3 = c_5$, and $K_4 = c_6$.

Now consider a finite element of length h, with nodes places at the end points (like in the CBT beam element); let the origin of the x-coordinate be at node 1 (left end of the element). Then express the four constants, K_1 through K_6, in terms of the following four nodal variables:

$$\Delta_1 = w(0), \quad \Delta_2 = \phi_x(0), \quad \Delta_3 = w(h), \quad \Delta_4 = \phi_x(h), \tag{9}$$

and express $w(x)$ and $\phi(x)$ in the form (which requires rewriting K_1–K_4 in terms of Δ_1–Δ_4)

$$w(x) \approx \sum_{j=1}^{4}\varphi_j^{(1)}\Delta_j, \quad \phi_x(x) \approx \sum_{j=1}^{4}\varphi_j^{(2)}\Delta_j. \tag{10}$$

In particular, show that $\varphi_i^{(1)}$ and $\varphi_i^{(2)}$ are given by

$$\varphi_1^{(1)} = \frac{1}{\mu}\left(\mu - 12\Omega\xi - 3\xi^2 + 2\xi^3\right),$$

$$\varphi_2^{(1)} = -\frac{h}{\mu}(1-\xi)\xi\left(1 + 6\Omega - \xi\right),$$

$$\varphi_3^{(1)} = \frac{1}{\mu}\left(12\Omega\xi + 3\xi^2 - 2\xi^3\right),$$

$$\varphi_4^{(1)} = \frac{h}{\mu}(1-\xi)\xi\left(6\Omega + \xi\right), \tag{11}$$

$$\varphi_1^{(2)} = \frac{6}{h\mu}(1-\xi)\xi,$$

$$\varphi_2^{(2)} = \frac{1}{\mu}\left[\mu - 4(1+3\Omega)\xi + 3\xi^2\right],$$

$$\varphi_3^{(2)} = -\frac{6}{h\mu}(1-\xi)\xi,$$

$$\varphi_4^{(2)} = \frac{1}{\mu}\left[-2(1-6\Omega)\xi + 3\xi^2\right]. \tag{12}$$

Here η is the non-dimensional local coordinate

$$\xi = \frac{x}{h}, \quad \mu = 1 + 12\Omega, \quad \Omega = \frac{D_{xx}}{S_{xz}h^2}. \tag{ix}$$

9.6 (*Continuation of Problem 9.5*) The displacement finite element model of the TBT is constructed using the principle of minimum total potential energy, or equivalently, using the weak form over the element,

$$0 = \int_0^h \left[D_{xx}\frac{d\delta\phi_x}{dx}\frac{d\phi_x}{dx} + S_{xz}\left(\delta\phi_x + \frac{d\delta w}{dx}\right)\left(\phi_x + \frac{dw}{dx}\right)\right] dx$$

$$- \int h_0 q(x)\delta w\, dx - Q_1\delta w(0) - Q_3\delta w(h) - Q_2\delta\phi_x(0) - Q_4\delta\phi_x(h), \tag{1}$$

where

$$Q_1 \equiv -N_{xz}(0) = -\left[S_{xz}\left(\frac{dw}{dx} + \phi_x\right)\right]_{x=0},$$

$$Q_3 \equiv -M_{xx}(0) = -\left[D_{xx}\frac{d\phi_x}{dx}\right]_{x=0},$$

$$Q_2 \equiv N_{xz}(h) = \left[S_{xz}\left(\frac{dw}{dx} + \phi_x\right)\right]_{x=h},$$

$$Q_4 \equiv M_{xx}(h) = \left[D_{xx}\frac{d\phi_x}{dx}\right]_{x=h}, \tag{2}$$

Substitute the approximation (10) of **Problem 9.5** into the weak form in Eq. (1) and show that the finite element model is of the form

$$\mathbf{K}\boldsymbol{\Delta} = \mathbf{q} + \mathbf{Q}, \tag{3}$$

where

$$K_{ij} = \int_0^h \left[S_{xz}\left(\varphi_i^{(2)} + \frac{d\varphi_i^{(1)}}{dx}\right)\left(\varphi_j^{(2)} + \frac{d\varphi_j^{(1)}}{dx}\right)\right.$$

$$\left. + D_{xx}\frac{d\varphi_i^{(2)}}{dx}\frac{d\varphi_j^{(2)}}{dx}\right] dx, \tag{4}$$

$$q_i = \int_0^h \varphi_i^{(1)} q(x)\, dx. \tag{5}$$

9.7 (*Continuation of Problems 9.6*) Show that Eq. (3) of **Problem 9.6** has the explicit form

$$
\frac{2D_{xx}}{\mu h^3}
\begin{bmatrix}
6 & -3h & -6 & -3h \\
-3h & 2h^2\lambda & 3h & h^2\kappa \\
-6 & 3h & 6 & 3h \\
-3h & h^2\kappa & 3h & 2h^2\lambda
\end{bmatrix}
\begin{Bmatrix}
\Delta_1 \\ \Delta_2 \\ \Delta_3 \\ \Delta_4
\end{Bmatrix}
=
\begin{Bmatrix}
q_1 \\ q_2 \\ q_3 \\ q_4
\end{Bmatrix}
+
\begin{Bmatrix}
Q_1 \\ Q_2 \\ Q_3 \\ Q_4
\end{Bmatrix},
\tag{6}
$$

where

$$
\lambda = 1 + 3\Omega, \qquad \kappa = 1 - 6\Omega.
\tag{7}
$$

9.8 Develop the beam finite element model based on the Levinson beam theory.

It is really quite amazing by what margins competent but conservative scientists and engineers can miss the mark, when they start with the preconceived idea that what they are investigating is impossible. When this happens, the most well-informed men become blinded by their prejudices and are unable to see what lies directly ahead of them.

Arthur C. Clarke

10 Finite Element Analysis of Circular Plates

There does not exist a category of science to which one can give the name applied science. There are science and the applications of science, bound together as the fruit of the tree which bears it.

Louis Pasteur

10.1 INTRODUCTORY REMARKS

This chapter is devoted to the finite element modeling of axisymmetric circular plates based on various plate theories. In particular, we will consider linear and nonlinear bending analysis of homogeneous and through-thickness functionally graded circular plates with different boundary conditions. Some of the finite element models presented herein cannot be found in the literature at the time of this writing.

In particular, we develop the displacement finite element models of the classical (CPT-D), first-order (FST-D), and third-order (TST-D) plate theories, and a mixed model of the classical plate theory (CPT-M). These finite element models account for the geometric nonlinearity, material gradation through the thickness, and the couple stress effect. As is made clear in the previous chapters, the domain of analysis for all axisymmetric circular plate problems is a radial line. In the polar coordinate system, the domain can be represented as a line between points $r = b$ and $r = a$, with $b = 0$ for solid circular plates (see Fig. 10.1.1). In the finite element

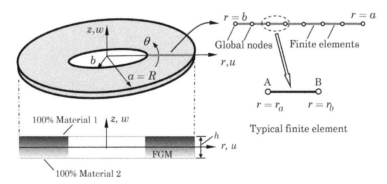

Figure 10.1.1: Axisymmetric bending of an annular plate of inner radius b and outer radius a; the computational domain and its mesh are highlighted in red.

DOI: 10.1201/9781003240846-10

analysis, a typical finite element occupies the domain between two points A and B with coordinates $r = r_a$ and $r = r_b$, respectively [i.e., $\Omega^e = (r_a, r_b)$]. The generalized primary and secondary variables are not shown on the element because they are different for different finite element model. Many of the ideas of the finite element analysis presented in Chapter 9 are also valid here.

10.2   DISPLACEMENT MODEL OF THE CPT

10.2.1   WEAK FORMS

From Hamilton's statement in Eq. (6.2.5), we have the following variational statement for the CPT:

$$
0 = \int_0^T \int_b^a \left[m_0 \left(\ddot{u}\, \delta u + \ddot{w}\, \delta w \right) - m_1 \left(\frac{\partial \ddot{w}}{\partial r} \delta u + \ddot{u}\, \frac{\partial \delta w}{\partial r} \right) + m_2 \frac{\partial \ddot{w}}{\partial r} \frac{\partial \delta w}{\partial r} \right.
$$
$$
+ N_{rr} \left(\frac{\partial \delta u}{\partial r} + \frac{\partial w}{\partial r} \frac{\partial \delta w}{\partial r} \right) + N_{\theta\theta} \left(\frac{\delta u}{r} \right) - M_{rr} \frac{\partial^2 \delta w}{\partial r^2}
$$
$$
\left. - \frac{1}{r} M_{\theta\theta} \frac{\partial \delta w}{\partial r} + P_{r\theta} \left(-\frac{\partial^2 \delta w}{\partial r^2} + \frac{1}{r} \frac{\partial \delta w}{\partial r} \right) - q\delta w \right] r\, dr\, dt. \quad (10.2.1)
$$

The statement in Eq. (10.2.1) is equivalent to the following two semi-discrete variational statements when specialized to a typical finite element $\Omega^e = (r_a, r_b)$:

$$
0 = \int_{r_a}^{r_b} \left[m_0 \ddot{u}\, \delta u - m_1 \frac{\partial \ddot{w}}{\partial r} \delta u + N_{rr} \frac{\partial \delta u}{\partial r} + N_{\theta\theta} \left(\frac{\delta u}{r} \right) \right] r\, dr
$$
$$
- Q_1 \delta u(r_a, t) - Q_4 \delta u(r_b, t), \quad (10.2.2)
$$
$$
0 = \int_{r_a}^{r_b} \left[m_0 \ddot{w}\delta w - m_1 \ddot{u} \frac{\partial \delta w}{\partial r} + m_2 \frac{\partial \ddot{w}}{\partial r} \frac{\partial \delta w}{\partial r} + N_{rr} \frac{\partial w}{\partial r} \frac{\partial \delta w}{\partial r} - M_{rr} \frac{\partial^2 \delta w}{\partial r^2} \right.
$$
$$
\left. - \frac{1}{r} M_{\theta\theta} \frac{\partial \delta w}{\partial r} + P_{r\theta} \left(-\frac{\partial^2 \delta w}{\partial r^2} + \frac{1}{r} \frac{\partial \delta w}{\partial r} \right) - q\delta w \right] r\, dr
$$
$$
- [Q_2 \delta w(r_a, t) + Q_5 \delta w(r_b, t) + Q_3 \delta\theta(r_a, t) + Q_6 \delta\theta(r_b, t)], \quad (10.2.3)
$$

where θ denotes the slope $\theta = -(\partial w/\partial r)$, Q_i are the generalized forces at the nodes of the element for a circular plate [see Fig. 10.2.2(a)]

$$
Q_1(t) \equiv -[rN_{rr}]_{r_a}, \qquad\qquad Q_4(t) \equiv [rN_{rr}]_{r_b}
$$
$$
Q_2(t) \equiv -\left[r\hat{V}_r\right]_{r_a}, \qquad\qquad Q_5(t) \equiv \left[r\hat{V}_r\right]_{r_b} \qquad (10.2.4)
$$
$$
Q_3(t) \equiv -\left[r\hat{M}_{rr}\right]_{r_a}, \qquad\qquad Q_6(t) \equiv \left[r\hat{M}_{rr}\right]_{r_b},
$$

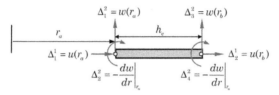

(a) Secondary variables

(b) Primary variables

Figure 10.2.2: Secondary and primary variables of a typical displacement finite element model for the CPT when linear interpolation of u Hermite cubic interpolation of w is assumed.

and the corresponding generalized displacements are: (u, w, θ) at the two nodes of the element. The stress resultants and N_{rr}, $N_{\theta\theta}$, M_{rr}, $M_{\theta\theta}$, and $P_{r\theta}$ are known in terms of u and w through Eqs. (6.2.21a)–(6.2.21f) as follows:

$$N_{rr} = A_{rr}\left[\frac{\partial u}{\partial r} + \frac{1}{2}\left(\frac{\partial w}{\partial r}\right)^2 + \nu\frac{u}{r}\right] - B_{rr}\left(\frac{\partial^2 w}{\partial r^2} + \frac{\nu}{r}\frac{\partial w}{\partial r}\right), \qquad (10.2.5a)$$

$$N_{\theta\theta} = A_{rr}\left[\frac{u}{r} + \nu\frac{\partial u}{\partial r} + \frac{\nu}{2}\left(\frac{\partial w}{\partial r}\right)^2\right] - B_{rr}\left(\nu\frac{\partial^2 w}{\partial r^2} + \frac{1}{r}\frac{\partial w}{\partial r}\right), \qquad (10.2.5b)$$

$$M_{rr} = B_{rr}\left[\frac{\partial u}{\partial r} + \frac{1}{2}\left(\frac{\partial w}{\partial r}\right)^2 + \nu\frac{u}{r}\right] - D_{rr}\left(\frac{\partial^2 w}{\partial r^2} + \frac{\nu}{r}\frac{\partial w}{\partial r}\right), \qquad (10.2.5c)$$

$$M_{\theta\theta} = B_{rr}\left[\frac{u}{r} + \nu\frac{\partial u}{\partial r} + \frac{\nu}{2}\left(\frac{\partial w}{\partial r}\right)^2\right] - D_{rr}\left(\nu\frac{\partial^2 w}{\partial r^2} + \frac{1}{r}\frac{\partial w}{\partial r}\right), \qquad (10.2.5d)$$

$$P_{r\theta} = S_{r\theta}\left(-\frac{\partial^2 w}{\partial r^2} + \frac{1}{r}\frac{\partial w}{\partial r}\right), \qquad (10.2.5e)$$

$$r\hat{V}_r = \frac{\partial}{\partial r}(r\hat{M}_{rr}) - \hat{M}_{\theta\theta} + rN_{rr}\frac{\partial w}{\partial r} - m_1\frac{\partial u}{\partial t^2} + m_2\frac{\partial^3 w}{\partial r\partial t^2}, \qquad (10.2.5f)$$

$$\hat{M}_{rr} = M_{rr} + P_{r\theta}, \quad \hat{M}_{\theta\theta} = M_{\theta\theta} + P_{\theta\theta}, \qquad (10.2.5g)$$

where A_{rr}, B_{rr}, D_{rr}, and $S_{r\theta}$ are defined in Eq. (6.2.22). It is clear that the above relations contain the second derivatives of the transverse deflection w with respect to r. Consequently, Hermite cubic interpolation of w is necessary.

The two variational statements in Eqs. (10.2.2) and (10.2.3) form the basis of the displacement finite element model of the CPT, as described next. We note that the two statements are coupled (i.e., the dependent unknowns u and w appear in both equations) when geometric nonlinearity for or functionally graded material is accounted in the formulation.

10.2.2   FINITE ELEMENT MODEL

An examination of the duality variables listed in Eq. (6.2.14) show that the Lagrange interpolation of u and Hermite interpolation of w is necessary. Let

$$u(r,t) = \sum_{j=1}^{2} \Delta_j^1(t)\psi_j(r), \quad w(r,t) = \sum_{J=1}^{4} \Delta_J^2(t)\varphi_J(r), \tag{10.2.6}$$

where $\psi_j(r)$ are the linear polynomials, $\varphi_J(r)$ are the Hermite cubic polynomials (they are the same as those listed in Section 9.1.2 with x replaced with r), (Δ_1^1, Δ_2^1) are the nodal values of $u(r)$ at r_a and r_b, respectively, and Δ_J^2 $(J = 1, 2, 3, 4)$ are the nodal values associated with $w(r)$ and its first derivative [see Fig. 10.2.2(b)]:

$$\Delta_1^1(t) = u(r_a, t), \quad \Delta_2^1(t) = u(r_b, t), \quad \Delta_1^2(t) = w(r_a, t),$$

$$\Delta_3^2(t) = w(r_b, t), \quad \Delta_2^2(t) = -\frac{\partial w}{\partial r}\bigg|_{r_a, t}, \quad \Delta_4^2(t) = -\frac{\partial w}{\partial r}\bigg|_{r_b, t}. \tag{10.2.7}$$

Substitution of the approximations in Eq.(10.2.6) into Eqs. (10.2.2) and (10.2.3), we obtain the following finite element model:

$$\begin{bmatrix} \mathbf{M}^{11} & \mathbf{M}^{12} \\ \mathbf{M}^{21} & \mathbf{M}^{22} \end{bmatrix} \begin{Bmatrix} \ddot{\boldsymbol{\Delta}}^1 \\ \ddot{\boldsymbol{\Delta}}^2 \end{Bmatrix} + \begin{bmatrix} \mathbf{K}^{11} & \mathbf{K}^{12} \\ \mathbf{K}^{21} & \mathbf{K}^{22} \end{bmatrix} \begin{Bmatrix} \boldsymbol{\Delta}^1 \\ \boldsymbol{\Delta}^2 \end{Bmatrix} = \begin{Bmatrix} \mathbf{F}^1 \\ \mathbf{F}^2 \end{Bmatrix}, \tag{10.2.8}$$

where mass and stiffness coefficients $M_{ij}^{\alpha\beta}$ and $K_{ij}^{\alpha\beta}$ and force coefficients F_i^α $(\alpha, \beta = 1, 2)$ are defined as follows:

$$M_{ij}^{11} = \int_{r_a}^{r_b} m_0 \psi_i \psi_j \, r dr, \qquad M_{iJ}^{12} = -\int_{r_a}^{r_b} m_1 \psi_i \frac{d\varphi_J}{dr} \, r dr,$$

$$M_{Ij}^{21} = -\int_{r_a}^{r_b} m_1 \frac{d\varphi_I}{dr} \psi_j \, r dr, \quad M_{IJ}^{22} = \int_{r_a}^{r_b} \left(m_0 \varphi_I \varphi_J + m_2 \frac{d\varphi_I}{dr} \frac{d\varphi_J}{dr} \right) r dr,$$

$$K_{ij}^{11} = \int_{r_a}^{r_b} A_{rr} \left[\frac{d\psi_i}{dr} \left(\frac{d\psi_j}{dr} + \frac{\nu}{r}\psi_j \right) + \frac{1}{r}\psi_i \left(\frac{1}{r}\psi_j + \nu\frac{d\psi_j}{dr} \right) \right] r dr,$$

$$K_{iJ}^{12} = \int_{r_a}^{r_b} \left\{ \frac{d\psi_i}{dr} \left[A_{rr} \frac{1}{2} \frac{\partial w}{\partial r} \frac{d\varphi_J}{dr} - B_{rr} \left(\frac{d^2\varphi_J}{dr^2} + \frac{\nu}{r} \frac{d\varphi_J}{dr} \right) \right] \right.$$

$$\left. + \frac{1}{r} \psi_i \left[\frac{1}{2} A_{rr} \nu \frac{\partial w}{\partial r} \frac{d\varphi_J}{dr} - B_{rr} \left(\nu \frac{d^2\varphi_J}{dr^2} + \frac{1}{r} \frac{d\varphi_J}{dr} \right) \right] \right\} r\, dr,$$

$$K_{Ij}^{21} = \int_{r_a}^{r_b} \left[\left(A_{rr} \frac{\partial w}{\partial r} \frac{d\varphi_I}{dr} - B_{rr} \frac{d^2\varphi_I}{dr^2} \right) \left(\frac{d\psi_j}{dr} + \frac{\nu}{r} \psi_j \right) \right.$$

$$\left. - B_{rr} \frac{1}{r} \frac{d\varphi_I}{dr} \left(\frac{1}{r} \psi_j + \nu \frac{d\psi_j}{dr} \right) \right] r\, dr,$$

$$K_{IJ}^{22} = \int_{r_a}^{r_b} \left\{ \frac{d\varphi_I}{dr} \left[\frac{1}{2} A_{rr} \left(\frac{\partial w}{\partial r} \right)^2 \frac{d\varphi_J}{dr} - B_{rr} \frac{\partial w}{\partial r} \left(\frac{d^2\varphi_J}{dr^2} + \frac{\nu}{r} \frac{d\varphi_J}{dr} \right) \right] \right.$$

$$- \frac{d^2\varphi_I}{dr^2} \left[\frac{1}{2} B_{rr} \frac{\partial w}{\partial r} \frac{d\varphi_J}{dr} - D_{rr} \left(\frac{d^2\varphi_J}{dr^2} + \frac{\nu}{r} \frac{d\varphi_J}{dr} \right) \right]$$

$$- \frac{1}{r} \frac{d\varphi_I}{dr} \left[\frac{1}{2} B_{rr} \nu \frac{\partial w}{\partial r} \frac{d\varphi_J}{dr} - D_{rr} \left(\nu \frac{d^2\varphi_J}{dr^2} + \frac{1}{r} \frac{d\varphi_J}{dr} \right) \right]$$

$$\left. + S_{r\theta} \left(-\frac{d^2\varphi_I}{dr^2} + \frac{1}{r} \frac{d\varphi_I}{dr} \right) \left(-\frac{d^2\varphi_J}{dr^2} + \frac{1}{r} \frac{d\varphi_J}{dr} \right) \right\} r\, dr,$$

$$F_i^1 = Q_1 \psi_i(r_a) + Q_4 \psi_i(r_b)$$

$$F_I^2 = \int_{r_a}^{r_b} q\varphi_I\, r\, dr + Q_2 \varphi_I(r_a) + Q_5 \varphi_I(r_b)$$

$$+ Q_3 \left[-\frac{d\varphi_I}{dx} \right]_{r_a} + Q_6 \left[-\frac{d\varphi_I}{dx} \right]_{r_b}. \tag{10.2.9}$$

10.3 MIXED MODEL OF THE CPT

10.3.1 WEAK FORMS

Here we present the mixed finite element model only for the static case without the effect of the couple stress terms because of the difficulty in formulating the mixed model with time derivative terms. The weak forms of Eqs. (6.2.40)–(6.2.42), which are valid only for the static case, are

$$0 = \int_{r_a}^{r_b} \frac{d\delta u}{dr} \left\{ \bar{A}_{rr} \left[r \frac{du}{dr} + \frac{r}{2} \left(\frac{dw}{dr} \right)^2 + \nu u \right] + r\bar{B}_{rr} M_{rr} \right\} dr$$

$$+ \int_{r_a}^{r_b} \left\{ \bar{A}_{rr} \left[\nu \, \delta u \frac{du}{dr} + \nu \frac{1}{2} \delta u \left(\frac{dw}{dr} \right)^2 + \frac{1}{r} \delta u\, u \right] + \nu \bar{B}_{rr} \delta u\, M_{rr} \right\} dr$$

$$+ \int_{r_a}^{r_b} (1 - \nu^2) B_{rr} \frac{1}{r} \left(\bar{B}_{rr} \delta u\, u - \delta u \frac{dw}{dr} \right) dr - Q_1\, \delta u(r_a) - Q_4\, \delta u(r_b),$$

$$\tag{10.3.1}$$

$$0 = \int_{r_a}^{r_b} \frac{d\delta w}{dr} \left[r \frac{dM_{rr}}{dr} + (1 - \nu) M_{rr} - (1 - \nu^2) \frac{D_{rr}}{r} \left(\bar{B}_{rr} u - \frac{dw}{dr} \right) \right] dr$$

$$+ \int_{r_a}^{r_b} \left\{ \bar{A}_{rr} \frac{dw}{dr} \frac{d\delta w}{dr} \left[r \frac{du}{dr} + \frac{r}{2} \left(\frac{dw}{dr} \right)^2 + \nu u \right] + r \bar{B}_{rr} \frac{d\delta w}{dr} \frac{dw}{dr} M_{rr} \right\} dr$$

$$- \int_{r_a}^{r_b} \delta w \, r \, q - Q_2 \, \delta w(r_a) - Q_5 \, \delta w(r_b), \tag{10.3.2}$$

$$0 = \int_{r_a}^{r_b} \left\{ r \frac{\partial \delta M_{rr}}{dr} \frac{dw}{dr} + (1 - \nu) \delta M_{rr} \frac{dw}{dr} + \bar{B}_{rr} \delta M_{rr} \left[r \frac{du}{dr} + \frac{r}{2} \left(\frac{dw}{dr} \right)^2 + \nu u \right] \right.$$

$$\left. - \frac{r}{D_{rr}} \delta M_{rr} \, M_{rr} \right\} dr - Q_3 \, \delta M_{rr}(r_a) - Q_6 \, \delta M_{rr}(r_b), \tag{10.3.3}$$

where [see Eq. (7.1.16) for the definition of A_{rr}, B_{rr}, D_{rr}, S_{rz} and $S_{r\theta}$]

$$D_{rr}^* = D_{rr} \, A_{rr} - B_{rr}^2, \quad \bar{A}_{rr} = \frac{D_{rr}^*}{D_{rr}}, \quad \bar{B}_{rr} = \frac{B_{rr}}{D_{rr}}, \tag{10.3.4}$$

and Q_i are the generalized forces [i.e., secondary variables; see Fig. 10.3.1(a)] at the two nodes of the element and they can be expressed in terms of the stress resultants as

$$Q_1 = -[r N_{rr}]_{r=r_a}, \qquad Q_4 = [r N_{rr}]_{r=r_b},$$

$$Q_2 = - \left[\frac{d}{dr} (r M_{rr}) - M_{\theta\theta} + \frac{d}{dr} \left(r N_{rr} \frac{dw}{dr} \right) \right]_{r=r_a},$$

$$Q_5 = \left[\frac{d}{dr} (r M_{rr}) - M_{\theta\theta} + \frac{d}{dr} \left(r N_{rr} \frac{dw}{dr} \right) \right]_{r=r_b}, \tag{10.3.5}$$

$$Q_3 = \left[r \frac{dw}{dr} \right]_{r=r_a}, \qquad Q_6 = \left[-r \frac{dw}{dr} \right]_{r=r_b}.$$

10.3.2   FINITE ELEMENT MODEL

An examination of Eqs. (10.3.1)–(10.3.3) (without the modified couple stress terms) indicate that the Lagrange interpolation of (u, w, M_{rr}) is admissible. Thus, one may use linear and higher-order Lagrange interpolation of these variables. The general form of the finite element interpolation is given by

$$u(r) = \sum_{j=1}^{m} u_j \psi_j^{(1)}(r), \quad w(r) = \sum_{j=1}^{n} w_j \psi_j^{(2)}(r), \quad M_{rr}(r) = \sum_{j=1}^{p} m_j \psi_j^{(3)}(r),$$

$$\tag{10.3.6}$$

where $\psi_j^{(1)}$, $\psi_j^{(2)}$, and $\psi_j^{(3)}$ are the Lagrange polynomials of different degree (i.e., $m \neq n \neq p$) used for u, w, and M_{rr}, respectively. However, when using different degree of interpolation, one must be careful

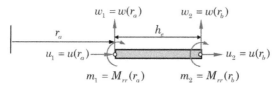

(a) Secondary variables

$w_1 = w(r_a)$ $w_2 = w(r_b)$

h_e

r_a

$u_1 = u(r_a)$ ──   ── $u_2 = u(r_b)$

$m_1 = M_{rr}(r_a)$ $m_2 = M_{rr}(r_b)$

(b) Primary variables

Figure 10.3.1: Secondary and primary variables of a typical mixed finite element model of the CPT when linear interpolation of u, w, and M_{rr} is assumed; $r\hat{V}_r = d(rM_{rr})/dr - M_{\theta\theta} + d(rN_{rr}(dw/dr))/dr$.

in selecting the values of m, n, and p for compatibility of these fields with each other. For simplicity, we take equal order interpolation (i.e., $m = n = p$) in implementing the mixed finite element model in a computer program. Figure 10.3.1 shows the secondary and primary degrees of freedom of the mixed finite element model when all variables are approximated using linear interpolation.

Substitution of the approximations in Eq. (10.3.6) into the weak forms in Eqs. (10.3.1)–(10.3.3) gives the following mixed finite element model of the CPT:

$$\begin{bmatrix} \mathbf{K}^{11} & \mathbf{K}^{12} & \mathbf{K}^{13} \\ \mathbf{K}^{21} & \mathbf{K}^{22} & \mathbf{K}^{23} \\ \mathbf{K}^{31} & \mathbf{K}^{32} & \mathbf{K}^{33} \end{bmatrix}^{(e)} \begin{Bmatrix} \mathbf{u} \\ \mathbf{w} \\ \mathbf{m} \end{Bmatrix}^{(e)} = \begin{Bmatrix} \mathbf{F}^1 \\ \mathbf{F}^2 \\ \mathbf{F}^3 \end{Bmatrix}^{(e)} , \qquad (10.3.7)$$

where $(\mathbf{u}, \mathbf{w}, \mathbf{m})$ denote the vectors of displacements u and w and moment M_{rr} at the nodes of the element. The non-zero coefficients of the above equations are given by

$$K_{ij}^{11} = \int_{r_a}^{r_b} \left[\bar{A}_{rr} r \frac{d\psi_i^{(1)}}{dr} \frac{d\psi_j^{(1)}}{dr} + \bar{A}_{rr}\nu \left(\frac{d\psi_i^{(1)}}{dr} \psi_j^{(1)} + \psi_i^{(1)} \frac{d\psi_j^{(1)}}{dr} + \frac{1}{r} \psi_i^{(1)} \psi_j^{(1)} \right) \right.$$

$$\left. + (1-\nu^2) B_{rr} \bar{B}_{rr} \frac{1}{r} \psi_i^{(1)} \psi_j^{(1)} \right] dr,$$

$$K_{ij}^{12} = \frac{1}{2} \int_{r_a}^{r_b} \bar{A}_{rr} \frac{dw}{dr} \left(r \frac{d\psi_i^{(1)}}{dr} \frac{d\psi_j^{(2)}}{dr} + \nu \psi_i^{(1)} \frac{d\psi_j^{(2)}}{dr} \right) dr$$

$$- \int_{r_a}^{r_b} B_{rr}(1-\nu^2) \frac{1}{r} \psi_i^{(1)} \frac{d\psi_j^{(2)}}{dr} \, dr,$$

$$K_{ij}^{13} = \int_{r_a}^{r_b} \bar{B}_{rr} \left(r \frac{d\psi_i^{(1)}}{dr} \psi_j^{(3)} + \nu \psi_i^{(1)} \psi_j^{(3)} \right) dr,$$

$$K_{ij}^{21} = - \int_{r_a}^{r_b} B_{rr}(1-\nu^2) \frac{1}{r} \frac{d\psi_i^{(2)}}{dr} \psi_j^{(1)} \, dr$$

$$+ \int_{r_a}^{r_b} \bar{A}_{rr} \frac{dw}{dr} \left(r \frac{d\psi_i^{(2)}}{dr} \frac{d\psi_j^{(1)}}{dr} + \nu \frac{d\psi_i^{(2)}}{dr} \psi_j^{(1)} \right) dr,$$

$$K_{ij}^{22} = \int_{r_a}^{r_b} D_{rr}(1-\nu^2) \frac{1}{r} \frac{d\psi_i^{(2)}}{dr} \frac{d\psi_j^{(2)}}{dr} \, dr + \frac{1}{2} \int_{r_a}^{r_b} \bar{A}_{rr} \left(\frac{\partial w}{\partial r} \right)^2 \frac{d\psi_i^{(2)}}{dr} \frac{d\psi_j^{(2)}}{dr} r \, dr,$$

$$K_{ij}^{23} = \int_{r_a}^{r_b} \left[r \frac{d\psi_i^{(2)}}{dr} \frac{d\psi_j^{(3)}}{dr} + (1-\nu) \frac{d\psi_i^{(2)}}{dr} \psi_j^{(3)} \right] dr + \int_{r_a}^{r_b} \bar{B}_{rr} \frac{dw}{dr} \frac{d\psi_i^{(2)}}{dr} \psi_j^{(3)} r \, dr,$$

$$K_{ij}^{31} = \int_{r_a}^{r_b} \bar{B}_{rr} \left(r \psi_i^{(3)} \frac{d\psi_j^{(1)}}{dr} + \nu \psi_i^{(3)} \psi_j^{(1)} \right) dr,$$

$$K_{ij}^{32} = \int_{r_a}^{r_b} \left[r \frac{d\psi_i^{(3)}}{dr} \frac{d\psi_j^{(2)}}{dr} + (1-\nu) \psi_i^{(3)} \frac{d\psi_j^{(2)}}{dr} \right] dr + \frac{1}{2} \int_{r_a}^{r_b} \bar{B}_{rr} \frac{\partial w}{\partial r} \psi_i^{(3)} \frac{d\psi_j^{(2)}}{dr} r \, dr,$$

$$K_{ij}^{33} = - \int_{r_a}^{r_b} \frac{1}{D_{rr}} \psi_i^{(3)} \psi_j^{(3)} r \, dr, \quad F_i^1 = \psi_i^{(1)}(r_a)Q_1 + \psi_i^{(1)}(r_b)Q_4,$$

$$F_i^2 = \int_{r_a}^{r_b} qr \, dr + \psi_i^{(2)}(r_a)Q_2 + \psi_i^{(2)}(r_b)Q_5,$$

$$F_i^3 = \psi_i^{(3)}(r_a)Q_3 + \psi_i^{(3)}(r_b)Q_6. \tag{10.3.8}$$

When $\psi_i^{(1)} = \psi_i^{(2)} = \psi_i^{(3)}$, i, j take the values of $1, 2, \ldots, n$, n being the number of nodes in the element.

10.4   DISPLACEMENT MODEL OF THE FST

10.4.1   WEAK FORMS

The virtual work statement in Eq. (7.1.5), which includes the couple stress effect, for the FST is equivalent to the following three integral statements over a typical finite element $\Omega^e = (r_a, r_b)$:

$$0 = \int_{r_a}^{r_b} \left(N_{rr} \frac{\partial \delta u}{\partial r} + N_{\theta\theta} \frac{\delta u}{r} + m_0 \ddot{u}\,\delta u + m_1 \ddot{\phi}_r\,\delta u \right) r\,dr$$
$$- Q_1\,\delta u(r_a, t) - Q_5\,\delta u(r_b, t), \tag{10.4.1}$$

$$0 = \int_{r_a}^{r_b} \left[N_{rr} \frac{\partial w}{\partial r} \frac{\partial \delta w}{\partial r} + N_{rz} \frac{\partial \delta w}{\partial r} + \tfrac{1}{2} P_{r\theta} \left(-\frac{\partial^2 \delta w}{\partial r^2} + \frac{1}{r} \frac{\partial \delta w}{\partial r} \right) \right.$$
$$\left. + m_0 \ddot{w}\,\delta w - q\delta w \right] r\,dr - Q_2\,\delta w(r_a, t) - Q_6\,\delta w(r_b, t)$$
$$- Q_3\,\delta\theta_r(r_a, t) - Q_7\,\delta\theta_r(r_b, t), \tag{10.4.2}$$

$$0 = \int_{r_a}^{r_b} \left[M_{rr} \frac{\partial \delta\phi_r}{\partial r} + \frac{1}{r} M_{\theta\theta}\delta\phi_r + N_{rz}\delta\phi_r + m_1 \ddot{u}\,\delta\phi_r + m_2 \ddot{\phi}_r\,\delta\phi_r \right.$$
$$\left. + \tfrac{1}{2} P_{r\theta} \left(\frac{\partial \delta\phi_r}{\partial r} - \frac{1}{r}\delta\phi_r \right) \right] r\,dr - Q_4\,\delta\phi_r(r_a) - Q_8\,\delta\phi_r(r_b), \tag{10.4.3}$$

where θ_r denotes the slope $\theta_r = -(\partial w/\partial r)$, Q_i are the generalized forces at the nodes of the element for a circular plate [see Fig. 10.4.1(b)]:

$$
\begin{aligned}
Q_1(t) &\equiv -\left[r N_{rr} \right]_{r_a}, & Q_5(t) &\equiv \left[r N_{rr} \right]_{r_b}, \\
Q_2(t) &\equiv -\left[r \bar{V}_r \right]_{r_a}, & Q_6(t) &\equiv \left[r \bar{V}_r \right]_{r_b}, \\
Q_3(t) &\equiv -\tfrac{1}{2}\left[r P_{r\theta} \right]_{r_a}, & Q_7(t) &\equiv \tfrac{1}{2}\left[r P_{r\theta} \right]_{r_b}, \\
Q_4(t) &\equiv -\left[r \bar{M}_{rr} \right]_{r_a}, & Q_8(t) &\equiv \left[r \bar{M}_{rr} \right]_{r_b},
\end{aligned}
\tag{10.4.4}
$$

and N_{rr}, $N_{\theta\theta}$, M_{rr}, $M_{\theta\theta}$, and $P_{r\theta}$ are known in terms of u and w as

$$N_{rr} = A_{rr}\left[\frac{\partial u}{\partial r} + \frac{1}{2}\left(\frac{\partial w}{\partial r} \right)^2 + \nu \frac{u}{r} \right] + B_{rr}\left(\frac{\partial \phi_r}{\partial r} + \frac{\nu}{r}\phi_r \right), \tag{10.4.5a}$$

$$N_{\theta\theta} = A_{rr}\left[\frac{u}{r} + \nu \frac{\partial u}{\partial r} + \frac{\nu}{2}\left(\frac{\partial w}{\partial r} \right)^2 \right] + B_{rr}\left(\nu \frac{\partial \phi_r}{\partial r} + \frac{1}{r}\phi_r \right), \tag{10.4.5b}$$

$$M_{rr} = B_{rr}\left[\frac{\partial u}{\partial r} + \frac{1}{2}\left(\frac{\partial w}{dr} \right)^2 + \nu \frac{u}{r} \right] + D_{rr}\left(\frac{\partial \phi_r}{\partial r} + \frac{\nu}{r}\phi_r \right), \tag{10.4.5c}$$

$$M_{\theta\theta} = B_{rr}\left[\frac{u}{r} + \nu \frac{\partial u}{\partial r} + \frac{\nu}{2}\left(\frac{\partial w}{\partial r} \right)^2 \right] + D_{rr}\left(\nu \frac{\partial \phi_r}{\partial r} + \frac{1}{r}\phi_r \right), \tag{10.4.5d}$$

$$N_{rz} = S_{rz}\left(\phi_r + \frac{\partial w}{\partial r} \right), \tag{10.4.5e}$$

$$P_{r\theta} = \frac{1}{2}S_{r\theta}\left[\frac{\partial \phi_r}{\partial r} - \frac{\partial^2 w}{\partial r^2} - \frac{1}{r}\left(\phi_r - \frac{\partial w}{\partial r} \right) \right], \tag{10.4.5f}$$

$$\bar{M}_{rr} = M_{rr} + 0.5 P_{r\theta}, \quad r\bar{V}_r = N_{rz} + N_{rr}\frac{\partial w}{\partial r}. \tag{10.4.5g}$$

10.4.2   FINITE ELEMENT MODEL

An examination of the weak forms in Eqs. (10.4.1)–(10.4.3) show that the Lagrange interpolation of (u, ϕ) and Hermite interpolation of w (only because of the modified couple stress terms) are required. Let

$$u(r,t) = \sum_{j=1}^{m} u_j(t)\psi_j^{(1)}(r), \quad w(r,t) = \sum_{j=1}^{4} \Delta_j(t)\psi_j^{(2)}(r)$$

$$\phi_r(r,t) = \sum_{j=1}^{n} s_j(t)\psi_j^{(3)}(r), \tag{10.4.6}$$

where $\psi_j^{(1)}$ and $\psi_j^{(3)}$ are the Lagrange polynomials of different degree used for u and ϕ_r, respectively, and $\psi_j^{(2)}(r) = \varphi_j(r)$ are the Hermite cubic polynomials when the couple stress effect is included. Figure 10.4.1 shows the secondary and primary variables of the displacement finite element model of the FST when linear interpolation of u and ϕ_r and Hermite cubic interpolation of w are used. If the effect of the modified couple stress is not considered, one may use Lagrange interpolation of w. In that case, we can use equal degree interpolation of all variables, $\psi_j^{(1)} = \psi_j^{(2)} = \psi_j^{(3)}$, and they can be linear or higher order.

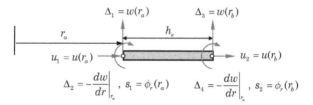

(a) Primary variables

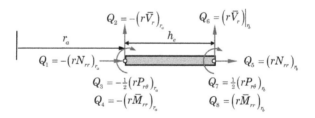

(b) Secondary variables

Figure 10.4.1: Primary and secondary variables of a typical displacement finite element model of the FST when linear interpolation of u and ϕ_r and Hermite cubic approximation of w are assumed.

Substitution of the approximations from Eq. (10.4.6) into the variational statements of Eqs. (10.4.1)–(10.4.3), we obtain the following finite element equations:

$$\begin{bmatrix} M^{11} & M^{12} & M^{13} \\ M^{21} & M^{22} & M^{23} \\ M^{31} & M^{32} & M^{33} \end{bmatrix} \begin{Bmatrix} \ddot{u} \\ \ddot{\Delta} \\ \ddot{s} \end{Bmatrix} + \begin{bmatrix} K^{11} & K^{12} & K^{13} \\ K^{21} & K^{22} & K^{23} \\ K^{31} & K^{32} & K^{33} \end{bmatrix} \begin{Bmatrix} u \\ \Delta \\ s \end{Bmatrix} = \begin{Bmatrix} F^1 \\ F^2 \\ F^3 \end{Bmatrix}. \quad (10.4.7)$$

The nonzero mass and stiffness coefficients, $M_{ij}^{\alpha\beta}$ and $K_{ij}^{\alpha\beta}$, and force coefficients F_i^{α} $(\alpha, \beta = 1, 2, 3)$ are defined as follows:

$$M_{ij}^{11} = \int_{r_a}^{r_b} m_0 \psi_i^{(1)} \psi_j^{(1)} \, r\,dr, \quad M_{ij}^{13} = \int_{r_a}^{r_b} m_1 \psi_i^{(1)} \psi_j^{(3)} \, r\,dr = M_{ji}^{31},$$

$$M_{ij}^{22} = \int_{r_a}^{r_b} m_0 \psi_i^{(2)} \psi_j^{(2)} \, r\,dr, \quad M_{ij}^{33} = \int_{r_a}^{r_b} m_2 \psi^{(3)} \psi^{(3)} \, r\,dr,$$

$$K_{ij}^{11} = \int_{r_a}^{r_b} A_{rr} \left[\frac{d\psi_i^{(1)}}{dr} \left(\frac{d\psi_j^{(1)}}{dr} + \frac{\nu}{r}\psi_j^{(1)} \right) + \frac{1}{r}\psi_i^{(1)} \left(\frac{1}{r}\psi_j^{(1)} + \nu\frac{d\psi_j^{(1)}}{dr} \right) \right] r\,dr,$$

$$K_{ij}^{12} = \frac{1}{2} \int_{r_a}^{r_b} A_{rr} \frac{\partial w}{\partial r} \left(\frac{d\psi_i^{(1)}}{dr} + \frac{\nu}{r}\psi_i^{(1)} \right) \frac{d\psi_j^{(2)}}{dr} \, r\,dr,$$

$$K_{ij}^{21} = \int_{r_a}^{r_b} A_{rr} \frac{\partial w}{\partial r} \frac{d\psi_i^{(2)}}{dr} \left(\frac{d\psi_j^{(1)}}{dr} + \frac{\nu}{r}\psi_j^{(1)} \right) r\,dr,$$

$$K_{ij}^{13} = \int_{r_a}^{r_b} B_{rr} \left[\frac{d\psi_i^{(1)}}{dr} \left(\frac{d\psi_j^{(3)}}{dr} + \frac{\nu}{r}\psi_j^{(3)} \right) + \frac{1}{r}\psi_i^{(1)} \left(\nu\frac{d\psi_j^{(3)}}{dr} + \frac{1}{r}\psi_j^{(3)} \right) \right] r\,dr,$$

$$K_{ij}^{22} = \int_{r_a}^{r_b} \left[\frac{1}{2} A_{rr} \left(\frac{\partial w}{\partial r} \right)^2 \frac{d\psi_i^{(2)}}{dr} \frac{d\psi_j^{(2)}}{dr} + S_{rz} \frac{d\psi_i^{(2)}}{dr} \frac{d\psi_j^{(2)}}{dr} \right.$$
$$\left. + \frac{1}{4} S_{r\theta} \left(-\frac{d^2\psi_i^{(2)}}{dr^2} + \frac{1}{r}\frac{d\psi_i^{(2)}}{dr} \right) \left(-\frac{d^2\psi_j^{(2)}}{dr^2} + \frac{1}{r}\frac{d\psi_j^{(2)}}{dr} \right) \right] r\,dr,$$

$$K_{ij}^{23} = \int_{r_a}^{r_b} \left[B_{rr} \frac{\partial w}{\partial r} \frac{d\psi_i^{(2)}}{dr} \left(\frac{d\psi_j^{(3)}}{dr} + \frac{\nu}{r}\psi_j^{(3)} \right) + S_{rz} \frac{d\psi_i^{(2)}}{dr}\psi_j^{(3)} \right.$$
$$\left. + \frac{1}{4} S_{r\theta} \left(-\frac{d^2\psi_i^{(2)}}{dr^2} + \frac{1}{r}\frac{d\psi_i^{(2)}}{dr} \right) \left(\frac{d\psi_j^{(3)}}{dr} - \frac{1}{r}\psi_j^{(3)} \right) \right] r\,dr,$$

$$K_{ij}^{31} = \int_{r_a}^{r_b} B_{rr} \left[\frac{d\psi_i^{(3)}}{dr} \left(\frac{d\psi_j^{(1)}}{dr} + \frac{\nu}{r}\psi_j^{(1)} \right) + \frac{1}{r}\psi_i^{(3)} \left(\nu\frac{d\psi_j^{(1)}}{dr} + \frac{1}{r}\psi_j^{(1)} \right) \right] r\,dr,$$

$$K_{ij}^{32} = \int_{r_a}^{r_b} \left[\frac{1}{2} B_{rr} \frac{\partial w}{\partial r} \left(\frac{d\psi_i^{(3)}}{dr} + \frac{\nu}{r} \psi_i^{(3)} \right) \frac{d\psi_j^{(2)}}{dr} + S_{rz} \psi_i^{(3)} \frac{d\psi_j^{(2)}}{dr} \right.$$

$$\left. + \frac{1}{4} S_{r\theta} \left(\frac{d\psi_i^{(3)}}{dr} - \frac{1}{r} \psi_i^{(3)} \right) \left(-\frac{d^2 \psi_j^{(2)}}{dr^2} + \frac{1}{r} \frac{d\psi_j^{(2)}}{dr} \right) \right] r \, dr,$$

$$K_{ij}^{33} = \int_{r_a}^{r_b} \left\{ D_{rr} \left[\frac{d\psi_i^{(3)}}{dr} \left(\frac{d\psi_j^{(3)}}{dr} + \frac{\nu}{r} \psi_j^{(3)} \right) + \frac{1}{r} \psi_i^{(3)} \left(\nu \frac{d\psi_j^{(3)}}{dr} + \frac{1}{r} \psi_j^{(3)} \right) \right] \right.$$

$$+ \frac{1}{4} S_{r\theta} \left(\frac{d\psi_i^{(3)}}{dr} - \frac{1}{r} \psi_i^{(3)} \right) \left(\frac{d\psi_j^{(3)}}{dr} - \frac{1}{r} \psi_j^{(3)} \right)$$

$$\left. + S_{rz} \psi_i^{(3)} \psi_j^{(3)} \right\} r \, dr,$$

$$F_i^1 = Q_1 \psi_i^{(1)}(r_a) + Q_5 \psi_i^{(1)}(r_b), \quad F_i^3 = Q_4 \psi_i^{(3)}(r_a) + Q_8 \psi_i^{(3)}(r_b),$$

$$F_i^2 = \int_{r_a}^{r_b} q \psi_i^{(2)} r \, dr + Q_2 \psi_i^{(2)}(r_a) + Q_6 \psi_i^{(2)}(r_b) + Q_3 \psi_i^{(2)}(r_a) + Q_7 \psi_i^{(2)}(r_b).$$

$$(10.4.8)$$

10.5   DISPLACEMENT MODEL OF THE TST

10.5.1   VARIATIONAL STATEMENTS

In this section, we present the displacement finite element model for static case of the TST for circular plates. This model may not be readily available in the literature. Extension to the dynamic case, once the equilibrium case is known, is straightforward.

The variational statements required to derive the finite element model can be obtained from Eq. (8.1.5) by omitting the time derivative terms and time integral:

$$0 = \int_b^a \left[N_{rr} \left(\frac{d\delta u}{dr} + \frac{dw}{dr} \frac{d\delta w}{dr} \right) + \frac{1}{r} N_{\theta\theta} \, \delta u + \bar{M}_{rr} \frac{d\delta\phi_r}{dr} + \frac{1}{r} \bar{M}_{\theta\theta} \, \delta\phi_r \right.$$

$$- \alpha P_{rr} \frac{d^2 \delta w}{dr^2} - \alpha \frac{1}{r} P_{\theta\theta} \frac{d\delta w}{dr} + \bar{N}_{rz} \left(\delta\phi_r + \frac{d\delta w}{dr} \right)$$

$$- q \delta w + c_f w \, \delta w + \tfrac{1}{2} \bar{P}_{r\theta} \left(\frac{d\delta\phi_r}{dr} - \frac{1}{r} \delta\phi_r \right)$$

$$\left. - \tfrac{1}{2} \tilde{P}_{r\theta} \left(\frac{d^2 \delta w}{dr^2} - \frac{1}{r} \frac{d\delta w}{dr} \right) \right] r \, dr, \qquad (10.5.1)$$

where the stress resultants N_{rr}, $N_{\theta\theta}$, $\bar{M}_{rr}$, $\bar{M}_{\theta\theta}$, P_{rr}, $P_{\theta\theta}$, $\bar{P}_{r\theta}$, $\tilde{P}_{r\theta}$, and $\bar{N}_{rz}$ are defined in Eqs. (8.1.24a)–(8.1.25d) in terms of the generalized displacements

(u, w, ϕ_r) and plate stiffness coefficients A_{rr}, B_{rr}, D_{rr}, and so on. They are repeated here for convenience of the reader:

$$\bar{M}_{rr} = M_{rr} - \alpha\, P_{rr}, \quad \bar{M}_{\theta\theta} = M_{\theta\theta} - \alpha\, P_{\theta\theta}, \quad \bar{N}_{rz} = N_{rz} - \beta\, P_{r\theta} - \beta Q_{\theta z},$$

$$(10.5.2a)$$

$$\bar{P}_{r\theta} = P_{r\theta} - \beta R_{r\theta} = \frac{1}{2}\hat{A}_{r\theta}\left(\frac{d\phi_r}{dr} - \frac{1}{r}\phi_r\right) + \frac{1}{2}\tilde{A}_{r\theta}\left(-\frac{d^2 w}{dr^2} + \frac{1}{r}\frac{dw}{dr}\right),$$

$$(10.5.2b)$$

$$\tilde{P}_{r\theta} = P_{r\theta} + \beta R_{r\theta} = \frac{1}{2}\tilde{A}_{r\theta}\left(\frac{d\phi_r}{dr} - \frac{1}{r}\phi_r\right) + \frac{1}{2}\tilde{A}_{r\theta}\left(-\frac{d^2 w}{dr^2} + \frac{1}{r}\frac{dw}{dr}\right).$$

$$(10.5.2c)$$

$$N_{rr} = A_{rr}\left[\frac{du}{dr} + \frac{1}{2}\left(\frac{dw}{dr}\right)^2 + \nu\frac{u}{r}\right] + \bar{B}_{rr}\left(\frac{d\phi_r}{dr} + \frac{\nu}{r}\phi_r\right)$$
$$- \alpha E_{rr}\left(\frac{\nu}{r}\frac{dw}{dr} + \frac{d^2 w}{dr^2}\right),$$

$$(10.5.2d)$$

$$N_{\theta\theta} = A_{rr}\left[\frac{u}{r} + \nu\frac{du}{dr} + \frac{\nu}{2}\left(\frac{dw}{dr}\right)^2\right] + \bar{B}_{rr}\left(\nu\frac{d\phi_r}{dr} + \frac{1}{r}\phi_r\right)$$
$$- \alpha E_{rr}\left(\frac{1}{r}\frac{dw}{dr} + \nu\frac{d^2 w}{dr^2}\right),$$

$$(10.5.2e)$$

$$M_{rr} = B_{rr}\left[\frac{du}{dr} + \frac{1}{2}\left(\frac{dw}{dr}\right)^2 + \nu\frac{u}{r}\right] + \bar{D}_{rr}\left(\frac{d\phi_r}{dr} + \frac{\nu}{r}\phi_r\right)$$
$$- \alpha F_{rr}\left(\frac{\nu}{r}\frac{dw}{dr} + \frac{d^2 w}{dr^2}\right),$$

$$(10.5.2f)$$

$$M_{\theta\theta} = B_{rr}\left[\frac{u}{r} + \nu\frac{du}{dr} + \frac{\nu}{2}\left(\frac{dw}{dr}\right)^2\right] + \bar{D}_{rr}\left(\nu\frac{d\phi_r}{dr} + \frac{1}{r}\phi_r\right)$$
$$- \alpha F_{rr}\left(\frac{1}{r}\frac{dw}{dr} + \nu\frac{d^2 w}{dr^2}\right),$$

$$(10.5.2g)$$

$$P_{rr} = E_{rr}\left[\frac{du}{dr} + \frac{1}{2}\left(\frac{dw}{dr}\right)^2 + \nu\frac{u}{r}\right] + \bar{F}_{rr}\left(\frac{d\phi_r}{dr} + \frac{\nu}{r}\phi_r\right)$$
$$- \alpha H_{rr}\left(\frac{\nu}{r}\frac{dw}{dr} + \frac{d^2 w}{dr^2}\right),$$

$$(10.5.2h)$$

$$P_{\theta\theta} = E_{rr}\left[\frac{u}{r} + \nu\frac{du}{dr} + \frac{\nu}{2}\left(\frac{dw}{dr}\right)^2\right] + \bar{F}_{rr}\left(\nu\frac{d\phi_r}{dr} + \frac{1}{r}\phi_r\right)$$
$$- \alpha H_{rr}\left(\frac{1}{r}\frac{dw}{dr} + \nu\frac{d^2 w}{dr^2}\right),$$

$$(10.5.2i)$$

$$N_{rz} = \bar{A}_{rz}\left(\phi_r + \frac{dw}{dr}\right), \quad R_{rz} = \bar{D}_{rz}\left(\phi_r + \frac{dw}{dr}\right), \tag{10.5.3a}$$

$$P_{r\theta} = \frac{1}{2}\bar{A}_{r\theta}\left(\frac{d\phi_r}{dr} - \frac{1}{r}\phi_r\right) + \frac{1}{2}\tilde{A}_{r\theta}\left(-\frac{d^2w}{dr^2} + \frac{1}{r}\frac{dw}{dr}\right), \tag{10.5.3b}$$

$$R_{r\theta} = \frac{1}{2}\bar{D}_{r\theta}\left(\frac{d\phi_r}{dr} - \frac{1}{r}\phi_r\right) + \frac{1}{2}\tilde{D}_{r\theta}\left(-\frac{d^2w}{dr^2} + \frac{1}{r}\frac{dw}{dr}\right), \tag{10.5.3c}$$

$$Q_{z\theta} = -\beta\,D_{z\theta}\left(\phi_r + \frac{dw}{dr}\right). \tag{10.5.3d}$$

The stiffness coefficients A_{rr}, B_{rr}, D_{rr}, E_{rr}, F_{rr}, H_{rr}, $A_{r\theta}$, $D_{r\theta}$, $F_{r\theta}$ are the extensional, extensional-bending, bending, and higher-order stiffness coefficients defined in Eq. (8.1.26), and stiffness coefficients with over bars are defined in (8.1.27):

$$\begin{aligned}
&\bar{B}_{rr} = B_{rr} - \alpha E_{rr}, \quad \bar{D}_{rr} = D_{rr} - \alpha F_{rr}, \quad \bar{F}_{rr} = F_{rr} - \alpha H_{rr}, \\
&\bar{A}_{rz} = A_{rz} - \beta D_{rz}, \quad \bar{D}_{rz} = D_{rz} - \beta F_{rz}, \\
&\bar{A}_{r\theta} = A_{r\theta} - \beta D_{r\theta}, \quad \bar{D}_{r\theta} = D_{r\theta} - \beta F_{r\theta}, \quad \hat{A}_{r\theta} = \bar{A}_{r\theta} - \beta \bar{D}_{r\theta}, \\
&\tilde{A}_{r\theta} = A_{r\theta} + \beta D_{r\theta}, \quad \tilde{D}_{r\theta} = D_{r\theta} + \beta F_{r\theta}, \quad D_{z\theta} = D_{r\theta}, \\
&\tilde{\tilde{A}}_{r\theta} = A_{r\theta} + 2\beta D_{r\theta} + \beta^2 F_{r\theta}, \quad \bar{\bar{A}}_{r\theta} = A_{r\theta} - \beta^2 F_{r\theta}, \\
&\alpha = \frac{4}{3h^2}, \quad \beta = 3\alpha.
\end{aligned} \tag{10.5.4}$$

The variational statement in Eq. (10.5.1) can be split into three independent variational statements (by virtue of the fact that δu, δw, and $\delta\phi_r$ are arbitrary variations independent of each other) over a typical finite element $\Omega^e = (r_a, r_b)$:

$$0 = \int_{r_b}^{r_a}\left(N_{rr}\frac{d\delta u}{dr} + \frac{1}{r}N_{\theta\theta}\,\delta u\right)r\,dr - Q_1\delta u(r_a) - Q_5\delta u(r_b), \tag{10.5.5}$$

$$\begin{aligned}
0 = \int_{r_b}^{r_a}&\left[N_{rr}\frac{dw}{dr}\frac{d\delta w}{dr} - \alpha P_{rr}\frac{d^2\delta w}{dr^2} - \alpha\frac{1}{r}P_{\theta\theta}\frac{d\delta w}{dr} + \bar{N}_{rz}\frac{d\delta w}{dr}\right. \\
&\left. - \frac{1}{2}\tilde{P}_{r\theta}\left(\frac{d^2\delta w}{dr^2} - \frac{1}{r}\frac{d\delta w}{dr}\right) - q\delta w + c_f\,w\,\delta w\right]r\,dr \\
&- Q_2\,\delta w(r_a) - Q_6\,\delta w(r_b) - Q_3\,\delta\theta_r(r_a) - Q_7\,\delta\theta_r(r_b), \tag{10.5.6}
\end{aligned}$$

$$\begin{aligned}
0 = \int_{r_b}^{r_a}&\left[\bar{M}_{rr}\frac{d\delta\phi_r}{dr} + \frac{1}{r}\bar{M}_{\theta\theta}\,\delta\phi_r + \bar{N}_{rz}\,\delta\phi_r\right. \\
&\left. + \frac{1}{2}\tilde{P}_{r\theta}\left(\frac{d\delta\phi_r}{dr} - \frac{1}{r}\delta\phi_r\right)\right]r\,dr - Q_4\,\delta\phi_r(r_a) - Q_8\,\delta\phi_r(r_b), \tag{10.5.7}
\end{aligned}$$

where $\theta_r = -dw/dr$. The secondary variables Q_i are defined by

$$Q_1 \equiv -\left[rN_{rr}\right]_{r_a}, \qquad\qquad Q_5 \equiv \left[rN_{rr}\right]_{r_b},$$

$$Q_2 \equiv -\left[r\hat{V}_r\right]_{r_a}, \qquad\qquad Q_6 \equiv \left[r\hat{V}_r\right]_{r_b},$$

$$Q_3 \equiv -\left[\alpha r P_{rr} + 0.5r\tilde{P}_{r\theta}\right]_{r_a}, \qquad Q_7 \equiv \left[\alpha r P_{rr} + 0.5r\tilde{P}_{r\theta}\right]_{r_b}, \qquad (10.5.8)$$

$$Q_4 \equiv -\left[r\bar{M}_{rr} + 0.5r\bar{P}_{r\theta}\right]_{r_a}, \qquad Q_8 \equiv \left[r\bar{M}_{rr} + 0.5r\bar{P}_{r\theta}\right]_{r_b},$$

$$\hat{V}_r = \bar{N}_{rz} + N_{rr}\frac{dw}{dr} + \alpha\left[\frac{d}{dr}(rP_{rr}) - P_{\theta\theta}\right] + \tfrac{1}{2}\left[\frac{d}{dr}\left(r\tilde{P}_{r\theta}\right) + \tilde{P}_{r\theta}\right].$$

and the corresponding generalized displacements are: $(u, w, -dw/dr, \phi_r)$ at the two nodes of the element, as shown in Fig. 10.5.1.

10.5.2   FINITE ELEMENT MODEL

An examination of the variational statements in Eqs. (10.5.4)–(10.5.6) show that the Lagrange interpolation of (u, ϕ_r) and Hermite interpolation of w are required. Thus, the FST model with couple stress effect and the TST finite element models have the same degrees of freedom (see Fig. 10.5.1). Let

$$u(r) = \sum_{j=1}^{m} u_j \psi_j^{(1)}(r), \quad w(r) = \sum_{j=1}^{4} \Delta_j \psi_j^{(2)}(r), \quad \phi_r(r) = \sum_{j=1}^{n} s_j \psi_j^{(3)}(r), \quad (10.5.9)$$

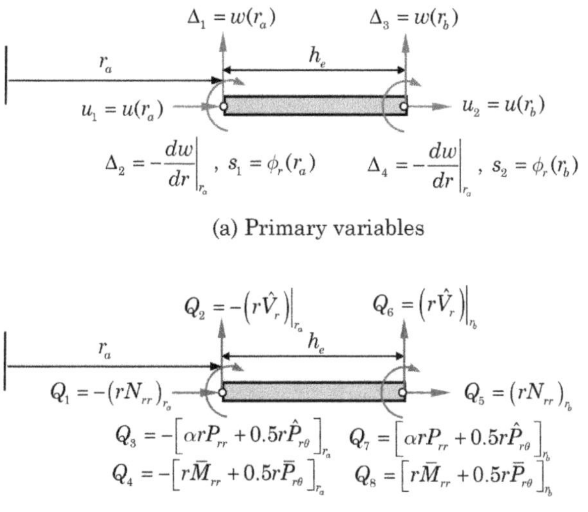

$$\Delta_1 = w(r_a) \qquad\qquad \Delta_3 = w(r_b)$$

$$u_1 = u(r_a) \qquad\qquad u_2 = u(r_b)$$

$$\Delta_2 = -\frac{dw}{dr}\bigg|_{r_a}, \ s_1 = \phi_r(r_a) \qquad \Delta_4 = -\frac{dw}{dr}\bigg|_{r_b}, \ s_2 = \phi_r(r_b)$$

(a) Primary variables

$$Q_2 = -\left(r\hat{V}_r\right)\big|_{r_a} \qquad\qquad Q_6 = \left(r\hat{V}_r\right)\big|_{r_b}$$

$$Q_1 = -(rN_{rr})_{r_a} \qquad\qquad Q_5 = (rN_{rr})_{r_b}$$

$$Q_3 = -\left[\alpha r P_{rr} + 0.5r\hat{P}_{r\theta}\right]_{r_a} \quad Q_7 = \left[\alpha r P_{rr} + 0.5r\hat{P}_{r\theta}\right]_{r_b}$$

$$Q_4 = -\left[r\bar{M}_{rr} + 0.5r\bar{P}_{r\theta}\right]_{r_a} \quad Q_8 = \left[r\bar{M}_{rr} + 0.5r\bar{P}_{r\theta}\right]_{r_b}$$

(b) Secondary variables

Figure 10.5.1: Primary and Secondary variables of a typical displacement finite element model of the TST when linear interpolation of u and ϕ_r and Hermite cubic approximation of w are assumed.

where $\psi_j^{(1)}$ and $\psi_j^{(3)}$ are the Lagrange polynomials of different degree used for u and ϕ_r, respectively, and $\psi_j^{(2)}$ are the Hermite cubic polynomials.

Substitution of the approximations in Eq. (10.5.9) into Eqs. (10.5.5)–(10.5.7), with the stress resultants known in terms of the displacements through Eqs. (10.5.2a)–(10.5.3d), we obtain:

$$
\begin{bmatrix} \mathbf{K}^{11} & \mathbf{K}^{12} & \mathbf{K}^{13} \\ \mathbf{K}^{21} & \mathbf{K}^{22} & \mathbf{K}^{23} \\ \mathbf{K}^{31} & \mathbf{K}^{32} & \mathbf{K}^{33} \end{bmatrix} \begin{Bmatrix} \mathbf{u} \\ \boldsymbol{\Delta} \\ \mathbf{s} \end{Bmatrix} = \begin{Bmatrix} \mathbf{F}^1 \\ \mathbf{F}^2 \\ \mathbf{F}^3 \end{Bmatrix}.
\tag{10.5.10}
$$

The nonzero stiffness coefficients $K_{ij}^{\alpha\beta}$ and force coefficients F_i^α $(\alpha, \beta = 1, 2, 3)$ are defined as follows $(\hat{\bar{A}}_{rz} = \hat{A}_{rz} + \beta^2 D_{z\theta})$:

$$
K_{ij}^{11} = \int_{r_a}^{r_b} A_{rr} \left[\frac{d\psi_i^{(1)}}{dr} \left(\frac{d\psi_j^{(1)}}{dr} + \frac{\nu}{r}\psi_j^{(1)} \right) + \frac{1}{r}\psi_i^{(1)} \left(\frac{1}{r}\psi_j^{(1)} + \nu\frac{d\psi_j^{(1)}}{dr} \right) \right] r\,dr,
$$

$$
K_{ij}^{12} = \frac{1}{2}\int_{r_a}^{r_b} A_{rr} \frac{dw}{dr} \left(\frac{d\psi_i^{(1)}}{dr} + \frac{\nu}{r}\psi_i^{(1)} \right) \frac{d\psi_j^{(2)}}{dr} r\,dr
$$
$$
- \alpha \int_{r_a}^{r_b} E_{rr} \left[\frac{d\psi_i^{(1)}}{dr} \left(\frac{\nu}{r}\frac{d\psi_j^{(2)}}{dr} + \frac{d^2\psi_j^{(2)}}{dr^2} \right) + \psi_i^{(1)} \left(\frac{1}{r^2}\frac{d\psi_j^{(2)}}{dr} + \frac{\nu}{r}\frac{d^2\psi_j^{(2)}}{dr^2} \right) \right] r\,dr,
$$

$$
K_{ij}^{21} = \int_{r_a}^{r_b} A_{rr} \frac{dw}{dr}\frac{d\psi_i^{(2)}}{dr} \left(\frac{d\psi_j^{(1)}}{dr} + \frac{\nu}{r}\psi_j^{(1)} \right) r\,dr
$$
$$
- \alpha \int_{r_a}^{r_b} E_{rr} \left[\frac{d^2\psi_i^{(2)}}{dr^2} \left(\frac{d\psi_j^{(1)}}{dr} + \frac{\nu}{r}\psi_j^{(1)} \right) + \frac{d\psi_i^{(2)}}{dr} \left(\frac{1}{r^2}\psi_j^{(1)} + \frac{\nu}{r}\frac{d\psi_j^{(1)}}{dr} \right) \right] r\,dr,
$$

$$
K_{ij}^{13} = \int_{r_a}^{r_b} \bar{B}_{rr} \left[\frac{d\psi_i^{(1)}}{dr} \left(\frac{d\psi_j^{(3)}}{dr} + \frac{\nu}{r}\psi_j^{(3)} \right) + \frac{1}{r}\psi_i^{(1)} \left(\nu\frac{d\psi_j^{(3)}}{dr} + \frac{1}{r}\psi_j^{(3)} \right) \right] r\,dr,
$$

$$
K_{ij}^{31} = \int_{r_a}^{r_b} \bar{B}_{rr} \left[\frac{d\psi_i^{(3)}}{dr} \left(\frac{d\psi_j^{(1)}}{dr} + \frac{\nu}{r}\psi_j^{(1)} \right) + \frac{1}{r}\psi_i^{(3)} \left(\nu\frac{d\psi_j^{(1)}}{dr} + \frac{1}{r}\psi_j^{(1)} \right) \right] r\,dr,
$$

$$
K_{ij}^{22} = \int_{r_a}^{r_b} \left\{ \frac{1}{2}A_{rr} \left(\frac{dw}{dr} \right)^2 \frac{d\psi_i^{(2)}}{dr}\frac{d\psi_j^{(2)}}{dr} - \alpha E_{rr}\frac{dw}{dr}\frac{d\psi_i^{(2)}}{dr} \left(\frac{\nu}{r}\frac{d\psi_j^{(2)}}{dr} + \frac{d^2\psi_j^{(2)}}{dr^2} \right) \right.
$$
$$
\left. - \alpha\frac{d^2\psi_i^{(2)}}{dr^2} \left[\frac{1}{2}E_{rr}\frac{dw}{dr}\frac{d\psi_j^{(2)}}{dr} - \alpha H_{rr} \left(\frac{\nu}{r}\frac{d\psi_j^{(2)}}{dr} + \frac{d^2\psi_j^{(2)}}{dr^2} \right) \right] \right\},
$$

$$- \alpha \frac{1}{r} \frac{d\psi_i^{(2)}}{dr} \left[\frac{\nu}{2} E_{rr} \frac{dw}{dr} \frac{d\psi_j^{(2)}}{dr} - \alpha H_{rr} \left(\frac{1}{r} \frac{d\psi_j^{(2)}}{dr} + \nu \frac{d^2\psi_j^{(2)}}{dr^2} \right) \right]$$

$$+ \frac{1}{4} \tilde{\bar{A}}_{r\theta} \left(-\frac{d^2\psi_i^{(2)}}{dr^2} + \frac{1}{r} \frac{d\psi_i^{(2)}}{dr} \right) \left(-\frac{d^2\psi_j^{(2)}}{dr^2} + \frac{1}{r} \frac{d\psi_j^{(2)}}{dr} \right)$$

$$+ \hat{\bar{A}}_{rz} \frac{d\psi_i^{(2)}}{dr} \frac{d\psi_j^{(2)}}{dr} + c_f \psi_i^{(2)} \psi_j^{(2)} \Bigg\} r dr,$$

$$K_{ij}^{23} = \int_{r_a}^{r_b} \Bigg\{ \bar{B}_{rr} \frac{dw}{dr} \frac{d\psi_i^{(2)}}{dr} \left(\frac{d\psi_j^{(3)}}{dr} + \frac{\nu}{r} \psi_j^{(3)} \right) + \hat{\bar{A}}_{rz} \frac{d\psi_i^{(2)}}{dr} \psi_j^{(3)}$$

$$+ \frac{1}{4} \tilde{\bar{A}}_{r\theta} \left(-\frac{d^2\psi_i^{(2)}}{dr^2} + \frac{1}{r} \frac{d\psi_i^{(2)}}{dr} \right) \left(\frac{d\psi_j^{(3)}}{dr} - \frac{1}{r} \psi_j^{(3)} \right)$$

$$- \alpha \bar{F}_{rr} \left[\frac{d^2\psi_i^{(2)}}{dr^2} \left(\frac{d\psi_j^{(3)}}{dr} + \frac{\nu}{r} \psi_j^{(3)} \right) + \frac{1}{r} \frac{d\psi_i^{(2)}}{dr} \left(\nu \frac{d\psi_j^{(3)}}{dr} + \frac{1}{r} \psi_j^{(3)} \right) \right] \Bigg\} r dr,$$

$$K_{ij}^{32} = \int_{r_a}^{r_b} \Bigg\{ \frac{d\psi_i^{(3)}}{dr} \left[\frac{1}{2} \bar{B}_{rr} \frac{dw}{dr} \frac{d\psi_j^{(2)}}{dr} - \alpha \bar{F}_{rr} \left(\frac{\nu}{r} \frac{d\psi_j^{(2)}}{dr} + \frac{d^2\psi_j^{(2)}}{dr^2} \right) \right]$$

$$+ \frac{1}{r} \psi_i^{(3)} \left[\frac{\nu}{2} \bar{B}_{rr} \frac{dw}{dr} \frac{d\psi_j^{(2)}}{dr} - \alpha \bar{F}_{rr} \left(\frac{1}{r} \frac{d\psi_j^{(2)}}{dr} + \nu \frac{d^2\psi_j^{(2)}}{dr^2} \right) \right] + \tilde{\bar{A}}_{rz} \psi_i^{(3)} \frac{d\psi_j^{(2)}}{dr}$$

$$+ \frac{1}{4} \tilde{\bar{A}}_{r\theta} \left(\frac{d\psi_i^{(3)}}{dr} - \frac{1}{r} \psi_i^{(3)} \right) \left(-\frac{d^2\psi_j^{(2)}}{dr^2} + \frac{1}{r} \frac{d\psi_j^{(2)}}{dr} \right) \Bigg\} r dr,$$

$$K_{ij}^{33} = \int_{r_a}^{r_b} \Bigg\{ \hat{D}_{rr} \left[\frac{d\psi_i^{(3)}}{dr} \left(\frac{d\psi_j^{(3)}}{dr} + \frac{\nu}{r} \psi_j^{(3)} \right) + \frac{1}{r} \psi_i^{(3)} \left(\nu \frac{d\psi_j^{(3)}}{dr} + \frac{1}{r} \psi_j^{(3)} \right) \right]$$

$$+ \frac{1}{4} \hat{A}_{r\theta} \left(\frac{d\psi_i^{(3)}}{dr} - \frac{1}{r} \psi_i^{(3)} \right) \left(\frac{d\psi_j^{(3)}}{dr} - \frac{1}{r} \psi_j^{(3)} \right) + \hat{\bar{A}}_{rz} \psi_i^{(3)} \psi_j^{(3)} \Bigg\} r dr,$$

$$F_i^1 = Q_1 \psi_i^{(1)}(r_a) + Q_5 \psi_i^{(1)}(r_b), \quad F_i^3 = Q_4 \psi_i^{(3)}(r_a) + Q_8 \psi_i^{(3)}(r_b),$$

$$F_i^2 = \int_{r_a}^{r_b} q \psi_i^{(2)} r dr + Q_2 \psi_i^{(2)}(r_a) + Q_6 \psi_i^{(2)}(r_b)$$

$$+ Q_3 \left[-\frac{d\psi_i^{(2)}}{dr} \right]_{r_a} + Q_7 \left[-\frac{d\psi_i^{(2)}}{dr} \right]_{r_b}, \tag{10.5.11}$$

10.6   TANGENT STIFFNESS COEFFICIENTS

10.6.1   PRELIMINARY COMMENTS

For the form of the finite element equations at hand, the tangent stiffness matrix $\mathbf{T}$ is assumed to be of the same form as the direct stiffness matrix $\mathbf{K}$ for each model. Then the coefficients of the submatrices $\mathbf{T}^{\alpha\beta}$ of the tangent stiffness matrix $\mathbf{T}$ can be computed using the definition

$$T_{ij}^{\alpha\beta} \equiv \frac{\partial R_i^\alpha}{\partial \Delta_j^\beta} = K_{ij}^{\alpha\beta} + \sum_{\gamma=1}^{\Lambda} \sum_{k=1}^{n_\gamma} \frac{\partial K_{ik}^{\alpha\gamma}}{\partial \Delta_j^\beta} \Delta_k^\gamma - \frac{\partial F_i^\alpha}{\partial \Delta_j^\beta}$$

$$= K_{ij}^{\alpha\beta} + \sum_{k=1}^{n_\gamma} \frac{\partial K_{ik}^{\alpha 1}}{\partial \Delta_j^\beta} \Delta_k^1 + \sum_{k=1}^{n_\gamma} \frac{\partial K_{ik}^{\alpha 2}}{\partial \Delta_j^\beta} \Delta_k^2 + \cdots - \frac{\partial F_i^\alpha}{\partial \Delta_j^\beta}, \quad (10.6.1)$$

where Λ takes the value of 2, 3, or 4, depending on the model. The reader is advised to revisit Section 9.9 for additional details for the computation of the tangent stiffness coefficients.

10.6.2   THE DISPLACEMENT MODEL OF THE CPT

The tangent stiffness coefficients for this case are given by

$$T_{ij}^{11} = K_{ij}^{11}, \quad T_{Ij}^{21} = K_{Ij}^{21},$$

$$T_{iJ}^{12} = K_{iJ}^{12} + \int_{r_a}^{r_b} \frac{1}{2} A_{rr} \frac{\partial w}{\partial r} \left(\frac{d\psi_i}{dr} \frac{d\varphi_J}{dr} + \frac{\nu}{r} \psi_i \frac{d\varphi_J}{dr} \right) r\,dr = T_{Ji}^{21},$$

$$T_{IJ}^{22} = K_{IJ}^{22} + \int_{r_a}^{r_b} \left[A_{rr} \left\{ \frac{d\varphi_I}{dr} \frac{d\varphi_J}{dr} \left[\left(\frac{\partial w}{\partial r} \right)^2 + \frac{\partial u}{\partial r} + \nu \frac{u}{r} \right] \right\} \right.$$

$$- B_{rr} \frac{d\varphi_I}{dr} \frac{d\varphi_J}{dr} \left(\frac{\partial^2 w}{\partial r^2} + \frac{\nu}{r} \frac{\partial w}{\partial r} \right)$$

$$\left. - \frac{1}{2} B_{rr} \frac{\partial w}{\partial r} \left(\frac{d^2 \varphi_I}{dr^2} \frac{d\varphi_J}{dr} + \frac{\nu}{r} \frac{d\varphi_I}{dr} \frac{d\varphi_J}{dr} \right) \right] r\,dr. \quad (10.6.2)$$

The tangent stiffness matrix is symmetric (i.e., $T_{ij}^{\alpha\beta} = T_{ji}^{\beta\alpha}$).

10.6.3   THE MIXED MODEL OF THE CPT

The tangent stiffness coefficients for this case are given by

$$T_{ij}^{11} = K_{ij}^{11}, \quad T_{ij}^{21} = K_{ij}^{21}, \quad T_{ij}^{13} = K_{ij}^{13}, \quad T_{ij}^{23} = K_{ij}^{23},$$

$$T_{ij}^{12} = K_{ij}^{12} + \int_{r_a}^{r_b} \frac{1}{2} A_{rr} \frac{dw}{dr} \left(r \frac{d\psi_i^{(1)}}{dr} \frac{d\psi_j^{(2)}}{dr} + \nu \psi_i^{(1)} \frac{d\psi_j^{(2)}}{dr} \right) dr = T_{ji}^{21},$$

$$T_{ij}^{22} = K_{ij}^{22} + \int_{r_a}^{r_b} \left\{ \bar{A}_{rr} \left[\left(\frac{\partial w}{\partial r} \right)^2 + \frac{\partial u}{\partial r} + \frac{\nu}{r} u \right] + \bar{B}_{rr} M_{rr} \right\} \frac{d\psi_i^{(2)}}{dr} \frac{d\psi_j^{(2)}}{dr} r dr,$$

$$T_{ij}^{32} = K_{ij}^{32} + \frac{1}{2} \int_{r_a}^{r_b} \bar{B}_{rr} \frac{\partial w}{\partial r} \psi_i^{(3)} \frac{d\psi_j^{(2)}}{dr} r dr,$$

$$T_{ij}^{31} = K_{ij}^{31}, \quad T_{ij}^{33} = K_{ij}^{33}. \tag{10.6.3}$$

The tangent stiffness matrix is symmetric (i.e., $T_{ij}^{\alpha\beta} = T_{ji}^{\beta\alpha}$).

10.6.4 THE DISPLACEMENT MODEL OF THE FST

The tangent stiffness coefficients for the FST are given by

$$T_{ij}^{11} = K_{ij}^{11}, \quad T_{ij}^{12} = 2K_{ij}^{12}, \quad T_{ij}^{13} = K_{ij}^{13}, \quad T_{ij}^{21} = K_{ij}^{21},$$

$$T_{ij}^{31} = K_{ij}^{31} = T_{ji}^{13}, \quad T_{ij}^{23} = K_{ij}^{23}, \quad T_{ij}^{33} = K_{ij}^{33},$$

$$T_{ij}^{22} = K_{ij}^{22} + \int_{r_a}^{r_b} \left\{ A_{rr} \left[\left(\frac{\partial w}{\partial r} \right)^2 + \frac{\partial u}{\partial r} + \nu \frac{u}{r} \right] \right.$$

$$\left. + B_{rr} \left(\frac{\partial \phi}{\partial r} + \frac{\nu}{r} \phi \right) \right\} \frac{d\psi_i^{(2)}}{dr} \frac{d\psi_j^{(2)}}{dr} r dr,$$

$$T_{ij}^{32} = K_{ij}^{32} + \frac{1}{2} \int_{r_a}^{r_b} B_{rr} \frac{\partial w}{\partial r} \left(\frac{d\psi_i^{(3)}}{dr} + \frac{\nu}{r} \psi_i^{(3)} \right) \frac{d\psi_j^{(2)}}{dr} r dr. \tag{10.6.4}$$

The tangent stiffness matrix of the FST element is also symmetric.

10.6.5 THE DISPLACEMENT MODEL OF THE TST

The tangent stiffness coefficients for the TST are given by

$$T_{ij}^{11} = K_{ij}^{11}, \quad T_{ij}^{21} = K_{ij}^{21}, \quad T_{ij}^{13} = K_{ij}^{13}, \quad T_{ij}^{31} = K_{ij}^{31}, \quad T_{ij}^{23} = K_{ij}^{23},$$

$$T_{ij}^{12} = K_{ij}^{12} + \frac{1}{2} \int_{r_a}^{r_b} A_{rr} \frac{dw}{dr} \left(\frac{d\psi_i^{(1)}}{dr} + \frac{\nu}{r} \psi_i^{(1)} \right) \frac{d\psi_j^{(2)}}{dr} r dr$$

$$T_{ij}^{22} = K_{ij}^{22} + \int_{r_a}^{r_b} \left\{ A_{rr} \left[\frac{du}{dr} + \frac{\nu}{r} u + \left(\frac{dw}{dr} \right)^2 \right] - \alpha E_{rr} \left(1.5 \frac{\nu}{r} \frac{dw}{dr} + \frac{d^2w}{dr^2} \right) \right.$$

$$\left. + \bar{B}_{rr} \left(\frac{d\phi_r}{dr} + \frac{\nu}{r} \phi_r \right) \right\} \frac{d\psi_i^{(2)}}{dr} \frac{d\psi_j^{(2)}}{dr} r dr$$

$$- \frac{1}{2} \alpha \int_{r_a}^{r_b} E_{rr} \frac{dw}{dr} \frac{d^2\psi_i^{(2)}}{dr^2} \frac{d\psi_j^{(2)}}{dr} r dr$$

$$T_{ij}^{32} = K_{ij}^{32} + \frac{1}{2} \int_{r_a}^{r_b} \bar{B}_{rr} \frac{dw}{dr} \left(\frac{d\psi_i^{(3)}}{dr} + \frac{\nu}{r} \psi_i^{(3)} \right) \frac{d\psi_j^{(2)}}{dr} r dr. \tag{10.6.5}$$

10.7 NUMERICAL RESULTS

10.7.1 PRELIMINARY COMMENTS

The evaluation of the element stiffness and force coefficients defined for various finite element models is carried out using numerical integration, which is necessary especially due to the nonlinear terms. Numerical integration also facilitates the evaluation of the transverse shear and nonlinear terms in the stiffness coefficients to avoid shear and membrane locking.

In this section, we present numerical results obtained with various finite element (FEM) models developed in the preceding sections. Numerical results obtained with the FEM are compared in all cases. We use four models of the FEM, as designated here:

CPT(D) – Displacement finite element model of the CPT
CPT(M) – Mixed finite element model of the CPT
FST(D) – Displacement finite element model of the FST
TST(D) – Displacement finite element model of the TST

The CPT(D) model uses the Hermite cubic interpolation of $w(r)$ and linear interpolation of $u(r)$ and has 3 degrees of freedom (DoF) per node; the CPT(M) also has 3 DoF, namely, (u, w, M_{rr}). The FST(D) model without the inclusion of the couple stress effect uses Lagrange interpolation of u, w, and ϕ_r, and has 3 DoF per node; when the couple stress effect is included, the FST(D) element has 4 DoF per node, with linear interpolation of u, Hermite cubic interpolation of w, and linear interpolation of ϕ_r. The TST(D) element has 4 DoF, just as the FST(D).

In general, there are three categories of expressions in the stiffness matrices of these elements: (1) linear terms except for the terms containing transverse shear coefficient A_{rz}, (2) shear terms containing transverse shear coefficient A_{rz} (only the FST), and (3) nonlinear terms due to the inclusion of the von Kámán nonlinearity. The CPT(D) and CPT(M) has only expressions of type 1 and 3, while all other models have the three categories of expressions.

The number of Gauss points used for the evaluation of the first category of terms is dictated by the highest degree polynomial. In the FST, it is dictated by the term $r(\phi + dw/dr)$, which is fourth order. Hence, the number of Gauss points used is 3. For the second category of terms (shear), as already discussed, the number of Gauss points is 2. Finally, for the nonlinear terms, the number of Gauss points used is 2. Thus, full integration with 3 points and reduced integration with 2 points is needed. However, 3×2 integration rule with refined meshes also gives results very close to those obtained with the 3×2 integration rule.

In obtaining the numerical solutions, we shall consider functionally graded circular plates of (outside) radius $a = R = 10$ in (25.4 cm) and thickness $h = 0.1$ in (0.254 cm), and subjected to uniformly distributed load of intensity

q_0 lb/in (1 lb/in = 175 N/m). The FGM plate is made of two materials with the following values of the moduli, Poisson's ratio, and shear correction coefficient:

$$E_1 = 30 \times 10^6 \text{ psi } (207 \text{ GPa}), \quad E_2 = 3 \times 10^6 \text{ psi } (21 \text{ GPa}),$$
$$\nu = 0.3, \quad K_s = 5/6 \text{ (used in FST models)}. \tag{10.7.1}$$

We shall investigate the parametric effects of the power-law index, n, and boundary conditions on the linear and nonlinear transverse deflections and bending moments.

10.7.2 LINEAR ANALYSIS

Here we consider functionally graded circular plates which are either pinned or clamped at the (outer) edge. The boundary conditions on the primarily variables in various models for the pinned case are as follows:

Displacement models :

$$u(0) = 0, \text{ and } \frac{dw}{dr} = 0 \text{ or } \phi_r(0) = 0 \text{ at } r = 0;$$
$$u(R) = w(R) = 0, \tag{10.7.2a}$$

Mixed models :

$$u(0) = 0, \quad u(R) = w(R) = M_{rr}(R) = 0. \tag{10.7.2b}$$

The exact solution for the transverse deflection of pinned (at the outer edge) functionally graded circular plates according to the FST, with the power-law given in Eq. (6.1.8) $[B_{rr} \neq 0; D_{rr}^* = D_{rr} A_{rr} - B_{rr}^2]$, are given by $(\xi = r/R)$

$$u(r) = \frac{B_{rr} q_0 R^3}{16 D_{rr}^*} \left(-\xi + \xi^3\right), \tag{10.7.3a}$$

$$w(r) = \frac{A_{rr} q_0 R^4}{64 D_{rr}^*} \left[\xi^4 - 2\left(\frac{3+\nu}{1+\nu}\right)\xi^2 + \frac{5+\nu}{1+\nu} - \frac{4B_{rr}^2}{D_{rr} A_{rr}(1+\nu)}\left(1-\xi^2\right)\right]$$
$$+ \frac{q_0 R^2}{4 A_{rz}}\left(1-\xi^2\right), \tag{10.7.3b}$$

$$\phi_r(r) = \frac{A_{rr} q_0 R^3}{16 D_{rr}^*}\left[-\frac{2B^2}{D_{rr} A_{rr}(1+\nu)}\xi + \frac{(3+\nu)}{(1+\nu)}\xi - \xi^3\right], \tag{10.7.3c}$$

$$M_{rr}(r) = \frac{(3+\nu)q_0 R^2}{16}\left(1-\xi^2\right), \tag{10.7.3d}$$

where $\xi = r/R$ and q_0 is the intensity of the uniform distributed load (UDL). The CPT solutions are given by setting $1/A_{rz}$ to zero, and the solutions for homogeneous plates are obtained by setting $B_{rr} = 0$.

The boundary conditions on the primary variables in various models for the clamped circular plate are as follows (replace $-dw/dr$ with ϕ_r for the FST):

Displacement models :

$$u(0) = 0, \quad \frac{dw}{dr}(0) = 0, \quad u(R) = w(R) = \frac{dw}{dr}(R) = 0. \qquad (10.7.4a)$$

Mixed models : $u(0) = 0, \quad u(R) = w(R) = 0.$ \qquad (10.7.4b)

The exact solution for the transverse deflection of functionally graded clamped circular plate according to the FST is $(\xi = r/R)$

$$u(r) = \frac{B_{rr}q_0 R^3}{16 D_{rr}^*} \left(-\xi + \xi^3 \right), \quad \phi_r(r) = \frac{A_{rr}q_0 R^3}{16 D_{rr}^*} \left(\xi - \xi^3 \right), \qquad (10.7.5a)$$

$$w(r) = \frac{A_{rr}q_0 R^4}{64 D_{rr}^*} \left(1 - \xi^2 \right)^2 + \frac{q_0 R^2}{4 A_{rz}} \left(1 - \xi^2 \right), \qquad (10.7.5b)$$

$$M_{rr}(r) = \frac{q_0 R^2}{16} \left[(1 + \nu) - (3 + \nu)\xi^2 \right]. \qquad (10.7.5c)$$

Extensive numerical studies have been carried out with various models, including mesh independence and the effect of the power-law index, and post-computation of the stress resultants (either the bending moments or the rotations). All finite element models, with 16-element uniform mesh [linear elements in CPT(M), FST(M), and FST(D), with 2 full integration points and 1 reduced integration point], yield results that are indistinguishable in graphical presentations. Most interestingly, it is found that the post-computed rotations (in the mixed model) and bending moments (in the displacement models) are very accurate (one cannot distinguish between the exact and numerical solutions), except at $r = 0$. Based on the numerical studies, it is found that the nodal generalized displacements predicted by CPT(D) match the exact CPT solutions for the pinned and clamped plates. The data provided in Eq. (10.7.1) is used in obtaining the numerical results.

Table 10.7.1 contains a comparison of the numerical values of the center deflection and bending moment (post-computed at the center of each element in the displacement models and at the nodes in the mixed models) obtained using various finite element models with the analytical solution for $n = 0, 1, 5$ (the bending moment is independent of n) (a load value of $q_0 = 1$ lb/in is used). It is clear that all models predict solutions that are very close to each other; the bending moment predicted by the mixed models at $r = 0$ are found to be inaccurate (the exact values are $M_{rr}(0) = 20.625$ lb-in for the pinned plates and $M_{rr}(0) = 8.125$ lb-in and $M_{rr}(R) = -12.5$ lb-in for the clamped plates) but they are very accurate at other points.

Figure 10.7.1 contains plots of the deflections $w(0)$ predicted for the pinned plates and clamped plates as a function of the normalized radial coordinate, r/R. The deflections predicted (shown by symbols) by all FEM models are essentially the same (i.e., the differences cannot be seen in the graph) and match with the analytical solutions (shown as lines); this also indicates that the effect of shear deformation is negligible (because $R/h = 100$, a thin plate).

Table 10.7.1: Comparison of the center deflection $w(0)$ (in) and post-computed bending moment $M_{rr}(0.03125R)$ (lb-in) predicted by various finite element models with the exact solutions of pinned and clamped circular plates (16 elements).

n	CPT(D)	CPT(M)	FST(D)	TST(D)	Exact-CPT	Exact-FST
		Pinned circular plates, $w(0)$				
0.0	0.23187	0.23024	0.23171	0.23225	0.23187	0.23190
1.0	0.45124	0.44725	0.45101	0.45192	0.45129	0.45134
5.0	0.74905	0.74092	0.74887	0.74972	0.74924	0.74934
		Pinned circular plates, $M_{rr}(0.03125R)$				
	20.598	– – –	20.681	20.694	20.605	20.605
		Clamped circular plates, $w(0)$				
0.0	0.05687	0.05646	0.05694	0.05724	0.05687	0.05690
1.0	0.13305	0.13173	0.13219	0.13372	0.13311	0.13315
5.0	0.25905	0.25575	0.25951	0.25965	0.25924	0.25934
		Clamped circular plates, $-M_{rr}(0.96875R)$				
	11.237	13.105	11.270	11.272	11.231	11.213

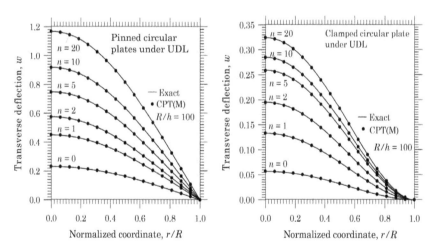

Figure 10.7.1: Plots of the center deflection of pinned and clamped (at $r = R$) circular plates as functions of the normalized radial coordinate, r/R, for various value of the power-law index, n [see Eq. (10.7.1) for the data]. The solid lines correspond to the analytical solutions and symbols to the finite element solutions.

The post-computed values in FST(D) and nodal values in CPT(M) of the bending moment M_{rr} are plotted as functions of the normalized radial coordinate (r/R) in Fig. 10.7.2. Except for the value at $r = 0$, the results match closely with the exact solution for both pinned and clamped circular plates. The numerical solutions predicted by different finite models are indistinguishable from each other in the graphs. The results are independent of the power-law index n.

Figure 10.7.3 shows the center deflection $w(0)$ and rotation $-dw/dr$ at $r = 0.5265\,R$ as functions of the power-law index n for the pinned and clamped circular plates. The solutions predicted by various finite element models are not distinguishable in the graphs. We note that the rate of increase of the deflection has two different regions; the first region has a rapid increase of the deflection while the second region is marked with a relatively slow increase. This is primarily because of the fact that the coupling coefficient B_{xx} varies with n rapidly for the smaller values of n followed by a slow decay after $n > 3$. The rate of increase in the deflection or slope in the second part is less for clamped plates than for the pinned plates.

10.7.3 NONLINEAR ANALYSIS WITHOUT COUPLE STRESS EFFECT

The resulting nonlinear equations are solved using the Newton iteration scheme. The initial solution vector is chosen to be $\mathbf{\Delta}^0 = \mathbf{0}$ so that the first iteration of the first load step yields the linear solution for the first load.

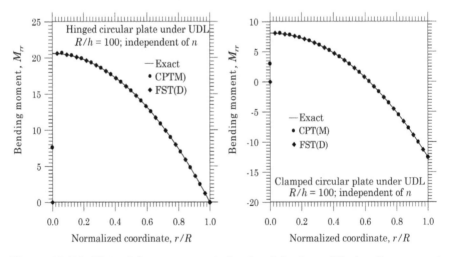

Figure 10.7.2: Plots of the post-computed and nodal values of the bending moment M_{rr} of pinned and clamped (at $r = R$) circular plates as functions of the normalized radial coordinate, r/R [see Eq. (10.7.1) for the data]. The solid lines correspond to the analytical solutions and symbols to the finite element solutions.

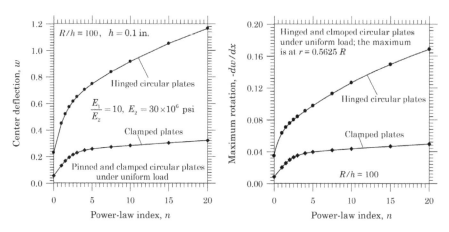

Figure 10.7.3: Plots of the center deflection $w(0)$ and $-dw/dr$ at $r = 0.5625\,R$ of pinned and clamped circular plates as a function of the power-law index, n.

The effect of the number of Gauss points used on the nonlinear deflection is also investigated. It is found that there is no significant difference between the results obtained with 3×2 and 2×1 ($M \times N$ means M integration points for linear terms and N integration points for shear as well as nonlinear terms). The FST element (when couple stress effect is considered) with Hermite cubic interpolation of w required 3×2 integration rule.

The geometric and material parameters given in Eq. (10.7.1) with a mesh of 16 elements and convergence tolerance of $\epsilon = 10^{-3}$ are used to obtain the numerical results for homogeneous plates ($n = 0$); Load increments of $q_0 = 0.15, 0.15, 0.3, \ldots, 0.3$, which correspond to $P = (q_0 R^4 / E_1 h^4) = 0.5, 1.0, 2.0, \ldots$ are used. The nonlinear analysis results for dimensionless center deflection $\bar{w} = w(0)/h$ and post-computed bending moment for different values of the load parameter $P = q_0 R^4 / E_1 h^4$ obtained by various models are presented in Tables 10.7.2 and 10.7.3, respectively, for the pinned plates. Similar results are presented in Tables 10.7.4 and 10.7.5 for the clamped plates. The results were obtained using the 2×1 integration rule with CPT(D), FST(D) with 3DoF, CPT(M), and TST(D) elements.

Plots of $\bar{w}$ versus $P = q_0 R^4 / E_1 h^4$ and M_{rr} versus $P = q_0 R^4 / E_1 h^4$ are shown in Figs. 10.7.4 and 10.7.5, respectively, for pinned and clamped plates. The results predicted by the TST(D) are slightly larger for larger loads, indicating that the TST represents plates as slightly flexible than the CPT and FST. The results predicted by the CPT(D), TST(D), and CPT(M) are indistinguishable in the graphs. The TST(D) also has convergence issues for FGM plates.

Table 10.7.2: Comparison of the center deflection $w(0)/h$ versus the load parameter $P = q_0 R^4/E_1 h^4$ predicted by various finite element models with the exact solutions of pinned circular plates (meshes of 16 elements is used; $n = 0$).

P	CPT(D)	FST(D)	CPT(M)	TST(D)
0.5	0.2975	0.2975	0.2959	0.2985
1.0	0.4808	0.4809	0.4788	0.4836
2.0	0.7040	0.7042	0.7016	0.7109
4.0	0.9649	0.9652	0.9622	0.9805
6.0	1.1378	1.1380	1.1348	1.1624
8.0	1.2714	1.2713	1.2680	1.3048
10.0	1.3819	1.3816	1.3782	1.4244
12.0	1.4773	1.4767	1.4732	1.5287
14.0	1.5617	1.5608	1.5573	1.6220
16.0	1.6378	1.6366	1.6330	1.7071
18.0	1.7074	1.7058	1.7022	1.7857
20.0	1.7716	1.7698	1.7662	1.8590

Table 10.7.3: Comparison of the post-computed bending moment $M_{rr}(0.03125\,R)$ versus the load parameter P predicted by various finite element models with the exact solutions of pinned circular plates (meshes of 16 elements is used; $n = 0$).

P	CPT(D)	FST(D)	TST(D)
0.5	2.6104	2.6221	2.6271
1.0	4.1403	4.1596	4.1761
2.0	5.8754	5.9019	5.9479
4.0	7.7172	7.7474	7.8568
6.0	8.8426	8.8726	9.0449
8.0	9.6724	9.7010	9.9343
10.0	10.3410	10.3670	10.6600
12.0	10.9070	10.9320	11.2820
14.0	11.4030	11.4260	11.8330
16.0	11.8480	11.8690	12.3310
18.0	12.2530	12.2720	12.7890
20.0	12.6260	12.6440	13.2150

Plots of dimensionless center deflection $\bar{w} = w(0)/h$ versus the load parameter P are shown in Figs. 10.7.6 and 10.7.7 for pinned and clamped plates, respectively, for various values of the power-law index. Figure 10.7.8 contains plots of the normalized deflection $\bar{w} = w(b)/h$ versus the load parameter P for circular plates and annular plates, both pinned at the outer edge; annular plates are assumed to be free at the inner edge (i.e., $r = b = 0.2\,R$). A uniform mesh of 20 elements is used for the circular plate while a uniform mesh of 16 elements is used for annular plates (so that the element lengths in each case are the same). The results obtained by CPT(D), FST(D), and CPT(M) are indistinguishable in the plots.

Table 10.7.4: Comparison of the center deflection $w(0)/h$ versus the load parameter $P = q_0 R^4/E_1 h^4$ for clamped circular plates predicted by various finite element models (meshes of 16 elements is used).

P	CPT(D)	FST(D)	CPT(M)	TST(D)
0.5	0.0850	0.0851	0.0844	0.0855
1.0	0.1680	0.1682	0.1669	0.1692
2.0	0.3229	0.3233	0.3208	0.3256
4.0	0.5769	0.5779	0.5742	0.5841
6.0	0.7705	0.7719	0.7676	0.7834
8.0	0.9246	0.9263	0.9218	0.9437
10.0	1.0526	1.0545	1.0499	1.0784
12.0	1.1625	1.1645	1.1598	1.1950
14.0	1.2591	1.2611	1.2564	1.2985
16.0	1.3455	1.3474	1.3428	1.3920
18.0	1.4239	1.4258	1.4212	1.4776
20.0	1.4959	1.4976	1.4931	1.5568

Table 10.7.5: Comparison of the post-computed bending moment $M_{rr}(0.03125\,R)$ versus the load parameter $P = q_0 R^4/E_1 h^4$ for clamped circular plates predicted by various finite element models (meshes of 16 elements is used).

P	CPT(D)	FST(D)	TST(D)
0.5	1.2086	1.2234	1.2230
1.0	2.3816	2.4109	2.4111
2.0	4.5201	4.5756	4.5827
4.0	7.7978	7.8911	7.9352
6.0	10.0340	10.1480	10.2480
8.0	11.6340	11.7570	11.9220
10.0	12.8420	12.9680	13.2010
12.0	13.7970	13.9220	14.2230
14.0	14.5790	14.7000	15.0700
16.0	15.2370	15.3530	15.7910
18.0	15.8040	15.9140	16.4190
20.0	16.3010	16.4050	16.9760

The deflections of the annular plates (shown in broken lines) are larger than the circular plates (shown in solid lines), as can be seen from Fig. 10.7.8, due to the fact that annular plates have less stiffness compared to the solid circular plates for the same boundary conditions.

Plots of the normalized deflection $\bar{w} = w(b)/h$ for annular plates with internal edge clamped and outer edge (i.e., $r = a = R$) free are presented in Fig. 10.7.9. A uniform mesh of 16 elements is used. As the internal radius of the annular plate increases from $b = 0.2\,R$ to $b = 0.5\,R$, the deflection of the outer edge decreases substantially because of the reduction in free span of the plate (see Fig. 10.7.9).

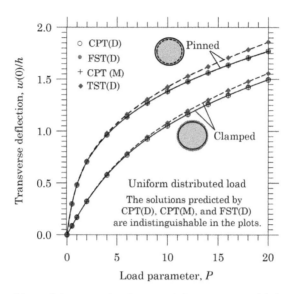

Figure 10.7.4: Plots of the normalized center deflection $\bar{w} = w(0)/h$ versus the load parameter $P = q_0 R^4 / E_1 h^4$ for homogeneous pinned and clamped circular plates using all four finite element models.

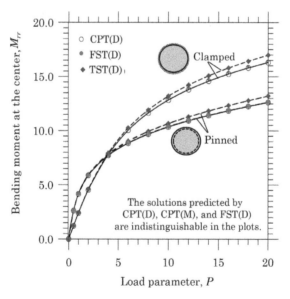

Figure 10.7.5: Plots of the post-computed bending moment $M_{rr}(0.03125\,R)$ versus the load parameter $P = q_0 R^4 / E_1 h^4$ for homogeneous pinned and clamped circular plates using CPT(D), FST(D), and TST(D).

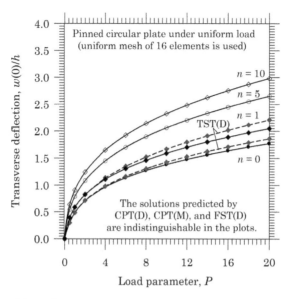

Figure 10.7.6: Plots of the normalized center deflection $\bar{w}$ versus the load parameter $P = q_0 R^4 / E_1 h^4$ for pinned circular plates, for various value of the power-law index, n.

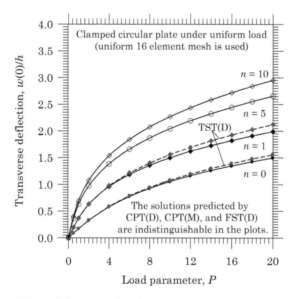

Figure 10.7.7: Plots of the normalized center deflection $\bar{w}$ versus the load parameter $P = q_0 R^4 / E_1 h^4$ for clamped circular plates, for various value of the power-law index, n.

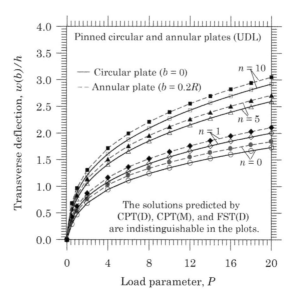

Figure 10.7.8: Plots of the normalized deflection $\bar{w}$ versus the load parameter $P = q_0 R^4 / E_1 h^4$ for annular plates with outer edge, $r = a = R$, pinned, for various value of the power-law index, n; $b = 0$ for solid circular plates and $b = 0.2\,R$ for annular plates.

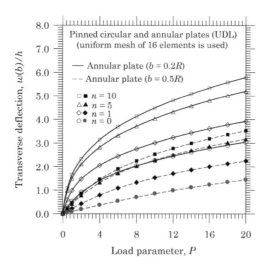

Figure 10.7.9: Plots of the normalized deflection $\bar{w}$ versus the load parameter $P = q_0 R^4 / E_1 h^4$ for annular plates clamped at the inner edge ($b = 0.2R, 0.5R$), for various value of the power-law index, n.

10.7.4 NONLINEAR ANALYSIS WITH COUPLE STRESS EFFECT

Next, we consider several examples of circular and annular plates with pinned and clamped boundary conditions to study the parametric effects of the power-law index n and the length scale parameters ℓ from the modified couple stress theory on the nonlinear load-deflection behavior.

The following parameters are used in obtaining the numerical results (no specific units are used) in the examples:

$$h = 0.1, \quad a = R = 10h, \quad \nu = 0.25, \quad E_1 = 10^6, \quad E_2 = 10^5, \quad K_s = 5/6. \quad (10.7.6)$$

Convergence studies were carried out with different numerical integration rules (to eliminate shear and membrane locking) and number of elements used in the mesh to verify the linear and nonlinear solution of clamped, homogeneous, solid circular plates under uniformly distributed transverse load by comparing the finite element solutions with the exact solution for the linear case. We note that the CPT(D) element has three $(u, w, -dw/dx)$ while the FST(D) (when the couple stress effect is considered) and TST(D) elements have four $(u, w, -dw/dx, \phi)$ degrees of freedom per node (see Reddy, Romanoff, and Loya [82] for additional discussion).

The boundary conditions used in the three theories are as follows $(a = R)$:

$$\text{CPT:} \quad u = \frac{dw}{dx} = 0 \ \text{at} \ r = 0; \qquad u = w = \frac{dw}{dx} = 0 \ \text{at} \ r = R$$

$$\text{FST:} \quad u = \frac{dw}{dx} = \phi_r = 0 \ \text{at} \ r = 0; \qquad u = w = \frac{dw}{dx} = \phi_r = 0 \ \text{at} \ r = R$$

$$\text{TST:} \quad u = \frac{dw}{dx} = \phi_r = 0 \ \text{at} \ r = 0; \qquad u = w = \frac{dw}{dx} = \phi_r = 0 \ \text{at} \ r = R$$

Table 10.7.6 summarizes the effect of the Gauss-point rule $M \times N$, with M points for full integration of the linear terms and N points for reduced integration (recall the discussion of Section 10.7.1) used for shear and nonlinear terms on the center transverse deflection of clamped, homogeneous, circular plates under a UDL of intensity q_0. The linear solutions at $q_0 = 100$ due the CPT and FST are $w(0) = 1.7578 \times 10^{-2}$ and $w(0) = 1.8341 \times 10^{-2}$, respectively. The analytical solutions are

$$w^{\text{CPT}}(0) = \frac{q_0 R^4}{64D} = 1.7578 \times 10^{-2}, \tag{10.7.7}$$

$$w^{\text{FST}}(0) = \frac{q_0 R^4}{64D} \left[1 + \frac{8}{3(1-\nu)K_s} \frac{h^2}{R^2}\right] = 1.8328 \times 10^{-2}, \tag{10.7.8}$$

where $D = Eh^3/12(1-\nu^2)$. It is clear that CPT(D) gives essentially the same results for all integration rules, while FST(D) gives the best results for 2×1 integration rule. The linear solutions predicted at $q_0 = 100$ (with $\Delta q_0 = 100$ or $P \equiv (q_0 R^4/E_1 h^4) = 1$) for the homogeneous plates using a mesh of 16 elements of the CPT(D) and FST(D) elements and 2×1 integration rule are $w(0) = 1.7578 \times 10^{-2}$ and $w(0) = 1.8341 \times 10^{-2}$, respectively. On the basis of this investigation, it is determined that NGP×LGP= 2×1 is a suitable Gauss rule for CPT(D) as well as FST(D). The difference between the results obtained using 2×1 and 3×2 is not significant in graphical presentation of the results.

Table 10.7.6: Dimensionless center transverse deflection $\bar{w} \times 10$ [$\bar{w} = w(0)/h$] of clamped circular plates under a uniform load of intensity q_0. The first row corresponds to the linear solution and the second row corresponds to the converged nonlinear solution at load $q_0 = 100$, where the convergence tolerance is taken to be $\epsilon = 10^{-3}$; the numbers in the parenthesis refer to the number of iterations taken for convergence.

Exact	2×2	3×3	2×1	3×2
		CPT(D)		
1.7578	1.7578	1.7578	1.7578	1.7578
	1.7298 (3)	1.7299 (3)	1.7300 (3)	1.7299 (3)
		FST(D)		
1.8328	1.7166	1.7166	1.8341	1.7166
	1.6923 (3)	1.6923 (3)	1.8021 (3)	1.6923 (3)

The load parameter $P = (q_0 R^4/E_1 h^4)$ versus the dimensionless deflections $w(0)/h$ for clamped solid circular plate under UDL of intensity q_0 are presented in Fig. 10.7.10 for various values of the power-law index n and the ratio of the length scale to the plate thickness ℓ/h [the length scale ℓ enters the calculation through $S_{r\theta}$ appearing in the finite element stiffness coefficients; see Eq. (6.2.22) for the definition of $S_{r\theta}$]. The load increment used was $\Delta q_0 = 100$, and a mesh of 16 elements is used. The values selected for n and $l = \ell/h$ are only to determine the parametric effects and they do not necessarily correspond to any specific physical system.

In Figs. 10.7.10 and 10.7.11, the results obtained with the CPT(D) are shown with symbols and solid lines while those obtained with the FST(D) are shown with broken lines. The results predicted by the two theories are very close, as can be seen from the plots. The deflections predicted by the FST(D) are slightly higher than those predicted by the CPT(D) for any n with $l = \ell/h = 0$. This increase in the deflection is due to the effect of transverse shear deformation. For any n and $l = \ell/h \neq 0$, the deflections predicted by the FST(D) are slightly lower than those predicted by the CPT(D).

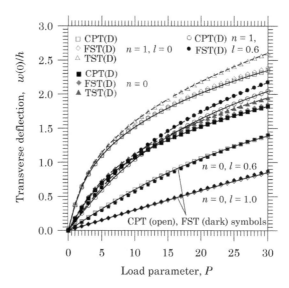

Figure 10.7.10: Load–deflection curves for clamped solid circular plates for $n = 0, 1$ and $l/h = 0.6$.

In general, irrespective of the theory used, the effect of the material length scale entering through the couple stress term is to stiffen the plates.

Plots of the load parameter $P = (q_0 R^4 / E_1 h^4)$ versus the normalized deflection $w(0)/h$ for circular plates pinned at $r = a$ are shown in Fig. 10.7.11 for $n = 0, 1, 5$ and $l = \ell/h = 0.6$. It is found that both CPT(D) and FST(D) give essentially the same results (i.e., the difference cannot be seen in the plots). The difference between $n = \ell/h = 0$ and $n \geq 0$ but $\ell/h = 0.6$ is small for circular plates.

Next, we study annular plates under UDL with different boundary conditions; the hole diameter is taken as $b = 0.25R$, and a mesh of 16 elements is used. Both the CPT(D) and FST(D) gave almost identical results; only results based on the CPT(D) are included in the figures, unless stated otherwise.

First, we consider the case of an annular plate pinned at $r = a = R$ and free at $r = b = 0.25R$. Plots of the load parameter P versus normalized center deflection $w(b)/h$ are shown in Fig. 10.7.12 for $n = 0, 5$ and $\ell/h = 0, 0.6$. Once again, we see the stiffening effect due to the couple stress.

Next, we consider the same problem with outer edge clamped. Unlike the case of pinned circular plates, the clamped circular plate load–deflection curves with $\ell/h = 0.6$ cross over those curves with $\ell/h = 0$ at certain loads, as can be seen from Fig. 10.7.13. Also, the effect of the couple stress is more significant in the clamped circular plates.

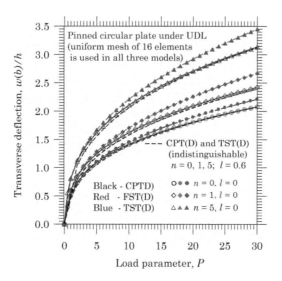

Figure 10.7.11: Load–deflection curves for clamped plates.

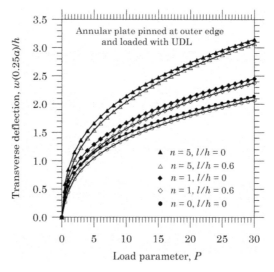

Figure 10.7.12: Load–deflection curves for annular plates under UDL and pinned at the outer edge.

Lastly, we consider annular plates clamped at the inner edge and subjected to UDL. The edge at $r = a = R$ is assumed to be free while edge $r = b$ is clamped. Figure 10.7.14 contains the load–deflection curves for the problem. Even for this cantilevered annular plate with inner edge clamped, the effect of the material length scale is to stiffen the plate (but to a smaller extent). The FST(D) and CPT(D) results are very close, indicating that the effect of the transverse shear deformation is negligible.

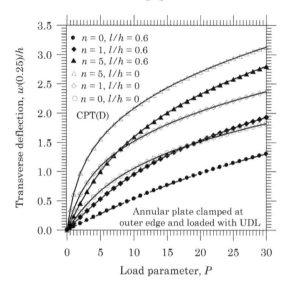

Figure 10.7.13: Load–deflection curves for annular plates under UDL and clamped at the outer edge.

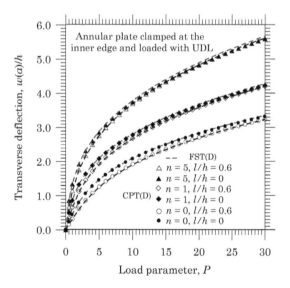

Figure 10.7.14: Load–deflection curves for annular plates under UDL and clamped at the inner edge.

10.8 CHAPTER SUMMARY

In this chapter, the displacement and mixed models of the CPT, the displacement model of the FST, and the displacement model of the TST are presented. In each case, the variational statements (or weak forms) are developed, finite element equations are derived, and the tangent stiffness coefficient of each model are summarized.

Numerical results of the linear and nonlinear bending of circular plates and annular plates for various boundary conditions and UDL are presented. In addition, the effect of the material variation (n) and couple stress (ℓ/h) are investigated.

Although only UDL is used in the numerical examples, a variety of loads, including point load at the center, linearly varying load with r, can be used.

SUGGESTED EXERCISES

10.1 verify the stiffness and force coefficients in Eq. (10.2.9) for the displacement model of the CPT.

10.2 verify the stiffness and force coefficients in Eq. (10.3.8) for the mixed model of the CPT.

10.3 verify the stiffness and force coefficients in Eq. (10.4.8) for the displacement model of the FST.

10.4 verify the stiffness and force coefficients in Eq. (10.5.11) for the displacement model of the TST.

10.5 verify the tangent stiffness coefficients in Eq. (10.6.2) for the displacement model of the CPT.

10.6 verify the tangent stiffness coefficients in Eq. (10.6.3) for the mixed model of the CPT.

10.7 verify the tangent stiffness coefficients in Eq. (10.6.4) for the displacement model of the FST.

10.8 verify the tangent stiffness coefficients in Eq. (10.6.5) for the displacement model of the TST.

10.9 The general solution $w(r)$ and ϕ_r of isotropic circular plates according to the FST, in the absence of any loads, is available from Eqs. (7.3.22) and (7.3.18) as

$$w(r) = -\frac{1}{D_{rr}}\left[-\frac{r^2}{4}(\log r - 1)c_1 + c_2\frac{r^2}{4} + c_3\log r + c_6\right] - \frac{1}{S_{rz}}c_1\log r$$
$$\equiv K_1 + K_2 r^2 + K_3\log r + K_4 r^2\log r, \tag{1}$$

$$\phi_r(r) = \frac{1}{D_{rr}}\left[\frac{r}{4}(2\log r - 1)c_1 + c_2\frac{r}{2} + \frac{c_3}{r}\right]$$
$$\equiv -2K_2 r - \frac{K_3}{r} - K_4\left[r(1 + 2\log r) + \frac{1}{r}\Gamma\right], \tag{2}$$

where $\Gamma = (4D_{rr}/S_{xz})$, and c_i are constants of integration. The CPT solution is obtained from (1) and (2) by setting $\Gamma = 0$ in Eq. (2).

Now consider a typical finite element located between $r_a \leq r \leq r_b$. Let the generalized displacements at nodes 1 and 2 of the element be defined as

$$w(r_a) = \Delta_1, \quad \phi_r(r_a) \equiv \Delta_2,$$
$$w(r_b) = \Delta_3, \quad \phi_r(r_b) \equiv \Delta_4, \tag{3}$$

where ϕ_r denotes the slope (positive clockwise), which has different meanings in different theories, as defined below:

$$\phi_r = \begin{cases} -\frac{dw}{dr}, & \text{for CPT,} \\ \\ \phi_r, & \text{for FST.} \end{cases} \qquad (4)$$

Next, let Q_1 and Q_3 denote the shear forces (i.e., values of rQ_r) at nodes 1 and 2, respectively, and Q_2 and Q_4 the bending moments (i.e., values of rM_{rr}) at nodes 1 and 2, respectively.

Using Eqs. (1) and (2), relate the nodal degrees of freedom Δ_i defined in Eqs. (3) and (4) to the constants K_i and then determine the finite element stiffness matrix.

Happiness is when what you think, what you say, and what you do are in harmony.
                                               Mahatma (Mohandas K.) Gandhi

There is only one happiness in this life, to love and be loved.   George Sand

References

The references cited in this book meant for additional reading.

1. S.P. Timoshenko, *A Course of Elasticity Theory. Part 2: Rods and Plates,* St. Petersburg: AE Collins Publishers, 1916; 2nd ed., Kiev: Naukova Dumka, 1972 (in Russian) (Apparently in this book Timoshenko and Ehrenfest jointly developed the beam theory that incorporates both rotary inertia and shear deformation).

2. S.P. Timoshenko, "On the Correction for Shear of the Differential Equation for Transverse Vibrations of Prismatic Bars," *Philosophical Magazine,* Ser. 6, **41**(245), 744–746, 1921; also see Timoshenko, S. P. *The Collected Papers,* McGraw-Hill, New York, 288–290, 1953 (this paper seems to be the one cited more often than his 1916 paper).

3. Isaac Elishakoff, "Who developed the so-called Timoshenko beam theory?," *Mathematics and Mechanics of Solids,* **25**(1), 97–116, 2020 (the author has written several papers on history of Timoshenko and Timoshenko's connections to other scientists).

4. J.N. Reddy, "A review of the literature on finite-element modeling of laminated composite plates," *The Shock and Vibration Digest,* **17**(4), 3–8, 1985.

5. J.N. Reddy and A.R. Srinivasa, "Misattributions and misnomers in mechanics: why they matter in the search for insight and precision of thought," *Vietnam Journal of Mechanics,* **42**(3), 283–291, 2020.

6. J.N. Reddy, "A simple higher-order theory for laminated composite plates," *Journal of Applied Mechanics,* **51**, 745–752, 1984.

7. J.N. Reddy, "A refined nonlinear theory of plates with transverse shear deformation," *International Journal of Solids and Structures,* **20**, 881–896, 1984.

8. J.N. Reddy, *Mechanics of Laminated Plates and Shells, Theory and Analysis*, 2nd ed., CRC Press, Boca Raton, FL, 2004.

9. J.N. Reddy, *Energy Principles and Variational Methods in Applied Mechanics*, 3rd ed., John Wiley & Sons, New York, NY, 2017.

10. P.R. Heyliger and J.N. Reddy, "A higher order beam finite element for bending and vibration problems," *Journal of Sound and Vibration,* **126**(2), 309–326, 1988.

11. J.N. Reddy, *Introduction to Continuum Mechanics with Applications*, 2nd ed., Cambridge University Press, New York, NY, 2013.

12. D.P.H. Hasselman and G.E. Youngblood, "Enhanced Thermal Stress Resistance of Structural Ceramics with Thermal Conductivity Gradient," *Journal of the American Ceramic Society,* **61**(1,2), 49–53, 1978.

13. M. Koizumi, "The Concept of FGM," *Ceramic Transactions, Functionally Gradient Materials,* **34**, 3–10, 1993.

14. V. Birman and L.W. Byrd, "Modelling and analysis of functionally graded materials and structures," *Applied Mechanics Reviews,* **60**, 195–216, 2007.

15. N. Noda and T. Tsuji, "Steady thermal stresses in a plate of functionally gradient material with temperature-dependent properties," *Transactions of Japan Society of Mechanical Engineers Series A,* **57**, 625–631, 1991.

16. Y. Obata, N. Noda, and T. Tsuji, "Steady thermal stresses in a functionally gradient material plate. Transactions of Japan Society of Mechanical Engineers," **58**, 1689–1695, 1992.

17. F. Erdogan, "Fracture mechanics of functionally graded materials," *Composites Engineering,* **5**, 753–770, 1995.

18. J.N. Reddy and C.D. Chin, "Thermomechanical Analysis of Functionally Graded Cylinders and Plates," *Journal of Thermal Stresses,* **26**(1), 593–626, 1998.

19. G.N. Praveen and J.N. Reddy, "Nonlinear Transient Thermoelastic Analysis of Functionally Graded Ceramic-Metal Plates," *Journal of Solids and Structures,* **35**(33), 4457–4476, 1998.

20. G.N. Praveen, C.D. Chin, and J.N. Reddy, "Thermoelastic Analysis of Functionally Graded Ceramic-Metal Cylinder," *Journal of Engineering Mechanics, ASCE,* **125**(11), 1259–1266, 1999.

21. J.N. Reddy, C.M. Wang, and S. Kitipornchai, "Axisymmetric bending of functionally graded circular and annular plates," *European Journal of Mechanics, A/SOlids,* **18**, 185–199, 1999.

22. J.N. Reddy, "Analysis of Functionally Graded Plates," *International Journal for Numerical Methods in Engineering,* **47**, 663–684, 2000.

23. Z.Q. Cheng and R.C. Batra, "Three-dimensional thermoelastic deformations of a functionally graded elliptic plate," *Composites, Part B,* **31**, 97–106, 2000.

24. J.N. Reddy and Z.-Q. Cheng, "Three-Dimensional Thermomechanical Deformations of Functionally Graded Rectangular Plates," *European Journal of Mechanics, A/Solids,* **20**(5), 841–860, 2001.

25. S.S. Vel and R.C. Batra, "Exact solution for thermoelastic deformations of functionally graded thick rectangular plates," *AIAA Journal,* **40**, 1421–1433, 2002.

26. B.V. Sankar and J.T Tzeng, "Thermal stresses in functionally graded beams," *AIAA Journal,* **40**, 1228–1232, 2002.

27. J. Yang and H.S. Shen, "Vibration characteristics and transient response of shear-deformable functionally graded plates in thermal environments," *Journal of Sound and Vibration,* **225**, 579–602, 2002.

28. J. Yang and H.S. Shen, "Non-linear bending analysis of shear deformable functionally graded plates subjected to thermo-mechanical loads under various boundary conditions," *Composites, Part B,* **34**, 103–115, 2003.

29. J. Kim, G.H. Paulino, "Finite element evaluation of mixed mode stress intensity factors in functionally graded materials," *International Journal for Numerical Methods in Engineering,* **53**, 1903–1935, 2002.

30. Ch. Zhang, A. Savaidis, G. Savaidis, H. Zhu, "Transient dynamic analysis of a cracked functionally graded material by a BIEM," *Computational Materials Science,* **26**, 167–174, 2003.

31. M. Kashtalyan, "Three-dimensional elasticity solution for bending of functionally graded rectangular plates," *European Journal of Mechanics, A/Solids* **23**, 853–864, 2004.

32. L.F. Qian, R.C. Batra and L.M. Chen, "Static and dynamic deformations of thick functionally graded elastic plates by using higher-order shear and normal deformable plate theory and meshless local Petrov–Galerkin method," *Composites, Part B,* **35**, 685–697, 2004.

33. S. Kitipornchai, J. Yang, and K.M. Liew, "Semi-analytical solution for nonlinear vibration of laminated FGM plates with geometric imperfections," *International Journal of Solids and Structures,* **41**, 305–315, 2004.

34. J.W. Aliaga and J.N. Reddy, "Nonlinear thermoelastic analysis of functionally graded plates using the third-order shear deformation theory," *International Journal of Computational Engineering Science,* **5**(4), 753–779, 2004.

35. A.J.M. Ferreira, R.C. Batra, C.M.C. Rouque, L.F. Qian and P.A.L.S. Martins, "Static analysis of functionally graded plates using third-order shear deformation theory and a meshless method," *Composite Structures,* **69**, 449–457, 2005.

36. I. Elishakoff and C. Gentilini, "Three-dimensional flexure of rectangular plates made of functionally graded materials," *Journal of Applied Mechanics*, **72**, 788–791, 2005.

37. H. Matsunaga, "Stress analysis of functionally graded plates subjected to thermal and mechanical loadings," *Composite Structures*, **87**, 344–357, 2009.

38. Sh. Hosseini-Hashemi, H. Rokni Damavandi Taher, H. Akhavan and M. Omidi, "Free vibration of functionally graded rectangular plates using first-order shear deformation plate theory," *Applied Mathematical Modelling,* **34**, 1276–1291, 2010.

39. M. Talha and B.N. Singh, "Static response and free vibration analysis of FGM plates using higher order shear deformation theory," *Applied Mathematical Modeling,* **34**(12), 3991–4011, 2010.

40. M.E. Golmakani and M. Kadkhodayan, "Nonlinear bending analysis of annular FGM plates using higher-order shear deformation plate theories," *Composite Structures,* **93**, 973–982, 2011.

41. M.K. Singha, T. Prakash, and M. Ganapathi, "Finite Element analysis of functionally graded plates under transverse load," *Finite Elements in Analysis and Design,* **47**, 453–460, 2011.

42. C.F. Lü, C.W. Lim, and W.Q. Chen, "Size-dependent elastic behavior of FGM ultrathin films based on generalized refined theory," *International Journal of Solids and Structures,* **46**, 1176–1185, 2009.

43. M.H. Kahrobaiyan, M. Asghari, M. Rahaeifard, and M.T. Ahmadian, "Investigation of the size-dependent dynamic characteristics of atomic force microscope microcantilevers based on the modified couple stress theory," *International Journal of Engineering Science,* **48**, 1985–1994, 2010.

44. J. Zhang and Y. Fu, "Pull-in analysis of electrically actuated viscoelastic microbeams based on a modified couple stress theory," *Meccanica,* **47**, 1649–1658, 2012.

45. B.S. Shariat, Y. Liu, and G. Rio, "Modelling and experimental investigation of geometrically graded NiTi shape memory alloys," *Smart Materials and Structures,* **22**, 025030, 2013.

46. X. Li, B. Bhushan, K. Takashima, C.W. Baek, Y.K. Kim, "Mechanical characterization of micro/nanoscale structures for MEMS/NEMS applications using nanoindentation techniques," *Ultramicroscopy,* **97**, 481–494, 2003.

47. B. Klusemann and B. Svendsen, "Homogenization methods for multiphase elastic composites: comparisons and benchmarks," *Technische Mechanik,* **30**(4), 374–386, 2010.

48. J.R. Zuiker, "Functionally graded materials : choice of micro mechanics model and limitations in property variation," *Composites Engineering,* **5**(7), 807–819, 1995.

49. Z. Hashin and S. Shtrikman, "On some variational principles in anisotropic and non homogeneous elasticity," *Journal of the Mechanics and Physics of Solids,* **10**(4), 335–342, 1962.

50. T. Mori and K. Tanaka, "Average stress in matrix and average elastic energy of materials with misfitting inclusions," *Acta Metallurgica,* **21**, 571–574, 1973.

51. J.R. Willis, "Bounds and self-consistent estimates for the overall properties of anisotropic composites," *Journal of the Mechanics and Physics of Solids,* **25**(3), 185–202, 1977.

52. Hui-Shen Shen and Zhen-Xin Wang, "Assessment of Voigt and Mori-Tanaka models for vibration analysis of functionally graded plates," *Composite Structures,* **94**, 2197–2208, 2012.

53. Y. Benveniste, "A new approach to the application of Mori-Tanaka's theory in composite materials," *Mechanics of Materials,* **6**, 147–157, 1987.

54. R. Hill, "The elastic behavior of a crystalline aggregate," *Proceedings of Physical Society A,* **65**, 349–354, 1952.

55. R.D. Mindlin, "Influence of couple-stresses on stress concentrations," *Experimental Mechanics,* **3**, 1–7, 1963.

56. W.T. Koiter, "Couple-stresses in the theory of elasticity: I and II," *Koninklijke Nederlandse Akademie van Wetenschappen* (Royal Netherlands Academy of Arts and Sciences), **B67**, 17–44, 1964.

57. R.A. Toupin, "Theories of elasticity with couple-stress," *Archive for Rational Mechanics and Analysis,* **17**, 85–112, 1964.

58. R.D. Mindlin, "Second gradient of strain and surface-tension in linear elasticity," *International Journal of Solids and Structures,* **1**, 217–238, 1965.

59. F. Yang, A.C.M. Chong, D.C.C. Lam, and P. Tong, "Couple stress based strain gradient theory for elasticity," *International Journal of Solids and Structures,* **39**, 2731–2743, 2002.

60. D.C.C. Lam, F. Yang, A.C.M. Chong, J. Wang, and P. Tong, "Experiments and theory in strain gradient elasticity," *Journal of the Mechanics and Physics of Solids,* **51**, 1477–1508, 2003.

61. W. Chen, M. Xu, and L. Li, "A model of composite laminated Reddy plate based on new modified couple stress theory," *Composite Structures,* **94**, 2012, 2143–2156, 2012.

62. A.R. Srinivasa and J.N. Reddy, "A model for a constrained, finitely deforming, elastic solid with rotation gradient dependent strain energy, and its specialization to von Kármán plates and beams," *Journal of the Mechanics and Physics of Solids,* **61**, 873–885, 2013.

63. J.N. Reddy and A.R. Srinivasa, "Nonlinear theories of beams and plates accounting for moderate rotations and material length scales," *International Journal of Non-Linear Mechanics,* **66**, 43–53, 2014.

64. M. Cadek, J.N. Coleman, K.P. Ryan, V. Nicolosi, G. Bister, A. Fonseca, J.B. Nagy, K. Szostak, F. Beguin, and W.J. Blau, "Reinforcement of polymers with carbon nanotubes: the role of nanotube surface area," *Nano Letters,* **4**, 353–356, 2004.

65. W.X. Chen, J.P. Tu, L.Y. Wang, H.Y. Gan, Z.D. Xu, and X.B. Zhang, "Tribological application of carbon nanotubes in a metal-based composite coating and composites," *Carbon.* **41**, 215–222, 2003.

66. S.K. Park and X.L. Gao, "Bernoulli-Euler beam model based on a modified couple stress theory," *Journal of Micromechanics and Microengineering,* **16**, 2355–2359, 2006.

67. S.K. Park and X.L. Gao, "Variational formulation of a modified couple stress theory and its application to a simple shear problem," *Zeitschrift für angewandte Mathematik und Physik (ZAMP),* **59**, 904–917, 2008.

68. H.M. Ma, X.L. Gao, J.N. Reddy, "A microstructure-dependent Timoshenko beam model based on a modified couple stress theory," *Journal of the Mechanics and Physics of Solids,* **56**, 3379–3391, 2008.

69. H.M. Ma, X.L. Gao, J.N. Reddy, "A nonclassical Reddy-Levinson beam model based on a modified couple stress theory," *International Journal of Multiscale Computational Engineering,* **8**(2), 167–180, 2010.

70. H.M. Ma, X.L. Gao, J.N. Reddy, "A non-classical Mindlin plate model based on a modified couple stress theory," *Acta Mechanica,* **220**, 217–235, 2011.

71. R. Aghababaei and J.N. Reddy, "Nonlocal third-order shear deformation plate theory with application to bending and vibration of plates," *Journal of Sound and Vibration,* **326**, 277–289, 2009.

72. J.V. Araújo dos Santos and J.N. Reddy, "Vibration of Timoshenko beams using non-classical elasticity theories," *Shock and Vibration,* **19**(3), 251–256, 2012.

73. J. V. Araujo dos Santos and J.N. Reddy, "Free vibration and buckling analysis of beams with a modified couple-stress theory," *International Journal of Applied Mechanics,* **4**(3), 1250026 (28 pages), 2012.

74. J.N. Reddy, "Microstructure-dependent couple stress theories of functionally graded beams," *Journal of the Mechanics and Physics of Solids,* **59**, 2382–2399, 2011.

75. J.N. Reddy, "A general nonlinear third-order theory of functionally graded plates," *International Journal of Aerospace and Lightweight Structures,* **1**(1), 1–21, 2011.

76. C.M.C. Roque, A.J.M. Ferreira, and J.N. Reddy, "Analysis of Timoshenko nanobeams with a nonlocal formulation and meshless method," *International Journal of Engineering Science,* **49**, 976–984, 2011.

77. J.N. Reddy and Jessica Berry, "Modified couple stress theory of axisymmetric bending of functionally graded circular plates," *Composites Structures,* **94**, 3664–3668, 2012.

78. J.N. Reddy and J. Kim, "A nonlinear modified couple stress-based third-order theory of functionally graded plates," *Composite Structures,* **94**, 1128–1143, 2012.

79. J. Kim and J.N. Reddy, "Analytical solutions for bending, vibration, and buckling of FGM plates using a couple stress-based third-order theory," *Composite Structures,* **103**, 86–98, Sep. 2013.

80. A. Arbind and J.N. Reddy, "Nonlinear analysis of functionally graded microstructure-dependent beams," *Composite Structures,* **98**, 272–281, 2013.

81. A. Arbind, J.N. Reddy, and A. Srinivasa, "Modified couple stress-based third-order theory for nonlinear analysis of functionally graded beams," *Latin American Journal of Solids and Structures,* **11**, 459–487, 2014.

82. J.N. Reddy, Jani Romanoff, and Jose Antonio Loya, "Nonlinear finite element analysis of functionally graded circular plates with modified couple stress theory," *European Journal of Mechanics - A/Solids,* **56**, 92–104, 2016.

83. J.N. Reddy and J. Kim, "A nonlinear modified couple stress-based third-order theory of functionally graded plates," *Composite Structures,* **94**, 1128–1143, 2012.

84. J. Kim and J.N. Reddy, "A general third-order theory of functionally graded plates with modified couple stress effect and the von Kármán nonlinearity: theory and finite element analysis," *Acta Mechanica,* **226**(9), 973–2998, 2015.

85. S. M. Mousavi, J. Paavola, and J.N. Reddy, "Variational approach to dynamic analysis of third-order shear deformable plates within gradient elasticity," *Meccanica,* **50**(6), 1537–1550, 2015.

86. J. Pei, F. Tian, and T. Thundat, "Glucose biosensor based on the microcantilever," *Analytical Chemistry,* **76**, 292–297, 2004.

87. E. Kröner, "Elasticity theory of materials with long range cohesive forces," *International Journal of Solids and Structures,* **3**(5), 731–742, 1967.

88. I.A. Kunin, "The theory of elastic media with microstructure and the theory of dislocations," in *Mechanics of Generalized Continua,* E. Kröner (Ed.), Springer, Berlin Heidelberg, 321–329, 1968.

89. A.C. Eringen, "Nonlocal polar elastic continua," *International Journal of Engineering Science,* **10**(1), 1–16, 1972.

90. A.C. Eringen, "Linear theory of nonlocal elasticity and dispersion of plane waves," *International Journal of Engineering Science,* **10**(5), 425–435, 1972.

91. A.C. Eringen, and D.G. Edelen, "On nonlocal elasticity," *International Journal of Engineering Science,* **10**(3), 233–248, 1972.

92. A.C. Eringen, "On differential equations of nonlocal elasticity and solutions of screw dislocation and surface waves," *Journal of Applied Physics,* **54**(9), 4703–4710, 1983.

93. A.C. Eringen, *Nonlocal Continuum Field Theories,* Springer, New York, NY, 2002.

94. J. Peddieson, G.R. Buchanan, and R.P. McNitt, "Application of nonlocal continuum models to nanotechnology," *International Journal of Engineering Science,* **41**(3–5), 305–312, 2003.

95. P. Lu, H.P. Lee, C. Lu, and P.Q. Zhang, "Dynamic properties of flexural beams using a nonlocal elasticity model," *Journal of Applied Physics,* **99**(7), article 073510, 2006.

96. J.N. Reddy, "Nonlocal theories for bending, buckling and vibration of beams," *International Journal of Engineering Science,* **45**, 288–307, 2007.

97. J.N. Reddy and S. D. Pang, "Nonlocal continuum theories of beams for the analysis of carbon nanotubes," *Applied Physics Letters,* **103**, 023511-1 to 023511-16, 2008.

98. J.N. Reddy and S. El-Borgi, "Eringen's nonlocal theories of beams accounting for moderate rotations," *International Journal of Engineering Science* **82**, 159–177, 2014.

99. J.N. Reddy, Sami El-Borgi, and Jani Romanoff, "Non-linear analysis of functionally graded microbeams using Eringen's nonlocal differential model," *International Journal of Non-Linear Mechanics,* **67**, 308–318, 2014.

100. Parisa Khodabakhshi and J.N. Reddy, "A unified integro-differential nonlocal model," *International Journal of Engineering Science,* **95**, 60–75, 2015.

101. Jose Fernandez-Saez, R. Zaera, J.A. Loya, and J.N. Reddy, "Bending of Euler–Bernoulli beams using Eringen's integral formulation: a paradox resolved," *International Journal of Engineering Science,* **99**, 107–116, 2016.

102. J.N. Reddy and M.L. Rasmussen, *Advanced Engineering Analysis,* John Wiley & Sons, New York, NY, 1982; reprinted by Krieger, Malabar, FL, 1991.

103. J.N. Reddy, *Applied Functional Analysis and Variational Methods in Engineering,* McGraw–Hill, New York, NY, 1984; reprinted by Krieger, Malabar, FL, 1991.

104. S.P. Timoshenko, "On the correction for shear of the differential equation for transverse vibrations of prismatic bar," *Philosophical Magazine,* **41**, 744–746, 1921.

105. S.P. Timoshenko, *Strength of Materials,* 2nd ed., D. Van Nostrand Company, New York, NY, 1940.

106. R.D. Mindlin and H. Deresiewicz, "Timoshenko's Shear Coefficient for Flexural Vibrations of Beams," Technical Report No. 10, ONR Project NR064-388, 1953.

107. G.R. Cowper, "The shear coefficient in Timoshenko's beam theory," *Journal of Applied Mechanics, ASME,* **33**(2), 335–340, 1966.

108. T. Kaneko, "On Timoshenko's correction for shear in vibrating beams," *Journal of Physics, D: Applied Physics,* **8**, 1927–1936, 1975.

109. J.J. Jensen, "On the shear coefficient in Timoshenko's beam theory," *Journal of Sound and Vibration* **87**, 621–635, 1983.

110. J.R. Hutchinson, "Shear coefficients for Timoshenko beam theory," *Journal of Applied Mechanics, ASME,* **68**, 87–92, 2001.

111. F. Gruttmann and W. Wagner, "Shear correction factors in Timoshenko's beam theory for arbitrary shaped cross-sections," *Computational Mechanics,* **27**, 199–-207, 2001.

112. S.B. Dong, C. Albdogan, and E. Taciroglu, "Much ado about shear correction factors in Timoshenko beam theory," *International Journal of Solids and Structures,* **47**(13), 1651–1665, 2010.

113. J.N. Reddy, *Theory and Analysis of Elastic Plates and Shells,* 2nd ed., CRC Press, Boca Raton, FL, 2007.

114. W. Ritz, "Über eine neue methode zur lösung gewisser randwertaufgeben," *Göttingener Nachsichten Mathematisch-Physikalische Klasse,* 236, 1908.

115. B. G. Galerkin, "Series solutions in rods and plates," *Vestnik Inzhenerov i Tekhnikov,* **19**, 1915.

116. C.M. Wang, "Timoshenko beam-bending solutions in terms of Euler–Bernoulli solutions", *Journal of Engineering Mechanics, ASCE*, **121**(6), 763–765, 1995.

117. C.M. Wang, T.Q. Yang, and K.Y. Lam, "Viscoelastic Timoshenko beam bending solutions from viscoelastic Euler–Bernoulli solutions," *Journal of Engineering Mechanics, ASCE*, **123**(7), 746–748, 1997.

118. J.N. Reddy and C.M. Wang, "Bending, buckling, and frequency relationships between the Euler–Bernoulli and Timoshenko non-local beam theories," *Asian Journal of Civil Engineering (Building and Housing)*, **10**(3), 265–281, 2009.

119. J.N. Reddy, "Canonical relationships between bending solutions of classical and shear deformation beam and plate theories," *Annals of Solid and Structural Mechanics*, **1**(1), 9–27, 2010.

120. C.M. Wang, J.N. Reddy, and K.H. Lee, *Shear Deformation Theories of Beams and Plates (Relationships with Classical Solutions)*, Elsevier, U.K., 2000.

121. J.N. Reddy, C.M. Wang, G.T. Lim, and K.H. Ng, "Bending Solutions of the Levinson beams and plates in terms of the classical theories," *Int. J. Solids & Structures*, **38**, 4701–4720, 2001.

122. J.N. Reddy and Archana Arbind, "Bending relationships between the modified couple stress-based functionally graded Timoshenko beams and homogeneous Bernoulli-Euler beams," *Annals of Solid and Structural Mechanics*, **3**(1), 15–26, 2012.

123. R.T. Fenner and J.N. Reddy, *Mechanics of Solids and Structures*, 2nd ed., CRC Press, Boca Raton, FL, 2012.

124. M. Levinson, "An accurate, simple theory of the static and dynamics of elastic plates," *Mechanics Research Communications*, **7**(6), 343–350, 1980.

125. M. Levinson, "A new rectangular beam theory," *International Journal of Solids and Structures*, **74**, 81–87, 1981.

126. W.B. Bickford, "A consistent higher order beam theory," in *Developments in Theoretical and Applied Mechanics*, **11**, 137–150, 1982.

127. G.I.N. Rozvany and Z. Mróz, "Column design: optimization of support conditions and segmentation," *Journal of Structural Mechanics*, **5**, 279–290, 1977.

128. N. Olhoff and B. Akesson, "Minimum stiffness of optimally located supports for maximum value of the column buckling loads," *Structural Optimization*, **3**, 163–175, 1991.

129. Ali Reza Daneshmehr, Mostafa Mohammad Abadi, and Amir Rajabpoor, "Thermal effect on static bending, vibration and buckling of Reddy beam based on modified couple stress theory," *Applied Mechanics and Materials*, **332**, 331–338, 2013.

130. Guanghui He and Xiao Yang, "Finite element analysis for buckling of two-layer composite beams using Reddy's higher order beam theory," *Finite Elements in Analysis and Design*, **83**, 49–57, 2014.

131. Z. Sun, L. Yang, and Y. Gao, "The displacement boundary conditions for Reddy higher-order shear cantilever beam theory," *Acta Mechanica*, **226**, 1359–1367, 2015.

132. Castrenze Polizzotto, "From the Euler–Bernoulli beam to the Timoshenko one through a sequence of Reddy-type shear deformable beam models of increasing order," *European Journal of Mechanics, A/Solids*, **53**, 62–74, 2015.

133. G. Jin, C. Yang, and Z, Liu, "Vibration and damping analysis of sandwich viscoelastic-core beam using Reddy's higher-order theory," *Composite Structures*, **140**, 390–409, 2016.

134. Reza Nazemnezhad and Mojtaba Zare, "Nonlocal Reddy beam model for free vibration analysis of multilayer nanoribbons incorporating interlayer shear effect," *European Journal of Mechanics, A/Solids*, **55**, 234–242, 2016.

135. A. Nadai, *Die Elastichen Platten*, Springer, Berlin, 1925.

136. H.M. Westergaard, "Stresses in concrete pavements computed by theoretical analysis," *Public Roads*, U.S. Dept. of Agriculture, Bureau of Public Roads, **7**(2), 1926.

137. J.R. Roark and W.C. Young, *Formulas for Stress and Strain*, McGraw–Hill, New York, NY, 1975.

138. N. McLachlan, *Bessel Functions for Engineers*, Oxford University Press, UK, 1948.

139. C. L. Kantham, "Bending and Vibration of Elastically Restrained Circular Plates," *Journal Franklin Institute*, **265**(6), 483–491, 1958.

140. A. W. Leissa, *Vibration of Plates*, NASA SP–160, Washington, D.C., 1969.

141. G. Kirchhoff, "Über die Gleichungen des Gleichgewichts eines elastischen Körpes bei nicht unendlich kleinen Verschiebungen seiner Theile," *Sitzungsber. Akad. Wiss. Wien (Vienna)*, **9**, 762–773, 1852.

142. R. Szilard, *Theories and Applications of Plate Analysis*, John Wiley & Sons, Hoboken, NJ, 2004.

143. J.N. Reddy and C.M. Wang, "Relationships between classical and shear deformation theories of axisymmetric bending of circular plates," *AIAA Journal*, **35** (12): 1862–1868, 1997.

144. C.M. Wang, and K.H. Lee, "Buckling load relationship between Reddy and Kirchhoff circular plates," *Journal of Franklin Institute*, **335**B(6), 989–995, 1998.

145. C.M. Wang, "Buckling of polygonal and circular sandwich plates", *AIAA Journal*, **33**(5), 962–964, 1995.

146. C.M. Wang and K.H. Lee, "Deflection and stress resultants of axisymmetric Mindlin plates in terms of corresponding Kirchhoff solutions," *International Journal of Mechanical Sciences*, **38**(11), 1179–1185, 1996.

147. C.M. Wang, "Relationships between Mindlin and Kirchhoff bending solutions for tapered circular and annular Plates", *Engineering Structures*, **19**(3), 255–258, 1997.

148. H. Marcus, *Die Theorie elastischer Gewebe und ihre Anwendung auf biegsame Platten*, Springer-Verlag, Berlin, Germany, 1924.

149. J. Ye, "Axisymmetric buckling of homogeneous and laminated circular plates," *Journal of Structural Engineering, ASCE*, **121**(8), 1221–1224, 1995.

150. C.M. Wang, Y. Xiang, and Q. Wang, "Axisymmetric buckling of Reddy circular plates on Pasternak foundation," *Journal of Engineering Mechanics*, **127**(3), 254–259, 2001.

151. Sh. Hosseini-Hashemi, M. Eshaghi, and H. Rokni Damavandi Taher, "An exact analytical solution for freely vibrating piezoelectric coupled circular/annular thick plates using Reddy plate theory," *Composite Structures*, **92**, 1333–1351, 2010.

152. H. Bisadi, M. Eshaghi, H. Rokni, and M. Ilkhani, "Benchmark solution for transverse vibration of annular Reddy plates," *International Journal of Mechanical Sciences*, **56**, 35–49, 2012.

153. M. Rahmat Talabi and A.R. Saidi, "An explicit exact analytical approach for free vibration of circular/annular functionally graded plates bonded to piezoelectric actuator/sensor layers based on Reddy's plate theory," *Applied Mathematical Modelling*, **37**, 7664–7684, 2013.

154. M. Fadaee, "A novel approach for free vibration of circular/annular sector plates using Reddy's third order shear deformation theory," *Meccanica*, **50**, 2325–2351, 2015.

155. J. N. Reddy, *An Introduction to the Finite Element Method*, 4th ed., McGraw–Hill, New York, NY, 2019.

156. J.N. Reddy, *An Introduction to Nonlinear Finite Element Analysis*, 2nd ed., Oxford University Press, Oxford, UK, 2015.

157. K.S. Surana and J.N. Reddy, *The Finite Element Method for Initial Value Problems, Mathematics and Computations*, CRC Press, Boca Raton, FL, 2018.

158. J.C. Houbolt, "A recurrence matrix solution for the dynamic response of elastic aircraft," *Journal of Aeronautical Science*, **17**, 540–550, 1950.

159. N.M. Newmark, "A method of computation for structural dynamics," *Journal of Engineering Mechanics, ASCE*, **85**, 67–94, 1959.

160. Wooram Kim and J.N. Reddy, "A new family of higher-order time integration algorithms for the analysis of structural dynamics," *Journal of Applied Mechanics, ASME* **84**, 071008-1 to 071008-17, July 2017.

Strength does not come from physical capacity. It comes from an indomitable will.
Mahatma (Mohandas K.) Gandhi

What we need is a strong education system that allows creativity to grow and encourages students to be interested in science and technology.
Frances Arnold

Coauthored Papers on Beams and Circular Plates

I've learned that people will forget what you said, people will forget what you did, but people will never forget how you made them feel.   Maya Angelou

1. P. R. Heyliger and J.N. Reddy, "A higher-order beam finite element for bending and vibration problems," *Journal of Sound and Vibration,* Vol. 126, No. 2, pp. 309–326, 1988.
2. D.H. Robbins and J.N. Reddy, "Analysis of piezoelectrically actuated beams using a layer-wise displacement theory," *Computers & Structures,* Vol. 41, No. 2, pp. 265–279, 1991.
3. J.N. Reddy, C.M. Wang, and K.Y. Lam, "Unified finite elements based on the classical and shear deformation theories of beams and axisymmetric circular plates," *Communications in Numerical Methods in Engineering,* Vol. 13, pp. 495–510, 1997.
4. C.M. Wang, J.N. Reddy, and K.H. Lee, "Relationship between bending solutions of classical and shear deformation beam theories," *International Journal of Solids & Structures,* Vol. 34, No. 26, pp. 3373–3384, 1997.
5. J.N. Reddy, "On locking-free shear deformable beam finite elements," *Computer Methods in Applied Mechanics and Engineering,* Vol. 149, pp. 113–132, 1997.
6. J.N. Reddy and C.M. Wang, "Relationships between classical and shear deformation theories of axisymmetric bending of circular plates," *AIAA Journal,* Vol. 35, No. 12, pp. 1862–1868, 1997.
7. A.A. Khdeir and J.N. Reddy, "An exact solution for the bending of thin and thick cross-ply laminated beams," *Composite Structures,* Vol. 37, pp. 195–203, 1997.
8. S.K. Kassegne and J.N. Reddy, "A layerwise shell stiffener and stand-alone curved beam element," *Asian Journal of Structural Engineering,* Vol. 2 Nos. 1 and 2, pp. 1–14, 1997.
9. A.A. Khdeir and J.N. Reddy, "Jordan canonical form solution for thermally induced deformations of cross-ply laminated composite beams," *Journal of Thermal Stresses,* Vol. 22(3), pp. 331–346, 1999.
10. J.N. Reddy, C. M. Wang, and S. Kitipornchai, "Axisymmetric bending of functionally graded circular and annular plates," *European Journal of Mechanics,* Vol. 18, pp. 185–199, 1999.
11. J.N. Reddy and J. I. Barbosa, "On vibration suppression of magnetostrictive beams," *Smart Materials and Structures,* Vol. 9, pp. 49–58, 2000.
12. K.K. Ang, J.N. Reddy and C.M. Wang, "Displacement control of Timoshenko beams via induced strain actuators," *Smart Materials and Structures,* Vol. 9, pp. 1–4, 2000.
13. A. Yavari and S. Sarkani and J.N. Reddy, "General solutions of beams with jump discontinuities on elastic foundation," *Archive of Applied Mechanics,* Vol. 71, No. 9, pp. 625–639, 2001.
14. J.N. Reddy, C.M. Wang, G.T. Lim, and K.H. Ng, "Bending solutions of the Levinson beams and plates in terms of the classical theories," *International Journal of Solids & Structures,* Vol. 38, pp. 4701–4720, 2001.
15. A. Yavari, S. Sarkani and J.N. Reddy, "On non-uniform Euler-Bernoulli and Timoshenko beams with jump discontinuities: application of distribution theory," *International Journal of Solids and Structures,* Vol. 38, pp. 8389–8406, 2001.
16. S. Marfia, E. Sacco, and J.N. Reddy, "Superelastic and shape memory effects in laminated shape-memory-alloy beams," *AIAA Journal,* Vol. 41, pp. 100–109, January 2003.
17. A. Chakraborty, S. Gopalakrishnan, and J.N. Reddy, "A new beam finite element for the analysis of functionally graded materials," *International Journal of Mechanical Sciences,* Vol. 45, No. 3, pp. 519–539, 2003.

535

18. A. Laulusa and J.N. Reddy, "On shear and extensional locking in nonlinear composite beams," *Engineering Structures,* Vol. 26, No. 2, pp. 151–170, 2004.
19. M.C. Ray and J.N. Reddy, "Effect of delamination on the active constrained layer damping of laminated composite beams," *AIAA Journal,* Vol. 42, No. 6, 1219–1226, 2004.
20. N. Murgude and J.N. Reddy, "Nonlinear analysis of microbeam under electrostatic loading," *Mechanics of Advanced Materials and Structures,* Vol. 13, No. 1, pp. 13–32, 2006.
21. J.N. Reddy and S.D. Pang, "Nonlocal continuum theories of beams for the analysis of carbon nanotubes," *Applied Physics Letters,* Vol. 103, pp. 023511-1 to 023511-16, 2008.
22. H.M. Ma, X.-L. Gao, and J.N. Reddy, "A microstructure-dependent Timoshenko beam model based on a modified couple stress theory," *Journal of the Mechanics and Physics of Solids,* Vol. 56, pp. 3379–3391, 2008.
23. J.N. Reddy and C.M. Wang, "Bending, buckling, and frequency relationships between the Euler–Bernoulli and Timoshenko non-local beam theories," *Asian Journal of Civil Engineering (Building and Housing),* Vol. 10, No. 3, 2009.
24. C.M. Wang, Y.Y. Zhang, Y. Xiang, and J.N. Reddy, "Recent studies on buckling of carbon nanotubes," *Applied Mechanics Reviews,* vol. 63, 030804-1 to 030804-18, May 2010.
25. G.S. Payette and J.N. Reddy, "Nonlinear quasi-static finite element formulations for viscoelastic Euler-Bernoulli and Timoshenko beams," *International Journal for Numerical Methods in Biomedical Engineering,* Vol. 26, No. 12, pp. 1736–1755, 2010.
26. H.M. Ma, X.-L. Gao, and J.N. Reddy, "A non-classical Reddy–Levinson beam model based on a modified couple stress theory," *International Journal for Multiscale Computational Engineering,* Vol. 8, No. 2, pp. 167–180, 2010.
27. J.V. Araujo dos Santos and J.N. Reddy, "A model for free vibration analysis of Timoshenko beams with couple stress," *Acta Mechanica Solida Sinica,* Vol. 23, pp. 240–248, December 2010.
28. Wooram Kim and J.N. Reddy, "A comparative study of least-squares and the weak-form Galerkin finite element models for the nonlinear analysis of Timoshenko beams," *Journal of Solid Mechanics,* Vol. 2, No.2, pp. 101–114, 2010.
29. R. Gunes, M. Aydin, M. K. Apalak, and J.N. Reddy, "The elasto-plastic impact analysis of functionally graded circular plates under low-velocities," *Composite Structures,* Vol. 93, No. 2, pp. 860–869, 2011.
30. Yiping Liu and J.N. Reddy, "A non-local curved beam model based on a modified couple stress theory," *International Journal of Structural Stability and Dynamics,* Vol. 11, No. 3, pp. 495–512, 2011.
31. C.M.C. Roque, A.J.M. Ferreira, and J.N. Reddy, "Analysis of Timoshenko nanobeams with a nonlocal formulation and meshless method," *International Journal of Engineering Science,* Vol. 49, pp. 976–984, 2011.
32. Wooram Kim and J.N. Reddy, "Nonconventional finite element models for nonlinear analysis of beams," *International Journal of Computational Methods,* Vol. 8, No. 3, pp. 349–368, 2011.
33. M.K. Apalak, R. Gunes, M. Aydin, and J.N. Reddy, "Impact performance of Al/SiC functionally graded circular plates," *International Journal of Materials and Product Technology,* Vol. 42, Nos. 1-2, pp. 56–65, 2011.
34. J.N. Reddy and Archana Arbind, "Bending relationships between the modified couple stress-based functionally graded Timoshenko beams and homogeneous Bernoulli–Euler beams" *Annals of Solid and Structural Mechanics,* Vol. 3, No. 1, pp. 15-26, 2012.
35. V.P. Vallala, G.S. Payette, and J.N. Reddy, "Spectral/hp finite element formulation for viscoelastic beams based on an higher-order beam theory," *International Journal of Applied Mechanics,* Vol. 4, No. 1, pp. 1–28, 2012.
36. J.N. Reddy and Jessica Berry, "Modified couple stress theory of axisymmetric bending of functionally graded circular plates," *Composites Structures,* Vol. 94, pp. 3664–3668, 2012.
37. J.V. Araujo dos Santos and J.N. Reddy, "Vibration of Timoshenko beams using non-classical elasticity theories," *Shock and Vibration,* Vol. 19, No. 3, pp. 251–256, 2012.

38. J.V. Araujo dos Santos and J.N. Reddy, "Free vibration and buckling analysis of beams with a modified couple-stress theory," *International Journal of Applied Mechanics,* Vol. 4, No. 3, pp. 1250056-1 (28 pages), 2012.

39. G.S. Payette and J.N. Reddy "A nonlinear finite element framework for viscoelastic beams based on the higher-order Reddy beam theory," *Journal of Engineering Materials and Technology,* Vol. 135, No. 1, pp. 011005-1 to 011005-11, 2013.

40. J.N. Reddy and Patrick Mahaffey, "Generalized beam theories accounting for von Kármán nonlinear strains with application to buckling and post-buckling," *Journal of Coupled Systems and Multiscale Dynamics,* Vol. 1, No.1, pp. 120–134, 2013.

41. P. Ghosh, J.N. Reddy, and A.R. Srinivasa, "Development and implementation of a beam theory model for shape memory polymers," *International Journal of Solids and Structures,* Vol. 50, Nos. 3-4, pp. 595–608, Feb 2013.

42. C.M.C. Roque, D.S. Fidalgo, A.J.M. Ferreira, and J.N. Reddy, "A study of a microstructure-dependent composite laminated Timoshenko beam using a modified couple stress theory and a meshless method," *Composite Structures,* Vol. 96, pp. 532–537, Feb 2013.

43. Md. Ishaquddin, P. Raveendranath, and J.N. Reddy, "Coupled polynomial field approach for elimination of flexure and torsion locking phenomena in the Timoshenko and Euler–Bernoulli curved beam elements," *Finite Elements in Analysis and Design,* Vol. 65, pp. 17–31, Mar 2013.

44. A.R. Srinivasa and J.N. Reddy, "A model for a constrained, finitely deforming, elastic solid with rotation gradient dependent strain energy, and its specialization to von Karman plates and beams," *Journal of the Mechanics and Physics of Solids,* Vol. 61, No. 3, pp. 873–885, Mar 2013.

45. M. Şimşek and J.N. Reddy, "Bending and vibration of functionally graded microbeams using a new higher order beam theory and the modified couple stress theory," *International Journal of Engineering Science,* Vol. 64, pp. 37–53, Mar 2013.

46. Archana Arbind and J.N. Reddy, "Nonlinear analysis of functionally graded microstructure-dependent beams," *Composite Structures,* Vol. 98, pp. 272–281, 2013.

47. M. Şimşek and J.N. Reddy, "A unified higher order beam theory for buckling of a functionally graded microbeam embedded in elastic medium using modified couple stress theory," *Composite Structures,* Vol. 101, pp. 47–58, Jul 2013.

48. M. Komijani, J.N. Reddy, M. R. Eslami, and M. Bateni, "An analytical approach for thermal stability analysis of two-layer Timoshenko beams," *International Journal of Structural Stability and Dynamics,* Vol. 13, No. 8, pp. 1350036-1 to 1350036-14, Dec 2013.

49. Archana Arbind, J.N. Reddy, and A. Srinivasa, "Modified couple stress-based third-order theory for nonlinear analysis of functionally graded beams," *Latin American Journal of Solids and Structures,* Vol 11, No 3, pp. 459–487, 2014.

50. B.G. Sinir, B.B. Özhan, and J.N. Reddy, "Buckling configurations and dynamic response of buckled Euler-Bernoulli beams with non-classical supports," *Latin American Journal of Solid Mechanics,* Vol. 11, No. 14, pp. 2516–2536, 2014.

51. M. Komijani, J.N. Reddy, and R. Eslami, "Nonlinear analysis of microstructure-dependent functionally graded piezoelectric material actuators," *Journal of the Mechanics and Physics of Solids,* Vol. 63, pp. 214–227, Feb 2014.

52. R. Gunes, Murat Aydin, M.K. Apalak, and J.N. Reddy, "Experimental and numerical investigations of low velocity impact on functionally graded circular plates," *Composites Part B,* Vol. 59, pp. 21–32, Mar 2014.

53. Jani Romanoff and J.N. Reddy, "Experimental validation of the modified couple stress Timoshenko beam theory for web-core sandwich panels," *Composite Structures,* Vol. 111, pp. 130–137, May 2014.

54. M. Komijani, S.E. Esfahani, J.N. Reddy, Yiping Liu, and M.R. Eslami, "Nonlinear thermal stability and vibration of pre/post-buckled temperature- and microstructure-dependent functionally graded beams resting on elastic foundation," *Composite Structures,* Vol. 112, pp. 292–307, Jun 2014.

55. H. Soltani, G.S. Payette and J.N. Reddy, "Vibration of elastic beams in presence of an inviscid fluid medium," *International Journal of Structural Stability and Dynamics*, Vol. 14, No. 6, 1450022-1 to 1450022-29 (29 pages), Aug 2014.

56. Ehsan Taati, Masoud Molaei Najafabadi, and J.N. Reddy, "Size-dependent generalized thermoelasticity model for Timoshenko micro-beams based on strain gradient and non-Fourier heat conduction theories," *Composite Structures*, Vol. 116, pp. 595–611, 2014.

57. Noël Challamel, Zhen Zhang, C.M. Wang, J.N. Reddy, Q. Wang, Thomas Michelitsch, and Bernard Collet, "On nonconservativeness of Eringen's nonlocal elasticity in beam mechanics: correction from a discrete-based approach," *Archives of Applied Mechanics*, Vol. 84 (9-11), pp. 1275–1292, 2014.

58. J.N. Reddy and A. R. Srinivasa, "Non-linear theories of beams and plates accounting for moderate rotations and material length scales," *International Journal of Non-Linear Mechanics*, Vol. 66, pp. 43–53, 2014.

59. M. Komijani, J.N. Reddy, and A. Ferreira, "Nonlinear stability and vibration of pre/post-buckled FGPM actuators," *Meccanica*, Vol. 49 (11), pp. 2729–2745, Nov 2014.

60. J.N. Reddy and S. El-Borgi, "Eringen's nonlocal theories of beams accounting for moderate rotations," *International Journal of Engineering Science*, Vol. 82, pp. 159–177, 2014.

61. J.N. Reddy, Sami El-Borgi, and Jani Romanoff, "Non-linear analysis of functionally graded microbeams using Eringen's nonlocal differential model," *International Jurnal of Non-Linear Mechanics*, Vol. 67, pp. 308–318, 2014.

62. Parisa Khodabakhshi and J.N. Reddy, "A unified integro-differential nonlocal model," *International Journal of Engineering Science*, Vol. 95, pp. 60–75, Oct 2015.

63. Sami El-Borgi, Ralston Fernandes, and J.N. Reddy, "Non-local free and forced vibrations of graded nanobeams resting on a non-linear elastic foundation," *International Journal of Non-Linear Mechanics*, Vol. 77, pp. 348–363, Dec 2015.

64. Jose Fernandez-Saez, R. Zaera, J.A. Loya, and J.N. Reddy, "Bending of Euler-Bernoulli Beams using Eringen's integral formulation: a paradox resolved," *International Journal of Engineering Science*, Vol. 99, pp. 107–116, Feb 2016.

65. J.N. Reddy, Jani Romanoff, and Jose Antonio Loya, "Nonlinear finite element analysis of functionally graded circular plates with modified couple stress theory," *European Journal of Mechanics – A/Solids*, Vol. 56, pp. 92–104, 2016.

66. M.E. Torki and J.N. Reddy, "Buckling of functionally-graded beams with partially delaminated piezoelectric layers," *International Journal of Structural Stability and Dynamics*, Vol. 16, No. 3, 1450104 (25 pages), Apr 2016.

67. Anssi T. Karttunen, Jani Romanoff, and J.N. Reddy, "Exact microstructure-dependent Timoshenko beam element," *International Journal of Mechanical Sciences*, Vol. 111, pp. 35–42, Jun 2016.

68. Saikat Sarkar and J.N. Reddy, "Exploring the source of non-locality in the Euler-Bernoulli and Timoshenko beam models," *International Journal of Engineering Science*, Vol. 104, pp. 110–115, Jul 2016.

69. Jani Romanoff, J.N. Reddy, and Jasmin Jelovica, "Using non-local Timoshenko beam theories for prediction of micro- and macro-structural responses," *Computers and Structures*, Vol. 156, pp. 410–420, Nov 2016.

70. Noël Challamel, J.N. Reddy, and C. M. Wang, "Eringen's stress gradient model for bending of nonlocal beams," *Journal of Engineering Mechanics*, ASCE, Vol. 142, No. 12, article number 04016095 (9 pages), Dec 2016.

71. Anssi T. Karttunen, J.N. Reddy, and Jani Romanoff, "Closed-form solution for circular microstructure-dependent Mindlin plates," *Acta Mechanica*, Vol. 228, pp. 323–331, 2017.

72. Parisa Khodabakhshi and J.N. Reddy, "A unified beam theory with strain gradient effect and the von Kŕmán nonlinearity," *ZAMM, Zeitschrift f'ur Angewandte Mathematik und Mechanik*, Vol. 97, No. 1, pp. 70–91, Jan 2017.

73. Bruno R. Goncalves, Anssi T. Karttunen, Jani Romanoff, and J.N. Reddy, "Buckling and free vibration of shear-flexible sandwich beams using a couple-stress-based finite element," *Composite Structures*, Vol. 165, pp. 233–241, Apr 2017.

74. Anssi T. Karttunen, Raimo von Hertzen, J.N. Reddy, and Jani Romanoff, "Exact elasticity-based finite element for circular plates," *Computers and Structures*, Vol. 182, pp. 219–226, Apr 2017.

75. Archana Arbind, J.N. Reddy, and A.R. Srinivasa, "Nonlinear analysis of beams with rotation gradient dependent potential energy for constrained micro-rotation," *European Journal of Mechanics, A/Solids*, Vol. 65, pp. 178–194, Sept. 2017.

76. Mohamed Trabelssi, Sami El-Borgi, Liao-liang Ke, J.N. Reddy, "Nonlocal free vibration of graded nanobeams resting on a nonlinear elastic foundation using DQM and LaDQM," *Composite Structures*, Vol. 176, pp. 736–747, Sept. 2017.

77. Rakesh Ranjan and J.N. Reddy, "Non uniform rational bspline (NURBS) based nonlinear analysis of straight beams with mixed formulations," *Journal of Solid Mechanics*, Vol. 10, No. 1, pp. 38–56, 2018.

78. P. Kasirajan, Amirtham Rajagopal, and J. N. Reddy, "Nonlocal nonlinear bending and free vibration analysis of a rotating laminated nano cantilever beam," *Mechanics of Advanced Materials and Structures*, Vol. 25, No. 5, pp. 439–450, 2018.

79. Anssi Karttunen, J.N. Reddy, and J. Romanoff, "Micropolar modelling approach for periodic sandwich beams," *Composite Structures*, Vol. 185, pp. 656–664, Feb 2018.

80. Z. Wang, A.R. Srinivasa, K.R. Rajagopal, and J.N. Reddy, "Simulation of inextensible elastic-plastic beams based on an implicit rate type model," *International Journal of Non-Linear Mechanics*, Vol. 99, pp. 165–172, Mar 2018.

81. Anssi Karttunen, J.N. Reddy, and J. Romanoff, "Two-scale constitutive modeling of a lattice core sandwich beam," *Composites Part B*, Vol. 160, pp. 66–75, 2019.

82. Praneeth Nampally, Anssi Karttunen, and J.N. Reddy, "Nonlinear finite element analysis of lattice core sandwich beams," *European Journal of Mechanics, A/Solids*, Vol. 74, pp. 431–439, 2019.

83. Mohammad Arefi, Elyas Mohammad-Rezaei Bidgoli, Rossana Dimitri, Francesco Tornabene, and J.N. Reddy, "Size-dependent vibration of functionally graded polymer composite curved nanobeams reinforced with graphene nanoplatelets," *Applied Sciences-Basel*, Vol. 9, No. 8, article 1580, 2019.

84. Recep Ekici, Vahdet Mesut Abaci, and J.N. Reddy, "3D Micro-structural modeling of vibration characteristics of smart particle-reinforced metal-matrix composite beams," *International Journal of Structural Stability and Dynamics*, Vol. 19, No. 7, 1950078, 2019.

85. K.S. Surana, D. Mysore, and J.N. Reddy, "Thermodynamic consistency of beam theories in the context of classical and nonclassical continuum mechanics and a thermodynamically consistent new formulation," *Continuum Mechanics and Thermodynamics*, Vol. 31, No. 5, pp. 1283–1312, 2019.

86. Shubhankar Roy Chowdhury and J.N. Reddy, "Geometrically exact micropolar Timoshenko beam and its application in modelling sandwich beams made of architected lattice core," *Composite Structures*, Vol. 226, 111228, 2019.

87. T.B. Nguyen, J.N. Reddy, J. Rungamornrat, J. Lawongkerd, T. Senjuntichai, and V.H. Luong, "Nonlinear analysis for bending, buckling and post-buckling of nano-beams with nonlocal and surface energy effects," *International Journal of Structural Stability and Dynamics*, Vol. 19, No. 11, 1950130, 2019.

88. J.N. Reddy and Praneeth Nampally, "A dual mesh finite domain method for the analysis of functionally graded beams," *Composite Structures*, Vol. 251, article 112648, 2020.

89. Anssi T. Karttunen and J.N. Reddy, "Hierarchy of beam models for lattice core sandwich structures," *International Journal of Solids and Structures*, Vol. 204, pp. 172–186, 2020.

90. J.N. Reddy, Praneeth Nampally, and A.R. Srinivasa, "Nonlinear analysis of functionally graded beams using the dual mesh finite domain method and the finite element method," *International Journal of Non-Linear Mechanics*, Vol. 127, 103575, Dec 2020.

91. Praneeth Nampally and J.N. Reddy, "Geometrically nonlinear Euler-Bernoulli and Timoshenko micropolar beam theories," *Acta Mechanica*, Vol. 231, pp. 4217–4242, 2020.

92. M. Di Paola, J.N. Reddy, and E. Ruocco, "On the application of fractional calculus for the formulation of viscoelastic Reddy beam," *Meccanica*, Vol. 55, pp. 1365–1378, 2020.

93. J.N. Reddy and A.R. Srinivasa, "Misattributions and misnomers in mechanics: why they matter in the search for insight and precision of thought," *Vietnam Journal of Mechanics,* Vol. 42, No. 3, pp. 283–291, 2020.

94. Praneeth Nampally and J.N. Reddy, "Bending analysis of functionally graded axisymmetric circular plates using the dual mesh finite domain method," *Latin American Journal of Solid Mechanics,* Vol. 17, No. 7, 24 pages, 2020.

95. E. Ruocco, J.N. Reddy, and C.M. Wang, "An enhanced Hencky bar-chain model for bending, buckling and vibration analyses of Reddy beams," *Engineering Structures,* Vol. 221, article 111056, 2020.

96. Piotr Jankowski, Krzysztof Kamil Żur, Jinseok Kim, and J.N. Reddy, "On the bifurcation buckling and vibration of porous nanobeams," *Composite Structures,* Vol. 250, article 112632, 2020.

97. J.N. Reddy, Praneeth Nampally, and Nam Phan, "Dual mesh control domain analysis of functionally graded circular plates accounting for moderate rotations," *Composite Structures,* Vol. 257, article 113153, 2021.

98. Piotr Jankowski, Krzysztof Kamil Żur, Jinseok Kim, C.W. Lim, and J.N. Reddy, "On the piezoelectric effect on stability of symmetric FGM porous nanobeams," *Compsite Structures,* Vol. 267, article 113880, 2021.

99. Eugenio Ruocco, J.N. Reddy, and E. Sacco, "Analytical solution for a 5-parameter beam displacement model," *International Journal of Mechanical Sciences,* Vol. 201, article 106496, 2021.

100. Kurt Soncco, Karl Nils Betancourt, Roman Arciniega, and J.N. Reddy, "Postbuckling analysis of nonlocal functionally graded beams," *Latin American Journal of Solids and Structures,* Vol. 18, No. 7, e400 (20 pages), 2021.

ANSWERS TO SELECTED PROBLEMS

> *If anything is worth doing, do it with all your heart.*   Gautama Buddha
>
> *You must be the change you wish to see in the world.*   Mahatma Gandhi
>
> *The best way to predict your future is to create it.*   Abraham Lincoln
>
> *The ones who are crazy enough to think that they can change the world are the ones who do.*   Steve Jobs

CHAPTER 1

1.3 Use the principle of balance of linear momentum and the following relations between the Cauchy stress tensor σ and the second Piola–Kirchhoff stress tensor $\mathbf{S}$, and between the area element $d\mathbf{a}$ in the deformed and the area element $d\mathbf{A}$ in the undeformed configurations of the material body:

$$\sigma = \frac{1}{J}\mathbf{F}\cdot\mathbf{S}\cdot\mathbf{F}^{\mathrm{T}}, \quad \hat{n}\,da = J\mathbf{F}^{-\mathrm{T}}\cdot\hat{\mathbf{N}}dA,$$

where $\hat{n}$ and $\hat{\mathbf{N}}$ are the unit normal vectors in the deformed and undeformed configurations. Also note that $\mathbf{F} = \mathbf{I} + \nabla\mathbf{u}$.

1.4 The inversion of the given strain–stress relation yields the required result:

$$\begin{Bmatrix} \sigma_1 \\ \sigma_2 \\ \sigma_6 \end{Bmatrix} = \begin{bmatrix} Q_{11} & Q_{12} & 0 \\ Q_{12} & Q_{22} & 0 \\ 0 & 0 & Q_{66} \end{bmatrix} \begin{Bmatrix} \varepsilon_1 \\ \varepsilon_2 \\ \varepsilon_6 \end{Bmatrix},$$

where

$$Q_{11} = \frac{E_1}{1 - \nu_{12}\nu_{21}}, \quad Q_{12} = \frac{\nu_{12}E_2}{1 - \nu_{12}\nu_{21}}, \quad Q_{22} = \frac{E_2}{1 - \nu_{12}\nu_{21}}, \quad Q_{66} = G_{12}.$$

1.5 We have $u_1 = u + x_3\,\phi$, $u_2 = 0$, and $u_3 = w$. The only nonzero component of the vector $\boldsymbol{\omega}$ is

$$\omega_2 = \frac{1}{2}\left(\frac{\partial u_1}{\partial x_3} - \frac{\partial u_3}{\partial x_1}\right) = \frac{1}{2}\left(\phi_1 - \frac{dw}{dx_1}\right).$$

Then, the only nonzero component of χ is χ_{12}:

$$\chi_{12} = \frac{1}{2}\frac{\partial\omega_2}{\partial x_1} = \frac{1}{4}\left(\frac{d\phi_1}{dx_1} - \frac{d^2w}{dx_1^2}\right).$$

CHAPTER 2

2.1 $V_E = (m_1 + m_2)gL_1(1 - \cos\theta_1) + m_2gL_2(1 - \cos\theta_2)$

2.2 $U = \int_0^L \frac{EA}{2}\left(\frac{du}{dx}\right)^2 dx + \frac{1}{2}k[u(L)]^2$ and $V = -\left[\int_0^L f(x)u\,dx + Pu(L)\right]$

2.3 (a) $U = \frac{19}{25}AKv\sqrt{v}$ (b) $U^* = \left(\frac{50}{57}\right)^2\frac{P^3}{3K^2A^2}$.

CHAPTER 3

3.3 The Lagrangian is $L = K - U - V_E$ with

$$K = \frac{1}{2}\int_0^T \int_0^L \left[m_0\left(\frac{\partial u}{\partial t}\right)^2 + m_0\left(\frac{\partial w}{\partial t}\right)^2 + m_2\left(\frac{\partial^2 w}{\partial x \partial t}\right)^2 - 2m_1 \frac{\partial u}{\partial t}\frac{\partial^2 w}{\partial x \partial t}\right] dx dt, \quad (1)$$

$$U = \frac{1}{2}\int_0^T \int_0^L \left\{ \bar{A}_{xx}\left[\frac{\partial u}{\partial x} + \frac{1}{2}\left(\frac{\partial w}{\partial x}\right)^2\right]^2 + \bar{B}_{xx}\bar{M}_{xx}\left(\frac{\partial w}{\partial x}\right)^2 \right.$$

$$\left. + 2\bar{M}_{xx}\frac{\partial w}{\partial x} + 2\bar{B}_{xx}\frac{\partial u}{\partial x}\bar{M}_{xx} - \frac{1}{D_{xx}}(\bar{M}_{xx})^2 + c_f w^2 \right\} dx dt, \quad (2)$$

$$V_E = \int_0^T \int_0^L (uf + qw)dx dt. \quad (3)$$

3.4 The solution is given by

$$\hat{D}_{xx}^* u(x) = B_{xx}\frac{q_0 L^3}{24}\left[-1 + 3\frac{x}{L} - 2\frac{x^2}{L^2} - \left(1 - \frac{x}{L}\right)^4\right], \quad (1)$$

$$\hat{D}_{xx}^* w(x) = \frac{\hat{D}_{xx}^*}{\hat{D}_{xx}}\frac{M_{xx}^T L^2}{2}\left(\frac{x^2}{L^2} - \frac{x}{L}\right) - \frac{B_{xx}^2}{\hat{D}_{xx}}\frac{q_0 L^4}{48}\left(\frac{x^2}{L^2} - \frac{x}{L}\right)$$

$$+ A_{xx}\frac{q_0 L^4}{120}\left[-6\left(\frac{x}{L}\right) + 10\left(\frac{x}{L}\right)^3 - 5\left(\frac{x}{L}\right)^4 + \left(\frac{x}{L}\right)^5\right]. \quad (2)$$

3.6 The maximum deflections at the center of a simply-supported beam due to uniform load q_0 (downward) and point load (upward) at the center (according to CBT) are, respectively:

$$w^q(0.5L) = -\frac{5}{384}\frac{q_0 L^4}{EI}, \quad w^q(0.5L) = \frac{FL^3}{48EI}.$$

The deflection at the free end of a clamped beam due to a point load F_0 (downward) at the free end is $w_A = F_0 L^3/3EI$. The required answers are: (a) $F_{\rm rod} = 3315.26$ lb (up), (b) $w_C = -0.8951$ in., and (c) $M_{\rm max}^{DE} = 7694.81$ lb-ft at $x = 6.0533$ ft.

3.7 Follow the following steps:

Step 1: Write the equilibrium of vertical forces and moments:

$$A_z + F_B + C_z - 12 \times 30 \times 10^3 = 0, \quad (1)$$

$$6F_B + 12R_C - 6\left(12 \times 30 \times 10^3\right) = 0, \quad (2)$$

where R_A and R_C denote the vertical forces at the ends of the beam, A and C, respectively, and F_B is the force in the elastic post at point B.

Step 2: Use the method of superposition by treating the given beam as two beams superposed: (1) a simply supported beam under uniform distributed load q_0 and (2) a simply supported beam with a point load F_B at the center point B. Then the transverse deflections at point B of the two beams are (from a book on mechanics of materials, e.g., see Fenner and Reddy [123])

$$(w_B)_1 = -\frac{5q_0 L^4}{384D_{xx}}, \quad (w_B)_2 = \frac{F_B L^3}{48D_{xx}}, \quad (3)$$

where $D_{xx} = EI$ is the bending stiffness of the beam. Thus, the net deflection of the beam at point B due to q_0 and F_B is

$$w_B = (w_B)_1 + (w_B)_2 = -\frac{5q_0 L^4}{384D_{xx}} + \frac{F_B L^3}{48D_{xx}}. \quad (4)$$

Step 3: The (geometric) compatibility requires that w_B be equal to the gap plus the elongation in the post:

$$(w_B)_1 + (w_B)_2 = -0.2 \times 10^{-3} - \frac{F_B L_P}{A_P E_P}. \tag{5}$$

Equations (4) and (5) together give F_B.

Step 4: Use Eqs. (1) and (2) to determine the vertical reaction forces R_A and R_C. Then write the expression for the $M(x)$ as a function of x, and determine the transverse deflection $w(x)$ using the equation $EI(d^2w/dx^2) = -M(x)$.

The answers are: (a) $F_B = 219.78$ kN, (b) $R_A = R_C = 70.11$ kN, and (c) $w(3) = -1.8963$ mm and $w(6) = -1.0745$ mm.

3.8 The general solution to this buckling problem is given by Eq. (3.7.62) with $\lambda^2 = \frac{N_{xx}^0}{D_{xx}^e}$. The boundary conditions for clamped-hinged beam are

$$w(0) = 0, \quad w'(0) = 0, \quad w(a) = 0, \quad M_{xx}(a) = 0,$$

where the origin of the coordinate system is taken at the clamped end. These boundary conditions yield the characteristic polynomial

$$\sin \lambda L - \lambda L \cos \lambda L = 0.$$

3.9 The general solution to this natural vibration problem is given by Eq. (3.7.83) with λ and μ defined by Eq. (3.7.84). The characteristic equation is

$$\sin \lambda L \cosh \lambda L - \cos \lambda L \sinh \lambda L = 0.$$

CHAPTER 4

4.1 Integrate the first equation and obtain

$$N_{xx} = -\int f(x)\, dx + c_1.$$

Integration of the second equation yields,

$$N_{xz} + \tfrac{1}{2}\frac{dP_{xy}}{dx} = -\int q(x)\, dx + c_2.$$

The third equation results in

$$M_{xx} + P_{xy} = -\int \int^x q(\xi)\, d\xi\, dx - c_2 x + c_3.$$

Then follow the procedure in Eqs. (4.6.8)–(4.6.16) to determine the generalized displacements.

4.5 The boundary conditions at $x = 0$ give

$$c_4 = c_5 = 0, \quad c_6 = \Omega L^2 \, M_{xx}^C(0), \quad \Omega = \frac{D_{xx}^*}{A_{xx} S_{xz} L^2}.$$

The hinged boundary condition at $x = L$ gives

$$c_1 = 0, \quad c_3 = -c_2 L, \quad c_2 = \frac{3\Omega}{L(1 + 3\Omega)} M_{xx}^C(0).$$

Thus, we have

$$u^F(x) = u^C(x) - \frac{B_{xx}}{2D_{xx}^*}(x^2 - 2Lx) - \frac{3\Omega}{L(1 + 3\Omega)} M_{xx}^C(0),$$

$$\phi^F(x) = -\frac{dw^C}{dx} + \frac{3\Omega}{L(1 + 3\Omega)}\frac{A_{xx}}{2D_{xx}^*}(x^2 - 2Lx),$$

$$w^F(x) = w^C(x) - \frac{A_{xx}}{D_{xx}^*}\left[\frac{3\Omega}{6L(1 + 3\Omega)}(x^3 - 3LX^2) + \Omega L^2\right] M_{xx}^C(0) + \frac{1}{S_{xz}} M^C(x),$$

$$M_{xx}^F = M_{xx}^C + \frac{3\Omega}{L(1 + 3\Omega)} M_{xx}^C(0)\,(x - L).$$

4.6 See the steps described for **Problem 3.7**. You also need to write the expression for the shear force $V(x)$ and relate to the shear strain $\gamma_{xz} = \phi_x + dw/dx$. The answers are: (a) $F_B = 219.64$ kN, (b) $R_A = R_C = 70.18$ kN, and (c) $w(3) = -1.9343$ mm and $w(6) = -1.0739$ mm.

CHAPTER 5

5.4 The statement of the principle of virtual displacements is

$$0 = \int_0^L \left\{ N_x \frac{d\delta u}{dx} + M_x \left(c_0 \frac{d^2 \delta w}{dx^2} + c_1 \frac{d\delta\phi_x}{dx} \right) + c_2 L_x \frac{d\delta\psi_x}{dx} + \frac{c_3}{h^3} P_{xx} \left(\frac{d\delta\phi}{dx} + \frac{d^2 \delta w}{dx^2} \right) \right. $$

$$ + N_{xz} \left[c_1 \delta\phi_x + (1 + c_0) \frac{d\delta w}{dx} \right] + 2c_2 \delta\psi_x R_{xz} + \frac{3c_3}{h^3} S_{xz} \left(\delta\phi_x + \frac{d\delta w}{dx} \right) \right\} dx $$

$$ - \int_0^L (f\delta u + q\delta w)\, dx $$

$$ = \int_0^L \left\{ -\frac{dN_{xx}}{dx} \delta u - \frac{d}{dx} (c_1 M_{xx}) \delta\phi + \frac{d^2}{dx^2} (c_0 M_{xx}) \delta w - \frac{d}{dx} (c_2 L_{xx}) \delta\psi \right. $$

$$ - \frac{d}{dx} \left(\frac{c_3}{h^3} P_{xx} \right) \delta\phi_x + \frac{d^2}{dx^2} \left(\frac{c_3}{h^3} P_{xx} \right) \delta w $$

$$ + c_1 \delta\phi_x N_{xz} + 2c_2 \delta\psi_x R_{xz} - (1 + c_0) \frac{dN_{xz}}{dx} \delta w + \frac{3c_3}{h^3} \left(S_{xz} \delta\phi - \frac{dS_{xz}}{dx} \delta w \right) \right\} dx $$

$$ - \int_0^L (f\delta u + q\delta w)\, dx + \left[N_{xx} \delta u + c_1 M_{xx} \delta\phi_x + \frac{c_3}{h^3} P_{xx} \left(\delta\phi_x + \frac{d\delta w}{dx} \right) \right. $$

$$ + c_2 L_{xx} \delta\psi_x - \frac{c_3}{h^3} \frac{dP_{xx}}{dx} \delta w + c_0 M_{xx} \frac{d\delta w}{dx} - c_0 \frac{dM_{xx}}{dx} \delta w + \frac{c_3}{h^3} P_{xx} \delta w $$

$$ + (1 + c_0) N_{xz} \delta w + \frac{3c_3}{h^3} S_{xz} \delta w \Big]_0^L . $$

CHAPTER 6

6.3 For this special case, we have

$$M_{rr} = -D_{rr} \left(\frac{d^2 w}{dr^2} + \frac{\nu}{r} \frac{dw}{dr} \right), \quad M_{\theta\theta} = -D_{rr} \left(\nu \frac{d^2 w}{dr^2} + \frac{1}{r} \frac{dw}{dr} \right).$$

The using Eq. (6.3.3) in Eq. (6.2.25), the required result is obtained.

6.4 For a linearly distributed load of the type

$$q(r) = q_0 \left(1 - \frac{r}{a} \right),$$

we obtain [see Eqs. (6.3.8a) and (6.3.8b)]

$$F(r) = \frac{q_0 r^4}{64} + \left(-\frac{q_0}{a} \right) \frac{r^5}{225},$$

$$F'(r) = \frac{q_0 r^3}{16} + \left(-\frac{q_0}{a} \right) \frac{r^4}{45}, \tag{1}$$

$$F''(r) = \frac{3q_0 r^2}{16} + \left(-\frac{q_0}{a} \right) \frac{4r^3}{45}.$$

The constants c_2 and c_4 are given by ($c_1 = c_3 = 0$)

$$c_2 = -\frac{2}{(1+\nu)} \left(F''(a) + \frac{\nu}{a} F'(a) \right) = -\frac{q_0 a^2}{(1+\nu)} \left(\frac{71 + 29\nu}{360} \right),$$

$$c_4 = -\left(F(a) + \frac{a^2}{4} c_2 \right) = \frac{q_0 a^4}{(1+\nu)} \left(\frac{183 + 43\nu}{4800} \right).$$

6.5 We obtain

$$c_2 = -\frac{2q_1 a^2}{(1+\nu)}\left(\frac{4+\nu}{45}\right), \quad c_4 = \frac{q_1 a^4}{(1+\nu)}\left(\frac{6+\nu}{150}\right).$$

6.6 We have

$$F(r) = \frac{q_0 r^6}{576a^2}, \quad F'(r) = \frac{q_0 r^5}{96a^2}, \quad F''(r) = \frac{5q_0 r^4}{96a^2}$$

and $(c_1 = c_3 = 0)$

$$Dw(r) = \frac{q_0 r^6}{576a^2} + \frac{r^2}{4}c_2 + c_4,$$

$$D\frac{dw}{dr} = \frac{q_0 r^5}{96a^2} + \frac{r}{2}c_2.$$

Using the boundary conditions, we obtain

$$w(a) = 0: \quad \frac{q_0 a^4}{576} + \frac{a^2}{4}c_2 + c_4 = 0,$$

$$\frac{dw}{dr}(a) = 0: \quad \frac{q_0 a^3}{96} + \frac{a}{2}c_2 = 0,$$

$$c_2 = -\frac{q_0 a^2}{48} = -\frac{12q_0 a^2}{576}, \quad c_4 = \frac{q_0 a^4}{288} = \frac{2q_0 a^4}{576}.$$

6.7 We have

$$F(r) = \frac{q_0 a^4}{14400}\left(225\frac{r^4}{a^4} - 64\frac{r^5}{a^5}\right), \quad F'(r) = \frac{q_0 r^3}{16} - \frac{q_0 r^4}{45a},$$

$$c_2 = -\frac{q_0 a^2}{8} + \frac{2q_0 a^2}{45} = -\frac{29q_0 a^2}{360},$$

$$c_4 = -\frac{q_0 a^4}{64} + \frac{q_0 a^4}{225} + \frac{q_0 a^4}{32} - \frac{q_0 a^4}{90} = \frac{129q_0 a^4}{14400}.$$

6.9 The solution is

$$W_1(r) = \frac{q_1 a^4}{30(1+\nu)D}\left(1 - \frac{r^2}{a^2}\right), \quad (W_1)_{max} = c_1 = \frac{q_1 a^4}{30(1+\nu)D}.$$

6.10 Assume solution of the form (for $n = 0$)

$$W_1(r) = c_1\phi_1(r,\theta) = c_1 f_1(r), \quad f_1(r) = 1 - \frac{r^2}{a^2},$$

and obtain

$$A_{11} = \frac{8\pi(1+\nu)D}{a^2}, \quad M_{11} = \frac{2\pi m_0 a^2}{6}.$$

Hence, the frequency is given by $(m_0 = \rho h)$

$$\omega^2 = \frac{A_{11}}{M_{11}} = \frac{24(1+\nu)D}{m_0 a^4} \quad \text{or} \quad \omega = \frac{4.899}{a^2}\sqrt{\frac{(1+\nu)D}{\rho h}}.$$

CHAPTER 7

7.1 From Eqs. (7.2.3) and (7.2.4), we have

$$rN_{rz} = D_{rr}\left(r\frac{d^2\phi_r}{dr^2} + \frac{d\phi_r}{dr} - \frac{1}{r}\phi_r\right)$$

$$= D_{rr}\,r\left[\frac{d^2\phi_r}{dr^2} + \frac{d}{dr}\left(\frac{1}{r}\phi_r\right)\right]$$

$$= D_{rr}\,r\frac{d}{dr}\left[\frac{1}{r}\frac{d}{dr}(r\phi_r)\right].$$

7.2 From Eqs. (6.2.21c) and (6.2.21d) (for isotropic beams), we have

$$\mathcal{M}^C \equiv \frac{M_{rr}^C + M_{\theta\theta}^C}{1 + \nu} = -D_{rr}\left(\frac{d^2 w^C}{dr^2} + \frac{1}{r}\frac{dw^C}{dr}\right) = -D_{rr}\frac{1}{r}\frac{d}{dr}\left(r\frac{dw^C}{dr}\right).$$

Similarly, from Eqs. (7.2.3) and (7.2.4) (for isotropic beams), we have

$$\mathcal{M}^F \equiv \frac{M_{rr}^F + M_{\theta\theta}^F}{1 + \nu} = D_{rr}\left(\frac{d\phi_r^F}{dr} + \frac{1}{r}\phi_r^F\right) = D_{rr}\frac{1}{r}\frac{d}{dr}\left(r\phi_r^F\right).$$

7.3 Equation (1) follows from Eqs. (1) and (3) of **Problem 7.2**. Equation (2) follows from Eqs. (8) and (11) of **Problem 7.2** and Eq. (1).

CHAPTER 8

8.1 The result in Eq. (8.3.22) follows from the solution of **Problem 7.2**.

CHAPTER 9

9.1 For example, T_{ij}^{12} is computed as follows:

$$T_{ij}^{12} = K_{ij}^{12} + \sum_{k=1}^{m}\frac{\partial K_{ik}^{11}}{\partial w_j}u_k + \sum_{k=1}^{n}\frac{\partial K_{ik}^{12}}{\partial w_j}w_k + \sum_{k=1}^{p}\frac{\partial K_{ik}^{13}}{\partial w_j}M_k$$

$$= K_{ij}^{12} + 0 + \sum_{k=1}^{n}\frac{\partial K_{ij}^{12}}{\partial w_j}w_k + 0$$

$$= K_{ij}^{12} + \frac{1}{2}\sum_{k=1}^{n}\int_{x_a^e}^{x_b^e}\bar{A}_{xx}\frac{\partial^2 w}{\partial w_j \partial x}\frac{d\psi_i^{(1)}}{dx}\frac{d\psi_k^{(2)}}{dx}w_k\,dx$$

$$= K_{ij}^{12} + \frac{1}{2}\int_{x_a^e}^{x_b^e}\bar{A}_{xx}\frac{d\psi_j^{(2)}}{dx}\frac{d\psi_i^{(1)}}{dx}\left(\sum_{k=1}^{n}\frac{d\psi_k^{(2)}}{dx}w_k\right)dx$$

$$= K_{ij}^{12} + \frac{1}{2}\int_{x_a^e}^{x_b^e}\bar{A}_{xx}\frac{\partial w}{\partial x}\frac{d\psi_i^{(1)}}{dx}\frac{d\psi_j^{(2)}}{dx}\,dx.$$

9.3 Using the element and node numbering shown in Fig. P9.3, the connectivity matrix is

$$\mathbf{B} = \begin{bmatrix} 1 & 2 & 3 & 4 \\ 3 & 4 & 5 & 6 \\ 7 & 8 & 9 & 10 \end{bmatrix}.$$

Since the degrees of freedom, 1, 5, 9, and 10 are specified to be zero, the condensed matrix is 6×6, and is given by Hence the assembled matrix is

$$\begin{bmatrix}
K_{22}^{(1)} & K_{23}^{(1)} & K_{24}^{(1)} & 0 & 0 & 0 \\
K_{32}^{(1)} & K_{33}^{(1)} + K_{11}^{(2)} + k & K_{34}^{(1)} + K_{12}^{(2)} & K_{14}^{(2)} & -k & 0 \\
K_{42}^{(1)} & K_{43}^{(1)} + K_{21}^{(2)} & K_{44}^{(1)} + K_{22}^{(2)} & K_{24}^{(2)} & 0 & 0 \\
0 & K_{41}^{(1)} & K_{42}^{(1)} & K_{44}^{(1)} & 0 & 0 \\
0 & -k & 0 & 0 & K_{11}^{(3)} + k & K_{12}^{(3)} \\
0 & 0 & 0 & 0 & K_{21}^{(3)} & K_{22}^{(3)}
\end{bmatrix},$$

where $K_{ij}^{(e)}$ are element coefficients of element Ω^e and $k = (AE/L)$ is the stiffness of the rod connecting the two beams. The condensed force vector is

$$\left\{\frac{q_0 h_1^2}{12},\ \frac{q_0 h_1}{2} + \frac{q_0 h_2}{2},\ -\frac{q_0 h_1^2}{12} + \frac{q_0 h_2^2}{12},\ -\frac{q_0 h_2^2}{12},\ 0,\ 0\right\}^T.$$

Here $h_1 = h_2$ are the lengths of elements 1 and 2.

The data and numerical results are as follows:

```
Geometric and material data of the problem:
    Diameter of the rod, dp ........ =  0.2500E+00 in.
    Value of distributed load, Q0 ... =  0.4200E+03 lb/ft
    Youngs modulus of beam 1, YM1 ... =  0.3000E+08 psi
    Youngs modulus of beam 2, YM2 ... =  0.1500E+07 psi
    Moment of inertia of beam 1, SMA1 =  0.2200E+02 in^4
    Moment of inertia of beam 2, SMA2 =  0.4150E+03 in^4
    Length of beam 1, BL1 .......... =  0.6000E+01 ft
    Half length of beam 2, BL2 ...... =  0.1000E+02 ft
    Length of the steel rod, RL ..... =  0.1000E+02 ft
```

Load vector:

```
0.42000E+05 -0.42000E+04  0.00000E+00 -0.42000E+05  0.00000E+00  0.00000E+00
```

Stiffness matrix:

```
0.20750E+08  0.25938E+06  0.10375E+08  0.00000E+00  0.00000E+00  0.00000E+00
0.25938E+06  0.20918E+05  0.00000E+00 -0.25938E+06 -0.12272E+05  0.00000E+00
0.10375E+08  0.00000E+00  0.41500E+08  0.10375E+08  0.00000E+00  0.00000E+00
0.00000E+00 -0.25938E+06  0.10375E+08  0.20750E+08  0.00000E+00  0.00000E+00
0.00000E+00 -0.12272E+05  0.00000E+00  0.00000E+00  0.33491E+05 -0.76389E+06
0.00000E+00  0.00000E+00  0.00000E+00  0.00000E+00 -0.76389E+06  0.36667E+08
```

Solution:

```
0.13213E-01 -0.89511E+00 -0.40694E-18 -0.13213E-01 -0.62496E+00 -0.13020E-01
```

Cable force: 0.33153E+04 lb

CHAPTER 10

10.5 We have $\boldsymbol{\Delta} = \mathbf{HK}$,

$$\begin{Bmatrix} \Delta_1 \\ \Delta_2 \\ \Delta_3 \\ \Delta_4 \end{Bmatrix} = \begin{bmatrix} 1 & r_a^2 & \log r_a & r_a^2 \log r_a \\ 0 & -2r_a & -\frac{1}{r_a} & -r_a \left(1 + 2\log r_a\right) - \frac{1}{r_a}\Gamma \\ 1 & r_b^2 & \log r_b & r_b^2 \log r_b \\ 0 & -2r_b & -\frac{1}{r_b} & -r_b \left(1 + 2\log r_b\right) - \frac{1}{r_b}\Gamma \end{bmatrix} \begin{Bmatrix} K_1 \\ K_2 \\ K_3 \\ K_4 \end{Bmatrix}, \qquad (1)$$

or $\mathbf{K} = \mathbf{H}^{-1}\boldsymbol{\Delta}$. Similarly, we can relate the nodal forces Q_i to the constants K_i (all of the stress resultants are from the FST):

$$Q_1 \equiv -2\pi \left(r N_{rz}\right)_{r=r_a} \quad = 8\pi D_{rr} K_4$$

$$Q_2 \equiv 2\pi \left(-r M_{rr}\right)_{r=r_a}$$

$$= 2\pi D_{rr} \left\{ 2(1+\nu)r_a K_2 - \frac{(1-\nu)}{r_a} K_3 + \left[\Lambda_a - \frac{(1-\nu)}{r_a}\Gamma\right] K_4 \right\}$$

$$Q_3 \equiv 2\pi \left(r N_{rz}\right)_{r=r_b} \quad = -8\pi D_{rr} K_4$$

$$Q_4 \equiv 2\pi \left(r M_{rr}\right)_{r=r_b}$$

$$= -2\pi D_{rr} \left\{ 2(1+\nu)r_b K_2 - \frac{(1-\nu)}{r_b} K_3 + \left[\Lambda_b - \frac{(1-\nu)}{r_b}\Gamma\right] K_4 \right\},$$

or $\mathbf{Q} = \mathbf{GK}$, where

$$\Lambda_a = [2(1+\nu)\log r_a + (3+\nu)]\, r_a, \quad \Lambda_b = [2(1+\nu)\log r_b + (3+\nu)]\, r_b. \qquad (2)$$

Thus, we have

$$\mathbf{Q} = \mathbf{GK}, \quad \mathbf{\Delta} = \mathbf{HK}. \qquad (2)$$

Then the stiffness matrix $\mathbf{K}$ is ($\mathbf{Q} = \mathbf{GK} = \mathbf{GH}^{-1}\mathbf{\Delta} \equiv \mathbf{K\Delta}$

$$\mathbf{K} = \mathbf{GH}^{-1}, \qquad (3)$$

where

$$\mathbf{G} = 2\pi D_{rr}\begin{bmatrix} 0 & 0 & 0 & 4 \\ 0 & 2(1+\nu)r_a & -\frac{(1-\nu)}{r_a} & \left[\Lambda_a - \frac{(1-\nu)}{r_a}\Gamma\right] \\ 0 & 0 & 0 & -4 \\ 0 & -2(1+\nu)r_b & \frac{(1-\nu)}{r_b} & -\left[\Lambda_b - \frac{(1-\nu)}{r_b}\Gamma\right] \end{bmatrix}. \qquad (4)$$

The stiffness matrix of the CPT is obtained from $\mathbf{K} = \mathbf{GH}^{-1}$ by setting $\Gamma = 0$.

Index

A

Alternating (permutation) symbol, 5, 6
Analytical solution, 102, 106, 111, 124,
       131, 162, 171, 223, 224, 247,
       248, 252, 322, 336, 351, 407,
       472, 494, 509-511, 518
    CBT, 106, 111, 124
    CPT, 315, 322, 336, 351
    FST, 366
    RBT, 244, 248, 252
    TST, 407
    also see Exact solution
Approximate solution, 113–115, 119,
    292, 418, 441
Approximation:
    alpha–family of, 140
    finite element, 415, 418-422, 425,
       432, 437
    functions, 113, 118, 125, 130, 292,
       294, 342, 415

    properties, 120, 142
    Galerkin, 28, 144, 344, 346
    Hermite cubic, 75, 135, 227,
       420-422, 425, 437, 455, 458
    linear, 417, 423
    Petrov–Galerkin, 143
    quadratic, 419, 420, 425, 460, 474
    Ritz, 115, 119, 121, 126, 132, 137,
       294, 345, 349, 352, 353
    variational methods of, 213, 293
    weighted-residual, 141
Axisymmetry, 303

B

Balance of:
    linear momentum, 18
    secondary variables, 417
Bars, 2, 113
Basis vectors, 4, 11, 62, 304
Beam theory:
    classical, see Euler–Bernoulli
    Euler–Bernoulli, 3, 61–68
    Bickford, 223, 231, 431
    Levinson, 223, 230, 241, 251
    Reddy, 223, 238, 244, 253, 254,
       259, 281, 299, 402

    Timoshenko, 3, 62, 151, 165, 187,
       191
Bending of:
    CBT beams, 61–68
    CPT circular plates, 303–311
    FST circular plates, 355–
    RBT beams, 231, 232–235
    TBT beams, 151–160
Bilinear functional, 116
Body couple, 187
Body force, 18, 38, 52, 222
Boundary conditions:
    essential (Dirichlet or geometric),
       36, 40, 46, 47, 50, 55
    natural (Neumann or force), 46,
       47, 49, 66
    simply supported, 82
Buckling:
    load, 93, 97, 175, 201, 276, 280

    critical, 93, 95, 175, 333, 352
Buckling of:
    beams, 92, 174, 199, 273
    circular plates, 332, 348, 409

C

Cantilever(ed) beam, 37, 88, 90, 129,
    170, 185, 271
Cantilevered annular plate, 504
Cartesian coordinate system, 4, 5, 14, 52
Cauchy's formula, 9, 52
Characteristic equation, 97, 100, 141
Circular plates:
    displacement field, 306, 355, 395
    governing equations, 308, 356, 397
    strains, 304–306, 355, 396
    stress resultants, 307, 356, 397
Classical beam theory, *see*
       Euler–Bernoulli beam theory
Classical plate theory (CPT), 303
Coefficient of thermal expansion, 19, 69,
    229, 235
Complementary energy, 35
Constitutive relations, 18, 19, 24, 69, 70,
    156, 229, 235, 236, 305, 310,
    357, 398
Couple stress, 23
    constitutive relation, 24, 69, 305

549

Cramer's rule, 123, 284
Curl of a vector, 10, 353
Curl theorem, 11
Curvature, 24, 68, 224, 306, 355, 396
Cylindrical bending, 78
Cylindrical coordinates, 12, 15, 17, 303

D

Deformation tensor, 13
Deformation gradient, 17
Del operator, 10
Dirac delta function, 112, 143
Directional derivative, 10
Dirichlet boundary condition, 47
Displacement finite element model, 423, 435
Displacement field of:
    CBT, 62
    CPT, 306
    FST, 355
    RBT, 223, 231
    TBT, 152, 204
    TST, 395
Displacement gradient, 13, 16
Divergence, 10, 11, 353
    theorem, 11, 57
Dot (scalar) product, 5, 10
Double–dot product, 33
Duality pairs, 45–47
Duality pairs of:
    CBT, 68, 74, 82, 102, 119, 125, 134
    CPT, 306, 309, 315, 316, 326, 479
    FST, 361, 368
    LBT, 242
    RBT, 227, 228, 234, 254, 262, 281, 437
    TBT, 156, 164, 427
    TST, 397
Dummy index, 5, 13
Dynamic (transient) response, 109, 136, 449, 482

E

Eigenvalues, 97, 101, 210, 338
Eigenvector, see Mode shape
Elastic foundation, 63, 67, 80, 91, 109, 171–173, 233, 281, 335, 340, 351
Elasticity tensor, 19
Energetically conjugate, 17
Engineering constants, 19
Equations of:
    motion of:
        CBT, 64, 66, 68, 72

        elasticity, 57
        CPT, 308
        FST, 356
        RBT, 226, 228
        TBT, 156
        TST, 397
    equilibrium of:
        CBT, 73, 74
        elasticity, 50, 222
        CPT, 309
        FST, 359
        RBT, 235
        TBT, 160
        TST, 359
Essential boundary condition(s), 47, 55, 70, 82, 119, 143, 342, 345, 459

F

Finite element model of:
    CBT, 423, 424
    CPT, 487, 490
    FST, 495
    RBT, 433
    TBT, 427, 430
    TST, 499
First variation, 40
Force (natural) boundary conditions, 47, 49, 66, 119, 121, 143, 227, 293, 343
Fourth-order:
    differential equation, 74, 127, 259, 312, 315, 403
    elasticity tensor, 19
Functional, 41–46
    bilinear, 116
    quadratic, 41, 115
    potential energy, 48, 51–55, 59, 115, 125, 129, 143, 213, 243, 293, 484
Functionally graded:
    beams, 19–23, 84, 148, 177, 182, 198, 207, 230, 462–466
    circular plates, 305, 379, 513
Fundamental frequency, 99, 109, 210–212, 290, 336, 351
Fundamental lemma of variational calculus, 42–48, 54, 67, 137, 226, 234

G

Galerkin's method, 143, 144, 213, 340, 344, 345–348
Gauss:

rule, 448, 503
theorem, 50, 53
Gradient:
theorem, 11, 45
vector, 10
Green strain tensor, 14–17
Green–Gauss theorem, 11
Green–Lagrange strain tensor, *see*
Green strain tensor

H

Hamilton's principle, 54–58, 61, 66, 113,
137, 154, 225, 233, 307, 340,
356, 396, 417
Hermite cubic polynomials, 75, 135, 227,
420-422, 425, 437, 455, 458
Heterogeneous, 20
Homogeneous:
beams, 20, 83, 109, 146, 155, 159,
166, 187, 189–194, 258, 273
boundary conditions, 37, 44, 50,
119, 120, 143, 293, 342, 345
plates, 494, 497, 503
Hooke's law, 18, 25, 33, 48
Hyperelastic material, 51

I

Inhomogeneous:
beams, 158
material, 114
Initial condition(s), 110, 139, 436
Integration by parts, 226
Internal energy, 29, 35
Interpolation functions, 416, 417–420
Hermite cubic, 75, 135, 227,
420-422, 425, 437, 455, 458
Lagrange family, 419, 421, 429,
431, 437, 455, 490, 493, 496,
502, 507
Isotropic, 51, 52, 69, 77, 250

J

Jacobian, 18

K

Kinematic relations, 36, 54, 62, 152,
223, 231, 232, 241, 304, 355
Kinetic energy, 66, 225
virtual, 66, 225
Kronecker Delta, 5, 6
Kirchhoff stress tensor, *see*
Piola–Kirchhoff stress tensor

L

Laplace operator, 10
Laplace transform, 111–113
Least-squares method, 143, 421
Levinson beam theory (LBT), 223, 230,
241, 251, 255, 285
Linear momentum, 18
Linearly independent, 61, 102, 103, 113,
118, 120, 122, 141, 142, 221,
342

M

Material coordinates, 18, 19
Material stiffness coefficients, 71, 237
Mode shape, 97, 108, 176, 202, 285, 289,
336–339
Modified couples stress, *see* couple stress
constitutive relation, 305
Moment of inertia, 462

N

Natural boundary conditions, 47, 66,
119, 123, 143, 227, 293, 340,
343
Navier's solutions, 102, 203, 280, 285
Neumann boundary condition, 47
Newmark's method (or scheme), 139,
434–437
Non-homogeneous equation, 19
Numerical solution, 195, 493, 496

O

Orthogonal, 141
components, 9
Orthogonality, 296

P

Permutation symbol, 6
Plane stress, 18, 19, 25
Piola–Kirchhoff stress tensor, 17, 25, 49,
64
Potential energy, 27–29, 36, 48, 51–53,
113–116, 125, 130, 213, 231,
243, 293, 475
Power-law:
exponent (or index), 22, 70, 85-86,
106, 130, 195, 199, 207, 230,
289, 308, 330, 353, 386, 398,
450, 453, 461, 467, 493, 502
model, 22
Primary variable, 44–46, 68, 75, 119,
131, 164, 417–419, 423, 428,

431, 437, 490, 493, 497, 502, 508
Principle of:
  minimum total potential energy, 27, 48, 51–54, 114, 121, 124, 135, 213, 214, 231, 243, 294, 475
  virtual displacements, 4, 28, 49–54, 63, 113, 124, 231, 299, 417

Q

Quadratic:
  elements, 417, 418, 419, 460, 465, 468, 471, 473
  function(s), 134, 222, 231, 299, 418, 419, 420, 425
  functional, 41, 115, 118, 131

R

Rectangular cartesian coordinates, 5, 9, 14, 33, 49, 52, 62, 64
Reddy beam theory, 232, 244, 280
Reduced integration, 447, 452, 493, 503
Ritz method, 114–141, 292–296, 340, 345, 348, 416

S

Second Piola–Kirchhoff stress, 14, 17, 49, 64, 222
Secondary variable, 45, 46, 68, 119, 417, 423, 428, 432, 477, 479, 481, 485, 488
Shear correction coefficient (or factor), 3, 151–157, 206, 213, 219, 251, 254, 263, 269–272, 280, 297, 301, 355, 381, 386, 388, 390, 395, 409, 507
Simply supported beam, 77, 82, 91, 95, 99, 101, 106–109, 124–129, 144, 173, 191, 194–198, 203–213, 217, 268, 281, 285–293, 295, 319, 450, 456, 524
Simply supported circular plate, 319–322, 333, 339, 353, 364, 376–378, 388, 407, 411–413
Spring constant,
  rotational, 335, 341, 411, 412
  extensional, 30, 48, 88, 95, 114, 129, 170, 181, 185, 341, 483
Stability:
  criterion, 141, 436, 437
  numerical, 140, 437

structural (also, *see* buckling), 92, 333–335
Stiffness coefficients:
  finite element, 424, 427, 429, 433, 482, 485, 489
  material, 71, 237
  tangent, 443, 491
Strain energy, 24, 31,
  complementary, 31–33, 35, 39, 59
  due to transverse shear, 154, 155
  virtual, 66, 225
Strain energy of:
  CBT, 124
  CPT, 305
  FST, 356
  RBT, 233
  TBT, 155
  TST, 396
Strain–displacement relations of:
  CBT, 63
  CPT, 306
  FST, 355
  RBT, 231–233
  TBT, 152
  TST, 395
Stress resultants of:
  CBT, 65
  CPT, 307
  FST, 356
  RBT, 226, 234
  TBT, 153
  TST, 397
Stress–strain relations, 18, 33, 65, 69, 228, 305
Summation convention, 5, 440

T

Thermal coefficients of expansion, 19, 69, 229, 235
Third-order beam theory, 4, 23, 62, 221–235
Timoshenko beam theory, 62, 151–164

U

Uniform load, 88, 104, 107, 111, 120, 173, 207, 286, 297, 316, 320, 363, 374, 459, 511, 516, 519

V

Variables:
  primary, *see Primary variables*
  secondary, *Secondary variables*

Variation:
    definition of, 40
    first, 40–42
    second, 43
Variational:
    calculus, 42, 44, 48
    formulation(s), 41
    methods, 28, 113, 213, 292
    operator, 39
    solution, 146, 218, 340–352,
Vibration, 92, 97–101, 108, 176, 202,
    207, 210, 285, 332, 335, 350,
    441
Vibration of:
    beams, 101, 202
    circular plates, 339, 348
Virtual:
    displacements, 38, 49–51, 54, 61,
    63, 66, 113, 124, 231, 299, 417
    forces, 39
    kinetic energy, 66, 225
    strain, 38, 50, 58, 155
    strain energy, 38, 50, 66, 68, 154,
    225
    work, 28, 36–39, 50, 61, 66, 225,
    417
Voigt scheme, 20–22

W

Weak forms, 119, 120, 143, 340, 417,
    419, 421, 423, 427, 431,
    434–436
Weight function(s), 141
Weighted-residual method(s), 141–143,
    344
    collocation method, 143
    Galerkin's method, 143, 144, 213,
    293, 340, 345
    least-squares method, 143
    Petrov–Galerkin method, 143
    subdomain method, 423
Work done, 28–30
    external, 29, 48
    internal, 30–33
    virtual, 37, 50, 225

Y

Young's modulus, 21, 48, 69, 114, 229,
    305

Z

Zero vector, 443

For Product Safety Concerns and Information please contact our EU
representative GPSR@taylorandfrancis.com
Taylor & Francis Verlag GmbH, Kaufingerstraße 24, 80331 München, Germany

www.ingramcontent.com/pod-product-compliance
Ingram Content Group UK Ltd.
Pitfield, Milton Keynes, MK11 3LW, UK
UKHW021114180425
457613UK00005B/86